Theoretical Concepts in Physics

An Alternative View of Theoretical Reasoning in Physics

A highly original, novel and integrated approach to theoretical reasoning in physics. This book illuminates the subject from the perspective of real physics as practised by research scientists. It is intended to be a supplement to the final years of an undergraduate course in physics and assumes that the reader has some grasp of university physics. By means of a series of seven case studies, the author conveys the excitement of research and discovery, highlighting the intellectual struggles to attain understanding of some of the most difficult concepts in physics. The case studies comprise the origins of Newton's law of gravitation, Maxwell's equations, linear and non-linear mechanics and dynamics, thermodynamics and statistical physics, the origins of the concept of quanta, special relativity, and general relativity and cosmology. The approach is the same as that in the highly acclaimed first edition, but the text has been completely revised and many new topics introduced.

Malcolm Longair graduated in electronic physics from the University of St Andrews in 1963. He completed his Ph.D. in the Radio Astronomy Group of the Cavendish Laboratory, University of Cambridge, in 1967. From 1968 to 1969 he was a Royal Society Exchange Visitor to the Lebedev Institute, Moscow. He has been an exchange visitor to the USSR Space Research Institute on six subsequent occasions and has held visiting professorships at institutes and observatories throughout the USA. From 1980 to 1990, he held the joint posts of Astronomer Royal for Scotland, Regius Professor of Astronomy of the University of Edinburgh and Director of the Royal Observatory, Edinburgh. He was Deputy Head of the Cavendish Laboratory with special responsibility for the teaching of physics from 1991 to 1997 and has been Head of the Cavendish Laboratory since 1997. He is also a Professorial Fellow of Clare Hall, Cambridge. Professor Longair has received many awards, including the first Britannica Award for the Dissemination of Learning and the Enrichment of Life in February 1986. In December 1990, he delivered the series of Royal Institution Christmas Lectures for Young People on television on the topic 'The origins of our universe'. He was made a CBE in the 2000 Millennium honours list. Professor Longair's primary research interests are in the fields of high energy astrophysics and astrophysical cosmology. He has written numerous books and over 250 journal articles on his research work.

Theoretical Concepts in Physics

An Alternative View of Theoretical Reasoning in Physics

MALCOLM S. LONGAIR

CAMBRIDGE
UNIVERSITY PRESS

PUBLISHED BY THE PRESS SYNDICATE OF THE UNIVERSITY OF CAMBRIDGE
The Pitt Building, Trumpington Street, Cambridge, United Kingdom

CAMBRIDGE UNIVERSITY PRESS
The Edinburgh Building, Cambridge CB2 2RU, UK
40 West 20th Street, New York, NY 10011-4211, USA
477 Williamstown Road, Port Melbourne, VIC 3207, Australia
Ruiz de Alarcón 13, 28014 Madrid, Spain
Dock House, The Waterfront, Cape Town 8001, South Africa

http://www.cambridge.org

First published 1984
Second edition published 2003

Printed in the United Kingdom at the University Press, Cambridge

Typefaces Times New Roman MT 10/13 pt and Frutiger   *System* LaTeX 2_ε [TB]

A catalogue record for this book is available from the British Library

Library of Congress Cataloguing in Publication data

Longair, M.S., 1941–
Theoretical concepts in physics: an alternative view of theoretical reasoning in physics /
Malcolm S. Longair – [2nd ed.].
p. cm.
Includes bibliographical references and index
ISBN 0 521 82126 6 – ISBN 0 521 52878 X (paperback)
1. Mathematical physics. I. Title.
QC20 .L64 2003 530.1–dc21 2002073612

ISBN 0 521 82126 6 hardback
ISBN 0 521 52878 X paperback

For
Deborah

Contents

Preface and acknowledgements

The inspiration for this book was a course of lectures which I delivered between 1977 and 1980 to undergraduates about to enter their final year in Physics and Theoretical Physics at Cambridge. The aim of the course was to provide a survey of the nature of theoretical reasoning in physics, which would put them in a receptive frame of mind for the very intensive courses of lectures on all aspects of physics in the final year. The objectives of the course are described in the first chapter and concern issues about which I feel very strongly: students can go through an undergraduate course in physics without gaining an understanding of the insights, approaches and techniques which are the tools of the professional physicist, let alone an impression of the intellectual excitement and beauty of the subject. The course was intended as an alternative to the normal mode of presentation and was entitled *Theoretical Concepts in Physics*.

An important feature of the course was that it was entirely optional and strictly non-examinable. The lectures were delivered at 9 am every Monday, Wednesday and Friday during a four-week period in July and August, the old Cambridge Summer Term, prior to the final year of the physics course. Despite the timing of the lectures, the fact that the course was not examinable, and the alternative attractions of Cambridge during the summer months, the course was very well attended. I was very gratified by the positive response of the students and this encouraged me to produce a published version of the course with the same title, but with a health warning in the subtitle, *An alternative view of theoretical reasoning in physics for final-year undergraduates*. I was not aware of any other book which covered the material in quite the same way.

The first edition of the book was published in 1984, and by then it had expanded to include other aspects of my experience of teaching physics and theoretical physics. By that time, I was in Edinburgh and responsible for running the Royal Observatory, Edinburgh and the Department of Astronomy. I returned to Cambridge in 1991 and became deeply involved in the revision of the physics syllabus, which led to the present three- or four-year course structure. For the last four years, I have delivered an updated version of the old course, now renamed *Concepts in Physics*. I have continued to expand the range of the material discussed – many of these recent additions are included in this new edition.

Many of the warnings which I issued in the first edition are still relevant. This book is a highly individual approach to physics and theoretical physics. In no way is it a substitute for the systematic exposition of physics and theoretical physics as taught in the standard undergraduate physics course. The contents of this book should be regarded as a complementary approach, which illuminates and reinforces the material from the viewpoint of how the physics actually came about, and how real physicists and theoretical physicists operate. If I succeed in even marginally improving students' appreciation of physics as professional physicists know and love it, the book will have achieved its aims.

In the first edition, I purposely maintained the first person singular to a much greater extent than would be appropriate in a conventional textbook. My intention was to emphasise the individuality of every physicist's approach to the subject and to feel free to express my own opinions and experiences of how physics is actually carried out. Twenty years later, I find that my style of writing has changed. My earlier writings now seem much 'bouncier' and 'uninhibited' than my present style of writing. Undoubtedly, part of this more cautious approach is the result of the experience of sometimes not having got the arguments quite right and needing to change the emphasis as a result of deeper understanding. Have no fear, however – there is just as much passion in the writing as there was in the first edition, but it is written necessarily from a more experienced perspective. As a result, I have rewritten the whole book from scratch, attempting to make the use of language as precise as possible, whilst maintaining the vitality of the earlier writing.

The views expressed in the text are obviously all my own, but many of my Cambridge and Edinburgh colleagues have played a major role in formulating and clarifying my ideas. The idea of the original course came from discussions with Alan Cook, Volker Heine and John Waldram. I inherited the Examples Class in Mathematical Physics from Volker Heine and the late J.M.C. Scott. Developing that class helped enormously in clarifying many of my own ideas. In later years, Brian Josephson helped with the course and provided many startling insights. The course in thermodynamics was given in parallel with one by Archie Howie and I learned a great deal from discussions with him. As part of the reforms which were introduced in the 1990s, Archie delivered the course *Concepts in Physics* and I have enjoyed exploring and extending many of his innovations.

In Edinburgh, Peter Brand, John Peacock and Alan Heavens contributed in important ways to my understanding. In Cambridge, many members of the Department have been very supportive of my endeavours to bring physics alive for undergraduates. I am particularly grateful to John Waldram and David Green for innumerable discussions concerning the courses we have shared. I also acknowledge invaluable discussions with Steve Gull and Anthony Lasenby. Sanjoy Mahajan kindly took a special interest in the section on dimensional methods and critically reviewed what I have written – I am most grateful for his help and insights. A special debt of gratitude is due to Peter Harman, who kindly read some of my writings on Maxwell and made helpful suggestions.

Two committees have continued to provide valuable insight into physics. First, there is the Department of Physics Teaching Committee. I have often thought that a video recording of some of the heated discussions about how to teach physics and theoretical physics would have taught students more about physics than a whole course of lectures. Second, the Staff–Student Consultative Committee for Physics is the forum where the organisers of the physics courses face a highly intelligent and articulate set of consumers at all stages in their physics education. The participation of the students in these discussions has greatly helped the exposition of much of this material.

I must also acknowledge the stimulation provided over the years by the many generations of undergraduates who attended this and the other courses I have given. Their comments and enthusiasm were largely responsible for the fact that the first edition of the book appeared at all. The same remark applies to this new edition – Cambridge students are a phenomenal resource, which makes lecturing and teaching an enormous privilege and real pleasure.

Perhaps the biggest debts I owe in my education as a physicist are to the late Martin Ryle and the late Peter Scheuer, who supervised my research work in the Radio Astronomy Group during the 1960s. I learned more from them about real physics than from anyone else. Almost as great has been the influence of the late Yakov Borisevich Zeldovich and my colleague Rashid Sunyaev. The year I spent in Moscow in 1968–9 was a revelation in opening up new ways of thinking about physics and astrophysics. Another powerful influence was Brian Pippard, whose penetrating understanding of physics was a profound inspiration. Although he and I have very different views of physics, there is virtually no aspect of physics which we have discussed in which his insight has not added immensely to my understanding.

Grateful thanks are due to innumerable people who have helped in the preparation of this book. In preparing the first edition in Edinburgh, the bulk of the text was expertly typed by Janice Murray and Susan Hooper. The line drawings were drawn by Marjorie Fretwell and many of these have been redrawn for the second edition. The reduction of the diagrams to a size suitable for publication and the production of all the photographs in the first edition was the work of Brian Hadley and his colleagues in the Photolabs at the Royal Observatory, Edinburgh. The staff of the Royal Observatory Library were very helpful in locating references and also in releasing for photographing the many treasures in the Crawford Collection of old scientific books.

In preparing the new edition, Judith Andrews performed wonders in converting much of the text of the first edition into LaTeX. Equally important, in acting as my secretary and personal assistant she ensured that, despite the task of running the Laboratory, time was made available to enable the book to be rewritten.

As in all my endeavours, the debts I owe to my wife, Deborah, and our children, Mark and Sarah, cannot be adequately expressed in words.

1 Introduction

1.1 An explanation for the reader

This book is for students who love physics and theoretical physics. It arises from the dichotomy which, in my view, pervades most attempts to teach the ideal course in physics. On the one hand, there is the way in which university teachers present the subject in lecture courses and examples classes. On the other hand, there is the way in which we actually practise the discipline as professional physicists. In my experience, there is often little relation between these activities. This is a great misfortune because students are then rarely exposed to their lecturers when they are practising their profession as physicists.

There are good reasons, of course, why the standard lecture course has evolved into its present form. First of all, physics and theoretical physics are not particularly easy subjects and it is important to set out the fundamentals in as clear and systematic a manner as possible. It is absolutely essential that students acquire a firm grounding in the basic techniques and concepts of physics. But we should not confuse this process with that of doing real physics. Standard lecture courses in physics and its associated mathematics are basically 'five-finger' exercises, designed to develop technique and understanding. But such exercises are very different from a performance of the *Hammerklavier* sonata at the Royal Festival Hall. You are only doing physics or theoretical physics when the answers *really* matter – when your reputation as a scientist hangs upon being able to reason correctly in a research context or, in more practical terms, when your ability in undertaking original research determines whether you are employable, or whether your research grant is renewed. This is a quite different process from working through drill exercises, for which answers are available at the back of the book.

Second, there is so much material which lecturers feel they have to include in their courses that all physics syllabuses are seriously overloaded. There is generally little time left for sitting back and asking 'What is this all about?' Indeed, the technical aspects of the subject, which are themselves fascinating, can become so totally absorbing that it is generally left to the students to find out for themselves many essential truths about physics.

Let me list some aspects of the practice of physics which can be missed in our teaching but which, I believe, are essential aspects of the way in which we carry it out as professionals.

(i) A series of lecture courses is by its nature a modular exercise. It is only too easy to lose a *global view* of the whole subject. Professionals use the whole of physics in tackling problems and there is no artificial distinction between thermal physics, optics, mechanics, electromagnetism, quantum mechanics and so on.

(ii) A corollary of this is that in physics any problem can normally be tackled and solved in a variety of different ways. Often *there is no single 'best way' of solving a problem*; much deeper insights into how the physics works can be obtained if the problem is approached from very different standpoints, for example, from thermodynamics, electromagnetism, quantum theory and so on.

(iii) How problems are tackled and how one thinks about physics are rather personal matters. No two professional physicists think in exactly the same way because we all have different experiences of using the tools of physics in a research context. When we come to write down the relevant equations and solve them, however, we should come to the same answers. The *individual physicist's response to the subject* is an integral part of the way in which physics is taught and practised, to a much greater extent than students or the lecturers themselves would like to believe. But it is the diversity of different lecturers' approaches to physics which provides insight into the nature of the mental processes by which they understand their subject. I remember vividly a splendid lecture by my colleague Douglas Gough summarising a colloquium in Vienna entitled *Inside the Stars*, in which he concluded with the following wonderful paragraph:

> 'I believe that one should never approach a new scientific problem with an unbiased mind. Without prior knowledge of the answer, how is one to know whether one has obtained the right result? But with prior knowledge, on the other hand, one can usually correct one's observations or one's theory until the outcome is correct... However, there are rare occasions on which, no matter how hard one tries, one cannot arrive at the correct result. Once one has exhausted all possibilities for error, one is finally forced to abandon a prejudice, and redefine what one means by 'correct'. So painful is the experience that one does not forget it. That subsequent replacing of the old prejudice by a new one is what constitutes a gain in real knowledge. And that is what we, as scientists, continually pursue.'[1]

In fact, Douglas's dictum is the foundation of the process of discovery in research. All of us have different prejudices and personal opinions about what the solutions to problems might be and it is this diversity of approach which leads to new understandings.

(iv) Another potential victim of the standard lecture course is an appreciation of what it feels like to be involved in *research at the frontiers of knowledge*. Lecturers are always at their best when they reach the part of the course where they can slip in the things which excite them in their research work. For a few moments, the lecturer is transformed from a teacher into a research scientist and then the students see the real physicist at work.

(v) It is often difficult to convey the *sheer excitement of the processes of research and discovery in physics* and yet these are the very reasons that most of us get so enthusiastic about our research; once you are into a challenging research problem, it will not go away. The caricature of the 'mad' scientist is not wholly a myth in that, in carrying out frontier research, it is almost essential to become at times totally absorbed in the problems to the virtual exclusion of the cares of normal life. The biographies of many of the greatest scientists illustrate the extraordinary powers of concentration which they possessed – the examples of Newton and Faraday spring immediately to mind as physicists who, once embarked upon a fertile seam of research, would work unrelentingly until the inspiration was exhausted. All professional physicists have experience of this total intellectual commitment at much more modest levels of achievement and it is only later that, on reflection, we regard these as among our best research experiences. Yet some students complete a physics course without really being aware of what it is that drives us on.

(vi) Much of this excitement can be conveyed through examples selected from the *history* of some of the great discoveries in physics and yet these seldom appear in our courses. The reasons are not difficult to fathom. First of all, there is just not time to do justice to the material. Second, it is not a trivial matter to establish the relevant historical material – physics has created its own mythologies as much as any other subject. Third, nowadays the history and philosophy of science are generally taught as wholly separate disciplines from physics and theoretical physics. My view is that an appreciation of some historical case studies can provide invaluable insight into the processes of research and discovery in physics and of the intellectual framework within which they took place. In these historical case studies, we recognise parallels with our own research experience.

(vii) In these historical examples, key factors familiar to all professional physicists are the central roles of *hard work*, *experience* and, perhaps most important of all, *intuition*. Many of the most successful physicists depend very heavily upon intuition gained through their wide experience and a great deal of hard work in physics and theoretical physics. It would be marvellous if experience could be taught, but I am convinced that it is something which can only be achieved by dedicated hard work. We all remember our mistakes and the blind alleys we have entered and these teach us as much about physics as our successes. Intuition is potentially a dangerous tool because one can make some very bad blunders by relying on it too heavily in frontier areas of physics. Yet it is certainly the source of many of the greatest discoveries in physics. These were not achieved using five-finger exercise techniques, but involved leaps of the imagination which transcended known physics.

(viii) These considerations bring us close to what I regard as the central core of our experience as physicists and theoretical physicists. There is an essential element of *creativity* which is not so different from creativity in the arts. The leaps of imagination involved in discovering, say, Newton's laws of motion, Maxwell's equations, relativity and quantum theory are not so different in essence from the creations of the greatest artists, musicians, writers and so on. The basic differences are that physicists must be creative within a very strict set of rules and that their theories should be testable by confrontation with experiment and observation. Very few of us indeed attain the almost superhuman level of intuition involved in discovering a wholly new physical theory, but we are driven by the same creative urge. Each small step we make contributes to the sum of our understanding of the nature of our physical universe. All of us in our own way tread in regions where no one has passed before.

(ix) The imagination and creativity involved in the very best experimental and theoretical physics result unquestionably in a real sense of *beauty*. The great achievements of physics evoke in me, at least, the same type of response that one finds with great works of art. I suspect that many of us feel the same way about physics but are generally too embarrassed to admit it. This is a pity because the achievements of experimental and theoretical physics rank among the very peaks of human endeavour. I think it is important to tell students when I find a piece of physics particularly beautiful – and there are many examples of this.

When I teach such topics, I experience the same process of rediscovery as on listening to a familiar piece of classical music – one's umpteenth hearing of the *Eroica* symphony or of *Le Sacre du printemps*. I am sure students should know about this.

(x) Finally, physics is *great fun*. The standard lecture course with its concentration on technique can miss so much of the enjoyment and stimulation of the subject. It is essential to convey our enthusiasm for physics. Although physics finds practical application in a myriad of different areas, I am quite unashamed about promoting it for its own sake – if any apologia for this position is necessary, it is that in coming to a real understanding of our physical world our intellectual and imaginative powers are stretched to their very limits.

In this book, I adopt a very different approach to theoretical reasoning in physics from that of the standard textbook. The emphasis is upon the genius and excitement of the discovery of new insights into the laws of physics, much of it through a careful analysis of historical case studies. But my aims are more than simply attempting to redress the balance in the way in which physics is presented. Some of these further aims can be appreciated from the history of how this book came about.

1.2 How this book came about

The origin of this book can be traced to discussions in the Cambridge Physics Department in the mid-1970s among those who were involved in teaching theoretically biased undergraduate courses. There was a feeling that the syllabuses lacked coherence from the theoretical perspective and that the students were not quite clear about the scope of *physics* as opposed to *theoretical physics*. Are they really such different topics?

As our ideas evolved, it became apparent that a discussion of these ideas would be of value to all final-year students. A course entitled 'Theoretical concepts in physics' was therefore designed, to be given in the summer term in July and August to undergraduates entering their final year. It was to be strictly non-examinable and entirely optional. Students obtained no credit from having attended the course beyond an increased appreciation of physics and theoretical physics. I was invited to give the first presentation of this course of lectures, with the considerable challenge of attracting students to 9.00 a.m. lectures on Mondays, Wednesdays and Fridays during the most glorious summer months in Cambridge.

We agreed that the course should contain discussion of the following elements:

(a) *the interaction between experiment and theory*. Particular stress would be laid upon the importance of experiment and, in particular, novel technology in leading to theoretical advances;
(b) the importance of having available the *appropriate mathematical tools for tackling theoretical problems*;
(c) *the theoretical background to the basic concepts of modern physics*, emphasising underlying themes such as *symmetry, conservation, invariance and so on*;
(d) *the role of approximations and models in physics*;

(e) *the analysis of real scientific papers in theoretical physics*, providing insight into how professional physicists tackle real problems.

I decided to approach these topics through a series of case studies designed to illuminate these different aspects of physics and theoretical physics. We also had the following aim:

(f) to *consolidate and revise* many of the basic physical concepts which all final-year undergraduates can reasonably be expected to have at their fingertips.

Finally, I wanted the course

(g) *to convey my own personal enthusiasm for physics and theoretical physics*. My own research is in high-energy astrophysics and astrophysical cosmology, but I remain a physicist at heart: my own view is that astronomy, astrophysics and cosmology are no more than subsets of physics, but applied to the Universe on the large scale. My own enthusiasm results from being involved in astrophysical and cosmological research at the very limits of our understanding of the Universe. I am one of the very lucky generation who began research in astrophysics in the early 1960s and who have witnessed the amazing revolutions which have taken place in our understanding of all aspects of the physics of the Universe. But similar sentiments could be expressed about all areas of physics. The subject is not a dead, pedagogic discipline, the only object of which is to provide examination questions for students. It is an active, extensive subject in a robust state of good health.

After giving the course for four summers, I moved to Edinburgh where the first edition of this book was written. I returned to Cambridge in 1991 and, from 1998, have presented the course, now called '*Concepts in physics*,' to the third-year undergraduates. In this second edition, I have introduced new case studies and elaborated many of the ideas which stimulated the original course. To make the coverage more complete and enhance its usefulness to students, I have included material from examples classes in mathematical physics as well as material arising from my experience of lecturing on essentially the whole of physics. Further explanations of areas in which it is my experience that students find help valuable are included in chapter appendices.

1.3 A warning to the reader

The reader should be warned of two things. First, this is necessarily a *personal view of the subject*. It is intentionally designed to emphasise items (i) to (x) and (a) to (g) – in other words, to emphasise all those aspects which tend to be squeezed out of physics courses because of lack of time.

Second, and even more important, this set of case studies is not a textbook. It is certainly *not* a substitute for the systematic development of these topics through standard physics and mathematics courses. You should regard this book as a supplement to the standard courses, but one which I hope may enhance your understanding, appreciation and enjoyment of physics.

1.4 The nature of physics and theoretical physics

Let us begin by making a formal statement about the basis of our scientific endeavour. The natural sciences aim to give a logical and systematic account of natural phenomena and to enable us to predict from our past experience to new circumstances. *Theory* is the formal basis for such arguments; it need not necessarily be expressed in mathematical language, but the latter gives us the most powerful and general method of reasoning we possess. Therefore, wherever possible we attempt to secure *data* in a form that can be handled *mathematically*. There are two immediate consequences for theory in physics.

The first consequence is that the basis of all physics is *experimental data* and the necessity that these data be in *quantified form*. Some would like to believe that the whole of theoretical physics could be produced by pure reason, but they are doomed to failure from the outset. The great achievements of theoretical physics have been solidly based upon the achievements of experimental physics, which provides powerful constraints upon physical theory. Every theoretical physicist should therefore have a good and sympathetic understanding of the methods of experimental physics, not only so that theory can be confronted with experiment in a meaningful way but also so that new experiments can be proposed which are realisable and which can discriminate between rival theories.

The second consequence, as stated earlier, is that we must have adequate *mathematical tools* with which to tackle the problems we need to solve. Historically, the mathematics and the experiments have not always been in step. Sometimes the mathematics has been available but the experimental methods needed to test the theory have been unavailable. In other cases, the opposite has been true – new mathematical tools have had to be developed to describe the results of experiment.

Mathematics is central to reasoning in physics but we should beware of treating it as the whole physical content of theory. Let me reproduce some words from the reminiscences of Paul Dirac about his attitude to mathematics and theoretical physics. Dirac sought mathematical beauty in all his work. For example, on the one hand he writes:

> Of all the physicists I met, I think Schrödinger was the one that I felt to be most closely similar to myself... I believe the reason for this is that Schrödinger and I both had a very strong appreciation of mathematical beauty and this dominated all our work. It was a sort of act of faith with us that any equations which describe fundamental laws of Nature must have great mathematical beauty in them. It was a very profitable religion to hold and can be considered as the basis of much of our success.[2]

On the other hand, earlier he writes:

> I completed my [undergraduate] course in engineering and I would like to try to explain the effect of this engineering training on me. Previously, I was interested only in exact equations. It seemed to me that if one worked with approximations there was an intolerable ugliness in one's work and I very much wanted to preserve mathematical beauty. Well, the engineering training which I received did teach me to tolerate approximations and I was able to see that even theories based upon approximations could have a considerable amount of beauty in them.
>
> There was this whole change of outlook and also another, which was perhaps brought on by the theory of relativity. I had started off believing that there were some exact laws of Nature and that all we had to do was to work out the consequences of these exact laws. Typical of these were Newton's

laws of motion. Now, we learned that Newton's laws of motion were not exact, only approximations, and I began to infer that maybe all the laws of nature were only approximations . . .

I think that if I had not had this engineering training, I should not have had any success with the kind of work I did later on because it was really necessary to get away from the point of view that one should only deal with exact equations and that one should deal only with results which could be deduced logically from known exact laws which one accepted, in which one had implicit faith. Engineers were concerned only in getting equations which were useful for describing nature. They did not very much mind how the equations were obtained. . . .

And that led me of course to the view that this outlook was really the best outlook to have. We wanted a description of nature. We wanted the equations which would describe nature and the best we could hope for was, usually,* approximate equations and we would have to reconcile ourselves to an absence of strict logic.[3]

These are very important and profound sentiments which should be familiar to the reader. There is really no strictly logical way in which we can formulate theory – we are continually approximating and using experiment to keep us on the right track. Note that Dirac was describing theoretical physics at its very highest level – concepts like Newton's laws of motion, special and general relativity, Schrödinger's equation and the Dirac equation are the *very summits of achievement of theoretical physics* and very few can work creatively at that level. The same sentiments apply, however, in their various ways to all aspects of research as soon as we attempt to model quantitatively the natural world.

Most of us are concerned with applying and testing known laws to physical situations in which their application has not previously been possible, or foreseen, and we often have to make numerous approximations to make the problem tractable. The essence of our training as physicists is to develop confidence in our physical understanding of physics so that, when we are faced with a completely new problem, we can use our experience and intuition to recognise the most fruitful ways forward.

1.5 The influence of our environment

1.5.1 The international scene

It is important to realise not only that all physicists are individuals with their own prejudices but also that these prejudices are strongly influenced by the tradition within which they have studied physics. I have had experience of working in a number of different countries, particularly in the USA and the former Soviet Union, and the different scientific traditions can be appreciated vividly in the marked difference in approach of physicists to research problems. This has added greatly to my understanding and appreciation of physics.

An example of a distinctively British feature of physics is the tradition of *model building*, to which we will return on several occasions. Model building seems to have been an especially British trait during the nineteenth and early twentieth centuries. The works of Faraday and Maxwell are full of models, as we will see, and at the beginning of the twentieth century, the variety of models for atoms was quite bewildering. The J.J. Thomson

* Editorial commas.

'plum-pudding' model of the atom is perhaps one of the more famous examples, but it is just the tip of the iceberg. Thomson was quite straightforward about the importance of model building:

> The question as to which particular method of illustration the student should adopt is for many purposes of secondary importance provided that he does adopt one.[4]

Thomson's assertion is splendidly illustrated by Heilbron's *Lectures on the History of Atomic Physics 1900–1920*.[5] The modelling approach is very different from the continental European tradition of theoretical physics – we find Poincaré remarking that 'The first time a French reader opens Maxwell's book, a feeling of discomfort, and often even of mistrust, is at first mingled with his admiration . . . '.[6] According to Hertz, Kirchhoff was heard to remark that he found it painful to see atoms and their vibrations wilfully stuck in the middle of a theoretical discussion.[7] It was reported to me after a lecture in Paris that one of the senior professors had commented that my presentation had not been 'sufficiently Cartesian'. I believe the British tradition of model-building is alive and well. I can certainly vouch for the fact that, when I think about some topic in physics or astrophysics, I generally have some picture, or model, in my mind rather than an abstract or mathematical idea.

I believe the development of *physical insight* is an integral part of the model-building tradition. The ability to guess correctly what will happen in a new physical situation without having to write down all the mathematics is a very useful talent and most of us develop it with time. It must be emphasised, however, that having physical insight is no substitute for producing exact mathematical answers. If you want to claim to be a theoretical physicist, you must be able to give the rigorous mathematical solution as well.

1.5.2 *The local scene*

The influence of our environment applies to different physics departments, as well as to different countries. If we consider the term 'theoretical physics', there is a wide range of opinion as to what constitutes theoretical physics as opposed to physics. It is a fact that in the Cavendish Laboratory in Cambridge, most of the lecture courses are strongly theoretically biased. By this I mean that these courses aim to provide students with a solid foundation in basic theory and its development and relatively less attention is paid to matters of experimental technique. If experiments are alluded to, the emphasis is generally upon the results rather than the experimental ingenuity by which the experimental physicists came to their answers. Although we now give courses on the fundamentals of experimental physics, we expect students to acquire most of their experimental training through practical experiments. This is in strong contrast to the nature of the Cambridge physics courses in the early decades of the twentieth century, which were strongly experimental in emphasis.

Members of departments of theoretical physics or applied mathematics would claim, however, that they teach much 'purer' theoretical physics than we do. In their undergraduate teaching, I believe this is the case. There is by definition a strong mathematical bias in the teaching of these departments, and they are often much more concerned about rigor in their use of mathematics than we are. In other physics departments, the bias is often towards experiment rather than theory. I find it amusing that some members of the Cavendish Laboratory who are considered to be 'experimentalists' within the department are regarded as 'theorists' by other physics departments in the UK!

The reason for discussing this issue of the local environment is that it can produce a somewhat biased view of what we mean by physics and theoretical physics. My own perspective is that 'physics' and 'theoretical physics' are part of a continuum of approaches to physical understanding – they are different ways of looking at the same body of material. This is one of the reasons our final-year courses are entitled 'Experimental and theoretical physics'. In my opinion, there are great advantages in developing mathematical models in the context of the experiments, or at least in an environment where day-to-day contact occurs naturally with those involved in the experiments.

1.6 The plan of the book

This book consists of seven *case studies*, each designed to cover major areas of physics and key advances in theoretical understanding. The case studies are entitled:

I The origins of Newton's laws of motion and of gravity
II Maxwell's equations
III Mechanics and dynamics – linear and non-linear
IV Thermodynamics and statistical physics
V The origins of the concept of quanta
VI Special relativity
VII General relativity and cosmology

These topics have a very familiar ring, but they are treated from a rather different perspective as compared with the standard textbooks – that is why the subtitle of this book is *An alternative view of theoretical reasoning in physics*. My aim is not just to explore the content of the topics but also to recreate the intellectual background to some of the greatest discoveries in theoretical physics.

At the same time, we can gain from such historical case studies important insights into the process of how real physics and theoretical physics are carried out. Such insights can convey some of the excitement and intense intellectual struggle involved in achieving new levels of physical understanding. In a number of these case studies, we will follow the processes of discovery by the same routes followed by the scientists themselves, using only the mathematical techniques available to scientists at the time. For example, we cannot cut corners by assuming we can represent electromagnetic waves by photons until after the discovery of quanta.

In considering each case study, we will also revise many of the basic concepts of physics with which you should be familiar. There are numerous appendices designed to help in areas in which I find students often value additional insight. Finally, each case study is prefaced by a short essay explaining the approach taken and the objectives, which are all somewhat different and designed to illustrate different aspects of physics and theoretical physics.

1.7 Apologies and words of encouragement

Let me emphasise at the outset that I am not a historian or philosopher of science. I use the history of science very much for my own purpose, which is to illuminate my own experience of how real physicists think and behave. The use of historical case studies is simply a device for conveying something of the reality and excitement of physics. I therefore apologise unreservedly to historians and philosophers of science for using the fruits of their researches, for which I have the most profound respect, to achieve my pedagogical goals. My hope is that students will gain an enhanced appreciation and respect for the works of professional historians of science from what they read in this book.

Establishing the history by which scientific discoveries were made is a hazardous and difficult business; even in the recent past it is often difficult to disentangle what really happened. In my background reading, I have relied heavily upon standard biographies and histories. For me, they have provided vivid pictures of how science actually works and I can relate them to my own research experience. If I have erred in some places, my exculpation can only be the words attributed to Giordano Bruno, 'Si non e vero, e molto ben trovato' (if it is not true, it is a very good invention).

My intention is that all advanced undergraduates in physics should be able to profit from this book, whether or not they are planning to become professional theoretical physicists. Although experimental physics can be carried out without a deep understanding of theory, that point of view misses so much of the beauty and stimulation of the subject. Remember, however, the case of Stark, who made it a point of principle to reject almost all theories on which his colleagues had reached a consensus. Contrary to their view, he showed that spectral lines could be split by an electric field, the Stark effect, for which he won the Nobel prize.

Finally, I hope you enjoy this material as much as I do. One of my aims is to put in context all the physics you have met so far and put you into a receptive frame of mind for appreciating the final years of your undergraduate lecture courses. I particularly want to convey a real appreciation of the great discoveries of physics and theoretical physics. These are achievements as great as any in any field of human endeavour.

1.8 References

1 Gough, D.O. (1993). In *Inside the Stars*, eds. W.W. Weiss and A. Baglin, IAU Colloquium No. 137, p. 775. San Francisco: Astron. Soc. Pacific Conf. Series, Vol. 40.
2 Dirac, P.A.M. (1977). In *History of Twentieth Century Physics, Proc. International School of Physics 'Enrico Fermi'*, Course 57, p. 136. New York and London: Academic Press.
3 Dirac, P.A.M. (1977). *Op. cit.*, p. 112.

4 Thomson, J.J. (1893). *Notes on Recent Researches in Electricity and Magnetism*, vi. Oxford: Clarendon Press. (Quoted by J.L. Heilbron in reference 5 below, p. 42.)
5 Heilbron, J.L. (1977). In *History of Twentieth Century Physics, Proc. International School of Physics 'Enrico Fermi'*, Course 57, p. 40. New York and London: Academic Press.
6 Duhem, P. (1991 reprint). *The Aim and Structure of Physical Theory*, p. 85. Princeton: Princeton University Press.
7 Heilbron, J.L. (1977). *Op. cit.*, p. 43.

Case Study I

The origins of Newton's laws of motion and of gravity

Our first case study encompasses essentially the whole of what can be considered the modern scientific process. Unlike the other case studies, it requires little mathematics but a great deal in terms of intellectual imagination. For me, it is a heroic tale of scientists of the highest genius lying the foundations of modern science. Everything is there – the rôles of brilliant experimental skill, of imagination in the interpretation of observational and experimental data and of the remarkable leaps of the imagination which were to lay the foundations for the Newtonian picture of the world. This achievement may not at first sight seem so remarkable to the twenty-first-century reader, but closer inspection shows that in fact it is immense. As expressed by Herbert Butterfield in his *Origins of Modern Science*,[1] the understanding of motion was one of the most difficult steps that scientists have ever undertaken. In the quotation by Douglas Gough in Chapter 1, he expresses eloquently the 'pain' experienced on being forced to discard a cherished prejudice in the sciences. How much more difficult must have been the process of laying the foundations of modern science, when the concept that the laws of nature can be written in mathematical form had not yet been formulated.

How did our modern appreciation of the nature of our physical Universe come about? I make no apology for starting at the very beginning. In Chapter 2, the first of three chapters that address Case Study I, we set the scene for the subsequent triumphs, and tragedies, of two of the greatest minds of modern science – Galileo Galilei and Isaac Newton. Their achievements were firmly grounded in the remarkable observational advances of Tycho Brahe and in Galileo's skill as an experimental physicist and astronomer. Galileo and his trial by the Inquisition are considered in some detail in Chapter 3, the emphasis being upon the scientific aspects of this controversial episode in the history of science. The issues involved can be considered as the touchstone for the modern view of the nature of scientific enquiry. Then, with the deep insights of Kepler and Galileo established, Newton's extraordinary achievements are placed in their historical context in Chapter 4.

It may seem somewhat strange to devote so much space at the beginning of this text to what many will consider to be ancient history, a great deal of which we now understand to be wrong and misleading. Having been through this material, I feel very differently about it. It is a gripping story and full of resonances about the way we practice science today. There are, in addition, other aspects to this story which I believe are important. The great

Figure I.1: Tycho Brahe with the instruments he constructed for accurate observations of the positions of the stars and planets. He is seen seated within the 'great mural quadrant', which produced the most accurate measurements of the positions of the stars and planets at that time. (After *Astronomiae Instauratae Mechanica*, 1602, p. 20, Nürnberg. From the Crawford Collection, Royal Observatory, Edinburgh.)

scientists involved in this case study had complex personalities and, to provide a rounder picture of their characters and achievements, it is helpful to understand their intellectual perspectives as well as their contributions to fundamental physics.

I.1 Reference

1 Butterfield, H. (1950). *The Origins of Modern Science.* London: G. Bell, New York: Macmillan (1951).

2 From Ptolemy to Kepler – the Copernican revolution

2.1 Ancient history

The first of the great astronomers of whom we have knowledge is *Hipparchus*, who was born in Nicaea in the second century BC. Perhaps his greatest achievement was his catalogue of the positions and brightnesses of 850 stars in the northern sky. The catalogue was completed in 127 BC and represented a quite monumental achievement. A measure of his skill as an astronomer is that he compared his positions with those of Timocharis made in Alexandria 150 years earlier and discovered the *precession of the equinoxes*, the very slow change in direction of the Earth's axis of rotation relative to the frame of reference of the fixed stars. We now know that this precession is caused by tidal torques due to the Sun and Moon acting upon the slightly non-spherical Earth. At that time, however, the Earth was assumed to be stationary and so the precession of the equinoxes had to be attributed to a movement of the 'sphere of fixed stars'.

The most famous of the ancient astronomical texts is the *Almagest* of *Claudius Ptolomeaus*, or *Ptolemy*, who lived in the second century AD. The word 'Almagest' is a corruption of the Arabic translation of the title of his book, the *Megelé Syntaxis* or *Great Composition*, which in Arabic becomes *al-majisti*. It consisted of 13 volumes and provided a synthesis of all the achievements of the Greek astronomers and, in particular, leant heavily upon the work of Hipparchus. Within the *Almagest*, Ptolemy set out what became known as the *Ptolemaic system of the world*, which was to dominate astronomical thinking for the next 1500 years.

How did the Ptolemaic system work? It is apparent to everyone that the Sun and Moon appear to move in roughly circular paths about the Earth. Their trajectories are traced out against the *sphere of the fixed stars*, which also appears to rotate about the Earth once per day. In addition, five planets are observable by the naked eye, Mercury, Venus, Mars, Jupiter and Saturn. The Greek astronomers understood that the planets did not move in simple circles about the Earth, but had somewhat more complex motions. Figure 2.1 shows Ptolemy's observations of the motion of Saturn in AD 133 against the background of the fixed stars. Rather than move in a smooth path across the sky, the path of the planet doubles back upon itself.

The challenge to the Greek astronomers was to work out mathematical schemes which could describe these motions. As early as the third century BC, a few astronomers had suggested that these phenomena could be explained if the Earth rotated on its axis and

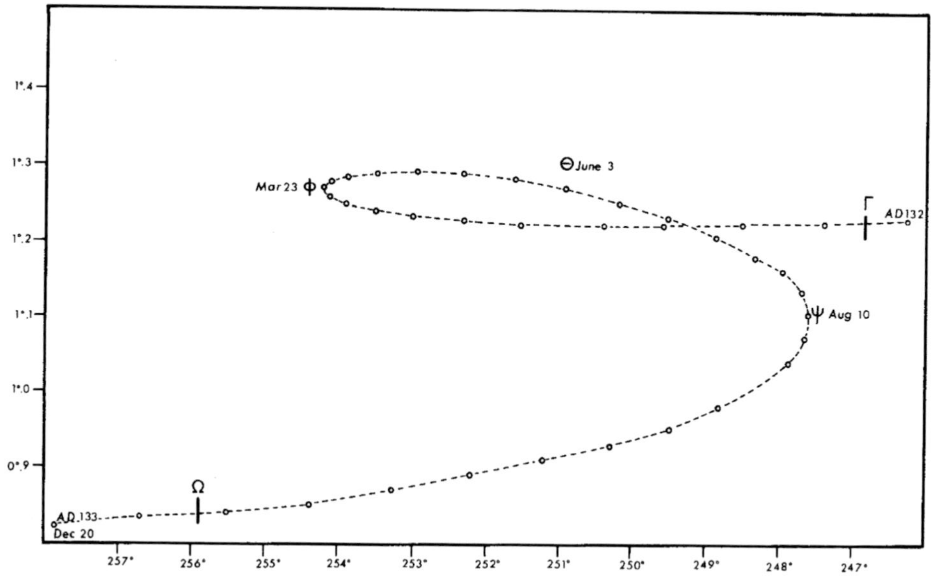

Figure 2.1: The motion of Saturn from 5 December AD 132 to 20 December AD 133 as observed by Ptolemy against the background of the fixed stars. (From O. Pedersen and M. Pihl, 1974, *Early Physics and Astronomy*, p. 71, London: McDonald and Co.)

even that the planets orbit the Sun. Heracleides of Pontus described a geo-heliocentric system which we will meet again in the work of Tycho Brahe. Most remarkably, Aristarchos proposed that the planets move in circular orbits about the Sun. In *The Sand Reckoner*, Archimedes wrote to King Gelon,

> You are not unaware that by the universe most astronomers understand a sphere the centre of which is at the centre of the Earth.... However, Aristarchos of Samos has published certain writings on the [astronomical] hypotheses. The presuppositions found in these writings imply that the universe is much greater than we mentioned above. Actually, he begins with the hypothesis that the fixed stars and the Sun remain without motion. As for the Earth, it moves around the Sun on the circumference of a circle with centre in the Sun.[1]

These ideas became the inspiration for Copernicus roughly eighteen centuries later. They were rejected at the time of Aristarchos for a number of reasons. Probably the most serious was the opposition of the upholders of Greek religious beliefs. According to Pedersen and Pihl (1974),[1]

> Aristarchos had sinned against deep-rooted ideas about Hestia's fire, and the Earth as a Divine Being. Such religious tenets could not be shaken by abstract astronomical theories incomprehensible to the ordinary man.[2]

From our perspective, the physical arguments against the heliocentric hypothesis are of equal interest. First, the idea that the Earth rotates about an axis was rejected. If the Earth rotated then when an object is thrown up in the air it would not come down again in the same spot – the Earth would have moved, because of its rotation, before the object landed. No one had ever observed this to be the case and so the Earth could not be rotating. The

Figure 2.2: The basic Ptolemaic system of the world showing the celestial bodies from the Earth in the order, Moon, Mercury, Venus, Sun, Mars, Jupiter, Saturn and the sphere of fixed stars. (From Andreas Cellarius, 1661, *Harmonia Macrocosmica* Amsterdam. Courtesy of F. Bertola, from *Imago Mundi*, 1995, Biblios, Padova.)

second problem resulted from the observation that if objects are not supported they fall under gravity. Therefore, if the Sun were the centre of the Universe rather than the Earth, everything ought to be falling towards that centre. Now, if objects are dropped they fall towards the centre of the Earth and not towards the Sun. It follows that the Earth must be located at the centre of the Universe. Thus, religious belief was supported by scientific rationale.

According to the Ptolemaic geocentric system of the world, the Earth is stationary at the centre of the Universe and the principal orbits of the other celestial objects are circles in the order Moon, Mercury, Venus, Sun, Mars, Jupiter, Saturn and finally the sphere of the fixed stars (Fig. 2.2). The problem with the elementary Ptolemaic system was that it could not account for the details of the motions of the planets, such as the retrograde motion shown in Fig. 2.1, and so the model had to become more complex. There was one central concept from Greek mathematics which played a key role in refining the Ptolemaic system. Part of the basic philosophy of the Greeks was that the only allowable motions were uniform motion in a straight line and uniform circular motion. Ptolemy himself stated that uniform circular motion was the only kind of motion 'in agreement with the nature of

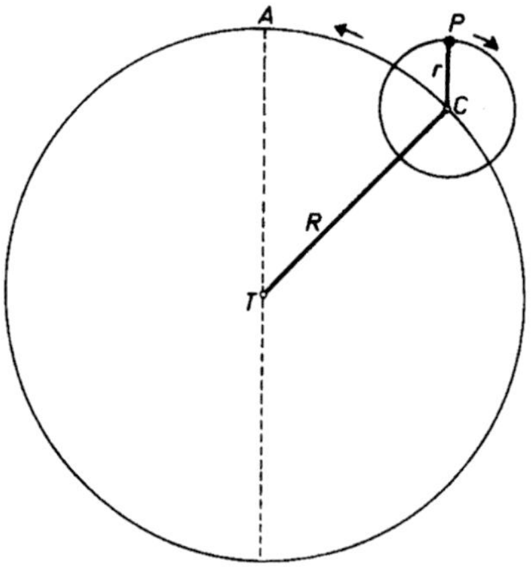

Figure 2.3: Illustrating circular epicyclic motion about a circular orbit according to the epicyclic model of Appolonios. (From O. Pedersen and M. Pihl, 1974, *Early Physics and Astronomy*, p. 83, London: McDonald and Co.)

Divine Beings'. Therefore, it was supposed that, in addition to their circular orbits about the Earth, the planets, as well as the Sun and Moon, had circular motions about the principal circular orbit (Fig. 2.3); the circles superimposed upon the main circular orbit were known as *epicycles*. It can be readily understood how the type of orbit shown in Fig. 2.1 can be reproduced by selecting suitable speeds for the motions of the planets in their epicycles.

One of the basic rules of *astrometry*, meaning the accurate measurement of the positions and movements of bodies on the celestial sphere, is that the accuracy with which their orbits are determined improves the longer the time span over which the observations are made. As a result, the simple epicyclic picture had to become more and more complex, the longer the time base of the observations. To improve the accuracy of the Ptolemaic model, the centre of the circle of a planet's principal orbit was allowed to differ from the position of the Earth, each circular component of the motion remaining uniform. As a consequence, it was found necessary to assume that the centre of the circle about which the epicycles took place also differed from the position of the Earth (Fig. 2.4). An extensive terminology was used to describe the details of the orbits, but there is no need to enter into these complexities here (see Pedersen and Pihl (1974) for more details[1]). The key point is that, by considerable geometrical ingenuity, Ptolemy and later generations of astronomers were able to give a good account of the observed motions of the Sun, Moon and the planets, but the models involved a considerable number of more or less arbitrary geometrical decisions. Although complicated, the Ptolemaic model was used in the preparation of almanacs and in the determination of the dates of religious festivals until after the Copernican revolution.

2.2 The Copernican revolution

By the sixteenth century, the Ptolemaic system was becoming more and more complicated as a tool for predicting the positions of celestial bodies. *Nicolaus Copernicus* revived the

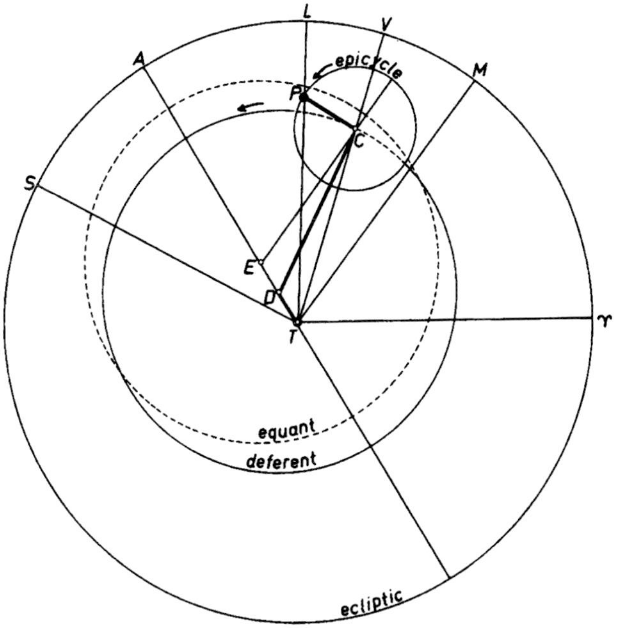

Figure 2.4: Illustrating the complexity of the Ptolemaic theory of the motion of the outer planets. (From O. Pedersen and M. Pihl, 1974, *Early Physics and Astronomy*, p. 94, London: McDonald and Co.)

idea of Aristarchus that a simpler model, in which the Sun is at the centre of the Universe, might provide a simpler description of the motions of the planets. Copernicus, born in Torun in Poland in 1473, first attended the University of Kraków and then went to Bologna, where his studies included astronomy, Greek, mathematics and the writings of Plato. In the early 1500s, he spent four years at Padua where he also studied medicine. By the time he returned to Poland, he had mastered all the astronomical and mathematical sciences. Copernicus made some observations himself and these were published between 1497 to 1529.

His great works were, however, his investigations of whether a heliocentric Universe could provide a simpler account of the motions of the planets. When he worked out the mathematics of this model, he found that it gave a remarkably good description. Again, however, he restricted the motions of the Moon and the planets to uniform circular orbits, according to the precepts of Aristotelian physics. In 1514 he circulated his ideas privately in a short manuscript called '*De hypothesibus motuum coelestium a se constitutis commentariolus*' (*A commentary on the theory of the motion of the heavenly objects from their arrangements*). The ideas were presented to Pope Clement VII in 1533, who approved of them and who in 1536 made a formal request that the work be published. Copernicus hesitated, but eventually wrote his great treatise summarising what is now known as the Copernican model of the universe in *De Revolutionibus Orbium Coelestium* (*On the Revolutions of the Heavenly Spheres*).[3] The publication of the work was delayed, but eventually it was published by Osiander in 1543. It is said that the first copy was brought to Copernicus on his death-bed on 24 May 1543. Osiander had inserted his own foreword into the treatise

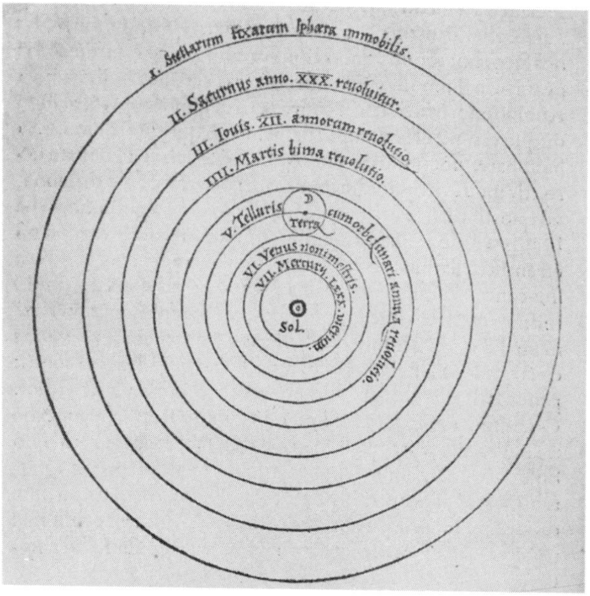

Figure 2.5: The Copernican Universe from Copernicus' treatise *De Revolutionibus Orbium Celestium*, 1543, opposite p. 10, Nurnberg. (From the Crawford collection, Royal Observatory, Edinburgh.)

stating that the Copernican model was no more than a calculating device for simplifying the predictions of planetary motions, but it is clear from the text itself that Copernicus was in no doubt that the Sun really was the centre of the Universe, and not the Earth.

Figure 2.5 shows the famous picture which appears opposite p. 10 of Copernicus' treatise, showing the planets in their familiar order with the Moon orbiting the Earth and the six planets orbiting the Sun. Beyond these lies the sphere of the fixed stars. The implications of the Copernican picture were profound, not only for science but also for the understanding of our place in the Universe. The scientific implications were twofold. First, the size of the Universe was vastly increased as compared with the Ptolemaic model. If the fixed stars were relatively nearby then they ought to display parallaxes, apparent motions relative to more distant background stars, because of the Earth's motion about the Sun. No such stellar parallax had ever been observed, and so the fixed stars must be very distant indeed.

In England, these ideas were enthusiastically adopted by the most important astronomer of the reign of Queen Elizabeth I, Thomas Digges, who was also the translator of large sections of *De Revolutionibus* into English. In his version of the Copernican model, the Universe is of infinite extent and the stars are scattered throughout space (Fig. 2.6). This is a remarkably prescient picture and one which Newton was to adopt, but it leads to some tricky cosmological problems, as we will see.

The second fundamental implication of the Copernican picture was that something was wrong with the Aristotelian concept that all objects fall towards the centre of the Universe,

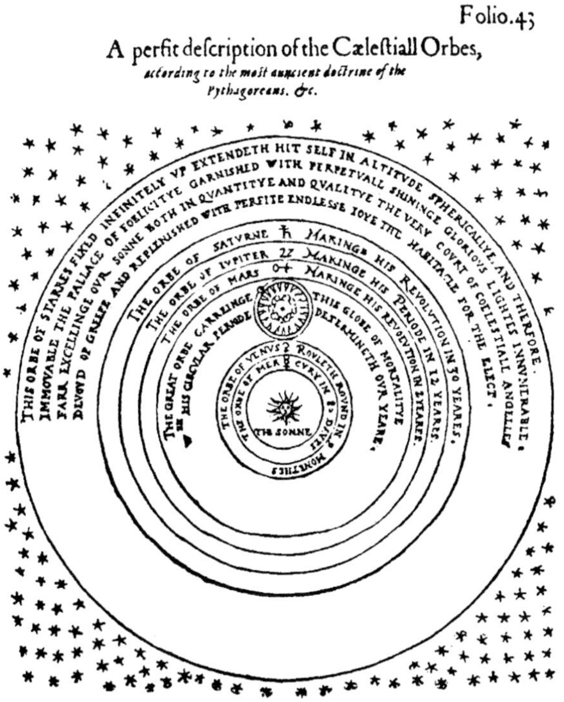

Figure 2.6: Thomas Digges' version of the Copernican picture of the world showing the solar system embedded in an infinite distribution of stars. (From Thomas Digges, 1576, *A Perfit Description of the Caelestiall Orbes*, London.)

which is now occupied by the Sun. This problem was only resolved with Newton's discovery of the nature of the law of gravity, namely, that it is an inverse- square law.

2.3 Tycho Brahe – the lord of Uraniborg

Copies of Copernicus' *De Revolutionibus Orbium Coelestium* circulated remarkably quickly thoughout Europe. One of the motivations behind Copernicus' researches was to produce simpler mathematical procedures for working out the motions of the Sun, Moon and the planets for the purposes of determining the exact date of the vernal equinox. These were needed to establish the correct dates for religious festivals and this was perhaps one reason for the favourable reception of Copernicus' heliocentric model by Pope Clement VII.

Until 1543, the predictions of the motions of the celestial bodies had been taken from what were known as the Alphonsine Tables, which were derived from the Ptolemaic system as refined by the Arabic astronomers. These tables had been prepared by the Rabbi Isaac ben Sid of Toledo and published in manuscript form in the *Libros del Sabre de Astronomica* in 1277 under the patronage of Alfonso X of Castile, also known as Alfonso the Wise. The tables were copied in manuscript form and were quickly disseminated throughout Europe. They were only published in the modern sense in 1483.

Modern scholarship has suggested that, in fact, predictions using the Copernican model were often not much better than those using the Alphonsine Tables. Predictions using the data in *De Revolutionibus* were made by Erasmus Reinhold, who produced what became known as the Prutenic, or Prussian, Tables, giving the positions of the stars and planets. These were published in 1551, no more than eight years after the first publication of *De Revolutionibus*.

The next hero of our story is *Tycho Brahe*, who was born into a noble family in 1546 at the family home of Knudstrup at Skåne, Denmark. He formed an early passion for astronomy. To prepare him for the life of a nobleman, however, he was sent to the University of Leipzig in March 1562 to study law, but he kept up his interest in astronomy, making observations secretly at night while his tutor slept. He also spent all the money he could save on astronomical books, tables and instruments. At this time he acquired his own copies of the Alphonsine and Prutenic Tables. The inspiration for his future work came from the predicted conjunction of Saturn and Jupiter in 1563. He found that the predictions of both tables were in error, by about a month if he used the Alphonsine Tables and by a few days if he used the Prutenic Tables. The need to improve the accuracy with which the planetary motions were known was one of the prime motivations for the monumental series of observations which he began in the late 1570s.

Once established in Denmark, he gave lectures at the University of Copenhagen where he discussed the Copernican theory.

> Tycho spoke of the skill of Copernicus, whose system, although not in accord with physical principles, was mathematically admirable and did not make the absurd assumptions of the ancients.[4]

Tycho continued his international collaboration with Landgrave William IV, whom he visited in Kassel in 1575 and, during that visit, he first became aware of the importance of the effect of *refraction* by the Earth's atmosphere on astronomical observations. Tycho was the first astronomer to take account of refraction in working out accurate stellar and planetary positions.

Tycho was determined to carry out a programme of measurement of the positions of the stars and planets to the very highest accuracy achievable. Frederick II of Denmark was persuaded that Tycho was an outstanding scientist who would bring honour to Denmark, and so, in 1576, to prevent Tycho 'brain-draining' to Germany made him an offer which he could not refuse. In the words of Frederick, he provided Tycho with

> ... our land of Hven with all our and the Crown's tenants and servants who thereon live, with all the rent and duty which comes from that ... as long as he lives and likes to continue and follow his *studia mathematica* ...[5]

The island of Hven lies halfway between Denmark and Sweden and Tycho considered this peaceful island retreat ideal for his astronomical ambitions. He was allowed to use the rents for the upkeep and the running of the observatories he was to build. In addition, he received regular supplementary funds from Frederick II to enable him to build his great observatories and to construct the instruments needed for making astronomical observations. In Tycho's own words, he believed that the total enterprise had cost Frederick 'more than a tun of gold'.[6] Victor Thoren estimates that Tycho received an annual income of about 1% of the Crown's revenue throughout his years at Hven.[7] This was the first example of 'big science' in the modern era.

The first part of the project was to build the main observatory, which he named Uraniborg, or the Heavenly Castle. Besides adequate space for all his astronomical instruments, he built alchemical laboratories on the ground floor where he carried out chemical experiments. The building also included a papermill and printing press, so that the results of the observations could be published promptly.

Uraniborg was completed by about 1580 and Tycho then began the construction of a second observatory, 'Stjerneborg', which was located at ground level. He now realised the value of constructing very solid foundations for the astronomical instruments, and they were clustered in a much more compact area in the new observatory. These astronomical instruments were the real glory of the observatory. They predated the telescope; all observations were made with the naked eye and consisted of measuring as accurately as possible the relative positions of the stars and planets in a systematic way. Two of the instruments seen in Figure I.1 are worthy of special mention. The first is the Great Globe, which can be seen on the ground floor of Uraniborg in Fig. I.1. On it, Tycho marked the positions of all the stars he recorded, thus forming a permanent record of the positions of, in the end, 777 stars.

The second was the great mural quadrant, also seen in Fig. I.1 being pointed out by Tycho himself. The mural quadrant is fixed in one position and had radius $6\frac{3}{4}$ feet. Observations were made by observing stars through a hole in the wall at the centre of the circle described by the quadrant. Observation of the position of a star consisted of measuring the angle from the horizon. Because of the large radius of the quadrant, very accurate positions could be measured. To keep an accurate track of time, Tycho had four clocks so that he would notice if any one of them was not keeping proper time.

Tycho's technical achievements were quite remarkable. He was the first scientist known to have understood the crucial importance of taking account of *systematic errors* in his observations. There are two beautiful examples of these. We have already mentioned the first of these, regarding atmospheric refraction. The second arises because large instruments bend under gravity, producing a systematic error in the sense that if the instrument is pointing vertically it does not bend whereas if it is horizontal the ends bend downwards and give incorrect angles. Tycho understood the necessity of eliminating these types of systematic error.

Another important advance was his understanding of the need to estimate how precise the observations were once the effects of systematic errors had been removed, in other words, the magnitude of the *random errors* in the observations. Tycho was also probably the first scientist to work out accurately the random errors in his observations. They turned out to be about ten times smaller than those of any earlier observations. This is a quite enormous increase in precision. A further key feature of the observations was that they were *systematic*, in the sense that they were made continuously over a period of about 20 years from 1576 to 1597 and systematically analysed by Tycho and his team of assistants. Throughout that period, he measured precisely the positions of the Sun, Moon, planets and the fixed stars. His

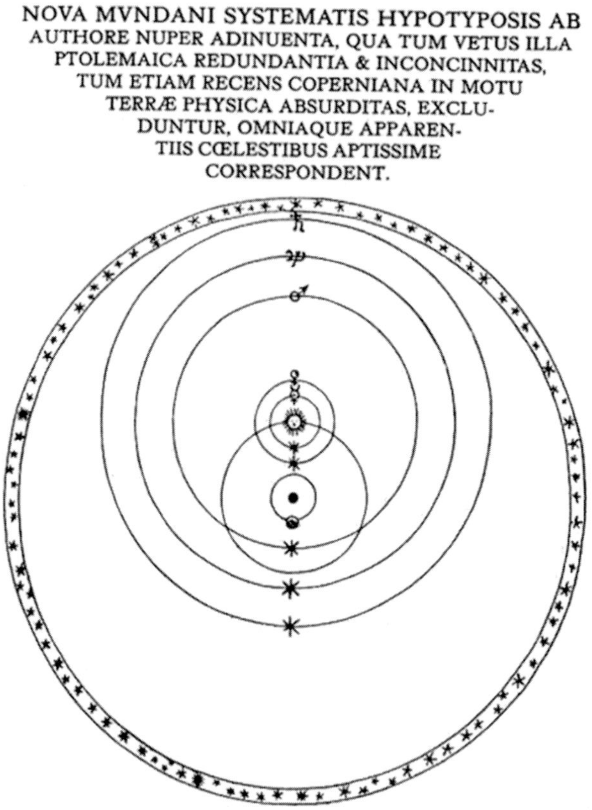
NOVA MVNDANI SYSTEMATIS HYPOTYPOSIS AB
AUTHORE NUPER ADINUENTA, QUA TUM VETUS ILLA
PTOLEMAICA REDUNDANTIA & INCONCINNITAS,
TUM ETIAM RECENS COPERNIANA IN MOTU
TERRÆ PHYSICA ABSURDITAS, EXCLU-
DUNTUR, OMNIAQUE APPAREN-
TIIS CŒLESTIBUS APTISSIME
CORRESPONDENT.

Figure 2.7: The Tychonic system of the world. (From V.E. Thoren, 1990, *The Lord of Uraniborg*, p. 252, Cambridge: Cambridge University Press.)

final catalogue contained the positions of 777 stars measured with a precision of about 1 to 2 minutes of arc. The crucial importance of knowing this figure will become apparent in the next section.

This was the summit of Tycho's achievement. He created the greatest observatory of its day, and Denmark and Hven became the centre of astronomy in Europe. After Frederick II's death in 1588, however, the support for pure science waned under his successor, Christian IV. In addition, Tycho was mismanaging the island. Matters came to a head when he left for exile in 1597, taking with him his observations, instruments and printing press. He eventually settled outside Prague under the patronage of Emperor Rudolf II at the castle of Benatek, where he began setting up again his magnificent set of instruments. During the remaining years of his life, a priority was the analysis of the mass of data he had secured and he had one stroke of great good fortune: one of his last acts, in 1600, was to employ Johannes Kepler to reduce the data from his observations of the planet Mars.

Tycho developed his own cosmology, which he regarded as a synthesis of the Copernican and Ptolemaic views of the Universe, in about 1583 (Fig. 2.7). In this model, the Earth is the centre of the Universe and the Moon and Sun orbit the Earth, while all the other planets orbit the Sun. Tycho was justly proud of his model, which bears more than a passing resemblance to the model of Heracleides mentioned in Section 2.1. Until one understands centrifugal forces and the law of gravity, there is not much wrong, but Tycho's own observations were to lead to the demolition of his theory and the establishment of the law of gravity.

Tycho's achievements rank among the greatest in observational and experimental science. We recognise in his work all the very best features of modern experimental science, and yet the rules had not been written in his day. The achievement is all the greater when we recall that the idea of quantitative scientific measurement scarcely existed at that time. Experiment and precise measurement did not exist – astronomical observation was by far the most exact of all the physical sciences. It should come as no surprise that Tycho's legacy played a central rôle in the Newtonian revolution.

2.4 Johannes Kepler and heavenly harmonies

We now introduce a very different character, *Johannes Kepler*. He was born in December 1571 in the Swabian city of Weil der Stadt. Kepler was a weak boy but a talented student, who entered the University of Tübingen in 1589. His first encounter with astronomy was through Michael Maestlin, who taught him mathematics, Euclid and trigonometry. In 1582, Maestlin had published his treatise *Epitome Astronomiae*, which contained a description of Copernicus's heliocentric Universe. Maestlin was, however, very cautious and, under the influence of the church, remained a Ptolemaian. In contrast, Kepler had no doubts. In his words:

> Already in Tübingen, when I followed attentively the instruction of the famous magister Michael Maestlin, I perceived how clumsy in many respects is the hitherto customary notion of the structure of the universe. Hence, I was so very delighted by Copernicus, whom my teacher very often mentioned in his lectures, that I not only repeatedly advocated his views in disputations of the candidates [students], but also made a careful disputation about the thesis that the first motion [the revolution of the heaven of fixed stars] results from the rotation of the Earth.[8]

Thus, from the very beginning, Kepler was a passionate and convinced Copernican. His instruction at the university included both astronomy and astrology and he became an expert at casting horoscopes, which was to stand him in very good stead.

Kepler's serious studies began in 1595 when he asked some basic questions about the Copernican picture of the Solar System. Why are there only six planets in orbit about the Sun? Why are their distances so arranged? Why do they move more slowly if they are further away from the Sun? Kepler's approach is summarised in two quotations from Max Casper. First,

> He was possessed and enchanted by the idea of harmony.[9]

Second,

> Nothing in this world was created by God without a plan; this was Kepler's principal axiom. His undertaking was no less than to discover this plan of creation, to think the thoughts of God over again . . .[10]

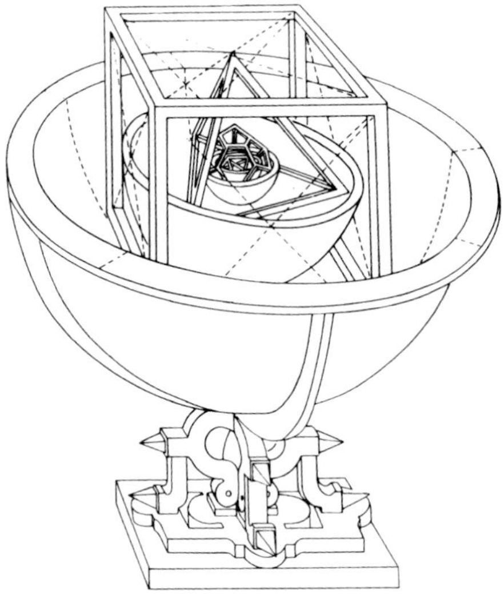

Figure 2.8: Kepler's model of nested polyhedra which he developed to account for the number of planets and their radial distances from the Sun. (From *Dictionary of Scientific Biography*, Vol. VII, 1973, p. 292, Charles Scribner's Sons, © 1970–80. Reprinted by permission of the Gale Group. After Kepler's original drawing in his *Mysterium Cosmographicum*, 1597.)

It was during his teaching of mathematics and geometry that the germ of an idea was planted. The moment of revelation is described in his book *Mysterium Cosmographicum* in his own inimitable style.

> Behold, reader, the invention and whole substance of this little book! In memory of the event, I am writing down for you the sentence in the words from that moment of conception: the Earth's orbit is the measure of all things: circumscribe around it a dodecahedron and the circle containing it will be Mars; circumscribe around Mars a tetrahedron, and the circle containing this will be Jupiter; circumscribe about Jupiter a cube and the circle containing this will be Saturn. Now inscribe within the Earth an icosahedron and the circle contained in it will be Venus; inscribe within Venus an octahedron, and the circle contained in it will be Mercury. You now have the reason for the number of planets.[11]

How does this work? It is a well-known fact of solid geometry that there are only five regular solids, the *five platonic solids*, in which all the edges are lines of equal length and in which the faces are all identical regular figures. Remarkably, by his choice of the order of the platonic solids Kepler was able to account for the radii of the planetary orbits to an accuracy of about 5%. In 1596, he explained his model of the solar system to the Duke of Würtemburg and designs were made to construct a real model of the solar system (Fig. 2.8), but it was never built. But Kepler went further. Copernicus had given no special physical significance to the Sun as the centre of the Solar System, but Kepler argued that the Sun was the origin of the forces which held the planets in their orbits.

Now many of Kepler's speculations are wrong and irrelevant to what follows. I seem to have devoted rather a lot of space to them, but there are two reasons for this. The first is that Kepler was seeking *physical* causes for the phenomena discovered by Copernicus. No-one had attempted to make this leap of the imagination before. The second is that this model is the first example of Kepler's fascination with harmonic and geometric principles.

With unbounded confidence and enthusiasm, Kepler published these ideas in 1597 in his *Mysterium Cosmographicum* (*The Mystery of the Universe*). He sent copies to many distinguished scientists of the day, including Tycho Brahe and Galileo. Galileo simply acknowledged receipt of the book, while Tycho Brahe was cautiously positive and encouraging in his response: he invited Kepler to come and work with him in Benatek.

The *Mysterium Cosmographicum* made a considerable impact upon astronomical thinking. Looking back much later, Kepler wrote of his book

> ...nearly all astronomical books which I published since that time have been related to some one of the main chapters in this little book, presenting themselves as its more detailed argument or perfection... The success which my book has had in the following years loudly testifies that no one ever produced a first book more deserving of admiration, more auspicious and, as far as its subject is concerned, more worthy.[12]

Kepler realised that what he needed to test his theory was much more accurate data on the orbits of the planets. The only person who had access to such data was Tycho Brahe. After various toings and froings, he ended up in Tycho's employ in 1600. There was a very great difference in outlook between Tycho and Kepler. When Kepler moved to Benatek, Tycho was 53 years old and Kepler 28. Tycho was the greatest astronomer of his time and of noble origin; Kepler was the greatest mathematician in Europe and of humble origin. Tycho wanted Kepler to work on the 'Tychonic' theory of the Solar System, whereas Kepler was already an ardent Copernican.

Just before Tycho died in 1601, he set Kepler to work on the problem of the orbit of Mars. On his deathbed, Tycho urged Kepler to complete a new set of astronomical tables to replace the Prutenic Tables. These were to be known as the 'Rudolphine Tables' in honour of the emperor Rudolph II, who had provided Tycho with the castle of Benatek as well as an enormous salary, 3000 gulden. Within two days of Tycho's death, Kepler was appointed Imperial Mathematician and the greatest period of his life's work began.

At first, Kepler assumed that the orbit of Mars was circular, as in the standard Copernican picture. His first discovery was that the motion of Mars could not be described by this model if it was referred to the centre of the Earth's orbit. Rather, the motion had to be referred to the true position of the Sun. This was an important advance. Kepler carried out an enormous number of calculations to try to fit the observed orbit of Mars to circular orbits, again following implicitly the precept that only circular motions should be used to describe the orbits of the planets. After a great deal of trial and error, the best orbits he could find still disagreed with the observations of Tycho Brahe by an error of 8 minutes of arc. This is where the knowledge of the errors in Tycho's observations were critical. As Kepler stated:

> Divine Providence granted to us such a diligent observer in Tycho Brahe that his observations convicted this Ptolemaic calculation of an error of 8 minutes of arc; it is only right that we should accept God's gift with a grateful mind... Because these 8 minutes of arc could not be ignored, they alone have led to a total reformation of astronomy.[13]

In other words, the random errors in Tycho's final determinations of the planetary orbits amounted to only about 1 to 2 minutes of arc, whereas the minimum discrepancy which

Kepler could find was at least four times this observational error. Before the time of Tycho, the random errors were about ten times greater and therefore Kepler would have had no problem in fitting these earlier observations to models involving circular orbits.

To paraphrase Kepler's more eloquent words, this disagreement was unacceptable and so he had to start again from the beginning. His next attempts to describe the solar system were based upon the use of ovoids (egg-shaped figures), in conjunction with a magnetic theory for the origin of the forces which hold the planets in their orbits. He found that it was very complicated and tedious to work out the orbits according to this magnetic theory, and so he adopted intuitively an alternative approach in which the motions of the planets are such that they sweep out equal areas in equal times. Whatever the actual shape of the orbit, the result is that the planet must move faster when it is closer to the Sun so that the area swept out by the line from the Sun to the planet is the same in equal time intervals. It turned out that this theory gives excellent predictions of the longitudes of the planets about the Sun and also of the Earth's orbit about the Sun. This great discovery is what we now know as *Kepler's second law of planetary motion*. Formally the statement is as follows:

Equal areas are swept out by the line from the Sun to a planet in equal times.

Kepler proceeded with the mammoth task of fitting ovoids and the areal law to the motion of Mars, but he could not obtain exact agreement, the minimum discrepancy amounting to about 4 minutes of arc, still outside the limits of Tycho's observational errors. In parallel with these researches, he was writing his treatise *A Commentary on the Motion of Mars* and he reached Chapter 51 before he realised that what he needed was a figure intermediate between an ovoid and a circle, an ellipse. He soon arrived at the key result that the orbit of Mars and indeed those of the other planets are ellipses with the Sun lying at one focus. The treatise on Mars was renamed *Astronomia Nova*, or *The New Astronomy*, with a subtitle *Based on Causes, or Celestial Physics*. It was published in 1609, four years after he had made the discovery of this law, which we now know as *Kepler's first law of planetary motion*. To state it formally,

The planetary orbits are ellipses with the Sun in one focus.

Notice the imaginative leap needed to place the Sun at the foci of the ellipses. Kepler already knew that the motion of Mars could not be referred to the centre of the Earth's orbit and so the focus is the next most obvious place. Kepler had discovered a crucial fact about the orbits of the planets, but he had no physical explanation for it. It turned out to be one of the key discoveries for the proper understanding of the law of gravity but it had to await the genius of Isaac Newton before its full significance was appreciated.

The next development was due to Galileo. We have to anticipate the great discoveries which he made in 1609 with his astronomical telescope. These were published in his book, the *Sidereus Nuncius* or the *Sidereal Messenger*, in 1610. Galileo was aware of the publication of Kepler's *Astronomia Nova* in the previous year and sent a copy of the *Sidereus Nuncius* to the Tuscan Ambassador at the Imperial Court in Prague, asking for a written opinion from Kepler. Kepler replied in a long letter on 19 April 1610, which he then published

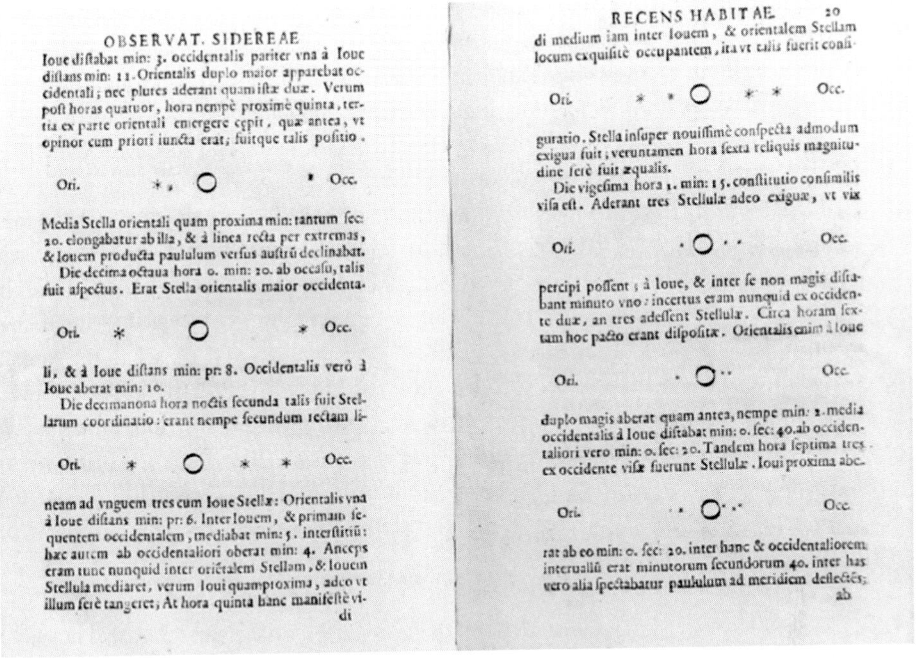

OBSERVAT. SIDEREAE

Ioue diſtabat min: 3. occidentalis pariter vna à Ioue diſtans min: 11. Orientalis duplo maior apparebat occidentali; nec plures aderant quam iſtæ duæ. Verum poſt horas quatuor, hora nempè proximè quinta, tertia ex parte orientali emergere cępit, quæ antea, vt opinor cum priori iuncta erat; fuitque talis poſitio.

Ori. Occ.

Media Stella orientali quam proxima min: tantum ſec: 20. elongabatur ab illa, & à linea recta per extremas, & Iouem producta paululum verſus auſtrũ declinabat.

Die decima octaua hora 0. min: 20. ab occaſu, talis fuit aſpectus. Erat Stella orientalis maior occidentali,

Ori. Occ.

li, & à Ioue diſtans min: pr: 8. Occidentalis verò à Ioue aberat min: 10.

Die decimanona hora noctis ſecunda talis fuit Stellarum coordinatio: erant nempe ſecundum rectam li-

Ori. Occ.

neam ad vnguem tres cum Ioue Stellæ: Orientalis vna à Ioue diſtans min: pr: 6. Inter Iouem, & primam ſequentem occidentalem, mediabat min: 5. interſtitiũ: hæc autem ab occidentaliori oberat min: 4. Anceps eram tunc nunquid inter orĩtalem Stellam, & Iouem Stellula mediaret, verum Ioui quamproxima, adeo vt illum ferè tangeret; At hora quinta hanc manifeſtè vidi

RECENS HABITAE. 20

di medium iam inter Iouem, & orientalem Stellam locum exquiſitè occupantem, ita vt talis fuerit confi-

Ori. Occ.

guratio. Stella inſuper nouiſſimè conſpecta admodum exigua fuit; veruntamen hora ſexta reliquis magnitudine ferè fuit æqualis.

Die vigeſima hora 1. min: 15. conſtitutio conſimilis viſa eſt. Aderant tres Stellulæ adeo exiguæ, vt vix

Ori. Occ.

percipi poſſent; à Ioue, & inter ſe non magis diſtabant minuto vno: incertus eram nunquid ex occidente duæ, an tres adeſſent Stellulæ. Circa horam ſextam hoc pacto erant diſpoſitæ. Orientalis enim à Ioue

Ori. Occ.

duplo magis aberat quam antea, nempe min: 2. media occidentalis à Ioue diſtabat min: 0. ſec: 40. ab occidentaliori vero min: 0. ſec: 20. Tandem hora ſeptima tres ex occidente viſæ fuerunt Stellulæ. Ioui proxima abe-

Ori. Occ.

rat ab eo min: 0. ſec: 20. inter hanc & occidentaliorem interuallũ erat minutorum ſecundorum 40. inter has vero alia ſpectabatur paululum ad meridiem deflectẽs; ab

Figure 2.9: Two pages from Galileo's *Sidereus Nuncius*, showing his drawings of the movements of the four Galilean satellites. (Courtesy of the Royal Observatory, Edinburgh.)

in May under the title *Dissertatio cum Nuncio Siderio* or *Conversation with the Sidereal Messenger*. As might be imagined, Kepler was wildly enthusiastic and, while Galileo presented his observations with some cautious attempts at interpretation, Kepler gave full rein to his imagination.

The most important discoveries for Kepler were the moons of Jupiter. These fitted in beautifully with the Copernican picture. Here was a miniature solar system: the four brightest moons of Jupiter – Io, Europa, Ganymede and Callisto – were seen orbiting the planet. In the *Sidereus Nuncius*, Galileo shows the motions of the four satellites – examples of the diagrams appearing in the *Sidereus Nuncius* are shown in Fig. 2.9. Kepler was in full flow. Here is the part of his letter explaining why Jupiter has moons:

The conclusion is quite clear. Our moon exists for us on Earth, not for the other globes. Those four little moons exist for Jupiter, not for us. Each planet in turn, together with its occupants, is served by its own satellites. From this line of reasoning we deduce with the highest degree of probability that Jupiter is inhabited.[14]

What can one say? It is intriguing that this form of extreme lateral thinking was part of the personality of the greatest mathematician in Europe at the time.

Kepler's third law was deeply buried in his treatise, *Harmonices Mundi* or *The Harmony of the World*. According to Max Casper, this work, published in 1619, was his crowning achievement, in which he attempted to synthesise all his ideas into one harmonious picture of the Universe. His harmonic theory was to encompass geometry, music, architecture, metaphysics, psychology, astrology and astronomy, as can be seen from the contents page of the five books which make up the treatise (Fig. 2.10). In modern terms, this was his 'grand unified theory', a concept which has haunted much of physics throughout the centuries.

By 1619, Kepler was no longer satisfied with an accuracy of 5% in the comparison of the radii of the planetary orbits with his harmonic theory. He now had much more accurate mean radii for the orbits of the planets and suddenly, at the time he had reached the writing of Book V, Chapter III, Eighth Division, of the *Harmony of the World*, he discovered what is now known as *Kepler's third law of planetary motion*:

> The period of a planetary orbit is proportional to the three-halves power of the mean distance of the planet from the Sun.

This was the crucial discovery which eventually led to Newton's law of gravity. Notice that this law is somewhat different from the solution he had proposed in the *Mysterium Cosmographicum*. It is not in conflict with it, however, since there is nothing in Kepler's third law to tell us why the planets have to lie at particular distances from the Sun.

It is important to note that Kepler's discoveries of the law of equal areas and of the ellipses which describe the paths of the planets, as well as the third law, were intuitive leaps of the imagination rather than the result of following some prescribed set of standard mathematical techniques.

It might be thought that the support of Kepler, a passionate Copernican, would have been invaluable to Galileo during his prosecution for advocating the Copernican picture of the world. As we will see, matters were not quite as simple as this. Galileo was rather cautious about Kepler's support and we can identify at least two aspects of his concerns. Two years after Kepler's death on 15 November 1630, Galileo wrote:

> I do not doubt but that the thoughts of Landsberg and some of Kepler's tend rather to the diminution of the doctrine of Copernicus than to its establishment as it seems to me that these (in the common phrase) have wished too much for it . . .[15]

We have already quoted Kepler's purple prose, which indicates part of the worry. It is difficult to be taken seriously as a research scientist if you spend time worrying about spoon-bending, unidentified flying objects, corn-circles and the like. In addition, the strong mystical streak in the *Harmony of the World* would not have appealed to Galileo, who was attempting to set the whole of natural philosophy on a secure mathematical foundation.

Another cause of Galileo's concerns is of the greatest interest. Again quoting his words:

> It seems to me that one may reasonably conclude that for the maintenance of perfect order among the parts of the Universe, it is necessary to say that movable bodies are movable only circularly.[16]

Kepler's assertion that the orbits of the planets were ellipses rather than circles was intellectually repugnant to Galileo. We can recognise again that this was no more than the

Ioannis Keppleri

HARMONICES
MVNDI

LIBRI V. QVORVM

Primus GEOMETRICVS, De Figurarum Regularium, quæ Proportiones Harmonicas conſtituunt, ortu & demonſtrationibus.

Secundus ARCHITECTONICVS, ſeu ex GEOMETRIA FIGVRATA, De Figurarum Regularium Congruentia in plano vel ſolido:

Tertius propriè HARMONICVS, De Proportionum Harmonicarum ortu ex Figuris; deque Naturâ & Differentiis rerum ad cantum pertinentium, contra Veteres:

Quartus METAPHYSICVS, PSYCHOLOGICVS & ASTROLOGICVS, De Harmoniarum mentali Eſſentiâ earumque generibus in Mundo; præſertim de Harmonia radiorum, ex corporibus cœleſtibus in Terram deſcendentibus, eiuſque effectu in Natura ſeu Anima ſublunari & Humana:

Quintus ASTRONOMICVS & METAPHYSICVS, De Harmoniis abſolutiſſimis motuum cœleſtium, ortuque Eccentricitatum ex proportionibus Harmonicis.

Appendix habet comparationem huius Operis cum Harmonices Cl. Ptolemæi libro III. cumque Roberti de Fluctibus, dicti Flud. Medici Oxonienſis ſpeculationibus Harmonicis, operi de Macrocoſmo & Microcoſmo inſertis.

Cum S. C. M^tis. Priuilegio ad annos XV.

Lincii Auſtriæ,

Sumptibus GODOFREDI TAMPACHII Bibl. Francof.
Excudebat IOANNES PLANCVS.

ANNO M. DC. XIX.

Figure 2.10: The table of contents of Kepler's treatise *The Harmony of the World*, 1619, Linz. (From the Crawford Collection, Royal Observatory, Edinburgh.)

legacy of Aristotelian physics, according to which the only allowable motions are uniform linear and circular motions. We would now say that this view was an unexamined prejudice on the part of Galileo, but we should not disguise the fact that we all make similar judgements in our own work.

Eventually Kepler completed the Rudolphine Tables, and they were published in September 1627. These set a new standard of accuracy in the prediction of solar, lunar and planetary positions. It is interesting that, in order to simplify the calculations, Kepler had invented his own form of logarithms. He had seen John Napier's *Mirifici Logarithmorum Canonis Descriptio* of 1614 as early as 1617 – this was the first set of tables of natural logarithms.

Although it was not recognised at the time, Kepler's three laws of planetary motion were to be crucial for Newton's great synthesis of the laws of gravity and celestial mechanics. But there was another giant, whose contributions were even more profound – Galileo Galilei. He has become a symbol of the birth of modern science and the struggle against received dogma. The trial of Galileo strikes right to the very heart of the modern concept of scientific method. But he contributed much more and these are the topics with which we will grapple in the next chapter.

2.5 References

1 Pedersen, O. and Pihl, M. (1974). *Early Physics and Astronomy*, p. 64. London: McDonald and Co.
2 Pedersen, O. and Pihl, M. (1974). *Op. cit.*, p. 65.
3 For a translation, see Duncan, A.M. (1976). *Copernicus: On the Revolutions of the Heavenly Spheres. A New Translation from the Latin*. London: David and Charles, New York: Barnes and Noble Books.
4 Hellman, D.C. (1970). *Dictionary of Scientific Biography*, Vol. 11, p. 403. New York: Charles Scribner's Sons.
5 Dreyer, J.L.E. (1890). *Tycho Brahe. A Picture of Scientific Life and Work in the Sixteenth Century*, pp. 86–7. Edinburgh: Adam and Charles Black.
6 Christianson, J. (1961). *Scientific American*, **204**, 118 (February issue).
7 Thoren, V.E. (1990). *The Lord of Uraniborg. A Biography of Tycho Brahe*, pp. 188–9. Cambridge: Cambridge University Press.
8 Casper, M. (1959). *Kepler*, trans. C. Doris Hellman, pp. 46–7. London and New York: Abelard-Schuman.
9 Casper, M. (1959). *Op. cit.*, p. 20.
10 Casper, M. (1959). *Op. cit.*, p. 62.
11 Kepler, J. (1596). From *Mysterium Cosmographicum*. See *Kepleri Opera Omnia*, ed. C. Frisch, Vol. 1, pp. 9ff.
12 Casper, M. (1959). *Op. cit.*, p. 71.
13 Kepler, J. (1609). From *Astronomia Nova*. See *Johannes Keplers Gesammelte Werke*, ed. M. Casper, Vol. III, p. 178, Munich: Beck (1937).

14 Kepler, J. (1610). In *Conversation with Galileo's Sidereal Messenger*, ed. and trans. E. Rosen (1965), p. 42. New York and London: Johnson Reprint Co.
15 Galilei, G. (1630). Quoted by Rupert Hall, A. (1970), *From Galileo to Newton 1630–1720. The Rise of Modern Science 2*, p. 41, London: Fontana Science.
16 Galilei, G. (1632). *Dialogues concerning the Two Chief Systems of the World*, trans. S. Drake, p. 32, Berkeley (1953).

3 Galileo and the nature of the physical sciences

3.1 Introduction

There are three separate but linked stories to be told. The first concerns Galileo as natural philosopher. Unlike Tycho Brahe the observer and Kepler the mathematician, Galileo was an experimental physicist whose prime concern was understanding the laws of nature in quantitative terms, from his earliest writings to his final great treatise *Discourse and Mathematical Demonstrations concerning Two New Sciences*.

The second story is astronomical, and occupies a relatively small, but crucial, period of Galileo's career, from 1609 to 1612, during which time he made a number of fundamental astronomical discoveries which had a direct impact upon his understanding of the physics of motion.

The third story, and the most famous of all, is his trial and subsequent house arrest, which continues to be the subject of considerable controversy. The scientific aspects of his censure and subsequent trial are of the greatest interest and strike right at the heart of the nature of the physical sciences. The widespread view is to regard Galileo as the hero and the Catholic Church as the villain of the piece, a source of conservative reaction and bigoted authority. From the methodological point of view Galileo made an logical error, but the church authorities made a much more disastrous blunder, which has resonated through science and religion ever since, and which was only officially acknowledged by Pope John Paul II in the 1980s.

My reasons for devoting a whole chapter to Galileo, his science and his tribulations are that it is a story which needs to be better known and which has resonances for the way in which physics as a scientific discipline is carried out today. Galileo's intellectual integrity and scientific genius are an inspiration – more than anyone else, he created the intellectual framework for the development of physics as we know it.

3.2 Galileo as an experimental physicist

Galileo Galilei was the son of Vincenzio Galileo, a distinguished musician and musical theorist, and was born in February 1564 in Pisa. In 1587, he was appointed to the chair of mathematics at the University of Pisa, where he was not particularly popular with his colleagues. One of the main causes was Galileo's opposition to Aristotelian physics, which remained the central pillar of natural philosophy. It was apparent to Galileo that Aristotle's physics was not in accord with the way in which matter actually behaves. For example, Aristotle's assertion concerning the fall of bodies of different weights reads as follows:

> If a certain weight moves a certain distance in a certain time, a greater weight will move the same distance in a shorter time, and the proportion which the weights bear to each other the times too will bear to one another; for example, if the half weight covers the distance in x, the whole weight will cover it in $x/2$.[1]

This is just wrong, as could have been demonstrated by a simple experiment – it seems unlikely that Aristotle ever tried the experiment himself. Galileo's objection is symbolised by the story of his dropping different weights from the Leaning Tower of Pisa. If different weights are dropped through the same height, they take the same time to reach the ground if the effects of air resistance are neglected, as was known to Galileo and earlier writers.

In 1592, Galileo was appointed to the chair of mathematics at Padua, where he was to remain until 1610. It was during this period that he produced his greatest work. Initially, he was opposed to the Copernican model of the solar system but, in 1595, he began to take it seriously in order to explain the origin of the tides in the Adriatic. He observed that the tides at Venice typically rise and fall by about five feet and therefore there must be quite enormous forces to cause this huge amount of water to be raised each half-day at high tide. Galileo reasoned that if the Earth rotated on its own axis and also moved in a circular orbit about the Sun then the changes in the direction of travel of a point in the surface of the Earth would cause the sea to slosh about and so produce the effect of the tides. This is not the correct explanation for the tides, but it led Galileo to favour the Copernican picture for physical reasons.

In Galileo's printed works, the arguments are given entirely in the abstract without reference in the conventional sense to experimental evidence. Galileo's genius as a pioneer scientist is described by Stillman Drake in his remarkable book *Galileo: Pioneer Scientist* (1990).[1] Drake deciphered Galileo's unpublished notes, which are not set down in any systematic way, and convincingly demonstrated that Galileo actually carried out the experiments to which he refers in his treatises with considerable experimental skill (Fig. 3.1).

Galileo's task was enormous – he disbelieved the basis of Aristotelian physics, but had no replacement for it. In the early 1600s, he undertook experimental investigations of the laws of free fall, the motion of balls rolling down slopes and the motion of pendulums; his results clarified the concept of acceleration for the first time.

A problem with physics up to the time of Galileo was that there was no way of measuring short time intervals accurately, and so he had to use considerable ingenuity in the design of his experiments. A very nice example is his experiment to investigate how a ball accelerates down a slope. He constructed a long shallow slope of length 2 metres at an angle of only 1.7° to the horizontal and cut a grove in it down which a heavy bronze ball could roll. He placed little frets on the slope so that there would be a little click as the ball passed over each fret. He then adjusted the positions of the frets along the slope so that the clicks occurred at equal time intervals (Fig. 3.2). Drake suggests that he could have equalised the time intervals to about 1/64 of a second by singing a rhythmic tune and making the clicks occur at equal beats in the bar. In view of Galileo's father's profession, this seems quite plausible.

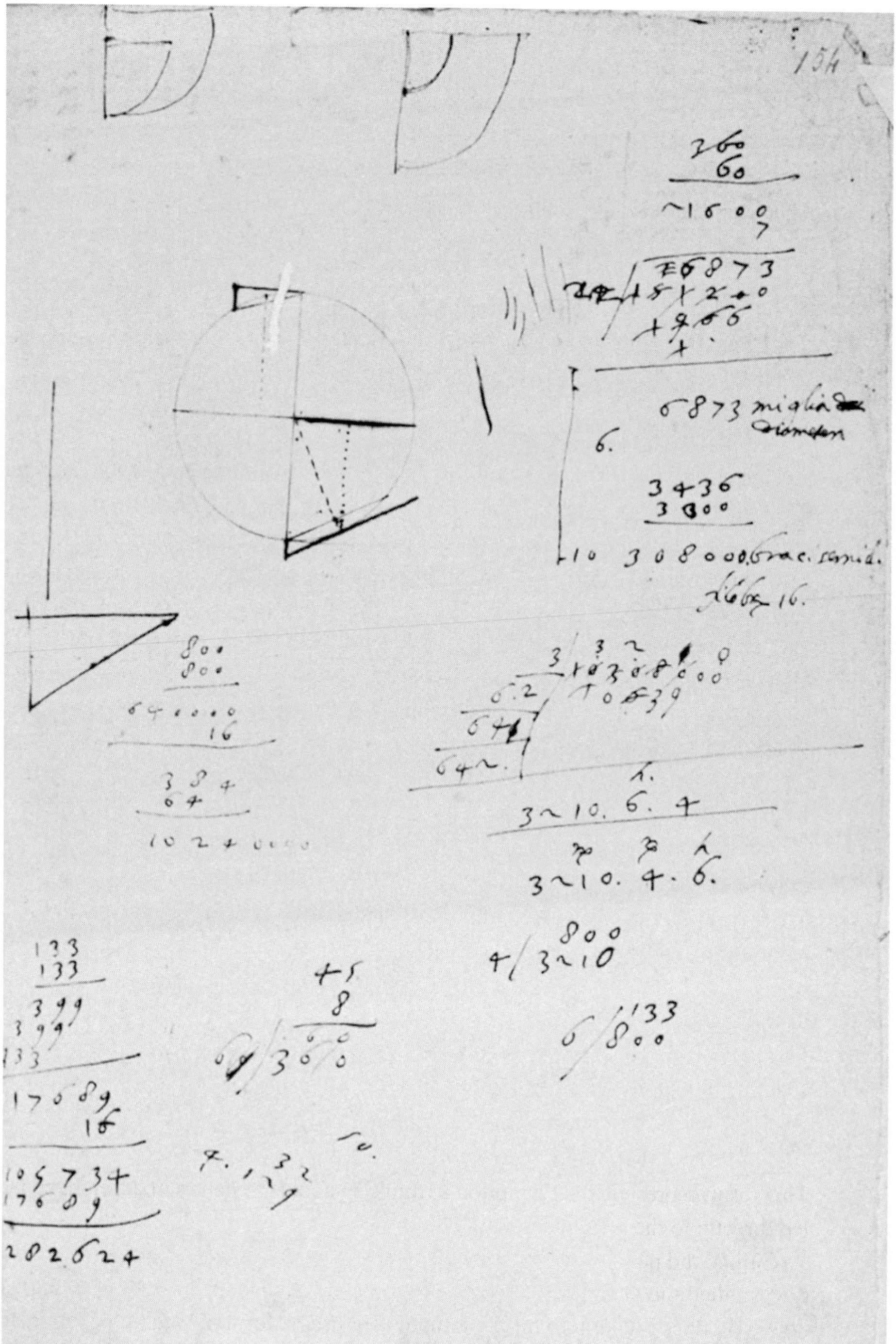

Figure 3.1: Part of Galileo's notes concerning the laws of the pendulum. (From S. Drake, 1990, *Galileo: Pioneer Scientist*, p. 19, Toronto: University of Toronto Press.)

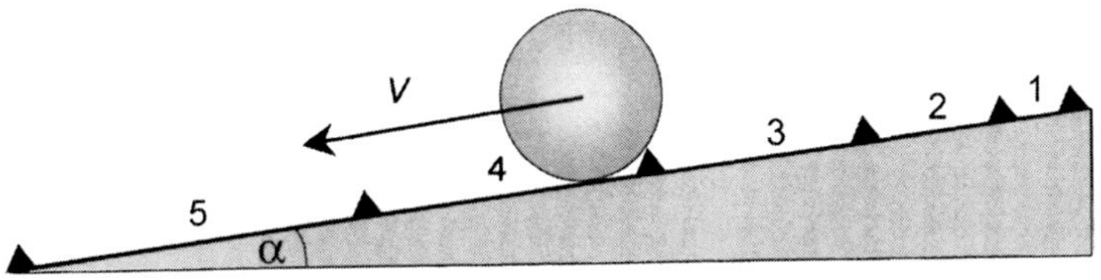

Figure 3.2: How Galileo established the law of motion under uniform acceleration. The numbers between the frets show their relative positions in order to produce a regular sequence of clicks.

By this means, he was able to measure the distance travelled as the ball rolled continuously down the slope and, by taking differences, he could work out the average speed between successive frets. He found that the speed increased as the odd numbers $1, 3, 5, 7, \ldots$ in equal time intervals.

Originally, Galileo had believed that, under constant acceleration, speed is proportional to distance travelled but, as a result of these precise experiments of 1604, he found, rather, that speed is proportional to time. He now had two relations: the first was the definition of speed, $x = vt$, and the second related speed to time under constant acceleration, $v = at$. Now, there is no algebra in Galileo's published works and the differential calculus had yet to be discovered. Suppose the speeds of a uniformly accelerated sphere are measured at times 0, 1, 2, 3, 4, 5 seconds (Fig. 3.2). Assume the sphere starts from rest at time 0. The speeds at the above times will be, say, $0, 1, 2, 3, 4, 5, \ldots$ cm s^{-1}, an acceleration of 1 cm s^{-2}. How far has the sphere travelled after 0, 1, 2, 3, 4, 5 seconds?

At zero time, no distance has been travelled. Between 0 and 1 s, the average speed is 0.5 cm s^{-1} and so the distance travelled must be 0.5 cm. In the next interval, between 1 and 2 s, the average speed is 1.5 cm s^{-1} and so the distance travelled in that interval is 1.5 cm; the total distance travelled from the position of rest is now $0.5 + 1.5 = 2$ cm. In the following interval, the average speed is 2.5 cm s^{-1}, the distance travelled is 2.5 cm and the total distance is 4.5 cm, and so on. We thus obtain a series of distances, 0, 0.5, 2, 4.5, 8, 12.5, ... cm, which can be written in cm as

$$\tfrac{1}{2}(0,\ 1,\ 4,\ 9,\ 16,\ 25, \ldots) = \tfrac{1}{2}(0,\ 1^2,\ 2^2,\ 3^2,\ 4^2,\ 5^2, \ldots). \tag{3.1}$$

This is Galileo's famous *time-squared law* for uniformly accelerated motion, expressed algebraically as

$$x = \tfrac{1}{2}at^2. \tag{3.2}$$

This result represented a revolution in thinking about the nature of accelerated motion and led directly to the Newtonian revolution.

Galileo did not stop there but went on to carry out two further brilliant experiments. He next studied the question of free fall, namely, if an object is dropped from a given height, how long does it take it to hit the ground? He used a form of water clock to measure time intervals accurately. Water was allowed to pour out of a tube at the bottom of a large vessel, kept full; the amount of water which flowed out was a measure of the time interval. By dropping objects from different heights, Galileo established that freely falling objects obey the time-squared law – in other words, when objects fall freely they experience a constant acceleration, the *acceleration due to gravity*.

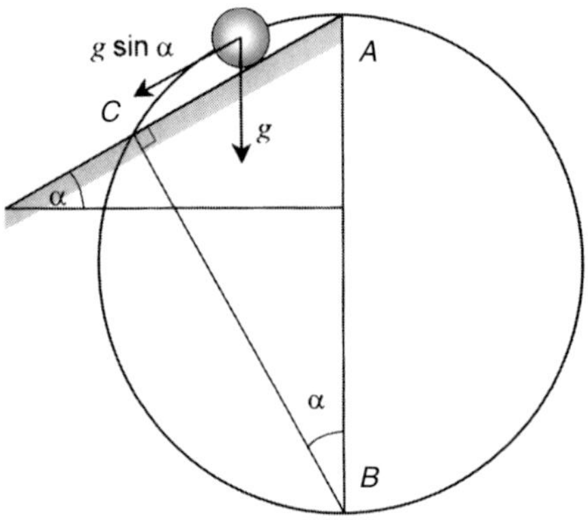

Figure 3.3: How Galileo established the theorem known by his name.

Having established these two results, he sought a relation between them. The desired relation, *Galileo's theorem*, is very beautiful. Suppose a body is dropped freely through a certain distance l, which is represented by the length AB in Fig. 3.3. Construct a circle whose diameter is AB. Now suppose the body slides without friction down an inclined plane and, for convenience, the top of the plane is placed at the point A. Galileo's theorem states that:

> The time it takes a body to slide down the slope from the point A to the point C, where the slope cuts the circle, is equal to the time it takes the body to fall freely from A to B.

In other words, the time it takes a body to fall along any *chord of a circle* is the same as the time it takes the body to fall freely down the diameter of the circle. The component of the acceleration due to gravity is $g \sin\alpha$ as the body slides down the slope; the component of acceleration perpendicular to the slope is zero (Fig. 3.3).

Now, any triangle constructed on the diameter of a circle and with its third point lying on the circle is a right-angled triangle. Therefore, we can equate angles, as shown in Fig. 3.3, from which it is apparent that AC/AB, the ratio of the distances travelled, is equal to $\sin\alpha$. Since, for equal times, the distance travelled is proportional to the acceleration, $x = \frac{1}{2}at^2$, this proves Galileo's theorem.

The next piece of genius was to recognise the relation between these deductions and the properties of swinging pendulums. As a youth, Galileo is said to have noticed that the period of the swing of a chandelier suspended in a church is independent of the *amplitude* of its swing. Galileo made use of his law of chords of a circle to explain this observation. If the pendulum is long enough, the arc AC described by the pendulum is almost exactly equal to the chord across the circle joining the extreme point of swing of the pendulum to the lowest point (Fig. 3.4). Inverting Fig. 3.3, it is therefore obvious why the period of the pendulum is independent of the amplitude of its swing – according to Galileo's theorem, the time to travel along any chord drawn to A will be the same as the time it takes the body to fall freely down twice the length of the pendulum. This is really brilliant physics.

What Galileo had achieved was to put into mathematical form the nature of acceleration under gravity. This had immediate practical application, because he could now work out the trajectories of projectiles. They travel with constant speed parallel to the ground and are accelerated by gravity in the vertical direction. For the first time, he was able to work out the parabolic paths of cannon balls and other projectiles (Fig. 3.5).

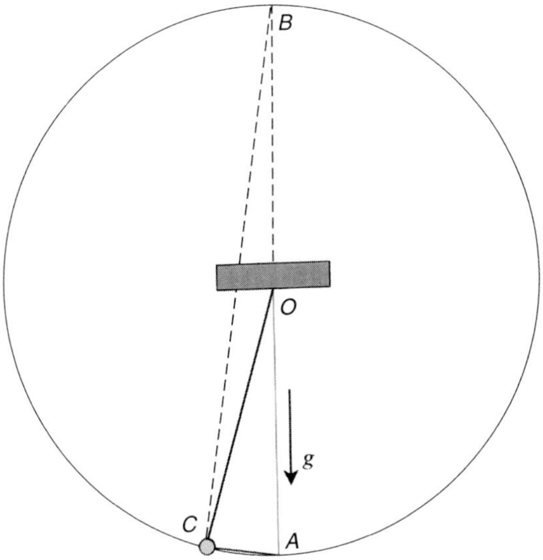

Figure 3.4: How Galileo showed that the period of a long pendulum is independent of the amplitude of the swing. Note the relation to Fig. 3.3.

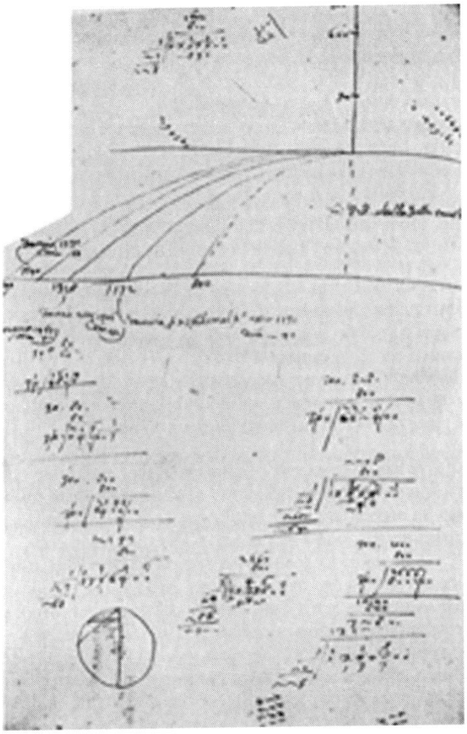

Figure 3.5: A page from Galileo's notebooks showing the trajectories of projectiles under the combination of acceleration under gravity and constant horizontal speed. (From S. Drake, 1990, *Galileo: Pioneer Scientist*, p. 107, Toronto: University of Toronto Press.)

Galileo began writing a systematic treatment of all these topics, showing how they could all be understood on the basis of the law of the constant acceleration; in 1610, in his own words, he was planning to write:

> ... three books on mechanics, two with demonstrations of its principles, and one concerning its problems; and though other men have written on the subject, what has been done is not one quarter of what I write, either in quantity or otherwise.[2]

Later he writes in the same vein:

> ... three books on local motion – an entirely new science in which no one else, ancient or modern, has discovered any of the most remarkable laws which I demonstrate to exist in both natural and violent movement; hence I may call this a new science and one discovered by me from its very foundations.[3]

The publication of these discoveries was delayed until the 1620s and 1630s. He was diverted from this task by news of the invention of the telescope. This was the beginning of his serious study of astronomy.

3.3 Galileo's telescopic discoveries

The invention of the telescope is attributed to the Dutch lens-grinder Hans Lipperhey, who in October 1608 applied to Count Maurice of Nassau for a patent for a device which could make distant objects appear closer. Galileo heard of this invention in July 1609 and set about building one for himself. By August, he had succeeded in constructing a telescope which magnified nine times, a factor three better than that patented by Lipperhey. This greatly impressed the Venetian Senate, who understood the importance of such a device for a maritime nation. Galileo was immediately given a lifetime appointment at the University of Padua at a vastly increased salary.

By the end of 1609, he had made a number of telescopes of increasing magnifying power, culminating in a telescope with a magnifying power of 30. In January 1610, he first turned his telescopes on the skies and immediately there came a flood of remarkable discoveries. These were rapidly published in March 1610 in his *Sidereus Nuncius* or *The Sidereal Messenger*.[4] In summary, the discoveries were:

(i) the Moon is mountainous rather than a perfectly smooth sphere (Fig. 3.6(*a*));
(ii) the Milky Way consists of vast numbers of stars rather than being a uniform distribution of light (Fig. 3.6(*b*));
(iii) Jupiter has four satellites, whose motions can be followed over several complete orbits in a matter of weeks (Fig. 2.9).

The book caused a sensation throughout Europe and Galileo won immediate international fame. These discoveries demolished a number of Aristotelian precepts which had been accepted over the centuries. For example, the resolution of the Milky Way into individual stars was quite contrary to the Aristotelean view. In the satellites of Jupiter, Galileo saw a prototype for the Copernican picture of the Solar System. The immediate effect of these

(*a*)

(*b*)

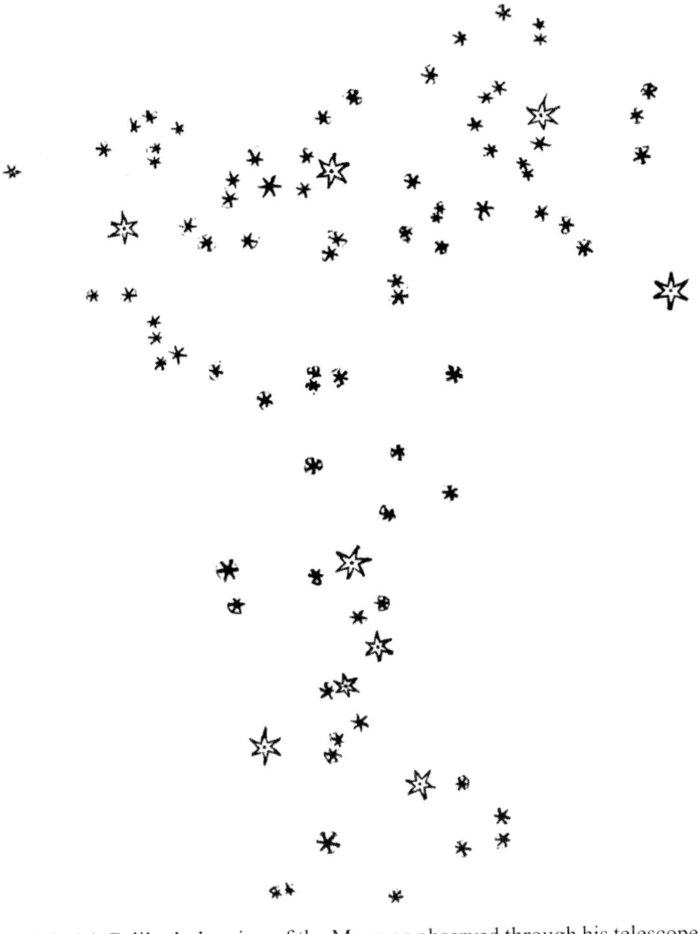

Figure 3.6: (*a*) Galileo's drawing of the Moon as observed through his telescope. (*b*) Galileo's sketch of the region of sky in the vicinity of Orion's belt, showing the resolution of the background light into faint stars. (From G. Galilei, 1610, *Sidereus Nuncius*, Venice. See also the translation by A. van Helden, 1989, Chicago: University of Chicago Press.)

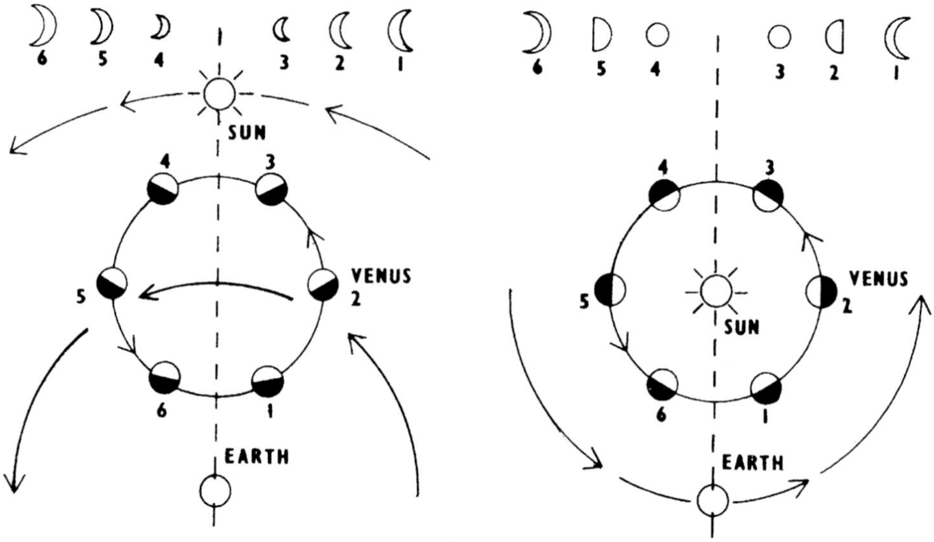

Figure 3.7: Illustrating the phases of Venus, according to the geocentric and heliocentric pictures of the structure of the Solar System. (From A. van Helden, 1989, *Sidereus Nuncius, or The Sidereal Messenger*, p. 108, Chicago: University of Chicago Press.)

discoveries was that Galileo was appointed Mathematician and Philosopher to the Grand Duke of Tuscany, Cosimo de Medici, to whom the *Sidereus Nuncius* was dedicated.

Later in 1610, he made two other crucial telescopic discoveries:

(iv) the rings of Saturn, which he took to be close satellites of the planet;
(v) the phases of the planet Venus.

This last discovery was of the greatest importance. When Venus is on the far side of its orbit with respect to the Earth, it appears circular but when it is on the same side of the Sun as the Earth, it looks like a crescent Moon. This was interpreted as evidence in favour of the Copernican picture because it is explained completely naturally if Venus and the Earth both orbit the Sun, the latter being the source of their illumination (Fig. 3.7). If, however, Venus moved on an epicycle about a circular orbit around the Earth and the Sun moved on a more distant sphere then the pattern of illumination relative to the Earth would be quite different, as illustrated in Fig. 3.7. In 1611 these great discoveries were presented by Galileo to the Pope and several cardinals, who were all favourably impressed by them. Galileo was elected to the Academia Lincei.

3.4 The trial of Galileo – the heart of the matter

Before recounting the events which led up to Galileo's appearance before the Inquisition and his conviction for the second most serious crime in the papal system of justice, let us summarise briefly some of the different facets of the debate between the Ptolemaeans, the

Copernicans and the church authorities; Finocchiaro provides an excellent summary in his documentary history *The Galileo Affair* (1989).[5] The established laws of physics remained in essence Aristotelian and only a few adventurous spirits doubted the basic correctness of the Ptolemaic system. There were problems with the Copernican picture and so Galileo *had* to become involved in these issues because they undermined his new-found understanding of the laws of motion.

3.4.1 *The issues*

The physical issues centred on these questions. First, does the Earth rotate on its axis with respect to the fixed stars? Second, do the Earth and the planets orbit the Sun? Specifically, is the Earth in motion? This latter concept was referred to as the *geokinetic hypothesis*. Finocchiaro summarises the pre-Galilean objections to this hypothesis under five headings.

(i) *The deception of the senses*. None of our senses gives us any evidence that the Earth is moving in an orbit about the Sun. If this were a fact of nature, surely it would be of such importance that our senses would make us aware of it.

(ii) *Astronomical problems*. First, the heavenly bodies were believed to be composed of different forms of matter from the material of the Earth. Second, Venus should show phases similar to the Moon if it were in orbit about the Sun. Third, if the Earth moved, why didn't the stars exhibit parallaxes?

(iii) *Physical arguments*. These were largely based upon Aristotelian physics and we have encountered some of them already.

(a) If the Earth moves, falling bodies should not fall vertically. Many counter-examples could be given – rain falls vertically, objects thrown vertically upwards fall straight down again, and so on. This was in contrast to the trajectory of an object dropped from the top of a ship's mast when a ship is in motion. In this case, the object does not fall vertically downwards, with respect to an observer on the shore.
(b) Projectiles fired in the direction of rotation of the Earth and in the opposite direction would have different trajectories. No such difference had been observed.
(c) Objects placed on a rotating potter's wheel are flung off if they are not held down. This was called the *extruding power of whirling*, what is now known as the *centrifugal force*. The same phenomenon should occur if the Earth is in a state of rotation, but we are not flung off the surface of the Earth.
(d) Next, there were purely philosophical arguments. Two forms of motion, uniform motion in a straight line and uniform circular motion, were thought to be the only 'natural' motions which objects could have. Objects must either fall in a straight line to the centre of the Universe or be in a state of uniform circular motion. We have already discussed the question whether objects fall towards the centre of the Earth or towards the Sun. Furthermore, according to Aristotelian physics, a simple body could have only one natural motion. But, according to Copernicus, objects dropped on Earth have *three* motions – downward motion under free fall, motion due to the rotation of the Earth on its axis and motion in a circular orbit about the Sun.
(e) Finally, if Aristotelian physics was to be rejected, what was there to replace it? The Copernicans had to provide a better theory and none was available.

(iv) *The Authority of the Bible*. There are no absolutely unambiguous statements in the Bible that assert that the Earth is stationary at the centre of the Universe. According to Finocchiaro, the most relevant statements[6] are as follows:

(a) Psalm 104:5. 'O Lord my God . . . who laid the foundations of the Earth, that it should not be removed forever.'
(b) Ecclesiastes 1:5. 'The Sun also riseth, and the Sun goeth down, and hasteth to the place where he ariseth.'
(c) Joshua 10:12,13. 'Then spake Joshua to the Lord in the day when the Lord delivered up the Amorites before the children of Israel, and he said in the sight of Israel, "Sun, stand thou still upon Gibeon; and thou, Moon, in the valley of Ajalon." And the Sun stood still, and the Moon stayed, until the people had avenged themselves upon their enemies.'

These are rather oblique references and it is intriguing that the Protestants were much more virulently anti-Copernican than the Catholics because of their belief in the literal truth of the Bible. The Catholic theologians took a more sophisticated and flexible interpretation of holy writ. However, the concept that the Earth is stationary at the centre of the Universe had also been the conclusion of the Church Fathers – the saints, theologians and churchmen who codified Christianity. To quote Finocchiaro,

> The argument claimed that all Church Fathers were unanimous in interpreting relevant Biblical passages . . . in accordance with the geostatic view; therefore the geostatic system is binding on all believers, and to claim otherwise (as Copernicus did) is heretical.[7]

(v) The most interesting argument from our perspective concerns the *hypothetical nature of Copernican theory*. It strikes at the very heart of the nature of the natural sciences. The crucial point is how we express statements concerning the success of the Copernican model. A correct statement is: '*If* the Earth rotates on its axis and moves in a circular orbit about the Sun, and if the other planets also orbit the Sun, *then* we can describe simply and elegantly the observed motions of the Sun, Moon and planets on the celestial sphere.' What we *cannot* do logically is to reverse the argument and state that *because* the planetary motions are explained simply and elegantly by the Copernican hypothesis *therefore* the Earth must rotate and move in a circular orbit about the Sun. This is an elementary error of logic, because there might well be quite different reasons why the Copernican model was successful. The key point is the difference between *induction* and *deduction*. Owen Gingerich[8] gives a pleasant example. A *deductive* sequence of arguments might run:

(a) If it is raining, the streets are wet.
(b) It is raining.
(c) Therefore, the streets are wet.

There is no problem here. But, now reverse (b) and (c) and we get into trouble.

(a) If it is raining, the streets are wet.
(b) The streets are wet.
(c) Therefore, it is raining.

This second line of reasoning is obviously false since the streets could be the streets of Venice, or could have been newly washed. In other words, you cannot prove anything about the absolute truth of statements in this way. This type of reasoning, in which we attempt to find general laws from specific pieces of evidence is called *induction*. All the physical sciences are to a greater or lesser extent based on induction, and so physical laws necessarily have a provisional, hypothetical nature. This was seen as in marked contrast to the absolute certainty of God's word as contained in the holy scriptures and its interpretation as dogma by the Church Fathers. According to Owen Gingerich,[8] this was the issue of substance which led to the trial and censure of Galileo – the juxtaposition of the hypothetical world picture of Copernicus with the truth as revealed in the Bible.

3.4.2 *The Galileo affair*

Prior to his great telescopic discoveries of 1610–11, Galileo was at best a cautious Copernican, but it gradually became apparent to him that his new understanding of the nature of motion eliminated the *physical* and *astronomical problems* listed under (ii) and (iii) above. The new evidence was consistent with the Copernican model; specifically, there are mountains on the Moon, just as there are on Earth, suggesting that the Earth and the Moon are similar bodies and the phases of Venus are exactly as would expected according to the Copernican picture. Thus, the physical and astronomical objections to Copernicanism could be discarded, leaving only the logical and theological problems to be debated.

As the evidence began to accumulate in favour of Copernicanism, conservative scientists and philosophers had to rely more and more upon the theological, philosophical and logical arguments. In December 1613, the Grand Duchess Dowager Christina asked Castelli, one of Galileo's friends and colleagues, about the religious objections to the motion of the Earth. Castelli responded to the satisfaction of both the Duchess and Galileo, but Galileo felt the need to set out the arguments in more detail. He suggested that there were three fatal flaws in the theological argument. To quote Finoccharo:

> First, it attempts to prove a conclusion [the Earth's rest] on the basis of a premise [the Bible's commitment to the geostatic system] which can only be ascertained with a knowledge of that conclusion in the first place . . . the business of Biblical interpretation is dependent on physical investigation, and to base a controversial physical conclusion on the Bible is to put the cart before the horse. Second, the Biblical objection is a *non sequitur*, since the Bible is an authority only in matters of faith and morals, not in scientific questions . . . Finally, it is questionable whether the Earth's motion really contradicts the Bible.[9]

This letter to the Grand Duchess Christina circulated privately and came into the hands of the conservatives. Sermons were delivered attacking the heliocentric picture and accusing its proponents of heresy. In March 1615, the Dominican friar Tommaso Caccini, who had already preached against Galileo, laid a formal charge of *suspicion of heresy* against Galileo before the Roman Inquisition. This charge was less severe than that of *formal heresy*, but was still a serious one. The Inquisition manual stated that, 'Suspects of heresy are those who occasionally utter propositions that offend the listeners . . . Those who keep, write, read, or give others to read books forbidden in the Index . . .' Further, there were two types of suspicion of heresy, *vehement* and *slight* suspicion of heresy, the former being considerably more serious than the latter. Once an accusation was made, there was a formal procedure which had to be followed.

Galileo responded by seeking the support of his friends and patrons and circulated three long essays privately. One of these repeated the arguments concerning the validity of the theological arguments and became known as *Galileo's letter to the Grand Duchess Christina*; the revised version was expanded from eight to forty pages. By good fortune, a Neapolitan friar, Paolo Antonio Foscarini, published a book in the same year arguing in detail that a moving Earth was compatible with the Bible. In December 1615, after a delay due to illness, Galileo himself went to Rome to clear his name and to prevent Copernicanism being condemned.

So far as Cardinal Roberto Bellarmine, the leading Catholic theologian of the day, was concerned, the main problem concerned the hypothetical nature of the Copernican picture. Here are his words, written on 12 April 1615 to Foscarini, after the *Letter to Christina* was circulating in Rome.

> . . . it seems to me that Your Paternity [Foscarini] and Mr Galileo are proceeding prudently by limiting yourselves to speaking suppositionally* and not absolutely, as I have always believed that Copernicus spoke. For there is no danger in saying that, by *assuming* the earth moves and the sun stands still, one saves all the appearances better than by postulating eccentrics and epicycles is to speak well; and that is sufficient for the mathematicians. However, it is different to want to affirm that in reality the Sun is at the centre of the world and only turns on itself without moving from east to west, and the Earth is in the third heaven and revolves with great speed around the Sun; this is a very dangerous thing, likely not only to irritate all scholastic philosophers and theologians, but also to harm the Holy Faith by rendering Holy Scripture false.[10]

Behind these remarks is a perfectly valid criticism of Galileo's support for the Copernican picture. It is not correct to state, as Galileo did, that the observation of the phases of Venus proves that the Copernican picture is necessarily correct; for example, in Tycho's cosmology, in which the planets orbit the Sun but the Sun together with the planets orbit the Earth (Fig. 2.7), exactly the same phases of Venus would be observed as in the Copernican picture. According to Gingerich, this was Galileo's crucial logical error. Strictly speaking, he could only make a hypothetical statement.

The findings of the Inquisition were favourable to Galileo personally – he was acquitted of the charge of suspicion of heresy. However, the Inquisition also asked a committee of 11 consultants for an opinion on the status of Copernicanism. On 16 February 1616, it reported unanimously that Copernicanism was philosophically and scientifically untenable and theologically heretical. This erroneous judgement was the prime cause of the subsequent condemnation of Galileo. It seems that the Inquisition had misgivings about this outcome because it issued no formal condemnation. Instead, it issued two milder instructions. First, Galileo was given a private warning by Cardinal Bellarmine to stop defending the Copernican world picture. Exactly what was said is a matter of controversy, but Bellarmine reported back to the Inquisition that the warning had been issued and that Galileo had accepted it.

* This word is often translated *hypothetically*.

The second result was a public decree by the Congregation of the Index. First, it reaffirmed that the doctrine of the Earth's motion was heretical; second, Foscarini's book was condemned and prohibited by being placed on the Index; third, Copernicus's *De Revolutionibus* was suspended until a few offending passages were amended; fourth, all similar books were subject to the same prohibition.

Rumours circulated that Galileo had been tried and condemned by the Inquisition and, to counteract these, Bellarmine issued a brief statement to the effect that Galileo had neither been tried nor condemned, but that he had been informed of the Decree of Index and instructed not to hold or defend the Copernican picture. Although he had been personally exonerated, the result was a defeat for Galileo.

3.5 The trial of Galileo

For the next seven years, Galileo kept a low profile and complied with the Papal instruction. In 1623, Gregory XV died and his successor, Cardinal Maffeo Barbarini, was elected Pope Urban VIII. He was Florentine and took a more relaxed view of the interpretation of the scriptures than his predecessor. An admirer of Galileo, he adopted the position that Copernicanism could be discussed hypothetically and might well prove to be of great value in making astronomical predictions. Galileo had six conversations with Urban VIII in Spring 1624 and came to the conclusion that Copernicanism could be discussed, provided that it was only considered hypothetically.

Galileo returned to Florence and immediately set about writing the *Dialogue on the Two Chief World Systems, Ptolemaic and Copernican*. He believed he had made every effort to comply with the wishes of the censors. The preface was written jointly by Galileo and the censors and, after some delay, the great treatise was published in 1632. Galileo wrote the book in the form of a dialogue between three speakers, Simplicio defending the traditional Aristotelian and Ptolemaic positions, Salviati defending the Copernican position and Sagredo an uncommitted observer and man of the world. Consistently, Galileo argued that the purpose was not to make judgements, but to pass on information and enlightenment. The book was published with full papal authority.

The Two Chief World Systems was well received in scientific circles, but very soon complaints and rumours began to circulate in Rome. A document dated February 1616, almost certainly a fabrication, was found in which Galileo was specifically forbidden from discussing Copernicanism in any form. By now, Cardinal Bellarmine had been dead 11 years. In fact, in his book Galileo had not treated the Copernican model hypothetically at all, but rather as a fact of nature – Salviati is Galileo speaking his own mind. The Copernican system was portrayed in a much more favourable light than the Ptolemaic picture, contradicting Urban VIII's conditions for discussion of the two systems of the world.

The pope was forced to take action – papal authority was being undermined at a time when the Counter-reformation and the reassertion of that authority were paramount political considerations. Galileo, now 68 years old and in poor health, was ordered to come to Rome under the threat of arrest. The result of the trial was a foregone conclusion. In the end, Galileo pleaded guilty to a lesser charge on the basis that, if he had violated the conditions imposed

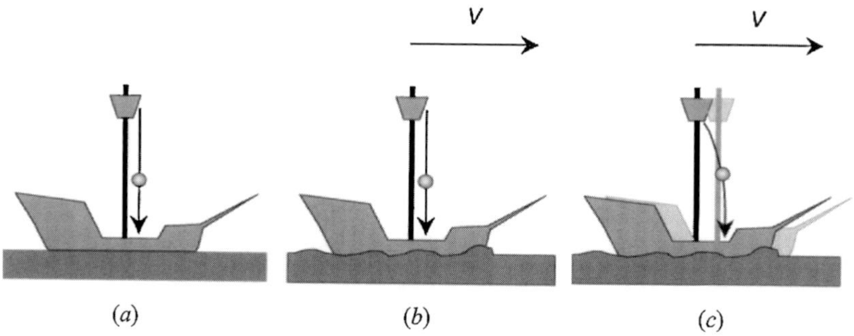

Figure 3.8: (*a*) Dropping an object from the top of a mast in a ship that is stationary in the frame of reference S. (*b*) Dropping an object from the top of a mast in a moving ship, viewed in the frame of reference S' of the ship. (*c*) Dropping an object from the top of a mast in a moving ship, as observed from the frame of reference S. The ship moves to the lighter grey position during the time the object falls.

upon him in 1616, he had done so inadvertently. The pope insisted upon interrogation under the formal threat of torture. On 22 June 1633, he was found guilty of 'vehement suspicion of heresy' and was forced to make a public abjuration, the proceedings being recorded in the Book of Decrees.

> I do not hold this opinion of Copernicus, and I have not held it after being ordered by injunction to abandon it. For the rest, here I am in your hands; do as you please.[11]

Galileo eventually returned to Florence where he remained under house arrest for the rest of his life – he died in Arcetri on 9 January 1642.

With indomitable spirit, Galileo set about writing his greatest work, *Discourses and Mathematical Demonstrations on Two New Sciences Pertaining to Mechanics and to Local Motion*, normally known as simply *Two New Sciences*. In this treatise, he brought together the understanding of the physical world which he had gained over a lifetime. The fundamental insights concern the second new science – the analysis of motion.

3.6 Galilean relativity

The ideas expounded in *Two New Sciences* had been in his mind since 1608. One of them is what is now called *Galilean relativity*. Relativity is often thought of as something invented by Einstein in 1905, but this does not do justice to Galileo's great achievement. Suppose an experiment is carried out on the shore and then on a ship moving at a constant speed. If the effect of air resistance is neglected, is there any difference in the outcome of any experiment? Galileo answers firmly, 'No, there is not.'

The relativity of motion is vividly illustrated as mentioned earlier, by dropping an object from the top of a ship's mast (Fig. 3.8). If the ship is stationary, the object falls vertically downwards. Now suppose the ship is moving. If the object is again dropped from the

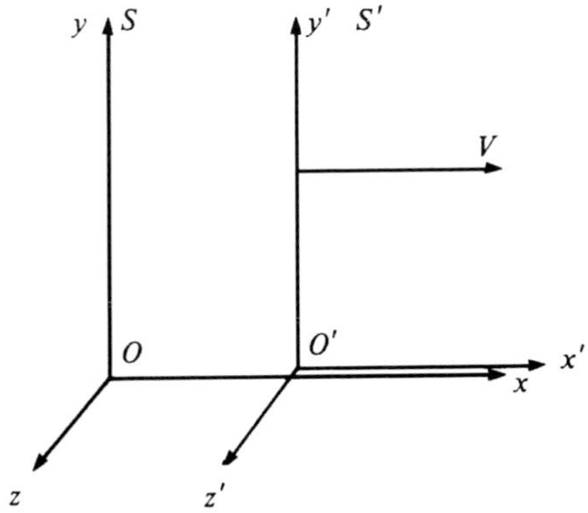

Figure 3.9: Illustrating two Cartesian frames of reference moving at relative velocity v in the direction of the positive x-axis in 'standard configuration'.

top of the mast, it again falls vertically downwards according to an observer on the ship. However, a stationary observer sitting on the shore notes that, relative to the shore, the path is curved (Fig. 3.8(c)). The reason is that, relative to the shore, the object has two separate components to its motion – vertical acceleration downwards due to gravity and uniform horizontal motion due to the motion of the ship.

This leads naturally to the concept of *frames of reference*. When the position of some object in three-dimensional space is measured, we can locate it by its coordinates in some rectangular coordinate system (Fig. 3.9). The point P has coordinates x, y, z in this stationary frame of reference, which we will call S. Now, suppose the ship moves along the positive x-axis at some speed v. We can set up another rectangular frame of reference S' on the ship, in which the coordinates of the point P are x', y', z'. It is now straightforward to relate the coordinates in these two frames of reference. If the object is stationary in the frame S then x is a constant but the value of x' changes as $x' = x - vt$, where t is the time, assuming that the origins of the two frames of reference are coincident at $t = 0$. The values of y and y' remain the same in S and S', as do z and z'. Also, time is the same in the two frames of reference. We have derived a set of relations between the coordinates of objects in the frames S and S':

$$\begin{aligned} x' &= x - vt, \\ y' &= y, \\ z' &= z, \\ t' &= t. \end{aligned} \tag{3.3}$$

These are known as the *Galilean transformations* between the frames S and S'. Frames of reference which move at constant relative speed to one another are called inertial frames of reference. Galileo's great insight can be summarised by stating that *the laws of physics are the same in every inertial frame of reference*. As a corollary of this insight, Galileo was the first to establish the *law of composition of velocities* – if a body has components of velocity in two different directions then the motion of the body can be found by adding together the separate effects of these motions. This was how he showed that the trajectories of cannon-balls and missiles are parabolae (Fig. 3.5).

In *Two New Sciences* Galileo described his discoveries concerning the nature of constant acceleration, the motion of pendulums and free fall under gravity. Finally, he stated his *law of inertia*, which asserts that a body moves at a constant velocity unless some impulse or force causes it to change that velocity – notice that it is now velocity rather than just speed which is constant, because the direction of motion does not change in the absence of forces. This is sometimes referred to as the *conservation of motion* – in the absence of forces, the separate components of the velocity remain unaltered. The word *inertia* is used here in the sense that it is a property of the body which resists change of motion. This law will become Newton's first law of motion. It can be appreciated why the Earth's motion caused Galileo no problems. Because of his understanding of Galilean relativity, he realised that the laws of physics would be the same whether the Earth were stationary or moving at a constant speed.

3.7 Reflections

We cannot leave this study without reflecting on the methodological and philosophical implications of the Galileo case. There is no question now that the Church made an error in condemning the new physics of Copernicus and Galileo. It was a further 350 years before Pope John Paul II admitted that an error had been made. In November 1979 on the occasion of the centenary of the birth of Albert Einstein, John Paul II stated that Galileo '... had to suffer a great deal – we cannot conceal the fact – at the hands of men and organisms of the Church.' He went on to assert that '... in this affair the agreements between religion and science are more numerous and above all more important than the incomprehensions which led to the bitter and painful conflict that continued in the course of the following centuries.'

For scientists, the central issue is the nature of scientific knowledge and the concept of truth in the physical sciences. Part of Cardinal Bellarmine's argument is correct. What Copernicus had achieved was a model that was much more elegant and economical for understanding the motions of the Sun, Moon and planets than the Ptolemaic picture, but in what sense was it the truth? If one were to put in enough effort, a Ptolemaic model of the Solar System could be created today which would replicate exactly the motions of the planets on the sky, but it would be of enormous complexity and provide little insight into the underlying physics which describes their motions. The value of the new model was not only that it provided a vastly improved framework for understanding the observed motions of the celestial bodies but also that, in the hands of Newton, it was to become the avenue for obtaining a very much deeper understanding of the laws of motion in general, leading to the unification of celestial physics, the laws of motion and the law of gravity. A scientifically satisfactory model has the capability not only of accounting economically for a large number of disparate observational and experimental phenomena but also of being extendable to make quantitative predictions about apparently unrelated phenomena.

Notice that I use the word *model* in describing this process rather than asserting that it is in any sense *truth*. Galileo's enormous achievement was to realise that the models to describe nature could be put on a rigorous mathematical basis. In one of his most famous remarks, he stated in his treatise *Il Saggiatore* (*The Assayer*) of 1624:

Philosophy is written in this very great book which always lies before our eyes (I mean the Universe), but one cannot understand it unless one first learns to understand the language and recognise the characters in which it is written. It is written in mathematical language and the characters are triangles, circles and other geometrical figures; without these means it is humanly impossible to understand a word of it; without these there is only clueless scrabbling around in a dark labyrinth.[12]

This is often abbreviated to the statement that

The Book of Nature is written in mathematical characters.

This was the great achievement of the Galilean revolution. The apparently elementary facts established by Galileo required an extraordinary degree of imaginative abstraction. Matter does not obey the apparently simple laws of Galileo – there is always friction, experiments can only be carried out with limited accuracy and often do not work. It needs deep insight and imagination to sweep away the unnecessary baggage and appreciate the basic simplicity in the way matter behaves. The modern approach to science is no more than the formalisation of the process begun by Galileo. It has been called the *hypothetico-deductive method*, whereby hypotheses are made and consequences deduced logically from them. A model is acceptable so long as it does not run significantly counter to the way matter is observed to behave. But models are only valid within well-defined regions of parameter space. Profesionals become very attached to them and the remarks by Dirac and Douglas Gough quoted in Chapter 1 describe the need to be satisfied with approximate theories and the 'pain' experienced on being forced to give up a cherished prejudice.

It is rare nowadays for religious dogma to impede progress in the physical sciences. However, *scientific prejudice* and *dogma* are the common currency of scientific debate. There is nothing particularly disturbing about this so long as we recognise what is going on. A scientific prejudice becomes embodied in a model, which provides a framework for carrying forward the debate and for suggesting experiments and calculations which can provide tests of the self-consistency of the model. We will find many examples throughout this book where the 'authorities' and 'received wisdom' were barriers to scientific progress. It takes a great deal of intellectual courage and perseverance to stand up to what is normally an overwhelming weight of conservative opinion. It is not just whimsy that leads us to use pontifical language to describe some of the bandwagons which can dominate areas of enquiry in the physical sciences. In extreme cases, through scientific patronage scientific dogma has attained an authority to the exclusion of alternative approaches. One of the most disastrous examples was the Lysenko affair in the USSR shortly after the Second World War, where Communist political philosophy strongly impacted the biological sciences, resulting in a catastrophe for these sciences in the Soviet Union.

Let me give two topical examples. It is intriguing how the idea of *inflation* during the very early stages of expansion of the Universe has attained the status of 'received dogma' among certain sections of the cosmological community. There are good reasons why this idea should be taken seriously, as will be discussed in Chapter 19. There is, however, no direct experimental evidence for the actual physics which could cause the inflationary expansion of the early Universe. Indeed, a common procedure is to work backwards and 'derive' the physics of inflation from the need to account for the features of the Universe as we observe it today. Then, theories of particle physics need to be found which can account for these forces. There is the obvious danger of ending up with bootstrapped self-consistency without any independent experimental check of the theory. Maybe this is the best one can do, but some of us will maintain a healthy scepticism until there are more independent arguments which support the conjecture of inflation.

The same methodology has occurred in the theory of elementary particles with the development of string theory. The creation of self-consistent quantum field theories involving one-dimensional objects rather than point particles has been a quite remarkable achievement. The latest versions of these theories involve the quantisation of gravity as an essential ingredient. Yet, they have not resulted in predictions which can be tested experimentally. Nonetheless, this is the area into which many of the most distinguished theorists have transferred all their efforts. It is taken as an article of faith that this is the most promising way of tackling these problems, despite the fact that it might well prove very difficult to find any experimental or observational tests of the theory in the foreseeable future.

3.8 References

1 Drake, S. (1990). *Galileo: Pioneer Scientist*, p. 63. Toronto: University of Toronto Press.
2 Drake, S. (1990). *Op. cit.*, p. 83.
3 Drake, S. (1990). *Op. cit.*, p. 84.
4 Galilei, G. (1610). *Sidereus Nuncius*, Venice. See the translation by A. van Helden (1989), *Sidereus Nuncius or The Sidereal Messenger*, Chicago: University of Chicago Press.
5 Finocchiaro, M.A. (1989). *The Galileo Affair. A Documentary History*. Berkeley: University of California Press.
6 Finocchiaro, M.A. (1989). *Op. cit.*, p. 24.
7 Finocchiaro, M.A. (1989). *Op. cit.*, p. 24.
8 Gingerich, O. (1982). *Scientific American*, **247**, 118.
9 Finocchiaro, M.A. (1989). *Op. cit.*, p. 28.
10 Finocchiaro, M.A. (1989). *Op. cit.*, p. 67.
11 Finocchiaro, M.A. (1989). *Op. cit.*, p. 287.
12 Sharratt, M. (1994). *Galileo: Decisive Innovator*, p. 140. Cambridge: Cambridge University Press.

4 Newton and the law of gravity

4.1 Introduction

Richard Westphal's monumental biography *Never at Rest* was the product of a lifetime's study of Isaac Newton's life and work. In the preface, he writes:

> The more I have studied him, the more Newton has receded from me. It has been my privilege at various times to know a number of brilliant men, men whom I acknowledge without hesitation to be my intellectual superiors. I have never, however, met one against whom I was unwilling to measure myself so that it seemed reasonable to say that I was half as able as the person in question, or a third or a fourth, but in every case a finite fraction. The end result of my study of Newton has served to convince me that with him there is no measure. He has become for me wholly other, one of the tiny handful of supreme geniuses who have shaped the categories of human intellect, a man not finally reducible to the criteria by which we comprehend our fellow beings.[1]

In the next paragraph, he writes:

> Had I known, when in youthful self-confidence I committed myself to the task, that I would end up in similar self-doubt, surely I would never have set out.[1]

Newton's impact upon science is so all pervasive that it is worthwhile filling in some of the background to his character and extraordinary achievements. The chronology which follows is that adopted in the Introduction of the volume *Let Newton Be*.[2]

4.2 Lincolnshire 1642–61

Newton was born in the hamlet of Woolsthorpe near Grantham on Christmas day 1642 (according to the old style Julian calendar). Newton's father was a successful farmer, who died three months before Newton's birth. When Isaac Newton was three years old, his mother married the Reverend Barnabas Smith and moved into his house, leaving Isaac behind to be brought up by his grandmother. Isaac hated his step-father, as is revealed by this entry in the list of sins he had committed up to the age of 19:

> Threatening my father and mother Smith to burn them and the house over them.[3]

It has been argued that this separation from his mother at an early age was a cause of his 'suspicious, neurotic, tortured personality'.[4] He was sent to the Free Grammar School at Grantham where he lodged with an apothecary. It was probably in these lodgings that he was introduced to chemistry and alchemy, which remained a lifelong passion. Newton himself claimed that he invented toys and did experiments as a teenager. He is reported to have constructed a model of a mill powered by a mouse, clocks, 'lanthorns' and fiery kites, which he flew to frighten the neighbours. He was already a loner who did not get on particularly well with his schoolmates. Following his school education, it was decided that he should return to the Free Grammar School at Grantham to prepare for entry to his uncle's old college, Trinity College, Cambridge.

4.3 Cambridge 1661–5

To begin with, Newton was a 'subsidiser', meaning that he earned his keep by serving the fellows and rich students at Trinity College. He took courses in Aristotelian philosophy, logic, ethics and rhetoric. His notebooks record that he was reading other subjects privately, the works of Thomas Hobbes, Henry More and René Descartes, as well as those of Kepler and Galileo. It became apparent to him that he was weak in mathematics and so he started to work 'ferociously' at the subject. By the end of this period, he had mastered whole areas of mathematics. He also continued to carry out experiments in a wide range of different topics. In 1664, he was elected a scholar of Trinity College, ensuring him a further four years of study without domestic duties. In 1665, he took his B.A. degree, and then the effects of the Great Plague began to spread north to Cambridge. The University was closed and Newton returned to his home at Woolsthorpe.

4.4 Lincolnshire 1665–7

During the next two years, Newton's burst of creative scientific activity must be one of the most remarkable ever recorded. The only comparable feat of which I am aware is the achievement of Albert Einstein in 1905. Over 50 years later, Newton wrote:

> In the beginning of 1665, I found the method of approximating series and the rule for reducing any dignity (power) of any binomial into such a series. The same year in May, I found the method of tangents of Gregory and Slusius and in November had the direct method of fluxions and the next year in January had the theory of colours and in May following I had entrance into the inverse method of fluxions. And in the same year I began to think of gravity extending to the orbit of the Moon and (having found out how to estimate the force with which a globe revolving within a sphere presses the surface of the sphere) from Kepler's rule of the periodical times of the planets being in sesquialternate proportion* of their distance from the centres of their Orbs, I deduced that the forces which keep the planets in their Orbs must [be] reciprocally as the squares of their distances from the centres about which they revolve: and thereby compared the force requisite to keep the Moon in her orb with the force of gravity at the surface of the Earth, and found them answer pretty nearly. All this was in the two plague years 1665–6. For in those days I was in the prime of my age for invention and minded mathematics and philosophy more than at any time since.[5]

* *Sesquialternate proportion* means 'to the power 3/2'.

It is a quite astonishing list – Newton had laid the foundations of three quite separate areas. In mathematics, he discovered the *binomial theorem* and the *differential* and *integral calculus*. In *optics*, he discovered the decomposition of light into its primary colours. He began his unification of *celestial mechanics* with the *theory of gravity*, which was to lead ultimately to his laws of motion and the theory of universal gravitation. Let us look into his achievements in the fields of optics and of the law of gravity in more detail.

4.4.1 Optics

Newton was a skilled experimenter and undertook a number of optical experiments using lenses and prisms while he was at Trinity College and at Woolsthorpe. Many investigators had noted that when white light is passed through a prism all the colours of the rainbow are produced. Descartes was the proponent of the generally accepted view that, when light passes through a prism, it is modified by the material of the prism and so becomes coloured.

In 1666, Newton, in his own words, 'applied myself to the grinding of Optick glasses of other figures than spherical' and among these he produced a triangular glass prism 'to try therewith the celebrated *Phaenomena of Colours*'. Sunlight was passed through a tiny hole onto the prism and, to Newton's surprise, the coloured light produced by the prism was of oblong form. This experiment led him to carry out his famous *experimentum crucis*, illustrated in Figs. 4.1(*a*), (*b*), the second picture showing Newton's own drawing of the experiment.

In this experiment, the coloured spectrum produced by the first prism was projected onto a plane board and a small hole was cut in the board to allow only one of the colours to pass through. A second prism was placed in the path of this second light beam, and Newton discovered that there was no further decomposition of this beam into more colours. The experiment was repeated for all the colours of the spectrum and consistently he found that there was no further splitting up of the colours, contrary to the expectations of Déscartes' theory. Newton concluded that:

> Light itself is a Heterogeneous mixture of differently refrangible rays.[6]

The word *refrangible* is what we would now call *refracted*. Newton had established that white light is a superposition of all the colours of the rainbow and that, when white light is passed through a prism, rays of different colour are deflected by different amounts.

This work was not presented in public until 1672, in a paper to the Royal Society, and Newton immediately found himself in the thick of a hot dispute. In addition to describing his new results, he regarded them as evidence in favour of his view that light can be considered to be made up of 'corpuscles', that is, of particles which travel from the source of light to the eye. Newton did not have a satisfactory theory of 'light corpuscles', or why they should be refracted in different ways by material media. His problems were exacerbated by the fact that the *experimentum crucis* was difficult to reproduce. The ensuing unpleasant debate with Huyghens and Hooke obscured the key result that white light is composed of all the colours of the spectrum.

This work led him to another important conclusion – that it is not possible to build a large refracting telescope of the type pioneeered by Galileo, because white light is split

(*a*)

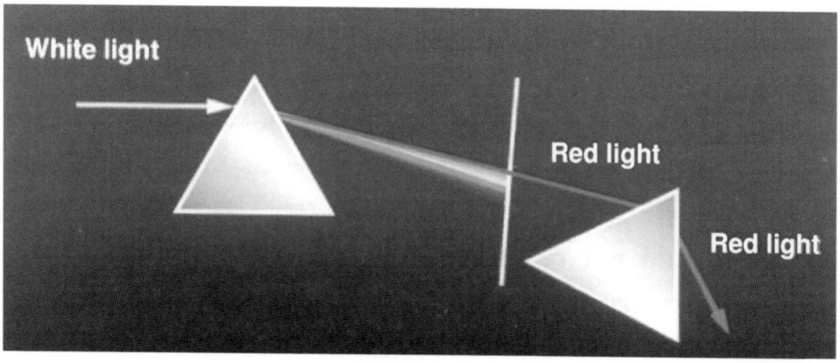

(*b*)

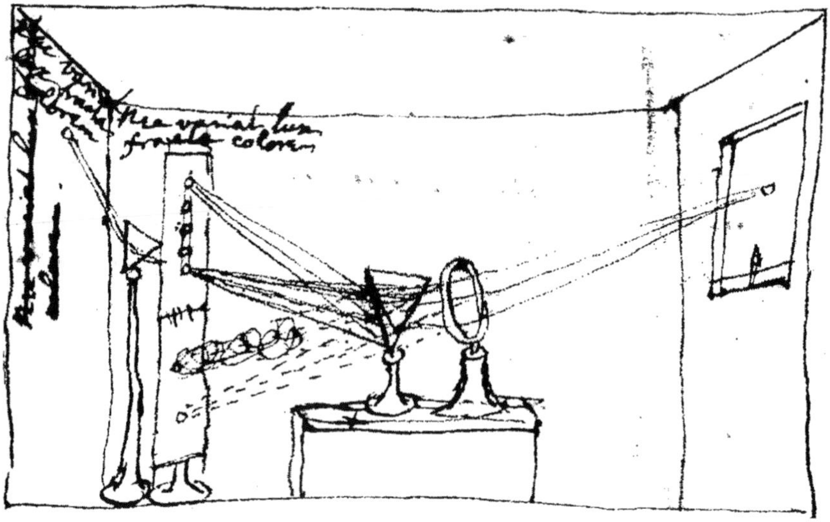

Figure 4.1: (*a*) A schematic diagram showing the decomposition of white light into the colours of the spectrum. On passing through the second prism, the colours are not split up into any further colours. (*b*) Newton's sketch of his *experimentum crucis*. (From C. Hakfoort, 1988, *Let Newton Be: a New Perspective on his Life and Works*, eds. J. Fauvel, R. Flood, M. Shortland, and R. Wilson, p. 87, Oxford: Oxford University Press. By kind permission of the Warden and Fellows, New College Oxford.)

up into its primary colours by a refracting lens and so different colours are focussed at different positions on the optical axis of the telescope, the phenomenon known as *chromatic aberration*. Because of this problem, Newton designed and built a new type of telescope, an *all-reflecting telescope*, which would not suffer from chromatic aberration. He ground the mirrors himself and constructed the mount for the telescope and mirrors. This configuration is now called a 'Newtonian' design and is shown in Fig. 4.2(*a*). Figure 4.2(*b*) shows a contemporary drawing of Newton's telescope as well as a comparison of the magnifying power of Newton's telescope with that of a refracting telescope, indicating the superiority of Newton's design. Nowadays, all large optical astronomical telescopes are reflectors, the descendants of Newton's invention.

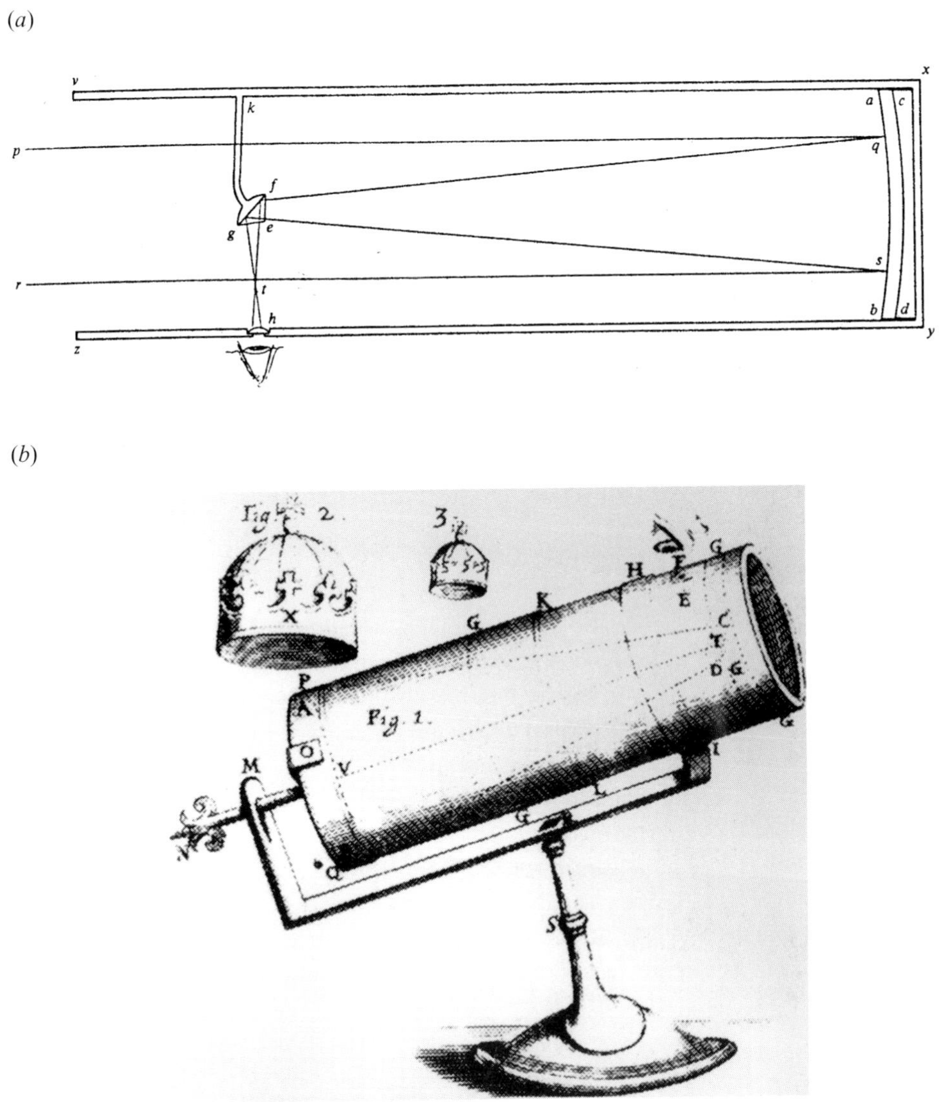

Figure 4.2: (*a*) The optical layout of Newton's reflecting telescope; *abcd* is the mirror made of speculum metal, *efg* is a prism and *h* is a lens. (From I.B. Cohen, 1970, *Dictionary of Scientific Biography*, Vol. 11, p. 57, New York: Charles Scribner's Sons © 1970–80. Reprinted by permission of the Gale Group.) (*b*) Newton's reflecting telescope, as illustrated in the *Philosophical Transactions of the Royal Society*, 1672. The two crowns show the improvement in magnification of Newton's telescope over the traditional refracting telescope, such as that used by Galileo. (From J. Fauvel, R. Flood, M. Shortland and R. Wilson eds., 1988, *Let Newton Be: a New Perspective on his Life and Works*, p. 16, Oxford: Oxford University Press.)

These were great experimental achievements, but Newton had an almost pathological dislike of writing up his work for publication. The systematic presentation of his optical work in his book *Opticks* was only published in 1704, long after his discoveries and inventions had been made.

4.4.2 The law of gravity

Newton's most famous achievement of his years at Woolsthorpe was the discovery of the law of gravity. The calculations carried out in 1665–6 were only the beginning of the story, but they contained in essence the theory which was to appear in its full glory in his *Principia Mathematica* of 1687. As Newton himself recounted, he was aware of Kepler's third law of planetary motion, which was deeply buried in Kepler's *Harmony of the World*. The volume was in the Trinity College library and it is likely that Isaac Barrow, the Lucasian Professor of Mathematics, drew Newton's attention to it.

In Newton's own words,

> the notion of gravitation [came to my mind] as I sat in contemplative mood [and] was occasioned by the fall of an apple.[7]

Newton asked whether the force of gravity, which makes the apple fall to the ground, is the same force which holds the Moon in its orbit about the Earth and the planets in their orbits about the Sun. To answer this question, he needed to know how the force of gravity varies with distance. He derived this relation from Kepler's third law by a simple argument. First, he needed an expression for *centripetal acceleration*, which he rederived for himself. Nowadays, this formula can be derived by a simple geometrical argument using vectors.

Figure 4.3 shows the velocity vectors of a particle moving at constant speed $|\boldsymbol{v}|$ in a circle with radius $|\boldsymbol{r}|$ at times t and $t + \Delta t$. In the time Δt, the angle subtended by the radius vector $\boldsymbol{r}$ changes by $\Delta\theta$. The change in the velocity vector in Δt is given by the vector triangle shown on the right. As $\Delta t \to 0$, the vector $\Delta\boldsymbol{v}$ continues to point towards the centre of the circle and its magnitude is $|\boldsymbol{v}|\Delta\theta$. Therefore, the centripetal acceleration is of magnitude

$$|\boldsymbol{a}| = \frac{|\Delta\boldsymbol{v}|}{\Delta t} = \frac{|\boldsymbol{v}|\Delta\theta}{\Delta t} = |\boldsymbol{v}|\omega = \frac{|\boldsymbol{v}|^2}{|\boldsymbol{r}|} \tag{4.1}$$

where ω is magnitude of the (constant) angular velocity of the particle. This is the famous expression for the *centripetal acceleration* a of an object moving in a circle of radius r at speed v.

In 1665, Newton was well aware of the fact that the orbits of the planets are actually ellipses, and not circles, but he assumed that it would be adequate to apply Kepler's third law to circular orbits, since the ellipticities of the orbits of the planets are generally very small.

He also knew that accelerations are proportional to the forces which cause them and so the force f which keeps a planet in its circular orbit must be proportional to its centripetal acceleration v^2/r, that is,

$$f \propto \frac{v^2}{r}; \tag{4.2}$$

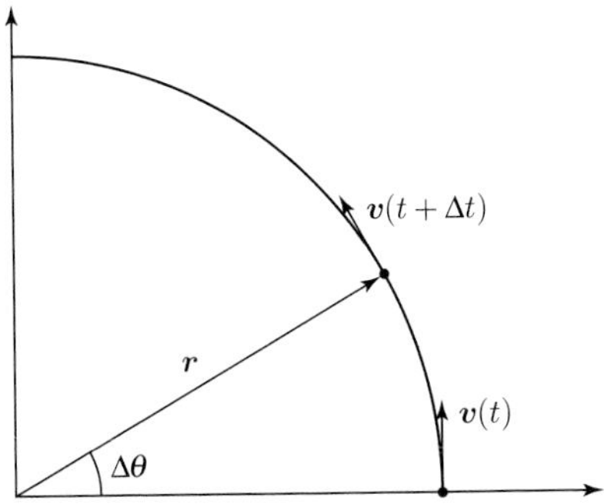

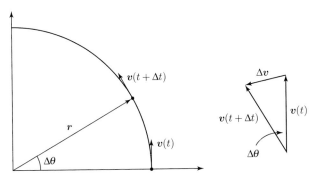

Figure 4.3: A vector diagram illustrating the origin of centripetal acceleration.

but Kepler's third law states that the period T of the planet's orbit is proportional to $r^{3/2}$. The speed of a planet in its orbit is $v = 2\pi r/T$ and therefore

$$v \propto \frac{1}{r^{1/2}}. \tag{4.3}$$

Now, we can substitute for v from (4.3) into (4.2) and so find

$$f \propto \frac{1}{r^2}. \tag{4.4}$$

This is the primitive form of Newton's *inverse square law of gravity* – the force of gravity decreases as the inverse square of the distance from the body.

This was the key result he was looking for – if gravity were truly universal then the force which causes apples to fall to the ground should be of exactly the same nature as the force which keeps the Moon in its orbit about the Earth. The only difference is that the acceleration of the apple should be greater than that of the Moon, because the Moon is 60 times further from the centre of the Earth than the apple. Thus, according to the universal theory of gravity, the centripetal acceleration of the Moon should be only $1/60^2 = 1/3600$ times the acceleration due to gravity on the surface of the Earth. Newton had all the data he needed to make this comparison. Using modern numbers, the acceleration due to gravity on the surface of the Earth is on average 9.80665 m s^{-2}. The period of revolution of the Moon about the Earth is 27.32 days and its mean distance from the Earth is $r = 384\,408\,000$ m. Putting these together, the mean speed of the Moon v is 1023 m s^{-1} and its centripetal acceleration v^2/r is 2.72×10^{-3} m s^{-2}. The ratio of this acceleration to the local acceleration due to gravity is $9.80665/2.72 \times 10^{-3} = 3600$, exactly as predicted.

Newton's calculations did not give quite such good agreement, but were sufficiently close to persuade him that the force which holds the Moon in its orbit about the Earth and the planets in their orbits about the Sun, is of exactly the same type as the force which causes the acceleration due to gravity on Earth. From this result, the general formula for the gravitational attraction between any two bodies of masses M_1 and M_2 follows:

$$f = \frac{GM_1M_2}{r^2}, \tag{4.5}$$

where G is the gravitational constant and r is the distance between the bodies. The force acts along the line joining the two bodies and is always an attractive force.

This work was not published, because there were a number of steps in the calculation which needed further elaboration.

(i) Kepler had shown that the orbits of the planets are ellipses and not circles – how did that affect the calculation?
(ii) Newton was uncertain about the influence of the other bodies in the Solar System upon each others' orbits.
(iii) He was unable to explain the details of the Moon's motion about the Earth, which, as we now know, is influenced by the fact that the Earth is not spherical.
(iv) Probably most important of all, there is a key assumption in the calculation, that all the mass of the Earth can be located at its centre in working out the acceleration due to gravity at its surface and its influence upon the Moon. The same assumption was made for all the bodies in the solar system. In his calculations of 1665–6, Newton regarded this step as an approximation. He was uncertain about its validity for objects close to the surface of the Earth.

Newton laid this work aside until 1679.

4.5 Cambridge 1667–96

The University reopened in 1667 and Newton returned to Trinity College in the summer of that year. He became a fellow of the college in the autumn of 1667 and two years later, at the age of 26, he was elected Lucasian Professor of Mathematics, a position which he held for the next 32 years. As Lucasian Professor, Newton's duties were not heavy. He was required to give at least one lecture per week in every term and to deposit his lectures, properly written up, in the University Library. In the period 1670 to 1672, he deposited lectures on optics, during 1673–83 lectures on arithmetic and algebra and in 1684–5 most of book I of what was to become the *Principia Mathematica*. In 1687, the lectures were entitled 'The system of the world' and this became part III of the *Principia*. There seems to be no record of his lectures for 1686, nor from 1688 until he left Cambridge in 1696.

His lectures were not particularly successful. His assistant Humphrey Newton, who was no relation, wrote of Newton during the years he was preparing the *Principia*:

> He seldom left his chamber except at term time, when he read [his lectures] in the [Old S]chools as being Lucasian Professor, where so few went to hear him, and fewer that understood him, that of times he did, in a manner, for want of hearers, read to the walls . . . [when he lectured he] usually stayed about half an hour; when he had no auditors, he commonly returned in a fourth part of that time or less.[8]

The first of Newton's publications appeared during this period. In late 1668, Nicholas Mercator published his book *Logarithmotechnica*, in which he described some of the techniques for the analysis of infinite series which Newton had already worked out in greater generality. Newton set about writing up his mathematical work so that his priority could be established. He wrote hastily the work known as *De Analysi* (*On Analysis*), which Isaac Barrow was allowed to show to a London mathematician, Mr Collins. Newton insisted, however, that the work remain anonymous. The results were communicated by letter among British and continental mathematicians and so the work became widely known.

Newton incorporated the most important parts of *De Analysi* into another manuscript, *Methodus Fluxionum et Serierum Infinitarum*, the *Method of Fluxions and Infinite Series*, which was not published until long afterwards in 1711 by William Jones. The manuscript was, however, read by Leibniz in October 1676 when he was on a visit to London. Although Leibniz scholars have established that he only copied out the sections on infinite series, this incident was the source of the later bitter accusations that Leibniz had plagiarised Newton's discovery of the differential and integral calculus.

Newton was prompted to return to his researches on the laws of motion and gravity in 1679 by an interchange of letters with Robert Hooke. Hooke challenged Newton to work out the curve followed by a particle falling in an inverse-square field of force, the law Newton had derived 14 years earlier. This stimulated Newton to derive two crucial results. As noted by Chandrasekhar in his remarkable commentary on the *Principia*,[9]

> Newton's interest in dynamics was revived sufficiently for him to realise for the first time the real meaning of Kepler's law of areas. And as he wrote, 'I found now that whatsoever was the law of force which kept the Planets in their Orbs, the area described by a radius drawn from them to the Sun would be proportional to the times in which they were described'; and he proved the two propositions that
>
> > all bodies circulating about a centre sweep out areas proportional to the time
>
> and that
>
> > a body revolving in an ellipse ... the law of attraction directed to a focus of the ellipse ... is inversely as the square of the distance.

The first remarkable discovery, that Kepler's second law, the law of areas, is correct whatever the nature of the force law, so long as it is a central force, is now recognised as a consequence of the law of conservation of angular momentum. Chandrasekhar noted that: 'The resurrection of Kepler's law of areas in 1679 was a triumphant breakthrough from which the *Principia* was later to flow'.[9] At this point, another dispute broke out with Hooke, who claimed that he was the originator of the inverse square law of gravity. Newton violently rejected this claim and would not communicate any of the results of his calculations to Hooke, or to anyone else.

In 1684, Edmund Halley travelled to Cambridge to ask Newton precisely the same question which had been posed by Hooke. Newton's immediate response was that the orbit is an ellipse, but the proof could not be found among his papers. In November 1684, Newton sent the proof to Halley. Halley returned to Cambridge where he saw an incomplete manuscript of Newton's entitled *De Motu Corporum in Gyrum*, or *On the Motion of Revolving Bodies*, which was eventually to be transformed into the first part of the *Principia*. With a great deal of persuasion, Newton agreed to set about systematising all his researches on motion, mechanics, dynamics and gravity.

Only in 1685, when he was working at white heat on the preparation of what was to become his great treatise, the *Philosophiae Naturalis Principia Mathematica*, or the

Principia for short, did he demonstrate that, for a spherically symmetric body, the gravitational attraction can be found exactly by placing all its mass at its centre. In the words of J.W.L. Glaisher, on the occasion of the bicentenary of the publication of the *Principia*,

> No sooner had Newton proved this superb theorem – and we know from his own words that he had no expectation of so beautiful a result till it emerged from his mathematical investigation – then all the mechanism of the universe at once lay spread before him. . . . We can imagine the effect of this sudden transition from approximation to exactitude in stimulating Newton's mind to still greater effort.[10]

Nowadays, this result is proved in a few lines using Gauss's theorem in vector calculus as applied to the inverse square law of gravity, but these techniques were not available to Newton. It is not generally appreciated what a crucial step this calculation was in the development of the *Principia*.

Humphrey Newton gives us a picture of Newton during the writing of the *Principia*.

> (He) ate sparing (and often) forgot to eat at all. (He rarely dined) in the hall, except on some public days (when he would appear) with shoes down at the heel, stockings untied, surplice on, and his head scarcely combed. (He) seldom went to the Chapel (but very often) went to St. Mary's Church, especially in the forenoon.[11]

The *Principia* is one of the greatest intellectual achievements of all time. The theory of statics, mechanics and dynamics, as well as the law of gravity, are developed entirely through mathematical relations and mechanistic interpretations of the physical origins of the forces are excluded. In the very beginning, what we now call *Newton's laws of motion* are set out in their definitive form – we will return to these in more detail in Chapter 7. Despite the fact that Newton had developed his own version of the integral and differential calculus, the *Principia* is written entirely in terms of geometrical arguments. Often these are difficult to follow for the modern reader, largely because the geometrical arguments used are no longer familiar to physicists. Chandrasekhar's remarkable reconstruction of the arguments in modern geometrical terms gives some impression of the methods which Newton must have used. As an example of the economy of expression used by Newton in the *Principia*, Fig. 4.4 shows a translation of the proof of the elliptical orbits under the influence of an inverse square field of force. Comparison with Chandrasekhar's much lengthier derivation shows how terse Newton's geometric proofs were.

4.6 Newton the alchemist

Throughout his time at Cambridge, Newton maintained a profound interest in alchemy and in the interpretation of ancient and biblical texts. These aspects of his work tend to be regarded, at best, as unfortunate aberrations, and yet they were of the greatest significance for Newton.

Newton studied alchemy with the same seriousness he devoted to his mathematics and physics. Whereas his great contributions to the latter disciplines became public knowledge, his alchemy remained very private, his papers containing over a million words on alchemical topics. From the late 1660s to the mid 1690s, he carried out extensive chemical experiments

PROPOSITION XI. PROBLEM VI

If a body revolves in an ellipse; it is required to find the law of the centripetal force tending to the focus of the ellipse.

Let S be the focus of the ellipse. Draw SP cutting the diameter DK of the ellipse in E, and the ordinate Qv in x; and complete the parallelogram $QxPR$. It is evident that EP is equal to the greater semiaxis AC: for drawing HI from the other focus H of the ellipse parallel to EC, because CS, CH are equal, ES, EI will be also equal; so that EP is the half-sum of PS, PI that is (because of the parallels HI, PR, and the equal angles IPR, HPZ), of PS, PH, which taken together are equal to the whole axis 2AC. Draw QT perpendicular to SP, and putting L for the principal latus rectum of the ellipse (or for $\frac{2BC^2}{AC}$), we shall have

$$L \cdot QR : L \cdot Pv = QR : Pv = PE : PC = AC : PC,$$

$$\text{also, } L \cdot Pv : Gv \cdot Pv = L : Gv, \text{ and, } Gv \cdot Pv : Qv^2 = PC^2 : CD^2.$$

By Cor. II, Lem. VII, when the points P and Q coincide, $Qv^2 = Qx^2$, and Qx^2 or $Qv^2 : QT^2 = EP^2 : PF^2 = CA^2 : PF^2$, and (by Lem. XII) $= CD^2 : CB^2$. Multiplying together corresponding terms of the four proportions, and simplifying, we shall have

$$L \cdot QR : QT^2 = AC \cdot L \cdot PC^2 \cdot CD^2 : PC \cdot Gv\, CD^2 \cdot CB^2 = 2PC : Gv,$$

since $AC \cdot L = 2BC^2$. But the points Q and P coinciding, 2PC and Gv are equal. And therefore the quantities $L \cdot QR$ and QT^2, proportional to these, will be also equal. Let those equals be multiplied by $\frac{SP^2}{QR}$, and $L \cdot SP^2$ will become equal to $\frac{SP^2 \cdot QT^2}{QR}$. And therefore (by Cor. I and V, Prop. VI) the centripetal force is inversely as $L \cdot SP^2$, that is, inversely as the square of the distance SP.Q.E.I.

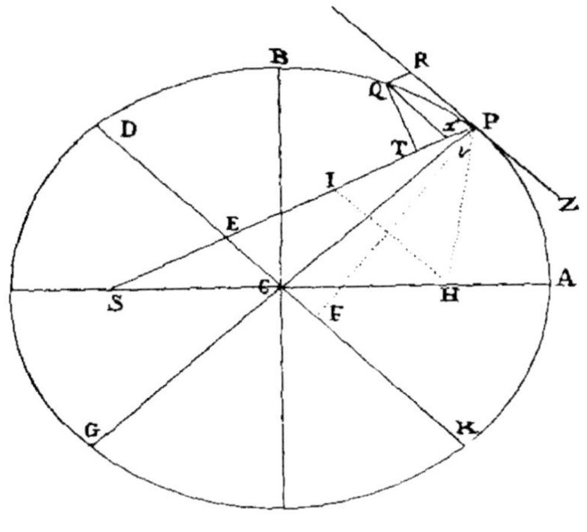

Figure 4.4: A translation from the Latin of Newton's proof that the orbits of the planets are ellipses under the influence of an inverse square law field of gravity. (From J. Roche, 1988, *Let Newton Be: a New Perspective on his Life and Works*, eds. J. Fauvel, R. Flood, M. Shortland and R. Wilson, p. 55, Oxford: Oxford University Press. Reprinted by permission of Oxford University Press.)

Figure 4.5: A seventeenth century engraving of the Great Gate and Chapel of Trinity College. The small shed seen in the bay of the chapel closest to the wall of Great Court may have been Newton's chemical laboratory. (Engraving by David Loggan, from his *Cantabrigia illustrated*, Cambridge 1690. Reprinted in J. Golinski, 1988, *Let Newton Be: a New Perspective on his Life and Works*, eds. J. Fauvel, R. Flood, M. Shortland, and R. Wilson, p. 152, Oxford: Oxford University Press. Courtesy of the Society of Antiquaries of London.)

in his laboratory at Trinity College. Figure 4.5 is a contemporary engraving of Trinity College; Newton's rooms were on the first floor to the right of the Gatehouse. It has been suggested that the shed which can be seen leaning against the Chapel outside the College was Newton's laboratory. He taught himself all the basic alchemical operations, including the construction of his own furnaces. Humphrey Newton, his assistant from 1685 to 1690, wrote:

> About six weeks at spring and six at the fall, the fire in the laboratory scarcely went out . . . what his aim might be, I was not able to penetrate into . . . Nothing extraordinary, as I can remember, happened in making his experiments, which if there did . . . I could not in the least discern it.[12]

The two Newtons kept the furnaces going continuously for six weeks at a time, taking turns to tend them overnight.

From the late 1660s to the mid 1690s, Newton devoted a huge effort to systematising everything he had read on chemistry and alchemy (Fig. 4.6). His first attempt to put some order into his understanding was included in his *Index Chemicus*, or *Chemical Dictionary*, of the late 1660s; it was followed over the next 25 years by successive versions. According to Golinski,[13] the final index cites more than 100 authors and 150 works. There are over 5000 page references under 900 separate headings. In addition, Newton devoted a huge effort to deciphering obscure allegorical descriptions of alchemical processes. It was part

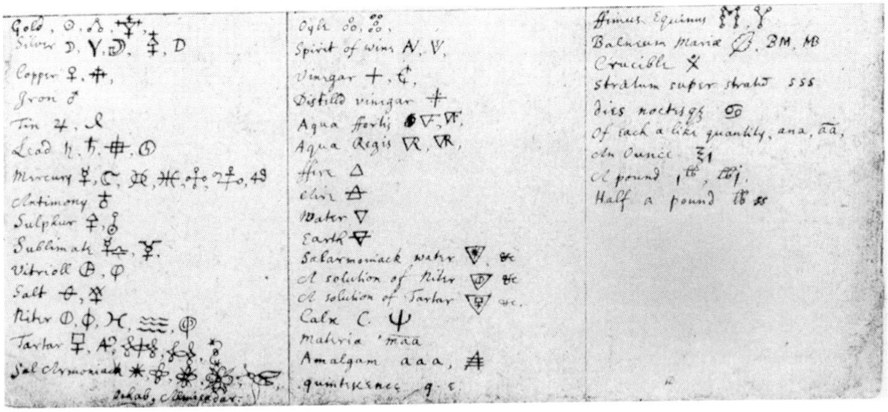

Figure 4.6: Newton's list of alchemical symbols. (From J. Golinski, 1988, *Let Newton Be: a New Perspective on his Life and Works*, eds. J. Fauvel, R. Flood, M. Shortland, and R. Wilson), p. 155, Oxford: Oxford University Press. With kind permission of the Provost, Fellows and Scholars of King's College, Cambridge.)

of the mystique of the alchemists that the fundamental truths should not be made known to the unworthy or the vulgar. The ultimate example of this was Newton's belief that the biblical account of the creation was actually an allegorical description of alchemical processes. In a manuscript note of the 1680s, Newton noted that

> Just as the world was created from dark chaos through the bringing forth of the light and through the separation of the airy firmament and of the waters from the Earth, so our work brings forth the beginning out of black chaos and its first matter through the separation of the elements and the illumination of matter.

What Newton was trying to achieve is illuminated by a manuscript dating from the 1670s entitled *Of Nature's Obvious Laws and Processes in Vegetation*. To quote Golinski,

> Newton distinguished 'vulgar chemistry' from a more sublime interest in the processes of vegetation... 'Nature's actions are either vegetable or purely mechanical' he wrote. The imitation of mechanical changes in nature would be common, or vulgar, chemistry, whereas the art of inducing vegetation was 'a more subtle, secret and noble way of working.'

Newton's objective was no less than to isolate the essence of what gave rise to growth and life itself. Such a discovery would make man like God and so had to be kept a secret from the vulgar masses.

4.7 The Interpretation of ancient texts and the scriptures

Newton devoted almost as much effort to interpreting the works of the ancients and biblical texts as he did to his alchemical studies. Newton convinced himself that all his great discoveries had in fact been known to the ancient Greek philosophers. In 1692, Nicholas Fatio de Duillier, Newton's protegé during the period 1689–93, wrote to Christiaan Huyghens that Newton had discovered that all the chief propositions of the *Principia* had been known to Pythagoras and Plato but that they had turned these discoveries into a 'great mystery'.

Newton believed that the ancients had kept this knowledge secret for the same reason he kept his alchemical work secret. These truths were of such great significance that they had to be preserved for those who could truly appreciate them. For this reason, the Greeks had to disguise their deep understandings in a coded language of symbolism, which only initiates such as Newton could penetrate.

Piyo Rattansi[16] suggests that Newton's objective was to give his great discoveries an ancient and honourable lineage. The *Principia Mathematica* had been well received on the continent as a text in mathematics, but not as physics. The continental physicists disliked the introduction of the 'unintelligible' force of attraction, which had to be introduced to account for the law of gravity. To them Newton's 'force of gravity' was a mystic force because they could not envisage any physical mechanism for causing gravitational attraction – all other physical forces could be explained by actions involving concrete bits of matter but what was it that caused gravity to act across the Solar System? The continental scientists interpreted the law of gravity as being the reintroduction of 'occult causes', which the whole of the revolution in science had been bent upon eliminating. Newton was claiming as ancient an authority for his discoveries as any branch of knowledge.

In addition, throughout his lifetime, he had a deep interest in the scriptures and their interpretation. His prime interest centred upon the near East and how a proper interpretation of the texts could be shown to predict retrospectively all the significant events of history. His aim was to show that the seat of all knowledge began in Israel and from there it travelled to Mesopotamia and Egypt. The results of these researches were published in two books, *The Chronology of the Ancient Kingdoms Amended* and *Observations upon the Prophecies of Daniel and the Apocalypse of St John*. The latter work runs for 323 pages and was published in twelve editions between 1733 and 1922.

Rattansi puts Newton's studies of the biblical texts in another revealing light:

> His biblical work was intended to vindicate the authority of the Bible against those Catholics who had tried to show that only by supplementing it with the tradition of the universal church could it be made authoritative. It served at the same time as a weapon against the free-thinkers who appealed to a purely 'natural' religion and thereby did away with the unique revelation enshrined in Christian religion. It did so by demonstrating that history was continually shaped according to providential design which could be shown, only after its fulfilment, to have been prefigured in the divine word.[17]

Later, he writes:

> Newton followed the Protestant interpretation of Daniel's visions and made the Roman church of later times the kingdom of the Antichrist which would be overthrown before the final victory of Christ's kingdom.[18]

Newton held the view that the scriptures had been deliberately corrupted in the fourth and fifth centuries. This view was at odds with the tenets of the Church of England and, as a fellow of Trinity College, he was expected to be a member of the Church. A major crisis was narrowly avoided by the arrival of a royal decree, exempting the Lucasian Professor from the necessity of taking holy orders.

In 1693, Newton had a nervous breakdown. After his recovery, he lost interest in academic studies and, in 1696, left Cambridge to take up the post of Warden of the Mint in London.

4.8 London 1696–1727

By the time Newton took up his position as Warden of the Mint, he was already recognised as the greatest living English scientist. Although the position was normally recognised as a sinecure, he devoted all his energies to the recoinage needed at that time to stabilise the currency. He was also responsible for prosecuting forgers of the coin of the realm, an offence which carried the death penalty. Apparently, Newton carried out this unpleasant aspect of his responsibility 'with grisly assiduity'. He was an effective manager and administrator and in 1700 was appointed Master of the Mint.

In 1703, following the death of Hooke, Newton was elected President of the Royal Society. He now held positions of great power, which he used to further his own interests. Halley entered the service of the Mint in 1696 and in 1707 David Brewster was appointed as general supervisor of the conversion of the Scottish coinage into British. Newton had secured the Savillian chair of Astronomy at Oxford for David Gregory in 1692 as well as the Savillian Chair of Geometry for Edmond Halley in the early 1700s. He also ensured that William Whiston was elected to his own Lucasian Chair in Cambridge from which he had eventually resigned in 1703.

In his old age, he did not mellow. An acrimonious quarrel arose with Flamsteed, the first Astronomer Royal, who had made very accurate observations of the Moon, which Newton wished to use in his analysis of its motion. Newton became impatient to use these observations before they were completed to Flamsteed's satisfaction. Newton and Halley took the point of view that, since Flamsteed was a public servant, the observations were national property and they eventually succeeded not only in obtaining the incomplete set of observations but also in publishing them in an unauthorised version without Flamsteed's consent in 1712. Flamsteed managed to recover about 300 of the unauthorised copies and had the pleasure of burning this spurious edition of his monumental work. He later published his own definitive set of observations in his *Historia Coelestis Brittanica*.

The infamous dispute with Leibniz was even nastier. It was fomented by Fatio de Duillier who brought the initial charge of plagiarism against Leibniz. Leibniz appealed to the Royal Society to set up an independent panel to pass judgement on the question of priority. Newton appointed a committee to look into the matter but then wrote the committee's report himself in his *Commercium Epistolicum*. He wrote it as if its conclusions were impartial findings which came down in Newton's favour. But he did not stop there. A review of the report was published, in the *Philosophical Transactions of the Royal Society*, which was also written by Newton, anonymously. The story is very unpleasant, to say the least, particularly since Leibniz had unquestionably made original and lasting contributions to the development of the differential and integral calculus. Indeed, the system of notation which is now used universally is that of Leibniz, not Newton.

Newton died on the 20 March 1727 at the age of eighty-five and was buried in Westminster Abbey.

4.9 References

1 Westphal, R.S. (1980). *Never at Rest: A Biography of Isaac Newton*, p. ix. Cambridge: Cambridge University Press.

2 Fauvel, J., Flood, R., Shortland, M. and Wilson, R. (eds.) (1988). *Let Newton Be: a New Perspective on his Life and Works*. Oxford: Oxford University Press. (The title alludes to Pope's epitaph intended for Newton: 'Nature, and Nature's Laws, Lay hid in night; God said, "Let Newton be", and all was light.')

3 Cohen, I.B. (1970). *Dictionary of Scientific Biography*, Vol. 11, p. 43. New York: Charles Scribner's Sons.

4 Fauvel, J. *et al.* (1988). *Op. cit.*, p. 12.

5 Fauvel, J. *et al.* (1988). *Op. cit.*, p. 14.

6 Cohen, I.B. (1970). *Op. cit.*, p. 53.

7 Stukeley, W. (1752). *Memoirs of Sir Isaac Newton's Life*, pp. 19–20 (ed. A.H. White, London, 1936).

8 Cohen, I.B. (1970). *Op. cit.*, p. 44.

9 Chandrasekhar, S. (1995). *Newton's Principia for the Common Reader*, p. 7. Oxford: Clarendon Press.

10 Chandrasekhar, S. (1995). *Op. cit.*, pp. 11–12.

11 Cohen, I.B. (1970). *Op. cit.*, p. 44.

12 Golinski, J. (1988). In Fauvel *et al.* eds. (1988), *Let Newton Be, op. cit.*, p. 153.

13 Golinski, J. (1988). *Op. cit.*, p. 156.

14 Golinski, J. (1988). *Op. cit.*, p. 160.

15 Golinski, J. (1988). *Op. cit.*, p. 151.

16 Rattansi, P. (1988). In Fauvel *et al.* eds. (1988), *Let Newton Be, op. cit.*, p. 185.

17 Rattansi, P. (1988). *Op. cit.*, p. 198.

18 Rattansi, P. (1988). *Op. cit.*, p. 200.

Appendix to Chapter 4: Notes on conic sections and central orbits

It may be useful to recall here some basic aspects of the geometry and algebra associated with conic sections and central forces.

A4.1 Equations for conic sections

The conic sections are the shapes obtained by slicing through a double cone at different angles. Their geometrical definition is as follows: they are the curves generated in a plane by the requirement that the ratio of the perpendicular distance of any point on the curve from a fixed straight line lying in the plane and the distance of that point on the curve from

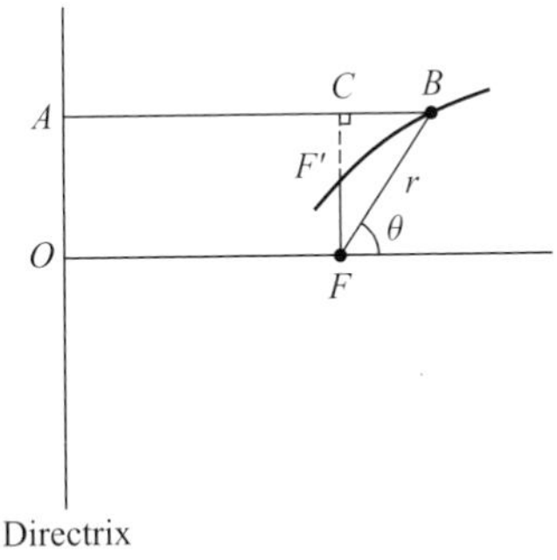

Figure A4.1

a fixed point be a constant. The fixed straight line is called the *directrix* and the fixed point the *focus*. From Fig. A4.1, this requirement can be written

$$\frac{AB}{BF} = \frac{AC + CB}{BF} = \text{constant} = k$$

that is,

$$\frac{AC + r\cos\theta}{r} = k \tag{A4.1}$$

where r and θ are polar coordinates with respect to F, the focus. Now AC and k are independent constants, and so, writing $AC = \lambda/e$ and $k = e^{-1}$, where λ and e are also constants, we can rewrite (A4.1) as

$$\frac{\lambda}{r} = 1 - e\cos\theta. \tag{A4.2}$$

We find immediately an interpretation for λ. When $\theta = \pi/2$ and $3\pi/2$ we find $r = \lambda$, that is, λ is the distance FF' in Fig. A4.1. Note that the curve is symmetric with respect to the line $\theta = 0$. The distance λ is known as the *semi-latus rectum*. The curves generated for different values of e are shown in Fig. A4.2. If $e < 1$, we obtain an ellipse, if $e = 1$, a parabola and if $e > 1$, a hyperbola. Notice that the hyperbola, $e > 1$, has two branches, the foci being referred to as the inner focus F_1 and the outer focus F_2 for the hyperbola branch on the right-hand side of the directrix, and vice versa for the other branch.

Equation (A4.2) can be written in different ways. Suppose instead we choose a Cartesian coordinate system with origin at the centre of the ellipse. In the polar coordinate system of (A4.2), the ellipse intersects the x-axis at $\cos\theta = \pm 1$, that is, at radii $r = \lambda/(1+e)$ and $r = \lambda/(1-e)$ from the focus on the x-axis. The semi-major axis of the ellipse therefore has length

$$a = \frac{1}{2}\left(\frac{\lambda}{1+e} + \frac{\lambda}{1-e}\right) = \frac{\lambda}{1-e^2}$$

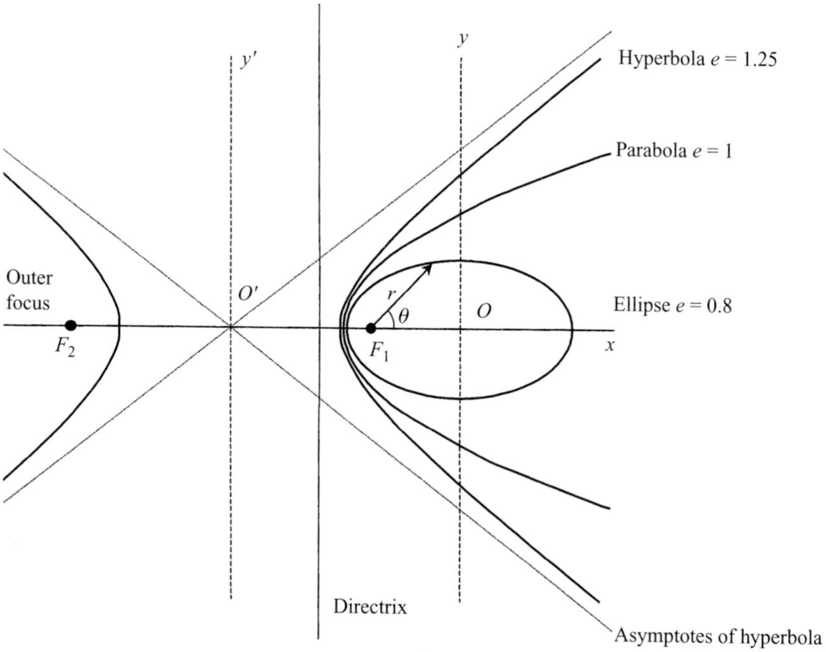

Figure A4.2: The conic sections: $e = 0$, circle (not shown); $0 < e < 1$, ellipse; $e = 1$, parabola; $e > 1$, hyperbola (which has two branches).

and the centre is at a distance $x = e\lambda/(1 - e^2)$ from F_1. In the new coordinate system, we can therefore write

$$x = r\cos\theta - e\lambda/(1 - e^2)$$
$$y = r\sin\theta.$$

With a bit of algebra, (A4.2) reduces to

$$\frac{x^2}{a^2} + \frac{y^2}{b^2} = 1 \tag{A4.3}$$

where $b = a(1 - e^2)^{1/2}$. Equation (A4.3) shows that b is the length of the semi-minor axis. The meaning of e also becomes apparent: if $e = 0$, the ellipse becomes a circle. It is therefore appropriate that e is called the *eccentricity* of the ellipse.

Exactly the same analysis can be carried out for hyperbolae, for which $e > 1$. The algebra is exactly the same, but now the Cartesian coordinate system is referred to an origin at O' in Fig. A4.2 and $b^2 = a^2(1 - e^2)$ is a negative quantity. We can therefore write

$$\frac{x^2}{a^2} - \frac{y^2}{b^2} = 1 \tag{A4.4}$$

where now $b = a(e^2 - 1)^{1/2}$.

One of the most useful ways of relating the conic sections to the orbits of test particles in central fields of force is to rewrite the equations in what is known as *pedal form*. In this form, the variable θ is replaced by a distance coordinate p, the perpendicular distance from

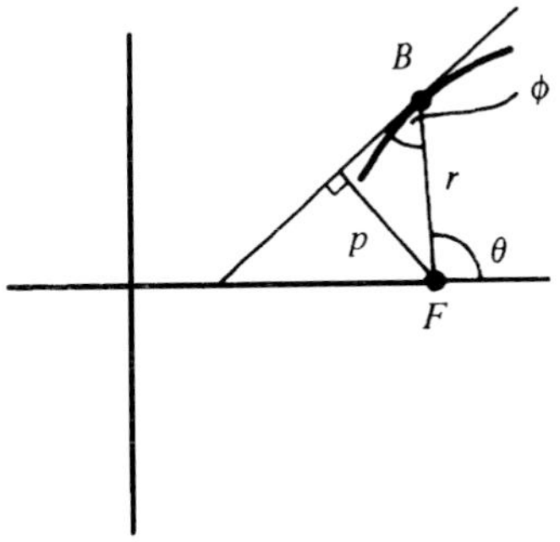

Figure A4.3

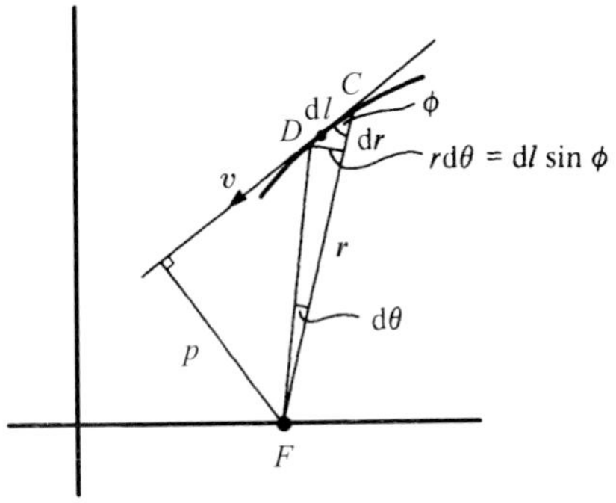

Figure A4.4

the tangent at a particular point on the curve to the focus, as illustrated in Fig. A4.3. From this diagram, it can be seen that $p = r \sin\phi$. We are interested in the tangent at the point B and so let us take the derivative of θ with respect to r. From (A4.2), we find

$$\frac{d\theta}{dr} = -\frac{\lambda}{r^2 e \sin\theta}. \tag{A4.5}$$

Figure A4.4 shows the changes $d\theta$ and dr as the point on the curve moves tangentially a distance dl. From the geometry of the small triangle defined by dl, dr and $r\,d\theta$, we see that

$$\tan\varphi = r\frac{d\theta}{dr}. \tag{A4.6}$$

We now have sufficient relationships to eliminate θ from (A4.2), because

$$p = r\sin\varphi \qquad \text{and} \qquad \tan\varphi = -\frac{\lambda}{re\sin\theta}. \tag{A4.7}$$

After a little algebra, we find that

$$\frac{\lambda}{p^2} = \frac{1}{A} + \frac{2}{r}, \tag{A4.8}$$

where $A = \lambda/(e^2 - 1)$. This the *pedal* or *pr equation* for conic sections. If A is positive, we obtain hyperbolae, if A is negative, ellipses and if A is infinite, parabolae. Notice that, in the case of hyperbolae, (A4.8) refers to values of p and r relative to the inner focus of the right-hand branch. If the hyperbola is referred to the outer focus of this branch then the equation becomes

$$\frac{\lambda}{p^2} = \frac{1}{A} - \frac{2}{r}.$$

We can now turn the whole exercise round and state: 'If we find equations of the form (A4.8) from a physical theory then the curves must be conic sections.'

A4.2 *Kepler's laws and planetary motion*

Let us recall Kepler's three laws of planetry motion, the origins of which were discussed in detail in Chapter 3.

K1: Planetary orbits are ellipses with the Sun at one focus;
K2: Equal areas are swept out by the radius vector from the Sun to a planet in equal times;
K3: The period T of a planetary orbit is proportional to the three-halves power of its mean distance r from the Sun, that is, $T \propto r^{3/2}$.

Consider first the motion of a test particle moving in a central field of force. For an isolated system, Newton's laws of motion lead directly to the *law of conservation of angular momentum*, as follows. If $\mathbf{\Gamma}$ is the total net torque acting on the system and $\boldsymbol{L}$ is the total angular momentum,

$$\mathbf{\Gamma} = \frac{\mathrm{d}\boldsymbol{L}}{\mathrm{d}t},$$

where $\boldsymbol{L} = m(\boldsymbol{r} \times \boldsymbol{v})$ for the case of a test mass m orbiting in the field of force. For a *central* field of force originating at the point from which $\boldsymbol{r}$ is measured, $\mathbf{\Gamma} = \boldsymbol{r} \times \boldsymbol{f} = 0$, since $\boldsymbol{f}$ and $\boldsymbol{r}$ are parallel or antiparallel vectors, and hence

$$\boldsymbol{L} = m(\boldsymbol{r} \times \boldsymbol{v}) = \text{constant vector}, \qquad \text{(A4.9)}$$

the law of conservation of angular momentum. Since there is no force acting outside the plane defined by the vectors $\boldsymbol{v}$ and $\boldsymbol{r}$, the motion is wholly confined to the vr-plane. The magnitude of $\boldsymbol{L}$ is constant and so (A4.9) can be rewritten

$$mrv \sin\varphi = \text{constant}. \qquad \text{(A4.10)}$$

But $r \sin\varphi = p$ and hence

$$pv = \text{constant} = h. \qquad \text{(A4.11)}$$

This calculation shows the value of introducing the geometrical quantity p. The *specific angular momentum* h is the angular momentum of the particle per unit mass.

Let us now work out the area swept out per unit time. The area of the triangle FCD in Fig. A4.4 is $\frac{1}{2} r\,\mathrm{d}l \sin\varphi$. Therefore, the area swept out per unit time is

$$\tfrac{1}{2} r \sin\varphi\, \mathrm{d}l/\mathrm{d}t = \tfrac{1}{2} rv \sin\varphi = \tfrac{1}{2} pv = \tfrac{1}{2} h = \text{constant}, \qquad \text{(A4.12)}$$

Kepler's second law; we can see that the latter is no more than a statement of the law of conservation of angular momentum in the presence of a central force. Notice that the result does not depend upon the radial dependence of the force – it is only the fact that it is a *central* force which is important, as was fully appreciated by Newton.

Kepler's first law can be derived from the law of conservation of energy in a gravitational field. Let us work out the orbit of a test particle of mass m in an inverse square field of force. Newton's law of gravity in vector form can be written

$$\boldsymbol{f} = -\frac{GmM}{r^2}\boldsymbol{i}_r,$$

where $\boldsymbol{i}_r$ is a unit vector in the direction of $\boldsymbol{r}$. Setting $\boldsymbol{f} = -m \,\mathrm{grad}\,\phi$, where ϕ is the gravitational potential energy per unit mass of the particle, we obtain $\phi = -GM/r$. Therefore, the expression for conservation of energy of the particle in the gravitational field is

$$\frac{mv^2}{2} - \frac{GmM}{r} = C, \qquad \text{(A4.13)}$$

where C is a constant of the motion. But we have shown that, because of the conservation of angular momentum, for *any* central field of force, $pv = h =$ constant. Therefore,

$$\frac{h^2}{p^2} = \frac{2GM}{r} + \frac{2C}{m} \qquad \text{(A4.14)}$$

or

$$\frac{h^2/(GM)}{p^2} = \frac{2}{r} + \frac{2C}{GMm}. \qquad \text{(A4.15)}$$

We recognise this equation as the pedal equation for conic sections, the exact form of the curve depending only on the sign of the constant C. Inspection of (A4.13) shows that if C is negative then the particle cannot reach $r = \infty$ and so takes up a bound elliptical orbit, but if C is positive then the orbits are hyperbolae. In the case $C = 0$, the orbit is parabolic. From the form of the equation, it is apparent that the origin of the force lies at the focus of the conic section.

To find the period of the particle in its elliptical orbit, we note that the area of the ellipse is πab and that the rate at which area is swept out is $\frac{1}{2}h$ (A4.12). Therefore the period of the elliptical orbit is

$$T = \frac{\pi ab}{h/2}. \qquad \text{(A4.16)}$$

Comparing (A4.8) and (A4.15), the semi-latus rectum is $\lambda = h^2/(GM)$ and, from the analysis of Section A4.1, a, b and λ are related by

$$b = a\left(1 - e^2\right)^{1/2} \quad \text{and} \quad \lambda = a\left(1 - e^2\right). \qquad \text{(A4.17)}$$

Substituting into A4.16, we find

$$T = \frac{2\pi}{(GM)^{1/2}}a^{3/2}; \qquad \text{(A4.18)}$$

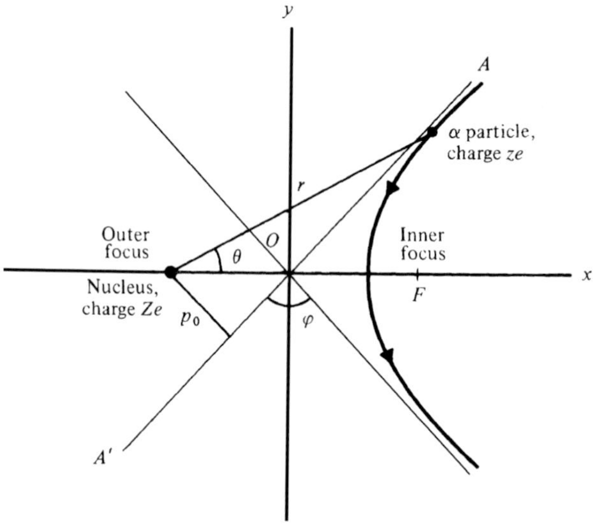

Figure A4.5

a is the semi-major axis of the ellipse and so is proportional to the mean distance of the particle from the focus. Consequently we have derived Kepler's law for the general case of elliptical orbits.

A4.3 *Rutherford scattering*

The scattering of α-particles by atomic nuclei was one of the great experiments in nuclear physics, carried out by Rutherford and his colleagues Geiger and Marsden in 1911. It established conclusively that the positive charge in atoms is contained within a point-like nucleus. The experiment involved firing α-particles at a thin gold sheet and measuring the angular distribution of the scattered α-particles. Rutherford was an experimenter of genius who was famous for his antipathy to theory. The calculation which follows was, however, carried out by Rutherford himself, consulting his theoretically minded colleagues to ensure that he had not made a blunder. Although it is now regarded as an elementary example of particle orbits in an inverse square field of force, no one had bothered to carry out the calculation before or realised its profound significance.

It is assumed that all the positive charge of an atom is contained in a compact nucleus. The deflection of an α-particle due to the inverse square law of electrostatic repulsion between the particle and the nucleus can be determined using the formalism of orbits under an inverse square field of force. Figure A4.5 shows the dynamics and geometry of the repulsion in both pedal and Cartesian forms. From the considerations of Section A.2, we can see that the trajectory must be a hyperbola, with the nucleus located at the outer focus. If there were no repulsion, the α-particle would travel along the diagonal AA' and pass by the nucleus at a perpendicular distance p_0, which is known as the *impact parameter*. The velocity of the α-particle at infinity is v_0.

The trajectory of the α-particle, (A4.4), is

$$\frac{x^2}{a^2} - \frac{y^2}{b^2} = 1,$$

from which we can find the asymptotes $x/y = \pm a/b$. Therefore, the scattering angle φ shown in Fig. A4.5 is given by

$$\tan\frac{\varphi}{2} = \frac{x}{y} = \frac{a}{b} = \frac{1}{(e^2-1)^{1/2}}. \tag{A4.19}$$

The pedal equation for the hyperbola with respect to its outer focus,

$$\frac{\lambda}{p^2} = \frac{1}{A} - \frac{2}{r}, \qquad A = \lambda/\sqrt{e^2-1},$$

may be compared with the equation for the conservation of energy of the α-particle in the field of the nucleus,

$$\frac{mv^2}{2} + \frac{Zze^2}{4\pi\epsilon_0 r} = \frac{mv_0^2}{2}. \tag{A4.20}$$

Since $pv = h$ is a constant, (A4.20) can be rewritten as

$$\left(\frac{4\pi\epsilon_0 m h^2}{Zze^2}\right)\frac{1}{p^2} = \frac{4\pi\epsilon_0 m v_0^2}{Zze^2} - \frac{2}{r}. \tag{A4.21}$$

Therefore $A = Zze/(4\pi\epsilon_0 m v_0^2)$ and $\lambda = 4\pi\epsilon_0 m h^2/(Zze^2)$. Recalling that $p_0 v_0 = h$, we can combine (A4.19) with the values of a and λ to find

$$\cot\frac{\varphi}{2} = \left(\frac{4\pi\epsilon_0 m}{Zze^2}\right) p_0 v_0^2. \tag{A4.22}$$

Thus, the number of α-particles scattered through an angle φ is directly related to their impact parameter p_0. If a parallel beam of α-particles is fired at a nucleus, the number of particles with impact parameters in the range p_0 to $p_0 + \mathrm{d}p_0$ is proportional to the area of an annulus of width $\mathrm{d}p_0$ at radius p_0:

$$N(p_0)\,\mathrm{d}p_0 \propto 2\pi p_0\,\mathrm{d}p_0.$$

Therefore, the number of particles scattered between φ to $\varphi + \mathrm{d}\varphi$ is

$$N(\varphi)\,\mathrm{d}\varphi \propto p_0\,\mathrm{d}p_0 \propto \left(\frac{1}{v_0^2}\cot\frac{\varphi}{2}\right)\left(\frac{1}{v_0^2}\csc^2\frac{\varphi}{2}\right)\mathrm{d}\varphi$$

$$= \frac{1}{v_0^4}\cot\frac{\varphi}{2}\csc^2\frac{\varphi}{2}\,\mathrm{d}\varphi. \tag{A4.23}$$

This result can also be written in terms of the *probability* $p(\varphi)$ that the α-particle is scattered through an angle φ. With respect to the incident direction, the number of particles scattered between angles φ and $\varphi + \mathrm{d}\varphi$ is

$$N(\varphi)\,\mathrm{d}\varphi = \tfrac{1}{2}\sin\varphi\, p(\varphi)\,\mathrm{d}\varphi. \tag{A4.24}$$

If the scattering is uniform over all solid angles then $p(\varphi) =$ constant. Equating the results (A4.23) and (A4.24), we find the famous result

$$p(\varphi) \propto \frac{1}{v_0^4} \csc^4 \frac{\varphi}{2}. \qquad \text{(A4.25)}$$

This was the probability law which Rutherford derived in 1911.[A1] He and his colleagues found that the α-particles scattered by a thin gold sheet followed exactly this relation for scattering angles between 5° and 150°, over which interval the function $\cot(\varphi/2)\csc^2(\varphi/2)$ varies by a factor of 40 000. From the known speeds of the α-particles, and the fact that the law was obeyed to large scattering angles, they deduced that the nucleus must be less than about 10^{-14} m in radius, that is, very much smaller than the size of atoms $\sim 10^{-10}$ m.

Pais has given an amusing account of Rutherford's discovery of the law.[A2] As he points out, Rutherford was lucky, in that he happened to use α-particles of the right energy to observe the distribution law of scattered particles. He also remarks that Rutherford did not mention this key result at the first Solvay conference held in 1911, nor was the full significance of the result appreciated for a few years. He first used the term 'nucleus' in his book on radioactivity in 1912, stating that

> The atom must contain a highly charged nucleus.

Only in 1914, during a Royal Society discussion, did he come out forcefully in favour of the nuclear model of the atom.

A4.4 Appendix references

A1 Rutherford, E. (1911). *Phil. Mag.*, **21**, 669.
A2 Pais, A. (1986). *Inward Bound*, pp.188–93. Oxford: Clarendon Press.

Case Study II
Maxwell's equations

Each case study has a different emphasis and this one is as extreme as they get. The central theme is the origin of Maxwell's equations, which might seem a much more straightforward story than some of the other case studies. This was my opinion until I understood how Maxwell actually arrived at his great discovery of the *displacement current*. It turns out to be as remarkable an example of model building as I have encountered anywhere in physics and strikes to the heart of the nature of electromagnetism. It is also a wonderful example of how fruitful it can be to work by analogy, provided one is constrained by experiment. The story culminates in the discovery that electromagnetic disturbances propagate at the speed of light, leading directly to the unification of light and electromagnetism and to Hertz's beautiful experiments, which fully vindicated Maxwell's theory.

Along the way, we pay tribute to Faraday's genius as an experimenter in discovering the phenomenon of *electromagnetic induction* (Figure II.1) and many other aspects of electromagnetic phenomena. His invention of the concept of *lines of force*, what I have called 'mathematics without mathematics', was crucial to the mathematisation of electromagnetism, and to Maxwell's theoretical studies. The key role of vector calculus in simplifying the mathematics of electromagnetism provides an opportunity for revising some of that material, and a number of useful results are included in the appendix to Chapter 5.

Then, we completely invert the process. In Chapter 6, we begin with the structure of Maxwell's equations and find out what we have to do to endow them with physical meaning, making the minimum number of assumptions. This may seem a somewhat contrived exercise, but it provides insight into the mathematical structure underlying the theory and is not so different from what has to be done in frontier areas of theoretical research, where it is crucial to relate the mathematics to what can be measured experimentally. It also serves to give the subject much greater coherence; classical electromagnetism can be fully encompassed by a set of four partial differential equations.

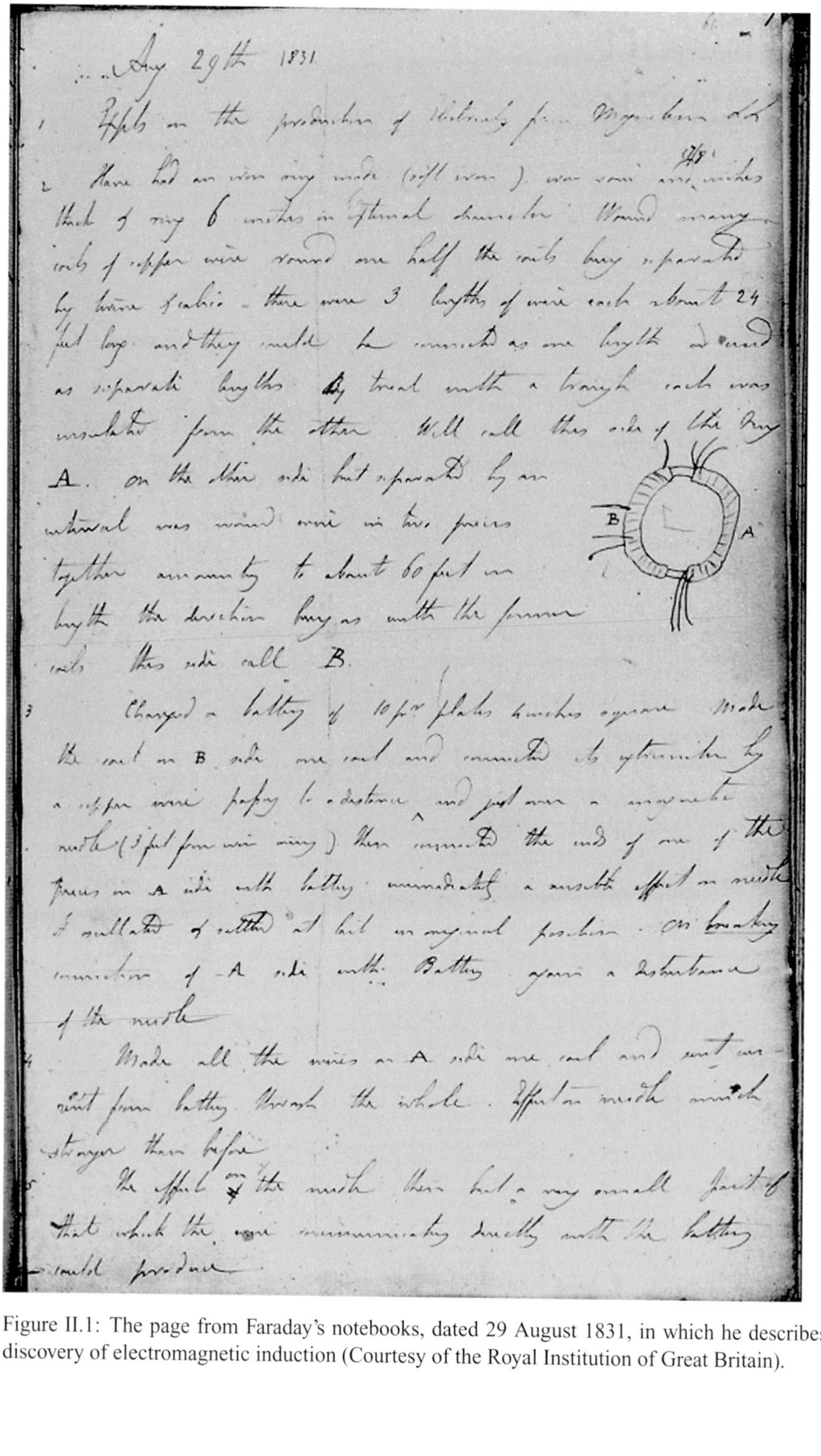

Figure II.1: The page from Faraday's notebooks, dated 29 August 1831, in which he describes the discovery of electromagnetic induction (Courtesy of the Royal Institution of Great Britain).

5 The origin of Maxwell's equations

5.1 How it all began

Electricity and magnetism have an ancient history. Magnetic materials are mentioned as early as 800 BC by the Greek writers, the word 'magnet' being derived from the mineral magnetite, which was known to attract iron in its natural state and which was mined in the Greek province of Magnesia in Thessaly. Magnetic materials were of special importance because of their use in compasses, and this is reflected in the English word for the mineral, lodestone, meaning leading stone. Static electricity was also known to the Greeks through the electrostatic phenomena observed when amber is rubbed with fur – the Greek work for amber is *elektron*. The first systematic study of magnetic and electric phenomena was published in 1600 by William Gilbert (1544–1603) in his treatise *De Magnete, Magneticisque Corporibus, et de Magno Magnete Tellure*. The main subject of the treatise was the Earth's magnetic field, which he showed was similar to that of a bar magnet. He also described the force between two bodies charged by friction and named it the *electric* force between them.

In addition to his famous experiments, in which he showed that lightning is an electrostatic discharge, Benjamin Franklin (1706–90) systematised the laws of electrostatics and defined the conventions for naming positive and negative electric charges. In the course of these studies, he also enunciated the *law of conservation of electric charge*. In 1767, Joseph Priestley (1733–1804) showed that inside a hollow conducting sphere there are no electric forces. From this, he inferred that the force law for electrostatics must be of inverse square form, just as in the case of gravity. A modern version of this experiment, carried out by Williams, Faller and Hall in 1971, established that the inverse square law holds good to better than one part in 3×10^{15}.

By the end of the eighteenth century, many of the basic experimental features of electrostatics and magnetostatics had been established. In the 1770s and 1780s, Charles-Augustin Coulomb (1736–1806) performed very sensitive electrostatic experiments, which established directly the inverse square law of electrostatics. He undertook similar experiments in magnetostatics, using very long magnetic dipoles so that the properties of each pole of the dipole could be considered separately. In SI notation, which we will use throughout this book, the laws can be written in scalar form as

$$f_e = \frac{q_1 q_2}{4\pi\epsilon_0 r^2}, \tag{5.1}$$

$$f_{\mathrm{m}} = \frac{\mu_0 p_1 p_2}{4\pi r^2}, \tag{5.2}$$

where q_1 and q_2 are the electric charges of two point objects separated by distance r and p_1 and p_2 their magnetic pole strengths. The constants $1/(4\pi\epsilon_0)$ and $\mu_0/(4\pi)$ are included in these definitions according to the SI convention. Purists might prefer the whole analysis of this chapter to be carried out in the original notation, but such adherence to historical authenticity might obscure the essence of the argument for the modern reader. In modern vector notation the directional dependence of the electrostatic force can incorporated explicitly:

$$\boldsymbol{f}_{\mathrm{e}} = \frac{q_1 q_2}{4\pi\epsilon_0 r^3}\boldsymbol{r} \quad \text{or} \quad \boldsymbol{f}_{\mathrm{e}} = \frac{q_1 q_2}{4\pi\epsilon_0 r^2}\boldsymbol{i}_r, \tag{5.3}$$

where $\boldsymbol{i}_r$ is the unit vector directed radially *away* from one charge in the direction of the other. A similar pair of expressions is found for magnetostatic forces:

$$\boldsymbol{f}_{\mathrm{m}} = \frac{\mu_0 p_1 p_2}{4\pi r^3}\boldsymbol{r} \quad \text{or} \quad \boldsymbol{f}_{\mathrm{m}} = \frac{\mu_0 p_1 p_2}{4\pi r^2}\boldsymbol{i}_r. \tag{5.4}$$

The late eighteenth and early nineteenth century was a period of extraordinary brilliance in French mathematics. Of special importance for this story are the works of Siméon-Denis Poisson (1781–1840), who was a pupil of Pierre–Simon Laplace (1749–1827), and Joseph-Louis Lagrange (1736–1813). In 1812, Poisson published his famous *Mémoire sur la distribution de l'électricité à la surface des corps conducteurs*, in which he demonstrated that many of the problems of electrostatics can be simplified by the introduction of the *electrostatic potential* V, which is the solution of Poisson's equation

$$\frac{\partial^2 V}{\partial x^2} + \frac{\partial^2 V}{\partial y^2} + \frac{\partial^2 V}{\partial z^2} = -\frac{\rho_{\mathrm{e}}}{\epsilon_0}, \tag{5.5}$$

where ρ_{e} is the electric charge density distribution. The electric field strength $\boldsymbol{E}$ is then given by*

$$\boldsymbol{E} = -\mathrm{grad}\, V. \tag{5.6}$$

In 1826, Poisson published the corresponding expressions for the magnetic flux density $\boldsymbol{B}$ in terms of the magnetostatic potential V_{m}:

$$\frac{\partial^2 V_{\mathrm{m}}}{\partial x^2} + \frac{\partial^2 V_{\mathrm{m}}}{\partial y^2} + \frac{\partial^2 V_{\mathrm{m}}}{\partial z^2} = 0, \tag{5.7}$$

where $\boldsymbol{B}$ is given by

$$\boldsymbol{B} = -\mu_0\, \mathrm{grad}\, V_{\mathrm{m}}. \tag{5.8}$$

* Inevitably, some symbols in physics are used to mean different physical quantities in different contexts. One obvious example is E for energy and E for electric field. To avoid confusion, E will nearly always mean energy; $\boldsymbol{E}$ will refer to the electric field vector, $|\boldsymbol{E}|$ to its magnitude and E_x, E_y, E_z to its components. So far as is practical we adhere to the recommended Royal Society conventions.

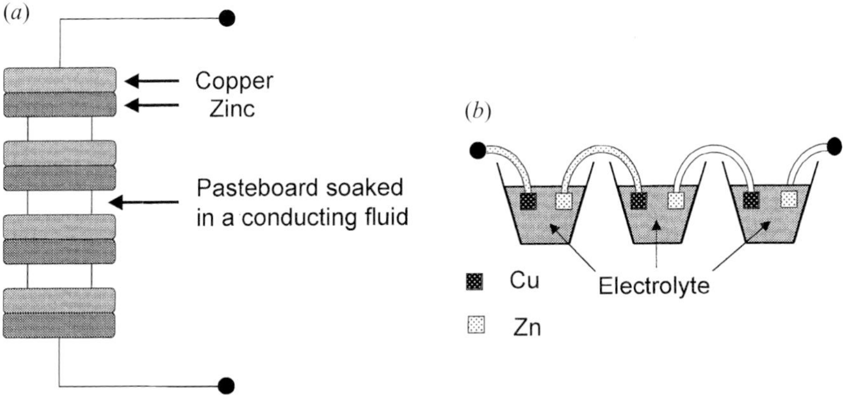

Figure 5.1: (*a*) Illustrating the construction of a voltaic pile. (*b*) Illustrating Volta's *crown of cups*, resembling a modern bank of batteries joined in series.

Until 1820, electrostatics and magnetostatics appeared to be quite separate phenomena, but this changed with the development of the science of *current electricity*. In parallel with the development of the laws of electrostatics and magnetostatics in the latter years of the eighteenth century, the Italian anatomist Luigi Galvani (1737–98) discovered that electrical effects could stimulate the muscular contraction of frogs' legs. In 1791, he showed that, when two dissimilar metals were used to make the connection between nerve and muscle, the same form of muscular contraction was observed. This was announced as the discovery of *animal electricity*. Alessandro Volta (1745–1827) suspected that the electric current was associated with the presence of different metals in contact with a moist body. In 1800, he demonstrated this by constructing what became known as a *voltaic pile*, which consisted of interleaved layers of copper and zinc separated by layers of pasteboard soaked in a conducting liquid (Fig. 5.1(*a*)). With this pile, Volta was able to demonstrate all the phenomena of electrostatics – the production of electric discharges, electric shocks, and so on. By far the most important aspect of Volta's experiments was, however, the fact that he had discovered a *controllable source* of electric current.

A problem with the voltaic cell was that it had a short lifetime because the pasteboard dried out. This led Volta to invent his *crown of cups*, in which the electrodes were placed in glass vessels (Fig. 5.1(*b*)) – these were the precursors of modern batteries. These discoveries were well known to the general public. It is intriguing that magnetism, along with mesmerism, was used by Despina in Mozart's *Cosi fan Tutti*, first performed in 1791, to revive the heavily disguised Ferrando and Guglielmo, and that 'galvanism' played an essential role in the inspiration for Mary Shelley's *Frankenstein*, written in 1816.

A key experimental advance was made in 1820 when Hans-Christian Ørsted (1777–1851) demonstrated that there is always a magnetic field associated with an electric current – this marked the beginning of the science of *electromagnetism*. As soon as his discovery was announced, the physicists Jean-Baptiste Biôt (1774–1862) and Félix Savart (1791–1841) set out to discover the dependence of the strength of the magnetic field at a position $\boldsymbol{r}$ from an element $d\boldsymbol{l}$ in which a current I is flowing. In the same year, they found the answer, the

Biôt–Savart law, which in modern vector notation can be written

$$\mathrm{d}\boldsymbol{B} = \frac{\mu_0 I \, \mathrm{d}\boldsymbol{l} \times \boldsymbol{r}}{4\pi r^3}. \tag{5.9}$$

Notice that the signs of the vectors are important in finding the correct direction of the field. The vector $\mathrm{d}\boldsymbol{l}$ is in the direction of the current I and $\boldsymbol{r}$ is measured from the current element $I\mathrm{d}\boldsymbol{l}$ to the point at vector distance $\boldsymbol{r}$.

Next, André-Marie Ampère (1775–1836) extended the Biot–Savart law to relate the current flowing though a closed loop to the integral of the component of the magnetic flux density around the loop. In modern vector notation, *Ampère's circuital law* in free space can be written

$$\oint_C \boldsymbol{B} \cdot \mathrm{d}\boldsymbol{s} = \mu_0 I_{\text{enclosed}}, \tag{5.10}$$

where I_{enclosed} is the total electric current flowing through the area enclosed by the loop C.

The story develops rapidly. In 1826, Ampère published his famous treatise *Theorie des phénomènes électro-dynamique, uniquement déduite de l'expérience*, which included a proof that the magnetic field of a current loop could be represented by an equivalent magnetic shell. In the treatise, he also formulated the equation for the force between two elements, $\mathrm{d}\boldsymbol{l}_1$ and $\mathrm{d}\boldsymbol{l}_2$, carrying currents I_1 and I_2:

$$\mathrm{d}\boldsymbol{F}_2 = \frac{\mu_0 I_1 I_2 \, \mathrm{d}\boldsymbol{l}_1 \times (\mathrm{d}\boldsymbol{l}_2 \times \boldsymbol{r})}{4\pi r^3}. \tag{5.11}$$

$\mathrm{d}\boldsymbol{F}_2$ is the force acting on the current element $I_2\mathrm{d}\boldsymbol{l}_2$, the vector $\boldsymbol{r}$ being drawn from $\mathrm{d}\boldsymbol{l}_1$ to $\mathrm{d}\boldsymbol{l}_2$. Ampère also demonstrated the relation between this law and the Biôt–Savart law.

In 1827, Georg Simon Ohm (1787–1854) formulated the relation between potential difference V and the current I, what is now known as *Ohm's law*, $V = RI$, where R is the resistance of the material through which the current flows. Sadly, this pioneering work was not well received by Ohm's colleagues in Cologne and he resigned from his post in disappointment. The central importance of his work was subsequently recognised, however, and he was awarded the Copley Medal of the Royal Society of London in 1841.

All the results described above were known by 1830 and comprise the whole of *static electricity*, namely, the forces between *stationary* charges, magnets and currents. An essential feature of Maxwell's equations is that they deal with *time-varying* phenomena as well. Over the succeeding 20 years, all the basic experimental features of time-varying electric and magnetic fields were established and the hero of this story is unquestionably Michael Faraday (1791–1867).

5.2 Michael Faraday – mathematics without mathematics

Michael Faraday was born into a poor family, his father being a blacksmith who moved with his family to London in early 1791. He began life as a bookbinder's apprentice working in a Mr Ribeau's shop and learned his early science by reading the books he had to bind.[1] These included the *Encyclopaedica Britannica*, his attention being particularly attracted by the article on electricity by James Tyler. He attempted to repeat some of the experiments himself and built a small electrostatic generator out of bottles and old wood.

In 1812, one of Mr Ribeau's customers gave Faraday tickets to Humphry Davy's lectures at the Royal Institution. Faraday sent a bound copy of his lecture notes to Davy, inquiring if there was any post which he might fill, but none was available then. In October of that same year, however, Humphry Davy was temporarily blinded by an explosion while working with the dangerous chemical 'nitrate of chlorine', and needed someone to write down his thoughts. Faraday was recommended for this task and subsequently, on 1 March 1813, he took up a permanent position as assistant to Davy at the Royal Institution, where he was to remain for the rest of his life.

Soon after Faraday's appointment, Davy decided to visit scientific institutions on continental Europe and took Faraday with him as his scientific assistant. During the next 18 months, they met most of the great scientists of the day – in Paris, they met Ampère, Humboldt, Gay-Lussac, Arago and many others. In Italy, they met Volta and, while in Genoa, observed experiments conducted on the torpedo, a fish capable of giving electric shocks.

Øersted's announcement of the discovery of a connection between electricity and magnetism in 1820 caused a flurry of scientific activity. Many articles were submitted to the scientific journals describing other electromagnetic effects and attempting to explain them. Faraday was asked to survey this mass of experiment and speculation by the editor of the *Philosophical Magazine* and, as a result, began his systematic study of electromagnetic phenomena.

Faraday proceeded to repeat all the experiments reported in the literature. In particular, he noted the movement of the poles of a small magnet in the vicinity of a current-carrying wire. It had already been noted by Ampère that the force acting upon the magnetic pole is such as to move it in a circle about the wire. Alternatively, if the magnet were kept fixed, then a current-carrying wire would feel a force, moving it in a circle about the magnet. Faraday proceeded to demonstrate these phenomena by two beautiful experiments (Fig. 5.2). In one of them, seen on the right-hand side in Fig. 5.2, a magnet was placed upright in a dish of mercury with one pole projecting above the surface. A wire was arranged so that one end of it was attached to a small cork, which floated on the mercury, and the other end was fixed above the end of the magnet. When a current was passed through the wire, the wire was found to rotate about the axis of the magnet, exactly as Faraday had expected. In the second experiment, illustrated on the left-hand side of Fig. 5.2, the current-carrying wire was fixed and the magnet free to rotate about the wire. These were the first *electric motors* ever constructed.

These experiments led Faraday to the crucial concept of *magnetic lines of force*, which sprang from observation of the patterns which iron filings take up about a magnet (Fig. 5.3). A magnetic line of force, or field line, represents the direction in which the force acting upon a magnetic pole acts when it is placed in a magnetic field. The greater the number of lines of force per unit area in the plane perpendicular to the field lines, the greater the force acting upon the magnetic pole. Faraday was to lay great emphasis upon the use of lines of force as a means of visualising the effects of stationary and time-varying magnetic fields.

Now, there was a problem with Faraday's picture. The force between two magnetic poles acts along the line between them. How could this behaviour be reconciled with the

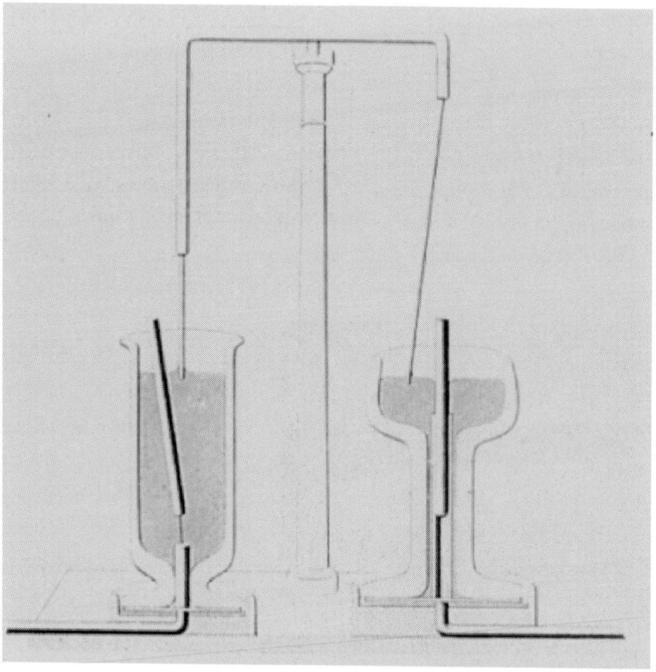

Figure 5.2: Faraday's experiments illustrating the forces acting between a current-carrying wire and a magnet. In the left-hand half of the diagram, the current-carrying wire is fixed and the magnet rotates about the vertical axis; in the right-hand half of the diagram, the magnet is fixed and the current-carrying wire rotates about the vertical axis. These were the first electric motors to be constructed. (Courtesy of the Royal Institution of Great Britain.)

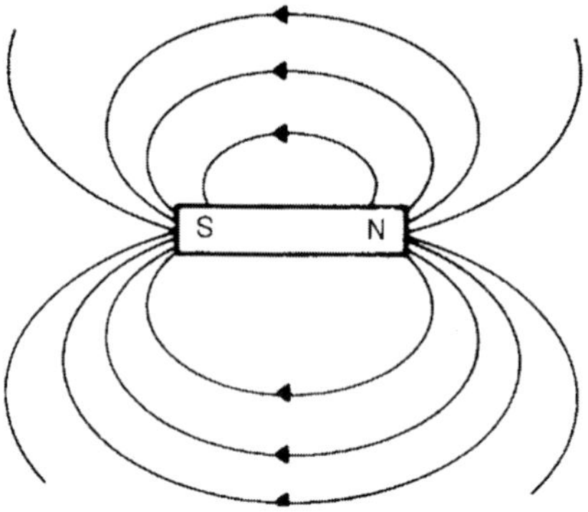

Figure 5.3: Illustrating Faraday's concept of the lines of force about a bar magnet.

circular lines of force observed about a current-carrying wire? Faraday showed how he could simulate all the effects of a magnet if the current-carrying wire were bent into a loop, as illustrated in Fig. 5.4. Using the concept of lines of force, he argued that the magnetic lines of force would be compressed within the loop with the result that one side of the loop would have one polarity and the other the opposite polarity. He proceeded to demonstrate experimentally that, indeed, all aspects of the forces associated with currents flowing in wires could be understood in terms of magnetic lines of force. In fact, all the laws concerning the

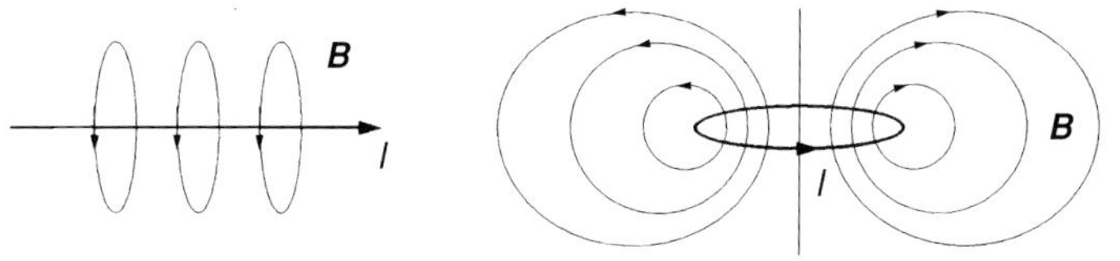

Figure 5.4: Illustrating Faraday's reasoning to illustrate the equivalence of a magnetic field and a bar magnet: the long straight wire on the left is bent into the loop on the right, compressing the field lines which lie within the loop.

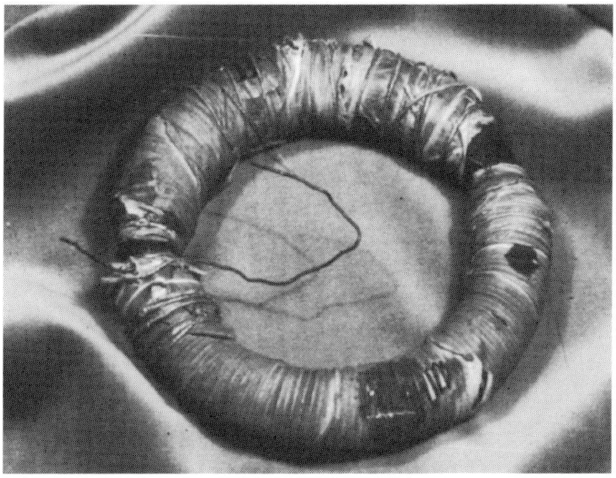

Figure 5.5: The apparatus with which Faraday first demonstrated electromagnetic induction. (Courtesy of the Royal Institution of Great Britain.)

forces between static magnets and currents can be derived from Faraday's deep insight of the exact equivalence of magnetic dipoles and current loops, as demonstrated in appendix section A5.7.

The great advance occurred in 1831. Believing firmly in the symmetry of nature, Faraday conjectured that since an electric current produced a magnetic field it must also be possible to generate an electric current from a magnetic field. In 1831, he learned of Joseph Henry's experiments in Albany, New York, in which very powerful electromagnets were used. Faraday immediately had the idea of observing the strain in the material of a electromagnet caused by the lines of force. He built a strong electromagnet by winding onto a thick iron ring, an insulating wire through which a current could be passed, thus creating a magnetic field within the ring. The effects of the strain were to be detected by another winding on the ring, which was attached to a galvanometer to measure the amount of electric current produced. A photograph of his original apparatus is shown in Fig. 5.5.

The experiment was conducted on 29 August 1831 and is recorded meticulously in Faraday's laboratory notebooks (Fig. II.1). The effect was not at all what Faraday might have expected. When the primary circuit was closed, there was a displacement of the galvanometer needle in the secondary winding – an electric current had been induced in the secondary wire through the medium of the iron ring. Deflections of the galvanometer were *only* observed when the current in the electromagnet was switched on and off – there was no effect when a steady current flowed in the electromagnet. In other words, the effect only seemed to be associated with changing currents and, consequently, changing magnetic fields. This was the discovery of *electromagnetic induction*.

Over the next few weeks, there followed a series of remarkable experiments in which the nature of electromagnetic induction was established. Faraday improved the sensitivity of his apparatus, and he also observed that the electric current produced in the second circuit was in opposite directions when the current was switched on and off. Next, he tried coils of different shapes and sizes and discovered that the iron bar was not needed to create the effect. On 17 October 1831, a new experiment was carried out in which an electric current was created by sliding a cylindrical bar magnet into a long coil (or solenoid) connected to a galvanometer. Then, in a famous experiment, demonstrated at the Royal Society of London on 28 October 1831, he showed how a continuous electric current could be generated by rotating a copper disc between the poles of the 'great horse-shoe magnet' which belonged to the Society. The axis and the edge of the disc were connected by a sliding contact to a galvanometer and, as the disc rotated, the needle was deflected. On 4 November 1831, he found that simply moving a copper wire between the poles of the magnet could create an electric current. Thus, within a period of four months, he had discovered the *transformer* and the *dynamo*.

As early as 1831, Faraday had established the qualitative form his law of induction in terms of the concept of lines of force – *the electromotive force induced in a current loop is directly related to the rate at which magnetic field lines are cut*, adding that

> By magnetic curves, I mean lines of magnetic force which would be depicted by iron filings.[2]

He now realised that the term 'electricity' could mean a number of different things. In addition to *magneto-electricity*, which he had just discovered, there was *static electricity*, which could be produced by friction, as had been known from ancient times. *Voltaic electricity* was associated with chemical effects in a voltaic pile. In *thermo-electricity*, a potential difference is created when materials of different types are placed in contact and the ends at which the joins are made are maintained at different temperatures. Finally, there is *animal electricity*, produced by fish such as torpedos and electric eels, which Faraday had seen on his travels with Davy. He asked the question which may seem obvious to us now, with hindsight, but which illustrates his deep insight at the time – are these different forms of electricity the same? In 1832, he performed an elegant series of experiments in which he showed that he could produce similar chemical, magnetic and other effects, no matter what the source of the electricity might be, including electric fish.

Although the law of induction began to emerge at an early stage, it took Faraday many years to complete all the necessary experimental work to demonstrate the general validity of the law – namely, that it is the rate of change of the total magnetic flux linking the circuit, whatever the origin of the flux, which determines the magnitude of the electromotive force induced in the circuit. In 1834, Lenz enunciated the law which cleared up the problem of the direction of the induced electromotive force in the circuit – *the electromotive force acts in such a direction as to oppose the change in magnetic flux*.

Faraday could not formulate his theoretical ideas mathematically, but he was convinced that the concept of lines of force provided the key to understanding electromagnetic phenomena. In 1846, he speculated in a discourse to the Royal Institution that light might be some form of disturbance propagating along magnetic field lines. He published these ideas in a paper entitled *Thoughts on Ray Vibrations*, but they were received with considerable scepticism. Faraday had, however, indeed hit upon the correct concept. As we will see in the next section, James Clerk Maxwell showed in 1864 that light is indeed a form of electromagnetic radiation. With his outstanding physical intuition and mathematical ability, he was able to put Faraday's discoveries into mathematical form and then to show that any electromagnetic wave propagating in a vacuum travels at the speed of light. As Maxwell himself acknowledged in his great paper 'A dynamical theory of the electromagnetic field', published in 1865,

> The conception of the propagation of transverse magnetic disturbances to the exclusion of normal ones is distinctively set forth by Professor Faraday in his *Thoughts on Ray Vibrations*. The electromagnetic theory of light as proposed by him is the same in substance as that which I have begun to develop in this paper, except that in 1846 there was no data to calculate the velocity of propagation.[3]

Although Faraday could not formulate his ideas mathematically, his deep feeling for the behaviour of electric and magnetic fields provided the essential insights needed by mathematicians such as Maxwell to develop the mathematical theory of the electromagnetic field. In Maxwell's words,

> As I proceeded with the study of Faraday, I perceived that his method of conceiving of phenomena was also a mathematical one, though not exhibited in the conventional form of mathematical symbols . . . I found, also, that several of the most fertile methods of research discovered by the mathematicians could be expressed much better in terms of ideas derived from Faraday than in their original form.[4]

I must confess that when I first learned electromagnetism lines of force were an obstacle to my understanding, largely because it was not explained clearly to me that they are only a *model* for what is going on. The things you actually measure in a laboratory experiment are vector forces at different points in space and the fictitious lines of force are conceptual models to represent these vector fields. We will return to this key point in the next section.

Before we leave Faraday, we must describe one further key discovery, which was to influence Maxwell's thinking about the nature of electromagnetic phenomena. Faraday had an instinctive belief in the unity of the forces of nature, and in particular that there should be a close relation between the phenomena of light, electricity and magnetism. In a series of experiments carried out towards the end of 1845, he attempted to find out whether the polarisation of light could be influenced by the presence of a strong electric field, but no effect was seen. Turning instead to magnetism, his experiments also showed no effect until he passed the light through lead borate glass in the presence of a strong magnetic field. He had cast these glasses himself in the years 1825 to 1830 as part of a project sponsored by the Royal Society of London to create superior optical glasses for use in astronomical instruments. These heavy glasses had the property of having large refractive indices. Faraday demonstrated the phenomenon now known as *Faraday rotation*, in which the plane of polarisation of linearly polarised light is rotated when the light rays travel along the magnetic field direction in the presence of a transparent dielectric. William Thomson, later Lord Kelvin, interpreted this phenomenon as evidence that the magnetic field caused a rotational motion of the electric charges in molecules. Following an earlier suggestion by Ampère, he envisaged magnetism as being essentially rotational in nature, and this was to influence strongly Maxwell's model for a magnetic field in free space.

Here we must leave Michael Faraday. He is an outstanding example of a meticulous experimenter of genius with no mathematical training, who was never able to express the results of his researches in mathematical form – there is not a single mathematical formula in his writings. He had, however, an intuitive genius for experiment and for devising empirical conceptual models to account for his results. These models embodied the mathematics necessary to formulate the theory of the electromagnetic field.

5.3 How Maxwell derived the equations for the electromagnetic field

James Clerk Maxwell (1831–79) was born and educated in Edinburgh. In 1850, he went up to Cambridge where he studied for the Mathematical Tripos with considerable distinction. As James David Forbes, Professor of Natural Philosophy at Edinburgh University, wrote to William Whewell, the Master of Trinity College, in April 1852,

> Pray do not suppose that . . . I am not aware of his exceeding uncouthness, as well Mathematical as in other respects; . . . I thought the Society and Drill of Cambridge the only chance of taming him and much advised his going.[5]

Allied to his formidable mathematical abilities was a physical imagination which could appreciate the empirical models of Faraday and give them mathematical substance. Peter Harman's brilliant study *The Natural Philosophy of James Clerk Maxwell*[6] is essential reading for understanding Maxwell's intellectual approach and achievement.

5.3.1 *'On Faraday's lines of force' (1856)*

A very distinctive feature of Maxwell's thinking was his ability to work by *analogy*. As early as 1856, he described his approach in an essay entitled *Analogies in Nature* written for the Apostles' Club at Cambridge. The technique is best illustrated by the examples given below, but its essence can be caught in the following passage from the essay.

> Whenever [men] see a relation between two things they know well, and think they see there must be a similar relation between things less known, they reason from one to the other. This supposes that, although pairs of things may differ widely from each other, the *relation* in the one pair may be the same as that in the other. Now, as in a scientific point of view the *relation* is the most important thing to know, a knowledge of the one thing leads us a long way towards knowledge of the other.[7]

In other words, the approach consists of recognising mathematical similarities between quite distinct physical problems and seeing how far one can go in applying the successes of one theory to different circumstances. In relation to electromagnetism, he found formal analogies between the mathematics of mechanical and hydrodynamical systems and the phenomena of electrodynamics. He acknowledged throughout this work his debt to William Thomson, who had made substantial steps in mathematising electric and magnetic phenomena. Maxwell's great contribution was not only to take this process very much further but also to give it real physical content.

In the same year, 1856, Maxwell published the first of his papers on electromagnetism, 'On Faraday's lines of force'.[8] In the preface to his *Treatise on Electricity and Magnetism* of 1873, he recalled:

> ... before I began the study of electricity I resolved to read no mathematics on the subject till I had first read through Faraday's *Experimental Researches in Electricity*.[9]

The first part of the paper enlarged upon the technique of analogy and drew particular attention to its application to incompressible fluid flow and magnetic lines of force. We will use the vector-operator expressions div, grad and curl in our development, although the use of vector methods was only introduced by Maxwell into the study of electromagnetism in a paper of 1870 entitled 'On the mathematical classification of physical quantities'[10] – he invented the terms 'slope' (now 'gradient'), 'curl' and 'convergence' (the opposite of divergence) to provide an intuitive feel for the meaning of these operators. In 1856, they were not available and the partial derivatives were written out in Cartesian form.

Let us recall the *continuity equation*, or *equation of conservation of mass*, for incompressible fluid flow. Consider a volume v bounded by a closed surface S. Using vector notation, the mass flow per unit time through a surface element $\mathrm{d}\boldsymbol{S}$ is $\rho \boldsymbol{u} \cdot \mathrm{d}\boldsymbol{S}$, where $\boldsymbol{u}$ is the fluid velocity and ρ its density distribution. Therefore, the total mass flux through the closed surface is $\int_S \rho \boldsymbol{u} \cdot \mathrm{d}\boldsymbol{S}$. This is equal to the rate of loss of mass from v, which is

$$-\frac{\mathrm{d}}{\mathrm{d}t}\int_v \rho \, \mathrm{d}v. \tag{5.12}$$

Therefore

$$-\frac{\mathrm{d}}{\mathrm{d}t}\int_v \rho \, \mathrm{d}v = \int_S \rho \boldsymbol{u} \cdot \mathrm{d}\boldsymbol{S}. \tag{5.13}$$

Now, applying the divergence theorem to the right-hand side of (5.13) we find

$$\int_S \rho \boldsymbol{u} \cdot \mathrm{d}\boldsymbol{S} = \int_v \mathrm{div}\,(\rho \boldsymbol{u})\, dv = -\int_v \frac{\partial \rho}{\partial t} \mathrm{d}v. \tag{5.14}$$

Applying the second equality to the volume element $\mathrm{d}v$,

$$\mathrm{div}\, \rho \boldsymbol{u} = -\frac{\partial \rho}{\partial t}. \tag{5.15}$$

If the fluid is incompressible, ρ does not depend on the time or space coordinates and hence

$$\mathrm{div}\, \boldsymbol{u} = 0. \tag{5.16}$$

Maxwell was very impressed by the concepts of lines and tubes of force as expounded by Faraday and drew an immediate analogy between the behaviour of magnetic field lines and the streamlines of incompressible fluid flow (Fig. 5.6). The velocity $\boldsymbol{u}$ is analogous to the magnetic flux density $\boldsymbol{B}$; for example, if the tubes of force, or streamlines, diverge, the

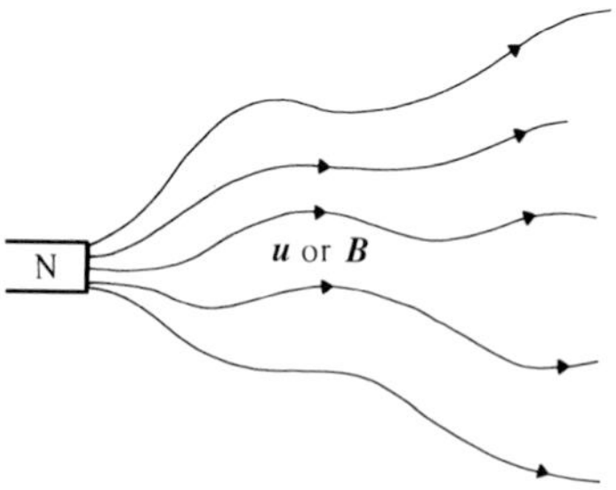

Figure 5.6: Illustrating the analogy between magnetic field lines and the streamlines in the flow of an incompressible fluid.

strength of the field decreases, as does the fluid velocity. This suggests that the magnetic field can be characterised by

$$\text{div}\,\boldsymbol{B} = 0. \tag{5.17}$$

In this paper of 1856, Maxwell recognised the important distinction between $\boldsymbol{B}$ and $\boldsymbol{H}$, associating $\boldsymbol{B}$ with magnetic fluxes and referring to it as the *magnetic induction*, and $\boldsymbol{H}$ with forces, calling it the *magnetic intensity*. The velocity $\boldsymbol{u}$ is associated with the flux density of an incompressible fluid through unit area of the surface, just as $\boldsymbol{B}$, the magnetic flux density, is the flux of a vector field; $\boldsymbol{H}$, the magnetic field strength, is associated with the force on a unit magnetic pole.

One of the great achievements of this paper was that all the relations between electromagnetic phenomena known at that time were expressed as a set of six 'laws', expressed in words rather than mathematics, very much akin to the style of Faraday.[11] Let us consider Maxwell's equations* in their modern guise.

Faraday's law of electromagnetic induction had been put into mathematical form by Neumann who, in 1845, wrote down explicitly the proportionality of the electromotive force $\mathcal{E}$ induced in a closed circuit C to the rate of change of magnetic flux, Φ,

$$\mathcal{E} = -\frac{\mathrm{d}\Phi}{\mathrm{d}t}, \tag{5.18}$$

where $\Phi = \int_S \boldsymbol{B} \cdot \mathrm{d}\boldsymbol{S}$ is the total magnetic flux through any surface S bounded by the circuit. This can be written in the form

$$\int_C \boldsymbol{E} \cdot \mathrm{d}\boldsymbol{s} = -\frac{\mathrm{d}}{\mathrm{d}t} \int_S \boldsymbol{B} \cdot \mathrm{d}\boldsymbol{S}. \tag{5.19}$$

The left-hand side is a path integral defining the electromotive force $\mathcal{E}$ induced in the circuit and the right-hand side contains the definition of Φ. Now Stokes' theorem can be applied

* Maxwell's equations in their modern vector form were first deduced by Oliver Heaviside,[12] and almost contemporaneously by Heinrich Hertz, in the 1880s.

to the left-hand side of (5.19) to give

$$\int_S \text{curl}\, \boldsymbol{E} \cdot \mathrm{d}\boldsymbol{S} = -\frac{\mathrm{d}}{\mathrm{d}t} \int_S \boldsymbol{B} \cdot \mathrm{d}\boldsymbol{S}. \tag{5.20}$$

Applying this result to the elementary surface area d$\boldsymbol{S}$, we find

$$\text{curl}\, \boldsymbol{E} = -\frac{\partial \boldsymbol{B}}{\partial t}. \tag{5.21}$$

Next, Maxwell rewrote the relation between the magnetic field produced by an electric current, Ampère's law, in vector form:

$$\oint_C \boldsymbol{H} \cdot \mathrm{d}\boldsymbol{s} = I_{\text{enclosed}}, \tag{5.22}$$

where I_{enclosed} is the current flowing through the surface bounded by C. Thus,

$$\oint_C \boldsymbol{H} \cdot \mathrm{d}\boldsymbol{s} = \int_S \boldsymbol{J} \cdot \mathrm{d}\boldsymbol{S}, \tag{5.23}$$

where $\boldsymbol{J}$ is the current density. Applying Stokes' theorem to (5.23),

$$\int_S \text{curl}\, \boldsymbol{H} \cdot \mathrm{d}\boldsymbol{S} = \int_S \boldsymbol{J} \cdot \mathrm{d}\boldsymbol{S}. \tag{5.24}$$

Therefore, for an elementary surface area d$\boldsymbol{S}$ we find

$$\text{curl}\, \boldsymbol{H} = \boldsymbol{J}. \tag{5.25}$$

We have already described Maxwell's reasoning in deriving div $\boldsymbol{B} = 0$ and, in the same way, he concluded that div $\boldsymbol{E} = 0$ in free space in the absence of electric charges. He knew, however, from Poisson's equation for the electrostatic potential (5.5), that in the presence of a charge density distribution ρ_e this equation must become

$$\text{div}\, \boldsymbol{E} = \frac{\rho_e}{\epsilon_0}. \tag{5.26}$$

It is convenient to gather together this *primitive* and *incomplete* set of Maxwell's equations:

$$\begin{aligned} \text{curl}\, \boldsymbol{E} &= -\frac{\partial \boldsymbol{B}}{\partial t}, \\ \text{curl}\, \boldsymbol{H} &= \boldsymbol{J}, \\ \text{div}\, \epsilon_0 \boldsymbol{E} &= \rho_e, \\ \text{div}\, \boldsymbol{B} &= 0. \end{aligned} \tag{5.27}$$

The final achievement of this paper was the formal introduction of what is now called the *vector potential* $\boldsymbol{A}$. Such a vector had already been introduced by Neumann, Weber and Kirchhoff in order to calculate induced currents and is defined by

$$\boldsymbol{B} = \text{curl}\, \boldsymbol{A}. \tag{5.28}$$

This definition is clearly consistent with (5.17), since div curl $A = 0$. Maxwell went further and showed how the *induced electric field* E could be related to A. Incorporating the definition (5.28) into (5.20), we obtain

$$\text{curl}\, \boldsymbol{E} = -\frac{\partial}{\partial t}(\text{curl}\, \boldsymbol{A}). \tag{5.29}$$

Interchanging the order of the time and spatial derivatives on the right-hand side,

$$\boldsymbol{E} = -\frac{\partial \boldsymbol{A}}{\partial t}. \tag{5.30}$$

5.3.2 *'On physical lines of force' (1861–2)*

These analyses gave formal coherence to the theory, but Maxwell still lacked a physical model for electromagnetic phenomena. He developed his solution to this problem in 1861–2 and these new and quite remarkable ideas were published in a series of papers entitled 'On physical lines of force'.[13] Since his earlier work on the analogy between $\boldsymbol{u}$ and $\boldsymbol{B}$, he had become more and more convinced that magnetism was essentially rotational in nature. His aim was to devise a model for the medium filling all space which could account for the stresses that Faraday had associated with magnetic lines of force – in other words, a mechanical model for the *aether*, which was assumed to be the medium through which light was propagated. In his intriguing book *Innovation in Maxwell's Electromagnetic Theory*,[14] David Siegel has carried out a detailed analysis of these papers, vividly demonstrating the richness of Maxwell's insights in drawing physical analogies between mechanical and electromagnetic phenomena.

The model was based upon the analogy between a rotating vortex tube and a tube of magnetic flux. The analogy follows from the following considerations. When left on their own, magnetic field lines expand apart, exactly as occurs in the case of a fluid vortex tube, if the rotational centrifugal forces are not balanced. The rotational kinetic energy of the vortices can be written

$$\int_v \rho \boldsymbol{u}^2 \, dv, \tag{5.31}$$

where ρ is the density of the fluid and $\boldsymbol{u}$ its rotational velocity. This expression is formally identical to that for the energy contained in a magnetic field distribution, $\int_v [\boldsymbol{B}^2/(2\mu_0)]\, dv$. Again $\boldsymbol{u}$ is analogous to $\boldsymbol{B}$ – the greater the velocity of rotation of the tube, the stronger the magnetic field. Maxwell postulated that everywhere the local magnetic flux density is proportional to the angular velocity of the vortex tube, so that the angular momentum vector $\boldsymbol{L}$ is parallel to the axis of the vortex and therefore parallel to the magnetic flux density vector $\boldsymbol{B}$.

Maxwell therefore began with a model in which all space is filled with vortex tubes (Fig. 5.7(*a*)). There is, however, an immediate mechanical problem. Friction between neighbouring vortices would lead to their disruption. Maxwell adopted the practical engineering solution of inserting 'idle wheels', or 'ball-bearings', between the vortices so that they could all rotate in the same direction without friction (Fig. 5.7(*b*)); his published picture of the vortices, represented by an array of rotating hexagons, is shown in Fig. 5.8. He then

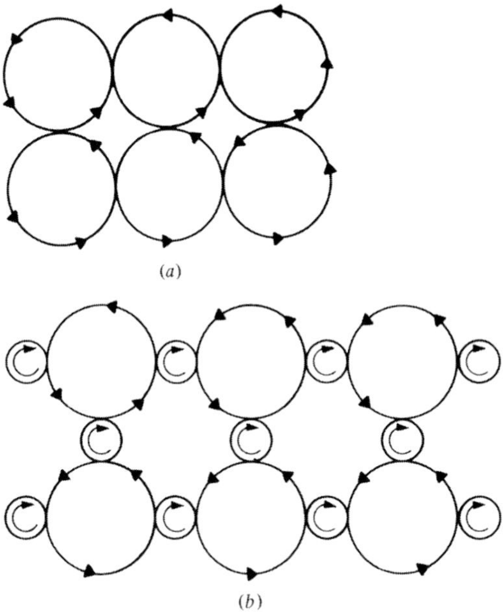

Figure 5.7: (*a*) Maxwell's original model of rotating vortices as a representation of a magnetic field. Friction at the points where the vortices touch would lead to dissipation of the rotational energy of the tubes. (*b*) Maxwell's model with 'idle wheels' or 'ball-bearings', which prevent dissipation of the rotational energy of the vortices. If these particles are free to move, they are identified with the particles which carry current in a conductor.

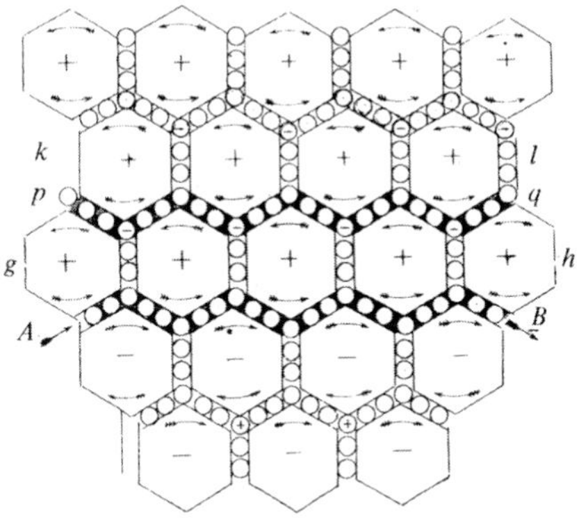

Figure 5.8: Maxwell's picture of the dynamic interaction of the vortices (represented by hexagons) and the current-carrying particles (after *Phil. Mag.*, 1861, Series 4, Vol. 21, Plate V, Fig. 2). A current of particles flows to the right along the shaded paths, for example, from A to B. Their motion causes the upper rows of vortices to rotate anticlockwise and the lower rows to rotate clockwise.

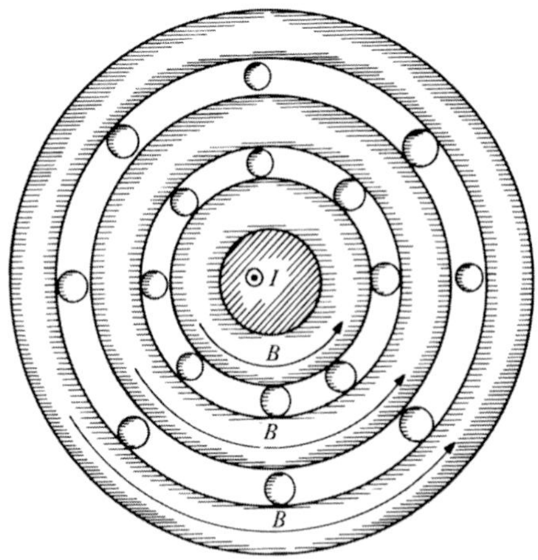

Figure 5.9: A representation of the magnetic field about a current-carrying wire according to Maxwell's model. The vortices become circular tori concentric with the axis of the wire.

identified the idle wheels with electric particles, which, if they were free to move, would carry an electric current as in a conductor. In insulators, *including free space*, they would not be free to move through the distribution of vortex tubes and so could not carry an electric current. It would not be surprising if this rather extreme example of the technique of working by analogy was a contributory cause of the 'feelings of discomfort, and often even of mistrust' to which Poincaré alludes on first encountering Maxwell's works (see subsection 1.5.1).

Remarkably, this mechanical model for the aether could account for all the known phenomena of electromagnetism. For example, consider the magnetic field produced by an electric current flowing in a wire. The current causes the vortices to rotate as an infinite series of vortex rings about the wire, as illustrated in Fig. 5.9. Correctly, the magnetic field lines form circular closed loops about the wire. Another example is the interface between two regions in which the magnetic flux densities are different but parallel. In the region of the stronger field, the vortices rotate with greater angular velocity and hence, in the interface, there must be a net force on the electric particles which drags them along the interface, as shown in Fig. 5.10, causing an electric current to flow. Notice that the direction of the current in the interface agrees with what is found experimentally.

As an example of induction, consider the effect of embedding a second wire in the magnetic field of Fig. 5.9, as shown in Fig. 5.11. If the current in the first wire is steady, there will be no current in the second wire. If, however, the current in the first wire changes, an impulse is communicated through the intervening idle wheels and vortices and a reverse current is induced in the second wire.

The last part of the paper contains the flash of genius which led to the discovery of the complete set of Maxwell's equations. He now considered how insulators store electrical energy. He made the reasonable assumption that in insulators the idle wheels, or electric particles, can be displaced from their equilibrium positions by the action of an electric field. Thus, he attributed the electrostatic energy in the medium to the elastic potential

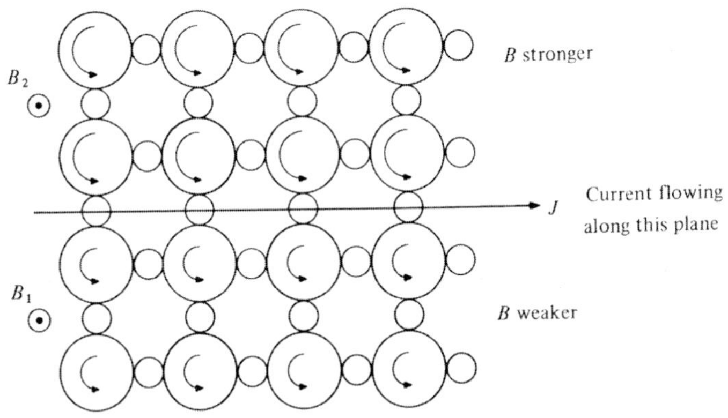

Figure 5.10: Illustrating how flow along a current sheet $\boldsymbol{J}$ is caused by a discontinuity in magnetic flux density according to Maxwell's model. The current-carrying particles flow along the line indicated. Because of friction, the lower set of vortices is slowed down whilst the upper set is speeded up. The direction of the discontinuity in $\boldsymbol{B}$ is seen to be correct according to Maxwell's equations.

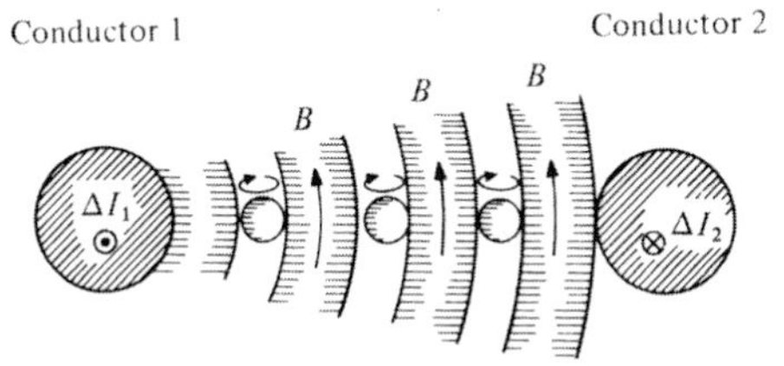

Figure 5.11: Illustrating the phenomenon of induction according to Maxwell's model. The induced current in conductor 2, ΔI_2, is in the opposite direction to ΔI_1, as can be seen from the sense of rotation of the current-carrying particles.

energy associated with the displacement of the electric particles. This had two immediate consequences. First, when the electric field applied to the medium is varying there are small changes in the positions of the electric particles in the insulating medium or vacuum, and so there are small currents associated with this motion of the particles. In other words, there is a current associated with the *displacement* of the electric particles from their equilibrium positions. Second, since the electric particles are bound elastically, any disturbance results in the propagation of waves though the medium. Armed with these ideas, Maxwell could then carry out a straightforward calculation to find the speed at which such disturbances are propagated through the insulator or vacuum, as follows.

In an linear elastic medium, the displacement of the electric particles is proportional to the electric field strength:

$$\boldsymbol{r} = \alpha \boldsymbol{E}. \tag{5.32}$$

When the strength of the field varies, the charges move, causing a current, as explained above; Maxwell called it the *displacement current*. If N is the number density of electric particles and q the charge on each of them, the *displacement current density* is

$$\boldsymbol{J}_\mathrm{d} = qN\dot{\boldsymbol{r}} = qN\alpha\dot{\boldsymbol{E}} = \beta\dot{\boldsymbol{E}}. \tag{5.33}$$

Maxwell argued that this displacement current density should be included in equation (5.25), which should now read

$$\mathrm{curl}\,\boldsymbol{H} = \boldsymbol{J} + \boldsymbol{J}_\mathrm{d} = \boldsymbol{J} + \beta\dot{\boldsymbol{E}}. \tag{5.34}$$

At this stage α and β are unknown constants to be determined from the known electric and magnetic properties of the medium.

First of all, let us work out the speed of propagation of a disturbance through the medium. Assuming that there are no real currents, $\boldsymbol{J} = 0$, (5.21) and (5.34) reduce to

$$\begin{aligned} \mathrm{curl}\,\boldsymbol{E} &= -\dot{\boldsymbol{B}}, \\ \mathrm{curl}\,\boldsymbol{H} &= \beta\dot{\boldsymbol{E}}. \end{aligned} \tag{5.35}$$

The *dispersion relation* for these waves, that is, the relation between the wave vector k and the angular frequency ω, can be found by the standard procedure. We seek wave solutions of the form $\mathrm{e}^{\mathrm{i}(\boldsymbol{k}\cdot\boldsymbol{r}-\omega t)}$, and so replace the vector operators by scalar and vector products according to the prescription

$$\begin{aligned} \mathrm{curl} &\rightarrow \mathrm{i}\boldsymbol{k}\times \\ \partial/\partial t &\rightarrow -\mathrm{i}\omega \end{aligned}$$

(see appendix section A5.6). Then equations (5.35) reduce to

$$\begin{aligned} \mathrm{i}(\boldsymbol{k}\times\boldsymbol{H}) &= -\mathrm{i}\omega\beta\boldsymbol{E}, \\ \mathrm{i}(\boldsymbol{k}\times\boldsymbol{E}) &= \mathrm{i}\omega\boldsymbol{B}. \end{aligned} \tag{5.36}$$

Eliminating $\boldsymbol{E}$ from equations (5.36),

$$\boldsymbol{k}\times(\boldsymbol{k}\times\boldsymbol{H}) = -\omega^2\beta\mu\mu_0\boldsymbol{H}, \tag{5.37}$$

where we have used the linear constitutive relation $\boldsymbol{B} = \mu\mu_0\boldsymbol{H}$, μ being the permeability of the medium. Using the vector relation

$$\boldsymbol{A}\times(\boldsymbol{B}\times\boldsymbol{C}) = \boldsymbol{B}(\boldsymbol{A}\cdot\boldsymbol{C}) - \boldsymbol{C}(\boldsymbol{A}\cdot\boldsymbol{B}),$$

we find that

$$\boldsymbol{k}(\boldsymbol{k}\cdot\boldsymbol{H}) - \boldsymbol{H}(\boldsymbol{k}\cdot\boldsymbol{k}) = -\omega^2\beta\mu\mu_0\boldsymbol{H}. \tag{5.38}$$

There is no solution for $\boldsymbol{k}$ parallel to $\boldsymbol{H}$, that is, for *longitudinal waves*, since the left-hand side of (5.37) is then zero. There exist solutions, however, for *transverse waves*, for which $\boldsymbol{k}\cdot\boldsymbol{H} = 0$. These represent plane transverse waves with the $\boldsymbol{E}$ and $\boldsymbol{H}$ vectors perpendicular to each other and to the direction of propagation of the waves. The dispersion relation for the waves is thus $k^2 = \omega^2\beta\mu\mu_0$. Since the speed of propagation of the waves is $c = \omega/k$, we find that

$$c^2 = 1/(\beta\mu\mu_0). \tag{5.39}$$

Notice that, because of the proportionality of k and ω, the phase velocity, c_p, and the group velocity of a wave packet, $c_g = d\omega/dk$, both have the same value, c, given by (5.39).

Now, Maxwell knew how to evaluate the constant β. The energy density stored in the dielectric is just the work done per unit volume in displacing the electric particles a distance $\boldsymbol{r}$, that is,

$$\text{work done} = \int \boldsymbol{f} \cdot d\boldsymbol{r} = \int Nq\boldsymbol{E} \cdot d\boldsymbol{r}. \tag{5.40}$$

But

$$\boldsymbol{r} = \alpha \boldsymbol{E} \tag{5.41}$$

and hence

$$d\boldsymbol{r} = \alpha\, d\boldsymbol{E}. \tag{5.42}$$

Therefore, the work done is

$$\int_0^E Nq\alpha \boldsymbol{E} \cdot d\boldsymbol{E} = \tfrac{1}{2}\alpha Nq|\boldsymbol{E}|^2 = \tfrac{1}{2}\beta|\boldsymbol{E}|^2. \tag{5.43}$$

But this is equal to the electrostatic energy density in the dielectric, which is $\frac{1}{2}\boldsymbol{D} \cdot \boldsymbol{E} = \frac{1}{2}\epsilon\epsilon_0|\boldsymbol{E}|^2$, where ϵ is the permittivity of the medium. Therefore $\beta = \epsilon\epsilon_0$. Inserting this value into (5.39), we find for the speed of the waves

$$c = (\mu\mu_0\epsilon\epsilon_0)^{-1/2}. \tag{5.44}$$

Notice that, even in a vacuum, for which $\mu = 1$ and $\epsilon = 1$, the speed of propagation of the waves is finite, $c = (\mu_0\epsilon_0)^{-1/2}$. Maxwell used Weber and Kohlrausch's experimental values for the product $\epsilon_0\mu_0$ and found to his amazement that c was almost exactly the speed of light: in his letters of 1861 to Michael Faraday and William Thomson, he showed that the values agreed within about 1%. In Maxwell's own words, with his own emphasis:

> The velocity of transverse modulations in our hypothetical medium, calculated from the electromagnetic experiments of MM. Kohlrausch and Weber, agrees so exactly with the velocity of light calculated from the optical experiments of M. Fizeau that we can scarcely avoid the inference that *light consists in the transverse modulations of the same medium which is the cause of electric and magnetic phenomena.*[15]

This remarkable calculation represented the unification of light with electricity and magnetism.

One cannot help but wonder that such pure gold came out of a specific mechanical model for the electromagnetic field in a vacuum. Maxwell was quite clear about the significance of the model:

> The conception of a particle having its motion connected with that of a vortex by perfect rolling contact may appear somewhat awkward. I do not bring it forward as a mode of connection existing in Nature . . . It is however a mode of connection which is mechanically conceivable and it serves to bring out the actual mechanical connections between known electromagnetic phenomena.[16]

Maxwell was well aware of the fact that the theory was 'awkward' and considered it only a 'provisional and temporary hypothesis'.[17] He therefore recast the theory on a much more abstract basis, without any special assumptions about the nature of the medium through which electromagnetic phenomena are propagated. In 1865, he published a further classic paper entitled 'A dynamical theory of the electromagnetic field'.[18] To quote Whittaker: 'In this, the architecture of his system was displayed, stripped of the scaffolding by aid of which it had been first erected'.[19]

In this paper, the equations appear in their final form as eight sets of *general equations for the electromagnetic field*. Maxwell's own view of the significance of this paper is revealed in what C.W.F. Everitt calls 'a rare moment of unveiled exuberance' in a letter to his cousin Charles Cay:

> I have also a paper afloat, containing an electromagnetic theory of light, which, till I am convinced to the contrary, I hold to be great guns.[20]

In their modern vector form, the equations read

$$\begin{aligned} \operatorname{curl} \boldsymbol{E} &= -\frac{\partial \boldsymbol{B}}{\partial t}, \\ \operatorname{curl} \boldsymbol{H} &= \boldsymbol{J} + \frac{\partial \boldsymbol{D}}{\partial t}, \\ \operatorname{div} \boldsymbol{D} &= \rho_{\mathrm{e}}, \\ \operatorname{div} \boldsymbol{B} &= 0. \end{aligned} \tag{5.45}$$

Notice how the inclusion of what is still called the *displacement current* term, $\partial \boldsymbol{D}/\partial t$, resolves a problem with the equation of continuity in current electricity. Let us take the divergence of the second equation of (5.45). Then, because the operator div curl gives zero for any vector,

$$\operatorname{div}(\operatorname{curl} \boldsymbol{H}) = \operatorname{div} \boldsymbol{J} + \frac{\partial}{\partial t}(\operatorname{div} \boldsymbol{D}) = 0. \tag{5.46}$$

Since $\operatorname{div} \boldsymbol{D} = \rho_{\mathrm{e}}$, we obtain

$$\operatorname{div} \boldsymbol{J} + \frac{\partial \rho_{\mathrm{e}}}{\partial t} = 0, \tag{5.47}$$

which is the continuity equation for the conservation of electric charge (see Section 6.2). Without the displacement term, the continuity equation would be nonsense and the primitive set of equations (5.27) would not be self-consistent.

5.4 Heinrich Hertz and the discovery of electromagnetic waves

The identification of light with electromagnetic radiation was a triumph, providing a physical foundation for the wave theory of light, which could successfully account for the phenomena of reflection, refraction, polarisation and so on. Although we now know that light is just one form of electromagnetic radiation, this was far from obvious at the time. In particular, there were no experiments which could measure the speed at which electromagnetic disturbances are propagated. Maxwell was well aware of the fact that the expression (5.44) for the speed of propagation of electromagnetic phenomena provided a key test of the theory. In the language of Maxwell's day, the speed of propagation of electromagnetic waves depended upon the ratio of the absolute units of electrostatics and electromagnetism, which, from (5.1) and (5.2), are $1/(4\pi\epsilon_0)$ and $\mu_0/(4\pi)$ respectively, in SI notation; their ratio $\nu = 1/(\epsilon_0\mu_0)$ is the square of the speed of light. In 1862, Léon Foucault established an improved estimate of the speed of light, 298 000 km s^{-1}, which was 4% below the value 310 000 km s^{-1} inferred from the experimentally determined value of ν – Maxwell noted the former value without comment in his paper of 1865. Maxwell devoted a considerable amount of ingenuity and experimental effort to improving the accuracy with which ν was known. In 1868, he obtained a new value for ν by balancing the electrostatic force of attraction of two oppositely charged discs against the electromagnetic force of repulsion between two current-carrying coils. The resulting estimate of ν gave a value of 288 000 km s^{-1} for the speed of light, 4% smaller than Foucault's measurement. Soon afterwards, M'Kichan, working in William Thomson's laboratory in Glasgow, made better estimates of ν, obtaining a value, 293 000 km s^{-1}, within 2% of Foucault's value of the speed of light. Following Maxwell's appointment to the Cavendish Chair of Experimental Physics at Cambridge in 1871, his programme of precise measurement of the fundamental units of electrostatics and electromagnetism remained a major preoccupation, which was continued by his successor John William Strutt, the third Lord Rayleigh.

Maxwell died in 1879 before direct experimental evidence was obtained for the existence of electromagnetic waves. The matter was finally laid to rest ten years after his death in a classic series of experiments by Heinrich Hertz, almost thirty years after Maxwell had identified light with electromagnetic radiation. I strongly recommend perusal of Hertz's great papers on electromagnetism included in his monograph *Electric Waves*,[21] which sets out beautifully his remarkable set of experiments.

Hertz found that he could detect the effects of electromagnetic induction at considerable distances from his apparatus. Examples of the types of emitter and detector which he used are shown in Fig. 5.12. Electromagnetic radiation was emitted when sparks were produced between the two large spheres by the application of a high voltage from an induction coil. The signal was emitted by a dipole antenna consisting of a pair of smaller spheres, attached to each of which is a long straight wire, these wires thus forming a dipole emitter. The method of detection of the radiation field was the observation of sparks in a narrow spark-gap which was attached to the ends of a dipole antenna of similar design to the emitter. In some of the experiments, rather than being in the form of a dipole, the detector was in the form of a loop or square aerial attached to the small spheres.

After a great deal of trial and error, Hertz was able to generate waves of relatively short wavelength. He made experiments with aerials of different dimensions and found that for particular arrangements there was a strong resonance, corresponding to the resonant frequencies of the emitter and detector. The frequency of the waves at resonance could be found from the resonant frequency of the emitter and detector, which he took to be $\omega = (LC)^{-1/2}$, where L and C are the inductance and capacitance of the dipole. He measured the wavelength of the radiation by placing a reflecting sheet at some distance from the spark-gap emitter, so that standing waves were set up along the line between the emitter and the sheet. By measuring the positions at which a minimum signal was observed, he was

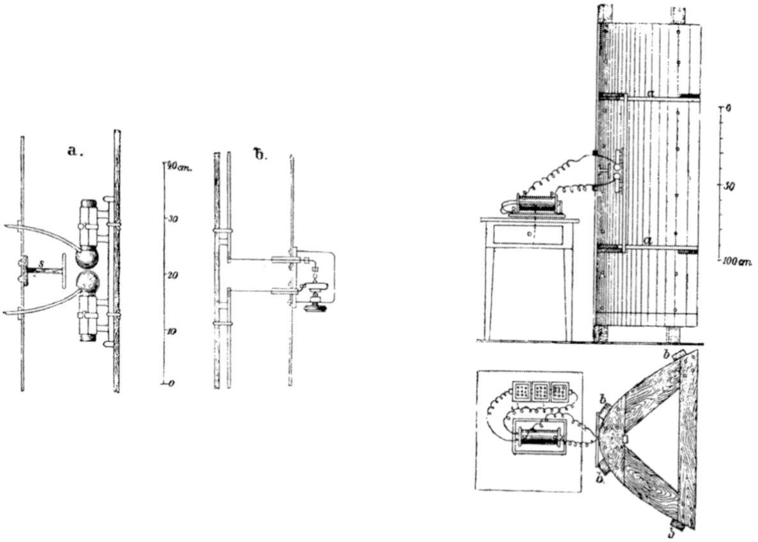

Figure 5.12: Hertz's apparatus for the generation and detection of electromagnetic radiation. The emitter *a* produced electromagnetic radiation in discharges between the spherical conductors. The detector *b* consisted of a similar device, the jaws of the detector being placed as close together as possible to achieve maximum sensitivity. The emitter was placed at the focus of a cylindrical paraboloid reflector to produce a directed beam of radiation. (From H. Hertz, 1893, *Electric Waves*, pp. 183–4, London: Macmillan and Co.)

able to measure the wavelength of the resonant waves, which was twice the distance between neighbouring minima. The speed of the waves could be found from the relation $c = \nu\lambda$.

The speed turned out to be almost exactly the speed of light in free space. Hertz then began a series of experiments which demonstrated conclusively that these waves behaved in all respects exactly like light. His great paper on the subject has the headings rectilinear propagation, polarisation, reflection, refraction. Some of the experiments were quite remarkable in their execution. To demonstrate refraction he constructed a prism weighing 12 cwt (about 0.6 tonnes) out of 'so-called hard pitch, a material like asphalt'. The experiments demonstrated convincingly that there existed electromagnetic waves of frequency about 1 GHz and wavelength 30 cm which behaved in all the above respects exactly like light. Notice that Hertz had to find the means of producing waves of frequency as high as 1 GHz or else the wavelength of the radiation would have been too long to fit into his laboratory. These momentous experiments were conclusive proof of the validity of Maxwell's equations.

5.5 Reflections

The story told above does scant justice to the complexity of the route by which the understanding of electromagnetic phenomena was gained. The story is much more complex than the 'streamlined' version which I have presented above. A vivid impression of what was really involved can be gained from Peter Harman's *The Natural Philosophy of James Clerk Maxwell*[6] and from Jed Buchwald's remarkable volume *From Maxwell to Microphysics: Aspects of Electromagnetic Theory in the Last Quarter of the Nineteenth Century*.[22] It is striking how long it took for Maxwell's deep insights to become generally accepted by the community of scientists. Maxwell himself was remarkably modest about his contribution. As Freeman Dyson has noted, when Maxwell was President of the Mathematical and Physical Sciences Section of the British Association for the Advancement of Science in 1870, his presidential lecture was a splendid opportunity for promoting his new ideas, and yet he scarcely mentioned his recent work on electromagnetism, simply referring in an off-hand way to 'Another theory of electricity which I prefer', without even mentioning that it was his own. As Dyson remarks, 'The moral of this story is that modesty is not always a virtue.'[23]

But the problems were deeper. Not only was Maxwell's theory complex; the discovery of the equations for the electromagnetic field also required a major shift in perspective for physicists of the late nineteenth century. It is worth quoting Dyson a little further.

> There were other reasons, besides Maxwell's modesty, why his theory was hard to understand. He replaced the Newtonian universe of tangible objects interacting with one another at a distance by a universe of fields extending through space and only interacting locally with tangible objects. The notion of a field was hard to grasp, because fields are intangible. The scientists of that time, including Maxwell himself, tried to picture fields as mechanical structures composed of a multitude of little wheels and vortices extending throughout space. These structures were supposed to carry the mechanical stresses that electric and magnetic fields transmitted between electric charges and currents. To make the fields satisfy Maxwell's equations, the system of wheels and vortices had to be extremely complicated. If you try to visualise the Maxwell theory with such mechanical models, it looks like a throwback to Ptolemaic astronomy with planets riding on cycles and epicycles in the sky. It does not look like the elegant astronomy of Newton.[23]

In fairness to Maxwell, the vortices and idle-wheels had disappeared by the time of his great paper of 1865, but his underlying reasoning was undoubtedly mechanical. Maxwell's route to the equations of the electromagnetic field is as extreme an example of model building in theoretical physics as any I know. Nonetheless, his technique of model building by analogy ended up with a set of equations which contained the essence of a description of physical phenomena in terms of fields rather than the classical laws of mechanics and dynamics. It was to have profound consequences. To quote Dyson again,

> Maxwell's theory becomes simple and intelligible only when you give up thinking in terms of mechanical models. Instead of thinking of mechanical objects as primary and electromagnetic stresses as secondary consequences, you must think of the electromagnetic field as primary and mechanical forces as secondary. The idea that the primary constituents of the universe are fields did not come easily to the physicists of Maxwell's generation. Fields are an abstract concept, far removed from the familiar world of things and forces. The field equations of Maxwell are partial differential equations. They cannot be expressed in simple words like Newton's law of motion, force equals mass times acceleration. Maxwell's theory had to wait for the next generation of physicists, Hertz and Lorentz and Einstein, to reveal its power and clarify its concepts. The next generation grew up with Maxwell's equations and was at home in a universe built out of fields. The primacy of fields was as natural to Einstein as the primacy of mechanical structures had been to Maxwell.[23]

In the 1880s, Heaviside and Hertz reduced Maxwell's equations to the simple form familiar to all physicists, equations (5.45). They possess a beautiful underlying symmetry and simplicity and can account for all the phenomena of classical electrodynamics. Their elegance and power are the subject of the next chapter.

5.6 References

1 For a delightful short biography of Faraday see Thomas, J.M. (1991), *Michael Faraday and the Royal Institution: the Genius of Man and Place*, Bristol: Adam Hilger, IoP Publications.
2 See Harman, P.M. (1998), *The Natural Philosophy of James Clerk Maxwell*, p. 74, Cambridge: Cambridge University Press.
3 Maxwell, J.C. (1865). In *The Scientific Papers of J. Clerk Maxwell* in two volumes (1890), ed. W.D. Niven, Vol. 1, p. 466. Cambridge: Cambridge University Press. (Reprinted 1965, New York: Dover Publications)
4 Maxwell, J.C. (1873). *Treatise on Electricity and Magnetism*, Vol. 1, p. ix. Oxford: Clarendon Press. (Reprint of third edition, 1998, Oxford Classics series.)
5 Harman, P.M. (1990). *The Scientific Letters and Paper of James Clerk Maxwell*, Vol. 1, p. 9. Cambridge: Cambridge University Press.
6 Harman, P.M. (1998). *The Natural Philosophy of James Clerk Maxwell*. Cambridge: Cambridge University Press.
7 Harman, P.M. (1990). *Op. cit.*, p. 381.
8 Maxwell, J.C. (1856). In *Scientific Papers*, Vol. 1, pp. 155–229.
9 Maxwell, J.C. (1873). *Op. cit.*, p. viii.
10 Maxwell, J.C. (1870). In *Scientific Papers*, Vol. 2, pp. 257–66.
11 Maxwell, J.C. (1856). In *Scientific Papers* Vol. 1, pp. 206–7.
12 There is a good account of Heaviside's equations by B.J. Hunt in *The Maxwellians*, pp. 119–28. Ithaca and London: Cornell University Press. See also P.J. Nahin (1988), *Oliver Heaviside: Sage in Solitude*, New York: IEEE Press.
13 Maxwell, J.C. (1861–2). In *Scientific Papers*, Vol. 1, pp. 459–512.
14 Siegel, D.M. (1991). *Innovation in Maxwell's Electromagnetic Theory. Molecular Vortices, Displacement Current, and Light*. Cambridge: Cambridge University Press.
15 Maxwell, J.C. (1861–2). In *Scientific Papers*, Vol. 1, p. 500.
16 Maxwell, J.C. (1861–2). In *Scientific Papers*, Vol. 1, p. 486.
17 Harman, P.M. (1998). *Op. cit.*, p. 113.
18 Maxwell, J.C. (1865). In *Scientific Papers*, Vol. 1, pp. 459–512.
19 Whittaker, E. (1951). *A History of the Theories of Aether and Electricity*, p. 255. London: Thomas Nelson and Sons.
20 Campbell, L. and Garnett, W. (1882). *The Life of James Clark Maxwell*, p. 342 (letter of 5 January 1865). London: Macmillan. See also P.M. Harman ed. (1995), *The Scientific Letters and Papers of James Clerk Maxwell*, Vol. 2, p. 203, Cambridge: Cambridge University Press.
21 Hertz, H. (1893). *Electric Waves*. London: Macmillan. (Reprinted 1962, New York: Dover Publications Inc.)

22 Buchwald, J.Z. (1985). *From Maxwell to Microphysics: Aspects of Electromagnetic Theory in the Last Quarter of the Nineteenth Century*. Chicago: Chicago University Press.
23 Dyson, F. (1999). In *James Clerk Maxwell Commemorative Booklet*. Edinburgh: James Clerk Maxwell Foundation.

Appendix to Chapter 5: Useful notes on vector fields

These notes are intended to reinforce what the reader has probably learned already about vector fields and to summarise some useful results which are used in Chapter 6.

A5.1 *The divergence theorem and Stokes' theorem*

The importance of these theorems is that they link expressions of the laws of physics in 'large-scale' form, in terms of integrals over finite volumes of space, to expressions in 'small-scale' form, in terms of the properties of the field in the vicinity of a particular point. The first form leads to integral equations and the second to differential equations. The fundamental theorems relating these two forms are as follows.

Divergence theorem:

$$\underbrace{\int_S \boldsymbol{A}\cdot \mathrm{d}\boldsymbol{S}}_{\text{large-scale}} = \int_v \underbrace{\mathrm{div}\,\boldsymbol{A}}_{\text{small-scale}}\,\mathrm{d}v \tag{A5.1}$$

where $\mathrm{d}\boldsymbol{S}$ is an element of surface area of the surface S which encloses volume v. The direction of the vector $\mathrm{d}\boldsymbol{S}$ is always taken to be directed *normally outwards*.

Stokes' theorem:

$$\underbrace{\int_C \boldsymbol{A}\cdot \mathrm{d}\boldsymbol{l}}_{\text{large-scale}} = \int_S \underbrace{\mathrm{curl}\,\boldsymbol{A}}_{\text{small-scale}}\cdot\mathrm{d}\boldsymbol{S}, \tag{A5.2}$$

where C is a closed curve and the integral on the left is taken round the closed curve. S is *any* open surface bounded by the loop C, $\mathrm{d}\boldsymbol{S}$ being the element of surface area; the sign of $\mathrm{d}\boldsymbol{S}$ is decided by the right-hand corkscrew convention. Notice that if $\boldsymbol{A}$ is a vector field of force then $\int \boldsymbol{A}\cdot\mathrm{d}\boldsymbol{l}$ is *minus* the external work needed to take a particle once round the circuit.

A5.2 *Results related to the divergence theorem*

Problem 1: By taking $\boldsymbol{A} = f\,\mathrm{grad}\,g$, derive two forms of *Green's theorem*,

$$\int_v [f\nabla^2 g + (\nabla f)\cdot(\nabla g)]\,\mathrm{d}v = \int_S f\nabla g\cdot\mathrm{d}\boldsymbol{S}, \tag{A5.3}$$

$$\int_v [f\nabla^2 g - g\nabla^2 f]\,\mathrm{d}v = \int_S (f\nabla g - g\nabla f)\cdot\mathrm{d}\boldsymbol{S}. \tag{A5.4}$$

Solution. Substitute $\boldsymbol{A} = f \operatorname{grad} g$ into the divergence theorem (A5.1):

$$\int_S f\nabla g \cdot \mathrm{d}\boldsymbol{S} = \int_v \operatorname{div}(f\nabla g)\,\mathrm{d}v.$$

Then, because $\nabla \cdot (a\boldsymbol{b}) = a\nabla \cdot \boldsymbol{b} + \nabla a \cdot \boldsymbol{b}$,

$$\int_S f\nabla g \cdot \mathrm{d}\boldsymbol{S} = \int_v (\nabla f \cdot \nabla g + f\nabla^2 g)\,\mathrm{d}v. \tag{A5.5}$$

Performing the same analysis for $g\nabla f$, we find

$$\int_S g\nabla f \cdot \mathrm{d}\boldsymbol{S} = \int_v (\nabla f \cdot \nabla g + g\nabla^2 f)\,\mathrm{d}v. \tag{A5.6}$$

Now subtracting (A5.6) from (A5.5) we obtain the second desired result,

$$\int_v (f\nabla^2 g - g\nabla^2 f)\,\mathrm{d}v = \int_S (f\nabla g - g\nabla f) \cdot \mathrm{d}\boldsymbol{S}. \tag{A5.7}$$

This result can be particularly useful if one of the functions is a solution of Laplace's equation, $\nabla^2 f = 0$.

Problem 2: Show that

$$\int_v \frac{\partial f}{\partial x_k}\,\mathrm{d}v = \int_S f\,\mathrm{d}S_k. \tag{A5.8}$$

Solution. Take the vector $\boldsymbol{A}$ to be $f\boldsymbol{i}_k$, where f is a scalar function of position and $\boldsymbol{i}_k$ is the unit vector in some direction. Substituting into the divergence theorem,

$$\int_S f\boldsymbol{i}_k \cdot \mathrm{d}\boldsymbol{S} = \int_v \operatorname{div} f\boldsymbol{i}_k\,\mathrm{d}v.$$

The divergence on the right-hand side is just $\partial f/\partial x_k$, and so

$$\int_S f\,\mathrm{d}S_k = \int_v \frac{\partial f}{\partial x_k}\,\mathrm{d}v.$$

This is Stokes' theorem projected in the $\boldsymbol{i}_k$-direction.

Problem 3: Show that

$$\int_v (\nabla \times \boldsymbol{A})\,\mathrm{d}v = \int_S (\mathrm{d}\boldsymbol{S} \times \boldsymbol{A}). \tag{A5.9}$$

Solution. Apply the result (A5.8) to the scalar quantities A_x and A_y, the components of the vector $\boldsymbol{A}$. Then

$$\int_v \frac{\partial A_y}{\partial x}\,\mathrm{d}v = \int_S A_y\,\mathrm{d}S_x, \tag{A5.10}$$

$$\int_v \frac{\partial A_x}{\partial y}\,\mathrm{d}v = \int_S A_x\,\mathrm{d}S_y. \tag{A5.11}$$

Subtracting (A5.11) from (A5.10),

$$\int_v \left(\frac{\partial A_y}{\partial x} - \frac{\partial A_x}{\partial y} \right) \mathrm{d}v = \int_S (A_y \, \mathrm{d}S_x - A_x \, \mathrm{d}S_y);$$

together with the corresponding equations in y, z and in z, x this amounts to

$$\int_v \operatorname{curl} \boldsymbol{A} \, \mathrm{d}v = \int_S (\mathrm{d}\boldsymbol{S} \times \boldsymbol{A}).$$

A5.3 Results related to Stokes' theorem

Problem: Show that an alternative way of writing Stokes' theorem is

$$\int_C f \, \mathrm{d}\boldsymbol{l} = \int_S (\mathrm{d}\boldsymbol{S} \times \operatorname{grad} f), \tag{A5.12}$$

where f is a scalar function.

Solution. Let us write $\boldsymbol{A} = f\boldsymbol{i}_k$, where $\boldsymbol{i}_k$ is a unit vector in the k-direction. Then Stokes' theorem

$$\int_C \boldsymbol{A} \cdot \mathrm{d}\boldsymbol{l} = \int_S \operatorname{curl} \boldsymbol{A} \cdot \mathrm{d}\boldsymbol{S}$$

becomes

$$\boldsymbol{i}_k \cdot \int_C f \, \mathrm{d}\boldsymbol{l} = \int_S \operatorname{curl} f\boldsymbol{i}_k \cdot \mathrm{d}\boldsymbol{S}.$$

We now need the expansion of curl $\boldsymbol{A}$, where $\boldsymbol{A}$ is the product of a scalar f and a vector $\boldsymbol{g}$:

$$\operatorname{curl} f\boldsymbol{g} = f \operatorname{curl} \boldsymbol{g} + (\operatorname{grad} f) \times \boldsymbol{g}. \tag{A5.13}$$

In the present case, $\boldsymbol{g}$ is the unit vector $\boldsymbol{i}_k$ and hence $\operatorname{curl} \boldsymbol{i}_k = 0$. Therefore,

$$\begin{aligned} \boldsymbol{i}_k \cdot \int_C f \, \mathrm{d}\boldsymbol{l} &= \int_S [(\operatorname{grad} f) \times \boldsymbol{i}_k] \cdot \mathrm{d}\boldsymbol{S} \\ &= \boldsymbol{i}_k \cdot \int_S \mathrm{d}\boldsymbol{S} \times \operatorname{grad} f, \end{aligned}$$

that is,

$$\int_C f \, \mathrm{d}\boldsymbol{l} = \int_S \mathrm{d}\boldsymbol{S} \times \operatorname{grad} f.$$

A5.4 Vector fields with special properties

A vector field $\boldsymbol{A}$ for which curl $\boldsymbol{A} = 0$ is called *irrotational* or *conservative*. More generally, a vector field is conservative if it satisfies any of the following conditions, all of which are equivalent:

(i) $\boldsymbol{A}$ can be expressed in the form $\boldsymbol{A} = -\operatorname{grad} \phi$, where ϕ is a scalar function of position, that is, ϕ depends only upon x, y and z;

(ii) curl $\boldsymbol{A} = 0$;
(iii) $\int_C \boldsymbol{A} \cdot \mathrm{d}\boldsymbol{l} = 0$ for a closed loop C;
(iv) $\int_A^B \boldsymbol{A} \cdot \mathrm{d}\boldsymbol{l}$ is independent of the path from A to B.

The conditions (ii) and (iii) are plainly identical because of Stokes' theorem

$$\int_C \boldsymbol{A} \cdot \mathrm{d}\boldsymbol{l} = \int_S \operatorname{curl} \boldsymbol{A} \cdot \mathrm{d}\boldsymbol{S} = 0.$$

If $\boldsymbol{A} = -\operatorname{grad} \phi$, then curl $\boldsymbol{A} = 0$ for any ϕ. Finally, we can write

$$\int_A^B \boldsymbol{A} \cdot \mathrm{d}\boldsymbol{l} = -\int_A^B \operatorname{grad} \phi \cdot \mathrm{d}\boldsymbol{l} = -(\phi_B - \phi_A),$$

showing that the integral is independent of the path between A and B. This last property is the reason for the name 'conservative field'. It does not matter what route one chooses between A and B. If $\boldsymbol{A}$ is a field of force then ϕ is the potential, equal to the work which needs to be expended to bring unit mass or charge to that point in the field from a point where the field is zero.

A vector field $\boldsymbol{A}$ for which div $\boldsymbol{A} = \nabla \cdot \boldsymbol{A} = 0$ is called a *solenoidal field*. If $\boldsymbol{B} = \operatorname{curl} \boldsymbol{A}$ then div $\boldsymbol{B} = 0$. Conversely, if div $\boldsymbol{B} = 0$ then $\boldsymbol{B}$ can be expressed as the curl of some vector $\boldsymbol{A}$; furthermore, we can add to $\boldsymbol{A}$ an arbitrary conservative vector field, which will always vanish when the curl of $\boldsymbol{A}$ is taken. Thus, if $\boldsymbol{A}' = \boldsymbol{A} - \operatorname{grad} \phi$,

$$\begin{aligned} \boldsymbol{B} = \operatorname{curl} \boldsymbol{A}' &= \operatorname{curl} \boldsymbol{A} - \operatorname{curl} \operatorname{grad} \phi \\ &= \operatorname{curl} \boldsymbol{A}. \end{aligned}$$

One of the most useful results in vector analysis is the identity

$$\operatorname{curl} \operatorname{curl} \boldsymbol{A} = \operatorname{grad} \operatorname{div} \boldsymbol{A} - \nabla^2 \boldsymbol{A},$$

perhaps more memorable as

$$\nabla \times (\nabla \times \boldsymbol{A}) = \nabla(\nabla \cdot \boldsymbol{A}) - \nabla^2 \boldsymbol{A}. \qquad \text{(A5.14)}$$

A5.5 *Vector operators in various coordinate systems*

It is useful to have at hand a list of the vector operators grad, div, curl and ∇^2 in rectangular (Cartesian), cylindrical and spherical polar coordinates. The standard books give the following.

grad:

Cartesian
$$\operatorname{grad} \Phi = \nabla \Phi = \boldsymbol{i}_x \frac{\partial \Phi}{\partial x} + \boldsymbol{i}_y \frac{\partial \Phi}{\partial y} + \boldsymbol{i}_z \frac{\partial \Phi}{\partial z}$$

cylindrical polar
$$\operatorname{grad} \Phi = \nabla \Phi = \boldsymbol{i}_r \frac{\partial \Phi}{\partial r} + \boldsymbol{i}_z \frac{\partial \Phi}{\partial z} + \boldsymbol{i}_\varphi \frac{1}{r} \frac{\partial \Phi}{\partial \varphi}$$

spherical polar
$$\operatorname{grad} \Phi = \nabla \Phi = \boldsymbol{i}_r \frac{\partial \Phi}{\partial r} + \boldsymbol{i}_\theta \frac{1}{r} \frac{\partial \Phi}{\partial \theta} + \boldsymbol{i}_\varphi \frac{1}{r \sin \theta} \frac{\partial \Phi}{\partial \varphi}$$

div:

Cartesian $$\operatorname{div}\boldsymbol{A} = \nabla\cdot\boldsymbol{A} = \frac{\partial A_x}{\partial x} + \frac{\partial A_y}{\partial y} + \frac{\partial A_z}{\partial z}$$

cylindrical polar $$\operatorname{div}\boldsymbol{A} = \nabla\cdot\boldsymbol{A} = \frac{1}{r}\left[\frac{\partial}{\partial r}(rA_r) + \frac{\partial}{\partial z}(rA_z) + \frac{\partial}{\partial\varphi}(A_\varphi)\right]$$

spherical polar $$\operatorname{div}\boldsymbol{A} = \nabla\cdot\boldsymbol{A} = \frac{1}{r^2\sin\theta}\left[\frac{\partial}{\partial r}(r^2\sin\theta\,A_r) + \frac{\partial}{\partial\theta}(r\sin\theta\,A_\theta) + \frac{\partial}{\partial\varphi}(rA_\phi)\right]$$

curl:

Cartesian $$\operatorname{curl}\boldsymbol{A} = \nabla\times\boldsymbol{A} = \left(\frac{\partial A_y}{\partial z} - \frac{\partial A_z}{\partial y}\right)\boldsymbol{i}_x + \left(\frac{\partial A_z}{\partial x} - \frac{\partial A_x}{\partial z}\right)\boldsymbol{i}_y + \left(\frac{\partial A_x}{\partial y} - \frac{\partial A_y}{\partial x}\right)\boldsymbol{i}_z$$

cylindrical polar $$\operatorname{curl}\boldsymbol{A} = \nabla\times\boldsymbol{A} = \frac{1}{r}\left[\frac{\partial}{\partial z}(rA_\varphi) - \frac{\partial}{\partial\varphi}(A_z)\right]\boldsymbol{i}_r + \frac{1}{r}\left[\frac{\partial}{\partial\varphi}(A_r) - \frac{\partial}{\partial r}(rA_\varphi)\right]\boldsymbol{i}_z + \left[\frac{\partial}{\partial r}(A_z) - \frac{\partial}{\partial z}(A_r)\right]\boldsymbol{i}_\varphi$$

spherical polar $$\operatorname{curl}\boldsymbol{A} = \nabla\times\boldsymbol{A} = \frac{1}{r^2\sin\theta}\left[\frac{\partial}{\partial\theta}(r\sin\theta\,A_\varphi) - \frac{\partial}{\partial\varphi}(rA_\theta)\right]\boldsymbol{i}_r + \frac{1}{r\sin\theta}\left[\frac{\partial}{\partial\varphi}(A_r) - \frac{\partial}{\partial r}(r\sin\theta\,A_\varphi)\right]\boldsymbol{i}_\theta + \frac{1}{r}\left[\frac{\partial}{\partial r}(rA_\theta) - \frac{\partial}{\partial\theta}(A_r)\right]\boldsymbol{i}_\varphi$$

Laplacian:

Cartesian $$\nabla^2\Phi = \frac{\partial^2\Phi}{\partial x^2} + \frac{\partial^2\Phi}{\partial y^2} + \frac{\partial^2\Phi}{\partial z^2}$$

cylindrical polar $$\nabla^2\Phi = \frac{1}{r}\frac{\partial}{\partial r}\left(r\frac{\partial\Phi}{\partial r}\right) + \frac{1}{r^2}\frac{\partial^2\Phi}{\partial\varphi^2} + \frac{\partial^2\Phi}{\partial z^2}$$

spherical polar
$$\nabla^2\Phi = \frac{1}{r^2}\frac{\partial}{\partial r}\left(r^2\frac{\partial\Phi}{\partial r}\right) + \frac{1}{r^2\sin\theta}\frac{\partial}{\partial\theta}\left(\sin\theta\frac{\partial\Phi}{\partial\theta}\right) + \frac{1}{r^2\sin^2\theta}\frac{\partial^2\Phi}{\partial\varphi^2}$$
$$= \frac{1}{r}\frac{\partial^2}{\partial r^2}(r\Phi) + \frac{1}{r^2\sin\theta}\frac{\partial}{\partial\theta}\left(\sin\theta\frac{\partial\Phi}{\partial\theta}\right) + \frac{1}{r^2\sin^2\theta}\frac{\partial^2\Phi}{\partial\varphi^2}$$

A5.6 *Vector operators and dispersion relations*

In finding three-dimensional solutions of wave equations, such as Maxwell's equations, it often simplifies the working considerably to use the relations $\nabla \to \mathrm{i}\boldsymbol{k}$, $\partial/\partial t \to -\mathrm{i}\omega$ obtained if the phase factor of the waves is written in the form $\exp[\mathrm{i}(\boldsymbol{k}\cdot\boldsymbol{r} - \omega t)]$, as we now show; the scalar product $\boldsymbol{k}\cdot\boldsymbol{r} = k_x x + k_y y + k_z z$ is also needed.

$$\begin{aligned}
\nabla \mathrm{e}^{\mathrm{i}\boldsymbol{k}\cdot\boldsymbol{r}} &= \operatorname{grad}\mathrm{e}^{\mathrm{i}\boldsymbol{k}\cdot\boldsymbol{r}} = \mathrm{i}\boldsymbol{k}\,\mathrm{e}^{\mathrm{i}\boldsymbol{k}\cdot\boldsymbol{r}},\\
\nabla\cdot(\boldsymbol{A}\,\mathrm{e}^{\mathrm{i}\boldsymbol{k}\cdot\boldsymbol{r}}) &= \operatorname{div}(\boldsymbol{A}\,\mathrm{e}^{\mathrm{i}\boldsymbol{k}\cdot\boldsymbol{r}}) = \mathrm{i}\boldsymbol{k}\cdot\boldsymbol{A}\,\mathrm{e}^{\mathrm{i}\boldsymbol{k}\cdot\boldsymbol{r}},\\
\nabla\times(\boldsymbol{A}\,\mathrm{e}^{\mathrm{i}\boldsymbol{k}\cdot\boldsymbol{r}}) &= \operatorname{curl}(\boldsymbol{A}\,\mathrm{e}^{\mathrm{i}\boldsymbol{k}\cdot\boldsymbol{r}}) = \mathrm{i}\boldsymbol{k}\times\boldsymbol{A}\,\mathrm{e}^{\mathrm{i}\boldsymbol{k}\cdot\boldsymbol{r}},
\end{aligned}$$

where $\boldsymbol{A}$ is a constant vector. Thus, it can be seen that $\nabla \to \mathrm{i}\boldsymbol{k}$ carries out the vector operations for wave solutions. In a similar way, it can be shown that $\partial/\partial t \to -\mathrm{i}\omega$.

When we attempt to solve a wave equation by this technique, we find that the exponential phase factor cancels out through the wave equation, resulting in a relation between $|\boldsymbol{k}|$ and ω known as a *dispersion relation*. Notice that the vectors $\boldsymbol{k}$ and $\boldsymbol{A}$ may themselves be complex. Here is an example.

A simple wave equation with a damping term $2\gamma\,\partial u/\partial t$ can be written as

$$\nabla^2 u - 2\gamma\frac{\partial u}{\partial t} - \frac{1}{v^2}\frac{\partial^2 u}{\partial t^2} = 0,$$

where v is the sound speed in the medium. Recalling that $\nabla^2 u = \nabla\cdot(\nabla u)$ and using the substitutions $\nabla \to \mathrm{i}\boldsymbol{k}$, $\partial/\partial t \to -\mathrm{i}\omega$, we find

$$(\mathrm{i}\boldsymbol{k})\cdot(\mathrm{i}\boldsymbol{k}) + 2\mathrm{i}\gamma\omega + \frac{\omega^2}{v^2} = 0, \qquad \text{so that} \qquad k^2 = \frac{\omega^2}{v^2} + 2\mathrm{i}\gamma\omega, \tag{A5.15}$$

the dispersion relation for the wave equation. If there is no damping $\gamma = 0$, the phase velocity of the waves is $v = \omega/k_0$ and the group velocity, $v_{\mathrm{g}} = \mathrm{d}\omega/\mathrm{d}k$, is also equal to v.

If $\gamma \neq 0$, the dispersion relation has an imaginary part, corresponding to the damping of the waves. If the damping is small, $\gamma v^2/\omega \ll 1$, (A5.15) reduces to

$$k = \frac{\omega}{v}\left(1 + \mathrm{i}\frac{\gamma v^2}{\omega}\right) = k_0\left(1 + \mathrm{i}\frac{\gamma v^2}{\omega}\right)$$

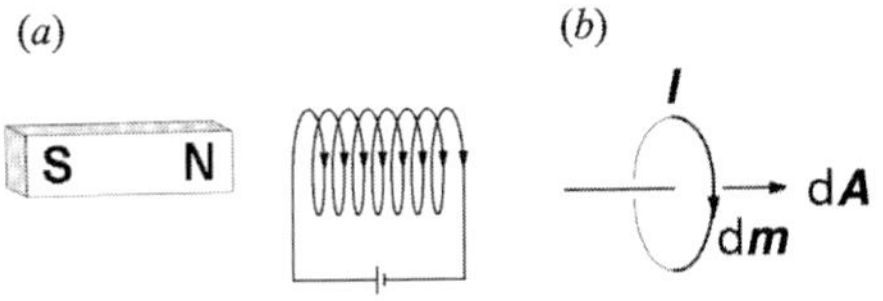

Figure A5.1: (*a*) Illustrating the equivalence of a bar magnet and a current loop. (*b*) Illustrating the definition of the magnetic moment of a current loop.

and so the expression for the propagation of the wave becomes

$$\begin{aligned} u(\boldsymbol{r}, t) &= u_0 \exp(-\gamma v r) \exp[\mathrm{i}(\boldsymbol{k} \cdot \boldsymbol{r} - \omega t)] \\ &= u_0 \exp(-\alpha t) \exp[\mathrm{i}(\boldsymbol{k} \cdot \boldsymbol{r} - \omega t)] \end{aligned}$$

where $\alpha = \gamma v^2$.

A5.7 How to relate the different expressions for the magnetic fields produced by currents

It is simplest to begin with Faraday's brilliant experiments of the 1820s (Section 5.2). Ørsted had shown that magnetic fields are produced by electric currents and Ampère that the magnetic field lines about a current-carrying wire are circular. Faraday reasoned that if the wire were bent into a circle then the magnetic field lines would be more concentrated inside the loop and less concentrated outside, as illustrated in Fig. 5.4. In an important series of experiments, he showed that *current loops produce identical magnetostatic effects to bar magnets*.

We therefore formally identify the *magnetic dipole moment* $\mathrm{d}\boldsymbol{m}$ associated with the current I flowing in an elementary loop of area $\mathrm{d}\boldsymbol{A}$ by the relation

$$\mathrm{d}\boldsymbol{m} = I\,\mathrm{d}\boldsymbol{A}. \tag{A5.16}$$

The direction of $\mathrm{d}\boldsymbol{m}$ is given by the right-hand cork-screw rule with the sense of rotation being given by the sense of the current flow (Fig. A5.1(*b*)). Therefore, the couple acting upon a current loop located in a magnetic field is

$$\mathrm{d}\boldsymbol{G} = \mathrm{d}\boldsymbol{m} \times \boldsymbol{B} = I(\mathrm{d}\boldsymbol{A} \times \boldsymbol{B}). \tag{A5.17}$$

Let us show how Ampère's circuital law can be derived from the definition of the magnetic moment of a current loop. There is one crucial and obvious difference between a magnet and a current loop – *you cannot pass through a bar magnet but you can pass through a current loop*. This may look obvious but it is rather profound. Suppose free magnetic poles existed. If we were to release one in the magnetic field of a bar magnet, it would crash into the magnet and stop there. In the case of a current loop, however, it would continually move in larger and larger orbits on passing through the current loop, gaining energy all the time from the source maintaining the current at a constant value. This does not happen because *there are no free magnetic poles*. If this experiment is performed with a magnetic dipole, it comes to rest inside the coil.

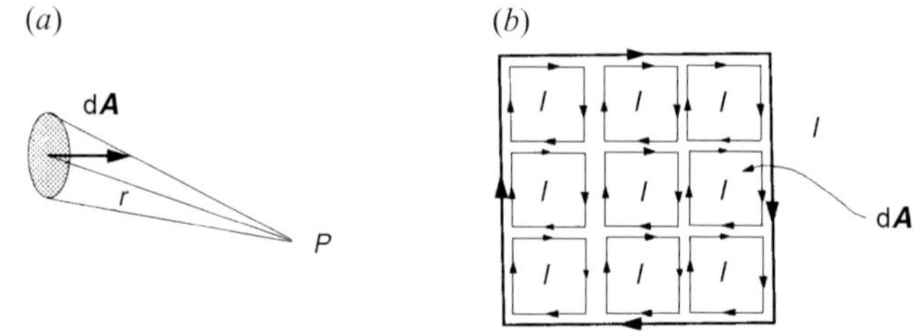

Figure A5.2: (*a*) Evaluating the magnetic potential V_{mag} of an elementary current loop. (*b*) Illustrating the use of Ampère's magnetic shell to evaluate the magnetostatic potential of a finite current loop.

Let us work out the magnetic flux density in the vicinity of an elementary current loop d$\boldsymbol{A}$ (Fig. A5.2(*a*)). The magnetostatic potential dV_{m} at any point P at distance $\boldsymbol{r}$ from the centre of the loop is given by

$$\mathrm{d}V_{\mathrm{m}} = \frac{\mathrm{d}\boldsymbol{m} \cdot \boldsymbol{i}_r}{4\pi r^2} = \frac{\mathrm{d}\boldsymbol{m} \cdot \boldsymbol{r}}{4\pi r^3}.$$

But d$\boldsymbol{m} = I\,\mathrm{d}\boldsymbol{A}$ and so

$$\mathrm{d}V_{\mathrm{m}} = I\,\frac{\mathrm{d}\boldsymbol{A} \cdot \boldsymbol{i}_r}{4\pi r^2}.$$

Now, $\mathrm{d}\boldsymbol{A} \cdot \boldsymbol{i}_r / r^2 = \mathrm{d}\Omega$ is the solid angle subtended by the current loop at the point P. Therefore,

$$\mathrm{d}V_{\mathrm{m}} = \frac{I\,\mathrm{d}\Omega}{4\pi}. \tag{A5.18}$$

Now, let us extend this result to a finite-sized current loop. We can split up the loop into many elementary loops, each carrying the current I (Fig. A5.2(*b*)) – this is the procedure described by Ampère in his treatise of 1825. Clearly, in the interior the oppositely flowing currents cancel out leaving only the current I flowing around the outside loop. Therefore, the magnetostatic potential at a point P is just the sum of the magnetostatic potentials due to all the elementary current loops:

$$V_{\mathrm{m}} = \sum \frac{I\,\mathrm{d}\Omega}{4\pi} = \frac{I\Omega}{4\pi}.$$

This is a remarkable simplification. We can now work out the magnetostatic potential V_{m} at any point in space and then find the magnetic flux density from (5.8),

$$V_{\mathrm{m}} = \frac{I\Omega}{4\pi}, \qquad \boldsymbol{B} = -\mu_0\,\mathrm{grad}\,V_{\mathrm{m}}.$$

Then, integrating, we find

$$\int_A^B \boldsymbol{B} \cdot \mathrm{d}\boldsymbol{s} = -\mu_0 \int_A^B \mathrm{grad}\,V_{\mathrm{m}} \cdot \mathrm{d}\boldsymbol{s} = -\mu_0[V_{\mathrm{m}}(B) - V_{\mathrm{m}}(A)].$$

Now, let us inspect what happens when we take a line integral from one side of a current loop to another, as shown in Fig. A5.3. The path starts at A, which is very close to the current loop but slightly on the nearer side of it. Therefore, the solid angle subtended by the loop

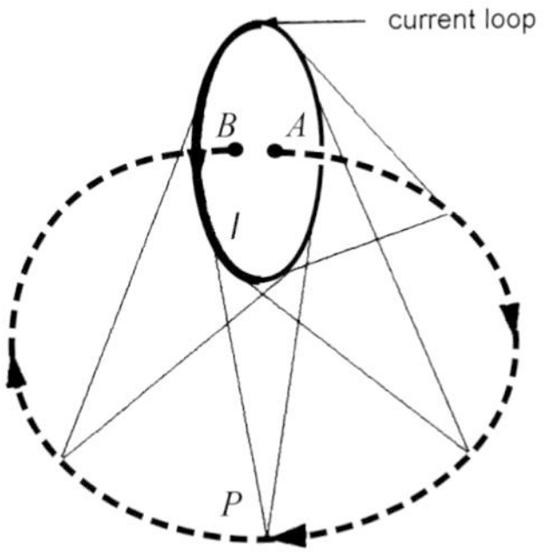

Figure A5.3: Evaluating the change in magnetic potential in passing from one side of a current loop to the other along the path shown.

is almost 2π steradians. As the point moves round the path, we see that the solid angle decreases and, at the point P, the solid angle reaches zero. Moving further round the path the solid angle becomes negative, and is almost -2π when it reaches B. Thus, the change in solid angle in going round the path from A to B is -4π steradians. Therefore

$$V_{\mathrm{m}}(B) - V_{\mathrm{m}}(A) = \frac{I\Delta\Omega}{4\pi} = -I.$$

If this were a magnet, this change in potential would be exactly compensated by the reverse field inside the magnet and that would be the end of the story. In the case of the current loop, however, we simply pass through the loop back to the point A where we started and, since we can let $B \to A$, we find

$$\oint \boldsymbol{B} \cdot \mathrm{d}\boldsymbol{s} = -\mu_0[V_{\mathrm{m}}(B) - V_{\mathrm{m}}(A)] = \mu_0 I. \tag{A5.19}$$

Remarkably, we have derived *Ampère's circuital theorem.*

Ampère's circuital theorem is an integral equation relating the integral of the magnetic flux density round a closed path to the enclosed current. Next we work out the magnetic flux density $\mathrm{d}\boldsymbol{B}$ at any point in space due to the current flowing in a line element $\mathrm{d}\boldsymbol{s}$ (Fig. A5.4), the *Biôt–Savart law* (5.9), which is the most general form of expression for the magnetic flux density produced by a current element $I\mathrm{d}\boldsymbol{s}$.

We begin with the expression for the magnetic flux density in terms of the magnetostatic potential:

$$\boldsymbol{B} = -\mu_0 \,\mathrm{grad}\, V_{\mathrm{m}}.$$

We showed above that if a current loop subtends a solid angle Ω at a point P then the magnetostatic potential at P is $V_{\mathrm{m}} = I\Omega/(4\pi)$. Taking the gradient of V_{m},

$$\boldsymbol{B} = -\frac{\mu_0 I}{4\pi}\,\mathrm{grad}\,\Omega.$$

The problem of finding the field $\mathrm{d}\boldsymbol{B}$ due to an element $\mathrm{d}\boldsymbol{s}$ of the current loop reduces to expressing grad Ω in a form that involves $\mathrm{d}\boldsymbol{s}$. To do this we consider how Ω changes when we move a vector distance $\mathrm{d}\boldsymbol{l}$ from P, as illustrated in Fig. A5.4(a). From the geometry

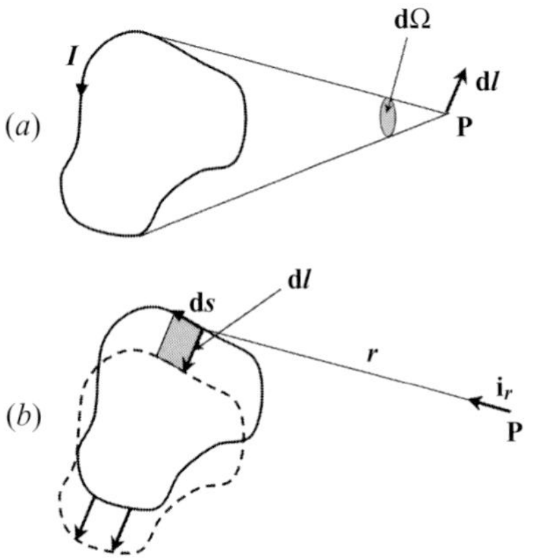

Figure A5.4: Evaluating the change in magnetic potential on moving a distance d$\boldsymbol{l}$ due to an arbitrarily shaped current loop.

of the diagram, we can write for the corresponding change in solid angle

$$d\Omega = (\text{grad } \Omega) \cdot d\boldsymbol{l}.$$

To find dΩ, the simplest thing to do is to shift the current loop by $-d\boldsymbol{l}$ and work out how the projected surface area changes as observed from P (Fig. A5.4(b)). The vector area formed by the parallelogram defined by d$\boldsymbol{s}$ and d$\boldsymbol{l}$ is

$$d\boldsymbol{A} = d\boldsymbol{s} \times d\boldsymbol{l}.$$

If $\boldsymbol{i}_r$ is the unit vector drawn in the direction of the element of area dA from P, then the projected area as observed from P is

$$dA_{\text{proj}} = \boldsymbol{i}_r \cdot (d\boldsymbol{s} \times d\boldsymbol{l}).$$

The change in solid angle contributed by the small parallelogram is therefore dA_{proj}/r^2, and so we find the total change in solid angle by integrating round the complete current loop:

$$d\Omega = \oint_C \frac{\boldsymbol{i}_r \cdot (d\boldsymbol{s} \times d\boldsymbol{l})}{r^2} = \oint_C \frac{\boldsymbol{r} \cdot (d\boldsymbol{s} \times d\boldsymbol{l})}{r^3}.$$

If now we recall the vector identity

$$\boldsymbol{A} \cdot (\boldsymbol{B} \times \boldsymbol{C}) = \boldsymbol{C} \cdot (\boldsymbol{A} \times \boldsymbol{B}),$$

we can write

$$d\Omega = \oint_C \frac{\boldsymbol{r} \cdot (d\boldsymbol{s} \times d\boldsymbol{l})}{r^3} = \oint_C \frac{d\boldsymbol{l} \cdot (\boldsymbol{r} \times d\boldsymbol{s})}{r^3} = d\boldsymbol{l} \cdot \oint_C \frac{\boldsymbol{r} \times d\boldsymbol{s}}{r^3}$$

since d$\boldsymbol{l}$ is a constant vector all round the current loop. Therefore

$$d\Omega = (\text{grad } \Omega) \cdot d\boldsymbol{l} = d\boldsymbol{l} \cdot \oint_C \frac{\boldsymbol{r} \times d\boldsymbol{s}}{r^3},$$

and so

$$\text{grad}\ \Omega = \oint_C \frac{\boldsymbol{r} \times \mathrm{d}\boldsymbol{s}}{r^3}.$$

Finally, since

$$\boldsymbol{B} = -\frac{\mu_0 I}{4\pi}\ \text{grad}\ \Omega$$

we obtain the expression

$$\boldsymbol{B} = -\frac{\mu_0 I}{4\pi} \oint_C \frac{\boldsymbol{r} \times \mathrm{d}\boldsymbol{s}}{r^3}.$$

This formula gives the total contribution to the field at P due to all the increments of path length round the circuit, and so we can remove the integral sign to find the contribution from the current element $\mathrm{d}\boldsymbol{s}$:

$$\mathrm{d}\boldsymbol{B} = \frac{\mu_0 I \mathrm{d}\boldsymbol{s} \times \boldsymbol{r}}{4\pi r^3}. \tag{A5.20}$$

Thus we have deduced the *Biôt–Savart law*. Notice, however, that Biôt and Savart deduced the law *experimentally*.

We have therefore four different ways of working out the magnetic field associated with currents. Which method is most appropriate depends upon the nature of the problem and the symmetry of the current distribution. The four ways are as follows.

(i) The magnetic moment of an elementary current loop $\mathrm{d}\boldsymbol{m} = I\ \mathrm{d}\boldsymbol{A}$ can be substituted into the relation between the magnetic moment of a dipole and the field it produces,

$$\mathrm{d}\boldsymbol{B} = \frac{\mu_0 |\mathrm{d}\boldsymbol{m}|}{4\pi r^3}(2\cos\theta\ \boldsymbol{i}_r + \sin\theta\ \boldsymbol{i}_\theta)$$

in polar coordinates.

(ii) The gradient of the magnetostatic potential can be taken, since

$$\boldsymbol{B} = -\mu_0\ \text{grad}\ V_\mathrm{m} = -\frac{\mu_0 I}{4\pi}\ \text{grad}\ \Omega.$$

(iii) Ampère's circuital theorem can be used:

$$\oint_C \boldsymbol{B} \cdot \mathrm{d}\boldsymbol{s} = \mu_0 I_{\text{enclosed}}.$$

(iv) The Biôt–Savart law can be used:

$$\mathrm{d}\boldsymbol{B} = \frac{\mu_0 I \mathrm{d}\boldsymbol{s} \times \boldsymbol{r}}{4\pi r^3}.$$

6 How to rewrite the history of electromagnetism

6.1 Introduction

Now that we have derived Maxwell's equations as he himself derived them, let us do everything backwards, starting with Maxwell's equations and regarding them simply as a set of vector equations relating the vector fields $\boldsymbol{E}$, $\boldsymbol{D}$, $\boldsymbol{B}$, $\boldsymbol{H}$ and $\boldsymbol{J}$. Therefore, initially we ascribe *no physical significance* to these fields. We then make a minimum number of postulates in order to give them physical significance and so derive from them all the experimentally established laws of electromagnetism. This approach is taken by Stratton in his book *Electromagnetic Theory*.[1]

We can then apply Maxwell's equations to further aspects of electromagnetic theory – the properties of electromagnetic waves, the emission of waves by accelerated charges, and so on – which provide tests of the theory that go far beyond the empirically derived laws from which Maxwell's equations were deduced. If the theory were to prove to be inconsistent with experiment then the interlocking nature of many of the results, as illustrated below, indicates how the whole edifice would have to be changed.

A number of my colleagues have objected strenuously to this approach to electromagnetism, principally on the grounds that historically it is most unlikely that anyone would have discovered Maxwell's equations by this route. I am not prepared to speculate about that. What I do know is that this procedure of starting with a mathematical structure, which is then given physical meaning, is found in other aspects of fundamental physics, for example in the theory of linear operators and quantum mechanics and in tensor calculus and the special and general theories of relativity. String theory is a contemporary example of the use of the same procedure, in an attempt to unify all the forces of nature. The mathematics gives formal coherence to the physical theory and enables predictions to be made about the behaviour of real systems in unexplored regions of parameter space. It also brings out the remarkable symmetries and elegance of the theory of the electromagnetic field.

This study began as an examples class in mathematical physics and it is instructive to maintain that format. Much of the analysis which follows is mathematically simple – the emphasis is upon the clarity with which we are able to make the correspondence between the mathematics and the physics.

6.2 Maxwell's equations as a set of vector equations

We start with the first two Maxwell's equations in the form

$$\operatorname{curl} \boldsymbol{E} = -\frac{\partial \boldsymbol{B}}{\partial t}, \tag{6.1}$$

$$\operatorname{curl} \boldsymbol{H} = \boldsymbol{J} + \frac{\partial \boldsymbol{D}}{\partial t}. \tag{6.2}$$

According to our procedure in this chapter, $\boldsymbol{E}$, $\boldsymbol{D}$, $\boldsymbol{B}$, $\boldsymbol{H}$ and $\boldsymbol{J}$ are to be regarded as unspecified vector fields which are functions of space and time coordinates and which are intended to describe the electromagnetic field. They are supplemented by a continuity equation for $\boldsymbol{J}$,

$$\operatorname{div} \boldsymbol{J} + \frac{\partial \rho_e}{\partial t} = 0. \tag{6.3}$$

We then make the first physical identification:

$$\rho_e \text{ is the electric charge density} \quad \text{and} \quad \text{charge is conserved.} \tag{6.4}$$

Problem: Show from (6.3) that $\boldsymbol{J}$ must be identified as a *current density*, that is, the rate of flow of charge through unit surface area.

Proof. Integrate (6.3) over some volume v bounded by a surface S:

$$\int_v \operatorname{div} \boldsymbol{J} \, dv = -\frac{\partial}{\partial t} \int_v \rho_e \, dv.$$

Now, according to the divergence theorem,

$$\int_v \operatorname{div} \boldsymbol{J} \, dv = \int_S \boldsymbol{J} \cdot d\boldsymbol{S}, \tag{6.5}$$

and so is equal to the rate at which charge is lost from the volume,

$$-\frac{\partial}{\partial t} \int_v \rho_e \, dv.$$

Thus

$$\int_S \boldsymbol{J} \cdot d\boldsymbol{S} = -\frac{\partial}{\partial t}(\text{total enclosed charge}). \tag{6.6}$$

Thus, $\boldsymbol{J}$ must represent the rate of flow, or flux, of charge per unit area through a plane perpendicular to $\boldsymbol{J}$.

6.3 Gauss's theorem in electromagnetism

Problem: Show that the fields $\boldsymbol{B}$ and $\boldsymbol{D}$ must satisfy the relations

$$\operatorname{div} \boldsymbol{B} = 0 \qquad \text{and} \qquad \operatorname{div} \boldsymbol{D} = \rho_e. \tag{6.7}$$

Derive the corresponding integral equations, which are different forms of *Gauss's theorem* in electromagnetism.

Proof. Take the divergences of (6.1) and (6.2). Since the divergence of a curl is always zero, we obtain

$$\begin{aligned} \text{div curl}\, \boldsymbol{E} &= -\frac{\partial}{\partial t}(\text{div}\, \boldsymbol{B}) = 0, \\ \text{div curl}\, \boldsymbol{H} &= \text{div}\, \boldsymbol{J} + \frac{\partial}{\partial t}(\text{div}\, \boldsymbol{D}) = 0. \end{aligned} \tag{6.8}$$

From (6.3), we find

$$\frac{\partial}{\partial t}(\text{div}\, \boldsymbol{D}) - \frac{\partial \rho_e}{\partial t} = \frac{\partial}{\partial t}(\text{div}\, \boldsymbol{D} - \rho_e) = 0. \tag{6.9}$$

Thus, the *partial derivatives* of div $\boldsymbol{B}$ and of div $\boldsymbol{D} - \rho_e$ with respect to time are both zero at all points in space. Therefore

$$\text{div}\, \boldsymbol{B} = \text{constant} \qquad \text{and} \qquad \text{div}\, \boldsymbol{D} - \rho_e = \text{constant}.$$

We have to establish the values of the constants. I have seen three approaches taken. (i) For *simplicity*, set both of the constants to zero and see if we obtain a self-consistent story. (ii) At some time, we believe we could so arrange charges and currents in the Universe to reduce div $\boldsymbol{B}$ and div $\boldsymbol{D} - \rho_e$ to zero. If we can do it for one moment, it must always be true. (iii) Look at the real world and see what the constants should be once we have a physical identification for the vector fields. We will find, whichever approach we take, that we obtain a self-consistent story if we take both constants to be zero, that is,

$$\text{div}\, \boldsymbol{B} = 0, \tag{6.10}$$

$$\text{div}\, \boldsymbol{D} - \rho_e = 0. \tag{6.11}$$

Note that this is an area in which we have to abandon strict logical deduction and adopt something which works.

We can now write these relations in integral form. Integrate (6.10) over any closed volume v and apply the divergence theorem, obtaining

$$\int_v \text{div}\, \boldsymbol{B}\, \mathrm{d}v = 0$$

and hence, using the divergence theorem (A5.1),

$$\int_S \boldsymbol{B} \cdot \mathrm{d}\boldsymbol{S} = 0. \tag{6.12}$$

Similarly, from (6.11),

$$\int_v \text{div}\, \boldsymbol{D}\, \mathrm{d}v = \int_v \rho_e\, \mathrm{d}v$$

and hence

$$\int_S \boldsymbol{D} \cdot \mathrm{d}\boldsymbol{S} = \int_v \rho_e\, \mathrm{d}v. \tag{6.13}$$

Equation (6.13) tells us that the field $\boldsymbol{D}$ originates on electric charges.

6.4 Time-independent fields as conservative fields of force

Problem: Show that, if the vector fields $\boldsymbol{E}$ and $\boldsymbol{B}$ are time independent, $\boldsymbol{E}$ must satisfy $\oint_C \boldsymbol{E} \cdot \mathrm{d}\boldsymbol{s} = 0$, that is, it is a conservative field and so can be written in the form $\boldsymbol{E} = -\operatorname{grad} \phi$, where ϕ is a scalar potential function. Prove that this is definitely not possible if the field $\boldsymbol{B}$ is time varying.

Proof. If $\partial \boldsymbol{B}/\partial t = 0$, (6.1) shows that

$$\operatorname{curl} \boldsymbol{E} = 0.$$

Therefore, from Stokes' theorem (A5.2), we find

$$\oint_C \boldsymbol{E} \cdot \mathrm{d}\boldsymbol{s} = 0,$$

where the line integral of $\boldsymbol{E}$ is taken around a closed contour C. This is one of the ways of defining a *conservative* field (see appendix section A5.4). Since curl grad $\phi = 0$, where ϕ is any scalar function, $\boldsymbol{E}$ can be derived from the gradient of a scalar function:

$$\boldsymbol{E} = -\operatorname{grad} \phi. \tag{6.14}$$

However, if $\boldsymbol{B}$ is time varying then we find that

$$\operatorname{curl} \boldsymbol{E} = -\frac{\partial \boldsymbol{B}}{\partial t} \neq 0.$$

Thus, $\boldsymbol{E}$ cannot be wholly expressed as $-\operatorname{grad} \phi$, if $\boldsymbol{B}$ is time varying.

6.5 Boundary conditions in electromagnetism

Across any sharp boundary between two media, the properties of the fields are expected to change discontinuously. We consider three problems.

Problem 1: From Section 6.3, show that the component of the vector field $\boldsymbol{B}$ perpendicular to the surface between the media, $\boldsymbol{B} \cdot \boldsymbol{n}$, is continuous at the boundary and likewise that $\boldsymbol{D} \cdot \boldsymbol{n}$ is continuous, if there are no surface charges on the boundary. If there are free surface charges, with surface charge density σ, show that $(\boldsymbol{D}_2 - \boldsymbol{D}_1) \cdot \boldsymbol{n} = \sigma$.

Proof. We erect a very short cylinder of cross-section S across the boundary (Fig. 6.1(a)) and then apply the rules which we know are true under all circumstances for the fluxes of the vector fields $\boldsymbol{B}$ and $\boldsymbol{D}$ through this volume. The diagram shows the $\boldsymbol{B}$ or $\boldsymbol{D}$ field passing through the boundary of the cylinder.

From (6.12), the flux of the vector field $\boldsymbol{B}$ through the surface of the cylinder is

$$\int_S \boldsymbol{B} \cdot \mathrm{d}\boldsymbol{S} = 0. \tag{6.15}$$

Now squash the cylinder until it is infinitesimally thin. Then the surface area round the edge of the cylinder $2\pi r\, \mathrm{d}y$ tends to zero as $\mathrm{d}y$ tends to zero, and only the contributions from the

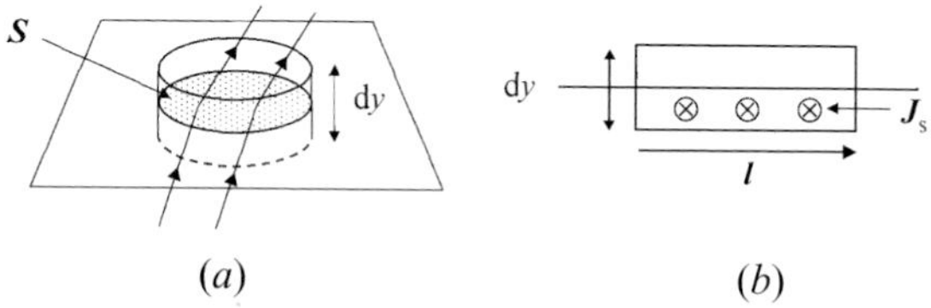

Figure 6.1: Illustrating the boundary conditions for (*a*) the normal components of $\boldsymbol{B}$ and $\boldsymbol{D}$, and (*b*) the components of $\boldsymbol{E}$ and $\boldsymbol{H}$ parallel to the surface.

upper and lower faces of the cylinder are left. If $\boldsymbol{n}$ is a unit vector normal to the interface, $\boldsymbol{B} \cdot \boldsymbol{n}$ is the flux of the vector field $\boldsymbol{B}$ per unit area. Taking medium 1 to be the lower layer and medium 2 the upper, the flux per unit area through the upper surface of the cylinder is $\boldsymbol{B}_2 \cdot \boldsymbol{n}$ and that entering through the lower surface is $-\boldsymbol{B}_1 \cdot \boldsymbol{n}$, recalling that $\mathrm{d}\boldsymbol{S}$ is always taken to point outwards through a closed surface. Therefore, (6.15) reduces to

$$(\boldsymbol{B}_2 - \boldsymbol{B}_1) \cdot \boldsymbol{n} = 0, \tag{6.16}$$

that is, the normal component of $\boldsymbol{B}$ is continuous.

In exactly the same way,

$$\int_S \boldsymbol{D} \cdot \mathrm{d}\boldsymbol{S} = \int_v \rho_\mathrm{e} \, \mathrm{d}v, \tag{6.17}$$

and when we squash the cylinder, the left-hand side becomes the difference in the values of $\int_S \boldsymbol{D} \cdot \mathrm{d}\boldsymbol{S}$ on either side of the surface and $\int_v \rho_\mathrm{e} \, \mathrm{d}v$ becomes the surface charge $\int_S \sigma \, \mathrm{d}S$, thus

$$(\boldsymbol{D}_2 - \boldsymbol{D}_1) \cdot \boldsymbol{n} = \sigma. \tag{6.18}$$

If there are no free surface charges, $\sigma = 0$ and the normal component of $\boldsymbol{D}$ is continuous.

Problem 2: If the fields $\boldsymbol{E}$ and $\boldsymbol{H}$ are static fields (and also $\boldsymbol{D}$, $\boldsymbol{B}$ and $\boldsymbol{J}$), show that (a) the tangential component of $\boldsymbol{E}$ is continuous at the boundary, and (b) the tangential component of $\boldsymbol{H}$ is continuous if there is no surface current density $\boldsymbol{J}_\mathrm{s}$. If there is, show that

$$\boldsymbol{n} \times (\boldsymbol{H}_2 - \boldsymbol{H}_1) = \boldsymbol{J}_\mathrm{s}.$$

Proof. (a) Create a long rectangular loop which crosses the surface between the two media as shown in Figure 6.1(*b*). The sides of the loop perpendicular to the surface, $\mathrm{d}y$, are very much shorter than the vector $\boldsymbol{l}$ defining the length of the horizontal arms. The fields are static and therefore, from (6.1) and the discussion in Section 6.4, curl $\boldsymbol{E} = 0$ or, applying Stokes' theorem, $\oint_C \boldsymbol{E} \cdot \mathrm{d}\boldsymbol{s} = 0$. We apply this result to the long rectangular loop.

Let $\boldsymbol{E}_1$ and $\boldsymbol{E}_2$ be the fields on the lower and upper sides respectively of the interface between the media. The loop can be squashed to infinitesimal thickness, so that the ends $\mathrm{d}y$ make no contribution to the line integral. Projecting the electric field $\boldsymbol{E}_1$ onto $\boldsymbol{l}$, the line integral along the upper arm of the rectangle is

$$\int \boldsymbol{E} \cdot \mathrm{d}\boldsymbol{s} = \boldsymbol{E}_2 \cdot \boldsymbol{l}.$$

Correspondingly, along the lower arm, the line integral is

$$\int \boldsymbol{E} \cdot \mathrm{d}\boldsymbol{s} = \boldsymbol{E}_1 \cdot (-\boldsymbol{l}).$$

Therefore, the line integral round the complete loop is

$$\oint_C \boldsymbol{E} \cdot \mathrm{d}\boldsymbol{s} = (\boldsymbol{E}_2 - \boldsymbol{E}_1) \cdot \boldsymbol{l} = 0.$$

We need to express this result in terms of the unit vector $\boldsymbol{n}$ normal to the interface. Let the unit vector parallel to $\boldsymbol{l}$ be $\boldsymbol{i}_l$ and the unit vector perpendicular to the loop be $\boldsymbol{i}_\perp$, such that $\boldsymbol{i}_l = \boldsymbol{i}_\perp \times \boldsymbol{n}$. Then

$$\begin{aligned}(\boldsymbol{E}_2 - \boldsymbol{E}_1) \cdot \boldsymbol{l} &= (\boldsymbol{E}_2 - \boldsymbol{E}_1) \cdot (\boldsymbol{i}_\perp \times \boldsymbol{n}) l = 0, \\ &= \boldsymbol{i}_\perp \cdot [(\boldsymbol{n} \times (\boldsymbol{E}_2 - \boldsymbol{E}_1)] l = 0. \end{aligned} \tag{6.19}$$

This result must be true for arbitrary values of l, and so

$$\boldsymbol{n} \times (\boldsymbol{E}_2 - \boldsymbol{E}_1) = 0, \tag{6.20}$$

that is, the tangential components of $\boldsymbol{E}$ are equal on either side of the interface.

(b) Now carry out the same analysis using (6.2) for the case of static magnetic fields, so that

$$\mathrm{curl}\, \boldsymbol{H} = \boldsymbol{J}.$$

Taking the scalar product with $\mathrm{d}\boldsymbol{S}$ and integrating across the area $\boldsymbol{S}$ of the loop, we obtain

$$\int_S \mathrm{curl}\, \boldsymbol{H} \cdot \mathrm{d}\boldsymbol{S} = \int_S \boldsymbol{J} \cdot \mathrm{d}\boldsymbol{S}.$$

Applying Stokes' theorem to the left-hand side gives

$$\oint_C \boldsymbol{H} \cdot \mathrm{d}\boldsymbol{s} = \int_S \boldsymbol{J} \cdot \mathrm{d}\boldsymbol{S}, \tag{6.21}$$

where the path integral is around the rectangular loop.

Now again squash the rectangular loop to zero thickness. Exactly as in (a), the line integral of $\boldsymbol{H}$ round the loop is

$$\oint_C \boldsymbol{H} \cdot \mathrm{d}\boldsymbol{s} = \boldsymbol{i}_\perp \cdot [\boldsymbol{n} \times (\boldsymbol{H}_2 - \boldsymbol{H}_1)] l. \tag{6.22}$$

The total surface current flowing through the rectangular loop, $\int_S \boldsymbol{J} \cdot \mathrm{d}\boldsymbol{S}$, can be rewritten as $(\boldsymbol{J}_\mathrm{s} \cdot \boldsymbol{i}_\perp) l$, where $\boldsymbol{J}_\mathrm{s}$ is the surface current density, that is, the current per unit length.

We can therefore rewrite (6.21) as

$$\boldsymbol{i}_\perp \cdot [\boldsymbol{n} \times (\boldsymbol{H}_2 - \boldsymbol{H}_1)] l = \boldsymbol{i}_\perp \cdot \boldsymbol{J}_\mathrm{s} l$$

giving

$$\boldsymbol{n} \times (\boldsymbol{H}_2 - \boldsymbol{H}_1) = \boldsymbol{J}_\mathrm{s}. \tag{6.23}$$

If there are no surface currents, $\boldsymbol{J}_\mathrm{s} = 0$, then $\boldsymbol{H} \times \boldsymbol{n}$, the tangential component of $\boldsymbol{H}$, is continuous.

At this stage, it is not obvious that the relations (6.20) and (6.23) are true in the presence of time-varying fields because at the outset of the problem we discarded the time-varying components. In fact, they are still true, as we now demonstrate in problem 3.

Problem 3: Prove that the statements (6.20) and (6.23) in problem 2 are correct even in the presence of time-varying fields.

Proof. This requires a little more care. First integrate (6.1) over a surface S and apply Stokes' theorem:

$$\int_S \operatorname{curl} \boldsymbol{E} \cdot \mathrm{d}\boldsymbol{S} = -\int_S \frac{\partial \boldsymbol{B}}{\partial t} \cdot \mathrm{d}\boldsymbol{S},$$

$$\oint_C \boldsymbol{E} \cdot \mathrm{d}\boldsymbol{s} = -\int_S \frac{\partial \boldsymbol{B}}{\partial t} \cdot \mathrm{d}\boldsymbol{S}.$$

Now apply this result to the little rectangle in Fig. 6.1(*b*). The analysis proceeds as before, except that there is a time-varying component on the right-hand side. Following the same analysis which led to (6.20),

$$\boldsymbol{i}_\perp \cdot [\boldsymbol{n} \times (\boldsymbol{E}_2 - \boldsymbol{E}_1)]l = -\frac{\partial \boldsymbol{B}}{\partial t} \cdot \boldsymbol{i}_\perp l\, \mathrm{d}y + \text{small end contributions.}$$

Now cancel out l on either side and let the height $\mathrm{d}y$ of the rectangle shrink to zero:

$$\boldsymbol{n} \times (\boldsymbol{E}_2 - \boldsymbol{E}_1) = -\left(\mathrm{d}y \frac{\partial \boldsymbol{B}}{\partial t}\right)_{\mathrm{d}y \to 0}.$$

$\partial \boldsymbol{B}/\partial t$ is a finite quantity in the case of time-varying fields but, in the limit in which $\mathrm{d}y$ tends to zero, the right-hand side makes no contribution, that is,

$$(\boldsymbol{E}_2 - \boldsymbol{E}_1) \times \boldsymbol{n} = 0$$

as before.

Exactly the same analysis can be carried out for time-varying $\boldsymbol{D}$ fields, that is,

$$\int_S \operatorname{curl} \boldsymbol{H} \cdot \mathrm{d}\boldsymbol{S} = \int_S \boldsymbol{J} \cdot \mathrm{d}\boldsymbol{S} + \int_S \frac{\partial \boldsymbol{D}}{\partial t} \cdot \mathrm{d}\boldsymbol{S},$$

$$\oint_C \boldsymbol{H} \cdot \mathrm{d}\boldsymbol{s} = \int_S \boldsymbol{J} \cdot \mathrm{d}\boldsymbol{S} + \int_S \frac{\partial \boldsymbol{D}}{\partial t} \cdot \mathrm{d}\boldsymbol{S}.$$

Now, as before, squash the rectangle to zero thickness, following the same reasoning as that which led to (6.23):

$$\boldsymbol{i}_\perp \cdot [\boldsymbol{n} \times (\boldsymbol{H}_2 - \boldsymbol{H}_1)]l = \boldsymbol{i}_\perp \cdot \boldsymbol{J}_\mathrm{s} l + \frac{\partial \boldsymbol{D}}{\partial t} \cdot \boldsymbol{i}_\perp l\, \mathrm{d}y$$
$$+ \text{ small end contributions,}$$

that is,

$$\boldsymbol{n} \times (\boldsymbol{H}_2 - \boldsymbol{H}_1) = \boldsymbol{J}_s + \left(\mathrm{d}y \frac{\partial \boldsymbol{D}}{\partial t} \right)_{\mathrm{d}y \to 0},$$

so that

$$\boldsymbol{n} \times (\boldsymbol{H}_2 - \boldsymbol{H}_1) = \boldsymbol{J}_s,$$

even in the presence of time-varying $\boldsymbol{D}$ fields.

6.6 Ampère's law

Problem: If $\boldsymbol{H}$ is not time varying (we will eventually identify this sort of $\boldsymbol{H}$ with a magnetostatic field), show that $\boldsymbol{H}$ can be expressed in the form $-\mathrm{grad}\, V_{\mathrm{mag}}$ when there are only permanent magnets and magnetisable materials but *no* currents. If there are steady currents show that $\oint_C \boldsymbol{H} \cdot \mathrm{d}\boldsymbol{s} = I_{\mathrm{enclosed}}$. This equation is often used even for time-varying fields. Why is this so?

Proof. In the absence of currents $\boldsymbol{J} = 0$, and so, from (6.2), in the absence of time-varying $\boldsymbol{D}$ fields curl $\boldsymbol{H} = 0$. $\boldsymbol{H}$ is therefore a conservative field and can be expressed as the gradient of a scalar field:

$$\boldsymbol{H} = -\mathrm{grad}\, V_{\mathrm{m}}.$$

If *steady* currents are present, curl $\boldsymbol{H} = \boldsymbol{J}$. Integrating over a surface S and applying Stokes' theorem,

$$\int_S \mathrm{curl}\, \boldsymbol{H} \cdot \mathrm{d}\boldsymbol{S} = \int_S \boldsymbol{J} \cdot \mathrm{d}\boldsymbol{S},$$

$$\oint_C \boldsymbol{H} \cdot \mathrm{d}\boldsymbol{s} = \text{total enclosed current} = I_{\mathrm{enclosed}}. \tag{6.24}$$

This equation can be used in the presence of time-varying $\boldsymbol{D}$ fields, provided that they are very slowly varying, that is, $\partial \boldsymbol{D}/\partial t \ll \boldsymbol{J}$. This is often a good approximation, but we have to check that it is valid when any of the fields or currents are time varying.

6.7 Faraday's law

Problem: Derive Faraday's law, $\oint_C \boldsymbol{E} \cdot \mathrm{d}\boldsymbol{s} = -\mathrm{d}\Phi/\mathrm{d}t$, where $\Phi = \int_S \boldsymbol{B} \cdot \mathrm{d}\boldsymbol{S}$ is the flux of the field $\boldsymbol{B}$ through the closed contour C. In this formulation, $\boldsymbol{E}$ is the *total* electric field, whereas the usual statement of Faraday's law refers only to the induced field. Explain this apparent discrepancy.

Proof. From (6.1), and using Stokes' theorem,

$$\int_S \operatorname{curl} \boldsymbol{E} \cdot \mathrm{d}\boldsymbol{S} = -\frac{\partial}{\partial t}\left(\int_S \boldsymbol{B} \cdot \mathrm{d}\boldsymbol{S}\right),$$

that is,

$$\oint_C \boldsymbol{E} \cdot \mathrm{d}\boldsymbol{s} = -\frac{\partial \Phi}{\partial t}. \tag{6.25}$$

Normally, Faraday's law refers only to the induced part of the field $\boldsymbol{E}$. There may also, however, be a component due to an electrostatic field, that is

$$\begin{aligned} \boldsymbol{E} &= \boldsymbol{E}_{\text{induced}} + \boldsymbol{E}_{\text{electrostatic}} \\ &= \boldsymbol{E}_{\text{induced}} - \operatorname{grad} V \\ \operatorname{curl} \boldsymbol{E} &= \operatorname{curl} \boldsymbol{E}_{\text{induced}} - \operatorname{curl} \operatorname{grad} V \\ &= -\frac{\partial \boldsymbol{B}}{\partial t}, \end{aligned}$$

which shows that $\boldsymbol{E}$ refers to the total field including the electrostatic part, but the latter does not survive curling.

6.8 The story so far

The above analysis has been based upon the mathematical properties of the set of vector equations introduced in Section 6.2. Although we have used words like magnets, electrostatic field and so on, the properties of the equations are independent of these physical identifications. We now need to give the fields $\boldsymbol{E}$, $\boldsymbol{D}$, $\boldsymbol{H}$ and $\boldsymbol{B}$ physical meaning.

We define $q\boldsymbol{E}$ to be the force acting on a stationary charge q. Let us see whether we have yet closed our set of equations. $\boldsymbol{J}$ is defined by (6.3) and (6.4) and $\boldsymbol{E}$ has just been defined. However, we still have to define $\boldsymbol{D}$, $\boldsymbol{B}$ and $\boldsymbol{H}$ and we have only two independent equations left, (6.1) and (6.2). We have not yet closed the set of equations. In order to do so, we have to introduce a further set of definitions based upon experimental evidence. First of all, we consider the behaviour of electromagnetic fields in a vacuum and make the further definitions, consistent with experiment, that

$$\text{in a vacuum,} \qquad \begin{aligned} \boldsymbol{D} &= \epsilon_0 \boldsymbol{E}, \\ \boldsymbol{B} &= \mu_0 \boldsymbol{H}, \end{aligned} \tag{6.26}$$

where ϵ_0 and μ_0 are constants. $\boldsymbol{D}$ is called the electric flux density or electric displacement, $\boldsymbol{E}$ is the electric field strength, $\boldsymbol{H}$ is the magnetic field strength and $\boldsymbol{B}$ the magnetic flux density or magnetic induction. The relations between $\boldsymbol{D}$ and $\boldsymbol{E}$ and between $\boldsymbol{B}$ and $\boldsymbol{H}$ are known as *constitutive equations* or *relations*.

Inside material media, these relations will not be correct, in general, and so we introduce vectors which describe the differences between the actual values and the vacuum definitions – these refer to the electric and magnetic *polarisation properties* of the material: thus,

in a material,
$$P = D - \epsilon_0 E, \tag{6.27}$$
$$M = \frac{B}{\mu_0} - H. \tag{6.28}$$

Granted these definitions, our objective is to find physical meanings for the 'polarisation vectors' P and M. Let us proceed with the mathematical analysis of the equations and see if these definitions lead to consistency with all the known laws of electricity and magnetism.

6.9 Derivation of Coulomb's law

Problem: The electrostatic field in a vacuum is due to a charge distribution ρ_e. Show that the field satisfies Poisson's equation $\nabla^2 V = -\rho_e/\epsilon_0$. From the definition of E in Section 6.8 and the results of Section 6.4, show that V is the work done in bringing unit charge from infinity to the point r in the field. Show that the solution of Poisson's equation is

$$V(r) = \int \frac{\rho_e(r')}{4\pi\epsilon_0 |r - r'|}\, d^3 r'.$$

Hence derive Coulomb's law, $F = q_1 q_2/(4\pi\epsilon_0 r^2)$.

Proof. We have shown that div $D = \rho_e$ and, in a vacuum, we have defined $D = \epsilon_0 E$. Therefore

$$\operatorname{div} \epsilon_0 E = \rho_e \qquad \text{and so} \qquad \operatorname{div}(\epsilon_0 \operatorname{grad} V) = -\rho_e,$$

that is,

$$\nabla^2 V = -\frac{\rho_e}{\epsilon_0}. \tag{6.29}$$

This is *Poisson's equation for the electrostatic potential V in a vacuum with a charge distribution ρ_e*.

The work done against the electrostatic force $f = qE$ in bringing the charge q from infinity to r is given by

$$\begin{aligned} \text{work done} &= -\int_\infty^r f \cdot dr = -q \int_\infty^r E \cdot dr = q \int_\infty^r \operatorname{grad} V \cdot dr \\ &= qV, \end{aligned}$$

where the electrostatic potential V is assumed to be zero at infinity. Thus, V measures the work done against an electrostatic field in bringing unit charge from infinity to that point in the field. Notice that qV is *not* the amount of energy needed to set up the electric field distribution as a whole. This will be evaluated in Section 6.12.

To find the solution of Poisson's equation, it is simplest to consider ρ_e as a distribution of point charges in a vacuum. The reason is that then, provided we are not at the actual position of any of these charges, we can use *Laplace's equation* $\nabla^2 V = 0$, which is a linear equation, for which the principle of superposition applies. Thus, if V_1 and V_2 are two independent solutions of Laplace's equation, $a_1 V_1 + a_2 V_2$ is also a solution. Therefore, the potential at any point is the superposition of the solutions of Laplace's equation, which can be related to the sources of the field, that is, to the charges q. Let us therefore consider as a source of the field an elementary volume of the charge distribution located at $\boldsymbol{r}' = 0$, that is, at the origin:

$$\rho_e(\boldsymbol{r}')\,\mathrm{d}^3\boldsymbol{r}' = q.$$

The proposed solution of Laplace's equation in this case is

$$V(r) = \frac{\rho_e(\boldsymbol{r}')\,\mathrm{d}^3\boldsymbol{r}'}{4\pi\epsilon_0 r} = \frac{q}{4\pi\epsilon_0 r}, \tag{6.30}$$

at a point P a radial distance r from the charge q at the origin. Let us test this vacuum solution away from the origin $r = 0$. We recall that Laplace's equation in spherical polar coordinates is

$$\frac{1}{r}\frac{\partial^2}{\partial r^2}(rV) + \frac{1}{r^2}\left[\frac{1}{\sin\theta}\frac{\partial}{\partial\theta}\left(\sin\theta\,\frac{\partial V}{\partial\theta}\right) + \frac{1}{\sin^2\theta}\frac{\partial^2 V}{\partial\varphi^2}\right] = 0 \tag{6.31}$$

(see appendix section A5.5) V is independent of θ and φ because of the spherical symmetry of the problem. Thus, substituting the proposed solution for $V(r)$ into the left-hand side of (6.31) gives

$$\frac{1}{r}\frac{\partial^2}{\partial r^2}(rV) = \frac{1}{r}\frac{\partial^2}{\partial r^2}\left(\frac{q}{4\pi\epsilon_0}\right) = 0$$

(provided $r \neq 0$). Thus, the proposed solution satisfies Laplace's equation away from the origin.

Now evaluate the charge Q enclosed by a sphere centred on the origin using Poisson's equation for ρ_e:

$$Q = \int_v \rho_e\,\mathrm{d}v = -\int_v \epsilon_0\nabla^2 V\,\mathrm{d}v$$

with $V = q/(4\pi\epsilon_0 r)$. The total enclosed charge is

$$Q = -\frac{q}{4\pi}\int_v \mathrm{div}\left(\mathrm{grad}\,\frac{1}{r}\right)\mathrm{d}v.$$

Using the divergence theorem,

$$Q = -\frac{q}{4\pi}\int_S\left(\mathrm{grad}\,\frac{1}{r}\right)\cdot\mathrm{d}\boldsymbol{S} = \frac{q}{4\pi}\int_S\frac{\mathrm{d}S}{r^2} = \frac{q}{4\pi}\int_S\mathrm{d}\Omega = q,$$

where $\mathrm{d}S$ is the element of area perpendicular to the radial direction and $\mathrm{d}\Omega$ is the element of solid angle. Thus, according to Poisson's equation, the correct charge is found at the origin from the proposed solution. If we move the charge away from $\boldsymbol{r}' = 0$, so that now r is replaced by $|\boldsymbol{r} - \boldsymbol{r}'|$, the solution (6.30) becomes

$$V(\boldsymbol{r}) = \frac{\rho_e(\boldsymbol{r}')\,\mathrm{d}^3\boldsymbol{r}'}{4\pi\epsilon_0|\boldsymbol{r} - \boldsymbol{r}'|}.$$

We conclude that this solution satisfies Poisson's equation in electrostatics.

Returning again to a single particle q_1 at the origin, $\nabla^2 V = -q_1/\epsilon_0$. The electrostatic force due to q_1 on another particle q_2 is $q_2 \boldsymbol{E}$, that is,

$$\begin{aligned} \boldsymbol{f} &= -q_2 \,\mathrm{grad}\, V = -\frac{q_1 q_2}{4\pi\epsilon_0}\,\mathrm{grad}\left(\frac{1}{r}\right) \\ &= \frac{q_1 q_2}{4\pi\epsilon_0 r^2}\boldsymbol{i}_r = \frac{q_1 q_2}{4\pi\epsilon_0}\frac{\boldsymbol{r}}{r^3}. \end{aligned} \tag{6.32}$$

This is the derivation of *Coulomb's inverse square law of electrostatics*. We have taken a rather circuitous route to reach the point at where most courses in electricity and magnetism begin. The point of interest is that we can start with Maxwell's equations and deduce Coulomb's law with a minimum number of assumptions.

6.10 Derivation of the Biôt–Savart law

Problem: Steady or slowly varying currents are flowing in a vacuum. What is 'slowly varying'? Show that curl $\boldsymbol{H} = \boldsymbol{J}$.

It can be shown that the solution of this equation is

$$\boldsymbol{H}(\boldsymbol{r}) = \int \frac{\boldsymbol{J}(\boldsymbol{r}') \times (\boldsymbol{r} - \boldsymbol{r}')}{4\pi\,|\boldsymbol{r} - \boldsymbol{r}'|^3}\,\mathrm{d}^3\boldsymbol{r}'. \tag{6.33}$$

Show that this leads to the Biôt–Savart law $\mathrm{d}\boldsymbol{H} = I\sin\theta\,(\mathrm{d}\boldsymbol{s} \times \boldsymbol{r})/(4\pi r^3)$ for the contribution to the magnetic field strength $\boldsymbol{H}$ at position vector $\boldsymbol{r}$ with respect to the current element $I\mathrm{d}\boldsymbol{s}$.

Proof. We have already discussed the issue of slowly varying fields in Section 6.6. 'Slowly varying' means that the displacement current term $\partial\boldsymbol{D}/\partial t \ll \boldsymbol{J}$. Then, from (6.2), we obtain curl $\boldsymbol{H} = \boldsymbol{J}$.

Just as in Section 6.9, where the solution of $\nabla^2 V = -\rho_e/\epsilon_0$ was required, so now we need the solution of curl $\boldsymbol{H} = \boldsymbol{J}$. The general solution (6.33) is obtained in the standard textbooks and need not be repeated here. Let us differentiate (6.33), so that the relation becomes one between a current element and the field at a distance $|\boldsymbol{r} - \boldsymbol{r}'|$:

$$\mathrm{d}\boldsymbol{H} = \frac{\boldsymbol{J}(\boldsymbol{r}') \times (\boldsymbol{r} - \boldsymbol{r}')}{4\pi\,|\boldsymbol{r} - \boldsymbol{r}'|^3}\,\mathrm{d}^3\boldsymbol{r}'. \tag{6.34}$$

Now we identify $\boldsymbol{J}(\boldsymbol{r}')\,\mathrm{d}^3\boldsymbol{r}'$ with the current element $I\mathrm{d}\boldsymbol{s}$. Then we obtain

$$\mathrm{d}\boldsymbol{H} = \frac{I\mathrm{d}\boldsymbol{s} \times \boldsymbol{r}}{4\pi r^3}, \qquad |\mathrm{d}\boldsymbol{H}| = \frac{I\sin\theta\,|\mathrm{d}\boldsymbol{s}|}{4\pi r^2}, \tag{6.35}$$

where $\boldsymbol{r}$ is the vector from the current element to the point in the field. This is the *Biôt–Savart law*.

6.11 The interpretation of Maxwell's equations in material media

Problem: Using the definitions in Section 6.8 and the results of Section 6.3, show that Maxwell's equations can be written in the form

$$\text{curl}\,\boldsymbol{E} = -\frac{\partial \boldsymbol{B}}{\partial t}, \tag{6.36a}$$

$$\text{div}\,\boldsymbol{B} = 0, \tag{6.36b}$$

$$\text{div}\,\epsilon_0 \boldsymbol{E} = \rho_e - \text{div}\,\boldsymbol{P}, \tag{6.36c}$$

$$\text{curl}\,\frac{\boldsymbol{B}}{\mu_0} = \left(\boldsymbol{J} + \frac{\partial \boldsymbol{P}}{\partial t} + \text{curl}\,\boldsymbol{M}\right) + \frac{\partial(\epsilon_0 \boldsymbol{E})}{\partial t}. \tag{6.36d}$$

Show that, in *electrostatics*, the field $\boldsymbol{E}$ may be calculated correctly everywhere by replacing a polarisable medium by a vacuum together with a volume charge distribution $-\text{div}\,\boldsymbol{P}$ and a surface charge density $\boldsymbol{P} \cdot \boldsymbol{n}$. Then write down the expression for the electrostatic potential V at any point in space in terms of these charges and show that the electric polarisation $\boldsymbol{P}$ represents the dipole moment per unit volume within the medium. You may use the fact that the electrostatic potential of an electric dipole of dipole moment $\boldsymbol{p}$ can be written as

$$V = \frac{1}{4\pi\epsilon_0}\,\boldsymbol{p} \cdot \nabla\left(\frac{1}{r}\right). \tag{6.37}$$

In *magnetostatics*, show that the field $\boldsymbol{B}$ may be correctly calculated by replacing magnetisable bodies by a current distribution curl $\boldsymbol{M}$ with surface current densities $-\boldsymbol{n} \times \boldsymbol{M}$. Then write down an expression for the magnetostatic vector potential $\boldsymbol{A}$ at any point in space in terms of these currents and show that the magnetisation $\boldsymbol{M}$ represents the magnetic dipole moment per unit volume within the medium. You may use the fact that the magnetostatic vector potential of a magnetic dipole of moment $\boldsymbol{m}$ at distance r can be written as

$$\boldsymbol{A} = \frac{\mu_0}{4\pi}\,\boldsymbol{m} \times \nabla\left(\frac{1}{r}\right). \tag{6.38}$$

Proof. I regard this as a particularly pleasant piece of analysis which gives insight into the mathematical structure of Maxwell's equations.

The first part of the problem is straightforward and simply involves rewriting Maxwell's equations. The general results (6.36a) and (6.36b) have already been discussed in Section 6.2:

$$\text{curl}\,\boldsymbol{E} = -\frac{\partial \boldsymbol{B}}{\partial t} \quad \text{and} \quad \text{div}\boldsymbol{B} = 0.$$

To derive (6.36c): since div $\boldsymbol{D} = \rho_e$ and $\boldsymbol{D} = \boldsymbol{P} + \epsilon_0 \boldsymbol{E}$, we obtain

$$\text{div}\,\epsilon_0 \boldsymbol{E} = \rho_e - \text{div}\,\boldsymbol{P}. \tag{6.39}$$

Similarly, from (6.2) and the definition $\boldsymbol{H} = (\boldsymbol{B}/\mu_0) - \boldsymbol{M}$,

$$\text{curl}\,\boldsymbol{H} = \boldsymbol{J} + \frac{\partial \boldsymbol{D}}{\partial t},$$

so that

$$\operatorname{curl}\left(\frac{\boldsymbol{B}}{\mu_0} - \boldsymbol{M}\right) = \boldsymbol{J} + \frac{\partial \boldsymbol{P}}{\partial t} + \frac{\partial(\epsilon_0 \boldsymbol{E})}{\partial t}$$

and finally

$$\operatorname{curl}\left(\frac{\boldsymbol{B}}{\mu_0}\right) = \left(\boldsymbol{J} + \frac{\partial \boldsymbol{P}}{\partial t} + \operatorname{curl} \boldsymbol{M}\right) + \frac{\partial(\epsilon_0 \boldsymbol{E})}{\partial t}. \tag{6.40}$$

Now let us consider the second part of the problem.

Electrostatics. Equation (6.36c) states that to calculate $\operatorname{div} \epsilon_0 \boldsymbol{E}$ at any point in space, we have to add to ρ_e the quantity $-\operatorname{div} \boldsymbol{P}$, which we may regard as an effective charge ρ_e^*.

Now consider the boundary conditions between the region containing the polarisable medium (medium 1) and a vacuum (medium 2). From Section 6.5, we know that, under all circumstances, the component of $\boldsymbol{D}$ normal to the boundary is continuous.

$$(\boldsymbol{D}_2 - \boldsymbol{D}_1) \cdot \boldsymbol{n} = \sigma \tag{6.41}$$

where σ is the surface charge density. If we suppose that there are no free charges on the surface, $\sigma = 0$. Therefore

$$(\boldsymbol{D}_2 - \boldsymbol{D}_1) \cdot \boldsymbol{n} = 0;$$

now using (6.28), we obtain

$$[\epsilon_0 \boldsymbol{E}_2 - (\epsilon_0 \boldsymbol{E}_1 + \boldsymbol{P}_1)] \cdot \boldsymbol{n} = 0,$$

and so

$$\epsilon_0 \boldsymbol{E}_2 - \epsilon_0 \boldsymbol{E}_1 = \boldsymbol{P}_1 \cdot \boldsymbol{n}. \tag{6.42}$$

Thus, when we replace the medium by a vacuum, there must also be an effective surface charge density $\sigma^* = \boldsymbol{P}_1 \cdot \boldsymbol{n}$ per unit area at the boundary.

Therefore, in calculating the $\boldsymbol{E}$ field anywhere in space, we find the correct answer if we replace the material medium by a charge distribution $\rho_e^* = -\operatorname{div} \boldsymbol{P}$ and a surface charge distribution $\sigma^* = \boldsymbol{P} \cdot \boldsymbol{n}$ per unit area. Let us check that this all makes sense. If we have replaced the material medium by *real charge distributions* ρ_e^* and σ^*, what is the field at any point in space?

The total electric field due to these charges is obtained from the gradient of the electrostatic potential V, where

$$\begin{aligned} V &= \frac{1}{4\pi\epsilon_0}\int_v \frac{\rho_e^*}{r}\,\mathrm{d}v + \frac{1}{4\pi\epsilon_0}\int_S \frac{\sigma^*}{r}\,\mathrm{d}S \\ &= -\frac{1}{4\pi\epsilon_0}\int_v \frac{\operatorname{div} \boldsymbol{P}}{r}\,\mathrm{d}v + \frac{1}{4\pi\epsilon_0}\int_S \frac{\boldsymbol{P}\cdot \mathrm{d}\boldsymbol{S}}{r}. \end{aligned}$$

Applying the divergence theorem to the second integral, we obtain

$$V = \frac{1}{4\pi\epsilon_0}\int_v \left[-\frac{\operatorname{div} \boldsymbol{P}}{r} + \operatorname{div}\left(\frac{\boldsymbol{P}}{r}\right)\right]\mathrm{d}v.$$

But

$$\operatorname{div}\left(\frac{\boldsymbol{P}}{r}\right)=\frac{\operatorname{div}\boldsymbol{P}}{r}+\boldsymbol{P}\cdot\operatorname{grad}\frac{1}{r},$$

and hence

$$V=\frac{1}{4\pi\epsilon_0}\int_v \boldsymbol{P}\cdot\operatorname{grad}\left(\frac{1}{r}\right)\mathrm{d}v. \tag{6.43}$$

This is the rather beautiful result we have been seeking. We recall that the potential at distance r from an electrostatic dipole is

$$V=\frac{1}{4\pi\epsilon_0}\boldsymbol{p}\cdot\operatorname{grad}\left(\frac{1}{r}\right)$$

where $\boldsymbol{p}$ is the electric dipole moment of the dipole. Thus, we can interpret (6.43) as showing that the quantity $\boldsymbol{P}$ is the *dipole moment per unit volume* within the material.

We can interpret the phenomenon of polarisation as the dipole moment per unit volume caused by placing the material in an electrostatic field $\boldsymbol{E}$. This is the origin of the term $\rho_e^* = -\operatorname{div}\boldsymbol{P}$. However, the dipoles at the surfaces of the material result in net positive charges on one side of the material and negative charges on the opposite side, so that the whole system remains neutral. It is these charges which give rise to the surface charge distribution $\rho_e^* = \boldsymbol{P}\cdot\boldsymbol{n}$.

Magnetostatics. For the third part of the problem, the analysis proceeds in exactly the same way. Equation (6.36d) states that we have to include a current density distribution $\boldsymbol{J}^* = \operatorname{curl}\boldsymbol{M}$ in the expression for $\boldsymbol{B}$ when we replace the medium by a vacuum. In the case of a surface current flowing in the boundary between two media, (6.23) indicates that

$$\boldsymbol{n}\times(\boldsymbol{H}_2-\boldsymbol{H}_1)=\boldsymbol{J}_\mathrm{s}.$$

If there are no surface currents, $\boldsymbol{J}_\mathrm{s}=0$ and so

$$\boldsymbol{n}\times(\boldsymbol{H}_2-\boldsymbol{H}_1)=0,$$

$$\boldsymbol{n}\times\left[\frac{\boldsymbol{B}_2}{\mu_0}-\left(\frac{\boldsymbol{B}_1}{\mu_0}-\boldsymbol{M}_1\right)\right]=0.$$

Thus,

$$\boldsymbol{n}\times\left[\frac{\boldsymbol{B}_2}{\mu_0}-\frac{\boldsymbol{B}_1}{\mu_0}\right]=-\boldsymbol{n}\times\boldsymbol{M}_1. \tag{6.44}$$

There must therefore be an effective surface current distribution $\boldsymbol{J}_\mathrm{s}^* = -(\boldsymbol{n}\times\boldsymbol{M}_1)$.

We now carry out an analysis similar to that which illustrated the meaning of the polarisation $\boldsymbol{P}$. The expression for the vector potential $\boldsymbol{A}$ due to the current I flowing in a current distribution is

$$\boldsymbol{A}=\frac{\mu_0}{4\pi}\int\frac{I\,\mathrm{d}\boldsymbol{s}}{r}.$$

Rewriting this expression in terms of the current density distribution $\boldsymbol{J}$, $I\,\mathrm{d}\boldsymbol{s} = \boldsymbol{J}\,\mathrm{d}\sigma\,\mathrm{d}l = \boldsymbol{J}\,\mathrm{d}v$ and so

$$\boldsymbol{A} = \frac{\mu_0}{4\pi}\int \frac{\boldsymbol{J}\,\mathrm{d}v}{r}.$$

The vector potential due to the current density distribution $\boldsymbol{J}^*$ and the surface current distribution $\boldsymbol{J}^*_{\mathrm{s}}$ is therefore

$$\begin{aligned}\boldsymbol{A} &= \frac{\mu_0}{4\pi}\int_v \frac{\boldsymbol{J}^*\,\mathrm{d}v}{r} + \frac{\mu_0}{4\pi}\int_S \frac{\boldsymbol{J}^*_{\mathrm{s}}\,\mathrm{d}S}{r} \\ &= \frac{\mu_0}{4\pi}\int_v \frac{\mathrm{curl}\,\boldsymbol{M}\,\mathrm{d}v}{r} - \frac{\mu_0}{4\pi}\int_S \frac{\mathrm{d}\boldsymbol{S}\times\boldsymbol{M}}{r}\end{aligned} \qquad (6.45)$$

since $\boldsymbol{n}\,\mathrm{d}S = \mathrm{d}\boldsymbol{S}$.

Now, from appendix section A5.2, problem 3, we know that

$$\int_v \mathrm{curl}\,\boldsymbol{a}\,\mathrm{d}v = \int_S (\mathrm{d}\boldsymbol{S}\times\boldsymbol{a}).$$

Therefore, setting $\boldsymbol{a} = \boldsymbol{M}/r$, we find

$$\int_v \mathrm{curl}\,\frac{\boldsymbol{M}}{r}\,\mathrm{d}v = \int_S \left(\mathrm{d}\boldsymbol{S}\times\frac{\boldsymbol{M}}{r}\right). \qquad (6.46)$$

Equation (6.46) then becomes

$$\boldsymbol{A} = \frac{\mu_0}{4\pi}\int_v \left(\frac{\mathrm{curl}\,\boldsymbol{M}}{r} - \mathrm{curl}\,\frac{\boldsymbol{M}}{r}\right)\mathrm{d}v. \qquad (6.47)$$

But,

$$\mathrm{curl}\,x\boldsymbol{a} = x\,\mathrm{curl}\,\boldsymbol{a} - \boldsymbol{a}\times\mathrm{grad}\left(\frac{1}{r}\right).$$

Therefore,

$$\boldsymbol{A} = \frac{\mu_0}{4\pi}\int_v \boldsymbol{M}\times\mathrm{grad}\left(\frac{1}{r}\right)\mathrm{d}v. \qquad (6.48)$$

This is the result we have been seeking. Comparison with (6.38) shows that $\boldsymbol{M}$ is the magnetic dipole moment per unit volume of the material.

6.12 The energy densities of electromagnetic fields

Problem: From the definition of the electric field $\boldsymbol{E}$ in Section 6.8, show that the rate at which batteries have to do work to push charges and currents against electrostatic fields, including emfs but neglecting ohmic heating, is

$$\int_v \boldsymbol{J}\cdot(-\boldsymbol{E})\,\mathrm{d}v$$

and hence show that the total energy of the system is

$$U = -\int_{\text{all space}} \mathrm{d}v \int_0^{\text{final fields}} \boldsymbol{J} \cdot \boldsymbol{E}\, \mathrm{d}t.$$

By taking the scalar products of (6.2) with $\boldsymbol{E}$ and (6.1) with $\boldsymbol{H}$, transform this expression into

$$U = \int_{\text{all space}} \mathrm{d}v \int_0^{\boldsymbol{D}} \boldsymbol{E} \cdot \mathrm{d}\boldsymbol{D} + \int_{\text{all space}} \mathrm{d}v \int_0^{\boldsymbol{B}} \boldsymbol{H} \cdot \mathrm{d}\boldsymbol{B}.$$

Proof. This is one of the classic pieces of analysis in electromagnetic theory. We start with the work done by the electromagnetic field on a particle of charge q. Work is only done by the electric field because, in the case of magnetic fields, no work is done on the particle; the magnetic force $\boldsymbol{f}$ acts perpendicular to the displacement of the particle since $\boldsymbol{f} = q(\boldsymbol{u} \times \boldsymbol{B})$. Thus, as the particle moves from $\boldsymbol{r}$ to $\boldsymbol{r} + \mathrm{d}\boldsymbol{r}$ in unit time, the work done by the electric field is $q\boldsymbol{E} \cdot \mathrm{d}\boldsymbol{r}$, that is,

$$\text{Work done in unit time} = q\boldsymbol{E} \cdot \boldsymbol{u}.$$

Therefore, per unit volume, the work done in unit time is $qN\boldsymbol{E} \cdot \boldsymbol{u} = \boldsymbol{J} \cdot \boldsymbol{E}$, where N is the number density of charges q. Now integrate over all space to find the total work done per unit time by the currents:

$$\int_v \boldsymbol{J} \cdot \boldsymbol{E}\, \mathrm{d}v.$$

The origin of the power to drive these currents is batteries, which must do work per unit time $\int_v (-\boldsymbol{J}) \cdot \boldsymbol{E}\, \mathrm{d}v$ on the system. Therefore, the total amount of energy supplied by the batteries is

$$U = -\int_{\text{all space}} \mathrm{d}v \int_0^t \boldsymbol{J} \cdot \boldsymbol{E}\, \mathrm{d}t. \tag{6.49}$$

The rest of the analysis is straightforward. We start by expressing $\boldsymbol{J} \cdot \boldsymbol{E}$ in terms of $\boldsymbol{E}$, $\boldsymbol{D}$, $\boldsymbol{H}$ and $\boldsymbol{B}$. From (6.2),

$$\boldsymbol{E} \cdot \boldsymbol{J} = \boldsymbol{E} \cdot \left(\operatorname{curl} \boldsymbol{H} - \frac{\partial \boldsymbol{D}}{\partial t}\right). \tag{6.50}$$

Now take the scalar product of (6.1) with $\boldsymbol{H}$ and add it to (6.45):

$$\boldsymbol{E} \cdot \boldsymbol{J} = \boldsymbol{E} \cdot \operatorname{curl} \boldsymbol{H} - \boldsymbol{H} \cdot \operatorname{curl} \boldsymbol{E} - \boldsymbol{E} \cdot \frac{\partial \boldsymbol{D}}{\partial t} - \boldsymbol{H} \cdot \frac{\partial \boldsymbol{B}}{\partial t}. \tag{6.51}$$

The vector relation $\nabla \cdot (\boldsymbol{a} \times \boldsymbol{b}) = \boldsymbol{b} \cdot (\nabla \times \boldsymbol{a}) - \boldsymbol{a} \cdot (\nabla \times \boldsymbol{b})$ enables us to write (6.51) as[2]

$$\boldsymbol{E} \cdot \boldsymbol{J} = -\operatorname{div}(\boldsymbol{E} \times \boldsymbol{H}) - \boldsymbol{E} \cdot \frac{\partial \boldsymbol{D}}{\partial t} - \boldsymbol{H} \cdot \frac{\partial \boldsymbol{B}}{\partial t}.$$

Therefore, integrating with respect to time, we find the total energy supplied to the system by the batteries per unit volume:

$$-\int_0^t \boldsymbol{J} \cdot \boldsymbol{E}\, \mathrm{d}t = \int_0^t \operatorname{div}(\boldsymbol{E} \times \boldsymbol{H})\, \mathrm{d}t + \int_0^{\boldsymbol{D}} \boldsymbol{E} \cdot \mathrm{d}\boldsymbol{D} + \int_0^{\boldsymbol{B}} \boldsymbol{H} \cdot \mathrm{d}\boldsymbol{B}. \tag{6.52}$$

Now integrate over all space, exchange the order of integration and apply the divergence theorem to the first integral on the right-hand side of (6.52). The latter becomes

$$\int_0^t \int_v \operatorname{div}(\boldsymbol{E} \times \boldsymbol{H})\, \mathrm{d}v\, \mathrm{d}t = \int_0^t \int_S (\boldsymbol{E} \times \boldsymbol{H}) \cdot \mathrm{d}\boldsymbol{S}\, \mathrm{d}t.$$

This integral represents the flux of energy per unit time through a surface S. The quantity $\boldsymbol{E} \times \boldsymbol{H}$ is known as the *Poynting vector* and is the rate of flow of energy in the direction normal to both $\boldsymbol{E}$ and $\boldsymbol{H}$ per unit area. Evidently, integrating the flux of $\boldsymbol{E} \times \boldsymbol{H}$ over the closed surface S represents the rate of loss of energy through the surface.

The other two terms in (6.52) represent the energies stored in the $\boldsymbol{E}$ and $\boldsymbol{H}$ fields per unit volume. We can therefore write the expression for the total energy in electric and magnetic fields:

$$U = \int_v \int_0^{\boldsymbol{D}} \boldsymbol{E} \cdot \mathrm{d}\boldsymbol{D}\, \mathrm{d}v + \int_v \int_0^{\boldsymbol{B}} \boldsymbol{H} \cdot \mathrm{d}\boldsymbol{B}\, \mathrm{d}v. \tag{6.53}$$

This is the answer we have been seeking. If the polarisable media are linear, $\boldsymbol{D} \propto \boldsymbol{E}$ and $\boldsymbol{B} \propto \boldsymbol{H}$, the total energy can be written

$$U = \int_v \tfrac{1}{2}(\boldsymbol{D} \cdot \boldsymbol{E} + \boldsymbol{B} \cdot \boldsymbol{H})\, \mathrm{d}v. \tag{6.54}$$

Problem: Transform the first term on the right-hand side of (6.53) into the form $\int \frac{1}{2}\rho_e V\, \mathrm{d}v$ in the case in which the polarisable media are linear, that is, $\boldsymbol{D} \propto \boldsymbol{E}$.

Proof. Let us do this one backwards. From (6.11),

$$\tfrac{1}{2}\int_v \rho_e V\, \mathrm{d}v = \tfrac{1}{2}\int_v (\operatorname{div} \boldsymbol{D})\, V \mathrm{d}v.$$

Now $\operatorname{div}(V\boldsymbol{D}) = \operatorname{grad} V \cdot \boldsymbol{D} + V \operatorname{div} \boldsymbol{D}$ and hence, substituting and using the divergence theorem,

$$\begin{aligned} \tfrac{1}{2}\int_v \rho_e V \mathrm{d}v &= \tfrac{1}{2}\int_v \operatorname{div}(V\boldsymbol{D})\, \mathrm{d}v - \tfrac{1}{2}\int_v \operatorname{grad} V \cdot \boldsymbol{D}\, \mathrm{d}v \\ &= \tfrac{1}{2}\int_S V\boldsymbol{D} \cdot \mathrm{d}\boldsymbol{S} + \tfrac{1}{2}\int_v \boldsymbol{E} \cdot \boldsymbol{D}\, \mathrm{d}v. \end{aligned} \tag{6.55}$$

We now have to deal with the surface integral in (6.55). We allow the integral to extend over a very large volume and ask how V and $\boldsymbol{D}$ vary as the radius r tends to infinity. If the system is isolated, the lowest-multipole electric field which could be present is associated with its net electric charge, for which $D \propto E \propto r^{-2}$ and $V \propto r^{-1}$. For example, if the system were neutral the lowest possible multipole of the field at a large distance would be a dipole field, for which $D \propto r^{-3}$, $V \propto r^{-2}$. Therefore, we need only consider the lowest-multipole case:

$$\tfrac{1}{2}\int_S V\boldsymbol{D} \cdot \mathrm{d}\boldsymbol{S} \propto \tfrac{1}{2}\int \frac{1}{r} \times \frac{1}{r^2} r^2\, \mathrm{d}\Omega \propto \frac{1}{r} \to 0 \text{ as } r \to \infty \tag{6.56}$$

where $\mathrm{d}\Omega$ is the element of solid angle. Thus, in the limit of infinite volume v, the first term on the right-hand side of (6.55) vanishes and

$$\tfrac{1}{2}\int_v \rho_e V\,\mathrm{d}v = \tfrac{1}{2}\int_v \boldsymbol{E}\cdot\boldsymbol{D}\,\mathrm{d}v. \qquad (6.57)$$

The right-hand side is exactly the same as $\iint \boldsymbol{E}\cdot \mathrm{d}\boldsymbol{D}\,\mathrm{d}v$, provided the media are linear.

Problem: Show that, in the absence of permanent magnets, the second term on the right-hand side of (6.54) becomes $\sum \frac{1}{2} I_n \phi_n$, where I_n is the current in the nth circuit and ϕ_n is the magnetic flux threading that circuit, again assuming the media are linear, $\boldsymbol{B} \propto \boldsymbol{H}$.

Proof. It is simplest to introduce the vector potential $\boldsymbol{A}$, which formed part of the story that led to Maxwell's discovery of his equations (see subsection 5.3.1). We define $\boldsymbol{A}$ as usual by $\boldsymbol{B} = \operatorname{curl} \boldsymbol{A}$. Now let us analyse the integral

$$\int_v \mathrm{d}v \int_0^B \boldsymbol{H}\cdot\mathrm{d}\boldsymbol{B}. \qquad (6.58)$$

Taking the differential of the definition of $\boldsymbol{A}$, $\mathrm{d}\boldsymbol{B} = \operatorname{curl}\mathrm{d}\boldsymbol{A}$. We use again the vector identity

$$\nabla\cdot(\boldsymbol{a}\times\boldsymbol{b}) = \boldsymbol{b}\cdot(\nabla\times\boldsymbol{a}) - \boldsymbol{a}\cdot(\nabla\times\boldsymbol{b});$$

therefore

$$\nabla\cdot(\boldsymbol{H}\times\mathrm{d}\boldsymbol{A}) = \mathrm{d}\boldsymbol{A}\cdot(\nabla\times\boldsymbol{H}) - \boldsymbol{H}\cdot(\nabla\times\mathrm{d}\boldsymbol{A}).$$

From this equation, substituting $\boldsymbol{H}\cdot\mathrm{d}\boldsymbol{B} = \boldsymbol{H}\cdot(\nabla\times\mathrm{d}\boldsymbol{A})$ into the integral (6.58) and applying Stokes' theorem to the second integral on the right-hand side, we obtain

$$\begin{aligned}\int \mathrm{d}v \int_0^B \boldsymbol{H}\cdot\mathrm{d}\boldsymbol{B} &= \int_v \mathrm{d}v \int_0^A (\nabla\times\boldsymbol{H})\cdot\mathrm{d}\boldsymbol{A} - \int_v \mathrm{d}v \int_0^A \nabla\cdot(\boldsymbol{H}\times\mathrm{d}\boldsymbol{A})\\ &= \int_v \mathrm{d}v \int_0^A (\nabla\times\boldsymbol{H})\cdot\mathrm{d}\boldsymbol{A} - \int_S \int_0^A (\boldsymbol{H}\times\mathrm{d}\boldsymbol{A})\cdot\mathrm{d}\boldsymbol{S}.\end{aligned}$$

We use the same argument which led to (6.56) to evaluate the surface integral at infinity in the second term on the right-hand side. Since the dependence of the magnetic field is no stronger than $A \propto r^{-2}$, $H \propto r^{-3}$ and there are no free magnetic poles, it follows that this term tends to zero as $r \to \infty$. Therefore

$$\int \mathrm{d}v \int_0^B \boldsymbol{H}\cdot\mathrm{d}\boldsymbol{B} = \int_v \mathrm{d}v \int_0^A (\nabla\times\boldsymbol{H})\cdot\mathrm{d}\boldsymbol{A}.$$

Assuming there are no displacement currents present, $\partial\boldsymbol{D}/\partial t = 0$ and hence $\nabla\times\boldsymbol{H} = \boldsymbol{J}$, that is,

$$\int_v \mathrm{d}v \int_0^B \boldsymbol{H}\cdot\mathrm{d}\boldsymbol{B} = \int_v \mathrm{d}v \int_0^A \boldsymbol{J}\cdot\mathrm{d}\boldsymbol{A}.$$

Now consider a closed current loop carrying a current I (Fig. 6.2). Consider a short section $\mathrm{d}\boldsymbol{l}$ of the current tube in which the current density is $\boldsymbol{J}$ and the cross-section $\mathrm{d}\sigma$. The advantage

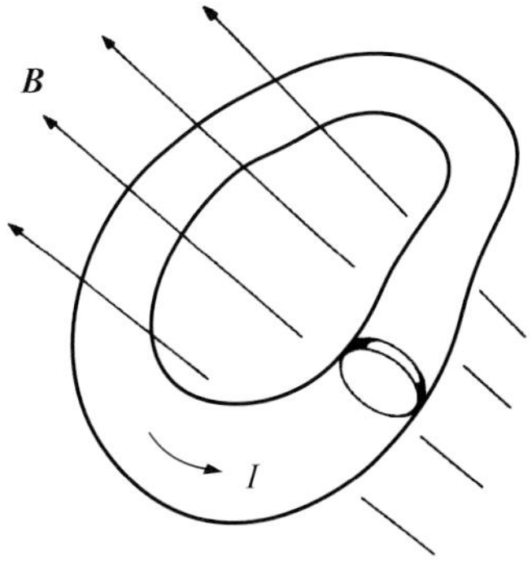

Figure 6.2: Illustrating magnetic flux lines passing through a closed current tube carrying a current I. The short section $\mathrm{d}\boldsymbol{l}$ of tube has cross-section $\mathrm{d}\boldsymbol{\sigma}$.

of using a closed tube is that the current I is constant, that is, $\boldsymbol{J}\cdot\mathrm{d}\boldsymbol{\sigma} = J\mathrm{d}\sigma = I = \text{constant}$. Therefore, since $\mathrm{d}v = \mathrm{d}\sigma\,\mathrm{d}l$,

$$\begin{aligned}\int_v \mathrm{d}v \int_0^A \boldsymbol{J}\cdot\mathrm{d}\boldsymbol{A} &= \int_l \int_0^A J\,\mathrm{d}\sigma\,\mathrm{d}\boldsymbol{A}\cdot\mathrm{d}\boldsymbol{l} \\ &= \int_l \int_0^A I\,\mathrm{d}\boldsymbol{A}\cdot\mathrm{d}\boldsymbol{l}.\end{aligned}$$

Now in taking the integral over $\mathrm{d}\boldsymbol{A}$ we notice that we are, in fact, working out an energy through the product $I\,\mathrm{d}A$ and therefore the energy needed to attain the value $\boldsymbol{A}$ is one half the product of I and $\boldsymbol{A}$, as occurs in electrostatics. We have disguised the dependence of the last integral on the vector field $\boldsymbol{J}$ by including it in the constant current I. Therefore,

$$\begin{aligned}\int_v \mathrm{d}v \int_0^A \boldsymbol{J}\cdot\mathrm{d}\boldsymbol{A} &= \tfrac{1}{2}I\int_l \boldsymbol{A}\cdot\mathrm{d}\boldsymbol{l} \\ &= \tfrac{1}{2}I\int_S \operatorname{curl}\boldsymbol{A}\cdot\mathrm{d}\boldsymbol{S} = \tfrac{1}{2}I\int_S \boldsymbol{B}\cdot\mathrm{d}\boldsymbol{S} = \tfrac{1}{2}I\Phi.\end{aligned}$$

Now we fill up the whole of space by a superposition of such current loops and flux linkages, so that

$$\int \boldsymbol{H}\cdot\mathrm{d}\boldsymbol{B} = \tfrac{1}{2}\sum_n I_n\Phi_n.$$

6.13 Concluding remarks

We are in great danger of writing a text book on vector fields in electromagnetism! It really is a remarkably elegant story. What is striking is the remarkable economy of Maxwell's equations in accounting for all the phenomena of classical electromagnetism. The formalism can be extended to treat much more complex systems involving anisotropic forces, for example those encountered in the propagation of electromagnetic waves in magnetised plasmas or in anisotropic material media. And it all originated in Maxwell's mechanical analogue for the properties of the vacuum through which electromagnetic phenomena are propagated.

6.14 References

1 Stratton, J.A. (1941). *Electromagnetic Theory*. London and New York: McGraw Hill.

2 The book by K.F. Riley, M.P. Hobson and S.J. Bence (2002), *Mathematical Methods for Physics and Engineering*, second edition, Cambridge University Press, can be warmly recommended as an excellent source for all the mathematics needed in this book. The chapter on vectors and vector operators is particularly helpful for this chapter.

Case Study III

Mechanics and dynamics – linear and non-linear

One of the key parts of any course in theoretical physics is the development of a wide range of procedures, which become more and more advanced, for treating problems in classical mechanics and dynamics. In one way or another, these are all extensions of the basic principles enunciated by Newton in the *Principia*, although some of them appear to bear only a distant resemblance to Newton's three laws of motion. As an example of the variety of ways in which the fundamentals of mechanics and dynamics can be expounded, here is a list of some of the different approaches which I found in the text book *Foundations of Physics*[1] by R.B. Lindsay and H. Margenau, which I read with some trepidation as a first-year undergraduate:

- Newton's laws of motion;
- D'Alembert's principle;
- the principle of virtual displacements;
- Gauss's principle of least constraint;
- Hertz's mechanics;
- Hamilton's principle and the principle of least action;
- Generalised coordinates and the methods of Lagrange;
- the canonical equations of Hamilton;
- the transformation theory of mechanics and the Hamilton–Jacobi equations.

This is not the place to go into these different approaches – this is more than adequately covered in standard texts such as Goldstein's *Classical Mechanics*.[2] Rather, in Chapter 7, I want to emphasise some of the features of these approaches which provide insights into different aspects of dynamical systems.

First of all, it is important to appreciate that these different approaches are *fully equivalent*. In principle, a given problem in mechanics or dynamics can be worked out by any of these techniques. They have been developed because Newton's laws of motion as they stand are not necessarily the simplest route to tackling any particular problem. Very often, other approaches provide a much more straightforward route. Furthermore, some of these procedures result in a much deeper appreciation of some of the basic features of dynamical systems. Among these, *conservation laws* and *normal modes of oscillation* are concepts of

(*a*)

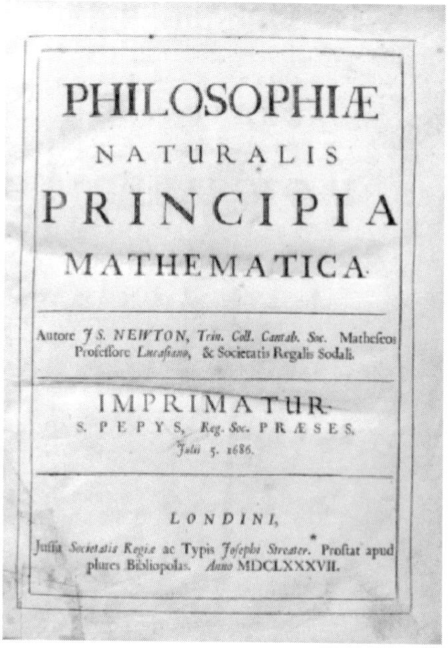

PHILOSOPHIÆ
NATURALIS
PRINCIPIA
MATHEMATICA.

Autore *JS. NEWTON*, *Trin. Coll. Cantab. Soc.* Matheſeos Profeſſore *Lucaſiano*, & Societatis Regalis Sodali.

IMPRIMATUR.
S. PEPYS, *Reg. Soc.* PRÆSES.
Julii 5. 1686.

LONDINI,

Juſſu *Societatis Regiæ* ac Typis *Joſephi Streater*. Proſtat apud plures Bibliopolas. *Anno* MDCLXXXVII.

(*b*)

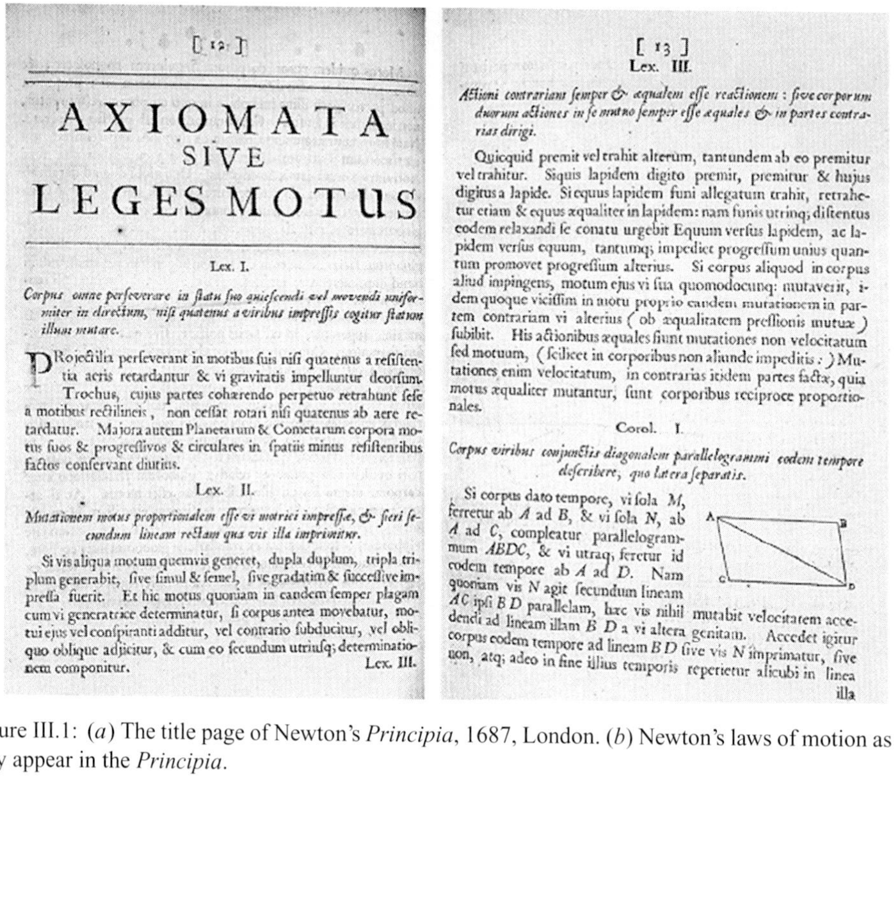

[12]

AXIOMATA
SIVE
LEGES MOTUS

Lex. I.

Corpus omne perſeverare in ſtatu ſuo quieſcendi vel movendi uniformiter in directum, niſi quatenus a viribus impreſſis cogitur ſtatum illum mutare.

PRojectilia perſeverant in motibus ſuis niſi quatenus a reſiſtentia aeris retardantur & vi gravitatis impelluntur deorſum. Trochus, cujus partes cohærendo perpetuo retrahunt ſeſe a motibus rectilineis, non ceſſat rotari niſi quatenus ab aere retardatur. Majora autem Planetarum & Cometarum corpora motus ſuos & progreſſivos & circulares in ſpatiis minus reſiſtentibus factos conſervant diutius.

Lex. II.

Mutationem motus proportionalem eſſe vi motrici impreſſæ, & fieri ſecundum lineam rectam qua vis illa imprimitur.

Si vis aliqua motum quemvis generet, dupla duplum, tripla triplum generabit, ſive ſimul & ſemel, ſive gradatim & ſucceſſive impreſſa fuerit. Et hic motus quoniam in eandem ſemper plagam cum vi generatrice determinatur, ſi corpus antea movebatur, motui ejus vel conſpiranti additur, vel contrario ſubducitur, vel obliquo oblique adjicitur, & cum eo ſecundum utriuſq; determinationem componitur.

Lex. III.

[13]

Lex. III.

Actioni contrariam ſemper & æqualem eſſe reactionem: ſive corporum duorum actiones in ſe mutuo ſemper eſſe æquales & in partes contrarias dirigi.

Quicquid premit vel trahit alterum, tantundem ab eo premitur vel trahitur. Siquis lapidem digito premit, premitur & hujus digitus a lapide. Si equus lapidem funi allegatum trahit, retrahetur etiam & equus æqualiter in lapidem: nam funis utrinq; diſtentus eodem relaxandi ſe conatu urgebit Equum verſus lapidem, ac lapidem verſus equum, tantumq; impediet progreſſum unius quantum promovet progreſſum alterius. Si corpus aliquod in corpus aliud impingens, motum ejus vi ſua quomodocunq; mutaverit, idem quoque viciſſim in motu proprio eandem mutationem in partem contrariam vi alterius (ob æqualitatem preſſionis mutuæ) ſubibit. His actionibus æquales fiunt mutationes non velocitatum ſed motuum, (ſcilicet in corporibus non aliunde impeditis:) Mutationes enim velocitatum, in contrarias itidem partes factæ, quia motus æqualiter mutantur, ſunt corporibus reciproce proportionales.

Corol. I.

Corpus viribus conjunctis diagonalem parallelogrammi eodem tempore deſcribere, quo latera ſeparatis.

Si corpus dato tempore, vi ſola *M*, ferretur ab *A* ad *B*, & vi ſola *N*, ab *A* ad *C*, compleatur parallelogrammum *ABDC*, & vi utraq; feretur id eodem tempore ab *A* ad *D*. Nam quoniam vis *N* agit ſecundum lineam *AC* ipſi *BD* parallelam, hæc vis nihil mutabit velocitatem accedendi ad lineam illam *BD* a vi altera genitam. Accedet igitur corpus eodem tempore ad lineam *BD* ſive vis *N* imprimatur, ſive non, atq; adeo in fine illius temporis reperietur alicubi in linea illa

Figure III.1: (*a*) The title page of Newton's *Principia*, 1687, London. (*b*) Newton's laws of motion as they appear in the *Principia*.

particular importance. Some of these procedures lead naturally to formalisms which can be taken over into quantum mechanics. We will recount how Dirac discovered his approach to quantum mechanics, as an illustration of how theoretical physicists work in practice.

These techniques are of great beauty and power, but there are many problems which are not soluble by analytic means. In Chapter 8, we go beyond the analytic to aspects of mechanical and dynamical systems which can only be tackled by the use of numerical methods using high-speed computers. Several different approaches to these problems are discussed – dimensional methods, the highly non-linear phenomena involved in chaotic dynamical systems and, most extreme of all, the study of systems which are so non-linear that only by computer modelling can we obtain insight into the origin of the regularities which appear within the disorder. Many of these areas of study are only in their infancy.

III.1 References

1 Lindsay, R.B. and Margenau, H. (1957). *Foundations of Physics*. London: Constable and Co. (Dover reprint).
2 Goldstein, H. (1950). *Classical Mechanics*. London: Addison-Wesley.

7 Approaches to mechanics and dynamics

7.1 Newton's laws of motion

In Case Study I, we outlined the history which led up to Newton's publication of his laws of motion in the *Principia Mathematica*, published by Samuel Pepys in 1687 (Fig. III.1). The *Principia* consists of three books, preceded by two short sections called 'Definitions' and 'Axioms, or the laws of motion'. The definitions are eight in number and describe concepts such as mass, momentum, impulse, impressed force, inertia and centrifugal force. The three axioms which follow are what we now call Newton's laws of motion. There has been considerable debate about the exact meanings of the definitions and axioms as they appear in the *Principia*, but in the analyses in Books I and III Newton uses these quantities unambiguously with their modern meanings. The three laws of motion appear in a form remarkably similar to the words used in all the standard modern textbooks.

- *Newton 1*: 'Every body continues in its state of rest, or of uniform motion in a straight line, except in so far as it is compelled by forces to change that state.' In vector notation,

$$\text{if} \qquad \boldsymbol{f} = 0 \qquad \text{then} \quad \frac{\mathrm{d}\boldsymbol{v}}{\mathrm{d}t} = 0. \tag{7.1}$$

- *Newton 2*: 'Change of motion [that is, momentum] is proportional to the force and takes place in the direction of the straight line in which the force acts.' In fact, Newton means that the rate of change of momentum is proportional to force, as is exemplified by his use of this law throughout the *Principia*. In modern notation,

$$\boldsymbol{f} = \frac{\mathrm{d}\boldsymbol{p}}{\mathrm{d}t}, \qquad \boldsymbol{p} = m\boldsymbol{v}, \tag{7.2}$$

  where $\boldsymbol{p}$ is the momentum of the body.
- *Newton 3*: 'To every action there is always an equal and contrary reaction; or, the mutual actions of any two bodies are always equal and oppositely directed along the same straight line.'

There are interesting logical problems about the definitions and the three laws of motion. Newton apparently regarded a number of the concepts as self-evident, for example, those of 'mass' and 'force'. Nowadays, it is preferable to regard the three laws of motion as providing the definitions of the quantities involved and at the same time reflecting our experimental experience.

As an example, simply using the idea that forces produce accelerations, we can define *mass* as follows. We devise a means by which a force of fixed magnitude can be applied to different bodies. By performing experiments, we find that the bodies suffer different accelerations under the action of this force. The *mass* of a body can therefore be defined as a quantity proportional to the inverse of its acceleration. Using Newton's third law, we can define a *relative* scale of mass. Suppose two point-like masses A and B interact with each other. Then, according to Newton's third law, $\boldsymbol{f}_A = -\boldsymbol{f}_B$. Notice that action and reaction always act on different bodies. If we measure their accelerations under their mutual interaction, we find that they are in a certain ratio a_{AB}/a_{BA}, which we can call M_{AB}. If we then compare C with A and B separately, we will measure acceleration ratios M_{BC} and M_{AC}. From experiment, we find that $M_{AB} = M_{AC}/M_{BC}$. This tells us that we can write M_{AB} as M_B/M_A, and so on. Therefore, we can define a mass scale by choosing one of M_A, M_B, M_C as the standard mass.

Now all of this may perhaps seem trivial, but it emphasises an important point. Even in something as apparently intuitive as Newton's laws, there are fundamental assumptions and results of experiment upon which the whole structure is based. In practice, the equations are a very good approximation to what actually happens in the real world. There is, however, no strictly logical way of setting up the foundations *ab initio*. This gives us no operational problem so long as we adhere to Dirac's dictum that 'We want the equations which describe nature . . . and have to reconcile ourselves to an absence of strict logic.'

We will not derive the conservation laws from Newton's laws of motion, a topic treated in the standard textbooks. We simply recall that these laws, which can be derived from Newton's laws in a straightforward manner, are:

- *the law of conservation of momentum*;
- *the law of conservation of angular momentum*;
- *the law of conservation of energy*, once suitable potential energy functions have been defined.

We will derive these from the Euler–Lagrange equation in Section 7.6.

One important concept is that of the *invariance* of Newton's laws of motion with respect to transformation between frames of reference in uniform relative motion. This subject is the key to the development of special relativity in Case Study VI. We discussed Galileo's discovery of inertial frames of reference and the invariance of the laws of physics under Galilean transformation of the coordinates in Section 3.6. Figure 3.9 shows frames of reference S and S' in uniform relative motion with relative velocity V. According to Galilean relativity, the coordinates in the frames of reference S and S' are related by

$$\begin{aligned} x' &= x - Vt, \\ y' &= y, \\ z' &= z, \\ t' &= t. \end{aligned} \tag{7.3}$$

Taking the second derivative of the first of the relations (7.3) with respect to time we obtain $\ddot{x}' = \ddot{x}$, that is, the acceleration and consequently the forces are the same in any inertial frame of reference. In relativistic language, the acceleration is an *invariant* between inertial frames of reference. Notice that time is absolute, in the sense that the coordinate t keeps a track of time in all inertial frames of reference. These transformations are implicit in the works of Galileo, who stated that there should be no difference in the laws of nature whether the observer is stationary or in a state of uniform rectilinear motion. For this reason, equations (7.3) are often referred to as *Galilean transformations*. It is intriguing that Galileo's motivation for advancing this argument was to demonstrate that there is nothing inconsistent with our view of the Universe if the Earth happens to be in motion rather than at rest at the centre of the Universe; that is, it was really part of his defence of the Copernican position.

7.2 Principles of 'least action'

Some of the most powerful alternative approaches to mechanics and dynamics involve finding the function which results in a minimum value of some other suitably defined function. In the formal development of mechanics, these procedures are stated *axiomatically* and the formalism for treating a wide range of problems is elaborated. As in so many aspects of basic theoretical physics, I find Feynman's approach in Chapter 19 of Vol. 2 of his *Lectures on Physics*,[1] entitled 'The principle of least action', quite brilliant in exposing the basic ideas behind this approach to dynamics. We will follow his exposition, but rather more briefly since we already have available most of the mathematical tools.

Let us consider the simple case of the dynamics of a particle in a conservative field of force, that is, one which can be derived from the derivative of a scalar potential. Such fields of force include the electrostatic force $\boldsymbol{f} = -q \,\mathrm{grad}\, \Phi_{\mathrm{e}}$ and the gravitational force $\boldsymbol{f} = -m \,\mathrm{grad}\, \Phi_{\mathrm{g}}$, where Φ_{e} and Φ_{g} are the electrostatic and gravitational potentials respectively. The potential energies of a test particle in these fields are $V = q\Phi_{\mathrm{e}}$ and $m\Phi_{\mathrm{g}}$.

Now let us introduce a set of axioms which enables us to work out the path of the particle subject to this force field. First, we define the *Lagrangian*

$$\mathcal{L} = \tfrac{1}{2} m \boldsymbol{v}^2 - V, \tag{7.4}$$

which is the difference between the kinetic and potential energies of the particle at any point in the field. Then, to derive the trajectory of the particle between two points in the field in a fixed time interval t_1 to t_2, we find the path that minimises the function

$$S = \int_{t_1}^{t_2} \left(\tfrac{1}{2} m \boldsymbol{v}^2 - V\right) \mathrm{d}t \tag{7.5}$$

$$= \int_{t_1}^{t_2} \left[\tfrac{1}{2} m \left(\frac{\mathrm{d}\boldsymbol{r}}{\mathrm{d}t}\right)^2 - V\right] \mathrm{d}t. \tag{7.6}$$

We will show that these statements are exactly equivalent to Newton's laws of motion or, rather, to what he accurately called his 'axioms'.

The quantity S does not have a simple official title, but Feynman calls it the 'action', so that minimising S can be considered as a principle of 'least action'. The problem with this

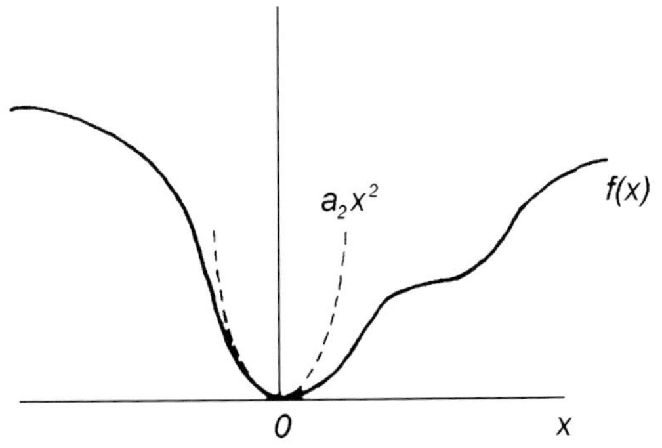

Figure 7.1: Illustrating how a function $f(x)$ can be approximated as a_2x^2 close to a minimum.

usage is that the word 'action' is used for something rather different from S in the formal development of mechanics. We will simply call it the function S.

We can readily see that this definition and procedure are certainly consistent with Newton's first law of motion. If there are no forces present, V = constant and hence we must minimise $\int_{t_1}^{t_2} \boldsymbol{v}^2\, \mathrm{d}t$. The minimum value corresponds to a constant velocity v between t_1 and t_2, as can be seen from the following argument. If the particle were instead to travel between the two points by accelerating and decelerating in such a way that t_1 and t_2 take the same value then the integral would have to be greater than that for constant velocity, because $\boldsymbol{v}^2$ appears in the integral and a basic rule of analysis tells us that $\langle \boldsymbol{v}^2 \rangle \geq \langle |\boldsymbol{v}| \rangle^2$. Thus, Newton's first law of motion is the solution corresponding to minimum S, that is, in the absence of forces $\boldsymbol{v}$ is constant.

To proceed further, we need the techniques of the *calculus of variations*. We will deal with some simple aspects of this in a moment but let us first analyse the case of motion in a scalar potential field in more detail. I find Feynman's analysis very attractive. Consider a function of a single variable, say $f(x)$. When we find a minimum value of this function, we find a point on the curve where $\mathrm{d}f(x)/\mathrm{d}x$ is zero; suppose that this occurs at $x = 0$ (Fig. 7.1). If we approximate the function in the region of the minimum at $x = 0$ by a power series,

$$f(x) = a_0 + a_1 x + a_2 x^2 + a_3 x^3 + \cdots ,$$

it is evident that the function can only have $\mathrm{d}f/\mathrm{d}x = 0$ at $x = 0$ if $a_1 = 0$. Thus, the function tends to a parabola as $x \to 0$, since the first non-zero coefficient in the above expansion of $f(x)$ about the minimum must be a_2, that is

$$f(x) = a_0 + a_2 x^2 + a_3 x^3 + \cdots .$$

In other words, for a small displacement x from the minimum, the change in the function $f(x)$ is only second order in x.

We now return to particle dynamics. Exactly the same principle is used to work out the trajectory of the particle by minimising S. If the minimal path of the particle is $\boldsymbol{x}_0(t)$ then another path between t_1 and t_2 is given by

$$\boldsymbol{x}(t) = \boldsymbol{x}_0(t) + \boldsymbol{\eta}(t), \tag{7.7}$$

where $\eta(t)$ describes the deviation of $\boldsymbol{x}(t)$ from the minimal path $\boldsymbol{x}_0(t)$. Just as we can define the minimum of the function $f(x)$ as occurring at the point at which there is no first-order dependence of $f(x)$ upon x, so we can define the minimum of the function S as occurring when there is no first-order dependence of S on η, that is, there should be no term in $\eta(t)$.

We substitute (7.7) into (7.6), obtaining

$$\begin{aligned} S &= \int_{t_1}^{t_2} \left[\frac{m}{2} \left(\frac{\mathrm{d}\boldsymbol{x}_0}{\mathrm{d}t} + \frac{\mathrm{d}\boldsymbol{\eta}}{\mathrm{d}t} \right)^2 - V(\boldsymbol{x}_0 + \boldsymbol{\eta}) \right] \mathrm{d}t \\ &= \int_{t_1}^{t_2} \left\{ \frac{m}{2} \left[\left(\frac{\mathrm{d}\boldsymbol{x}_0}{\mathrm{d}t} \right)^2 + 2 \frac{\mathrm{d}\boldsymbol{x}_0}{\mathrm{d}t} \cdot \frac{\mathrm{d}\boldsymbol{\eta}}{\mathrm{d}t} + \left(\frac{\mathrm{d}\boldsymbol{\eta}}{\mathrm{d}t} \right)^2 \right] - V(\boldsymbol{x}_0 + \boldsymbol{\eta}) \right\} \mathrm{d}t. \end{aligned} \tag{7.8}$$

Now, we are only interested in eliminating quantities to first order in η and hence we can drop the term $(\mathrm{d}\eta/\mathrm{d}t)^2$. Furthermore, we can expand $V(\boldsymbol{x}_0 + \boldsymbol{\eta})$ to first order in η by a Taylor expansion,

$$V(\boldsymbol{x}_0 + \boldsymbol{\eta}) = V(\boldsymbol{x}_0) + \nabla V \cdot \boldsymbol{\eta}. \tag{7.9}$$

Substituting (7.9) into (7.8) and preserving only quantities to first order in η, we find

$$S = \int_{t_1}^{t_2} \left[\frac{m}{2} \left(\frac{\mathrm{d}\boldsymbol{x}_0}{\mathrm{d}t} \right)^2 - V(\boldsymbol{x}_0) + m \frac{\mathrm{d}\boldsymbol{x}_0}{\mathrm{d}t} \cdot \frac{\mathrm{d}\boldsymbol{\eta}}{\mathrm{d}t} - \nabla V \cdot \boldsymbol{\eta} \right] \mathrm{d}t. \tag{7.10}$$

The first two terms inside the integral refer to the unknown minimal path and are constant. We are interested in ensuring that the last two terms have no dependence upon η, this being the condition for a minimum in S. Let us therefore concentrate upon these two terms; for convenience we write

$$S' \equiv \int_{t_1}^{t_2} \left(m \frac{\mathrm{d}\boldsymbol{x}_0}{\mathrm{d}t} \cdot \frac{\mathrm{d}\boldsymbol{\eta}}{\mathrm{d}t} - \boldsymbol{\eta} \cdot \nabla V \right) \mathrm{d}t. \tag{7.11}$$

We integrate the first term by parts so that η, and not $\mathrm{d}\eta/\mathrm{dt}$, appears in the integrand, that is,

$$S' = m \left[\boldsymbol{\eta} \cdot \frac{\mathrm{d}\boldsymbol{x}_0}{\mathrm{d}t} \right]_{t_1}^{t_2} - \int_{t_1}^{t_2} \left[\frac{\mathrm{d}}{\mathrm{d}t} \left(m \frac{\mathrm{d}\boldsymbol{x}_0}{\mathrm{d}t} \right) \cdot \boldsymbol{\eta} + \boldsymbol{\eta} \cdot \nabla V \right] \mathrm{d}t. \tag{7.12}$$

We know that the function η must be zero at t_1 and t_2 because that is where all paths must begin and end. Therefore, the first term in (7.12) is zero and, since the integral must be zero for all first-order perturbations η about the minimal solution $\boldsymbol{x}_0$ we can write

$$S' = - \int_{t_1}^{t_2} \left\{ \boldsymbol{\eta} \cdot \left[\frac{\mathrm{d}}{\mathrm{d}t} \left(m \frac{\mathrm{d}\boldsymbol{x}_0}{\mathrm{d}t} \right) + \nabla V \right] \right\} \mathrm{d}t = 0. \tag{7.13}$$

This must be true for arbitrary perturbations $\eta(t)$, and hence the term in square brackets must be zero, that is,

$$\frac{\mathrm{d}}{\mathrm{d}t} \left(m \frac{\mathrm{d}\boldsymbol{x}_0}{\mathrm{d}t} \right) = -\nabla V. \tag{7.14}$$

We recognise that we have recovered Newton's second law of motion, since $\boldsymbol{f} = -\nabla V$, that is,

$$\boldsymbol{f} = \frac{\mathrm{d}}{\mathrm{d}t}\left(m\frac{\mathrm{d}\boldsymbol{x}_0}{\mathrm{d}t}\right) = \frac{\mathrm{d}\boldsymbol{p}}{\mathrm{d}t}. \tag{7.15}$$

Thus, our alternative formulation of the laws of motion in terms of an action principle is exactly equivalent to Newton's statement of the laws of motion. There is plainly a great deal to be done in working out all the ramifications of this procedure.

It will have to suffice to say that we can generalise these procedures to take account of both conservative *and* non-conservative forces. The key point is that we have a prescription which involves writing down directly the kinetic and potential energies (T and V respectively) of a system and then forming the Lagrangian $\mathcal{L}$ and finding the minimum value of the function S. The big advantage of this procedure is that it is often a straightforward matter to write down these energies in some suitable system of coordinates. The next step is therefore clear. We need rules which tell us how to find the minimum value of S in any set of coordinates convenient for the problem in hand. These rules turn out to be the *Euler–Lagrange equations*.

An interesting point about action principles in physics should be noted. In general, we do not have a definite prescription for finding the Lagrangian. To quote Feynman,

> The question of what the action (S) should be for a particular case must be determined by some kind of trial and error. It is just the same problem as determining the laws of motion in the first place. You just have to fiddle around with the equations that you know and see if you can get them into the form of a principle of least action.[1]

An interesting and important example is that of a relativistic particle moving in an electromagnetic field. The appropriate Lagrangian is

$$\mathcal{L} = -\gamma^{-1} m_0 c^2 - q\Phi_{\mathrm{e}} + q\boldsymbol{v}\cdot\boldsymbol{A}, \tag{7.16}$$

where the Lorentz factor $\gamma = (1 - v^2/c^2)^{-1/2}$, Φ_{e} is the electrostatic potential and $\boldsymbol{A}$ the magnetic vector potential. Notice that, although $q\Phi_{\mathrm{e}}$ and $q\boldsymbol{v}\cdot\boldsymbol{A}$ have the familiar forms of potential energy, the term $-m_0c^2/\gamma$ is certainly not the relativistic kinetic energy. This term does, however, reduce correctly to the non-relativistic form for the kinetic energy, that is, $-m_0c^2/\gamma \to \frac{1}{2}m_0v^2 - m_0c^2$ as $v \to 0$; the constant term $-m_0c^2$ does not matter since we are only interested in minimising with respect to variables in the Lagrangian. We will demonstrate how this Lagrangian results in the relativistic expression for the Lorentz force, once we have derived the tools needed to analyse Lagrangians in general in the next section.

7.3 The Euler–Lagrange equation

For simplicity, we consider only forces which can be derived from a scalar potential V. As an example, suppose we consider a problem involving N particles interacting via a scalar potential energy function V. The positions of the N particles are given by the vector $[\boldsymbol{r}_1, \boldsymbol{r}_2, \boldsymbol{r}_3, \ldots, \boldsymbol{r}_N]$. Since we need three components to describe each position, for example, x_i, y_i, z_i, this vector has $3N$ coordinates. It may well prove more convenient to work in terms of a different set of position coordinates, which we can write as $[q_1, q_2, q_3, \ldots, q_{3N}]$. There will then be a set of relations between the two sets of coordinates of the form

$$q_i = q_i(\boldsymbol{r}_1, \boldsymbol{r}_2, \boldsymbol{r}_3, \ldots, \boldsymbol{r}_N), \tag{7.17}$$

and

$$r_i = r_i(q_1, q_2, q_3, \ldots, q_{3N}). \tag{7.18}$$

Notice that this is no more than a change of variables.

The aim of the procedure now is to write down equations for the dynamics of the particles, that is, an equation for each independent coordinate, in terms of the q_i rather than the r_i. We are guided by the analysis of the previous section on action principles to form the quantities T and V, the kinetic and potential energies respectively, in terms of the new set of coordinates and then to find the Lagrangian $\mathcal{L}$ and the stationary value of the 'action' S, $\int_{t_1}^{t_2} \mathcal{L}\,\mathrm{d}t$:

$$\mathcal{L} = T - V \qquad \text{and} \qquad \delta \int_{t_1}^{t_2} (T - V)\,\mathrm{d}t = 0. \tag{7.19}$$

This formulation is called *Hamilton's principle* and $\mathcal{L}$ as before is the *Lagrangian*. The key point is that Hamilton's principle does not say anything about the coordinate system in which we are working. It is the extension to more general coordinate systems of the arguments which we developed in Section 7.3 for Cartesian coordinates.

Now, the kinetic energy of the system is

$$T = \tfrac{1}{2} \sum_i m_i \dot{\boldsymbol{r}}_i^2. \tag{7.20}$$

In terms of our new coordinate system, we can write without loss of generality

$$\begin{aligned} r_i &= r_i(q_1, q_2, q_3, \ldots, q_{3N}, t), \\ \dot{r}_i &= \dot{r}_i(\dot{q}_1, \dot{q}_2, \dot{q}_3, \ldots, \dot{q}_{3N}, q_1, q_2, \ldots, q_{3N}, t). \end{aligned}$$

Notice that we have now included explicitly the time dependence of r_i and $\dot{r}_i$. Therefore, we can rewrite the kinetic energy as some function of the coordinates $\dot{q}_i$, q_i and t, that is, $T(\dot{q}_i, q_i, t)$, where we understand that all the values of i from 1 to $3N$ are included. Similarly, we can write the expression for the potential energy entirely in terms of the coordinates q_i and t, that is, $V(q_i, t)$, where i takes $3N$ values. We require a procedure for finding the stationary value of

$$S = \int_{t_1}^{t_2} [T(\dot{q}_i, q_i, t) - V(q_i, t)]\,\mathrm{d}t. \tag{7.21}$$

We now repeat our analysis of Section 7.2, in which we found the condition for S to be independent of first-order perturbations about the minimal path. In the same way as before, we let $q_{0i}(t)$ be the minimal solution and write the expression for another function $q_i(t)$ in the form

$$q_i(t) = q_{0i}(t) + \eta_i(t). \tag{7.22}$$

We rewrite S as

$$S = \int_{t_1}^{t_2} \mathcal{L}(\dot{q}_i, q_i, t)\,\mathrm{d}t \tag{7.23}$$

and now insert the trial solution (7.22) into (7.23). Then

$$S = \int_{t_1}^{t_2} \mathcal{L}[\dot{q}_{0i}(t) + \dot{\eta}_i(t), q_{0i}(t) + \eta_i(t), t]\,\mathrm{d}t.$$

Performing Taylor expansions to first order in the $\dot{\eta}_i(t)$ and the $\eta_i(t)$,

$$S = \int_{t_1}^{t_2} \mathcal{L}[\dot{q}_{0i}(t), q_{0i}(t), t]\,\mathrm{d}t + \int_{t_1}^{t_2} \left[\frac{\partial \mathcal{L}}{\partial \dot{q}_i}\dot{\eta}_i(t) + \frac{\partial \mathcal{L}}{\partial q_i}\eta_i(t)\right]\mathrm{d}t.$$

As before, we integrate the terms in $\dot{\eta}_i(t)$ by parts and then

$$S = S_0 + \left[\frac{\partial \mathcal{L}}{\partial \dot{q}_i}\eta_i(t)\right]_{t_1}^{t_2} - \int_{t_1}^{t_2} \left[\frac{\mathrm{d}}{\mathrm{d}t}\left(\frac{\partial \mathcal{L}}{\partial \dot{q}_i}\right)\eta_i(t) - \frac{\partial \mathcal{L}}{\partial q_i}\eta_i(t)\right]\mathrm{d}t. \tag{7.24}$$

Again, because the $\eta_i(t)$ must always be zero at the end points, the first term in brackets disappears and the result can be written

$$S = S_0 - \int_{t_1}^{t_2} \eta_i(t)\left[\frac{\mathrm{d}}{\mathrm{d}t}\left(\frac{\partial \mathcal{L}}{\partial \dot{q}_i}\right) - \frac{\partial \mathcal{L}}{\partial q_i}\right]\mathrm{d}t. \tag{7.25}$$

We require the integral to be zero for all first-order perturbations about the minimal solution. Therefore, we find the conditions

$$\frac{\partial \mathcal{L}}{\partial q_i} - \frac{\mathrm{d}}{\mathrm{d}t}\left(\frac{\partial \mathcal{L}}{\partial \dot{q}_i}\right) = 0 \tag{7.26}$$

for the minimisation of S. Equation (7.26) represents $3N$ equations for the time evolution of the $3N$ coordinates. This fundamental equation is known as the *Euler–Lagrange equation*. From our previous studies, it is apparent that this is no more than Newton's second law of motion written in the q_i coordinate system.

Let us use the Euler–Lagrange equation to derive the motion of a relativistic particle in electric and magnetic fields from the Lagrangian (7.16). Consider first the case of a particle in an electric field which is derivable from a scalar potential, so that the Lagrangian becomes

$$\mathcal{L} = -\frac{m_0 c^2}{\gamma} - q\Phi_{\mathrm{e}}. \tag{7.27}$$

Considering first the x-component of the motion of the particle,

$$\frac{\partial \mathcal{L}}{\partial x} - \frac{\mathrm{d}}{\mathrm{d}t}\left(\frac{\partial \mathcal{L}}{\partial v_x}\right) = 0, \tag{7.28}$$

and hence

$$-q\frac{\partial \Phi_{\mathrm{e}}}{\partial x} - \frac{\mathrm{d}}{\mathrm{d}t}\left[-m_0 c^2 \frac{\partial}{\partial v_x}\left(1 - \frac{v_x^2 + v_y^2 + v_z^2}{c^2}\right)^{1/2}\right] = 0,$$

giving

$$-q\frac{\partial \Phi_e}{\partial x} - \frac{d}{dt}(\gamma m_0 v_x) = 0.$$

Similar relations are found for the y- and z- components, and so we can add them together vectorially to write

$$q \text{ grad } \Phi_e + \frac{d\boldsymbol{p}}{dt} = 0 \qquad \text{or} \qquad \frac{d\boldsymbol{p}}{dt} = -q \text{ grad } \Phi_e = q\boldsymbol{E}, \tag{7.29}$$

where $\boldsymbol{p} = \gamma m_0 \boldsymbol{v}$ is the relativistic three-momentum. We have recovered the expression for the acceleration of a relativistic particle in an electric field $\boldsymbol{E}$.

A Lagrangian that includes the term $q\boldsymbol{v} \cdot \boldsymbol{A}$, (7.16), requires a little more work. Since we have already dealt with the terms $-m_0c^2\gamma^{-1} - q\Phi_e$, let us consider only

$$\mathcal{L} = q\boldsymbol{v} \cdot \boldsymbol{A} = q(v_x A_x + v_y A_y + v_z A_z). \tag{7.30}$$

Again, considering only the x-component of the Euler–Lagrange equation,

$$\begin{aligned} &\frac{\partial \mathcal{L}}{\partial x} - \frac{d}{dt}\left(\frac{\partial \mathcal{L}}{\partial v_x}\right) = 0, \\ &q\left(v_x\frac{\partial A_x}{\partial x} + v_y\frac{\partial A_y}{\partial x} + v_z\frac{\partial A_z}{\partial x}\right) - \frac{d}{dt}(qA_x) = 0. \end{aligned} \tag{7.31}$$

When we add together vectorially the components in the x-, y- and z- directions, the term in d/dt sums to

$$-q\frac{d\boldsymbol{A}}{dt}. \tag{7.32}$$

Still considering only the x-component of (7.30), the terms in the large parentheses in (7.31) can be reorganised as follows:

$$\begin{aligned} q\left(v_x\frac{\partial A_x}{\partial x} + v_y\frac{\partial A_y}{\partial x} + v_z\frac{\partial A_z}{\partial x}\right) &= q\left[\left(v_x\frac{\partial A_x}{\partial x} + v_y\frac{\partial A_x}{\partial y} + v_z\frac{\partial A_x}{\partial z}\right)\right. \\ &\quad \left. + v_y\left(\frac{\partial A_y}{\partial x} - \frac{\partial A_x}{\partial y}\right) + v_z\left(\frac{\partial A_z}{\partial x} - \frac{\partial A_x}{\partial z}\right)\right]. \end{aligned} \tag{7.33}$$

The right-hand side of (7.33) can be recognised as the x-component of the vector

$$q(\boldsymbol{v} \cdot \nabla)\boldsymbol{A} + q[\boldsymbol{v} \times (\nabla \times \boldsymbol{A})], \tag{7.34}$$

where the operator $\boldsymbol{v} \cdot \nabla$ is described in appendix section A7.1. Therefore, the result of applying the Euler–Lagrange formalism to the term $q\boldsymbol{v} \cdot \boldsymbol{A}$ is

$$-q\frac{d\boldsymbol{A}}{dt} + q(\boldsymbol{v} \cdot \nabla)\boldsymbol{A} + q[\boldsymbol{v} \times (\nabla \times \boldsymbol{A})]. \tag{7.35}$$

But, as shown in appendix section A7.1, the total and partial time derivatives are related by

$$\frac{d}{dt} = \frac{\partial}{\partial t} + (\boldsymbol{v} \cdot \nabla), \tag{7.36}$$

and hence the term (7.35) becomes

$$-q\frac{\partial \boldsymbol{A}}{\partial t} + q[\boldsymbol{v} \times (\nabla \times \boldsymbol{A})]. \tag{7.37}$$

This is the beautiful result we have been seeking. Following the reasoning which led to (5.30) in subsection 5.3.1, the *induced* electric field is given by

$$\boldsymbol{E}_{\text{induced}} = -\frac{\partial \boldsymbol{A}}{\partial t}, \tag{7.38}$$

and, by definition, $\boldsymbol{B} = \text{curl}\,\boldsymbol{A}$. Therefore, (7.37) reduces to

$$q\boldsymbol{E}_{\text{induced}} + q(\boldsymbol{v} \times \boldsymbol{B}). \tag{7.39}$$

We can now reassemble the complete Lagrangian from (7.29) and (7.39) and write the equation of motion of the particle as

$$\begin{aligned}\frac{\mathrm{d}\boldsymbol{p}}{\mathrm{d}t} &= q\boldsymbol{E} + q\boldsymbol{E}_{\text{induced}} + q(\boldsymbol{v} \times \boldsymbol{B}) \\ &= q\boldsymbol{E}_{\text{tot}} + q(\boldsymbol{v} \times \boldsymbol{B}),\end{aligned} \tag{7.40}$$

where $\boldsymbol{p} = \gamma m_0 \boldsymbol{v}$, and $\boldsymbol{E}_{\text{tot}}$ includes both the electrostatic and induced electric fields. We can see that we have recovered the complete expression for the equation of motion of a relativistic charged particle moving in combined electric and magnetic fields.

7.4 Small oscillations and normal modes

Let us give a simple example which illustrates the power of the Euler–Lagrange equation in analysing the dynamical behaviour of systems that undergo *small oscillations*. This leads naturally to the concept of the *normal modes* of oscillation of the system. A typical problem concerns a hollow cylinder, that is, a pipe with thin walls, suspended by strings of equal length at either end. The strings are attached to the circumference of the cylinder as shown in Fig. 7.2(*a*), so that it can wobble and swing at the same time. We consider first a swinging motion in which the strings and the flat ends of the cylinder remain in the same planes. An end-on view of the displaced cylinder is shown in Fig. 7.2(*b*). For the sake of definiteness, we take the length of the string to be $3a$ and the radius of the cylinder $2a$.

First of all, we choose a suitable set of coordinates to define completely the position of the cylinder when it undergoes a small displacement from the equilibrium position. It can be seen from Fig. 7.2(*b*) that θ and φ fulfil this role and are the natural coordinates to use for such pendulum-like motion. We now write down the Lagrangian for the system in terms of the coordinates $\dot{\theta}, \dot{\varphi}, \theta, \varphi$.

The *kinetic energy* T consists of two parts, one associated with the *translational motion* of the cylinder and the other with its *rotation*. In the analysis, we consider only small displacements from the equilibrium position $\theta = 0$, $\varphi = 0$. The horizontal displacement of the centre of mass of the cylinder from the equilibrium position is

$$x = 3a\sin\theta + 2a\sin\varphi \approx 3a\theta + 2a\varphi, \tag{7.41}$$

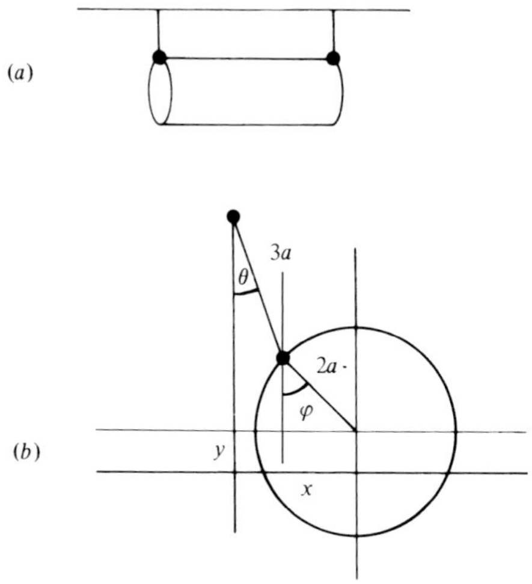

Figure 7.2: Illustrating (a) the swinging motion of a hollow cylinder suspended by strings at either end and (b) the coordinate system used to analyse its motion.

and consequently its translational motion is $\dot{x} = 3a\dot{\theta} + 2a\dot{\varphi}$. The linear kinetic energy is therefore $\frac{1}{2}ma^2(3\dot{\theta} + 2\dot{\varphi})^2$, where m is the mass of the cylinder. The rotational motion of the cylinder is $\frac{1}{2}I\dot{\varphi}^2$, where I is the moment of inertia of the cylinder about its horizontal axis. In this case, $I = 4a^2m$. Thus, the total kinetic energy of the cylinder is

$$\begin{aligned} T &= \tfrac{1}{2}ma^2(3\dot{\theta} + 2\dot{\varphi})^2 + 2a^2m\dot{\varphi}^2 \\ &= \tfrac{1}{2}ma^2(9\dot{\theta}^2 + 12\dot{\theta}\dot{\varphi} + 8\dot{\varphi}^2). \end{aligned} \tag{7.42}$$

The *potential energy* is entirely associated with the vertical displacement y of the centre of mass of the cylinder above its equilibrium position:

$$\begin{aligned} y &= 3a(1 - \cos\theta) + 2a(1 - \cos\varphi) \\ &= \tfrac{3}{2}a\theta^2 + a\varphi^2, \end{aligned} \tag{7.43}$$

for small values of θ and φ. Consequently, the potential energy relative to equilibrium is

$$V = \tfrac{1}{2}mg(3a\theta^2 + 2a\varphi^2). \tag{7.44}$$

The Lagrangian is therefore

$$\mathcal{L} = T - V = \tfrac{1}{2}\,ma^2(9\dot{\theta}^2 + 12\dot{\theta}\dot{\varphi} + 8\dot{\varphi}^2) - \tfrac{1}{2}mg(3a\theta^2 + 2a\varphi^2). \tag{7.45}$$

Notice that we have derived a Lagrangian which is quadratic in $\dot{\theta}$, $\dot{\varphi}$, θ and φ. The reasons are that the kinetic energy is a quadratic function of velocity and that the potential energy is evaluated relative to the equilibrium position, $\theta = \varphi = 0$, and consequently its expansion about the minimum is second order in the displacement (see Section 7.2).

We now use the Euler–Lagrange equation,

$$\frac{\partial\mathcal{L}}{\partial q_i} - \frac{\mathrm{d}}{\mathrm{d}t}\left(\frac{\partial\mathcal{L}}{\partial\dot{q}_i}\right) = 0, \tag{7.46}$$

to solve for the motion of the cylinder. Let us take the $(\theta, \dot{\theta})$ pair of coordinates first. Substituting (7.45) into (7.46) and taking $\dot{q}_i = \dot{\theta}$ and $q_i = \theta$, we find

$$\tfrac{1}{2}ma^2 \frac{\mathrm{d}}{\mathrm{d}t}(18\dot{\theta} + 12\dot{\varphi}) = -\tfrac{1}{2}mga\, 6\theta. \tag{7.47}$$

Similarly, for the coordinate pair $\dot{q}_i = \dot{\varphi}, q_i = \varphi$, we find

$$\tfrac{1}{2}ma^2 \frac{\mathrm{d}}{\mathrm{d}t}(12\dot{\theta} + 16\dot{\varphi}) = -\tfrac{1}{2}mga\, 4\varphi. \tag{7.48}$$

We end up with two differential equations:

$$\begin{aligned} 9\ddot{\theta} + 6\ddot{\varphi} &= -\frac{3g}{a}\theta, \\ 6\ddot{\theta} + 8\ddot{\varphi} &= -\frac{2g}{a}\varphi. \end{aligned} \tag{7.49}$$

The key to finding the normal modes of oscillation of the system is that, in a normal mode, all components oscillate at the same frequency; thus we seek oscillatory solutions of the form

$$\begin{aligned} \ddot{\theta} &= -\omega^2\theta, \\ \ddot{\varphi} &= -\omega^2\varphi. \end{aligned} \tag{7.50}$$

Now insert the trial solutions (7.50) into (7.49). Setting $\lambda = a\omega^2/g$,

$$\begin{aligned} (9\lambda - 3)\theta + 6\lambda\varphi &= 0, \\ 6\lambda\theta + (8\lambda - 2)\varphi &= 0. \end{aligned} \tag{7.51}$$

The condition that (7.51) be satisfied for all θ and φ is that the determinant of the coefficients be zero:

$$\begin{vmatrix} 9\lambda - 3 & 6\lambda \\ 6\lambda & 8\lambda - 2 \end{vmatrix} = 0. \tag{7.52}$$

Multiplying out this determinant,

$$6\lambda^2 - 7\lambda + 1 = 0, \tag{7.53}$$

which has solutions $\lambda = 1$ and $\lambda = \frac{1}{6}$. Therefore, since $\omega^2 = \lambda(g/a)$, the angular frequencies of oscillation are $\omega_1 = (g/a)^{1/2}$ and $\omega_2 = [g/(6a)]^{1/2}$, and so the ratio of the frequencies of oscillation of the normal modes is $6^{1/2} : 1$.

We can now find the physical nature of these modes by inserting the solutions $\lambda = \frac{1}{6}$ and $\lambda = 1$ into (7.51). The results are:

$$\begin{array}{llll} \text{for} \ \ \lambda = 1, & \omega_1 = (g/a)^{1/2}, & \theta_1 = A_1 \mathrm{e}^{\mathrm{i}(\omega_1 t + \psi_1)}, & \varphi_1 = -\theta_1; \\ \text{for} \ \ \lambda = \frac{1}{6}, & \omega_2 = (g/6a)^{1/2}, & \theta_2 = A_2 \mathrm{e}^{\mathrm{i}(\omega_2 t + \psi_2)}, & \varphi_2 = \frac{3}{2}\theta_2. \end{array} \tag{7.54}$$

Here ψ_1 and ψ_2 are the phases of the oscillations and A_1, A_2 are the amplitudes of oscillation.

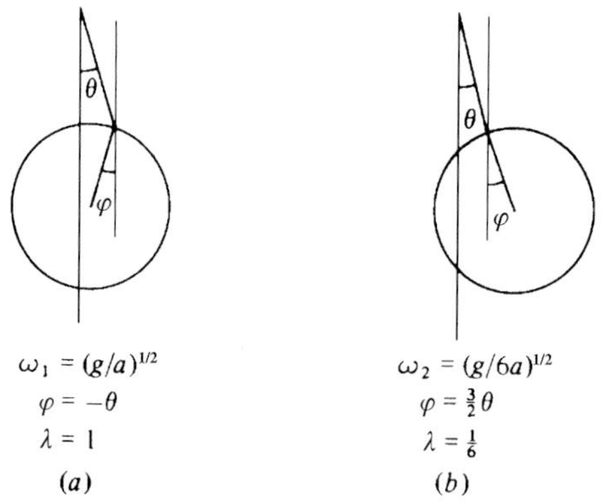

Figure 7.3: Illustrating the normal modes of oscillation of the cylinder about its equilibrium position.

These modes of oscillation are illustrated in Fig. 7.3. According to our analysis, if we set the cylinder oscillating in either of the modes (a) or (b) shown in Fig. 7.3, it will continue to do so for all time at frequencies ω_1 and ω_2 respectively.

We also note that we can represent any other such swinging motion by a superposition of modes 1 and 2, if we choose suitable amplitudes and phases for the normal modes. The x-displacements of the centre of mass of the cylinder in modes 1 and 2 are, from (7.41) and (7.54),

$$x_1(t) = 3a\theta + 2a\varphi = A_1 a e^{i(\omega_1 t + \psi_1)},$$

$$x_2(t) = 3a\theta + 2a\varphi = 6A_2 a e^{i(\omega_2 t + \psi_2)}.$$

For a general motion of the above type, the displacement of the centre of mass is

$$x(t) = x_1(t) + x_2(t) = A_1 a e^{i(\omega_1 t + \psi_1)} + 6A_2 a e^{i(\omega_2 t + \psi_2)}.$$

Then, if the general oscillation begins at $t = 0$ with some specific values of θ, φ, $\dot{\theta}$ and $\dot{\varphi}$, we can find suitable values for A_1, A_2, ψ_1 and ψ_2, since we have four initial conditions and four unknowns. The beauty of this procedure is that we have found the behaviour $x(t)$ of the system at *any* subsequent time.

This shows the fundamental importance of *normal modes*. Any configuration of a system at $t = 0$ can be represented by a suitable superposition of its normal modes at $t = 0$ and this enables us to predict the subsequent dynamical behaviour of the system.

Complete sets of *orthonormal functions* are of particular importance in the analysis of normal modes. The functions themselves are independent and normalised to unity so that a general function of these coordinates can be represented by a superposition of them. A simple example is the Fourier series

$$f(x) = \frac{a_0}{2} + \sum_{n=1}^{\infty} a_n \cos\frac{2\pi n x}{L} + \sum_{n=1}^{\infty} b_n \sin\frac{2\pi n x}{L},$$

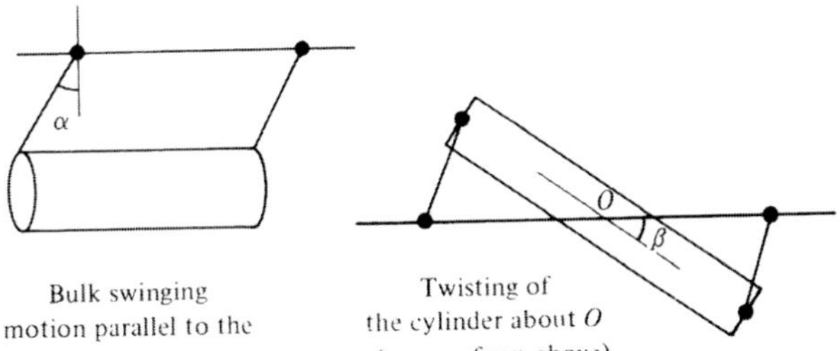

Figure 7.4: The two other normal models of oscillation of the suspended cylinder.

where an orthonormal set of sine and cosine functions is used to describe precisely the function $f(x)$ defined in the interval $0 < x < L$. The separate terms with coefficients a_0, a_n, b_n can be thought of as normal modes of oscillation with end points at 0 and L at which appropriate boundary conditions are applied. There are many different sets of orthonormal functions, which find application in a wide range of aspects of physics and theoretical physics. For example, in spherical polar coordinates, the associated Legendre polynomials provide a complete sets of orthogonal functions defined on a sphere.

In practice, normal modes are not completely independent: in real physical situations, there exist small higher-order terms in the Lagrangian which result in a coupling between them. These enable energy to be exchanged between the modes so that eventually energy is shared between them even if they were very different to begin with. This idea is the basis of the equipartition theorem, which will be discussed in more detail in Chapter 10.

Furthermore, if they are not maintained, the modes will eventually decay by dissipative processes. The time evolution of the system can be accurately determined by following the decay of each normal mode with time. In the example given above, the time evolution of the system would be given by

$$x(t) = A_1 a \mathrm{e}^{-\gamma_1 t}\, \mathrm{e}^{\mathrm{i}(\omega_1 t + \psi_1)} + 6A_2 a \mathrm{e}^{-\gamma_2 t}\, \mathrm{e}^{\mathrm{i}(\omega_2 t + \psi_2)},$$

where γ_1 and γ_2 are the damping constants for each mode.

One final point should be made about normal modes. It is evident that dynamical systems can become very complicated. Suppose, for example, the cylinder were allowed to move arbitrarily rather than being required to oscillate with the end of the cylinder and the string in one plane. The motion would become more complex, but we can guess what the other normal modes must be. These are indicated schematically in Fig. 7.4: a swinging motion along the axis of the cylinder (Fig. 7.4(*a*)) and torsional oscillation of the cylinder about O (Fig. 7.4(*b*)). Indeed, in many physical problems one can get quite a long way by guessing the forms of the normal modes by inspection: we reiterate the key point that in a normal mode all parts of the system must oscillate at the same frequency ω. We will have a great deal more to say about normal modes in our case studies on the origins of statistical mechanics and of the concept of quanta.

7.5 Conservation laws and symmetry

We recall that, in Newtonian mechanics, conservation laws are derived from the first integral of the equations of motion. In exactly the same way, we can derive a set of conservation laws from the first integral of the Euler–Lagrange equations. Approaching the problem from the perspective of these equations brings out clearly the close relations between symmetry and conservation laws. In fact, a great deal depends simply upon the form of the Lagrangian itself.

7.5.1 Lagrangian not a function of q_i

In this case $\partial\mathcal{L}/\partial q_i = 0$ for all the coordinates q_i, and hence the Euler–Lagrange equations read

$$\frac{\mathrm{d}}{\mathrm{d}t}\left(\frac{\partial\mathcal{L}}{\partial\dot{q}_i}\right) = 0, \qquad \frac{\partial\mathcal{L}}{\partial\dot{q}_i} = \text{constant}.$$

An example of such a Lagrangian is the motion of a particle in the absence of a field of force. Then $\mathcal{L} = \frac{1}{2}m\dot{q}_i^2$, $\dot{q}_i = \dot{x}_i$, and

$$\frac{\partial\mathcal{L}}{\partial\dot{q}_i} = \frac{\partial\mathcal{L}}{\partial\dot{x}} = m\dot{x} = \text{constant}, \tag{7.55}$$

that is, Newton's first law of motion.

This calculation is suggestive of how we can define a generalised momentum p_i. For an arbitrary system with coordinates q_i, we can define *conjugate momenta* as

$$p_i \equiv \frac{\partial\mathcal{L}}{\partial\dot{q}_i}; \tag{7.56}$$

p_i is not necessarily anything like a normal momentum, but it does have the property that if $\mathcal{L}$ does not depend upon q_i then it is a constant of the motion.

7.5.2 Lagrangian independent of time

This is obviously related to energy conservation. Let us perform a straightforward analysis. According to the chain rule,

$$\frac{\mathrm{d}\mathcal{L}}{\mathrm{d}t} = \frac{\partial\mathcal{L}}{\partial t} + \sum_i \dot{q}_i\frac{\partial\mathcal{L}}{\partial q_i}. \tag{7.57}$$

The Euler–Lagrange equation (7.26) can be written as

$$\frac{\mathrm{d}}{\mathrm{d}t}\left(\frac{\partial\mathcal{L}}{\partial\dot{q}_i}\right) = \frac{\partial\mathcal{L}}{\partial q_i}.$$

Therefore, substituting for $\partial\mathcal{L}/\partial q_i$ in (7.57),

$$\frac{\mathrm{d}\mathcal{L}}{\mathrm{d}t} - \frac{\mathrm{d}}{\mathrm{d}t}\left(\sum_i \dot{q}_i \frac{\partial\mathcal{L}}{\partial\dot{q}_i}\right) = \frac{\partial\mathcal{L}}{\partial t},$$

$$\frac{\mathrm{d}}{\mathrm{d}t}\left(\mathcal{L} - \sum_i \dot{q}_i \frac{\partial\mathcal{L}}{\partial\dot{q}_i}\right) = \frac{\partial\mathcal{L}}{\partial t}. \qquad (7.58)$$

But the Lagrangian does not have any explicit dependence upon time, that is, $\partial\mathcal{L}/\partial t = 0$. Therefore

$$\sum_i \dot{q}_i \frac{\partial\mathcal{L}}{\partial\dot{q}_i} - \mathcal{L} = \text{constant}. \qquad (7.59)$$

As will be shown below, this expression is exactly the same as the *law of conservation of energy* in Newtonian mechanics. The quantity which is conserved is known as the *Hamiltonian, H*:

$$H = \sum_i \dot{q}_i \frac{\partial\mathcal{L}}{\partial\dot{q}_i} - \mathcal{L}. \qquad (7.60)$$

We can write this in terms of the conjugate momentum $p_i = \partial\mathcal{L}/\partial\dot{q}_i$:

$$H = \sum_i p_i \dot{q}_i - \mathcal{L}. \qquad (7.61)$$

Notice that, in the case of Cartesian coordinates, the Hamiltonian becomes

$$\begin{aligned} H &= \sum_i (m_i \dot{\boldsymbol{r}}_i) \cdot \dot{\boldsymbol{r}}_i - \mathcal{L}, \\ &= 2T - (T - V), \\ &= T + V, \end{aligned}$$

which shows explicitly the relation to energy conservation. Notice that, formally, the conservation law arises from the invariance of the Lagrangian with respect to time.

7.5.3 *Lagrangian independent of the absolute positions of the particles*

By this statement, we mean that $\mathcal{L}$ depends only on $\boldsymbol{r}_1 - \boldsymbol{r}_2$ and not on the absolute value of $\boldsymbol{r}_1$. Suppose we change all the q_i by the same small amount ϵ. Then, the requirement that $\mathcal{L}$ remains unchanged on shifting all the coordinates of all the particles by ϵ becomes

$$\mathcal{L} + \sum_i \frac{\partial\mathcal{L}}{\partial q_i}\delta q_i = \mathcal{L} + \sum_i \epsilon \frac{\partial\mathcal{L}}{\partial q_i} = \mathcal{L}.$$

Invariance thus requires that

$$\sum_i \frac{\partial\mathcal{L}}{\partial q_i} = 0. \qquad (7.62)$$

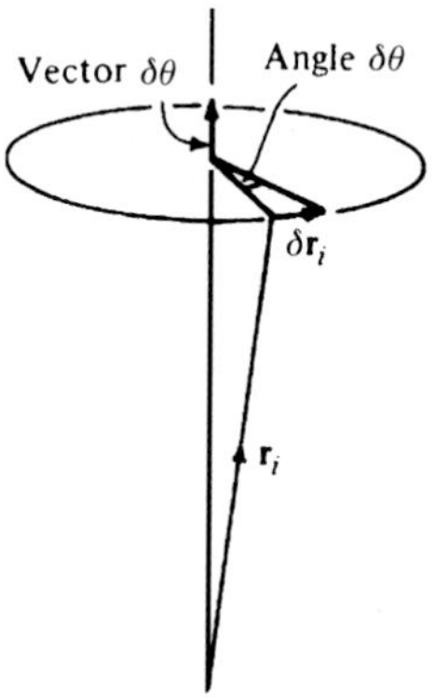

Figure 7.5: Illustrating the rotation of the system of coordinates through a small angle $\delta\boldsymbol{\theta}$.

Now, the Euler–Lagrange equations (7.26) add to give

$$\frac{\mathrm{d}}{\mathrm{d}t}\left(\sum_i \frac{\partial \mathcal{L}}{\partial \dot{q}_i}\right) = \sum_i \frac{\partial \mathcal{L}}{\partial q_i};$$

therefore, from (7.56) and (7.62) we find

$$\frac{\mathrm{d}}{\mathrm{d}t}\sum_i p_i = 0. \tag{7.63}$$

This is the law of *conservation of linear momentum* and results from the requirement that the Lagrangian be invariant with respect to spatial translations.

7.5.4 Lagrangian independent of the orientation of the system in space

By this statement we mean that $\mathcal{L}$ is invariant under rotations. If the system is rotated through a small angle $\delta\boldsymbol{\theta}$ about an arbitrary axis, as in Fig. 7.5, the position and velocity vectors, $\boldsymbol{r}_i$ and $\boldsymbol{v}_i$ respectively, change by

$$\delta\boldsymbol{r}_i = \delta\boldsymbol{\theta} \times \boldsymbol{r}_i \qquad \text{and} \qquad \delta\boldsymbol{v}_i = \delta\boldsymbol{\theta} \times \boldsymbol{v}_i .$$

We require that $\delta\mathcal{L} = 0$ under this rotation and hence that, for the complete system of particles,

$$\delta\mathcal{L} = \sum_i \left(\nabla_{r_i}\mathcal{L}\cdot\delta\boldsymbol{r}_i + \nabla_{v_i}\mathcal{L}\cdot\delta\boldsymbol{v}_i\right) = 0.$$

Hence we obtain

$$\sum_i \left[\frac{\mathrm{d}}{\mathrm{d}t}\nabla_{r_i}\mathcal{L}\cdot(\mathrm{d}\boldsymbol{\theta}\times\boldsymbol{r}_i) + \nabla_{v_i}\mathcal{L}\cdot(\mathrm{d}\boldsymbol{\theta}\times\boldsymbol{v}_i)\right] = 0. \tag{7.64}$$

Reordering the vector products, (7.64) becomes

$$\sum_i \left\{ \mathrm{d}\boldsymbol{\theta} \cdot \left[\boldsymbol{r}_i \times \frac{\mathrm{d}}{\mathrm{d}t} \nabla_{\boldsymbol{r}_i} \mathcal{L} \right] + \mathrm{d}\boldsymbol{\theta} \cdot \left(\boldsymbol{v}_i \times \nabla_{\boldsymbol{v}_i} \mathcal{L} \right) \right\} = 0,$$

$$\sum_i \mathrm{d}\boldsymbol{\theta} \cdot \left[\frac{\mathrm{d}}{\mathrm{d}t} \left(\boldsymbol{r}_i \times \nabla_{\boldsymbol{v}_i} \mathcal{L} \right) \right] = 0,$$

$$\mathrm{d}\boldsymbol{\theta} \cdot \sum_i \left[\frac{\mathrm{d}}{\mathrm{d}t} \left(\boldsymbol{r}_i \times \nabla_{\boldsymbol{v}_i} \mathcal{L} \right) \right] = 0. \tag{7.65}$$

Therefore, if the Lagrangian is to be independent of orientation,

$$\sum_i \left(\boldsymbol{r}_i \times \nabla_{\boldsymbol{v}_i} \mathcal{L} \right) = \text{constant}, \tag{7.66}$$

that is, using a generalisation of (7.56),

$$\sum_i (\boldsymbol{r}_i \times \boldsymbol{p}_i) = \text{constant}. \tag{7.67}$$

This is the *law of conservation of angular momentum* and results from the requirement that the Lagrangian be invariant under rotations.

7.6 Hamilton's equations and Poisson brackets

The next extension is to express the equations of motion in terms of p_i and q_i rather than in terms of $\dot{q}_i$ and q_i, that is, we use the canonical momentum p_i defined by $\partial\mathcal{L}/\partial\dot{q}_i$. Remember that this does not necessarily correspond to what we mean by momentum in Newtonian mechanics. Equation (7.61) relating the Hamiltonian to the Lagrangian may be written

$$H = \sum_i p_i \dot{q}_i - \mathcal{L}(q, \dot{q}). \tag{7.68}$$

It looks as though H depends upon p_i, $\dot{q}_i$ and q_i but, in fact, we can rearrange the equation so that H is a function of only p_i and q_i. Let us take the total differential of H in the usual way, assuming that $\mathcal{L}$ is time independent. Then

$$\mathrm{d}H = \sum_i p_i \,\mathrm{d}\dot{q}_i + \sum_i \dot{q}_i \,\mathrm{d}p_i - \sum_i \frac{\partial \mathcal{L}}{\partial \dot{q}_i} \mathrm{d}\dot{q}_i - \sum_i \frac{\partial \mathcal{L}}{\partial q_i} \mathrm{d}q_i. \tag{7.69}$$

Since $p_i \equiv \partial\mathcal{L}/\partial\dot{q}_i$, the first and third terms on the right-hand side are equal and hence cancel. Therefore,

$$\mathrm{d}H = \sum_i \dot{q}_i \,\mathrm{d}p_i - \sum_i \frac{\partial \mathcal{L}}{\partial q_i} \mathrm{d}q_i. \tag{7.70}$$

This differential depends only on the increments $\mathrm{d}p_i$ and $\mathrm{d}q_i$, and hence we can compare $\mathrm{d}H$ with its formal expansion in terms of p_i and q_i:

$$\mathrm{d}H = \sum_i \frac{\partial H}{\partial p_i} \mathrm{d}p_i + \sum_i \frac{\partial H}{\partial q_i} \mathrm{d}q_i.$$

It follows immediately that

$$\frac{\partial H}{\partial q_i} = -\frac{\partial \mathcal{L}}{\partial q_i} \quad \text{and} \quad \frac{\partial H}{\partial p_i} = \dot{q}_i .$$

The Euler–Lagrange equation can be written

$$\frac{\partial \mathcal{L}}{\partial q_i} = \frac{\mathrm{d}}{\mathrm{d}t}\left(\frac{\partial \mathcal{L}}{\partial \dot{q}_i}\right) = \frac{\mathrm{d}p_i}{\mathrm{d}t},$$

and so

$$\frac{\partial H}{\partial q_i} = -\dot{p}_i .$$

The equations of motion have therefore been reduced to the relations

$$\dot{q}_i = \frac{\partial H}{\partial p_i}, \qquad \dot{p}_i = -\frac{\partial H}{\partial q_i}. \tag{7.71}$$

This pair of equations is known as *Hamilton's equations*. They are first-order differential equations for each of the $3N$ coordinates. The p_i and the q_i are now treated on the same footing.

The last comment I want to make about these techniques concerns the objects known as *Poisson brackets*. In conjunction with Hamilton's equations of motion, they can be used to make the formalism even more compact. The Poisson bracket for the functions g and h is defined to be the quantity

$$[g, h] = \sum_{i=1}^{n} \left(\frac{\partial g}{\partial p_i}\frac{\partial h}{\partial q_i} - \frac{\partial g}{\partial q_i}\frac{\partial h}{\partial p_i} \right). \tag{7.72}$$

We may write in general

$$\dot{g} = \sum_{i=1}^{n} \left(\frac{\partial g}{\partial q_i}\dot{q}_i + \frac{\partial g}{\partial p_i}\dot{p}_i \right) \tag{7.73}$$

for the variation of any physical quantity g and hence, using Hamilton's equations, we can write

$$\dot{g} = [H, g]. \tag{7.74}$$

Therefore, Hamilton's equations can be written

$$\dot{q}_i = [H, q_i], \qquad \dot{p}_i = [H, p_i].$$

The Poisson brackets (7.72) have a number of very useful properties. If we identify g with q_i and h with q_j, we find that

$$[q_i, q_j] = 0.$$

Similarly,

$$[p_j, p_k] = 0,$$

and if $j \neq k$,

$$[p_j, q_k] = 0.$$

But if $g = p_k$ and $h = q_k$,

$$[p_k, q_k] = 1, \qquad [q_k, p_k] = -1.$$

Quantities with Poisson brackets which are zero are said to *commute*. Those with Poisson brackets equal to unity are said to be *canonically conjugate*. From the relation established in (7.74), we see that any quantity which commutes with the Hamiltonian H does not change with time. In particular, H itself is constant in time because it commutes with itself. Yet again we have returned to the conservation of energy.

You will find that Poisson brackets play an important role in the development of quantum mechanics, as demonstrated in Dirac's classic text *The Principles of Quantum Mechanics*.[2] There is a very nice story in Dirac's memoirs about how he came to realise their importance. In October 1925, Dirac was worried by the fact that, according to his formulation of quantum mechanics, the dynamical variables did not commute, that is, for two variables u and v, uv is not the same as vu. Dirac apparently had a strict rule about relaxing on Sunday afternoons by taking country walks. He writes:

> It was during one of the Sunday Walks in October 1925 when I was thinking very much about this $uv - vu$, in spite of my intention to relax, that I thought about Poisson brackets . . . I did not remember very well what a Poisson bracket was. I did not remember the precise formula for a Poisson bracket and only had some vague recollections. But there were exciting possibilities there and I thought that I might be getting on to some big new idea.
>
> Of course, I could not [find out what a Poisson bracket was] right out in the country. I just had to hurry home and see what I could then find about Poisson brackets. I looked through my notes and there was no reference there anywhere to Poisson brackets. The text books which I had at home were all too elementary to mention them. There was nothing I could do, because it was Sunday evening then and the libraries were all closed. I just had to wait impatiently through that night without knowing whether this idea was any good or not but still I think that my confidence grew during the course of the night. The next morning, I hurried along to one of the libraries as soon as it was open and then I looked up Poisson brackets in Whittaker's *Analytic Dynamics* and I found that they were just what I needed. They provided the perfect analogy with the commutator.[3]

This is a good example of how professionals work. They do not actually remember all the mathematics they have come across, but they keep their eyes and ears open to everything they hear, no matter how remote it might seem from their immediate interests; then, someday, these bits of information end up being important. They may not remember them exactly, but they know where to find them. These remarks apply equally aptly to our understanding of the whole of physics.

7.7 A warning

My approach to classical mechanics has intentionally been non-rigorous so that I can bring out certain specific features of different approaches to mechanics and dynamics. The subject often appears somewhat abstruse and mathematical and I have intentionally concentrated upon the simple parts which relate most directly to our understanding of Newton's laws. I would emphasise that the subject can be made rigorous and such approaches are found in books such as Goldstein's *Classical Mechanics*[4] or Landau and Lifshitz's *Mechanics*.[5] My aim has been to show that there are good physical reasons for developing these more complex approaches to mechanics and dynamics.

7.8 References

1 Feynman, R.P. (1964). *Lectures in Physics*, eds. R.P. Feynman, R.B. Leighton and M. Sands, Vol. 2, Chapter 19. London: Addison-Wesley.
2 Dirac, P.A.M. (1935). *The Principles of Quantum Mechanics*. Oxford: Clarendon Press.
3 Dirac, P.A.M. (1977). In *History of Twentieth Century Physics, Proc. International School of Physics 'Enrico Fermi'*, Course 57, p. 122. New York and London: Academic Press.
4 Goldstein, H. (1950). *Classical Mechanics*. London: Addison-Wesley.
5 Landau, L.D. and Lifshitz, E.M. (1960). *Mechanics*, Vol. 1 of *Course of Theoretical Physics*. Oxford: Pergamon Press.

Appendix to Chapter 7: The motion of fluids

The aim of this appendix is to revise some important aspects of the application of Newton's laws of motion to fluids. Fluid dynamics is a vast and important subject which can rapidly become very complex for reasons which will soon become apparent. The intention of these notes is to emphasise some of the basic differences encountered in dealing with fluids as opposed to systems of particles or solid bodies.

A7.1 The equation of continuity

First of all, we derive the equation which tells us that the fluid does not 'get lost'. By a procedure similar to that used in electromagnetism (Section 6.2), consider the net mass flux from a volume v bounded by a surface S. If $\mathrm{d}\boldsymbol{S}$ is an element of surface area, the direction of the vector being taken normally outwards, the mass flow through $\mathrm{d}\boldsymbol{S}$ is $\rho \boldsymbol{u} \cdot \mathrm{d}\boldsymbol{S}$ (Fig. A7.1). Integrating over the surface of the volume, the rate of outflow of mass must equal the rate of loss of mass from within v, that is,

$$\int_S \rho \boldsymbol{u} \cdot \mathrm{d}\boldsymbol{S} = -\frac{\mathrm{d}}{\mathrm{d}t}\int_v \rho \, \mathrm{d}v. \tag{A7.1}$$

Using the divergence theorem,

$$\int_v \mathrm{div} \rho \boldsymbol{u} \, \mathrm{d}v = -\frac{\mathrm{d}}{\mathrm{d}t}\int_v \rho \, \mathrm{d}v,$$

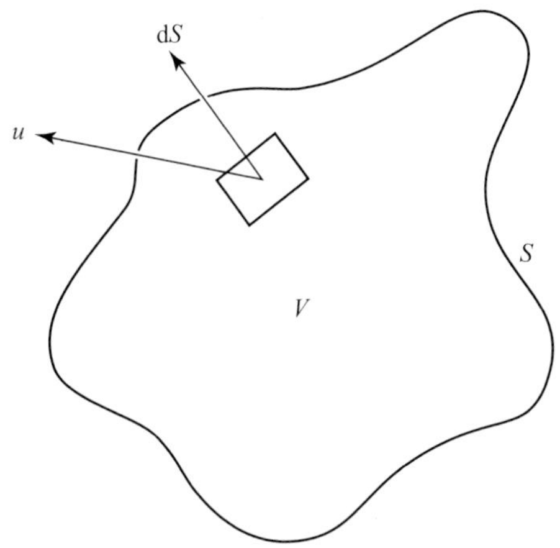

Figure A7.1: Illustrating the mass flux of fluid from a volume v.

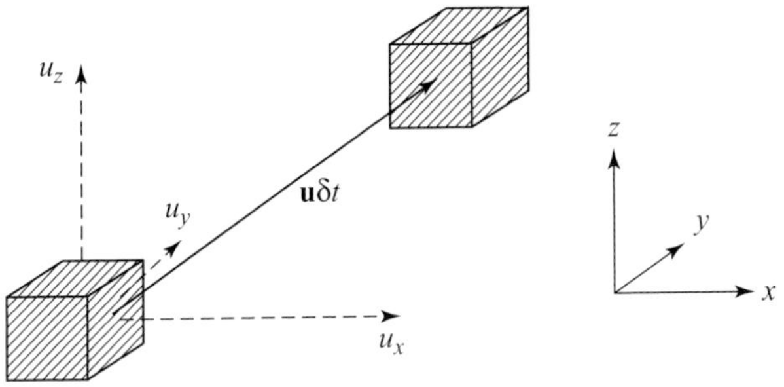

Figure A7.2: Illustrating the relation between the total and partial derivatives, $\mathrm{d}/\mathrm{d}t$ and $\partial/\partial t$ respectively.

that is,

$$\int_v \left(\mathrm{div}\,\rho\boldsymbol{u} + \frac{\mathrm{d}\rho}{\mathrm{d}t}\right)\mathrm{d}v = 0.$$

This result must be true for any elementary volume within the fluid and hence

$$\mathrm{div}\,\rho\boldsymbol{u} + \frac{\partial\rho}{\partial t} = 0. \qquad \text{(A7.2)}$$

This is the *equation of continuity*. Notice that we have changed from a total to a partial derivative between the last two equations. The reason is that, in microscopic form, we use partial derivatives to denote variations in the properties of the fluid *at a fixed point in space*. This is distinct from derivatives which follow the properties of a particular element of the fluid. The latter are denoted by total derivatives, for example, $\mathrm{d}\rho/\mathrm{d}t$, as can be seen from the following argument.

We define $\mathrm{d}\rho/\mathrm{d}t$ to be the rate of change of density of an element of the fluid which we follow as it moves through the fluid. This is shown schematically in Fig. A7.2, in which the motion of the element is followed over the time interval δt. If the velocity of the fluid at (x, y, z) is $\boldsymbol{u} = (u_x, u_y, u_z)$, it follows that $\mathrm{d}\rho/\mathrm{d}t$ is given by

$$\frac{\mathrm{d}\rho}{\mathrm{d}t} = \lim_{\delta t \to 0} \frac{1}{\delta t}[\rho(x + u_x\,\delta t, y + u_y\,\delta t, z + u_z\,\delta t, t + \delta t) - \rho(x, y, z, t)]. \qquad \text{(A7.3)}$$

Now perform a Taylor expansion of the first term in the brackets.

$$\frac{\mathrm{d}\rho}{\mathrm{d}t} = \lim_{\delta t \to 0} \frac{1}{\delta t} \left[\rho(x, y, z, t) + \frac{\partial \rho}{\partial x} u_x \, \delta t + \frac{\partial \rho}{\partial y} u_y \, \delta t + \frac{\partial \rho}{\partial z} u_z \, \delta t + \frac{\partial \rho}{\partial t} \, \delta t - \rho(x, y, z, t) \right]$$

$$= \frac{\partial \rho}{\partial t} + u_x \frac{\partial \rho}{\partial x} + u_y \frac{\partial \rho}{\partial y} + u_z \frac{\partial \rho}{\partial z}.$$

Thus

$$\frac{\mathrm{d}\rho}{\mathrm{d}t} = \frac{\partial \rho}{\partial t} + \boldsymbol{u} \cdot \operatorname{grad} \rho. \tag{A7.4}$$

Notice that equation (A7.4) is no more than the chain rule yet it relates our definitions of $\mathrm{d}/\mathrm{d}t$ and $\partial/\partial t$. Notice also that (A7.4) is a relation between differential operators:

$$\frac{\mathrm{d}}{\mathrm{d}t} = \frac{\partial}{\partial t} + \boldsymbol{u} \cdot \operatorname{grad} \tag{A7.5}$$

and this recurs throughout fluid dynamics. One can work in either of these frames of reference. If we work in the coordinate system which follows an element of the fluid, the coordinates are called *Lagrangian coordinates*. If we work in a fixed external reference frame, they are called *Eulerian coordinates*. Generally, Eulerian coordinates are used, although there are occasions when it is simpler to use a Lagrangian approach.

Equation (A7.4) enables us to rewrite the equation of continuity (A7.2) as follows:

$$\operatorname{div} \rho \boldsymbol{u} + \frac{\mathrm{d}\rho}{\mathrm{d}t} = \boldsymbol{u} \cdot \operatorname{grad} \rho.$$

Expanding $\operatorname{div} \rho \boldsymbol{u}$,

$$\boldsymbol{u} \cdot \operatorname{grad} \rho + \rho \operatorname{div} \boldsymbol{u} + \frac{\mathrm{d}\rho}{\mathrm{d}t} = \boldsymbol{u} \cdot \operatorname{grad} \rho.$$

Thus the equation of continuity becomes

$$\frac{\mathrm{d}\rho}{\mathrm{d}t} = -\rho \operatorname{div} \boldsymbol{u}. \tag{A7.6}$$

If the fluid is incompressible, then for any element of fluid $\rho =$ constant and the flow is described by $\operatorname{div} \boldsymbol{u} = 0$. If the flow is not time dependent, so that $\partial\rho/\partial t = 0$, and is irrotational, so that $\operatorname{curl} \boldsymbol{u} = 0$, then the velocity field $\boldsymbol{u}$ can be expressed in terms of a velocity potential ϕ defined by $\boldsymbol{u} = \operatorname{grad} \phi$. The velocity potential ϕ can then be found as the solution of Laplace's equation,

$$\nabla^2 \phi = 0, \tag{A7.7}$$

subject to satisfying the boundary conditions of the problem.

A7.2 The equation of motion for an incompressible fluid in the absence of viscosity

To derive the equation of motion, consider the forces acting upon a particular unit element of volume of the fluid. Newton's laws of motion can be applied in a Lagrangian system of coordinates. Then, neglecting viscous forces and assuming that the fluid is incompressible,

$$\rho \frac{\mathrm{d}\boldsymbol{u}}{\mathrm{d}t} = -\mathrm{grad}\, p - \rho\, \mathrm{grad}\, \phi, \tag{A7.8}$$

where p is pressure and ϕ the gravitational potential. Now we rewrite this equation in terms of Eulerian coordinates, that is, in terms of partial rather than total derivatives. The analysis proceeds exactly as in Section A7.1, but now we consider the vector quantity $\boldsymbol{u}$, rather than the scalar quantity ρ. This is only a minor complication because the three components of the vector $\boldsymbol{u}$ are scalar functions for which the relation (A7.5) applies. For example,

$$\frac{\mathrm{d}u_x}{\mathrm{d}t} = \frac{\partial u_x}{\partial t} + \boldsymbol{u} \cdot \mathrm{grad}\, u_x.$$

There are similar equations for u_y and u_z and therefore, adding all three together vectorially,

$$\begin{aligned} \boldsymbol{i}_x \frac{\mathrm{d}u_x}{\mathrm{d}t} + \boldsymbol{i}_y \frac{\mathrm{d}u_y}{\mathrm{d}t} + \boldsymbol{i}_z \frac{\mathrm{d}u_z}{\mathrm{d}t} &= \boldsymbol{i}_x \frac{\partial u_x}{\partial t} + \boldsymbol{i}_y \frac{\partial u_y}{\partial t} + \boldsymbol{i}_z \frac{\partial u_z}{\partial t} \\ &\quad + \boldsymbol{i}_x(\boldsymbol{u} \cdot \mathrm{grad}\, u_x) + \boldsymbol{i}_y(\boldsymbol{u} \cdot \mathrm{grad}\, u_y) + \boldsymbol{i}_z(\boldsymbol{u} \cdot \mathrm{grad}\, u_z). \end{aligned}$$

This can be written

$$\frac{\mathrm{d}\boldsymbol{u}}{\mathrm{d}t} = \frac{\partial \boldsymbol{u}}{\partial t} + (\boldsymbol{u} \cdot \mathrm{grad})\boldsymbol{u}. \tag{A7.9}$$

Note that in evaluating $(\boldsymbol{u} \cdot \mathrm{grad})\boldsymbol{u}$ we perform the operation $[u_x(\partial/\partial x) + u_y(\partial/\partial y) + u_z(\partial/\partial z)]$ on *all three of the components of the vector* $\boldsymbol{u}$, obtaining the component of $(\boldsymbol{u} \cdot \mathrm{grad})\boldsymbol{u}$ in each direction.

The equation of motion (A7.8) in Eulerian coordinates is therefore

$$\frac{\partial \boldsymbol{u}}{\partial t} + (\boldsymbol{u} \cdot \mathrm{grad})\, \boldsymbol{u} = -\frac{1}{\rho}\, \mathrm{grad}\, p - \mathrm{grad}\, \phi. \tag{A7.10}$$

This equation indicates clearly where the problems of fluid mechanics come from. The second term on the left-hand side introduces a nasty non-linearity in the velocity $\boldsymbol{u}$ and this causes all sorts of complications when trying to find exact solutions in fluid dynamical problems. It is obvious that the subject can rapidly become one of great mathematical complexity.

An interesting way of rewriting (A7.10) is to introduce the vector $\boldsymbol{\omega}$, which is called the *vorticity* of the flow, and is defined by $\boldsymbol{\omega} = \mathrm{curl}\, \boldsymbol{u} = \nabla \times \boldsymbol{u}$. Now

$$\boldsymbol{u} \times \boldsymbol{\omega} = \boldsymbol{u} \times \mathrm{curl}\, \boldsymbol{u} = \tfrac{1}{2}\, \mathrm{grad}\, \boldsymbol{u}^2 - (\boldsymbol{u} \cdot \mathrm{grad})\, \boldsymbol{u}, \tag{A7.11}$$

and so the equation of motion becomes

$$\frac{\partial \boldsymbol{u}}{\partial t} - (\boldsymbol{u} \times \boldsymbol{\omega}) = -\mathrm{grad}\left(\tfrac{1}{2} u^2 + \frac{p}{\rho} + \phi\right). \tag{A7.12}$$

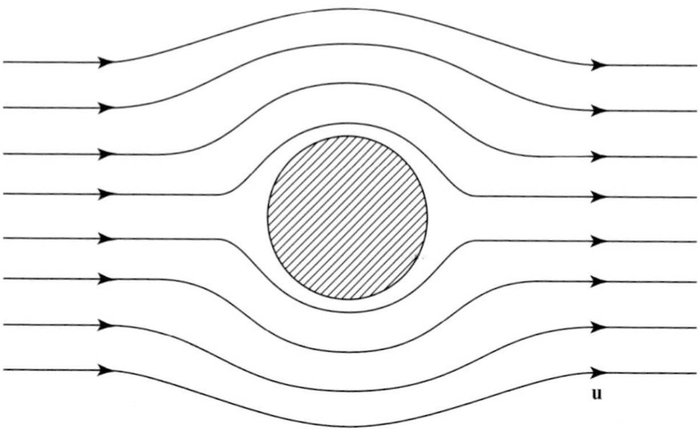

Figure A7.3: Illustrating the streamlines associated with the flow of an incompressible viscous fluid past a solid sphere in the limit of small Reynolds number Re $\ll 1$.

Equation (A7.12) is useful for finding specific solutions to certain classes of fluid-dynamical problems. For example, if we are concerned only with steady motion, $\partial \boldsymbol{u}/\partial t = 0$ and hence

$$\boldsymbol{u} \times \boldsymbol{\omega} = \text{grad}\left(\tfrac{1}{2}u^2 + \frac{p}{\rho} + \phi\right). \tag{A7.13}$$

If the flow is irrotational, $\text{curl}\,\boldsymbol{u} = 0$ and $\boldsymbol{u}$ can be derived from the gradient of a scalar potential (see appendix section A5.4). Since $\boldsymbol{\omega} = \text{curl}\,\boldsymbol{u}$ the right-hand side must be zero and hence

$$\tfrac{1}{2}u^2 + \frac{p}{\rho} + \phi = \text{constant}. \tag{A7.14}$$

Another important way in which this conservation law can be applied is to introduce the concept of *streamlines*, which are defined to be lines in the fluid that are everywhere parallel to the instantaneous velocity vector $\boldsymbol{u}$. For example, Fig. A7.3 shows the streamlines associated with the flow of an incompressible viscous fluid past a solid sphere. If we follow the flow along a streamline, the quantity $\boldsymbol{u} \times \boldsymbol{\omega}$ is perpendicular to $\boldsymbol{u}$. Hence, from (A7.13), in the direction of $\boldsymbol{u}$, that is, *along a streamline*, we again find

$$\tfrac{1}{2}u^2 + \frac{p}{\rho} + \phi = \text{constant}.$$

Notice that this result is true along a particular streamline even if $\boldsymbol{\omega} \neq 0$. We have derived *Bernoulli's theorem*.

A7.3 The equation of motion for an incompressible fluid including viscous forces

We have come almost as far as we can without writing down the stress tensor for the fluid and the equations of motion in tensor form. The proper derivation of the equations of motion including viscous forces needs the full tensor treatment and here we will do no more

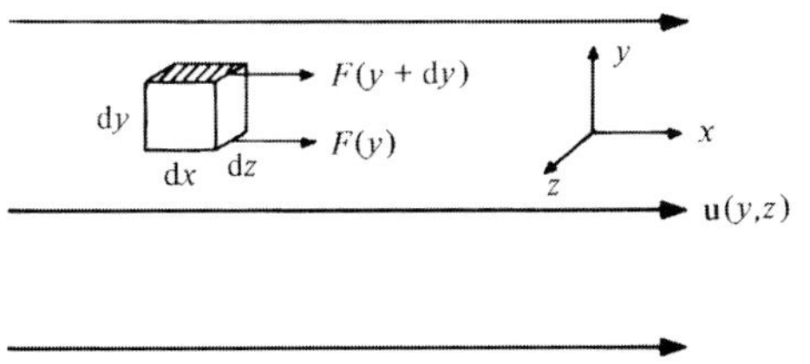

Figure A7.4: The action of viscous forces on a volume $dV = dx\ dy\ dz$ in unidirectional flow in the positive x-direction.

than rationalise the form the equations must have. Consider steady unidirectional flow of an incompressible fluid in the positive x-direction and work out the viscous forces on an element of the fluid of volume $dV = dx\,dy\,dz$ (Fig. A7.4). Consider first the viscous forces acting on the top and bottom surfaces of the volume at y and $y + dy$. The viscous force on the bottom surface is

$$\mu\, dx\, dz\, \frac{\partial u_x}{\partial y}(y),$$

where μ is the viscosity of the fluid. The force on the top surface is

$$\mu\, dx\, dz\, \frac{\partial u_x}{\partial y}(y + dy).$$

The net force on the element of fluid is the difference between these forces. Performing a Taylor expansion,

$$\mu\, dx\, dz \frac{\partial u_x}{\partial y}(y + dy) = \mu\, dx\, dz \left(\frac{\partial u_x}{\partial y}(y) + \frac{\partial^2 u_x}{\partial y^2}(y)\, dy \right).$$

Therefore, the net force acting on the volume dV due to the velocity gradient in the y-direction is

$$\mu\, dx\, dz\, \frac{\partial^2 u_x}{\partial y^2}(y)\, dy.$$

The same calculation can be performed for the viscous forces acting on the faces defined by $dx\ dy$. The net force is

$$\mu\, dx\, dy\, \frac{\partial^2 u_x}{\partial z^2}(y)\, dz.$$

The equation of motion of the element of fluid is therefore

$$\rho\, dV \frac{du_x}{dt} = \mu \left(\frac{\partial^2 u_x}{\partial y^2} + \frac{\partial^2 u_x}{\partial z^2} \right) dx\, dy\, dz,$$

that is,

$$\rho \frac{du_x}{dt} = \mu \left(\frac{\partial^2 u_x}{\partial y^2} + \frac{\partial^2 u_x}{\partial z^2} \right).$$

Since the flow is unidirectional in the x-direction, $\partial u_x/\partial x = 0$ and hence we can write

$$\rho \frac{\mathrm{d}u_x}{\mathrm{d}t} = \mu \nabla^2 u_x .$$

When this viscosity term, in vector form, is added to (A7.10) we obtain the equation of motion of incompressible fluid flow:

$$\frac{\partial \boldsymbol{u}}{\partial t} + (\boldsymbol{u} \cdot \mathrm{grad})\boldsymbol{u} = -\frac{1}{\rho}\,\mathrm{grad}\, p - \mathrm{grad}\,\phi + \frac{\mu}{\rho}\,\nabla^2 \boldsymbol{u}. \qquad \text{(A7.15)}$$

This is the *Navier–Stokes equation*. The viscous force appears in the form suggested by our analysis of steady unidirectional flow. For a full treatment, the reader should consult, for example, Batchelor's text *An Introduction to Fluid Dynamics*,[A1] Landau and Lifshitz's text book *Fluid Mechanics*[A2] or Faber's *Fluid Dynamics for Physicists*,[A3] which I particularly enjoy for its physical insight.

A7.4 Appendix references

A1 Batchelor, G.K. (1967). *An Introduction to Fluid Dynamics*. Cambridge: Cambridge University Press.

A2 Landau, L.D. and Lifshitz, E.M. (1959). *Fluid Mechanics*, Vol. 5 of *Course of Theoretical Physics*. Oxford: Pergamon Press.

A3 Faber, T.E. (1995). *Fluid Dynamics for Physicists*. Cambridge: Cambridge University Press.

8 Dimensional analysis, chaos and self-organised criticality

8.1 Introduction

The increasingly powerful mathematical tools described in Chapter 7 provided the means for tackling complex dynamical problems in classical physics. Despite these successes, in many areas of physics problems can become rapidly very complex and, although we may be able to write down the differential or integral equations which describe the behaviour of the system, often it is not possible to find analytic solutions.

The objective of this chapter is to study techniques developed to tackle these complex problems, some of them so non-linear that they seem quite beyond the scope of traditional analysis. First, we review the techniques of *dimensional analysis*. Used with care and insight, this approach is powerful and finds many applications in pure and applied physics. We will give as examples the non-linear pendulum, fluid flow, explosions, turbulence and so on.

Next, we briefly study *chaos*, the analysis of which became feasible only with the development of high-speed computers. The equations of motion are deterministic and yet the outcome is extremely sensitive to the precise initial conditions. Beyond these examples are even more extreme systems, in which so many non-linear effects come into play that it is impossible to predict the outcome of an experiment, in any conventional sense. And yet regularities are found in the form of scaling laws. There must be some underlying simplicity in the way in which the system behaves, despite the horrifying complexity of the many processes involved. These topics involve *fractals* and the burgeoning field of *self-organised criticality*.

Finally, we look briefly at what Roger Penrose has termed *non-computational* physics. Are there problems which even in principle we cannot solve by mathematical techniques, no matter how powerful our computers?

8.2 Dimensional analysis

Ensuring that equations are dimensionally balanced is part of our basic training as physicists. Dimensional analysis is, however, much deeper than simple consistency and can help us solve complicated problems without our having to solve complicated equations. Let us begin with a simple example and then extend it to illustrate the real power of dimensional analysis.

8.2.1 *The simple pendulum*

Consider a simple pendulum undergoing small oscillations. Its period τ can depend only on the length of the string l, the mass of the bob m and the gravitational acceleration g. The period of the pendulum can therefore be written

$$\tau = f(m, l, g). \tag{8.1}$$

Equation (8.1) must be dimensionally balanced and so let us write it in the form

$$\tau \sim m^{\alpha} l^{\beta} g^{\gamma}, \tag{8.2}$$

where α, β and γ are constants. We adopt the convention in this chapter that the units need not match if we use the proportional sign '$\propto$', but when we use the sign '$\sim$', the equation must be dimensionally balanced. In terms of dimensions, (8.1) reads

$$[\mathrm{T}] \equiv [\mathrm{M}]^{\alpha}[\mathrm{L}]^{\beta}[\mathrm{LT}^{-2}]^{\gamma}, \tag{8.3}$$

where [L], [T] and [M] mean the dimensions of length, time and mass respectively. The dimensions must be the same on either side of the equation and so, equating the powers of [T], [M] and [L], we find

$$\begin{aligned} [\mathrm{T}]: &\quad 1 = -2\gamma, &\quad \gamma = -\tfrac{1}{2}; \\ [\mathrm{M}]: &\quad 0 = \alpha, &\quad \alpha = 0; \\ [\mathrm{L}]: &\quad 0 = \beta + \gamma, &\quad \beta = +\tfrac{1}{2}. \end{aligned}$$

Therefore,

$$\tau \sim \sqrt{l/g}, \tag{8.4}$$

the correct expression for the simple pendulum – note that the period τ must be independent of the mass of the pendulum bob, m. This crude application of dimensional methods is found in many textbooks but does scant justice to its real power in pure and applied physics.

Let us expand the scope of dimensional analysis. We introduce a remarkable theorem enunciated by Edgar Buckingham in 1914, following some derogatory remarks about the method of dimensions by the distinguished theorist R.C. Tolman. Buckingham made creative use of his theorem in understanding how to scale from model ships in ship tanks to the real thing. The procedure is as follows.

- First, guess what the important quantities in the problem are, as above.
- Then apply the Buckingham Π theorem,* which states that a system described by n variables, built from r independent dimensions, is described by $n - r$ independent dimensionless *groups*.[1] Such a 'group' is a dimensionless quantity obtained by the multiplication or division of some or all of the variables, each raised to an appropriate power.

* I am indebted to Dr Sanjoy Mahajan for introducing me to the remarkable power of this theorem. His forthcoming book *Order of Magnitude Physics: the Art of Approximation in Science* (2003), written with Strel Phinney and Peter Goldreich, has my strongest recommendation for all students of physics.[2]

Table 8.1. *The non-linear pendulum*

Variable	Dimensions	Description
θ_0	–	angle of release
m	[M]	mass of bob
τ	[T]	period of pendulum
g	$[L][T]^{-2}$	acceleration due to gravity
l	[L]	length of pendulum

- Form from the important quantities all the simplest $n - r$ possible dimensionless groups.
- The most general solution to the problem can be written as a function of the $n - r$ independent dimensionless groups.

Let us repeat the pendulum problem, now allowing large oscillations – the non-linear case. First, as before, we guess what the independent variables are likely to be; Table 8.1 shows a list of the variables with their dimensions.

There are $n = 5$ variables and $r = 3$ independent dimensions and so, according to the Buckingham Π theorem, we can form just two independent dimensionless groups. θ_0 is already dimensionless and so let us choose $\Pi_1 = \theta_0$. Only one variable depends upon m and so *no dimensionless group can contain* m, that is, the period of the pendulum must be independent of the mass of the bob.

There are many possibilities for Π_2, for example, $(\tau^2 g/l)^2$, $\theta_0 l/(\tau^2 g)$ and so on. We require Π_2 to be independent of Π_1, and we would like it to be as simple as possible, that is, the group should contain the smallest possible number of factors or divisors. Then, all other dimensionless groups can be formed from these two independent examples, Π_1 and Π_2. The simplest independent group Π_2 we can form is, by inspection,

$$\Pi_2 = \frac{\tau^2 g}{l}.$$

Therefore, the general solution for the motion of the pendulum can be written as

$$f(\Pi_1, \Pi_2) = f\left(\theta_0, \frac{\tau^2 g}{l}\right) = 0, \tag{8.5}$$

where f is a function of θ_0 and $\tau^2 g/l$. Notice that we have not written down any differential equation.

Now, (8.5) might well turn out to be some complicated function, but we guess that it may often be possible to write the solution as some unknown functional dependence of one of the dimensionless groups upon the other. In this case, we could then write (8.5) as follows:

$$\Pi_2 = f_1(\Pi_1) \qquad \text{or} \qquad \tau = f_1(\theta_0)\sqrt{l/g}. \tag{8.6}$$

Now, we might not be able to determine $f_1(\theta_0)$ from theory, but we can determine *experimentally* how the period τ depends upon θ_0 for particular values of l and g. Then,

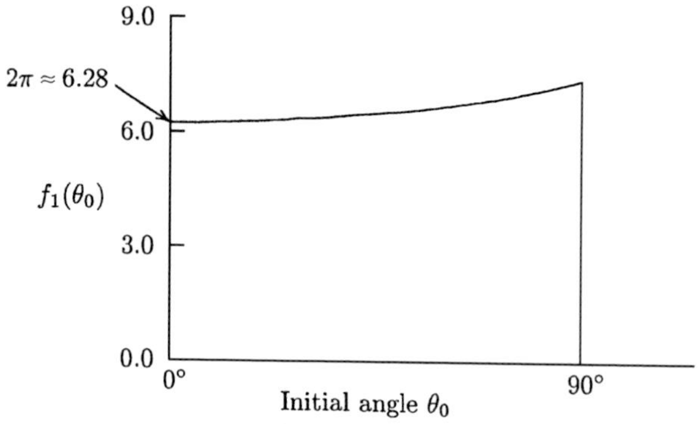

Figure 8.1: The dependence of the period of a non-linear pendulum upon the angle of release θ_0.

$f_1(\theta_0)$ is determined and will be the same for *all other simple pendulums*: $f_1(\theta_0)$ is *universal*. Notice that this solution is valid for *all* values of θ_0, not just for small values, $\theta_0 \ll 1$.

In this particular example, we can find the exact non-linear solution for an arbitrary value of θ_0. By the conservation of energy,

$$mgl(1 - \cos\theta) + \tfrac{1}{2}I\dot{\theta}^2 = \text{constant} = mgl(1 - \cos\theta_0), \tag{8.7}$$

where $I = ml^2$ and θ_0 is the maximum amplitude of the pendulum. Let us first recall the linear solution, which is found for small values of θ_0. The equation (8.7) becomes

$$\ddot{\theta} + \frac{g}{l}\theta = 0,$$

and so the angular frequency and period of oscillation are

$$\omega = \sqrt{\frac{g}{l}} \qquad \tau = \frac{2\pi}{\omega} = 2\pi\sqrt{\frac{l}{g}}.$$

Now returning to the exact differential equation (8.7), we can write the latter as

$$\frac{d\theta}{dt} = \sqrt{\frac{2g}{l}}(\cos\theta - \cos\theta_0)^{1/2}. \tag{8.8}$$

The period of the pendulum is four times the time $t(\theta_0)$ it takes to swing from 0 to θ_0 and so

$$\tau = 4\int_0^{t(\theta_0)} dt = 4\sqrt{\frac{l}{2g}}\int_0^{\theta_0} \frac{d\theta}{(\cos\theta - \cos\theta_0)^{1/2}}. \tag{8.9}$$

Notice that the integral depends only upon θ_0. Performing a numerical integration, we obtain the function $f_1(\theta_0)$ shown in Fig. 8.1.

We can therefore write

$$\tau = \sqrt{l/g}\, f_1(\theta_0). \tag{8.10}$$

Just as predicted by dimensional analysis, the result only depends upon a general function of the dimensionless parameters θ_0 and $\tau^2 g/l$.

Table 8.2. *G.I. Taylor's analysis of explosions*

Variable	Dimensions	Description
E	$[M][L]^2[T]^{-2}$	energy release
ρ_0	$[M][L]^{-3}$	external density
γ	–	ratio of specific heats
r_f	$[L]$	shock front radius
t	$[T]$	time

8.2.2 G.I. Taylor's analysis of explosions

In 1950, G.I. Taylor published a famous analysis of the dynamics of the shock waves associated with atomic explosions. A huge amount of energy E is released within a small volume, and this results in a strong spherical shock front, which propagates through the surrounding air. The internal pressure is enormous and very much greater than the pressure of the surrounding air. The dynamics of the shock front depend, however, upon the density ρ_0 of the surrounding air, which is swept up by the expanding shock front and causes its deceleration. The compression of the ambient gas by the shock front plays a role in its dynamics and depends upon the ratio of specific heats of air, which is $\gamma = 1.4$ for the molecular gases oxygen and nitrogen. The only other parameters in the problem are the radius of the shock front r_f and the time t. Table 8.2 shows the variables and their dimensions.

From the Buckingham Π theorem, there are five variables and three independent dimensions and so we can form two dimensionless groups. One of them can be $\Pi_1 = \gamma$. We find the second dimensionless quantity by elimination. From the quotient of E and ρ_0, we find

$$\left[\frac{E}{\rho_0}\right] = \frac{[M][L]^2[T]^{-2}}{[M][L]^{-3}} = \frac{[L^5]}{[T^2]} = \left[\frac{r_f^5}{t^2}\right].$$

Thus, the simplest choice for Π_2 is

$$\Pi_2 = \frac{Et^2}{\rho_0 r_f^5}.$$

We can therefore write the solution to the problem as $\Pi_2 = f(\Pi_1)$, that is,

$$r_f = A\left(\frac{E}{\rho_0}\right)^{1/5} t^{2/5} f(\gamma), \tag{8.11}$$

where A is a constant.

In 1941, Taylor had been asked by the UK government to carry out a theoretical study of very-high-energy explosions. His report was presented to the UK Civil Defence Research Committee of the Ministry of Home Security in that year and was only declassified in 1949. In his detailed analysis of the expansion of the very strong shocks associated with such explosions, he derived (8.11) and showed that $f(\gamma = 1.4) = 1.03$. He compared his results with high-speed films of chemical explosions available at that time and found that his theoretical predictions were in reasonable agreement with what was observed.[3] In 1949,

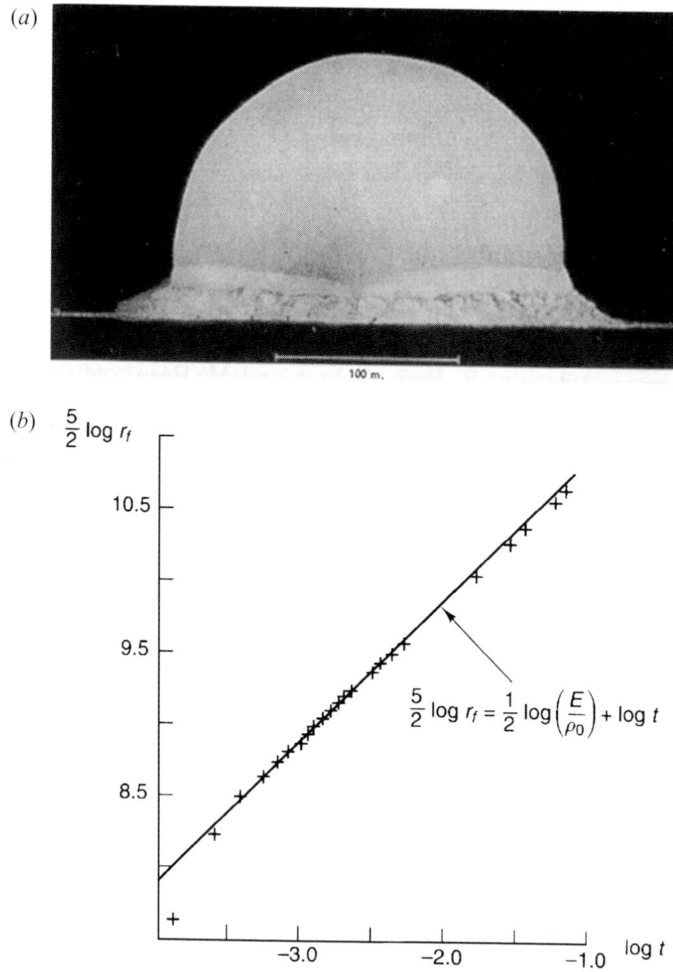

Figure 8.2: (*a*) A frame from Mack's film of a Nevada nuclear explosion taken 15 ms after ignition. (*b*) G.I. Taylor's analysis of the dynamics of the shock front of the nuclear explosion, showing that $r_{\rm f} \propto t^{2/5}$. (From G.I. Taylor, 1950, *Proc. Roy. Soc. A*, **201**, 175.)

he compared his calculations with the results of higher-energy explosions using TNT-RDX and again confirmed the validity of (8.11).[4]

In 1947, the US military released Mack's movie of the first atomic bomb explosion, which took place in the New Mexico desert in 1945, under the title 'Semi-popular motion picture record of the Trinity explosion'[5] (Fig. 8.2(*a*)). Taylor was able to determine, directly from the movie, the relation between the radius of the shock wave and time (Fig. 8.2(*b*)). To the annoyance of the US government, he was able to work out the energy E released in the explosion[4] and found it to be about 10^{14} J, equivalent to the explosion of about 20 kilotons of TNT. This work was published only in 1950, nine years after he had carried out his first calculations.

Table 8.3. *Steady horizontal motion of a sphere through a viscous fluid*

Variable	Dimensions	Description
F_d	$[M][L][T]^{-2}$	drag force
ρ_f	$[M][L]^{-3}$	density of fluid
R	$[L]$	radius of sphere
ν	$[L^2][T]^{-1}$	kinematic viscosity
v	$[L][T]^{-1}$	speed

8.2.3 *Fluid dynamics – drag in fluids*

Let us now apply the techniques of dimensional analysis to the flow of an incompressible fluid past a sphere of radius R. In the case in which viscous forces are dominant, the exact answer for the drag force f_d was derived by George Gabriel Stokes and is known as Stokes' formula, $F_d = 6\pi \nu R v$, where $\nu = \mu/\rho$ is the *kinematic viscosity* and v the speed of the flow. The Navier–Stokes equation for incompressible fluid flow was discussed in appendix section A7.3:

$$\frac{\partial \boldsymbol{u}}{\partial t} + (\boldsymbol{u} \cdot \nabla)\boldsymbol{u} = -\frac{1}{\rho}\nabla p + \nu \nabla^2 \boldsymbol{u}, \tag{8.12}$$

with

$$\nabla \cdot \boldsymbol{u} = 0.$$

The complexity of fluid-dynamical problems arises from the non-linear term $(\boldsymbol{u} \cdot \nabla)\boldsymbol{u}$ in (8.12). Let us develop the form of solution for the drag force on the sphere using dimensional arguments. It will turn out that there are two limiting cases, one in which *viscosity* mediates the drag force, the other in which the *inertia* of the fluid is dominant.

In applying dimensional analysis, we consider the physically equivalent case of a sphere moving horizontally through a static fluid at a speed v equal to the fluid speed at a large distance from the sphere in the static-sphere case. In the same way as before, we list in Table 8.3 the variables which are likely to be important in determining the horizontal motion of the sphere. They include the drag force on the sphere, the speed of the sphere and the density and viscosity of the fluid. The density of the sphere is not expected to have any effect on its steady horizontal motion, but this motion will depend on the surface area of the sphere and therefore on its radius. According to the Buckingham Π theorem, there are five variables and three independent dimensions and so we can form two independent dimensionless groups.

The last three entries in Table 8.3 form one dimensionless group,

$$\Pi_1 = \frac{vR}{\nu} = \mathsf{Re},$$

where Re is the *Reynolds number*, a dimensionless measure of the *speed* v of the sphere. The second dimensionless group must involve the top two entries in the table, to be independent of Π_1; the dimensionless quantity

$$\Pi_2 = \frac{F_\mathrm{d}}{\rho_\mathrm{f} R^2 v^2}$$

can be constructed. According to the theorem, the most general relation depends only on Π_1 and Π_2 and, since we want to find an expression for F_d, let us write

$$\Pi_2 = f(\Pi_1), \qquad \text{that is,} \qquad F_\mathrm{d} = \rho_\mathrm{f} R^2 v^2 \, f\left(\frac{vR}{\nu}\right).$$

Now, the drag force should be proportional to the kinematic viscosity ν. This means that the function $f(x)$ must be proportional to $1/x$. Therefore, the expression for the drag force is

$$F_\mathrm{d} = A \rho_\mathrm{f} R^2 v^2 \left(\frac{vR}{\nu}\right)^{-1} = A \nu \rho_\mathrm{f} R v, \tag{8.13}$$

where A is a constant to be found. We cannot find it using dimensional techniques. A full treatment[6] results in the answer $A = 6\pi$ and so

$$F_\mathrm{d} = 6\pi \, \nu \rho_\mathrm{f} R v = 6\pi \rho_\mathrm{f} \frac{R^2 v^2}{\mathsf{Re}}. \tag{8.14}$$

If now we imagine that the steady motion of the sphere is vertical, we can equate the drag force to the gravitational force on the sphere, $F_\mathrm{g} = (4\pi/3)\rho_\mathrm{sp} R^3 g$, where ρ_sp is the mean density of the sphere, and obtain the terminal speed v:

$$(4\pi/3)\rho_\mathrm{sp} R^3 g = 6\pi \, \nu \rho_\mathrm{f} R v,$$

that is,

$$v = \frac{2}{9}\left(\frac{gR^2}{\nu}\right)\left(\frac{\rho_\mathrm{sp}}{\rho_\mathrm{f}}\right). \tag{8.15}$$

So far, we have ignored the effect of buoyancy on the terminal velocity of the sphere. This enters through another dimensionless group,

$$\Pi_3 = \frac{\rho_\mathrm{f}}{\rho_\mathrm{sp}}.$$

The effect of buoyancy is to reduce the effective weight of the sphere or, equivalently, the magnitude of the acceleration due to gravity, so that

$$g \to g\left(1 - \frac{\rho_\mathrm{f}}{\rho_\mathrm{sp}}\right).$$

Notice the reasoning behind this buoyancy correction. The sphere would not fall under gravity if $\rho_\mathrm{f} = \rho_\mathrm{sp}$, but, if $\rho_\mathrm{f} \ll \rho_\mathrm{sp}$, the result (8.15) should be obtained. Therefore the corrected terminal speed is

$$v = \frac{2}{9}\left(\frac{gR^2}{\nu}\right)\left(\frac{\rho_\mathrm{sp}}{\rho_\mathrm{f}} - 1\right). \tag{8.16}$$

This was an important result historically, since it was used by Millikan in his famous oil-drop experiments to determine precisely the charge of the electron (see Sections 14.3 and 15.4).

In the analysis so far, it has been assumed implicitly that viscous forces play a dominant role in transferring momentum between the sphere and the fluid. Let us investigate the conditions under which this result is expected to apply. The terms in the Navier–Stokes equation which describe the response of the fluid to the action of pressure and gravitational potential gradients are

$$\frac{\partial \boldsymbol{u}}{\partial t} + (\boldsymbol{u} \cdot \nabla)\boldsymbol{u}, \qquad \nu \nabla^2 \boldsymbol{u}. \tag{8.17}$$

The acceleration of the fluid involves the term $(\boldsymbol{u} \cdot \nabla)\boldsymbol{u}$, which is associated with the *inertia* of the fluid, and the term $\nu \nabla^2 \boldsymbol{u}$ is associated with its viscosity. To estimate the relative importance of these two terms for an object of dimension R, let us make order-of-magnitude estimates of their relative magnitudes. Suppose that R is the scale over which momentum is transferred, either by inertia or by viscosity. Then, to order of magnitude, the partial differentials in (8.17) are given by

$$(\boldsymbol{u} \cdot \nabla)\boldsymbol{u} \sim \frac{v^2}{R}, \qquad \nu \nabla^2 \boldsymbol{u} \sim \frac{\nu v}{R^2}. \tag{8.18}$$

The relative importance of the two terms is given by the ratio

$$\frac{\text{inertial acceleration}}{\text{viscous acceleration}} \approx \frac{v^2/R}{\nu v/R^2} = \frac{vR}{\nu} = \mathsf{Re}, \tag{8.19}$$

where Re is the Reynolds number. In other words, the Reynolds number describes the relative importance of inertia and viscosity in a flow. The solution (8.13) is dominated by the viscous transport of momentum within the fluid. Hence it is applicable for flows in which the Reynolds number $\mathsf{Re} \ll 1$; the streamlines for this case are illustrated in Fig. 8.3(a).

Another way of looking at this result is in terms of the characteristic time t_ν for the diffusion of the viscous stresses through the fluid, which can be found to order of magnitude by approximating

$$\frac{\partial \boldsymbol{u}}{\partial t} = \nu \nabla^2 \boldsymbol{u} \qquad \text{as} \qquad \frac{v}{t_\nu} \approx \frac{\nu v}{R^2} \qquad \text{so that} \qquad t_\nu \approx \frac{R^2}{\nu}.$$

The time for the fluid to flow past the sphere is $t_v \sim R/v$. Thus, the condition that viscous forces have time to transmit momentum is

$$t_v \gg t_\nu, \qquad \text{so that} \qquad \frac{R}{v} \gg \frac{R^2}{\nu} \qquad \text{or} \qquad \frac{vR}{\nu} = \mathsf{Re} \ll 1.$$

In the high-Reynolds-number limit, $\mathsf{Re} = vR/\nu \gg 1$, the flow past the sphere becomes turbulent. Then the viscosity is unimportant and can be dropped from our table of relevant parameters (Table 8.4).

There are now four variables and three independent dimensions and so we can form only one dimensionless group, which we have found already:

$$\Pi_2 = \frac{F_\mathrm{d}}{\rho_\mathrm{f} R^2 v^2}.$$

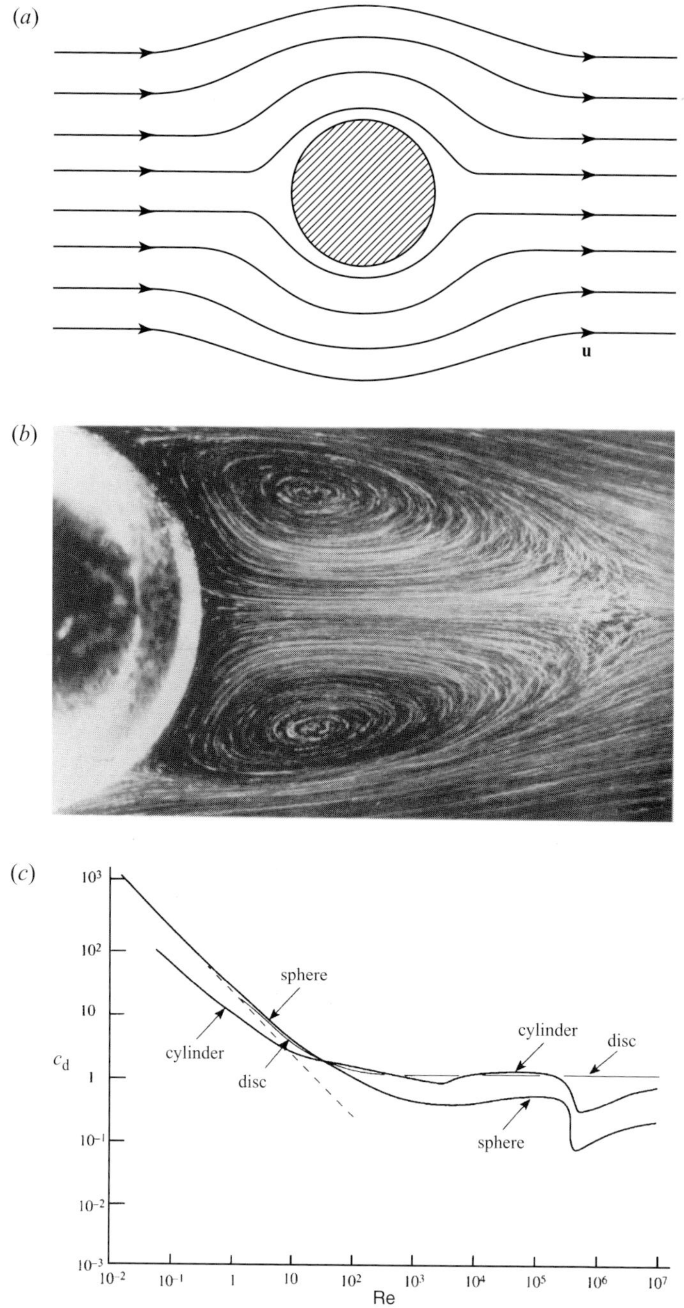

Figure 8.3: (*a*) Viscous flow past a sphere at low Reynolds number, $\mathrm{Re} \ll 1$. (*b*) Turbulent flow past a sphere at $\mathrm{Re} = 117$. (Courtesy of S. Taneda, from S. Taneda, 1956, *J. Phys. Soc.* (Japan), **11**, 1106, Photo 1g.) (*c*) The drag coefficient c_d for different objects as a function of Reynolds number Re. (After T.E. Faber, 1995, *Fluid Dynamics for Physicists*, pp. 258, 266, Cambridge: Cambridge University Press.)

Table 8.4. *Steady motion of a sphere in a fluid in the limit* Re $\gg 1$

Variable	Dimensions	Description
F_d	$[M][L][T]^{-2}$	drag force
ρ_f	$[M][L]^{-3}$	density of fluid
R	$[L]$	radius of sphere
v	$[L][T]^{-1}$	speed

Table 8.5. *Drag coefficient* c_d *for objects of different shapes*

Object	c_d
sphere	0.5
cylinder	1.0
flat plate	2.0
car	0.4

It can be seen that Π_2 contains all the variables relevant to this case and no others can be involved. Just as in the case of Taylor's analysis of atomic explosions, the simplest approach is to try first the case in which Π_2 is taken to be a constant:

$$F_d = C\rho_f v^2 R^2, \tag{8.20}$$

where C is a constant of order unity.

We can readily interpret the formula $f_d \propto \rho_f v^2 R^2$. The factor $\rho_f v^2$ is approximately the rate at which the momentum of the fluid per unit area arrives at the sphere and R^2 is the projected area of the sphere; thus, (8.20) is roughly the rate at which the momentum of the incident fluid is pushed away by the sphere per second. To find the limiting velocity of a sphere falling under gravity, and taking account of the buoyancy of the medium, we set F_d equal to the buoyancy force:

$$F_d = C\rho_f v^2 R^2 = \frac{4\pi R^3}{3}(\rho_{sp} - \rho_f)g,$$

and so

$$v \approx \sqrt{gR\frac{(\rho_{sp} - \rho_f)}{\rho_f}}.$$

Conventionally, the drag force is written

$$F_d = \tfrac{1}{2}c_d \rho_f v^2 A,$$

where A is the projected surface area of the object and c_d is the *drag coefficient*. Some values are given in Table 8.5.

In the design of Formula 1 racing cars, the aim is to make c_d as small as possible.

Table 8.6. *Homogeneous turbulence*

Variable	Dimensions	Description
$E(k)$	$[L]^3[T]^{-2}$	energy per unit wavenumber[a]
ε_0	$[L]^2[T]^{-3}$	rate of energy input[a]
k	$[L]^{-1}$	wavenumber

[a] Both per unit mass of fluid.

While there are analytic solutions for low Reynolds numbers, when viscosity is dominant, there are none in the inertial regime at high Reynolds number. Figure 8.3(*b*) shows the turbulent vortices which develop behind a sphere at high Reynolds numbers. The variation of the drag coefficient for a wide range of Reynolds numbers is shown in Fig. 8.3(*c*). Inspection of (8.14) shows that in the limit $\mathsf{Re} \ll 1$, $f_d \propto \mathsf{Re}^{-1}$, consistent with the relations shown in Fig. 8.3(*c*), but at high Reynolds numbers, the drag force is more or less independent of Re.

8.2.4 *Kolmogorov spectrum of turbulence*

Another remarkable example of the use of dimensional analysis in fluid dynamics is in the study of turbulence. As can be seen from Fig. 8.3(*b*), turbulent eddies form behind a sphere in a flow at high Reynolds number. One of the great problems of fluid dynamics is understanding such turbulent flows, which arise from the non-linear term $(\boldsymbol{u} \cdot \nabla)\boldsymbol{u}$ in the Navier–Stokes equation. Empirically, the features of turbulence are well established. The energy of the eddies is injected on large scales, say, on the scale of the sphere seen in Fig. 8.3(*b*). The large-scale eddies fragment into smaller-scale eddies, which in turn break up into even smaller eddies, and so on. In terms of energy transport, the energy injected on the largest scale cascades down though smaller and smaller eddies until the scales are sufficiently small that the energy is dissipated as heat by viscous forces at the molecular level. This process can be characterised by a *spectrum of turbulence*. This describes the amount of energy present on each scale, that is, the amount of energy per unit wavelength λ, in a steady state, when energy is continuously injected on a large scale and ultimately dissipated at the molecular level. The processes involved in the energy cascade are highly non-linear and there is no analytic theory of the spectrum of turbulence. Kolmogorov showed, however, that progress can be made by dimensional analysis.

In Kolmogorov's analysis, attention is concentrated upon the transfer of energy between the large scale l at which energy is injected and the scale $\lambda_{\rm visc}$ at which molecular dissipation is important. An important feature of the analysis is that there are no natural length scales between l and $\lambda_{\rm visc}$. The aim of the calculation is to determine the amount of kinetic energy in eddies on different length scales, and the natural way of describing this is in terms of $E(k)\,\mathrm{d}k$, the energy contained in eddies per unit mass in the wavenumber interval k to $k + \mathrm{d}k$, where $k = 2\pi/\lambda$.

Suppose the rate of supply of kinetic energy per unit mass is ε_0 on the large scale. As before we can form a table of relevant variables (Table 8.6). There are only three variables

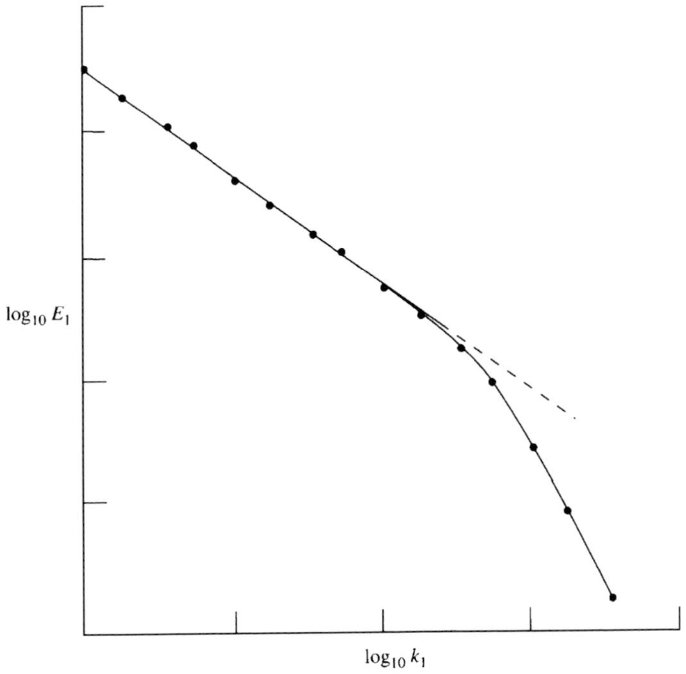

Figure 8.4: The experimental spectrum of homogeneous turbulence compared with the relation $E(k) \propto k^{-5/3}$ (broken line) derived from dimensional analysis. (From T.E. Faber, 1995, *Fluid Dynamics for Physicists*, p. 357, Cambridge: Cambridge University Press.)

in the problem and two independent dimensions and so we can form only one dimensionless group,

$$\Pi_1 = \frac{E^3(k)k^5}{\varepsilon_0^2}.$$

As before, the simplest assumption is that Π_1 is a constant of order unity and so

$$E(k) \sim \varepsilon_0^{2/3} k^{-5/3}.$$

This relation, known as the *Kolmogorov spectrum of turbulence*, turns out to be a remarkably good fit to the spectrum of turbulence on scales intermediate between those on which the energy is injected and those on which dissipation by molecular viscosity is important (Fig. 8.4).

These are only a few of the many uses of dimensionless analysis in fluid and gas dynamics. Many beautiful examples of the power of these techniques are given by G.I. Barenblatt in his excellent book *Scaling, Self-similarity, and Intermediate Asymptotics*.[7] A particularly appropriate example for the Cambridge students who attended the course on which this book is based is one in which it is shown that the speed of a rowing boat is proportional to $N^{1/9}$ where N is the number of rowers.

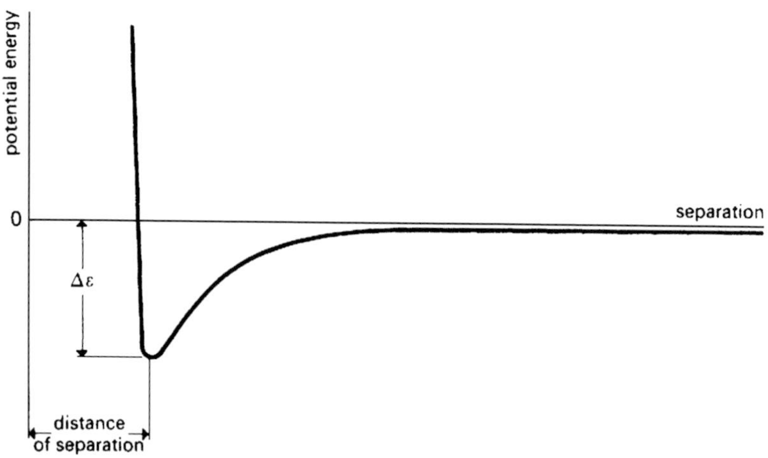

Figure 8.5: A schematic diagram showing the intermolecular potential energy between one molecule of a gas and a single neighbour as a function of their separation. The equilibrium distance of separation is σ. (From D. Tabor, 1991, *Gases, Liquids and Solids and Other States of Matter*, p. 125, Cambridge: Cambridge University Press.)

8.2.5 *The law of corresponding states*

Let us now study the application of dimensional techniques to the equation of state of imperfect gases, the aim being to derive the *law of corresponding states*. For n moles of a perfect gas, the pressure p, volume V and temperature T are related by the ideal gas law,

$$pV = nRT, \tag{8.21}$$

where R is the gas constant. This law provides an excellent description of the properties of all gases at very low pressures and high temperatures, specifically, when the gas is far from the transition temperature to a liquid or solid.

In the case of an *imperfect gas*, we need to take account of the finite size of the molecules and the intermolecular forces acting between them. Thus, we need to relate the macroscopic properties of the gas to the microscopic properties of the molecules. The simplest way of characterising the forces acting between molecules is in terms of the potential energy of the molecules as a function of their separation (Fig. 8.5) (see, for example, Tabor[8]). There is an attractive potential between the molecules, which becomes more negative as their separation decreases. This could, for example, be due to van der Waals forces. The potential continues to become more negative until at some scale σ the molecules are touching and then strong repulsive forces prevent them coming any closer together. Thus, σ is not only the distance of closest approach but is also a measure of the size of the molecules. As seen in the diagram, the properties of the intermolecular potential can be characterised in a simple way by the intermolecular separation σ and the depth of the attractive potential well $\Delta\varepsilon$.

As usual, we list all the variables which might be important in this problem including $\Delta\varepsilon$, σ and the mass m of a molecule of the gas.

Table 8.7. *The variables in the equation of state for an imperfect gas*

Variable	Dimensions	Description
p	$[M][L]^{-1}[T]^{-2}$	pressure
V	$[L]^3$	volume
kT	$[M][L]^2[T]^{-2}$	'temperature'
N	–	number of molecules
m	$[M]$	mass of molecule
σ	$[L]$	intermolecular spacing
$\Delta\varepsilon$	$[M][L]^2[T]^{-2}$	depth of attractive potential

From the Buckingham Π theorem, there should be $7 - 3 = 4$ independent dimensionless groups. Inspection of Table 8.7 reveals that, in fact, there are fewer quantities with independent dimensions than might appear at first sight. The quantities pV, kT and $\Delta\varepsilon$ all have the dimensions of energy, $[M][L]^2[T]^{-2}$. Furthermore, there is no way of creating a dimensionless quantity involving m alone, since there is no quantity involving just [T] which could be used along with m to eliminate [T] from the quantities with the dimensions of energy. Therefore, m cannot appear as a variable in the problem.

So, although we started with three independent dimensions, there are actually only two, [L] and $[M][L]^2[T]^{-2}$. The dimensions of the six independent quantities in Table 8.7 can now be constructed from [L] and $[M][L]^2[T]^{-2}$. The Buckingham Π theorem tells us that we can create $6 - 2 = 4$ independent dimensionless quantities from these as before.

Let us choose the following set of four dimensionless groups, to begin with:

$$\Pi_1 = N, \qquad \Pi_2 = \frac{kT}{\Delta\varepsilon}, \qquad \Pi_3 = \frac{V}{\sigma^3}, \qquad \Pi_4 = \frac{p\sigma^3}{\Delta\varepsilon}.$$

In this choice of the Π_i, we have been guided by the need to express the macroscopic quantities p, V and T in terms of the microscopic quantities σ and $\Delta\varepsilon$.

To make progress, we now need to include some physics in the analysis. The equation of state involves only the three variables p, V and T, and yet we have four independent dimensionless groups. Suppose the gas is in some state characterised by p, V and T. p and T are *intensive* variables, which are independent of the 'extent' of the system. Suppose we cut the system in half. The pressure and temperature remain unchanged, but the volumes of the two parts and the number of molecules in each part are both halved. Therefore, the equation of state can only involve the ratio $\Pi_5 = \Pi_3/\Pi_1 = V/(N\sigma^3)$.

We have reduced the problem to three dimensionless quantities, which have physical significance. Π_2 is the ratio of the typical thermal energy kT of the molecules to their binding energy $\Delta\varepsilon$ – this ratio describes the relative importance of the kinetic energy of the molecules of the gas in relation to their binding energies. Π_4 expresses the pressure in terms of that associated with the binding energy of the molecules and the volume over which it influences the properties of the gas. $\Pi_5 = V/(N\sigma^3)$ is the ratio of the volume V to the total volume occupied by the molecules – the volume $N\sigma^3$ is referred to as the *excluded volume*.

Table 8.8. *Values of $p^* V^*/(RT^*)$ for different gases*

Gas	$p^* V^*/(RT^*)$	Gas	$p^* V^*/(RT^*)$
He	0.327	Ar	0.291
Xe	0.277	H_2	0.306
O_2	0.292	N_2	0.292
Hg	0.909	H_2O	0.233
CO_2	0.277		

Equations of state traditionally relate the pressure to the volume and temperature, which suggests that the relation between the remaining dimensionless groups should have the form

$$\Pi_4 = f(\Pi_2, \Pi_5).$$

Substituting, we find

$$\frac{p\sigma^3}{\Delta\varepsilon} = f\left(\frac{V}{N\sigma^3}, \frac{kT}{\Delta\varepsilon}\right), \qquad \text{or} \qquad \frac{p}{p^*} = f\left(\frac{V}{V^*}, \frac{T}{T^*}\right),$$

where

$$p^* = \frac{\Delta\varepsilon}{\sigma^3}, \qquad V^* = N\sigma^3, \qquad T^* = \frac{\Delta\varepsilon}{k}. \tag{8.22}$$

With this choice of dimensionless quantities, $p^* V^*/(NkT^*) = 1$. The equation of state of the gas written in terms of p^*, V^* and T^* is called a *reduced equation of state*. The *law of corresponding states* is the statement that *the reduced equations of state are of the same form for all gases*.

Thus, if the shape of the potential function for gases were the same for all gases, we would expect $p^* V^*/(RT^*)$ to be constant for one mole of gas. Table 8.8 gives the values for a selection of gases. It can be seen that this procedure is reasonably successful for gases of the same molecular family. Examples of equations of state which have proved successful include the *van der Waals* and *Dieterici* equations of state, which can be written

$$\left(\pi + \frac{3}{\phi^2}\right)(3\phi - 1) = 8\theta \tag{8.23}$$

and

$$\pi(2\phi - 1) = \theta \,\exp\left[2\left(1 - \frac{1}{\theta\phi}\right)\right], \tag{8.24}$$

where $\pi = p/p^*$, $\phi = V/V^*$ and $\theta = T/T^*$. Notice that we gain physical insight into the meanings of the quantities p^*, V^* and T^*: (8.22) shows how these quantities provide information about the separation between molecules and about their binding energies.

This is only one of many examples which could be given of the use of dimensional analysis to obtain insight into physical processes where the detailed microphysics is non-trivial. Other pleasing examples include Lindemann's law of melting and critical phenomena in phase transitions, for example, the ferromagnetic transition. We will find an important application of the use of these techniques in our study of Wien's derivation of his displacement law (Section 11.3).

8.3 Introduction to chaos

One of the most remarkable developments in theoretical physics over the last 30 years has been the study of the chaotic dynamical behaviour found in systems which follow deterministic equations, such as Newton's laws of motion. These studies were foreshadowed by the work of mathematicians such as Poincaré, but only with the development of high-speed computers has the extraordinary richness of the subject become apparent. The character of chaotic systems can be appreciated by considering a dynamical system which is started twice, but from very slightly different initial conditions, say, due to a small error.

- In the case of a *non-chaotic system*, this small difference leads to an error in prediction which grows linearly with time.
- In a *chaotic system*, the difference grows exponentially with time, so that the state of the system is essentially unknown after a few characteristic times, despite the fact that the equations of motion are entirely deterministic.

Thus, chaos is quite different from randomness – the solutions are perfectly well defined, but the non-linearities present in the equations mean that the outcome is very sensitive indeed to the precise input parameters. The study of chaotic dynamical systems has rapidly become a heavy industry, but fortunately there are several excellent books on the subject. James Gleick's *Chaos: Making a New Science*[9] is a splendid example of a popular science book communicating the real essence of a technical subject with journalistic flair; a simple analytic introduction is provided by the attractive textbook by G.L. Baker and J.P. Gollub entitled *Chaotic Dynamics – An Introduction*.[10] My intention here is to describe briefly a few important aspects of these developments which indicate the richness of the field.

8.3.1 *The discovery of chaotic behaviour*

Chaotic behaviour in dynamical systems was discovered by Edward Lorenz of the Massachusetts Institute of Technology in the late 1950s. His interest was the study of weather systems and he derived a set of non-linear differential equations to describe their evolution. This was at the very dawn of the introduction of computers into the study of dynamical systems and, in 1959, he programmed his set of equations into his new Royal McBee electronic computer, which could carry out computations at a rate of 17 calculations per second. In his own words,

> The computer printed out the figures to three significant places, although they were carried in the computer to six, but we didn't think we cared about the last places at all. So I simply typed in what the computer printed out one time and decided to have it print it out in more detail, that is, more frequently. I let the thing run on while I went out for a cup of coffee and, a couple of hours later when it had simulated a couple of months of data, I came back and found it doing something quite different from what it had done before.

It became obvious that what had happened was that I had not started off with the same conditions – I had started off with . . . the original conditions plus extremely small errors. In the course of a couple of simulated months these errors had grown – they doubled about every four days. This meant that, in two months, they had doubled 15 times, a factor of 30 000 or so, and the two solutions were doing something entirely different.[11]

This is the origin of what became known popularly as the *butterfly effect*, namely, even a butterfly flapping its wings in one continent could cause hurricanes in some other part of the globe. This term can be regarded as a metaphor for the concept of *sensitive dependence upon initial conditions*. Lorenz simplified his equations to describe the interaction between temperature variations and convective motion:

$$\begin{aligned} \frac{\mathrm{d}x}{\mathrm{d}t} &= -\sigma x + \sigma y, \\ \frac{\mathrm{d}y}{\mathrm{d}t} &= xz + rx - y, \\ \frac{\mathrm{d}z}{\mathrm{d}t} &= xy - bz, \end{aligned} \tag{8.25}$$

where σ, r and b are constants. This system also displayed chaotic behaviour. Lorenz's computations marked the beginning of a remarkable story which involved a number of individuals working largely in isolation, who all came across different aspects of what we would now call chaotic behaviour. The systems studied included dripping taps, electronic circuits, turbulence, population dynamics, fractal geometry, the stability of Jupiter's great red spot, the dynamics of stellar orbits and the tumbling of Hyperion, one of the satellites of Saturn. This story is delightfully recounted in Gleick's book.

I particularly enjoy Baker and Gollub's short book, which describes very clearly many of the essential features of chaotic systems. As they discuss, the minimum conditions for chaotic dynamics are:

- the system has at least three dynamical variables;
- the equations of motion contain a non-linear term coupling several of the variables together.

These features are satisfied by Lorenz's equations (8.25) for convective motions – there are three variables x, y, z and the coupling between them takes place through the non-linear terms xz and xy on the right-hand sides of the last two equations. These conditions are necessary in order to obtain the following distinctive features of chaotic behaviour:

- the trajectories of the system in three dimensions diverge;
- the motion is confined to a finite region of the phase space of the dynamical variables;
- each trajectory is unique and does not intersect any other trajectory.

It is simplest to give a specific example of how a dynamical system can become chaotic – Baker and Gollub's analysis of a damped driven pendulum illustrates elegantly how chaotic motion comes about.

8.3.2 *The damped driven pendulum*

Consider a damped sinusoidally driven pendulum of length l with a bob of mass m and damping constant γ. The pendulum is driven at angular frequency ω_D. We have already written down an equation of motion for the non-linear pendulum, (8.7), which does not include damping or driving terms:

$$mgl(1 - \cos\theta) + \tfrac{1}{2}I\dot{\theta}^2 = \text{constant}. \tag{8.26}$$

Setting $I = ml^2$ and differentiating with respect to time,

$$ml\ddot{\theta} = -mg\sin\theta. \tag{8.27}$$

The damping force is proportional to the speed of the pendulum bob and so can be included as a term $-\gamma l\dot{\theta}$ on the right-hand side of (8.27). Likewise, the driving force to be included on the right-hand side can be written $A\cos\omega_D t$. The equation of motion is therefore

$$ml\frac{d^2\theta}{dt^2} + \gamma l\frac{d\theta}{dt} + mg\sin\theta = A\cos\omega_D t. \tag{8.28}$$

Changing variables, Baker and Gollub rewrite (8.28) in simplified form:

$$\frac{d^2\theta}{dt^2} + \frac{1}{q}\frac{d\theta}{dt} + \sin\theta = \alpha\cos\omega_D t, \tag{8.29}$$

where q is the *quality factor* of the pendulum. The equation has been normalised so that the natural angular frequency of small oscillations of the pendulum ω is 1. It is simplest to regard angular frequencies and time as dimensionless. In the cases we are about to consider, the amplitudes of the oscillations can be very large. The pendulum wire has to be regarded as stiff when the angular deviations exceed $\theta = \pm\pi/2$.

Equation (8.29) can be written as three independent first-order equations as follows:

$$\begin{aligned} \frac{d\omega}{dt} &= -\frac{1}{q}\omega - \sin\theta + \alpha\cos\varphi, \\ \frac{d\theta}{dt} &= \omega, \\ \frac{d\varphi}{dt} &= \omega_D. \end{aligned} \tag{8.30}$$

These equations contain three independent dynamical variables, ω, θ and φ:

- ω is the instantaneous angular velocity;
- θ is the angle of the pendulum with respect to its vertical equilibrium position;
- φ is the phase of the oscillatory driving force, which has constant angular frequency ω_D.

The non-linear couplings between these variables are introduced through the terms $\sin\theta$ and $\alpha\cos\varphi$ in the first of the three equations. Whether the motion is chaotic depends upon the values of α, ω_D and q.

One of the problems in representing chaotic behaviour is that the diagrams necessarily have to be three dimensional in order to represent the evolution of all three dynamical variables. Fortunately, in this case the phase of the driving oscillation is constant and so the dynamical behaviour can be followed in a simple phase diagram in which the instantaneous angular velocity ω is plotted against θ. Such diagrams for four values of α and q are shown in Fig. 8.6 and repay detailed study. The phase diagrams have all been redrawn to the same scale. The behaviour of the pendulums in real space is shown in the left-hand column, as well as the corresponding values of α and q.

Considering first the simplest example of a *moderately driven pendulum*, with $\alpha = 0.5$ and $q = 2$; the phase diagram shows simple periodic behaviour in a steady state, the pendulum swinging between $\theta = \pm 45^\circ$. Inspection of the phase diagram shows that, as required physically, at $\theta = \pm 45^\circ$ the angular velocity ω is zero. This closed-loop behaviour is familiar from the simple pendulum. In the terminology of dynamical systems, the origin of the phase diagram is called an *attractor*. This terminology can be understood from the observation that, if there were no driving term, the trajectory of the damped pendulum in the phase diagram would spiral in towards the origin. The trajectory shown in Fig. 8.6(a) is known as a *limit cycle*.

The three other diagrams show the periodic behaviour of the pendulum when the amplitude of the driving oscillator is large and the pendulum has reached a steady state.

- In Fig. 8.6(b), $\alpha = 1.07$ and $q = 2$, the pendulum follows a double-looped trajectory in the phase diagram, in one swing exceeding π radians from $\theta = 0$.
- At an even larger amplitude (Fig. 8.6(c)), $\alpha = 1.35$ and $q = 2$, the pendulum performs a complete loop about the origin in real space. In the phase diagram, the ends of the trajectory at $\pm\pi$ radians should be joined up.
- At an even greater driving amplitude (Fig. 8.6(d)), $\alpha = 1.45$ and $q = 2$, the pendulum performs two complete loops in real space in returning to its initial position in the steady state. Again, the end points at $\pm\pi$ radians in the phase diagram should be joined up. It is a useful exercise to trace the behaviour of the pendulum in real and phase space.

At some other values of α and q, there is, however, no steady state periodic behaviour. In Fig. 8.7, the phase diagrams for $q = 2$ and $\alpha = 1.15$ and 1.50 are shown. In both cases, the motion is chaotic in the sense that the trajectories are non-periodic and are very sensitive to the initial conditions.

There are other revealing ways of presenting the information given by the phase diagrams. One way of characterising the behaviour of the pendulum is to plot on the phase-space diagrams only the positions of the pendulum at times that are *multiples of the period of the forcing oscillation*, that is, we make *stroboscopic pictures* of the ω- and θ- coordinates at the forcing frequency ω_D. This form of presentation is called a *Poincaré section* and enables much of the complication of the dynamical motion to be eliminated.

For a simple pendulum, there is only one point on the Poincaré section. The examples shown in Fig. 8.8(a), (b) correspond respectively to the phase diagrams in Fig. 8.6(d) and Fig. 8.7(b). In Fig. 8.8(a) the phenomenon of period doubling is apparent. This is the type of non-linear phenomenon found, for example, in dripping taps.

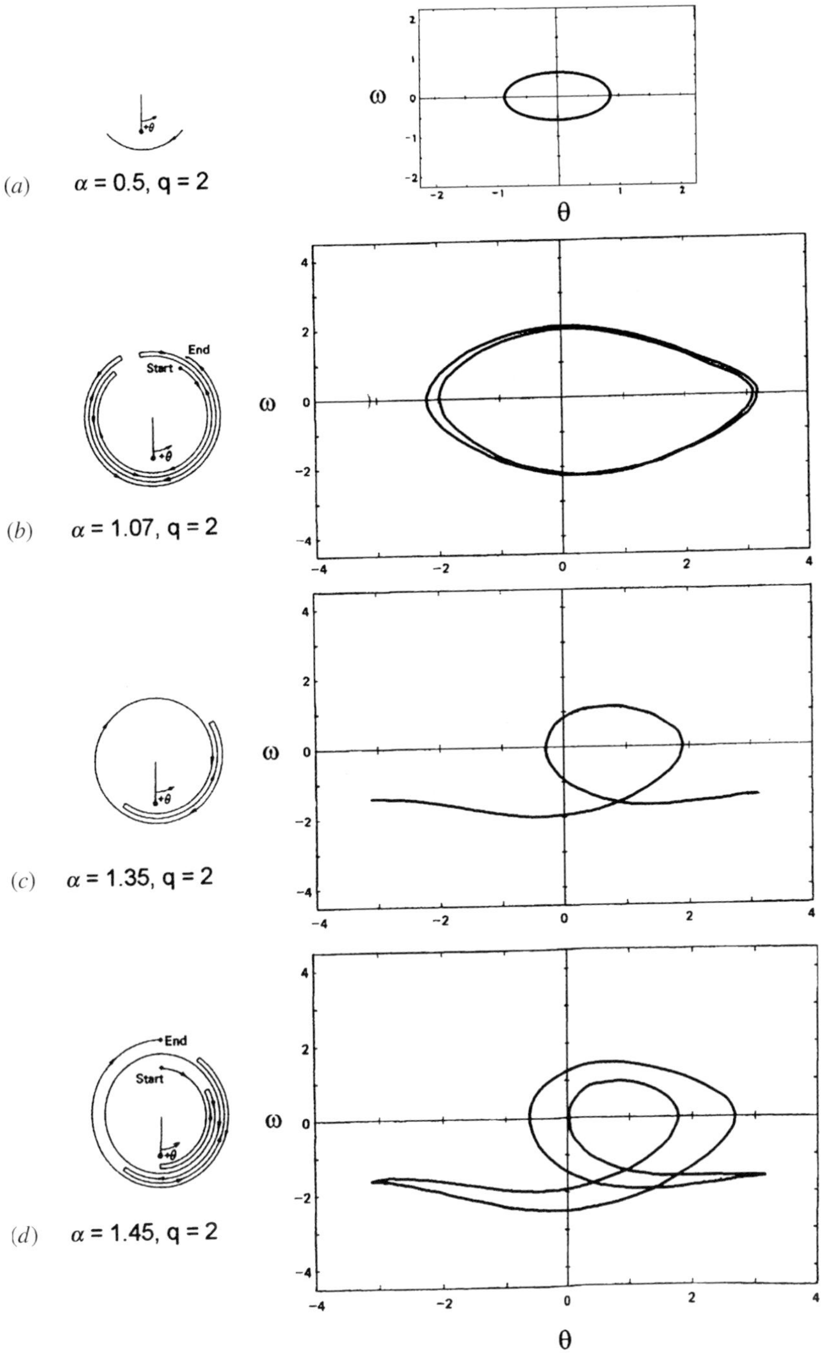

Figure 8.6: $\omega\theta$ phase diagrams for the non-linear pendulum for different values of the quality factor q and the amplitude α of the driving oscillation. (After G.L. Baker, and J.P. Gollub, 1990, *Chaotic Dynamics – An Introduction*, pp. 21–2, Cambridge: Cambridge University Press.)

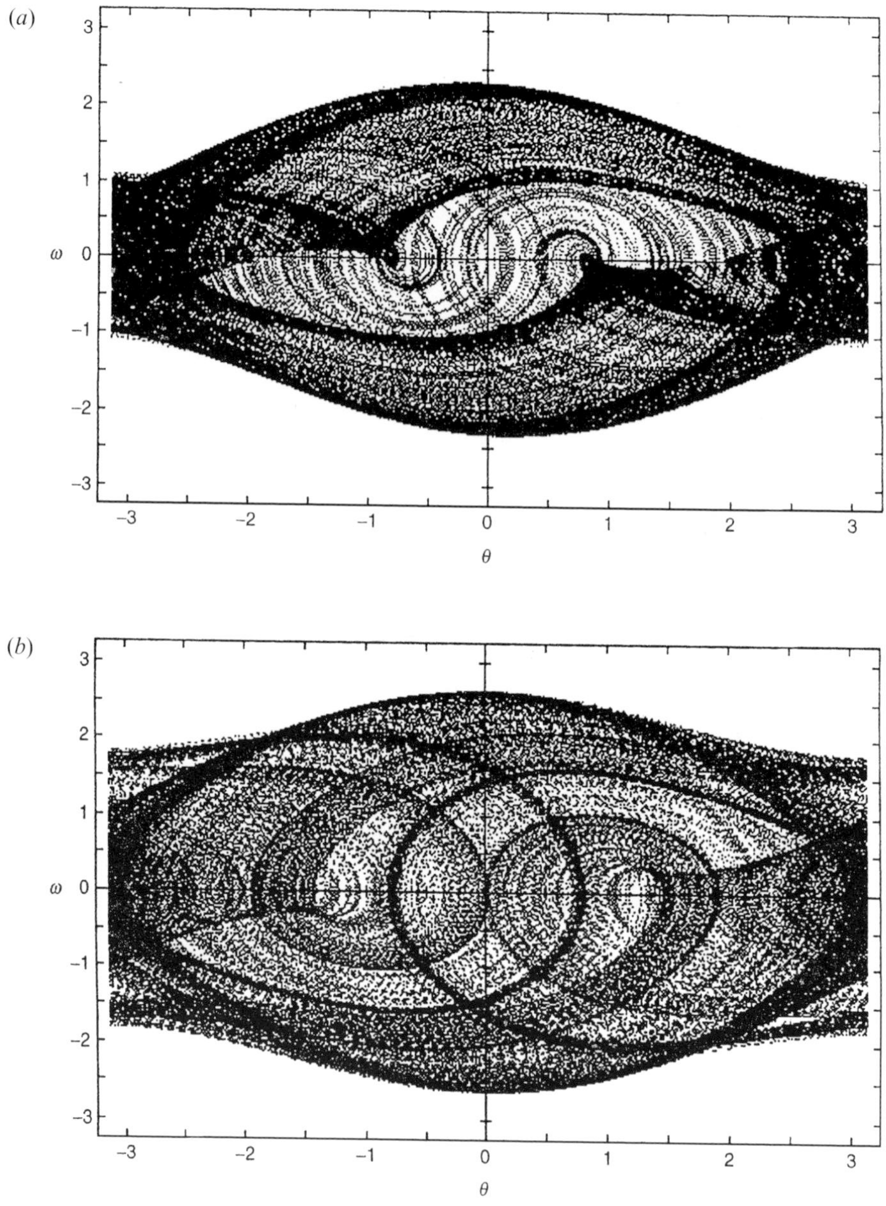

Figure 8.7: Phase diagrams for the non-linear pendulum for $q = 2$ and $(a)\ \alpha = 1.15$ and $(b)\ \alpha = 1.50$ are shown. In both cases, the motion is chaotic. (From G.L. Baker and J.P. Gollub, 1990, *Chaotic Dynamics – An Introduction*, pp. 49 and 53, Cambridge: Cambridge University Press.)

In the chaotic example, Fig. 8.8(*b*), the trajectories never cross the reference plane at the same point, but there is regular structure. For chaotic motion, the structure of the Poincaré sections is self-similar, that is, the same on different scales. As discussed by Baker and Gollub, the geometrical structure of the diagrams is now fractal and a fractional fractal

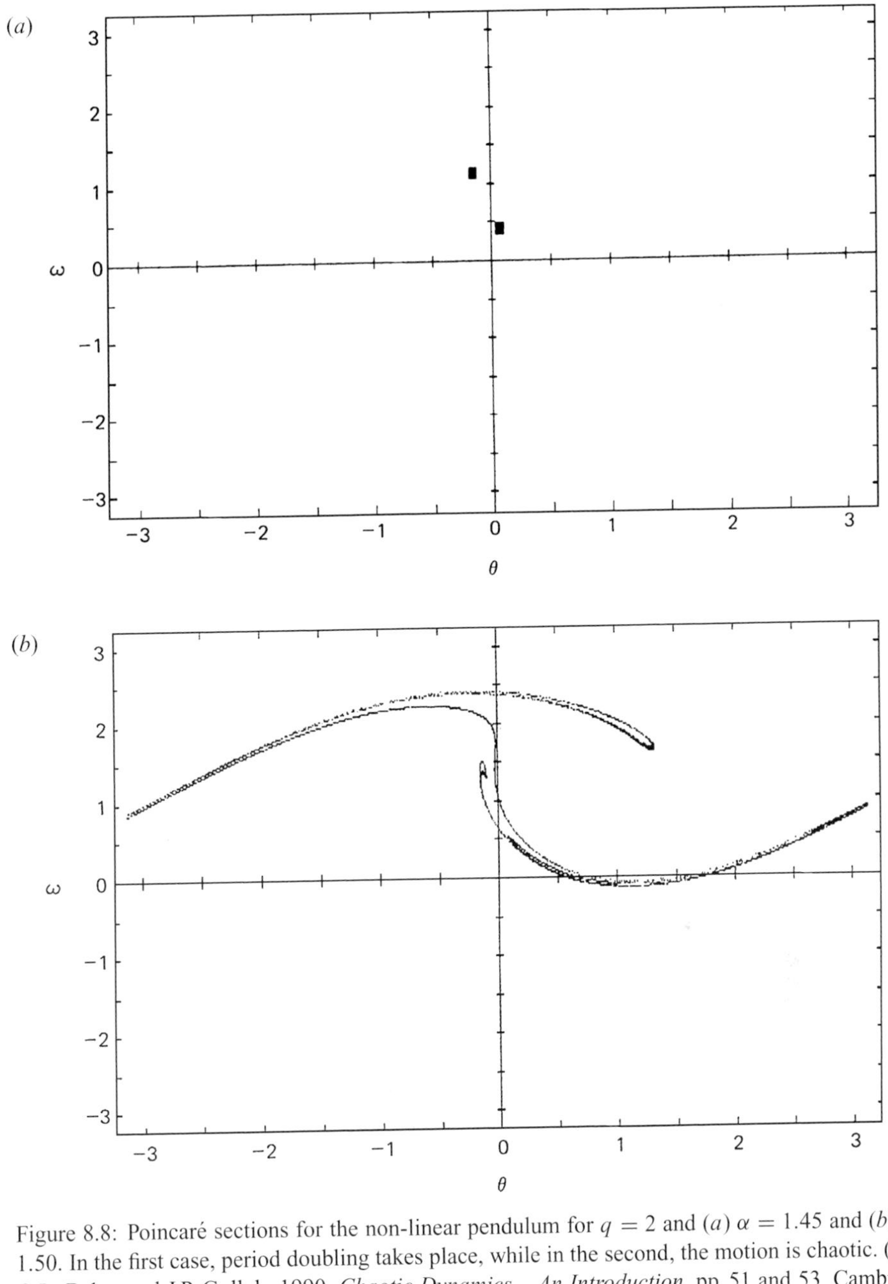

Figure 8.8: Poincaré sections for the non-linear pendulum for $q = 2$ and (a) $\alpha = 1.45$ and (b) $\alpha = 1.50$. In the first case, period doubling takes place, while in the second, the motion is chaotic. (From G.L. Baker and J.P. Gollub, 1990, *Chaotic Dynamics – An Introduction*, pp. 51 and 53, Cambridge: Cambridge University Press.)

dimension can be defined. They describe how the fractal dimension of the phase-space trajectories can be calculated.

Another powerful technique for analysing the motion of the pendulum is to take the *power spectrum* of its motion in the time domain. In the case of periodic motion, the power

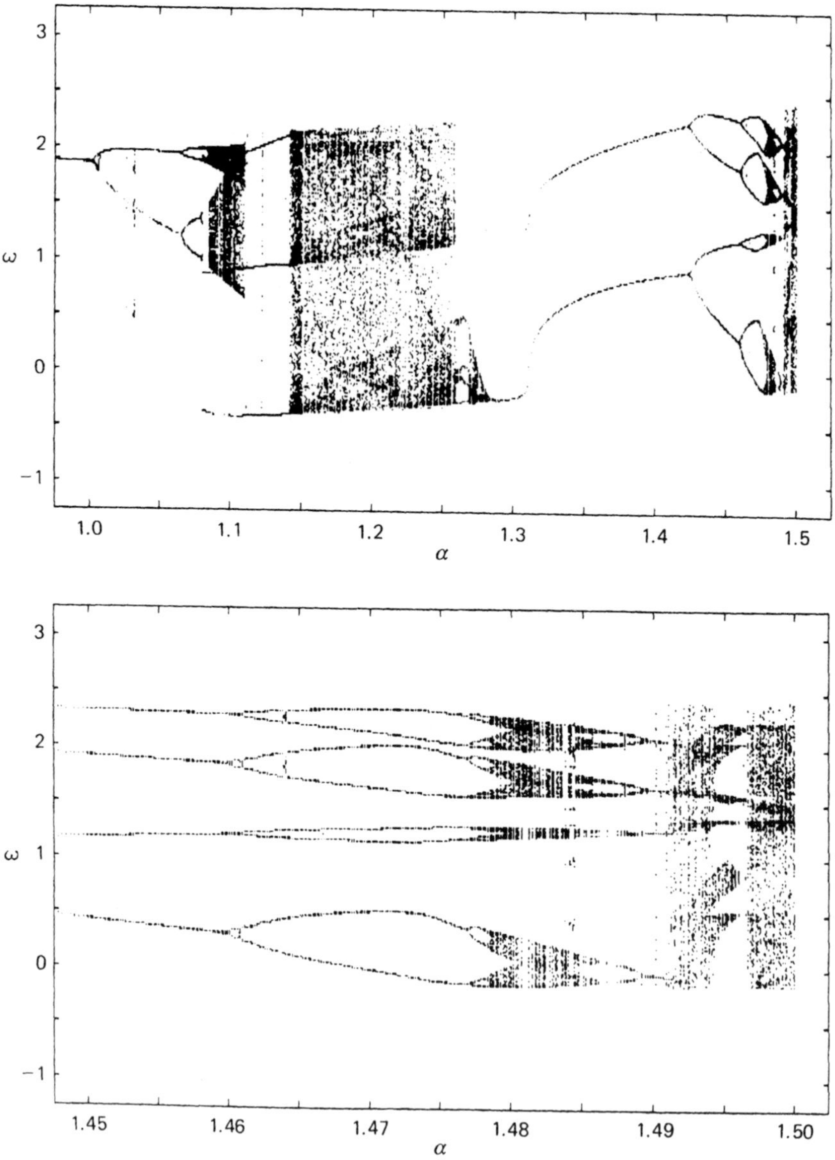

Figure 8.9: The bifurcation diagram for a damped driven pendulum with $q = 2$ for a range of values of the amplitude of the driving oscillation α. The values of ω are those at the beginning of each driving cycle. The lower diagram is an expanded version of the upper diagram for the parameter range $1.45 \leq \alpha \leq 1.50$. (From G.L. Baker and J.P. Gollub, 1990, *Chaotic Dynamics – An Introduction*, p. 69, Cambridge: Cambridge University Press.)

spectra display spikes corresponding to period doublings, quadruplings and so on. In the case of chaotic systems, broad-band power spectra are found.

Finally, the overall behaviour of the pendulum for different magnitudes of the forcing amplitude α can be displayed on a *bifurcation diagram* (Fig. 8.9). On the vertical axis, the magnitude of the pendulum's angular velocity at a fixed phase in each period of the driving

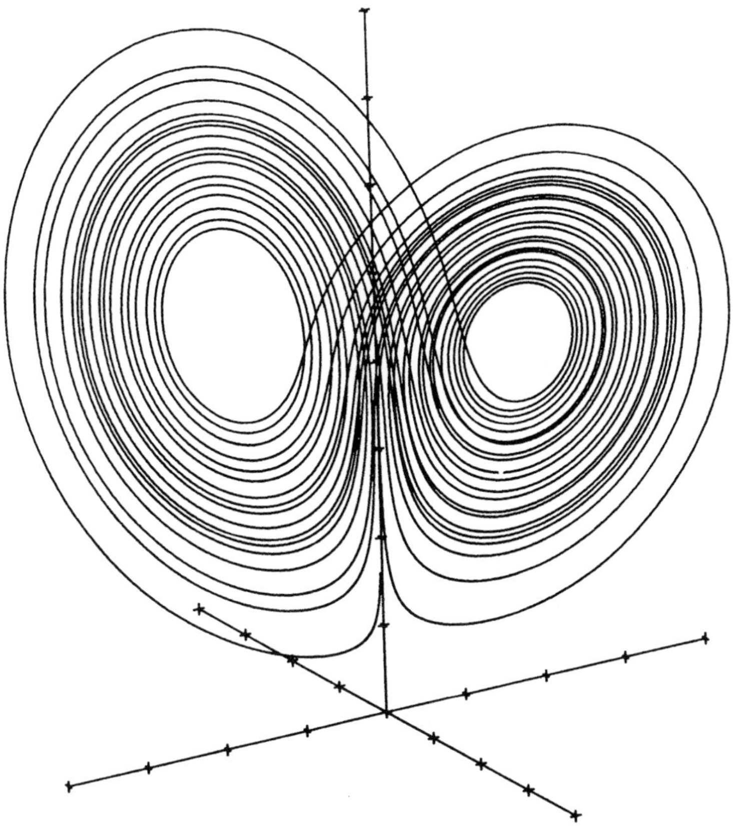

Figure 8.10: Part of a single trajectory on a Lorenz-attractor diagram, which is a three-dimensional representation of the evolution of a chaotic system with three independent variables. Each trajectory of the system is represented by a continuous line, which never intersects itself. Different trajectories never cross. Particles on neighbouring trajectories diverge rapidly resulting in chaotic motion. (From J. Gleick, 1987, *Chaos: Making a New Science*, p. 28, New York: Viking.)

frequency is shown. The bifurcation diagram shown in Fig. 8.9 (the lower diagram is an expansion of part of the upper diagram) displays the behaviour of the pendulum for the examples we have already studied with $q = 2$, but with a continuous range of values of α. It is useful to compare the behaviour of the pendulum in real and phase space in Figs. 8.6 and 8.7 with their location on the bifurcation diagram. Figure 8.9 (lower diagram) makes the point very clearly that the onset of chaotic behaviour of the pendulum proceeds through by a sequence of frequency doublings. This behaviour has also been observed in laboratory experiments on turbulence. Fig. 8.9 is a quite extraordinary diagram showing ranges of parameters in which the system has well-defined modes of periodic oscillation in between regions in which the behaviour is chaotic.

All the diagrams presented above are ways of representing on a two-dimensional sheet of paper what are at their simplest three-dimensional diagrams. It is instructive to study the dynamical motion of Lorenz's three-parameter model of convection in three dimensions (Fig. 8.10). For a given set of input parameters, a unique trajectory is traced out in the three-dimensional phase space, as illustrated by the continuous line in the diagram. For a slightly different set of initial conditions, a similar trajectory would be traced out initially, but it would rapidly diverge from that shown. Notice that the evolution of the system is confined to a finite region of parameter space. In this representation, there are two attractors and the system can pass from the sphere of influence of one attractor to the other. Each is an example of a *strange attractor*.

8.3.3 Logistic maps and chaos

Remarkably, chaotic behaviour was also discovered by an entirely separate line of development. Robert May was working on the problems of population growth and, in particular, how a biological population reaches a stable state. The nth generation with X_n members results in the $(n+1)$th generation with X_{n+1} members, and so we can write

$$X_{n+1} = f(X_n). \tag{8.31}$$

In a simple picture, we might imagine, for example, that the average number of offspring for a pair of human parents was the traditional 2.4 or some other number, which we will call μ. Then, we would expect that if $X_{n+1} = \mu X_n$, there would be unlimited, unsustainable growth in the size of the population. Therefore some form of cut-off to limit the size of the population to a finite value N is necessary. A mathematical way of introducing the cut-off would be to modify the recurrence relation to

$$x_{n+1} = \mu x_n(1 - x_n) \tag{8.32}$$

where x_n is a *population measure* defined by

$$x_n = X_n/N.$$

Equation (8.32) is a *difference equation*, similar to the differential equation introduced by P.F. Verhulst to model the development of populations:

$$\frac{dx}{dt} = \mu x(1 - x). \tag{8.33}$$

The behaviour of the population can be modelled on diagrams known as *logistic maps*, such as those illustrated in Fig. 8.11. In each of these diagrams, the abscissa represents the population measure x_n of the nth population and the ordinate that of the $(n+1)$th generation. This latter number becomes the next value on the abscissa, to which (8.32) is next applied. In the first case (Fig. 8.11(a)), the value of μ is taken to be 2 and the sequence starts with the value $x_1 = 0.2$. Then, the next generation has population measure $x_2 = 0.32$. Applying the formula (8.32), the successive generations have population measures $x_3 = 0.4352$, $x_4 = 0.4916$, $x_5 = 0.4999$, settling down to a stable value with $x_n = 0.5$.

In the second case, the value of μ is 3.3 and the sequence of generations shown in of Fig. 8.11(b) is obtained. Starting with the value $x_1 = 0.2$, the subsequent generations have population measures $x_2 = 0.528$, $x_3 = 0.8224$, $x_4 = 0.4820$, $x_5 = 0.8239$, $x_6 = 0.4787, \ldots$; the population measure x ultimately settles down to a value which

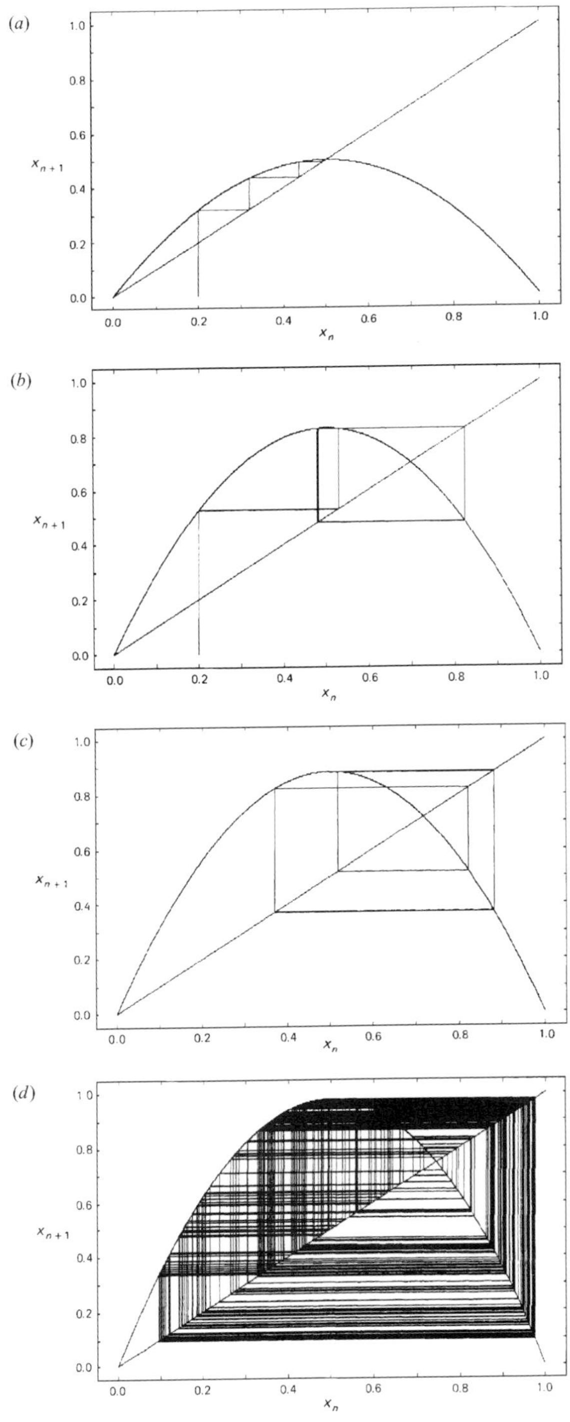

Figure 8.11: Logistic maps for the recurrence relation $x_{n+1} = \mu x_n(1 - x_n)$ for $\mu = 2, 3.3, 3.53$ and 3.9.

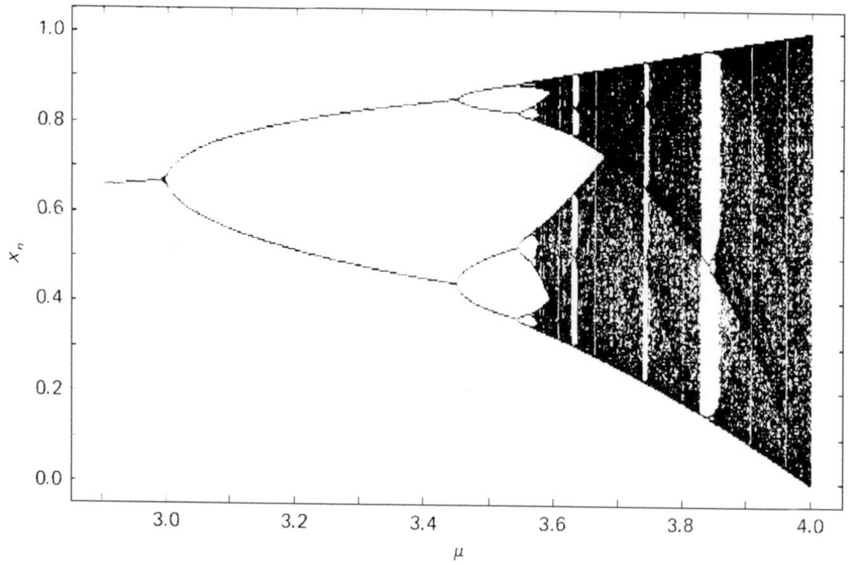

Figure 8.12: A bifurcation diagram for the logistic maps of $x_{n+1} = \mu x_n(1 - x_n)$. (From G.L. Baker, and J.P. Gollub, 1990, *Chaotic Dynamics – An Introduction*, p. 69, Cambridge: Cambridge University Press.)

oscillates between the values $x_n = 0.48$ and $x_{n+1} = 0.83$ for even n. In other words, the population undergoes a *limit cycle* between these values.

In the third example, $\mu = 3.53$ (Fig. 8.11(c)), the first six values of the population measure are $x_1 = 0.2, x_2 = 0.5648, x_3 = 0.8677, x_4 = 0.4053, x_5 = 0.8508, x_6 = 0.4480$. Eventually, the system settles down to a limit cycle between four stable states with values $x_n = 0.37$, $x_{n+1} = 0.52$, $x_{n+2} = 0.83$ and $x_{n+3} = 0.88$. We recognise in Figs. 8.3(b), (c) the phenomena of period doubling and quadrupling, as was found for the damped forced pendulum.

In the fourth example, with $\mu = 3.9$ (Fig. 8.11(d)), the system becomes chaotic. Just as in the case of the pendulum, we can create a bifurcation diagram to illustrate the stable values of the population measure (Fig. 8.12). It is remarkable that the bifurcation diagrams of Fig. 8.9 and 8.12 are so similar. Clearly, there are universal features of the processes of period doubling and the approach to chaotic behaviour in both examples.

In a remarkable analysis, Mitchell Feigenbaum discovered that for such maps there are universal regularities in the approach to chaotic behaviour through period doubling. He showed that the ratio of spacings between consecutive values of μ at bifurcations tends to a universal constant, the *Feigenbaum number*. If the first bifurcation occurs at μ_1, the second at μ_2 and so on, the Feigenbaum number is defined to be

$$\lim_{k\to\infty} = \frac{\mu_k - \mu_{k-1}}{\mu_{k+1} - \mu_k} = 4.669\,201\,609\,102\,990\,9\cdots$$

This number is a universal constant for the period-doubling route to chaos for maps with a quadratic maximum. It is an example of *universality* in chaotic dynamical systems, by which is meant that certain features of non-linear maps are independent of the particular form of map.

There are many applications of these ideas to complex physical systems. These include the tumbling motion of the satellite Hyperion in orbit about Saturn, turbulence, lasers, chemical systems, electronic circuits with non-linear elements, and so on. These are very dramatic developments. The ideas of chaos in non-linear systems now pervade many frontier areas of modern physics – it is no longer a mathematical curiosity. A key point is that there are universal features which underlie all chaotic phenomena and which are now an integral part of physics.

8.4 Scaling laws and self-organised criticality

The chaotic phenomena described in Section 8.3 reveal underlying regularities in non-linear dynamical systems. The examples studied were relatively simple and describable by a limited number of non-linear differential equations. There are, however, many examples in nature in which non-linearities of many different forms are present, and there is little hope that we can write down the appropriate equations, let alone solve them. Nonetheless, it is found that regularities appear. How does this come about? The emerging field of *self-organised criticality* may contain the seeds of a new type of physics which can begin to tackle these problems. Fortunately, a splendid book on this subject has been written by Per Bak, entitled *How Nature Works – The Science of Self-Organised Criticality*.[12] We will follow his presentation and some of his examples as a brief introduction to the key ideas.

8.4.1 *Scaling laws*

The germ of the idea for this new approach to phenomena that are multiply non-linear originated from the discovery that some remarkable *scaling laws* are found in complex systems. The common feature of all these examples is that there is no way in which we can write down the equations which might describe all relevant aspects of the behaviour of the system, and yet regularities appear. Let us list some of the more remarkable examples discussed by Bak.

(i) A typical example of a scaling law is the *Gutenberg–Richter law* for the probability of occurrence of earthquakes of different magnitude on the Richter scale. As an example, the earthquake statistics for the New Madrid zone in the south-east USA were studied by Arch Johnson and Susan Nava for the period 1974–83, using the Richter scale as a measure of the magnitude of the earthquake; the Richter scale is logarithmic in the energy release E of the earthquake. They confirmed the Gutenberg–Richter law that the annual probability of occurrence of an earthquake with magnitude greater than m is a power law in the energy E (Fig. 8.13). On the face of it, it is quite extraordinary that there should be any simple relation between the probability of an earthquake occurring and its magnitude. Consider all the immensely complicated phenomena which can affect the occurrence of an earthquake – my non-expert list would include plate tectonics, mountains, valleys, lakes, geological structures, and so on. Whatever actually happens, it is evident that the processes

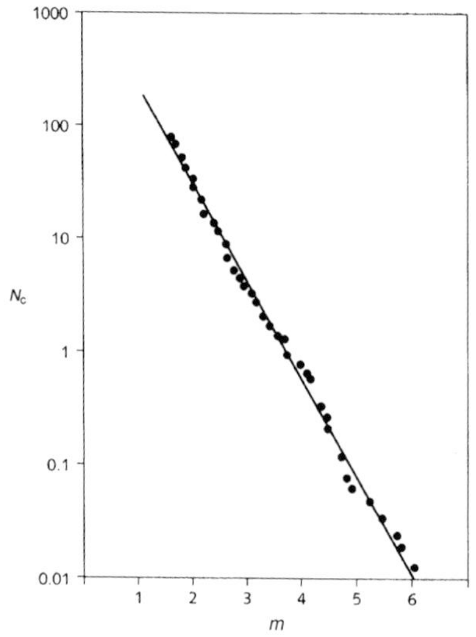

Figure 8.13: The number of earthquakes per year N_c with magnitude greater than m on the Richter scale, plotted against m. (From P. Bak, 1997, *How Nature Works – The Science of Self-Organised Criticality*, p. 13, Oxford: Oxford University Press.)

are extremely non-linear, and yet the linearity of the Gutenberg–Richter law suggests that, despite the complexity, there is some underlying systematic behaviour: large earthquakes are just extreme examples of smaller earthquakes.

(ii) In 1963, Benoit Mandelbrot analysed the *monthly variation in cotton prices* over a 30-month period and found that the distribution was very 'spikey'. He found that, when he measured the distribution of the *changes* in the cotton price from month to month, these variations followed a power-law distribution. Other commodities, and hence the stock market itself, also followed this pattern. This really is a horribly non-linear problem. Consider the many non-linear effects which determine the cotton price – supply, demand, rushes on the stock market, speculation, psychology, insider dealing, and so on. How could these result in a simple power-law behaviour?

(iii) John Sepkopski spent 10 years in libraries and museums studying *biological extinctions*. He made an enormous survey of all the marine fossil records available to him in order to determine the fractions of biological families present on the Earth which had become extinct over a period of 600 million years. The data were grouped into periods of 4 million years and he determined what fraction of all species disappeared during each of these periods. The data were again very 'spikey', with huge variations in the extinction rate. Nonetheless, David Raup pointed out that the probability distribution of the fraction of extinctions in each 4-million-year period followed a power-law distribution, with a few large extinction events, when many species disappeared, and larger numbers of smaller events. It is intriguing that the famous epoch of the extinction of the dinosaurs, which occurred about 65 million years before the present epoch, was by no means the largest extinction event in the fossil record. The extinction of the dinosaurs may well have been associated in some way with the collision of an asteroid with the Earth – the 180-km crater resulting from that collision was recently identified at Chicxulub on the Yutacán peninsula of the Gulf of Mexico.

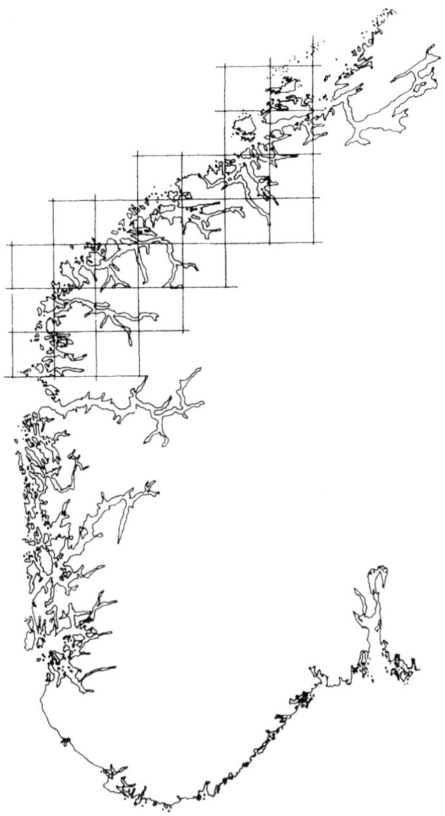

Figure 8.14: The coast of Norway showing how the fractal nature of its coastline can be quantified by determining the number of boxes of side δ required to enclose it. In this example, 26 boxes are needed to enclose the region shown. (From P. Bak, 1997, *How Nature Works – the Science of Self-Organised Criticality*, p. 13. Oxford: Oxford University Press. After J. Feder, 1988, *Fractals*, New York: Plenum Press. Courtesy of J. Feder and colleagues, University of Oslo.)

(iv) The term *fractal* was introduced by Benoit Mandelbrot to describe geometrical structures with *similar features on all length scales*. In his memoirs, he confesses that he has always been interested in describing mathematically the 'messiness' of the world rather than its obvious regularities. In his great book *The Fractal Geometry of Nature*,[13] he made the profound observation that the forms found in nature are generally of a fractal character. A classic example is the coastline of Norway. Being buffetted by the Atlantic and the severe northern climate, Norway has a very ragged shoreline containing fjords, fjords within fjords and so on. The simplest way of quantifying its fractal nature is to ask how many square boxes of side δ are needed to enclose the complete coastline (Fig. 8.14). Remarkably, the number of boxes $L(\delta)$ follows a power-law distribution

$$L(\delta) \propto \delta^{1-D},$$

where $D = 1.52 \pm 0.01$ is known as the *fractal* or *Hausdorff dimension* of the distribution. Notice an important feature of this distribution: because it is a power law there is no natural length scale with which to identify on what scale we might be looking at a photograph of the coastline. To put it another way, the coastline displays the same appearance of irregularity no matter what scale we select.

(v) A similar scaling law is found in the *large-scale distribution of galaxies in the Universe*. Galaxies have a strong tendency to be found in groups or clusters. These in turn are clustered into associations known as superclusters of galaxies. The simplest way of describing the clustering properties of galaxies is by means of a two-point correlation function defined by the relation

$$N(\theta)\,d\Omega = N_0[1 + w(\theta)]\,d\Omega,$$

where N_0 is a suitably defined average surface density and $w(\theta)$ describes the excess probability of finding a galaxy at angular position θ within the solid angle $d\Omega$ on the sky. Observations of large samples of galaxies have shown that the function $w(\theta)$ has the form

$$w(\theta) \propto \theta^{-0.8},$$

corresponding to a spatial correlation function

$$\xi(r) \propto r^{-1.8} \qquad \text{on physical scales } r \leq 8 \text{ Mpc}; \tag{8.34}$$

1 Mpc $\approx 3 \times 10^{22}$ m. Thus, although clusters of galaxies are prominent structures in the Universe, clustering occurs on all scales and there are no preferred scales, just as in a fractal distribution. On scales $r \gg 8$ Mpc, the strength of the clustering decreases more rapidly with increasing scale than (8.34) – on these scales the density contrast $\delta\rho/\rho \ll 1$, and so structures with these enormous dimensions are still in their linear stage of development. On scales less than 8 Mpc, the density contrasts are greater than unity and consequently involve non-linear gravitational interactions.[18]

(vi) $1/f$ *noise* is found in many different circumstances. It was originally discovered in electronic circuits, in which it manifests itself as voltage spikes far exceeding in size what would be expected from random Johnson noise (see the appendix to Chapter 15). As my colleague Adrian Webster once remarked, '$1/f$ noise is the spectrum of trouble'. Its origin has not been established. Two typical examples of $1/f$ noise are the global temperature variations since 1865, when such information was first available, and the time variability of the luminosities of quasars. They are qualitatively similar. Generally, the $1/f$ noise spectrum has the form

$$I(f) \propto f^{-\alpha} \qquad \text{where} \qquad 0 < \alpha < 2$$

and f is the frequency. In both cases, the observed behaviour is thought to be the result of a large number of non-linear processes which end up producing this systematic behaviour. In the case of electronic circuits, precautions must be taken so that the occasional presence of these large spikes does not upset the operation of the device.

(vii) In his book *Human Behaviour and the Principle of Least Effort*,[14] George Kingsley Zipf noted a number of remarkable scaling laws in the patterns of human endeavour. For example, he showed that there is systematic behaviour in the ranking of *city sizes*, just as there is in the case of earthquakes. The cities of the world can be ranked in order of their populations and it is found that city population obeys a power-law distribution similar to the Gutenberg–Richter law. Obviously, a large number of non-linear physical and social effects determine the population and hence the size of any city, and yet these all conspire to create a power-law distribution of size.

(viii) Zipf also showed that the frequency with which words are used in the English language follows a power law. Words in the English language can be ranked in order of their frequency of use. For example, 'the' is the most common, followed by 'of', 'and', 'to' and so on. What are the myriad of social and perceptual processes which result in a power-law rather than some other distribution?

(ix) In discussing these distributions with Benoit Mandelbrot, I was intrigued to learn that the classical music of Haydn, Mozart and Beethoven follows a $1/f$ noise distribution, while the music of Boulez and Stockhausen is not fractal!

In all these examples, massively non-linear phenomena are important and there appears at first sight to be little prospect of finding any analytic theory to account for the appearance of regular scaling laws in the real world about us. And yet there is hope from a somewhat unexpected direction.

8.4.2 *Sand piles and rice piles*

A clue to understanding these different examples of complex non-linear systems may be found in the behaviour of sand or rice piles. Every child is familiar with the problems of building sandcastles. If the walls of a sandcastle become too steep, avalanches of various sizes occur until its angle is reduced to some particular value. In a controlled experiment, dry sand is poured steadily onto the top of a pile and the sand pile maintains its regular conical shape by undergoing a continuous series of avalanches of different sizes. The remarkable result is that, as it grows, the angle of a particular sand pile remains the same. This example has become the archetype for the processes involved in what is called *self-organised criticality*. A key point is that it is not feasible to work out analytically the physics of the avalanches which occur in sand or rice piles. There are just too many non-linear phenomena coming into play simultaneously. Figure 8.15 shows how complex it would be to try to make any prediction for a rice pile.

Per Bak and his colleagues have constructed a remarkable computer simulation which may help to explain the behaviour of rice and sand piles. They illustrate the principles by a simple model consisting of a 5×5 grid of squares. Zero, 1, 2 or 3 grains are placed randomly on each square of the grid to begin with. Then, grains are added randomly to the grid with the following rules. When a square has four grains, these are redistributed, one to each of the four immediately neighbouring squares, and the number on the original square is set to zero. If this process results in any of the neighbouring squares having four grains then the process is repeated, and this sequence is repeated over and over again until there are no more than three grains on each square – at this point, the avalanche has stopped. Figure 8.16 illustrates this process for a specific initial configuration of grains. One grain is added to the central square and then the evolution takes place through the next seven iterations as shown. In this particular example, it can be seen that the single grain added to the centre results in nine toppling events in seven iterations of the algorithm. The solid area of eight squares in the centre of the grid at the bottom of Fig. 8.16 shows those squares which have suffered at least one toppling event.

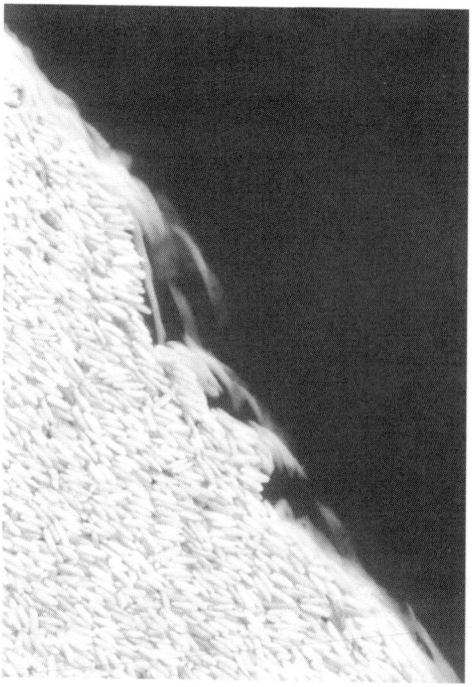

Figure 8.15: Illustrating the mini-avalanches involved in the formation of a rice pile. (From P. Bak, 1997, *How Nature Works – The Science of Self-Organised Criticality*, Plate 4, Oxford: Oxford University Press, from V. Frette *et al.* (1995). *Nature*, **379**, 49.)

Exactly the same procedure has been repeated for much larger arrays, such as the 50×50 grid shown in Fig. 8.17. The grey regions represents boxes in which at least one toppling event has taken place. The four frames (a), (b), (c) and (d) represent a single avalanche at different stages in its evolution.

Although this is a highly idealised model, it has a number of intriguing and suggestive features.

- If the simulations are run many times, the probability distribution of the sizes of the avalanches can be found. This distribution turns out to be of power-law form, of exactly the same form as the Gutenberg–Richter law for earthquakes.
- The rice pile becomes 'critical' without any fine-tuning of the parameters.
- The outline of the perimeter of the rice pile is very ragged and the same type of analysis as is used to show that the coastline of Norway is fractal shows that the outline of the rice pile is also fractal.
- The rules governing the redistribution of the rice grains can be changed quite significantly, but the same type of generic behaviour is found.
- It has not been possible to find an analytic model for any self-organised critical system.

These are quite remarkable results. The last bullet point is particularly interesting. Problems of this type could only be tackled once high-speed computers were available, so that very large numbers of simulations could be run, with computational tools able to analyse the data. Bak gives many other intriguing examples in his remarkable book. The same types of

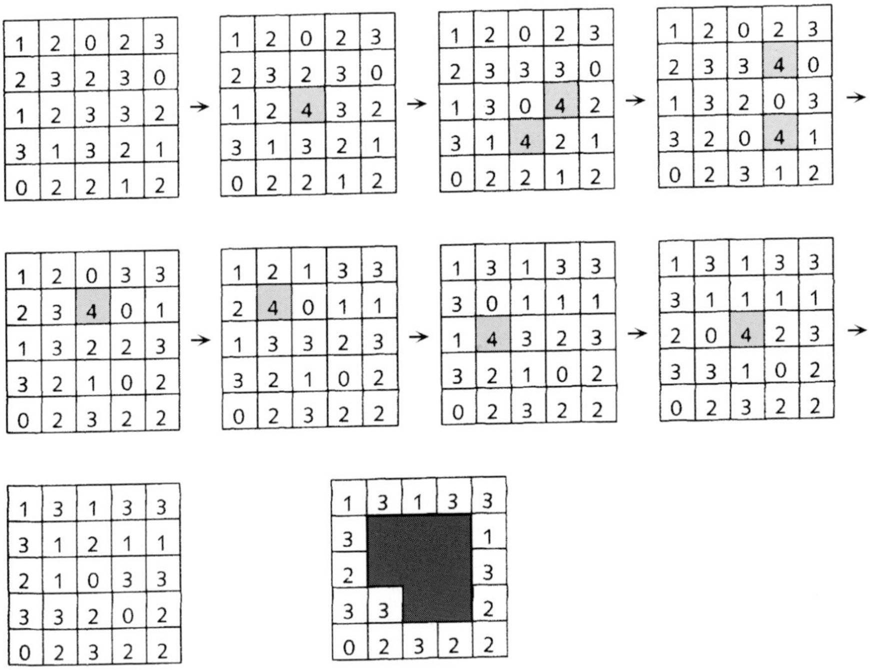

Figure 8.16: Illustrating the evolution of a mini-avalanche in a computer model of a rice pile. (From P. Bak, 1997, *How Nature Works – The Science of Self-Organised Criticality*, p. 53, Oxford: Oxford University Press. Reprinted by permission of Oxford University Press.)

analysis have been used to study the acoustic emission from volcanoes, crystal-cluster or colloid growth, forest fires, cellular automata and so on. The common characteristic of these self-organised critical phenomena is that they involve massively non-linear events and yet end up producing scaling laws with no preferred scales. Bak makes the case that this type of behaviour is central to the processes of evolution in nature.

8.5 Beyond computation

It might seem that this would be about as far as one could go in a physics text about the use of computation to solve complex problems; and yet there might be even more extreme forms of problem, which go beyond computation. These thoughts are inspired by Roger Penrose's controversial ideas expounded in his books *The Emperor's New Mind* and *Shadows of the Mind*.[15,16] The inspiration comes from a somewhat unexpected source and involves the problem of tiling an infinite plane by small numbers of tiles of fixed geometric shapes. It is obvious that an infinite plane can be tiled by simple tiles of regular geometry, for example, by squares, triangles, and so on. What is much less obvious is that there exist shapes which can tile an infinite plane without leaving any holes but which have the extraordinary property that the patterns created are *non-periodic*: this means that a pattern never repeats itself anywhere on an infinite plane. I find this a quite staggeringly non-intuitive result.

Penrose has been able to reduce the numbers of shapes required to tile an infinite plane non-periodically to only two. One example of a pair of tiles which have this property is

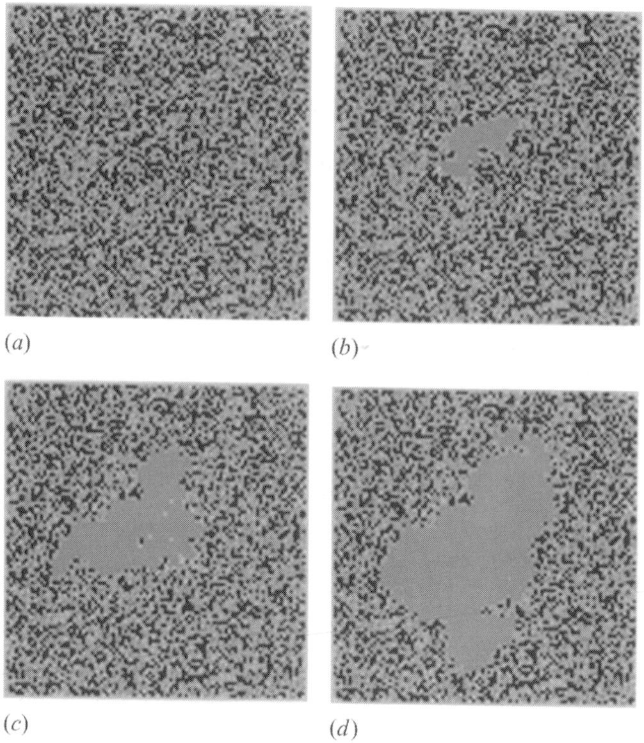

Figure 8.17: Illustrating the evolution of a mini-avalanche in a computer model of a rice pile on a 50×50 grid. The grey regions in these frames show the evolution of the area in which at least one toppling event has occurred. (From P. Bak, 1997, *How Nature Works – The Science of Self-Organised Criticality*, Plate 1, Oxford: Oxford University Press. Courtesy of J. Feder and colleagues, University of Oslo. Figure by M. Creutz.)

shown in Fig. 8.18. Penrose refers to the two shapes as 'kites' and 'darts'. The strange patterns seen in Fig. 8.18 continue to infinity and the configuration of the tilings cannot be predicted at any point in the plane since the patterns never recur in exactly the same way. Furthermore, how would a computer be able to establish whether the tiling was periodic or not? It might be that analogues of the tiling problem turn up in different aspects of physical problems. It will be intriguing to find out whether there are in fact problems in physics which end up being truly non-computational. A simple introduction to some of these ideas is contained is Penrose's short book *The Large, the Small and the Human Mind.*[17]

8.6 References

1 Buckingham, E. (1914). *Phys. Rev.*, **4**, 345.

2 Mahajan, S., Goldreich, P. and Phinney, S. (2003). *Order of Magnitude Physics: The Art of Approximation in Science*, see http://wol.ra.phy.cam.ac.uk/sanjoy.

3 Taylor, G.I. (1950). *Proc. Roy. Soc. A*, **201**, 159.

4 Taylor, G.I. (1950). *Proc. Roy. Soc. A*, **201**, 175.

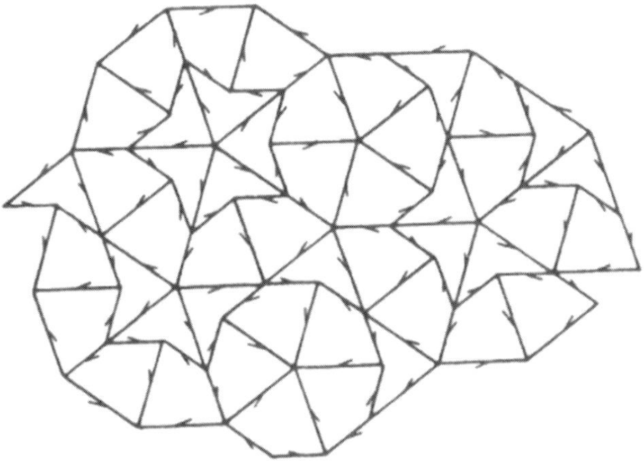

Figure 8.18: Illustrating tiling of a plane by two regular figures which Penrose calls 'kites' and 'darts'. An infinite plane can be tiled by these two shapes and the tiling is non-periodic. (From R. Penrose, 1986. In *M.C. Escher: Art and Science* (eds. H.S.M. Coxeter, M. Emmer, R. Penrose and M.L. Teuber), Amsterdam: Elsevier Science Publishers.)

5 Mack, J.E. (1947). Semi-popular motion picture record of the Trinity explosion, MDDC221, US Atomic Energy Commission.

6 Faber, T.E. (1995). *Fluid Dynamics for Physicists*. Cambridge: Cambridge University Press.

7 Barenblatt, G.I. (1995) *Scaling, Self-similarity, and Intermediate Asymptotics*. Cambridge: Cambridge University Press.

8 Tabor, D. (1991). *Gases, Liquids and Solids and Other States of Matter*. Cambridge: Cambridge University Press.

9 Gleick, J. (1987). *Chaos: Making a New Science*. New York: Viking.

10 Baker, G.L. and Gollub, J.P. (1990). *Chaotic Dynamics – An Introduction*. Cambridge: Cambridge University Press (second edition 1996).

11 From television interview with Edward Lorenz in the programme 'Chaos', produced by Chris Haws for Channel 4 by InCA (1988).

12 Bak, P. (1997). *How Nature Works – The Science of Self-Organised Criticality*. Oxford: Oxford University Press.

13 Mandelbrot, B. (1983). *The Fractal Geometry of Nature*. New York: Freeman.

14 Zipf, G.K. (1949). *Human Behaviour and the Principle of Least Effort*. Cambridge, Massachusetts: Addison-Wesley Publishers.

15 Penrose, R. (1989). *The Emperor's New Mind*. Oxford: Oxford University Press.

16 Penrose, R. (1994). *Shadows of the Mind*. Oxford: Oxford University Press.

17 Penrose, R. (1995). *The Large, the Small and the Human Mind*. Cambridge: Cambridge University Press.

18 Longair, M. (1998). *Galaxy Formation*. Berlin: Springer-Verlag.

Case Study IV

Thermodynamics and statistical physics

Thermodynamics is the science of how the properties of matter and systems change with temperature. A system may be viewed on the *microscopic scale*, in which case we study the interactions of the constituent particles or quanta and how these change with temperature. In this approach, we need to construct physical models for these interactions. The opposite approach is to study the system on the *macroscopic scale* and then the unique status of *classical thermodynamics* becomes apparent. In this approach, the behaviour of matter and radiation *in bulk* is studied and in effect we *deny that they have any* internal structure *at all*. In other words, the science of classical thermodynamics is solely concerned with relations between macroscopic measurable quantities such as pressure, volume and temperature.

Now this may seem to make classical thermodynamics a rather dull subject but, in fact, it is quite the opposite. In many physical problems, we may not know in detail the correct microscopic physics, and yet the thermodynamic approach can provide answers about the macroscopic behaviour of the system which are independent of the unknown detailed microphysics. Another way of looking at it is to think of classical thermodynamics as providing the boundary conditions which any microscopic model must satisfy. The thermodynamic arguments have absolute validity independent of the model adopted to explain any particular phenomenon.

It is remarkable that these profound statements can be made on the basis of the *first and second laws of thermodynamics*.* Let us state them immediately. The *first law of thermodynamics* is a statement about the conservation of energy:

Energy is conserved when heat is taken into account.

The *second law of thermodynamics* tells us how thermodynamic systems evolve with time. This can be deduced from the statement of the law due to Clausius:

No process is possible whose sole result is the transfer of heat from a colder to a hotter body.

* It is common practice to refer to four laws of thermodynamics altogether – the 'zeroth law' provides a definition of thermal equilibrium (see subsection 9.3.1) and the 'third law' states that the equilibrium entropies of all systems and the entropy changes in all reversible isothermal processes tend to zero as the temperature tends to zero.[4] The first and second laws are, however, very deep and contain the essence of thermodynamics.

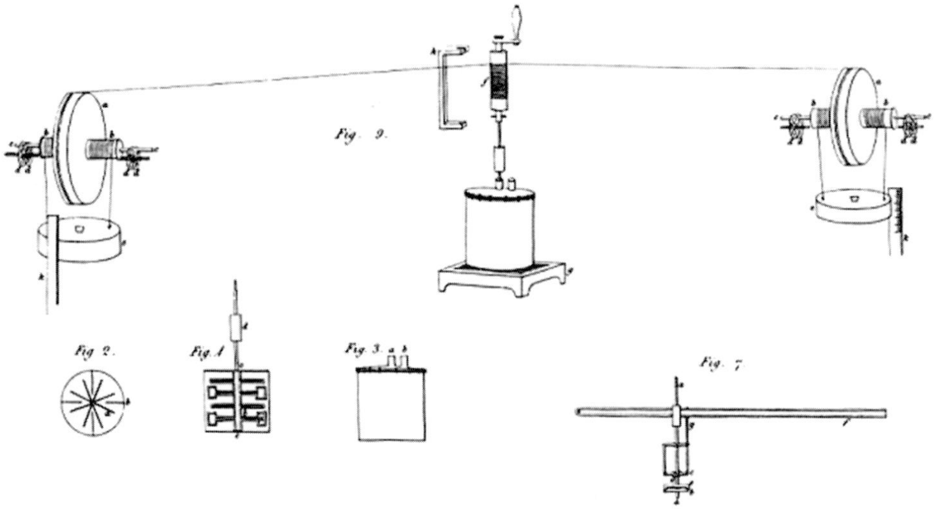

Figure IV.1: Joule's famous paddle-wheel experiment with which he determined the mechanical equivalent of heat. (From J.P. Joule, 1850, *Phil. Trans. Roy. Soc.*, **140**, opposite p. 64.)

From the perspective of classical thermodynamics, these laws are no more than reasonable hypotheses formulated on the basis of practical experience. They have proved, however, to be hypotheses of the greatest power. They have been applied to myriads of phenomena and time and again have proved to be correct. Examples include the extreme physical conditions such as matter in bulk at nuclear densities inside neutron stars and in the early stages of the Big Bang model of the Universe ($\rho \sim 10^{18}$ kg m^{-3}), and at ultra-low temperatures in laboratory experiments. There is no way in which the laws of classical thermodynamics can be proved – they are simply expressions of our common experience of the thermal properties of matter and radiation. It is intriguing that, when Max Planck was struggling to understand the nature of the spectrum of black-body radiation in 1900, he stated that:

> The two laws (of thermodynamics), it seems to me, must be upheld under all circumstances. For the rest, I was ready to sacrifice every one of my convictions about the physical laws.[1]

Indeed, the two laws of thermodynamics have never been found to be wanting in any physical problem.

There are two important themes running through the development of Chapter 9. The first is that the discovery of the laws of thermodynamics illuminates a number of conceptual problems which confronted the nineteenth-century pioneers. The resolution of these problems helped clarify concepts such as heat, energy, work, and so on. The second is that in our analysis of the second law and the concept of entropy, the early history of thermodynamics helps explain how Carnot came to his understanding of the importance of studying perfect heat engines. For much of the argument, we deal with the *ideal* behaviour of perfect machines or systems, and then contrast this with what happens in the real imperfect world.

We should distinguish this *thermodynamic* approach from that of *model building*, in which we seek to interpret the nature of thermodynamic laws in terms of microscopic processes. While Chapter 9 will remain strictly *model free*, in Chapter 10 we study two models, the *kinetic theory of gases* as enunciated by Clausius and Maxwell and the *statistical mechanics* of Boltzmann and Gibbs. These theories are explanatory in a sense that thermodynamics is not. For many students, thermodynamics is more accessible through statistical physics than through the subtleties of classical thermodynamics. In many applications, these statistical theories are very successful indeed and yet there remain problems which are too complicated to be treated by well-understood physics. A good example is the case of high-temperature superconductivity, in which the physics of strongly correlated electrons requires the development of new theoretical insights which are not yet available. Then the classical thermodynamic approach comes into its own.

It should come as no surprise to learn that, in fact, the science of classical thermodynamics developed in parallel with the atomic, or molecular, interpretation of the properties of matter. We find many of the same pioneers of classical thermodynamics – Kelvin, Clausius, Maxwell and Boltzmann – reappearing in the history of the microscopic physics of solids, liquids and gases. We will tell two stories in Chapter 10 – the kinetic theory of gases, and then the way in which the successes of that theory led to statistical mechanics. In fact, these two approaches form part of one single story because, despite the fact that there were good reasons why the kinetic theory was by no means readily accepted, it clearly suggested that the law of increase of entropy is in fact a statistical result, as first recognised by Maxwell.

The development of these concepts is admirably described by Peter Harman in his book *Energy, Force and Matter – The Conceptual Development of Nineteenth Century Physics.*[2] I should also confess to deriving enormous pleasure from an old textbook, which I picked up second-hand a long time ago when I was in my last year at school, by H.S. Allen and R.S. Maxwell entitled *A Textbook of Heat.*[3] The book is charming and full of splendid historical anecdotes, as well as presenting with great clarity the experimental ingenuity which went into establishing many of the results we now take for granted. It is a shame that books like this are no longer published.

IV.1 References

1 Planck, M. (1931). Letter from M. Planck to R.W. Wood. See Hermann, A. (1971), *The Genesis of Quantum Theory (1899–1913)*, pp. 23–4, Cambridge, Massachusetts: MIT Press.

2 Harman, P.M. (1982). *Energy, Force and Matter – The Conceptual Development of Nineteenth Century Physics*. Cambridge: Cambridge University Press.

3 Allen, H.S. and Maxwell, R.S. (1952). *A Textbook of Heat*, Parts I and II. London: Macmillan (first edition 1939).

4 Waldram, J.R. (1985). *The Theory of Thermodynamics.* Cambridge: Cambridge University Press.

9 Basic thermodynamics

9.1 Heat and temperature

Like so many aspects of the revolution which led to the birth of modern science, the origins of the scientific study of heat and temperature can be traced to the early years of the seventeenth century.[1] The quantitative study of heat and temperature depended upon the development of instruments which could give a quantitative measure of concepts such as 'the degree of hotness'. The first appearance of the word *thermomètre* occurred in 1624 in the volume *Récréaction Mathématique* by the French Jesuit Jean Leurechon. This was translated into English in 1633 by William Oughtred, under the title *Of the Thermometer, or an Instrument to Measure the Degrees of Heat or Cold in Aire.* The *Oxford English Dictionary* gives 1633 as the date of the first appearance of the word 'thermometer' in English literature.

There had been earlier descriptions of the use of the expansion of gases to measure 'the degree of hotness', by Galileo and others, but the first thermometers which resemble their modern counterparts were constructed in the 1640s to 1660s. Coloured alcohol within a glass bulb was used as the expanding substance and this liquid extended into the long stem of the thermometer, which was sealed off to prevent evaporative loss of the fluid. In 1701, Daniel Fahrenheit constructed an alcohol-in-glass thermometer in which the sealed tube was evacuated. Then, in 1714, he extended the temperature range of the thermometer by using mercury rather than fluids, such as alcohol, which boil at relatively low temperatures. In the period 1714 to 1724 he determined the fixed points of what became known as the *Fahrenheit temperature scale* as the boiling point of water at 212 °F, and the temperature of melting ice, at 32 °F. In 1742, Anders Celsius, Professor of Astronomy and founder of the Uppsala Observatory in Sweden, presented his mercury-in-glass thermometer to the Swedish Academy of Sciences, adopting a centigrade, now called Celsius, temperature scale, in which the temperature of melting ice was taken to be 0 °C and the boiling point of water 100 °C. On the Celsius scale, absolute zero is −273.15 °C. The kelvin temperature scale takes absolute zero as 0 K and has the same degree interval as the Celsius scale. The kelvin has been adopted internationally as the standard SI unit of temperature.

Joseph Black, who was appointed Professor of Chemistry and Medicine, first in Glasgow and then in Edinburgh, was a key figure in unravelling the nature of heat and temperature. It is intriguing that one of Black's assistants, who helped carry out his experiments in the 1760s, was James Watt, who will play a central role in the subsequent story. Black's great achievements included distinguishing quantitatively between the concepts of heat and temperature and defining what we now refer to as *specific heat capacities*. He also discovered and defined the concept of *latent heat*.

The experiments he undertook were simple but effective. To quantify the study of heat, Black used the *method of constant heat supply*, in which heat is supplied at a constant rate to a body and its temperature rise measured. By careful experimentation, Black showed that the heat Q supplied to a mass m of a substance is proportional to the temperature difference ΔT, that is, $Q = mc\Delta T$, where c is defined to be the *specific heat capacity* of the substance, the word *specific* meaning, as usual, per unit mass. His great realisation was that the specific heat capacities are different for different materials, and this provided the basis for the development of the mathematical theory of heat.

In his study of the latent heat associated with the melting of ice, he noted that the only effect of adding heat to the ice was to convert it into liquid water *at the same temperature*. As he expressed it, the heat that entered the water was

> . . . absorbed and concealed within the water, so as not to be discoverable by the application of a thermometer.

In one of his classic experiments, he set up in a large empty hall two glass vessels, each four inches in diameter. The first contained ice at its melting point and the other water at the same temperature. After half an hour, the water had increased in temperature by 7 °F, whereas the ice in the second vessel was only partially melted. In fact, the ice in the second vessel was only completely melted after ten and a half hours. These data can be used to show that the latent heat of melting of ice is 80 cal gm^{-1}. Similar ingenuity was used in determining the latent heat of steam as it condenses into water at its boiling point.

9.2 Heat as motion versus the caloric theory of heat

Towards the end of the eighteenth century, a major controversy developed concerning the nature of heat. The idea of heat as a form of motion can be traced back to the atomic theory of the Greek philosopher Democritus in the fifth century BC; he considered matter to be made up of indivisible particles whose motions gave rise to the myriad different forms in which matter is found in nature.

These ideas found resonances in the thoughts of scientists and philosophers during the seventeenth century. Francis Bacon wrote, for example:

> Heat itself, its essence and quiddity, is motion and nothing else.

This view was affirmed by Robert Boyle, who was persuaded of the kinetic, or dynamic, theory by the heat generated when a nail is hammered into a block of wood. He concluded that the action of the hammer impressed

> . . . a vehement and variously determined agitation of the small parts of the iron, which, being a cold body before, by that superinduced commotion of all its small parts, becomes in divers senses hot.

These opinions were shared by Thomas Hooke, who referred to fluidity as being

> nothing else but a certain pulse or shake of heat; for heat being nothing else but a very brisk and vehement agitation of the parts of a body, the parts of the body are thereby made so loose from one another, that they may easily move any way, and become a fluid.

Similar views were held by Isaac Newton, Gottfried Leibniz, Henry Cavendish, Thomas Young and Humphry Davy, as well as by the seventeenth-century philosophers Thomas Hobbes and John Locke.

During the eighteenth century, however, an alternative picture was developed, its origin being derived from the concept of *phlogiston*. The idea was that this hypothetical element could be combined with a body to make it combustible. Following the discovery of oxygen by Joseph Priestley in 1774, Antoine-Laurent Lavoisier, the father of modern chemistry, demonstrated that the process of combustion should rather be regarded as the process of *oxidation* and that it was unnecessary to involve 'phlogiston' in the process. Nonetheless, there was a strong body of opinion that heat should be considered a physical substance. Lavoisier and others named this 'matter of heat' *caloric*, which was to be thought of as an 'imponderable fluid', that is, a massless fluid, which was conserved when heat flowed from one body to another. Its flow was conceived as being similar to the flow of water in a water wheel, in which mass is conserved. To account for the diffusion of heat, caloric was assumed to be a self-repellent, elastic fluid, although it was attracted by matter. If one body were at a higher temperature than another and they were brought into thermal contact, caloric was said to flow from the hotter to the cooler body until they came into equilibrium at the same temperature. The hotter the body, the more caloric it contained and hence, because it was self-repellent, its volume would increase, accounting for its thermal expansion. In his *Traité Élémentaire de Chimie* of 1789, Lavoisier included caloric in his table of simple substances.[2]

There were, however, problems with the caloric theory – for example, when a warm body is brought into contact with ice, caloric flows from the warm body into the ice but, although ice is converted into water, the temperature of the ice–water mixture remains the same. It had to be supposed that caloric could combine with ice to form water. The theory attracted the support of a number of distinguished scientists, including Joseph Black, Pierre Simon Laplace, John Dalton, John Leslie, Claude Berthollet and Johan Berzelius. When Sadi Carnot came to formulate his pioneering ideas on the operation of perfect heat engines, he used the caloric theory to describe the flow of heat in a heat engine.

The two theories came into conflict at the end of the eighteenth century. Among the most important evidence against the caloric theory were the experiments of Count Rumford. Rumford was born Benjamin Thompson in the small rural town of Woburn in Massachusetts, USA, and had a quite extraordinary career, nicely summarised in the title of G.I. Brown's biography *Scientist, Soldier, Statesman, Spy – Count Rumford*.[2] In addition to these accomplishments, he was a founder of the Royal Institution of Great Britain and contracted an unhappy marriage to Lavoisier's widow. Despite his many practical contributions to society, he was not a well-liked figure. Brown quotes contemporary views of him as 'unbelievably cold-blooded, inherently egotistic and a snob', while another summarises his character as 'the most unpleasant personality in the history of science since Isaac Newton'.

Figure 9.1: A model of Rumford's cannon-boring experiments carried out at the arsenal at Munich in 1797–8. (Courtesy of the Deutsches Museum, Munich.)

Rumford's great contribution to physics was his realisation that heat could be created in unlimited quantities by the force of friction. This phenomenon was mostly vividly demonstrated by his experiments of 1798 in which cannons were bored with a blunt drill (Fig. 9.1). In his most famous experiment, he heated about 19 pounds (9 kg) of water in a box surrounding the cylinder of the cannon and after two and a half hours, the water actually boiled. In his own inimitable style, he remarked:

> It would be difficult to describe the surprise and astonishment expressed in the countenances of the bystanders, on seeing so large a quantity of water heated, and actually made to boil, without any fire . . .
>
> In reasoning on this subject, we must not forget *that most remarkable circumstance*, that the source of heat generated by friction in these experiments appeared evidently to be *inexhaustible*. It is hardly necessary to add, that anything which any *insulated* body or system of bodies can continue to furnish *without limitation* cannot possibly be a material substance; it appears to me to be extremely difficult, if not quite impossible, to form any distinct idea of anything capable of being excited and communicated in those experiments, except it be *MOTION*.[2]

The summit of Rumford's achievement was the first determination of the mechanical equivalent of heat, which he found from the amount of work done by the horses in rotating the drilling piece. In modern terms, he found a value of 5.60 J cal^{-1} in 1798 (the present-day value is 4.1868 J cal^{-1}). It was to be more than 40 years before Mayer and Joule returned to this problem.

Among the contributions which were to prove to be of prime importance in the understanding the nature of heat was Fourier's treatise *Analytical Theory of Heat*,[3] published in 1822. In this treatise, Fourier worked out the mathematical theory of heat transfer in the form of differential equations which did not require the construction of any specific model of the physical nature of heat. Fourier's methods were firmly based in the French tradition of rational mechanics and gave mathematical expression to the *effects* of heat without enquiring into its causes.

The idea of the conservation of energy for mechanical systems had been discussed in eighteenth-century treatises, in particular, in the work of Leibniz. He argued that 'living force' or *vis viva*, which we now call kinetic energy, is conserved in mechanical processes. By the 1820s, the relation of kinetic energy to the work done on a system had been clarified.

In the 1840s, a number of scientists independently came to the correct conclusion about the interconvertibility of heat and work, following the path pioneered by Rumford. One of the first was Julius Mayer, who in 1840 was appointed ship's doctor on the three-masted ship *Java* for a voyage to the Far East. He had plenty of time for reflection on scientific matters, among these being the production of heat in living organisms. At Surabaya, several of the crew fell ill and when he bled them, their venous blood was much redder in the tropics than it was in more moderate climes. The locals claimed that this was due to the need for less oxygen to maintain the body at its normal temperature than in colder climates. Mayer reasoned that the energy produced by the food one eats is partly converted into heat to maintain the body temperature and partly into the work done by one's muscles. On returning to Europe, Mayer proposed the equivalence of work and energy in 1842 on the basis of experiments on the work done in compressing and heating a gas. He elevated his results to the principle of the conservation of energy and from the adiabatic expansion of gases derived an estimate for the mechanical equivalent of heat of 3.58 J cal^{-1}. Mayer had great difficulty in getting his work published and his pioneering works were somewhat overshadowed by those of James Joule, who claimed that Mayer's work was 'an unsupported hypothesis, not an experimentally established quantitative law'. Joule proceeded to carry out meticulous experiments to test the hypothesis.

James Prescott Joule was born into a family which had grown wealthy through the foundation of a brewery by his grandfather. Joule's pioneering experiments were carried out in a laboratory, which his father had built for him next to the brewery. His genius was as a meticulous experimenter and the reason for singling out his name from among the other pioneers of the 1840s is that he put the science of heat on a firm experimental basis. Perhaps the most important aspect of his work was his ability to measure accurately very small temperature changes in his experiments.

It is a delight to track Joule's dogged progress towards a reliable estimate of the mechanical equivalent of heat.[4] For the sake of historical authenticity, let us depart from strict adherence to SI units and quote Joule's results in the units he used. Thus, the unit of thermal energy is the British thermal unit (BTU), which is the heat needed to raise the temperature of 1 pound of water by 1°F at STP. The unit of mechanical energy was the foot-poundal. The present-day value of the mechanical equivalent of heat, 4.1868 J cal^{-1}, corresponds to 778.2 foot-poundal BTU^{-1}.

1843 In Joule's dynamo experiments, the work done in driving a dynamo was compared with the heat generated by the electric current; they resulted in a publication entitled 'Production of heat by voltaic electricity'. In his earliest experiments, he established that the quantity of heat produced is proportional to RI^2, where R is the resistance and I the current. By 1843, he was able to derive from these electrical experiments a value for the mechanical equivalent of heat of 838 foot-poundal BTU^{-1}.

1843 In a postscript to his paper, Joule remarked that, in other experiments, he had shown that 'heat is evolved by the passage of water through narrow tubes' and quoted a result of 770 foot-poundal BTU^{-1}. These experiments hint at the origin of his most famous experiments, the paddle-wheel experiments.

1845 Joule carried out his own version of Mayer's experiment in which the change in temperature in an adiabatic expansion of air was measured. A volume of air was compressed and then allowed to expand. The change in internal energy, as measured by the fall in temperature of the gas on expansion, was then compared with the work done. By this means, he estimated the mechanical equivalent of heat to be 798 foot-poundal BTU^{-1}.

1847 In 1845, Joule realised that:

> If my views be correct, a fall [of a cascade] of 817 feet will of course generate one degree of heat, and the temperature of the river Niagara will be raised about one-fifth of a degree [Fahrenheit] by its fall of 160 feet.

The delightful sequel to this proposal was recounted by William Thomson after he had met Joule at the meeting of the British Association for the Advancement of Science in Oxford in 1847. In Thomson's words,

> About a fortnight later, I was walking down from Chamounix to commence the tour of Mont Blanc, and whom should I meet walking up but Joule, with a long thermometer in his hand, and a carriage with a lady in it not far off. He told me that he had married since we parted at Oxford and he was going to try for the elevation of temperature in waterfalls.[1]

Prospective spouses of physicists should be warned that this is by no means an isolated example of physicists doing physics during their honeymoons.

1845–1878 The first paddle-wheel experiment was carried out in 1845, a drawing of the experimental arrangement being published in 1850 (Fig. IV.1). The work done by the weights in driving the paddle wheel goes into heat through the frictional force between the water and the vanes of the paddle wheel. In 1845, Joule he presented his preliminary value, 890 foot-poundal BTU^{-1}. Improved estimates of 781-8 foot-poundal BTU^{-1} were reported in 1847. Yet more refined versions of the paddle-wheel experiments continued until 1878, when Joule gave his final result of 772.55 foot-poundal BTU^{-1}, corresponding to 4.13 J cal^{-1} – this can be compared with the present-day value of 4.1868 J cal^{-1}.

It was the results of the paddle-wheel experiments which greatly excited William Thomson (later, Lord Kelvin) in 1847, when he was only 22. At the same Oxford meeting referred to above, Thomson heard Joule describe the results of his recent experiments on the mechanical equivalent of heat. Up till that time, he had adopted the assumption made by Carnot that heat (or caloric) is conserved in the operation of heat engines. He found Joule's results astonishing and, as his brother James wrote, '[Joule's] views have a slight tendency to unsettle one's mind'. Joule's results became widely known in continental Europe and, by 1850, Helmholtz and Clausius had formulated what is now known as the *law of conservation of energy*, or the *first law of thermodynamics*. Helmholtz, in particular, was the first to express the conservation laws in a mathematical form in such a way as to incorporate mechanical and electrical phenomena, heat and work. In 1854, Thomson invented the word *thermo-dynamics* to describe the new science of heat to which he was to make such major contributions.

9.3 The first law of thermodynamics

My concern in this section is to describe the content of the first law of thermodynamics as clearly as possible. My approach has been strongly influenced by Brian Pippard's *The Elements of Classical Thermodynamics*,[5] which can be warmly recommended for the clarity and depth of its exposition.

9.3.1 The zeroth law and the definition of empirical temperature

To begin with, we need to establish what is meant by the term *empirical temperature* and how it follows from what is called the *zeroth law of thermodynamics*. Immediately, we encounter one of the distinctive features of thermodynamic arguments. In many of these, we make statements of the form 'It is a fact of experience that . . . ' and from such axioms, we derive the appropriate mathematical structures.

Let us consider first the thermal properties of fluids, meaning liquids or gases, because then the pressure is isotropic at all points in the fluid and also because then we can change the shape of the containing vessel without changing its volume, so that no work is done. At this stage, we have no definition of temperature. We now state our first axiom:

It is a fact of experience that the properties of a given number of moles of a particular fluid are entirely determined by only two properties, the pressure p and the volume V of the vessel.

It is assumed that the fluid is not subjected to other influences, such as an electric or magnetic field. If the system is wholly defined by two bulk properties, it is known as a two-coordinate system. Most of the systems we will deal with are two-coordinate systems, but it is relatively straightforward to develop the formalism for multi-coordinate systems.

Notice carefully what this assertion means. Suppose a certain volume of fluid is taken through a sequence of processes such that it ends up having pressure p_1 and volume V_1. Then, suppose we take another identical quantity of fluid and perform a totally different set of operations so that it also ends up with coordinates p_1 and V_1. We have asserted that these two fluids are totally indistinguishable in their physical properties.

Now take two isolated systems consisting of fluids with coordinates p_1, V_1 and p_2, V_2. They are brought into thermal contact and left for a very long time. In general, they will change their properties in such a way that they reach a state of *thermal equilibrium*, in which there is no net heat transfer between them. The concept of thermal equilibrium is crucial – all components which make up the system have been allowed to interact thermally until after a very long time no further changes in the bulk properties of the system are found. In attaining this state, in general heat is exchanged and work is done. In thermal equilibrium, the thermodynamic coordinates of the two systems are, say, p_1, V_1 and p_2, V_2. Now, it is clear that these four values cannot be arbitrary. It is a fact of experience that two fluids cannot have arbitrary values for p_1, V_1 and p_2, V_2 and be in thermal equilibrium. There must be some mathematical relation between the four quantities in thermal equilibrium, which can be expressed as

$$F(p_1, V_1, p_2, V_2) = 0.$$

This equation tells us the value of the fourth quantity if the other three are fixed.

So far we have not used the word *temperature*. We will find a suitable definition from another common fact of experience, which is so central to the subject that it has been raised to the status of a law of thermodynamics – the *zeroth law*. The formal statement is:

If two systems, 1 and 2, are separately in thermal equilibrium with a third, 3, then they must also be in thermal equilibrium with each another.

We can write this down mathematically. If system 1 is in thermal equilibrium with system 3, this means that

$$F(p_1, V_1, p_3, V_3) = 0,$$

or, expressing the pressure p_3 in terms of the other variables,

$$p_3 = f(p_1, V_1, V_3).$$

If system 2 is in equilibrium with system 3, this means, in the same way, that

$$p_3 = g(p_2, V_2, V_3).$$

Therefore, in thermal equilibrium,

$$f(p_1, V_1, V_3) = g(p_2, V_2, V_3). \tag{9.1}$$

But the zeroth law tells us that 1 and 2 must also be in thermal equilibrium, and consequently there must exist a function such that

$$h(p_1, V_1, p_2, V_2) = 0. \tag{9.2}$$

Equation (9.2) tells us that (9.1) must be of such a form that the dependence on V_3 cancels out on either side, that is, it must be a function of the form $f(p_1, V_1, V_3) = \phi_1(p_1, V_1)\zeta(V_3) + \eta(V_3)$. Therefore, substituting into (9.1), we find

$$\phi_1(p_1, V_1) = \phi_2(p_2, V_2) = \phi_3(p_3, V_3) = \theta = \text{constant}, \tag{9.3}$$

in thermal equilibrium.

If now we consider not three different systems but the same system in three different states that are in thermal equilibrium with each other, then $\phi_1 = \phi_2 = \phi_3$.

Thus for this particular system there exists a function ϕ of p and V which takes a definite value for these particular equilibrium states. The value of this function is defined to be the *empirical temperature* θ. Quantities which take definite values for a particular equilibrium state are called *functions of state*. In this example, p, V and θ are functions of state. Consequently, we have defined an *equation of state* for the system which relates p and V to the empirical temperature θ in thermodynamic equilibrium:*

$$\phi(p, V) = \theta. \tag{9.4}$$

* If a system is in thermal, mechanical and chemical equilibrium, it is said to be in thermodynamic equilibrium.

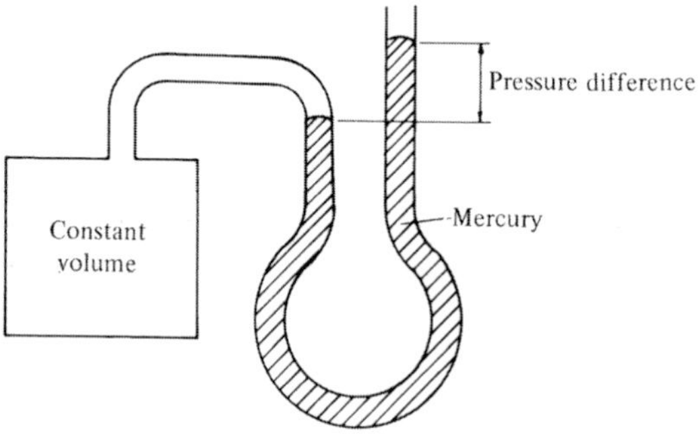

Figure 9.2: Illustrating the principle of the constant volume gas thermometer. The thermometer makes use of the fact that, at low pressures, the gas follows closely the ideal gas law $pV = RT$ and so, if the volume is fixed, the pressure difference provides a direct measure of the temperature.

We can now find from experiment all the combinations of p and V which correspond to a fixed value of the empirical temperature θ. Notice that we have three quantities, p, V and θ which describe the equilibrium state, but any two of them are sufficient to define that state completely. On a p, V diagram for the system, lines of constant θ are called *isotherms*.

At this stage, the empirical temperature θ looks nothing like what we customarily recognise as temperature and, in fact, we could use an expression of the type (9.4) to devise horribly complicated temperature scales. To put the subject on a firm experimental foundation, we need to choose some system, a coordinate of which will provide a *thermometric scale*. Once this scale is fixed then all other systems will have the same value of empirical temperature when they are in thermal equilibrium with the chosen system.

The subject of thermometric scales is a vast one. We will only note the importance of the *constant volume gas thermometer*, which is illustrated in Fig. 9.2. Although bulky and unwieldy, the constant volume gas thermometer is of special importance because it was found empirically that all gases have the same variation of pressure with volume at low pressures, that is, at a fixed temperature, the product pV is the same for one mole of all gases at low enough pressures. The relation between these quantities is the *perfect gas law*,

$$pV = RT, \tag{9.5}$$

where R is the universal gas constant for 1 mole of gas.

An *ideal*, or *perfect*, *gas* obeys (9.5) precisely. The temperature scale defined by (9.5) is called the *ideal gas temperature* scale. We will show in subsection 9.5.2 that the ideal gas temperature scale is identical to the thermodynamic temperature scale, which provides a rigorous theoretical basis for the study of heat; therefore, in the limit of low pressures, gas thermometers can be used to measure *thermodynamic temperatures* directly. Symbolically, we can write

$$T = \lim_{p \to 0} (pV)/R. \tag{9.6}$$

Notice that in general p and V are much more complicated functions of T, in particular, at high pressures and close to phase transitions.

9.3.2 The mathematical expression of the first law of thermodynamics

We have already stated the *first law of thermodynamics*:

Energy is conserved if heat is taken into account.

To give mathematical substance to the law, we must define precisely what we mean by the terms heat, energy and work. The last two are straightforward. The definition of the work done on a system by a force $\boldsymbol{f}$ is

$$W = \int_{\boldsymbol{r}_1}^{\boldsymbol{r}_2} \boldsymbol{f} \cdot \mathrm{d}\boldsymbol{r}. \tag{9.7}$$

When work is done on a system, its *energy* is increased. Let us note some of the different ways in which work can be done on a system.

- *Work done in compressing a fluid.* When a fluid is compressed under the action of an external pressure p, the work done is positive. Therefore, since the volume decreases, the amount of work done is

$$\mathrm{d}W = -p\,\mathrm{d}V. \tag{9.8}$$

  If the fluid itself does work by expanding, the work done on the surroundings is positive and so the external work done on the system is negative; thus the sign of the volume increment is important.
- *Work done in stretching a wire by* $\mathrm{d}\boldsymbol{l}$: $\mathrm{d}W = \boldsymbol{f} \cdot \mathrm{d}\boldsymbol{l}$.
- *Work done by an electric field* $\boldsymbol{E}$ *on a charge* q: $\mathrm{d}W = q\boldsymbol{E} \cdot \mathrm{d}\boldsymbol{r}$.
- *Work done against surface tension in increasing the surface area of a liquid by* $\mathrm{d}A$: $\mathrm{d}W = \gamma\,\mathrm{d}A$, where γ is the coefficient of surface tension.
- *Work done by a couple* $\boldsymbol{G}$ *in causing an angular displacement* $\mathrm{d}\boldsymbol{\theta}$: $\mathrm{d}W = \boldsymbol{G} \cdot \mathrm{d}\boldsymbol{\theta}$.
- *Work done on a dielectric per unit volume by an electric field* $\boldsymbol{E}$: $\mathrm{d}W = \boldsymbol{E} \cdot \mathrm{d}\boldsymbol{P}$, where $\boldsymbol{P}$ is the polarisation, that is, the dipole moment per unit volume (see Sections 6.11 and 6.12).
- *Work done by a magnetic field* $\boldsymbol{B}$ *per unit volume on a magnetisable medium*: $\mathrm{d}W = \boldsymbol{B} \cdot \mathrm{d}\boldsymbol{M}$, where $\boldsymbol{M}$ is the magnetisation of the medium, that is, its magnetic dipole moment per unit volume (see Sections 6.11 and 6.12).

Thus, the work done on a system is the scalar product of a generalised force $\boldsymbol{X}$ and a generalised displacement $\mathrm{d}\boldsymbol{x}$, $\mathrm{d}W = \boldsymbol{X} \cdot \mathrm{d}\boldsymbol{x}$. Notice also that the work done is always the product of an *intensive variable* $\boldsymbol{X}$, by which we mean a property defined at a particular point in the material, and an *extensive variable* $\mathrm{d}\boldsymbol{x}$ which describes the 'displacement' of the system under the action of the intensive variable.

Now consider an *isolated* system in which there is no thermal interaction with the surroundings. It is a fact of experience that if a certain amount of work is done on the system then the system attains the same new equilibrium state no matter how the work is done; for example, we may compress a volume of gas, or stir it with a paddle wheel, or pass an electric current through it. Joule's great contribution to thermodynamics was to demonstrate by precise experiment that this is indeed the case. The result is that, by these very different processes, the energy of the system is increased. We say that there is an increase in the *internal energy* U of the system by virtue of a given amount of work W having been done on it. Since it does not matter how the work is done, U must be a *function of state*. For this isolated system,

$$W = U_2 - U_1 \qquad \text{or} \qquad W = \Delta U. \tag{9.9}$$

Now suppose the system is not isolated: there is a thermal interaction between the surroundings and the system. Then the system will reach a new internal energy state that is determined not just by the work done on the system. We therefore define the *heat supplied* Q to be

$$Q = \Delta U - W. \tag{9.10}$$

This is our definition of *heat*. It may seem a rather roundabout way of proceeding, but it has the great benefit of logical consistency. It completely avoids specifying exactly what heat is, the problem which was at the root of many of the conceptual difficulties which existed, until about 1850, in incorporating heat into the conservation laws.

It is convenient to write (9.10) in differential form:

$$\mathrm{d}Q = \mathrm{d}U - \mathrm{d}W. \tag{9.11}$$

It is important to distinguish between those differentials which refer to functions of state and those which do not. As stated in subsection 9.3.1, p, V and T are functions of state, as is U. Each takes a definite value for a particular system in a particular state. $\mathrm{d}U$ is the differential of a function of state, as are $\mathrm{d}p$, $\mathrm{d}V$ and $\mathrm{d}T$, but $\mathrm{d}Q$ and $\mathrm{d}W$ are not because we can pass from U_1 to U_2 by adding together different amounts of Q and W. We therefore write the differentials of these quantities as $đQ$, $đW$. Written thus, (9.11) becomes

$$đQ = \mathrm{d}U - đW. \tag{9.12}$$

Equation (9.12) is the formal mathematical expression of the *first law of thermodynamics*. We are now in a position to apply the law of conservation of energy as expressed by (9.12) to all sorts of different problems.

9.3.3 *Some applications of the first law of thermodynamics*

(i) *Specific heat capacities.* U is a function of state and we have asserted that we can describe the properties of a gas entirely in terms of only two coordinates, that is, two functions of state. So, let us express U in terms of T and V: $U = U(V, T)$. Then the differential of U is

$$\mathrm{d}U = \left(\frac{\partial U}{\partial T}\right)_V \mathrm{d}T + \left(\frac{\partial U}{\partial V}\right)_T \mathrm{d}V. \tag{9.13}$$

Substituting into (9.12) and using (9.8),

$$đQ = \left(\frac{\partial U}{\partial T}\right)_V \mathrm{d}T + \left[\left(\frac{\partial U}{\partial V}\right)_T + p\right]\mathrm{d}V. \tag{9.14}$$

We can now define mathematically the concept of the *heat capacity* C, first described experimentally by Joseph Black in the 1760s. At *constant volume*, we define

$$C_V \equiv \left(\frac{\text{đ}Q}{\mathrm{d}T}\right)_V = \left(\frac{\partial U}{\partial T}\right)_V.$$

At *constant pressure*,

$$C_p \equiv \left(\frac{\text{đ}Q}{\mathrm{d}T}\right)_p = \left(\frac{\partial U}{\partial T}\right)_V + \left[\left(\frac{\partial U}{\partial V}\right)_T + p\right]\left(\frac{\partial V}{\partial T}\right)_p. \tag{9.15}$$

These expressions tell us the temperature rise for a given input of heat. Notice that these heat capacities do not refer to any particular volume or mass. It is conventional to use *specific heat capacities* or *specific heats*, where the word 'specific' takes its usual meaning of 'per unit mass'. Conventionally, specific quantitites are written in lower case letters, that is,

$$c_V = C_V/m, \qquad c_p = C_p/m. \tag{9.16}$$

Subtracting, we find

$$C_p - C_V = \left[\left(\frac{\partial U}{\partial V}\right)_T + p\right]\left(\frac{\partial V}{\partial T}\right)_p. \tag{9.17}$$

The interpretation of this equation is straightforward. The second term in the brackets describes the rate at which work is done by the system in pushing back its surroundings at constant p. The first term clearly has something to do with the internal properties of the gas because it describes how the internal energy changes with volume. It must be associated with the work done against different types of intermolecular force within the gas. Thus, $C_p - C_V$ provides information about $(\partial U/\partial V)_T$.

(ii) *The Joule expansion.* One way of finding out the relation between C_p and C_V is to perform what is known as a *Joule expansion*, the free expansion of a gas into a larger volume. The system is thermally isolated from its surroundings, so that there is no heat inflow or outflow. The walls of the vessels are fixed and therefore no work is done in the free expansion.

The apparatus used in careful experiments carried out by Joule in 1845 is shown in Fig. 9.3. Vessel A contained dry air at a pressure of 22 atmospheres and B was evacuated. C was a carefully constructed stopcock. In the first experiment, the whole experiment was enclosed in a water bath (Fig. 9.3(a)). When the stopcock was opened the air flowed from A into B, but although Joule could measure temperature changes as small as 0.005 °F, he could detect no change in the temperature of the water bath. In his words,

> ...no change of temperature [in the water bath] occurs when air is allowed to expand in such a manner as not to develop mechanical power.[1]

In the second experiment, A, B and C were surrounded by separate water baths (Fig. 9.3(b)). When the stopcock was opened, the temperatures of the water baths around B and C increased, while the temperature of the water bath around A decreased. This occurred because the gas in A does work in pushing the gas into B. If the gas in A, B and C was then allowed to come into thermal contact following the expansion and reach a new equilibrium state, no overall change in the temperature was observed in Joule's experiments.

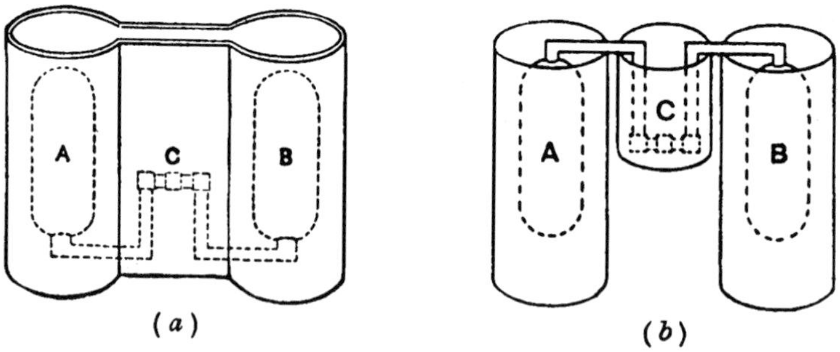

Figure 9.3: Illustrating Joule's experiments on the expansion of gases. In (*a*), the vessels A, B and the stopcock C are surrounded by a water bath. In (*b*), the vessels and the stopcock each have separate water baths. (From H.S. Allen and R.S. Maxwell, 1952, *A Text-book of Heat*, Vol. II, p. 575, London: Macmillan.)

This type of expansion is known as a *Joule expansion*. Now, $\Delta U = Q + W$. There is no heat transfer into or out of the system, $Q = 0$, and no external work has been done by the gas, $W = 0$. Therefore, there is no change in the internal energy U: $\Delta U = 0$.

We can now complete our definition of a *perfect gas* according to classical thermodynamics:

(a) its equation of state is the perfect gas law, $pV = RT$ for one mole of gas;
(b) in a Joule expansion, there is no change in temperature at constant internal energy.

For a perfect gas, there is a simple relation between C_p and C_V. From (b),

$$\left(\frac{\partial U}{\partial V}\right)_T = 0.$$

From (a),

$$\left(\frac{\partial V}{\partial T}\right)_p = \left[\frac{\partial}{\partial T}\left(\frac{RT}{p}\right)\right]_p = \frac{R}{p}.$$

Therefore, from (9.16) and (9.17), for one mole of a perfect gas

$$C_p - C_V = R. \tag{9.18}$$

It should be emphasised that the internal energy of a perfect gas is *a function only of temperature*: it is independent of volume and therefore must also be independent of pressure, because any given function of state of a gas, here the internal energy, is entirely determined by two other functions of state.

For real, imperfect, gases undergoing a Joule expansion there is, however, a change in T with volume. Because of the large heat capacities of his water baths, Joule could not detect the small overall change in temperature which would have occurred for non-ideal gases. Physically, the reason is that work is done against the intermolecular van der Waals forces. Also, at very high pressures there is an effective repulsive force because the molecules cannot be squeezed, what is known as hard-core repulsion. Thus, in a real gas there is a small change in temperature in a Joule expansion. The *Joule coefficient* $(\partial T/\partial V)_U$ measures

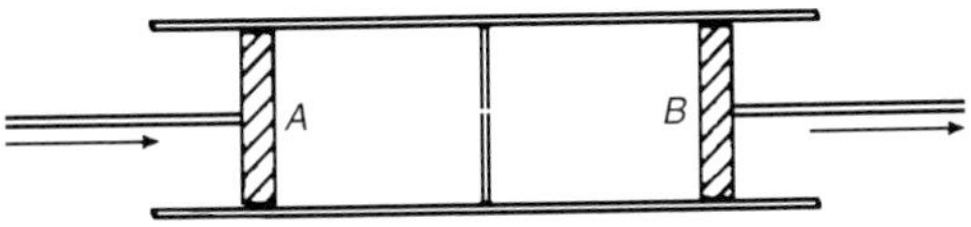

Figure 9.4: Illustrating the Joule–Kelvin experiment on the expansion of gases. In these experiments, the fluid was forced through one or more small holes or narrow tubes, often referred to as a porous plug, in such a way that the velocity of the fluid was very small on passing through the plug. Engineers refer to such a flow as a *throttled expansion*. (After H.S. Allen and R.S. Maxwell, 1952, *A Textbook of Heat*, Part II, p. 687, London: MacMillan and Co.)

the change in T as volume increases at constant internal energy U, and we can relate this to other properties of the gas.

(iii) *The enthalpy and the Joule–Kelvin expansion.* The heat capacity at constant volume C_V involves the derivative of a function of state, and so we may ask if there is a derivative of another function of state corresponding to C_p. Let us write $U = U(p, T)$ instead of $U(V, T)$, remembering that we need only two coordinates to specify the state of the gas; then

$$\mathrm{d}U = \left(\frac{\partial U}{\partial p}\right)_T \mathrm{d}p + \left(\frac{\partial U}{\partial T}\right)_p \mathrm{d}T.$$

Now proceed as before:

$$\begin{aligned} \text{đ}Q &= \mathrm{d}U + p\,\mathrm{d}V \\ &= \left(\frac{\partial U}{\partial p}\right)_T \mathrm{d}p + p\,\mathrm{d}V + \left(\frac{\partial U}{\partial T}\right)_p \mathrm{d}T; \end{aligned}$$

thus

$$\begin{aligned} \left(\frac{\text{đ}Q}{\mathrm{d}T}\right)_p &= p\left(\frac{\partial V}{\partial T}\right)_p + \left(\frac{\partial U}{\partial T}\right)_p \\ &= \left[\frac{\partial}{\partial T}(pV + U)\right]_p. \end{aligned} \tag{9.19}$$

The quantity $pV + U$ is entirely composed of functions of state and hence this new function must also be a function of state – it is known as the *enthalpy* H. Thus,

$$H = U + pV,$$

$$C_p = \left(\frac{\text{đ}Q}{\mathrm{d}T}\right)_p = \left(\frac{\partial H}{\partial T}\right)_p = \left[\frac{\partial}{\partial T}(U + pV)\right]_p. \tag{9.20}$$

The enthalpy often appears in flow processes and, in particular, in the type of expansion known as a *Joule–Kelvin expansion*. In this case, gas is transferred from one cylinder to another, the pressures in both cylinders being maintained at constant values p_1 and p_2 (Fig. 9.4). Suppose that a certain mass of gas is pushed through one or more small holes or narrow tubes, what is often referred to as a 'porous plug', from the left-hand to the right-hand cylinder. The gas is initially on the left-hand side, with internal energy U_1, volume V_1 and pressure p_1. Piston A pushes this gas at constant pressure p_1 through the plug, doing

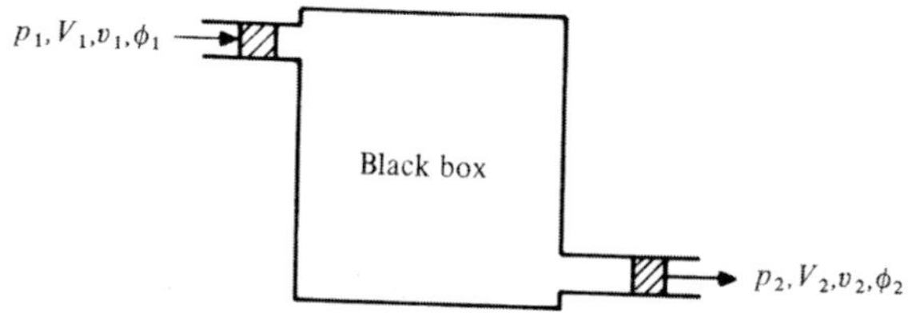

Figure 9.5: Illustrating the conservation of energy in fluid flow in the presence of a gravitational field.

work $p_1 V_1$ on the gas. The gas ends up in the right-hand cylinder with pressure p_2, volume V_2 and temperature T_2, and the work done by the gas acting on piston B is $p_2 V_2$. The system is thermally isolated and hence $Q = 0$. Therefore, from (9.10) the increase in internal energy equals the net work done on the gas, $\Delta U = W$, and so

$$U_2 - U_1 = p_1 V_1 - p_2 V_2.$$

Rearranging, we see that enthalpy is conserved:

$$p_1 V_1 + U_1 = p_2 V_2 + U_2 \qquad \text{or} \qquad H_1 = H_2. \tag{9.21}$$

For one mole of a perfect gas, $H = pV + U = RT + U(T)$. But $U(T) + RT$ is a unique function of temperature and hence T must be the same before and after the Joule–Kelvin expansion. Thus, for a perfect gas, there is no change in temperature in a Joule–Kelvin expansion. In real gases, however, there is a temperature change because of intermolecular forces. The temperature change may be either positive or negative depending upon the pressure and temperature: the Joule–Kelvin coefficient is defined to be $(\partial T/\partial p)_H$. The Joule–Kelvin experiment is a more sensitive method for determining deviations from perfect-gas-law behaviour than the Joule expansion.

We have now come very close to deriving a more general conservation equation, in which we take into account other contributions to the total energy, for example, the kinetic energy of bulk motion and the potential energy of the gas if it is in a gravitational field. Notice that the conservation of enthalpy is simply a version of the law of conservation of energy taking account of the work done by and on the gas in a throttling process. Let us consider flow through a 'black box', again with no heat loss, and add in these energies as well (Fig. 9.5).

We consider the steady flow of a given mass of gas or fluid m as it enters and leaves the black box; assuming there is no heat loss as the gas passes through the black box, the law of conservation of energy becomes

$$\begin{aligned} H_1 + \tfrac{1}{2}mv_1^2 + m\phi_1 &= H_2 + \tfrac{1}{2}mv_2^2 + m\phi_2, \\ p_1 V_1 + U_1 + \tfrac{1}{2}mv_1^2 + m\phi_1 &= p_2 V_2 + U_2 + \tfrac{1}{2}mv_2^2 + m\phi_2, \end{aligned} \tag{9.22}$$

that is,

$$\begin{aligned} \frac{p}{m/V} + \frac{U}{m} + \tfrac{1}{2}v^2 + \phi &= \text{constant}, \\ \frac{p}{\rho} + u + \tfrac{1}{2}v^2 + \phi &= \text{constant}, \end{aligned} \tag{9.23}$$

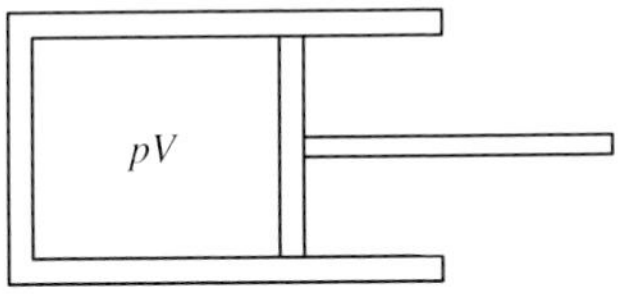

Figure 9.6: Illustrating an adiabatic expansion within a perfectly insulated system.

where u is the *specific energy density* and ρ is the density of the fluid. This is one of the equations of fluid flow. In particular, for an incompressible fluid $u_1 = u_2$ and so we obtain *Bernoulli's equation*,

$$\frac{p}{\rho} + \tfrac{1}{2}v^2 + \phi = \text{constant}, \tag{9.24}$$

which we derived from the perspective of fluid dynamics in appendix section A7.3.

Notice that we assumed that the additional terms present in Bernoulli's equation were absent in the Joule–Kelvin expansion. It was assumed that a Joule–Kelvin expansion takes place very slowly, so that the kinetic energy terms can be neglected and the two volumes are at the same gravitational potential.

(iv) *Adiabatic expansion.* In an adiabatic expansion, the volume of the gas changes without any thermal contact between the system and its surroundings. The classical method of illustration is the expansion or compression of the gas within a perfectly insulated cylinder (Fig. 9.6). A reversible adiabatic expansion takes place very slowly so that the system passes through an infinite number of *equilibrium states* between the initial and final thermodynamic coordinates. We will return to this key concept in our discussion of reversible processes in section 9.5.1. We can write for such a process

$$đQ = \mathrm{d}U + p\,\mathrm{d}V = 0. \tag{9.25}$$

If $C_V = (\partial U/\partial T)_V$ is taken to refer to 1 mole of a perfect gas, for which $(\partial U/\partial V)_T = 0$, then $\mathrm{d}U = nC_V\,\mathrm{d}T$ for n moles. During the expansion, the gas passes through an infinite series of equilibrium states, for which the perfect gas law applies, $pV = nRT$, and hence, from (9.25),

$$nC_V\,\mathrm{d}T + \frac{nRT}{V}\,\mathrm{d}V = 0,$$

so that

$$\frac{C_V}{R}\frac{\mathrm{d}T}{T} = -\frac{\mathrm{d}V}{V}. \tag{9.26}$$

Integrating,

$$\frac{V_2}{V_1} = \left(\frac{T_2}{T_1}\right)^{-C_V/R} \quad \text{or} \quad VT^{C_V/R} = \text{constant}. \tag{9.27}$$

Since $pV = nRT$ at all stages in the expansion, this result can also be written

$$pV^{\gamma} = \text{constant}$$

where

$$\gamma = 1 + \frac{R}{C_V}.$$

We have already shown that, for one mole of gas, $C_V + R = C_p$ and consequently

$$1 + \frac{R}{C_V} = \frac{C_p}{C_V} = \gamma; \qquad (9.28)$$

γ is the *ratio of specific heats*, or the *adiabatic index*. For a monatomic gas, $C_V = \frac{3}{2}R$ and consequently $\gamma = \frac{5}{3}$.

(v) *Isothermal expansion.* In this case, the same arrangement as in Fig. 9.6 is used except that now there is heat exchange with the surroundings, so that the gas within the cylinder remains at the same temperature, T = constant. For a perfect gas, this means that its internal energy is unchanged. Thus $\Delta U = 0$ in the expansion and so, from (9.10), $Q = -W$: the work done in pushing back the piston must be made up by a corresponding inflow of heat. The work done is

$$\int_{V_1}^{V_2} p\,\mathrm{d}V = \int_{V_1}^{V_2} \frac{RT}{V}\,\mathrm{d}V = RT\,\ln\frac{V_2}{V_1}. \qquad (9.29)$$

This is the amount of heat which must be supplied from the surroundings to maintain an isothermal expansion. This result will find important applications in the analysis of heat engines.

(vi) *Summary of expansion types.* Let us summarise the four different types of expansion described in this section.

- *Isothermal expansion*, $\Delta T = 0$. Heat must be supplied or removed from the system to maintain $\Delta T = 0$ (item (v)).
- *Adiabatic expansion*, $Q = 0$. No heat exchange takes place with the surroundings (item (iv)).
- *Joule expansion*, $\Delta U = 0$. The free expansion of a gas to a larger volume with fixed walls involves no change in internal energy (item (ii)). For a perfect gas there is no change in temperature.
- *Joule–Kelvin expansion*, $\Delta H = 0$. When a gas passes from one volume to another and the pressures in the two vessels are maintained at p_1 and p_2 during the transfer, enthalpy is conserved (item (iii)). For a perfect gas there is no change in temperature.

The basic principle behind these apparently different phenomena is simply the conservation of energy – they are no more than simple elaborations of the first law of thermodynamics.

9.4 The origin of the second law of thermodynamics

It always comes as a surprise that historically the first law of thermodynamics proved to be so much more troublesome than the second law. After 1850, the first law enabled a logically self-consistent definition of heat to be formulated and the law of conservation of energy to be placed on a firm conceptual basis. It was realised, however, that there must be other restrictions upon thermodynamic processes. In particular, we do not yet have any rules about the direction in which heat flows, or about the direction in which thermodynamic systems evolve. These are derived from the *second law of thermodynamics*, which can be expressed in the following form, presented by Rudolph Clausius:

> No process is possible whose sole result is the transfer of heat from a colder to a hotter body.

It is remarkable that such powerful results can be derived from this simple statement. Equally remarkable is the way in which the fundamentals of thermodynamics were derived from the operation of steam engines. As L.J. Henderson has remarked,

> Until 1850, the steam engine did more for science than science did for the steam engine.[6]

Let us illustrate the truth behind this perceptive statement.

9.4.1 *James Watt and the steam engine*

The invention of the steam engine was perhaps the most important of the developments which laid the foundations for the industrial revolution which swept through developed countries from about 1750 to 1850.[7] Before that time, the principal sources of power were water-wheels, which had to be located close to rivers and streams, and wind power, which was unreliable. Horses were used to provide power, but they were expensive. The importance of the steam engine was that power could be generated at the location where it was needed.

The first commercially viable and successful steam engine was built by Thomas Newcomen in 1712 to pump water out of coal mines and to raise water to power water-wheels. His steam engine (Fig. 9.7) operated at atmospheric pressure. The principle of operation was that the cylinder, fitted with a piston, was filled with steam, which was then condensed by injecting a jet of cold water. The resulting vacuum in the cylinder caused the piston to descend, resulting in a power stroke which was communicated to the pump by the mechanical arrangement shown in Fig. 9.7. When the power stroke was finished, the steam valve was opened and the weight of the pump rod pulled the piston back to the top of the cylinder. Through the next half century, the efficiency of Newcomen's design improved steadily thanks to the efforts of John Smeaton who, with better design and engineering, more than doubled the efficiency of Newcomen's original engine (Table 9.1).

One of the great heroes of this story is James Watt, who was trained as a scientific instrument maker in Glasgow. While repairing a model Newcomen steam engine in 1764, he was struck by the waste of steam and heat in its operation. In May 1765, he made the crucial innovation which led to the underpinning of the whole of thermodynamics. He realised that the efficiency of the steam engine would be significantly increased if the steam condensed in a separate chamber from the cylinder. This invention of the *condenser* was perhaps his greatest achievement. He took out his famous patent for this invention in 1768.

Figure 9.8 shows one of the steam engines built by Watt in 1788. The engine operated at atmospheric pressure. The key innovation was the separation of the condenser F from the cylinder and piston E. The cylinder was insulated and kept at a constant high temperature T_1 while the condenser was kept cool at a lower temperature T_2. The condenser was evacuated

Table 9.1. *The efficiencies of steam engines*[8]

Date	Builder	Duty	Percentage thermal efficiency
1718	Newcomen	4.3	0.5
1767	Smeaton	7.4	0.8
1774	Smeaton	12.5	1.4
1775	Watt	24.0	2.7
1792	Watt	39.0	4.5
1816	Woolf compound engine	68.0	7.5
1828	improved Cornish engine	104.0	12.0
1834	improved Cornish engine	149.0	17.0
1878	Corliss compound engine	150.0	17.2
1906	triple-expansion engine	203.0	23.0

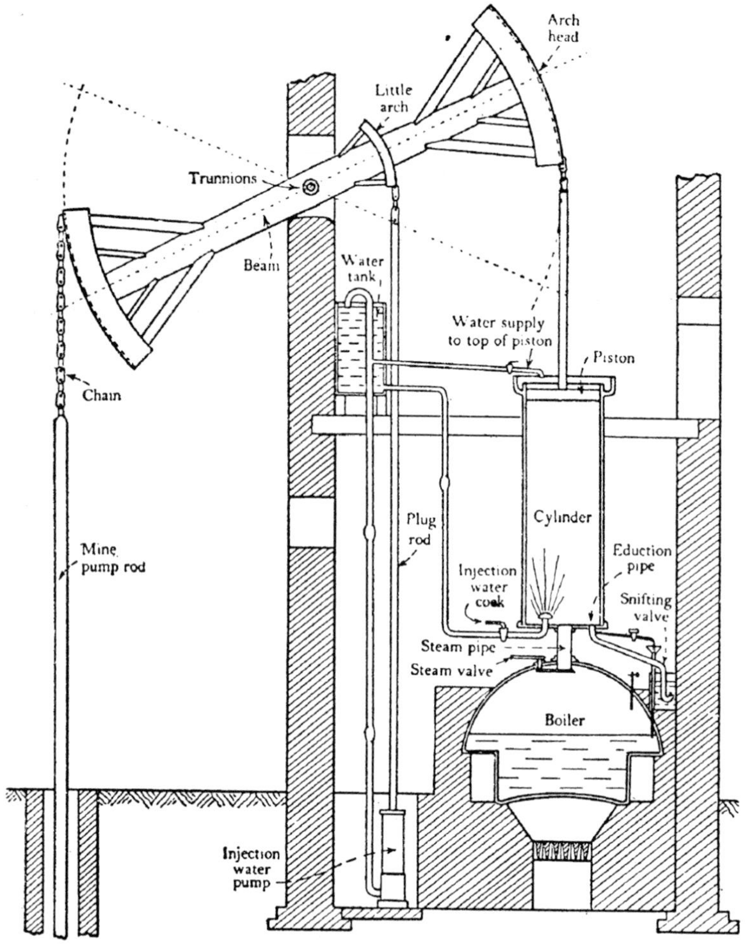

Figure 9.7: Newcomen's atmospheric steam engine of 1712. (From H.W. Dickenson, 1939, in *A Short History of the Steam Engine*, Fig. 7. Cambridge: Cambridge Unviersity Press.)

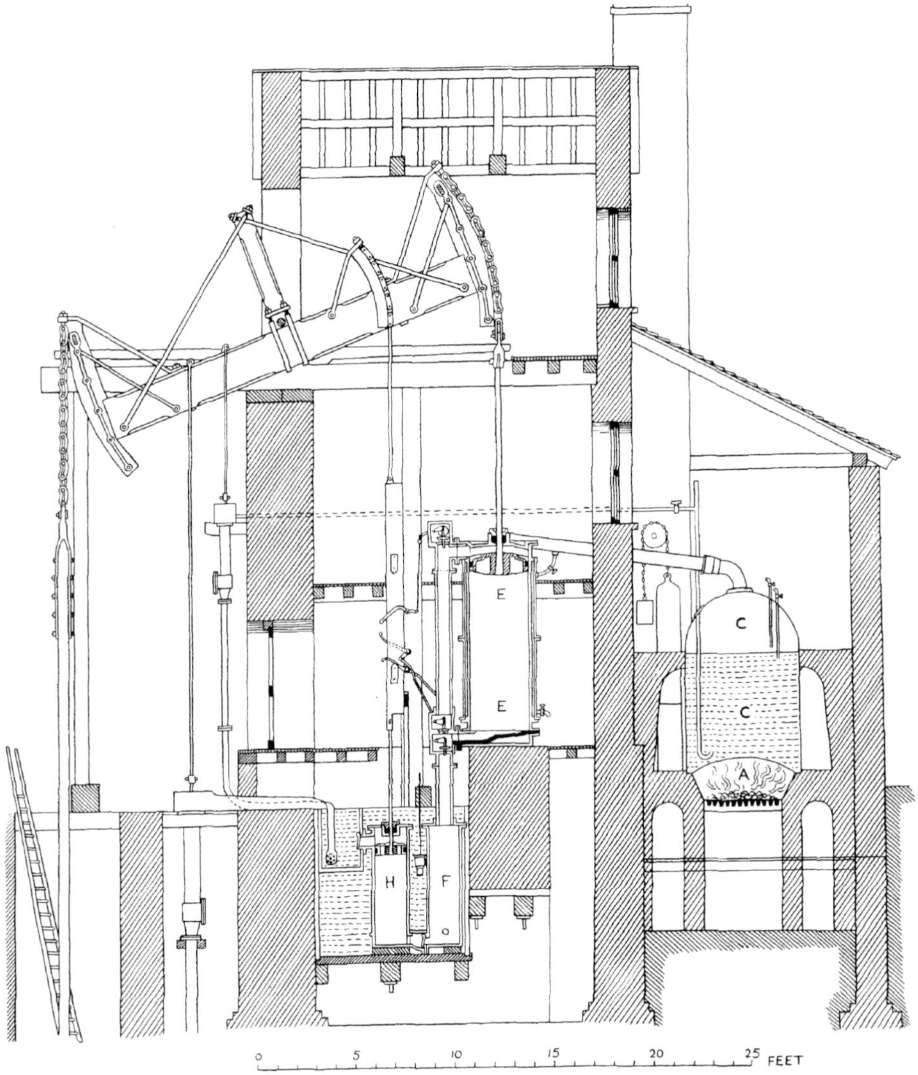

Figure 9.8: James Watt's single-acting steam engine of 1788. The boiler C is placed in an outhouse. The cylinder E is kept at a high temperature all the time. F is the separate condenser and H an air pump. (From H.W. Dickenson, 1958, in *A History of Technology*, Vol. IV, eds. C. Singer, E.J. Holmyard, A.R. Hall and T.I. Williams, p. 184, Oxford: Clarendon Press. Reprinted by permission of Oxford University Press.)

by the air pump H. The steam in E condensed when the condenser valve was opened and the power stroke took place. Steam was then allowed to fill the upper part of the cylinder and, once the stroke was finished, steam was allowed into the lower section of the cylinder, so that the weight of the pump rod restored the piston to its original position.

Pressure to increase the efficiency of steam engines continued from the mine owners, who were interested in reducing the costs of pumping operations. In 1782, Watt patented a double-action steam engine in which steam power was used in both the power and return

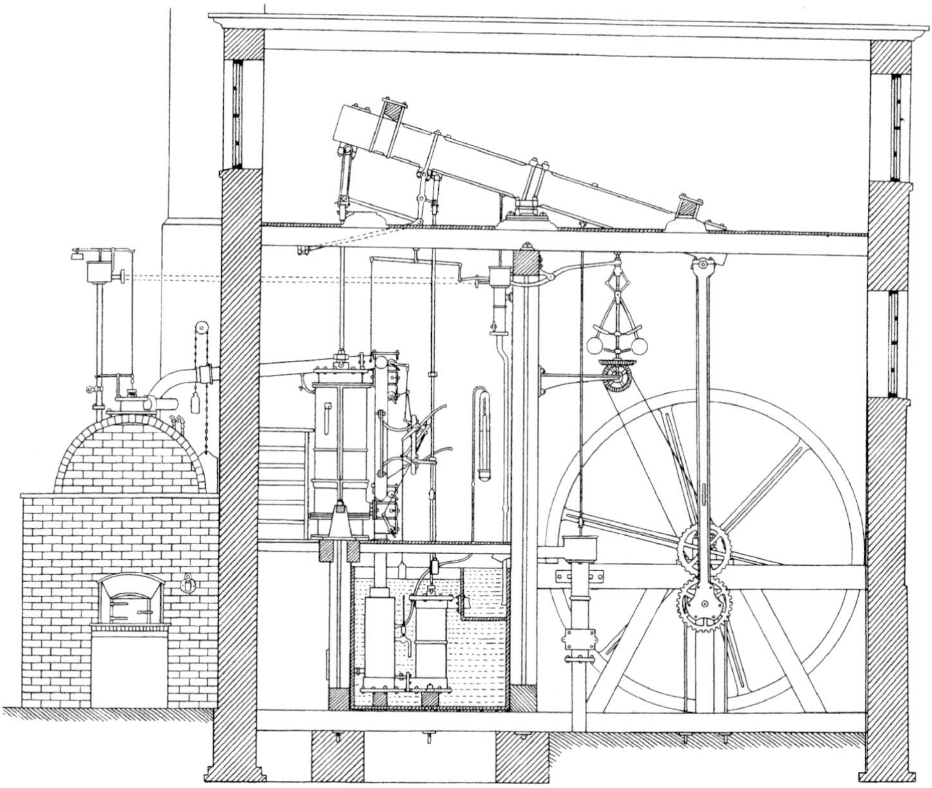

Figure 9.9: James Watt's double-acting rotative steam engine of 1784. The parallel motion bars are seen on the raised end of the beam. The sun-and-planet gear converts the reciprocating action of the engine into rotary motion. (From H.W. Dickenson, 1958, in *A History of Technology*, Vol. IV, eds. C. Singer, E.J. Holmyard, A.R. Hall and T.I. Williams, p. 187, Oxford: Clarendon Press. Reprinted by permission of Oxford University Press.)

strokes of the piston, resulting in twice the amount of work done. This required a new design of the parallel motion bars, so that the piston could push as well as pull. Watt referred to this as 'one of the most ingenious, simple pieces of mechanism I have contrived'. There was also a governor to regulate the supply of steam. In the version of the steam engine shown in Fig. 9.9, the reciprocating motion of the piston was converted into rotary motion by the sun-and-planet gear.

The next significant advance occurred in the early 1800s with Richard Trevithick's steam engine, which operated with high-pressure steam rather than at atmospheric pressure. As a consequence, more power could be generated from a smaller steam engine and these new engines were built into an early steam vehicle and an early version of the steam locomotive. Table 9.1 shows how the efficiencies of steam engines improved over the period from Newcomen to the beginning of the twentieth century.[2] It is amusing that the 'duty' of the engines was defined in units of one million foot-poundals of energy per bushel of coal, where one bushel is 84 pounds, surely the ultimate non-SI unit. Table 9.1 was of the greatest importance to the manufacturers and industrialists who were to drive through the industrial revolution. There was great interest in trying to understand what the maximum efficiency of a heat engine could be and these very practical considerations led to Sadi Carnot's interest in the efficiencies of heat engines.

9.4.2 *Sadi Carnot and the réflexions*

Carnot was the eldest son of Lazare Carnot, who was a member of the Directory after the French Revolution and later, during the Hundred Days in 1815, Napoleon's Minister of the Interior. After 1807, Lazare Carnot devoted much of his energies to the education of his sons. Sadi Carnot was educated at the elite École Polytechnique, where his teachers included Poisson, Gay-Lussac and Ampère. After a period as a military engineer, from 1819 onwards he was able to devote himself wholly to research. Carnot was concerned about the design of efficient steam engines, since he considered that France had fallen behind the United Kingdom in this technology. He was also impressed by the fact that although reliable estimates of the efficiencies of working heat engines had been established, those who had developed them were inventors and engineers, rather than physicists, and certainly not theorists. His great work *Réflexions sur la puissance motrice du feu et sur les machines propres à développer cette puissance*[9] was published in 1824; it is normally translated as *Reflections on the Motive Power of Fire*. The treatise concerned a theoretical analysis of the maximum efficiency of heat engines. In his approach, he was strongly influenced by his father's work on steam engines. The treatise is, however, of much greater generality and is an intellectual achievement of the greatest originality.

Earlier work on the maximum efficiency of steam engines had involved empirical studies, such as comparison of the fuel input with the work output, or else theoretical studies on the basis of specific models for the behaviour of the gases in heat engines. Carnot's aims were unquestionably practical in intent, but his basic perceptions were entirely novel. In my view, the imaginative leap involved is one of genius.

In seeking to derive a completely general theory of heat engines, he was guided by his father's premise, in the latter's study of steam engines, of the impossibility of perpetual motion. In the *Réflexions*, Sadi Carnot adopted the caloric theory of heat and assumed that caloric was conserved in the cyclic operation of heat engines. He postulated that it was the transfer of caloric from a hotter to a colder body which is the origin of the work done by a heat engine. The flow of caloric was envisaged as being analogous to the flow of fluid which, as in a water-wheel, can produce work when it falls down a potential gradient.

Carnot's basic insights into the operation of heat engines were twofold. The first was the recognition that a heat engine works most efficiently if the transfer of heat takes place as part of a cyclic process. The second was that the crucial factor in determining the amount of work which can be extracted from a heat engine is the temperature difference between the source of heat and the sink into which the caloric flows. It turns out that these basic ideas are independent of the particular model of the heat-flow process.

By another stroke of imaginative insight, he devised the cycle of operations which we now know as the *Carnot cycle*, as an idealisation of the behaviour of any heat engine. We will discuss the cycle in more detail in subsection 9.5.2. A key feature of the Carnot cycle is that all the processes are carried out *reversibly* so that, by reversing the sequence of operations, work can be done on the system and caloric transferred from a colder to a hotter body. By joining together an arbitrary heat engine and a reversed Carnot heat engine, he was able to demonstrate that no heat engine can ever produce more work than a Carnot heat engine. If it were otherwise, by joining the two engines together we could either transfer heat from a colder to a hotter body without doing any work or produce a net amount of work without any net heat transfer, both phenomena being in violation of common experience. The influence of Lazare Carnot's premise about the impossibility of perpetual motion is apparent. We will demonstrate these results formally in subsection 9.5.2.

It is tragic that Sadi Carnot died of cholera at the age of 36 in August 1832, before the profound significance of his work was appreciated by anyone. However, in 1834, Emile Clapeyron reformulated Carnot's arguments analytically and related the ideal Carnot engine to the standard pressure–volume indicator diagram. There the matter rested until William Thomson studied certain aspects of Clapeyron's paper and went back to Carnot's original version in the *Réflexions*. The big problem for Thomson and others at that time was to reconcile Carnot's work, in which caloric was conserved, with Joule's experiments which demonstrated the interconvertibility of heat and work. The matter was resolved by Rudolf Clausius, who showed that Carnot's theorem concerning the maximum efficiency of heat engines was correct but the assumption of no heat loss was wrong. In fact, there is a conversion of heat into work in the Carnot cycle. This reformulation by Clausius constitutes the bare bones of the second law of thermodynamics.[10] As we will show, however, the law goes far beyond the efficiency of heat engines: it serves not only to define the thermodynamic temperature scale but also to resolve the problem of how systems evolve thermodynamically. Let us now develop these concepts more formally, bringing out the basic assumptions made in formulating the second law mathematically.

9.5 The second law of thermodynamics

Let us first discuss the crucial distinction between *reversible* and *irreversible* processes, which has been implicit in a number of the arguments presented so far.

9.5.1 *Reversible and irreversible processes*

A reversible process is one which is carried out infinitely slowly so that, in passing from state A to state B, the system passes through an infinite number of equilibrium states. Such a process is also termed quasistatic. Since the process takes place infinitely slowly, there is no friction or turbulence, no gain in kinetic energy of the moving parts of the system and no sound waves are generated. At no stage are there unbalanced forces. At each stage, we make only an infinitesimal change to the system. The implication is that, by reversing the process precisely, we can get back to the point from which we started and nothing will have changed in either the system or its surroundings. Clearly, if there were frictional losses we could not get back to where we started without extracting some energy from the surroundings.

Let us emphasise this point by considering in detail how we could carry out a reversible isothermal expansion. Suppose we have a large heat reservoir at temperature T and a

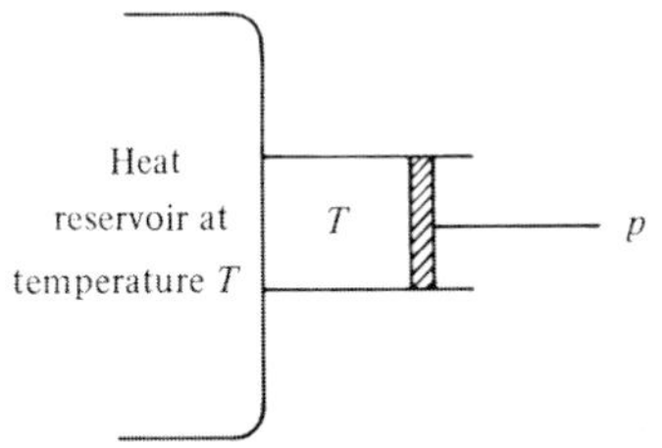

Figure 9.10: Illustrating a reversible isothermal expansion.

cylinder with gas in thermal contact with it, also at temperature T (Fig. 9.10). No heat flows if the two are at the same temperature. But if we allow an infinitesimally small movement of the piston outwards then the gas in the cylinder cools infinitesimally and so an infinitesimal amount of heat flows into the gas by virtue of the temperature difference. This small amount of energy brings the gas back to T. The system is reversible because, if we compress the gas at T slightly, it heats up and heat flows from the gas into the reservoir. Thus, provided we consider only infinitesimally slow changes, the heat flow process occurs reversibly.

Clearly, this is not possible if the reservoir and the piston are at different temperatures. In this case, we cannot reverse the direction of heat flow by making an infinitesimal change in the temperature of the cooler object. This makes the important point that, in reversible processes, the system must be able to evolve from one state to another by passing through an infinite set of equilibrium states which we join together by infinitesimal increments of work and energy flow.

To reinforce this point, let us repeat the argument for a reversible adiabatic expansion. The cylinder is completely thermally isolated from the rest of the Universe. Again, we perform each step infinitesimally slowly. There is no flow of heat in or out of the system and there is no friction. Therefore, since each infinitesimal step is reversible, we can perform the whole expansion by adding lots of them together.

Let us contrast this behaviour with the other two expansions we have described. In the Joule expansion, the gas expands suddenly into a large volume and this cannot take place without all sorts of non-equilibrium processes taking place. Unlike the adiabatic and isothermal expansions, there is no way in which we can design a series of equilibrium states through which the final state is reached. The Joule–Kelvin expansion is a case in which there is a discontinuity in the properties of the gas on passing into the second cylinder. We do not reach the final state by passing through a series of equilibrium states infinitely slowly.

This long introduction stresses the point that reversible processes are highly idealised, but they provide the norm against which all real processes can be compared. It is important to appreciate exactly what we mean by the term 'infinitely slowly', or quasistatically, in the above arguments. In practice, what this means is that the time scale over which the change in the system takes place must be very much longer than the time it takes for thermodynamic equilibrium to be re-established at each stage in the expansion by elementary physical processes within the system. If this is the case, the system can approach ideal behaviour, if enough care is taken.

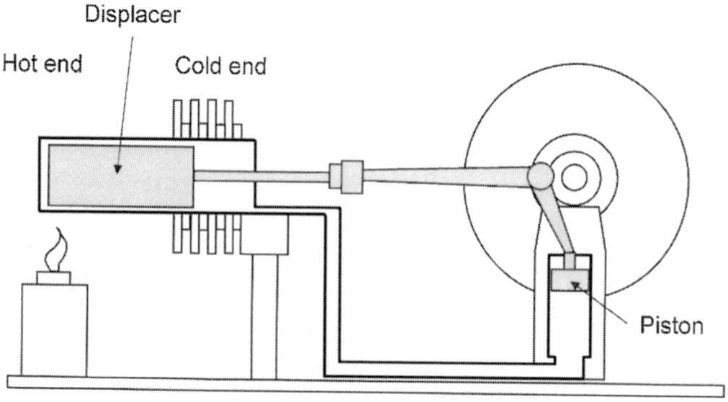

Figure 9.11: Illustrating the construction of a simple Stirling engine. (After the 'Super Vee' Stirling engine, available from the All Hot Air Company.)

9.5.2 *The Carnot cycle and the definition of thermodynamic temperature*

Clausius's statement of the second law is another 'fact of experience', which is asserted without proof:

No process is possible whose sole result is the transfer of heat from a colder to a hotter body.

The implication is that heat cannot be transferred from a cold to a hot body without some interaction with the surroundings of the system. Notice that the law assumes that we can define what we mean by the terms 'hotter' and 'colder' – all we have available so far is an empirical temperature scale defined on the basis of the properties of perfect gases. We will show that this is identical to the thermodynamic temperature scale.

It is a matter of common experience that it is easy to convert work into heat, for example, by friction, but it is much more difficult to devise an efficient means of converting heat into work, as is illustrated by the efficiencies of steam engines listed in Table 9.1. Carnot's deep insights into the operation of an ideal heat engine are worth repeating.

(i) In any efficient heat engine, there is a working substance, which is used *cyclically* in order to minimise the loss of heat.
(ii) In all real heat engines, heat enters the working substance at a high temperature T_1, the working substance does work and then returns heat to a heat sink at a lower temperature T_2.
(iii) The best we can achieve is to construct a heat engine in which all processes take place *reversibly*, that is, all the changes are carried out infinitely slowly so that there are no dissipative losses, such as friction or turbulence.

These are idealisations of the operation of real heat engines. A simple illustration of how these principles work in practice is provided by the *Stirling heat engine*, which was invented by the Rev. Dr Robert Stirling of Kilmarnock in Scotland in 1816 (Fig. 9.11). In Fig. 9.11 the working substance, air, fills the whole of the volume indicated by a heavy line. At the beginning of the cycle, the gas is raised to a high temperature by the heater at the 'hot end', and the piston is pushed upwards, resulting in the configuration shown in the diagram.

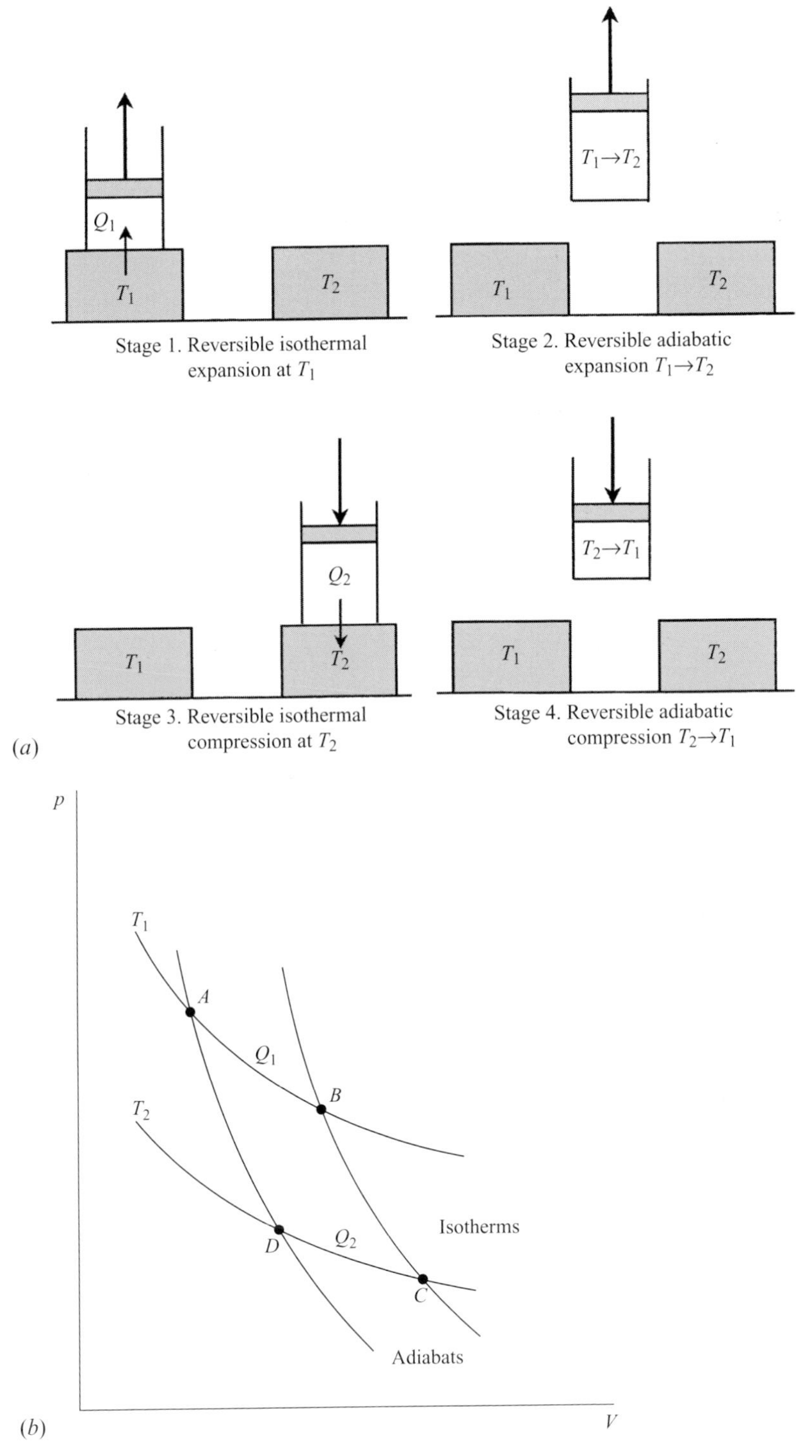

Figure 9.12: (*a*) Illustrating schematically the four stages in the Carnot cycle of an ideal engine. (*b*) The four stages of operation of an ideal Carnot engine indicated on a p–V or indicator diagram.

Stirling's brilliant innovation was to move the working substance from the heat source to a heat sink by means of a *displacer*, which is shown in Fig. 9.11. The hot gas comes into contact with the 'cold end' of the volume containing the displacer, which is maintained at room temperature by cooling fins or by a flow of water through cooling pipes. The hot gas gives up heat at the cold end so that the pressure drops and the return stroke of the piston takes place. The piston is attached to the displacer, which moves to the cold end of its cylinder, thus pushing the cooled gas to the hot end where it is reheated and the next power stroke can begin. By careful design, the efficiencies of such heat engines can approach the theoretical maximum values discussed below.

The sequence of operations is almost identical to that studied by Carnot in his paper of 1824. The working substance acquires heat at the hot end at temperature T_1, it does work on the piston and then gives up heat at the cold end at temperature T_2. The working substance is then reheated and the cycle begins again. There are many different types of Stirling engine and they are of the greatest interest as an ecologically friendly source of power. It is intriguing that Carnot did not know of Stirling's invention when he wrote the *Réflexions*.

Carnot's famous cycle of operation of a perfect heat engine is illustrated in Fig. 9.12(a). I strongly recommend Feynman's careful description of the cycle in Chapter 44 of Vol. 1 of his *Lectures on Physics*.[11] My exposition is modelled on his.

Two very large heat reservoirs, 1 and 2, are maintained at temperatures T_1 and T_2. The working substance is gas contained within a cylinder having a frictionless piston. We now carry out the following reversible sequence of operations, which simulates how heat engines do work. Recall that 'reversible' means that the operations are carried out infinitely slowly and without loss of heat through dissipative processes.

(i) The cylinder is placed in thermal contact with the reservoir at the higher temperature, T_1, and a reversible isothermal expansion is carried out. As described above, in a reversible isothermal expansion heat flows from the reservoir into the gas in the cylinder. At the end of the expansion, a quantity of heat Q_1 has been absorbed by the gas and a certain amount of work has been done by the working substance on the surroundings. Let us indicate this change as A to B on a p–V indicator diagram (Fig. 9.12(b)).

(ii) The cylinder is now isolated from the heat reservoirs and a reversible adiabatic expansion is performed so that the temperature of the working substance falls from T_1 to T_2, the temperature of the second reservoir. Again work is done on the surroundings.

(iii) The cylinder is now placed in thermal contact with the second reservoir at temperature T_2 and the gas in the cylinder is compressed, again reversibly at temperature T_2. In this process, the temperature of the gas is continuously increased infinitesimally above T_2 so that heat flows into the second reservoir. In this process, heat Q_2 is returned to the second reservoir and work is done on the working substance. The isothermal compression continues until the isotherm at T_2 intersects the adiabatic curve that will take the working substance back to its initial state.

(iv) The cylinder is again isolated from the heat reservoirs and a reversible adiabatic compression is carried out, bringing the gas back to its original state. Again work is done on

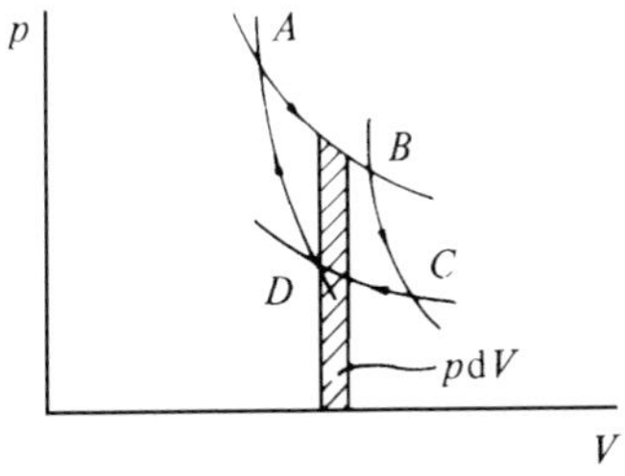

Figure 9.13: The area *ABCD* within the closed cycle in the indicator diagram represents the total work done, $W = \oint p\,\mathrm{d}V$.

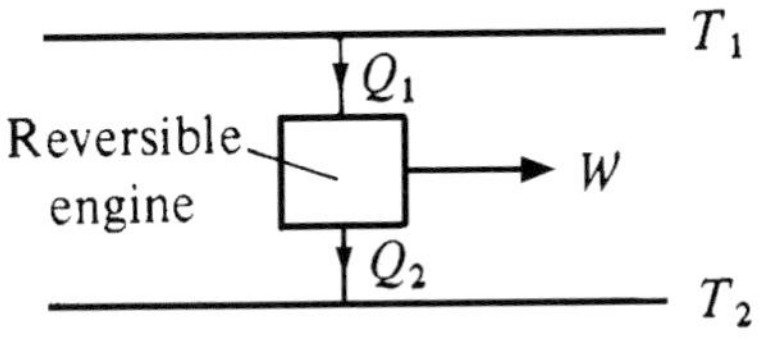

Figure 9.14: A representation of a reversible Carnot heat engine.

the working substance by the surroundings to increase its temperature under adiabatic conditions.

This is the sequence of operations of an ideal heat engine. The net effect of this cycle is that the working substance gains heat Q_1 from reservoir 1 at T_1 and returns heat Q_2 to reservoir 2 at T_2. In the process, the heat engine has done work. The amount of work done is just $W = \oint p\,\mathrm{d}V$ taken round the cycle. From the diagram it is obvious that the cyclic integral is equal to the area of the closed curve described by the cycle of the engine in the p–V indicator diagram (Fig. 9.13). We also know from the first law of thermodynamics that this work must be equal to the net heat supplied to the working substance, $Q_1 - Q_2$, that is,

$$W = Q_1 - Q_2 \tag{9.30}$$

Note incidentally that the adiabatic curves must be steeper than the isotherms because for the former $pV^\gamma = \text{constant}$, $\gamma > 1$, and for the latter $pV = \text{constant}$, for all gases. We can represent this machine schematically, as shown in Fig. 9.14.

The beautiful aspect of this argument is that all stages of the cycle are reversible and so we can run the whole sequence of actions backwards (Fig. 9.15). The *reversed Carnot cycle* is therefore:

(i) an adiabatic expansion from A to D, reducing the temperature of the working substance from T_1 to T_2;

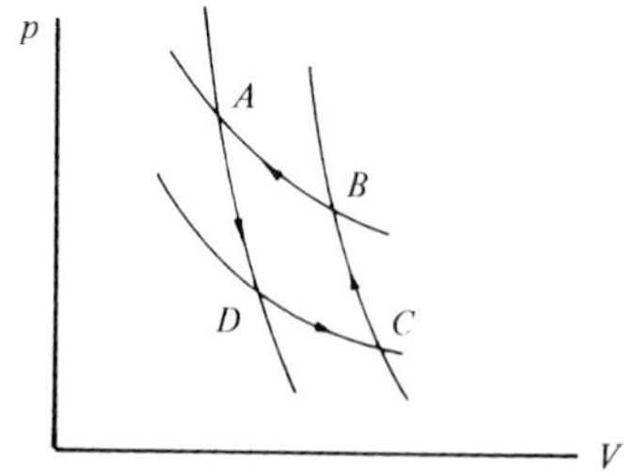

Figure 9.15: A reversed Carnot cycle representing a refrigerator or a heat pump.

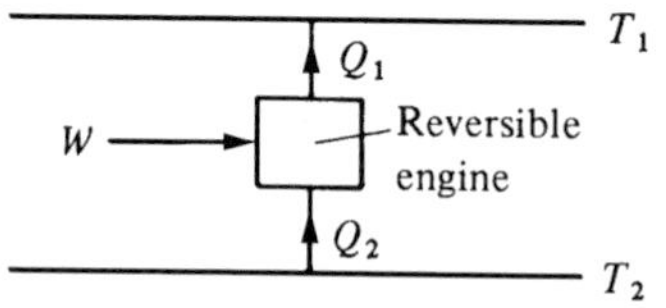

Figure 9.16: A representation of a reversed Carnot cycle acting as a refrigerator or heat pump.

(ii) an isothermal expansion at T_2 in which heat Q_2 is withdrawn from the reservoir;
(iii) an adiabatic compression from C to B, so that the working substance returns to temperature T_1;
(iv) an isothermal compression at T_1, so that the working substance gives up heat Q_1 to the reservoir at T_1.

The reversed Carnot cycle describes the operation of an ideal refrigerator or heat pump, in which heat is extracted from the reservoir at the lower temperature and passed to the reservoir at the higher temperature. This refrigerator or heat pump is shown schematically in Fig. 9.16. Notice that, in the reversed cycle, work has been done in order to extract the heat from T_2 and deliver it to T_1.

We can now define the *efficiencies* of heat engines, refrigerators and heat pumps. For the standard heat engine running forwards, the efficiency is defined to be

$$\eta = \frac{\text{work done in a cycle}}{\text{heat input}} = \frac{W}{Q_1} = \frac{Q_1 - Q_2}{Q_1}.$$

For a refrigerator,

$$\eta = \frac{\text{heat extracted from reservoir 2}}{\text{work done}} = \frac{Q_2}{W} = \frac{Q_2}{Q_1 - Q_2}.$$

For a heat pump, the cycle is run backwards but used to supply Q_1 at T_1 by doing work W:

$$\eta = \frac{\text{heat supplied to reservoir 1}}{\text{work done}} = \frac{Q_1}{W} = \frac{Q_1}{Q_1 - Q_2}.$$

We can now do three things – prove Carnot's theorem, show the equivalence of the Clausius and Kelvin statements of the second law of thermodynamics and derive the concept of thermodynamic temperature.

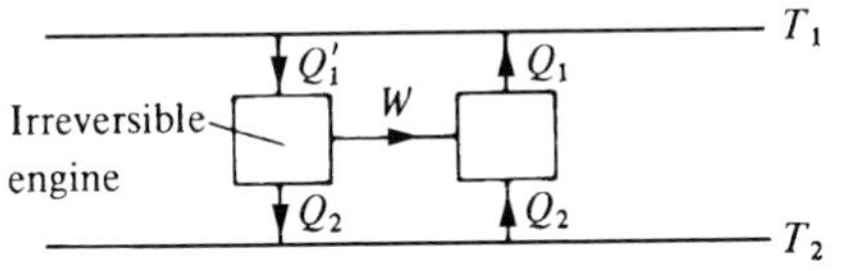

Figure 9.17: Illustrating the proof of Carnot's theorem.

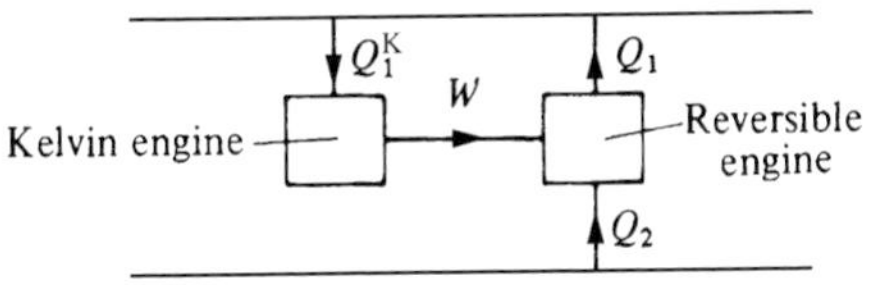

Figure 9.18: Illustrating the equivalence of Kelvin's and Clausius's statements of the second law using a hypothetical Kelvin engine which converts all the heat supplied at T_1 into work.

9.5.3 Carnot's theorem

Carnot's theorem states that

Of all heat engines working between two given temperatures, none can be more efficient than a reversible heat engine.

Suppose the opposite were true, namely, that we could construct an irreversible heat engine which has efficiency greater than that of a reversible engine working between the same temperatures. Then we could use the work produced by the irreversible engine to drive the reversible engine backwards (Fig. 9.17). Let us now regard the system consisting of the two combined engines as a single engine. All the work produced by the irreversible engine is used to drive the reversible engine backwards. Therefore, since our supposition is that $\eta_{\text{irr}} > \eta_{\text{rev}}$, we conclude that $W/Q'_1 > W/Q_1$, and consequently that $Q_1 - Q'_1 > 0$. Thus, overall, the only net effect of this combined engine is to produce a transfer of energy from the lower to the higher temperature without doing any work, which is forbidden by the second law. Therefore, no irreversible engine can have efficiency greater than that of a reversible engine operating between the same two temperatures. This is the theorem which Carnot published in his *Réflexions* of 1824, 26 years before the formal statement of the laws of thermodynamics by Clausius in 1850.

9.5.4 The equivalence of the Clausius and Kelvin statements of the second law

Kelvin's statement of the second law is

No process is possible whose sole result is the complete conversion of heat into work.

Suppose it were possible for an engine to convert completely heat Q_1 into work W. Again, we could use this work to drive a reversible Carnot engine backwards as a heat pump (Fig. 9.18). Then, regarded as a single system, no net work is done but the total amount of

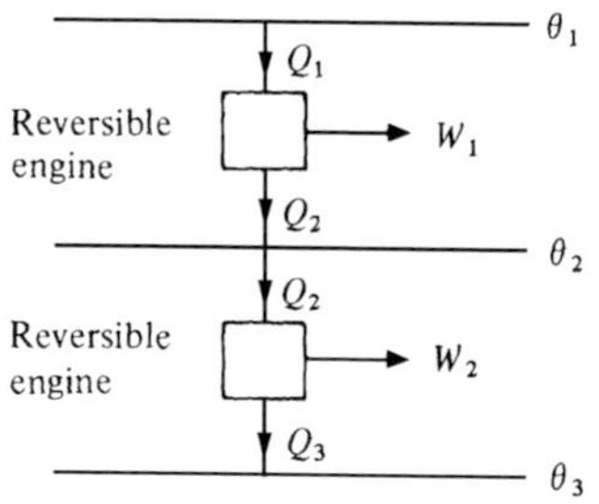

Figure 9.19: Illustrating the origin of the definition of thermodynamic temperature.

heat delivered to the reservoir at T_1 is

$$Q_1 - Q_1^{\mathrm{K}} = Q_1 - W = Q_1 - (Q_1 - Q_2) = Q_2,$$

where Q_1^{K} means the heat supplied to the hypothetical Kelvin engine, which is completely converted into work W. There is, therefore, a net transfer of heat Q_2 to the reservoir T_1 without any other change in the Universe and this is forbidden by the Clausius statement. Therefore, the Clausius and Kelvin statements are equivalent.

9.5.5 *Thermodynamic temperature*

At last we can define the *thermodynamic temperature*, the clue coming from Carnot's theorem. A reversible heat engine working between two temperatures has the maximum possible efficiency and this is a unique function of the two temperatures, which so far we have called rather sloppily T_1 and T_2. Expressing this result in terms of the heat input and output,

$$\eta = \frac{Q_1 - Q_2}{Q_1} \qquad \text{so that} \qquad \frac{Q_1}{Q_2} = \frac{1}{1-\eta}. \tag{9.31}$$

Let us now be somewhat more precise in the logic of the argument. Let the ratio Q_1/Q_2 be some function $f(\theta_1, \theta_2)$ of the empirical temperatures θ_1 and θ_2. Now join together two heat engines in series, as shown in Fig. 9.19. The engines are connected in such a way that the heat Q_2 delivered at temperature θ_2 is used to supply a second heat engine, which then delivers work W_2 and transfers heat Q_3 to the reservoir at θ_3. Thus,

$$\frac{Q_1}{Q_2} = f(\theta_1, \theta_2) \qquad \text{and} \qquad \frac{Q_2}{Q_3} = f(\theta_2, \theta_3). \tag{9.32}$$

However, we can consider the combined system as a single engine operating between θ_1 and θ_3, in which case

$$\frac{Q_1}{Q_3} = f(\theta_1, \theta_3). \tag{9.33}$$

Thus, since

$$\frac{Q_1}{Q_3} = \frac{Q_1}{Q_2} \times \frac{Q_2}{Q_3}$$

it follows that

$$f(\theta_1, \theta_3) = f(\theta_1, \theta_2) f(\theta_2, \theta_3). \tag{9.34}$$

Consequently, the function f must have the form

$$f(\theta_1, \theta_3) = \frac{g(\theta_1)}{g(\theta_3)} = \frac{g(\theta_1)}{g(\theta_2)} \times \frac{g(\theta_2)}{g(\theta_3)}. \tag{9.35}$$

We then adopt a definition of *thermodynamic temperature* T consistent with this requirement:

$$\frac{g(\theta_1)}{g(\theta_3)} \equiv \frac{T_1}{T_3}, \qquad \text{that is,} \qquad \frac{Q_1}{Q_3} = \frac{T_1}{T_3}. \tag{9.36}$$

Now we can show that this definition is identical to that corresponding to the perfect-gas temperature scale. We write the perfect-gas-law temperature scale as T^{P}, and hence $pV = RT^{\mathrm{P}}$. For the Carnot cycle with a perfect gas, from (9.29),

$$\begin{aligned} Q_1 &= \textstyle\int_A^B p\,\mathrm{d}V = RT_1^{\mathrm{P}} \ln \frac{V_B}{V_A}, \\ Q_2 &= \textstyle\int_D^C p\,\mathrm{d}V = RT_2^{\mathrm{P}} \ln \frac{V_C}{V_D}. \end{aligned} \tag{9.37}$$

Along the adiabatic legs of the cycle,

$$pV^{\gamma} = \text{constant} \qquad \text{and} \qquad T^{\mathrm{P}} V^{\gamma-1} = \text{constant}. \tag{9.38}$$

Therefore,

$$\begin{aligned} \left(\frac{V_B}{V_A}\right)^{\gamma-1} &= \frac{T_2^{\mathrm{P}}}{T_1^{\mathrm{P}}}. \\ \left(\frac{V_D}{V_C}\right)^{\gamma-1} &= \frac{T_1^{\mathrm{P}}}{T_2^{\mathrm{P}}}. \end{aligned} \tag{9.39}$$

Multiplying the two equations in (9.39) together, we find that

$$\frac{V_B V_D}{V_A V_C} = 1, \tag{9.40}$$

that is,

$$\frac{V_B}{V_A} = \frac{V_C}{V_D} \tag{9.41}$$

and, consequently, from the relations (9.37) and (9.36) we see that

$$\frac{Q_1}{Q_2} = \frac{T_1^{\mathrm{P}}}{T_2^{\mathrm{P}}} = \frac{T_1}{T_2}. \tag{9.42}$$

At last, we have derived a rigorous thermodynamic definition of temperature based upon the operation of perfect heat engines, the line of thought initiated by Carnot.

Now we can rewrite the maximum efficiencies of heat engines, refrigerators and heat pumps in terms of the thermodynamic temperatures between which they work:

$$\begin{aligned}
&\text{Heat engine,} \qquad \eta = \frac{Q_1 - Q_2}{Q_1} = \frac{T_1 - T_2}{T_1};\\
&\text{Refrigerator,} \qquad \eta = \frac{Q_2}{Q_1 - Q_2} = \frac{T_2}{T_1 - T_2};\\
&\text{Heat pump,} \qquad \eta = \frac{Q_1}{Q_1 - Q_2} = \frac{T_1}{T_1 - T_2}.
\end{aligned} \tag{9.43}$$

9.6 Entropy

You may have noticed something rather remarkable about the relation (9.42) which provides our definition of thermodynamic temperature. Taking the heats Q_1 and Q_2 to be positive quantities, (9.42) reads

$$\frac{Q_1}{T_1} - \frac{Q_2}{T_2} = 0. \tag{9.44}$$

This can be rewritten in the following form

$$\int_A^C \frac{đQ}{T} - \int_C^A \frac{đQ}{T} = 0, \tag{9.45}$$

since heat is supplied or removed only along the isothermal sections of the Carnot cycle. Since all sections of the cycle are reversible, this means that if we pass from A to C down either leg of the cycle,

$$\underbrace{\int_A^C \frac{đQ}{T}}_{\text{via } B} = \underbrace{\int_A^C \frac{đQ}{T}}_{\text{via } D}. \tag{9.46}$$

This strongly suggests that we have discovered another function of state, because the integral (9.46) implies that it does not matter how we get from A to C. This is confirmed by the following argument.

For any two-coordinate system, we can always move between any two points in the indicator diagram by a large number of mini-Carnot cycles, all of them reversible, as illustrated in Fig. 9.20. Whichever path we take, we will always find the same result. Mathematically, for any two points A and B,

$$\sum_A^B \frac{đQ}{T} = \text{constant}. \tag{9.47}$$

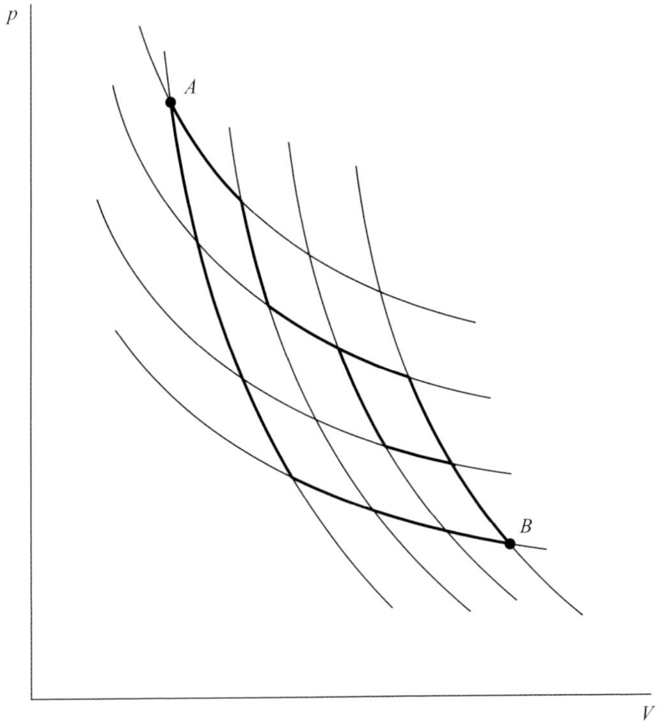

Figure 9.20: An indicator diagram illustrating different paths following isotherms and adiabats between A and B.

Writing this in integral form,

$$\int_A^B \frac{đQ}{T} = \text{constant} = S_B - S_A. \tag{9.48}$$

The new function of state S is called the *entropy* of the system. Notice that it is defined by means of any *reversible* process connecting the states A and B. By T, we mean the temperature at which the heat is supplied to the system, which in this case is the same as the temperature of the system itself, because the processes of heat exchange are performed reversibly.

An amusing footnote to the controversy about caloric and the nature of heat is that, in a sense, *entropy* was Carnot's *caloric fluid* since entropy, rather than heat, is conserved in the Carnot cycle.

Now, for any real engine, the efficiency must be less than that of an ideal Carnot engine working between the same temperatures, and therefore

$$\eta_{\rm irr} \leq \eta_{\rm rev},$$

$$\frac{Q_1 - Q_2}{Q_1} \leq \frac{Q_1(\text{rev}) - Q_2(\text{rev})}{Q_1(\text{rev})}.$$

Therefore,

$$\frac{Q_2}{Q_1} \geq \frac{Q_2(\text{rev})}{Q_1(\text{rev})} = \frac{T_2}{T_1},$$

$$\frac{Q_2}{T_2} \geq \frac{Q_1}{T_1}.$$

Thus, for this cycle,

$$\frac{Q_1}{T_1} - \frac{Q_2}{T_2} \leq 0.$$

This suggests that, in general, when we add together a large number of mini-Carnot cycles

$$\oint \frac{đQ}{T} \leq 0, \tag{9.49}$$

where the $đQ$s are the heat inputs and extractions in the real cycle. Notice that the equals sign applies only in the case of a reversible heat engine. Notice also the sign convention for heat: when heat enters the system, the sign is positive; when it is removed, the sign is negative.

The relation $\oint đQ/T \leq 0$ is fundamental to the whole of thermodynamics and is known as *Clausius's theorem*. This proof of Clausius's theorem is adequate for our present purposes. A limitation is that it only applies to two-coordinate systems and, in general, one needs to be able to deal with multi-coordinate systems. This question is treated nicely in Pippard's book,[5] where it is shown that in fact the same result is true for multi-coordinate systems.

9.7 The law of increase of entropy

Notice that the change in entropy (9.48) is defined by means of a reversible sequence of changes from the state A to the state B. When we take the system from state A to state B in the real imperfect world, we can never effect precisely reversible changes and so the heat involved is not directly related to the entropy difference between the initial and final states. Let us again compare what happens in a reversible and an irreversible change between two states. Suppose an irreversible change takes place from A to B (Fig. 9.21). Then we can complete the cycle by taking any *reversible* path back from B to A. Then, according to Clausius's theorem,

$$\oint \frac{đQ}{T} \leq 0 \qquad \text{or} \qquad \underbrace{\int_A^B \frac{đQ}{T}}_{\text{irrev}} + \underbrace{\int_B^A \frac{đQ}{T}}_{\text{rev}} \leq 0,$$

that is,

$$\underbrace{\int_A^B \frac{đQ}{T}}_{\text{irrev}} \leq \underbrace{\int_A^B \frac{đQ}{T}}_{\text{rev}}.$$

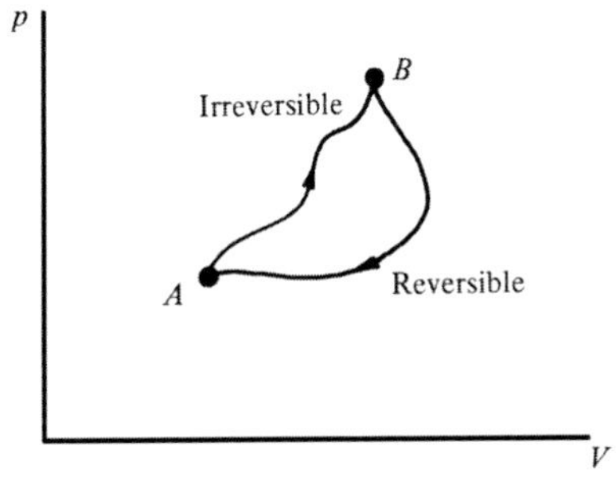

Figure 9.21: Reversible and irreversible paths between the points A and B in a indicator diagram.

But, because the second change is reversible,

$$\underbrace{\int_A^B \frac{đQ}{T}}_{\text{rev}} = S_B - S_A,$$

according to the definition of entropy. Consequently,

$$\underbrace{\int_A^B \frac{đQ}{T}}_{\text{irrev}} \leq S_B - S_A,$$

or, writing this for an irreversible differential heat flow,

$$\frac{đQ}{T} \leq dS. \tag{9.50}$$

Thus, we obtain the general result that, for any differential heat flow, $đQ/T \leq \mathrm{d}S$, where the equality holds only in the case of a reversible change.

This is a key result because at last we can see how we can find quantitatively what it is that determines the direction in which physical processes evolve. Notice that, if we are dealing with reversible processes, the temperature at which heat is supplied is that of the system itself. This is not necessarily the case for irreversible processes.

For an *isolated system*, there is no thermal contact with the surroundings and hence in the above inequality $đQ = 0$; therefore

$$\mathrm{d}S \geq 0. \tag{9.51}$$

Thus, the entropy of an isolated system cannot decrease. A corollary of this is that, in approaching equilibrium, the entropy of an isolated system must tend towards a maximum, and the final equilibrium configuration will be that for which the entropy is the greatest.

For example, consider two bodies at temperatures T_1 and T_2 and suppose that a small amount of heat $đQ$ is exchanged between them. Then heat flows from the hotter to the colder body and we find an entropy decrease $đQ/T_1$ of the hotter body and an entropy increase $đQ/T_2$ of the colder body, so that the total entropy change is $đQ/T_2 - đQ/T_1 > 0$, which is positive, meaning that we were correct in our inference about the direction of heat flow.

So far we have only dealt with isolated systems. Now let us consider the case in which there is thermal contact between the system and its surroundings. As a first example, take a system round a complete cycle of operations so that it ends up in exactly the same state as at the beginning of the cycle, which, in real systems, may well involve irreversible processes. Since the system ends up in exactly the same state in which it started, all the functions of state are exactly the same and hence, *for the system*, the total change in entropy is zero, $\Delta S = 0$. According to Clausius' theorem, this must be greater than or equal to $\oint đQ_{\rm sys}/T$, that is,

$$0 = \Delta S \geq \oint \frac{đQ_{\rm sys}}{T}.$$

In order to return to the initial state, heat must have been communicated between the system and its surroundings. The most efficient way of doing this is to transfer the heat reversibly to and from the surroundings. Then, in each stage of the cycle, $đQ_{\rm sys} = -đQ_{\rm surr}$ and hence

$$0 = \Delta S \geq \oint \frac{đQ_{\rm sys}}{T} = -\oint \frac{đQ_{\rm surr}}{T}.$$

Since the heat transfer with the surroundings is reversible, we can equate the last quantity $\oint đQ_{\rm surr}/T$ with $\Delta S_{\rm surr}$. Thus, we find

$$0 \leq -\Delta S_{\rm surr} \qquad \text{or} \qquad \Delta S_{\rm surr} \geq 0.$$

Thus, although there is no change in entropy of the system itself, there is an increase in the entropy of the surroundings so that, for the Universe as a whole, the entropy increases.

As a second example, consider an irreversible change taking a system from states 1 to 2. Again we assume that the heat transfer with the surroundings takes place reversibly. Then, as above,

$$\Delta S_{\rm sys} \geq \int_1^2 \frac{đQ_{\rm irr}}{T} = -\int \frac{đQ_{\rm surr}}{T} = -\Delta S_{\rm surr},$$
$$\Delta S_{\rm sys} + \Delta S_{\rm surr} \geq 0.$$

Notice the implication of these examples. When irreversible processes are present in the cycle, although the entropy of the system itself could stay the same or even decrease, the entropy of the Universe as a whole always increases. It is important to consider the surroundings as well as the system itself in this type of argument.

These examples illustrate the origin of the popular expression of the two laws of thermodynamics, due to Clausius, who in 1865 first used the word 'entropy', the Greek word for transformation.

(i) The energy of the Universe is constant.
(ii) The entropy of the Universe tends to a maximum.

The entropy change of the Universe need not necessarily involve heat exchange. It is a measure of the *irreversibility* of the process; the Joule expansion of a gas is a good example of an irreversible process in which there is no heat exchange with the surroundings.

We can obtain a general expression for the entropy change of a perfect gas by considering a *reversible* path from $A = (V_0, T_0)$ to $B = (V, T)$. For such a path, $\int_A^B đQ/T = \Delta S$. Now

$$đQ = dU + p\,dV,$$

and hence for 1 mole of gas

$$đQ = C_V dT + \frac{RT}{V} dV.$$

Therefore

$$\int_A^B \frac{đQ}{T} = C_V \int_A^B \frac{dT}{T} + R \int_A^B \frac{dV}{V},$$

that is,

$$S - S_0 = C_V \ln \frac{T}{T_0} + R \ln \frac{V}{V_0}. \tag{9.52}$$

Now, since the entropy change of a system depends *only* upon the endpoints, we can apply (9.52) to a perfect gas before and after a Joule expansion from V_0 to V. Since $T =$ constant in the Joule expansion of a perfect gas, (9.52) becomes

$$S - S_0 = R \ln \frac{V}{V_0}. \tag{9.53}$$

Thus, although there is no heat flow, the system has undergone an irreversible change and, as a result, there must be an increase in the entropy of the Universe in reaching the new equilibrium state from the old. In a reversible path between the same two endpoints the entropy change of the gas would have been cancelled by the entropy change of the surroundings.

Equation (9.53) is a key result which we will meet again in the study of the origin of quanta. Notice also that the increase in entropy of the gas is associated with a change in the volume of real space accessible to the system. We will find a corresponding result for the more general case of phase space in Chapter 10.

Finally, let us write down the entropy change of a perfect gas in an interesting form. From equation (9.52),

$$\begin{aligned} S - S_0 &= C_V \ln \frac{pV}{p_0 V_0} + R \ln \frac{V}{V_0} \\ &= C_V \ln \frac{p}{p_0} + (C_V + R) \ln \frac{V}{V_0} \\ &= C_V \ln \frac{pV^\gamma}{p_0 V_0^\gamma}. \end{aligned} \tag{9.54}$$

Thus, if the expansion is reversible and adiabatic, so that $pV^\gamma =$ constant, there is no entropy change. For this reason, reversible adiabatic expansions are often called *isentropic* expansions. Note also that this provides us with an interpretation of the entropy function – *isotherms* are curves of constant temperature while *adiabats* are curves of constant entropy.

9.8 The differential form of the combined first and second laws of thermodynamics

At last, we arrive at the equation which synthesises the physical content of the first and second laws of thermodynamics. For reversible processes, the entropy change of a system into which heat $đQ$ flows at temperature T is

$$dS = \frac{đQ}{T},$$

and hence, combining this with the relation $đQ = dU + p\,dV$,

$$T dS = dU + p\,dV. \tag{9.55}$$

More generally, if we write the work done on a system by a set of generalised forces X_i as

$$\sum_i X_i\, dx_i,$$

we find that

$$T dS = dU + \sum_i X_i\, dx_i. \tag{9.56}$$

The remarkable thing about this formula is that it combines the first and second laws entirely in terms of functions of state, and hence the relation must be true for all changes connecting two end-points each in thermodynamic equilibrium.

We shall not take this story any further, but we should note that the relations (9.55) and (9.56) are the origins of some very powerful results, which we will need later. In particular, for any gas, we can write

$$\left(\frac{\partial S}{\partial U}\right)_V = \frac{1}{T}, \tag{9.57}$$

that is, the partial derivative of the entropy with respect to U at constant volume defines the thermodynamic temperature T. This is of particular significance in statistical mechanics, because the concept of entropy S enters very early in the development of statistical mechanics and the internal energy U is one of the first quantities to be determined statistically. Thus, we already have the key relation needed to define the concept of temperature in statistical mechanics.

9.9 References

1 The textbook by Allen and Maxwell contains excellent historical material and quotations, used in this chapter, as well as illustrations of many of the experiments which laid the foundation for classical thermodynamics: Allen, H.S. and Maxwell, R.S. (1952), *A Textbook of Heat*, Parts I and II, London: Macmillan (first edition 1939).

2 Brown, G.I. (1999). *Scientist, Soldier, Statesman, Spy – Count Rumford*. Stroud, Gloucestershire, UK: Sutton Publishing.

3 Fourier, J.B.J. (1822). *Analytical Theory of Heat*, trans. A. Freeman (reprint edition, New York, 1955).
4 Joule, J.P. (1843). See *The Scientific Papers of James Prescott Joule*, two volumes, London 1884–7 (reprint London, 1963).
5 Pippard, A.B. (1966). *The Elements of Classical Thermodynamics*. Cambridge: Cambridge University Press.
6 Forbes, R.J. (1958). In *A History of Technology*, Vol. IV, eds. C. Singer, E.J. Holmyard, A.R. Hall and T.I. Williams, p. 165. Oxford: Clarendon Press.
7 For splendid discussions concerning the development of the steam engine during the industrial revolution, see the articles by R.J. Forbes and H.W. Dickenson in *A History of Technology* (1958), Vol. IV, eds. C. Singer, E.J. Holmyard, A.R. Hall and T.I. Williams. Oxford: Clarendon Press.
8 Forbes, R.J. (1958). Op. cit. p. 164.
9 Carnot, N.L.S. (1824). *Réflexions sur la puissance motrice du feu et sur les machine propres à développer cette puissance*. Paris: Bachelier.
10 Peter Harman's authoritative study provides key material about the history of the intellectual background to many of the theoretical issues discussed in this chapter: P.M. Harman (1982), *Energy, Force and Matter, The Conceptual Development of Nineteenth Century Physics*, Cambridge: Cambridge University Press.
11 Feynman, R.P (1963). *Lectures on Physics*, eds. R.P. Feynman, R.B. Leighton and M. Sands, Vol. 1, Chapter 44. Redwood City, California: Addison-Wesley.

Appendix to Chapter 9: Maxwell's relations and Jacobians

This appendix includes some mathematical results which are useful in dealing with problems in classical thermodynamics. There is no pretence at completeness or mathematical rigour in what follows.

A9.1 *Perfect differentials in thermodynamics*

The use of perfect differentials occurs naturally in classical thermodynamics. The reason is that, for much of the discussion, we have considered two-coordinate systems, in which the thermodynamic equilibrium state is completely defined by only two functions of state. When a system changes from one equilibrium state to another, the change in any function of state depends only upon the initial and final coordinates. Thus, if z is a function of state which is defined completely by two other functions of state, x and y, the differential change $\mathrm{d}z$ can be written

$$\mathrm{d}z = \left(\frac{\partial z}{\partial x}\right)_y \mathrm{d}x + \left(\frac{\partial z}{\partial y}\right)_x \mathrm{d}y. \tag{A9.1}$$

The fact that a given change $\mathrm{d}z$ does not depend upon the way in which we make the incremental steps in $\mathrm{d}x$ and $\mathrm{d}y$ enables us to set restrictions upon the functional dependence of z upon x and y. This relation is most simply demonstrated by two ways in which the

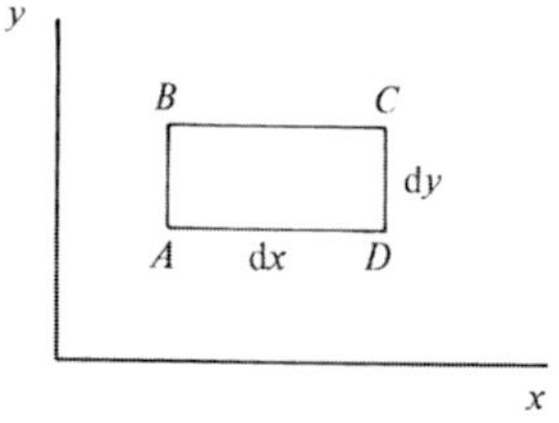

Figure A9.1: Illustrating two ways of reaching the state C from A.

differential change $\mathrm{d}z$ illustrated in Fig. A9.1 can be made. We can reach C from A either via D or B. Since z is a function of state, the change $\mathrm{d}z$ must be the same along the paths ABC and ADC. Along the path ADC, we first move a distance $\mathrm{d}x$ in the x-direction, and then a distance $\mathrm{d}y$ in the y-direction from the point $x + \mathrm{d}x$, so that

$$\begin{aligned} z(C) &= z(A) + \left(\frac{\partial z}{\partial x}\right)_y \mathrm{d}x + \left\{\frac{\partial}{\partial y}\left[z + \left(\frac{\partial z}{\partial x}\right)_y \mathrm{d}x\right]\right\}_x \mathrm{d}y \\ &= z(A) + \left(\frac{\partial z}{\partial x}\right)_y \mathrm{d}x + \left(\frac{\partial z}{\partial y}\right)_x \mathrm{d}y + \frac{\partial^2 z}{\partial y\,\partial x}\mathrm{d}x\mathrm{d}y. \end{aligned} \tag{A9.2}$$

where by $\partial^2 z/\partial y\partial x$ we mean

$$\left[\frac{\partial}{\partial y}\left(\frac{\partial z}{\partial x}\right)_y\right]_x.$$

Along the path ABC, the value of $z(C)$ is

$$\begin{aligned} z(C) &= z(A) + \left(\frac{\partial z}{\partial y}\right)_x \mathrm{d}x + \left\{\frac{\partial}{\partial x}\left[z + \left(\frac{\partial z}{\partial y}\right)_x \mathrm{d}y\right]\right\}_y \mathrm{d}x \\ &= z(A) + \left(\frac{\partial z}{\partial y}\right)_x \mathrm{d}y + \left(\frac{\partial z}{\partial x}\right)_y \mathrm{d}x + \frac{\partial^2 z}{\partial x\,\partial y}\mathrm{d}x\mathrm{d}y. \end{aligned} \tag{A9.3}$$

Notice that, in (A9.2) and (A9.3), the order in which the double partial derivatives is taken is important. Since if z is a function of state then equations (A9.2) and (A9.3) must be identical, the function z must have the property that

$$\frac{\partial^2 z}{\partial y\,\partial x} = \frac{\partial^2 z}{\partial x\,\partial y}, \qquad \text{that is,} \qquad \frac{\partial}{\partial y}\left(\frac{\partial z}{\partial x}\right) = \frac{\partial}{\partial x}\left(\frac{\partial z}{\partial y}\right), \tag{A9.4}$$

which is the mathematical condition for z to be a perfect differential of x and y.

A9.2 Maxwell's relations

We have introduced quite a number of functions of state: p, V, T, S, U, H. Indeed, we could go on to define an infinite number of functions of state by combinations of these functions. When differential changes in any function of state of a system are made, there are corresponding changes in all the other functions of state. The four most important of these differential relations are called *Maxwell's relations*. Let us derive two of them.

Consider first equation (9.55), which relates differential changes in S, U and V:

$$T\,\mathrm{d}S = \mathrm{d}U + p\,\mathrm{d}V \qquad \text{and so} \qquad \mathrm{d}U = T\,\mathrm{d}S - p\,\mathrm{d}V. \tag{A9.5}$$

Since $\mathrm{d}U$ is the differential of a function of state, it must be a perfect differential (see Section A9.1) and hence

$$\mathrm{d}U = \left(\frac{\partial U}{\partial S}\right)_V \mathrm{d}S + \left(\frac{\partial U}{\partial V}\right)_S \mathrm{d}V. \tag{A9.6}$$

Comparison with (A9.5) shows that

$$T = \left(\frac{\partial U}{\partial S}\right)_V, \qquad p = -\left(\frac{\partial U}{\partial V}\right)_S. \tag{A9.7}$$

Since $\mathrm{d}U$ is a perfect differential, we also know from (A9.4) that

$$\frac{\partial}{\partial V}\left(\frac{\partial U}{\partial S}\right) = \frac{\partial}{\partial S}\left(\frac{\partial U}{\partial V}\right). \tag{A9.8}$$

Substituting (A9.7) into (A9.8),

$$\left(\frac{\partial T}{\partial V}\right)_S = -\left(\frac{\partial p}{\partial S}\right)_V. \tag{A9.9}$$

This is the first of four similar relations between T, S, p and V. We can follow exactly the same procedure for the enthalpy H:

$$H = U + pV,$$

and so

$$\begin{aligned} \mathrm{d}H &= \mathrm{d}U + p\,\mathrm{d}V + V\,\mathrm{d}p, \\ &= T\,\mathrm{d}S + V\,\mathrm{d}p. \end{aligned} \tag{A9.10}$$

By exactly the same mathematical procedure, we find

$$\left(\frac{\partial T}{\partial p}\right)_S = \left(\frac{\partial V}{\partial S}\right)_p. \tag{A9.11}$$

We can repeat the analysis for two other important functions of state, the *Helmholtz free energy* $F = U - TS$ and the *Gibbs free energy* $G = U - TS + pV$, and find the other two Maxwell relations:

$$\left(\frac{\partial S}{\partial V}\right)_T = \left(\frac{\partial p}{\partial T}\right)_V, \tag{A9.12}$$

$$\left(\frac{\partial V}{\partial T}\right)_p = -\left(\frac{\partial S}{\partial p}\right)_T. \tag{A9.13}$$

The functions F and G are particularly useful in studying processes which occur at constant temperature and constant pressure respectively.

The set of Maxwell's relations (A9.9), (A9.11), (A9.12) and (A9.13) are useful in many thermodynamic problems and we will use them in subsequent chapters. It is helpful to have a mnemonic by which to remember them; Jacobians provide one way of remembering them.

A9.3 Jacobians in thermodynamics

Jacobians occur in the transformation of products of differentials between different systems of coordinates. In a simple example, suppose that the volume element $\mathrm{d}V = \mathrm{d}x\, \mathrm{d}y\, \mathrm{d}z$ is to be transformed into some other coordinate system in which the differentials of the coordinates are $\mathrm{d}u$, $\mathrm{d}v$ and $\mathrm{d}w$. Then, the volume element can be written

$$\mathrm{d}V_{uvw} = \left| \frac{\partial(x, y, z)}{\partial(u, v, w)} \right| \mathrm{d}u\, \mathrm{d}v\, \mathrm{d}w, \tag{A9.14}$$

where the Jacobian is defined to be

$$\frac{\partial(x, y, z)}{\partial(u, v, w)} \equiv \begin{vmatrix} \frac{\partial x}{\partial u} & \frac{\partial y}{\partial u} & \frac{\partial z}{\partial u} \\ \frac{\partial x}{\partial v} & \frac{\partial y}{\partial v} & \frac{\partial z}{\partial v} \\ \frac{\partial x}{\partial w} & \frac{\partial y}{\partial w} & \frac{\partial z}{\partial w} \end{vmatrix}, \tag{A9.15}$$

(see Riley, Hobson and Bence 2002).[A1]

We need only the two-coordinate version of these relations for use with Maxwell's equations. If the variables x, y and u, v are related by

$$\begin{aligned} x &= x(u, v), \\ y &= y(u, v), \end{aligned} \tag{A9.16}$$

the Jacobian is defined to be the determinant

$$\frac{\partial(x, y)}{\partial(u, v)} = \begin{vmatrix} \frac{\partial x}{\partial u} & \frac{\partial y}{\partial u} \\ \frac{\partial x}{\partial v} & \frac{\partial y}{\partial v} \end{vmatrix}. \tag{A9.17}$$

Then it is straightforward to show from the properties of determinants that:

$$\begin{gathered} \frac{\partial(x, y)}{\partial(x, y)} = -\frac{\partial(y, x)}{\partial(x, y)} = 1, \\ \frac{\partial(v, v)}{\partial(x, y)} = 0 = \frac{\partial(k, v)}{\partial(x, y)} \quad \text{if } k \text{ is a constant}, \\ \frac{\partial(u, v)}{\partial(x, y)} = -\frac{\partial(v, u)}{\partial(x, y)} = \frac{\partial(-v, u)}{\partial(x, y)} = \frac{\partial(v, -u)}{\partial(x, y)}, \end{gathered} \tag{A9.18}$$

and so on. There are similar rules for changes to the denominators. Notice, in particular, that

$$\frac{\partial(u, y)}{\partial(x, y)} = \left(\frac{\partial u}{\partial x} \right)_y, \tag{A9.19}$$

$$\frac{\partial(u, v)}{\partial(x, y)} = \frac{\partial(u, v)}{\partial(r, s)} \times \frac{\partial(r, s)}{\partial(x, y)} = \frac{1}{\partial(x, y)/\partial(u, v)}, \tag{A9.20}$$

and that

$$\left(\frac{\partial y}{\partial x}\right)_z \left(\frac{\partial z}{\partial y}\right)_x \left(\frac{\partial x}{\partial z}\right)_y = -1. \tag{A9.21}$$

Riley, Hobson and Bence (2002)[A1] give the general n-coordinate relations equivalent to (A9.17)–(A9.21).

The value of the Jacobian notation is that we can write all four Maxwell's relations in the compact form

$$\frac{\partial(T, S)}{\partial(x, y)} = \frac{\partial(p, V)}{\partial(x, y)}, \tag{A9.22}$$

where x, y represent two of the four variables T, S, p, V. Let us give an example of how this works. From (A9.19), the Maxwell relation

$$\left(\frac{\partial T}{\partial V}\right)_S = -\left(\frac{\partial p}{\partial S}\right)_V \tag{A9.23}$$

is exactly the same as

$$\frac{\partial(T, S)}{\partial(V, S)} = -\frac{\partial(p, V)}{\partial(S, V)} = \frac{\partial(p, V)}{\partial(V, S)}. \tag{A9.24}$$

We can generate all four Maxwell relations by introducing the four appropriate combinations of T, S, p and V into the denominators of (A9.22). By virtue of (A9.20), (A9.22) is exactly the same as

$$\frac{\partial(T, S)}{\partial(p, V)} = 1. \tag{A9.25}$$

The Jacobian (A9.25) is the key to remembering all four relations. We need only remember that the Jacobian is $+1$ if T, S, p and V are written in the above order. One way of remembering this is that the intensive variables, T and p, and the extensive variables, S and V, should appear in the same order in the numerator and denominator if the sign is positive.

Appendix reference

A1 Riley, K.F., Hobson, M.P. and Bence, S.J. (2002). *Mathematical Methods for Physics and Engineering*, second edition, pp. 148–51. Cambridge: Cambridge University Press.

10 Kinetic theory and the origin of statistical mechanics

10.1 The kinetic theory of gases

The controversy between the caloric and the kinetic, or dynamic, theories of heat was resolved in favour of the latter by James Joule's experiments of the 1840s and by the establishment of first and second laws of thermodynamics by Rudolph Clausius in the early 1850s. Before 1850, various kinetic theories of gases had been proposed, in particular, by John Herapath and John James Waterston, to whom we will return later. Joule had noted that the equivalence of heat and work could be interpreted in terms of the kinetic theory and, in his initial formulation of the first and second laws, Clausius described how they could also be interpreted within this framework, although he emphasised that the laws are quite independent of the particular microscopic theory.

The first systematic account of the kinetic theory was published by Clausius in 1857 in a classic paper entitled 'The nature of the motion which we call heat.'[1] This contained a simple derivation of the perfect gas law, assuming that the molecules of a gas can be considered to be elastic spheres which exert a pressure on the walls of the containing vessel by bouncing off them. Let us repeat Clausius's derivation, in a simple form.

10.1.1 Assumptions underlying the kinetic theory of gases

The objective of the kinetic theory is to apply *Newtonian physics* to large assemblies of atoms or molecules.* It turns out that many of the key results can be derived independently of the velocity distribution of the molecules and we will adopt that approach in this section.

The basic postulates of the kinetic theory are as follows.

- Gases consist of elementary entities, molecules, which are in constant motion. Each molecule has kinetic energy $\frac{1}{2}mv^2$ and their velocity vectors point in random directions.
- The molecules can be considered to be very small solid spheres of diameter a.
- The *long-range* forces between molecules are weak and undetectable in a Joule expansion and so they are taken to be zero. The molecules can collide, however, and when they collide they do so *elastically*, meaning that there is no loss of kinetic energy in any collision.

* For economy of expression, I will refer to both atoms and molecules as 'molecules' in this chapter.

- The origin of the *pressure* on the wall of a vessel is the force per unit area due to the elastic collisions of enormous numbers of molecules with the wall. These collisions also are assumed to be perfectly elastic – no kinetic energy is lost if the wall is stationary.
- The temperature is related to the average kinetic energy of the molecules of the gas. This makes sense in that, if work is done on the gas, the kinetic energies of the molecules are increased, thus increasing the internal energy and hence the temperature.

One key aspect of Clausius's model was the role of *collisions* between molecules. In his paper of 1857, he found that the typical speeds of the molecules of air were 461 and 492 m s^{-1} for oxygen and nitrogen respectively. The Dutch meteorologist Buys Ballot criticised this aspect of the theory, since it is well known that pungent odours take minutes to permeate a room, and not seconds. Clausius's response was to point out that the molecules of air collide with each other and therefore they must *diffuse* from one location to another, rather than travel in straight lines. In his response, published in 1858, Clausius introduced for the first time the concept of the *mean free path* of the molecules of the gas.[2] Thus, in the kinetic theory, it must be supposed that the molecules are continually colliding with each other.

An important consequence of these collisions is that there must inevitably be a *velocity dispersion* among the molecules of a gas. If a collision is head-on then all the kinetic energy is transferred from an incident to a stationary molecule. If it is a glancing collision, there is very little transfer of energy. As a result, we cannot consider all the molecules of the gas to have the same speed – random collisions inevitably lead to a *velocity dispersion*. The collisions perform another crucial function, central to Maxwell's analysis of molecular motion – they *randomise* the velocity vectors of the molecules and so the pressure of the gas is *isotropic*.

10.2 Kinetic theory of gases – first version

Let us begin with an elementary analysis, which gets straight to the heart of the matter although it has its dodgy moments. Consider the force acting on one wall of a smooth-walled cubical enclosure resulting from the impact of the molecules of a gas. According to Newton's second law of motion, the pressure is the rate of change in momentum per unit area of all the molecules striking the wall elastically per second.

Consider the rate of change of momentum due to the vast number of elastic collisions of the molecules with unit area of, say, the yz-plane. The pressure on the yz-plane is associated only with changes in the x-components of their momentum vectors, $2mv_x$. Now, all the molecules with positive values of v_x within a distance v_x of the wall arrive at the wall in one second. Therefore, the pressure p is found by summing over all the molecules within distance v_x of the wall which are moving towards it, that is,

$$p = \sum_i 2mv_x \times v_x = \sum_i 2mv_x^2 = n(+v_x)2m\overline{v_x^2}, \qquad (10.1)$$

where $n(+v_x)$ is the number density of molecules with v_x components in the positive x-direction and $\overline{v_x^2}$ is the mean square velocity component in the x-direction. At any time only half the molecules have positive values of v_x, the other half having negative values.

Therefore, if n is the total number density of molecules then $n(+v_x) = n/2$. Hence, $p = nm\overline{v_x^2}$. Because the distribution of velocities is isotropic,

$$\overline{v^2} = \overline{v_x^2} + \overline{v_y^2} + \overline{v_z^2} \qquad \text{and} \qquad \overline{v_x^2} = \overline{v_y^2} = \overline{v_z^2}.$$

Therefore, $\overline{v_x^2} = \frac{1}{3}\overline{v^2}$ and so

$$p = \tfrac{1}{3}nm\overline{v^2}. \tag{10.2}$$

We recognise that $\frac{1}{2}nm\overline{v^2}$ is just the total kinetic energy of the molecules of the gas per unit volume. Now, the perfect gas law for one mole of gas is $pV = RT$, or

$$p = (R/V)T = nkT, \tag{10.3}$$

where $k = R/N_0$ and N_0 is Avogadro's constant, the number of molecules per mole. The gas constant per molecule, k, is known as *Boltzmann's constant* and has the value 1.38×10^{-23} J K^{-1}. Comparing (10.2) and (10.3),

$$\tfrac{3}{2}kT = \tfrac{1}{2}m\overline{v^2}. \tag{10.4}$$

This equation provides a direct link between the temperature T of the gas and the *mean kinetic energy* $\frac{1}{2}m\overline{v^2}$ *of the molecules*. This is a key result.

10.3 Kinetic theory of gases – second version

Let us perform a more satisfactory calculation in which we integrate over all the angles at which molecules arrive at the walls of a cubical vessel. The determination of the flux of molecules arriving at the wall can be split into two parts.

10.3.1 *Probability of molecules arriving from a particular direction*

This calculation turns up in many different calculations in physics. Suppose we choose a particular direction in space and ask,

> What is the probability that a molecule has a velocity vector pointing within the range of angles θ to $\theta + \mathrm{d}\theta$ with respect to the chosen direction?

Because of collisions, the velocity vectors of the molecules point in random directions and so there is an equal probability that a velocity vectors will point in any direction. Therefore, the probability that a velocity vector lies within the solid angle $\mathrm{d}\Omega$ between the cones at θ and $\theta + \mathrm{d}\theta$ can be written in terms of the area $\mathrm{d}A$ subtended by $\mathrm{d}\Omega$ on the surface of a sphere of radius r:

$$\mathrm{d}P = \frac{\mathrm{d}A}{4\pi r^2}. \tag{10.5}$$

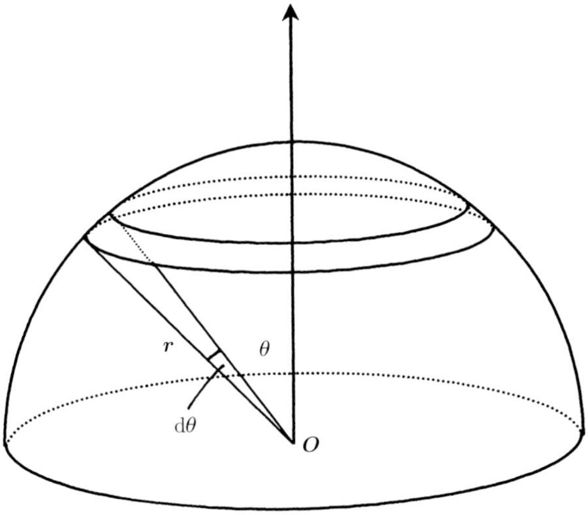

Figure 10.1: Evaluating the probability of molecules approaching the origin from directions in the range θ to $\theta + \mathrm{d}\theta$.

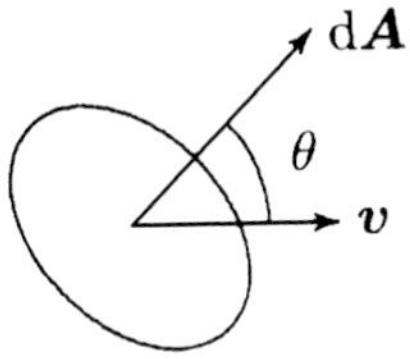

Figure 10.2: Illustrating the flux of molecules with velocity vectors $\boldsymbol{v}$ through an element of surface area $\mathrm{d}\boldsymbol{A}$. The convention is used of associating with the area a vector $\mathrm{d}\boldsymbol{A}$ which is normal to the plane of the area and has magnitude $|\mathrm{d}\boldsymbol{A}|$.

Thus the probability $\mathrm{d}P \equiv p(\theta)\,\mathrm{d}\theta$ that the velocity vector of a molecule lies within angles θ to $\theta + \mathrm{d}\theta$ of our chosen direction is given by substituting the surface area of the annulus shown in Fig. 10.1, $\mathrm{d}A = 2\pi r \sin\theta \times r\,\mathrm{d}\theta = 2\pi r^2 \sin\theta\,\mathrm{d}\theta$, into (10.5):

$$\mathrm{d}P = P(\theta)\,\mathrm{d}\theta = \frac{\mathrm{d}A}{4\pi r^2} = \tfrac{1}{2}\sin\theta\,\mathrm{d}\theta. \tag{10.6}$$

10.3.2 The flux of molecules arriving from a particular direction

The second part of the calculation involves finding the flux of molecules, all moving with velocity $\boldsymbol{v}$, at angle θ to an element of surface area $\mathrm{d}\boldsymbol{A}$, as illustrated in Fig. 10.2; the number density of the molecules is n. The component of velocity parallel to the plane of the elementary area is $|\boldsymbol{v}|\sin\theta$, but this results in no flow of molecules through $\mathrm{d}\boldsymbol{A}$. The component of velocity parallel to $\mathrm{d}\boldsymbol{A}$, that is, perpendicular to the surface, is $|\boldsymbol{v}|\cos\theta$ and so the *flux of molecules* through $\mathrm{d}\boldsymbol{A}$, that is, the number of molecules passing through $\mathrm{d}\boldsymbol{A}$ per second, is

$$J = n|\boldsymbol{v}||\mathrm{d}\boldsymbol{A}|\cos\theta = n\boldsymbol{v}\cdot\mathrm{d}\boldsymbol{A}. \tag{10.7}$$

10.3.3 *The flux of an isotropic distribution of molecules through unit area*

We now combine the results of subsections 10.3.1 and 10.3.2. One of the most important calculations is the total flux of molecules arriving at unit area of the vessel.

First, we evaluate the number of molecules arriving at the wall per second, assuming that the velocity distribution of the molecules is isotropic. From (10.7), the flux of molecules $J(v, \theta)$ with velocity $\boldsymbol{v}$ arriving at angle θ to an elementary area $\mathrm{d}\boldsymbol{A}$ is (see Fig. 10.2)

$$J(v, \theta) = n(v, \theta)\,\boldsymbol{v}\cdot \mathrm{d}\boldsymbol{A} = n(v, \theta)\, v \cos\theta\, \mathrm{d}A, \tag{10.8}$$

where $n(v, \theta)$ is the number density of molecules with speed v arriving at angle θ (Fig. 10.1). $\mathrm{d}\theta$ can be made as small as we like, so that we can consider all the molecules in the angular range θ to $\theta + \mathrm{d}\theta$ to be arriving from the direction θ. Therefore, if $n(v)$ is the total number density of molecules with speeds v, the number arriving at $\mathrm{d}\boldsymbol{A}$ within the angles θ to $\theta + \mathrm{d}\theta$ is, from (10.6),

$$n(v, \theta)\, \mathrm{d}\theta \equiv n(v)P(\theta)\, \mathrm{d}\theta = \tfrac{1}{2}n(v) \sin\theta\, \mathrm{d}\theta. \tag{10.9}$$

Therefore, substituting (10.9) into (10.8),

$$J(v, \theta)\, \mathrm{d}\theta = \tfrac{1}{2}n(v) \sin\theta\, \mathrm{d}\theta \times v \cos\theta\, \mathrm{d}A. \tag{10.10}$$

The total flux of molecules with speed v arriving at the area $\mathrm{d}A$ of the wall is therefore the integral of $J(v, \theta)$ from $\theta = 0$ to $\theta = \pi/2$, since those molecules with velocity vectors between $\pi/2$ and π are moving away from the wall. Therefore,

$$J(v) = \int_0^{\pi/2} \tfrac{1}{2}n(v)\, v\, \mathrm{d}A \sin\theta \cos\theta\, \mathrm{d}\theta = \tfrac{1}{4}n(v)\, v\, \mathrm{d}A. \tag{10.11}$$

Therefore, the flux of molecules with speed v per unit area is

$$J(v) = \tfrac{1}{4}n(v)\, v. \tag{10.12}$$

We now add together the contributions from molecules with all speeds and so the total flux arriving at the wall per unit area is

$$J = \sum_v \tfrac{1}{4}n(v)\, v = \tfrac{1}{4}n\, \overline{v}, \tag{10.13}$$

where $\overline{v}$ is the mean speed of the molecules,

$$\overline{v} = \frac{1}{n}\sum_v n(v)v, \tag{10.14}$$

and n their total number density. If instead the distribution of speeds were given by a continuous probability distribution $f(v)$, such that the probability that the speed lies in the range v to $v + \mathrm{d}v$ is $f(v)\, \mathrm{d}v$ and $\int_0^\infty f(v)\, \mathrm{d}v = 1$, the average speed would be

$$\overline{v} = \int_0^\infty vf(v)\, \mathrm{d}v. \tag{10.15}$$

10.3.4 The equation of state of a perfect gas

It is now straightforward to repeat the calculation of subsection 10.3.3 to find the pressure acting on the walls. A molecule arriving at the wall with speed v and at angle θ contributes a momentum change $2mv\cos\theta$ to the pressure. From (10.10), the flux of molecules arriving within the angles θ to $\theta + \mathrm{d}\theta$ is

$$J(v,\theta)\,\mathrm{d}\theta = \tfrac{1}{2}n(v)\,v\,\mathrm{d}A\sin\theta\cos\theta\,\mathrm{d}\theta. \tag{10.16}$$

Therefore, the contribution of these molecules to the rate of change in momentum *per unit area*, that is, their contribution $\mathrm{d}p(v,\theta)$ to the pressure acting on the walls of the vessel is

$$\begin{aligned}\mathrm{d}p(v,\theta) &= 2mv\cos\theta \times J(v,\theta)\,\mathrm{d}\theta \\ &= 2mv\cos\theta \times \tfrac{1}{2}n(v)\,v\sin\theta\cos\theta\,\mathrm{d}\theta \\ &= n(v)\,mv^2\sin\theta\cos^2\theta\,\mathrm{d}\theta.\end{aligned} \tag{10.17}$$

Now integrate from $\theta = 0$ to $\pi/2$ to find the pressure due to all the molecules with speed v:

$$p(v) = n(v)\,mv^2\int_0^{\pi/2}\sin\theta\cos^2\theta\,\mathrm{d}\theta = \tfrac{1}{3}n(v)\,mv^2. \tag{10.18}$$

Now sum over all speeds to find the total pressure of the gas:

$$p = \sum_v \tfrac{1}{3}n(v)\,mv^2 = \tfrac{1}{3}nm\overline{v^2} \tag{10.19}$$

where

$$\overline{v^2} = \frac{1}{n}\sum_v n(v)v^2 \tag{10.20}$$

is the *mean square speed of the molecules* and n is their total number density. This is a much more satisfactory derivation of (10.2). Again, if the distribution of molecular speeds were continuous and given by the probability distribution $f(v)\,\mathrm{d}v$, the mean square speed would be

$$\overline{v^2} = \int_0^\infty v^2 f(v)\,\mathrm{d}v. \tag{10.21}$$

As before, comparing (10.19) with the perfect gas law $p = nkT$, we find (10.4), $\frac{3}{2}kT = \frac{1}{2}m\overline{v^2}$.

If the vessel contained a mixture of different gases at the same temperature then each molecule would attain the same kinetic energy. Therefore, the total pressure from the different species i would be

$$p = \sum_i \tfrac{1}{3}n_i m_i\overline{v_i^2} = \sum p_i, \tag{10.22}$$

where p_i is the pressure contributed by each species. This is *Dalton's law of partial pressures*.

Clausius assumed that the internal energy of the gas is simply the sum of the kinetic energies of the molecules, that is, for one mole

$$U = \tfrac{1}{2}N_0 m\overline{v^2} = \tfrac{3}{2}N_0 kT, \tag{10.23}$$

where N_0 is Avogadro's constant. Therefore

$$U = \tfrac{3}{2}RT, \tag{10.24}$$

and it follows that the heat capacity at constant volume is

$$C_V = \left(\frac{\partial U}{\partial T}\right)_V = \frac{3}{2} \tag{10.25}$$

and the ratio of specific heats is

$$\gamma = \frac{C_V + R}{C_V} = \frac{5}{3}. \tag{10.26}$$

This remarkably elegant argument is so much taken for granted nowadays that one forgets what an achievement it represented in 1857. However, whilst accounting for the perfect gas law admirably, it did not give good agreement with the known values of γ for diatomic molecular gases, which are about 1.4. There must therefore exist other ways of storing energy within molecular gases which can increase the internal energy per molecule and hence C_V, an important point clearly recognised by Clausius in the last sentences of his paper.[1]

Two aspects of this story are worthy of further note. First, Clausius knew that there had to be a dispersion of speeds $f(v)$ of the molecules of the gas, but he had no way of knowing what it was and so worked in terms of mean values only. Second, the idea that the pressure of the gas should be proportional to $nm\overline{v^2}$ had been worked out 15 years earlier by Waterston. As early as 1843, Waterston wrote in a book published in Edinburgh,

> A medium constituted of elastic spherical molecules that are continually impinging against each other with the same velocity will exert against a vacuum an elastic force that is proportional to the square of this velocity and to its density.[3]

In December 1845, Waterston developed a more systematic version of his theory and gave the first statement of the equipartition theorem for a mixture of gases of different molecular species. He also worked out the ratio of specific heats, although his calculation contained a numerical error. Waterston's work was submitted to the Royal Society for publication in 1845, but was harshly refereed and remained unpublished until 1892, eight years after his death. The paper was discovered by Lord Rayleigh in the archives of the Royal Society of London and he then had it published with an introduction which he himself wrote. In it, Rayleigh wrote

> Impressed by the above passage [from a later work by Waterston] and with the general ingenuity and soundness of Waterston's views, I took the first opportunity of consulting the Archives, and saw at once that the memoir justified the large claims made for it, and that it makes an immense advance in the direction of the now generally received theory. The omission to publish it was a misfortune, which probably retarded the development of the subject by ten or fifteen years.[4]

Later he notes

> The character of the advance to be dated from this paper will be at once understood when it is realised that Waterston was the first to introduce into the theory the conception that heat and temperature are to be measured by 'vis viva' (that is, kinetic energy) . . . In the second section, the great feature is the

statement (VII) that in mixed media the mean square velocity is inversely proportional to the specific weight of the molecules. The proof which Waterston gave is doubtless not satisfactory; but the same may be said of that advanced by Maxwell fifteen years later.[5]

One final quotation is particularly memorable.

The history of this paper suggests that highly speculative investigations, especially by an unknown author, are best brought before the scientific world through some other channel than a scientific society, which naturally hesitates to admit into its printed records matter of uncertain value. Perhaps one may go further and say that a young author who believes himself capable of great things would usually do well to secure the favourable recognition of the scientific world by work whose scope is limited, and whose value is easily judged, before embarking on greater flights.[6]

What is one to say? The situation is rather tragic and yet it is so easy to see how it can come about. How often have we dismissed the ideas of an unknown scientist or student whose work we did not fully comprehend? I wish I could believe I had not, but I will probably never know for certain.

10.4 Maxwell's velocity distribution

Both of Clausius's papers were known to Maxwell when he turned to the study of the kinetic theory of gases in 1859. His paper entitled 'Illustrations of the dynamical theory of gases',[7] published in 1860, is characteristically novel and profound. The quite amazing achievement was that, in this single paper, he derived the correct formula for the velocity distribution $f(v)$ and introduced statistical concepts into the kinetic theory of gases and thermodynamics. Francis Everitt has written that this derivation of what we now know as Maxwell's velocity distribution marks the beginning of a new epoch in physics.[8] From it follow directly the statistical nature of the laws of thermodynamics, which is the key to Boltzmann's statistics, and the modern theory of statistical mechanics.

Maxwell's derivation of the distribution occupies no more than half a dozen short paragraphs. The problem is stated as Proposition IV of his paper,

To find the average number of particles whose velocities lie between given limits, after a great number of collisions among a great number of equal particles.

We will follow closely Maxwell's exposition with only minor changes in notation. The total number of molecules is N and the x-, y- and z- components of their velocities are v_x, v_y and v_z. Maxwell supposed that the velocity distribution in the three orthogonal directions must be the same after a great number of collisions, that is,

$$N f(v_x)\,\mathrm{d}v_x = N f(v_y)\,\mathrm{d}v_y = N f(v_z)\,\mathrm{d}v_z, \tag{10.27}$$

where f is the same function. Now the three perpendicular components of the velocity are entirely independent, and hence the number of molecules with velocities in the range v_x to $v_x + \mathrm{d}v_x$, v_y to $v_y + \mathrm{d}v_y$ and v_z to $v_z + \mathrm{d}v_z$ is

$$N f(v_x) f(v_y) f(v_z)\,\mathrm{d}v_x\,\mathrm{d}v_y\,\mathrm{d}v_z; \tag{10.28}$$

the total velocity of any molecule v is $v^2 = v_x^2 + v_y^2 + v_z^2$. Because large numbers of collisions have taken place, the probability that a particle has speed v, $\phi(v)$, must be isotropic and depend only on v, that is,

$$f(v_x)f(v_y)f(v_z) = \phi(v) = \phi[(v_x^2 + v_y^2 + v_z^2)^{1/2}], \tag{10.29}$$

where again we have normalised $\phi(v)$ so that

$$\int_{-\infty}^{\infty}\int_{-\infty}^{\infty}\int_{-\infty}^{\infty} \phi(v)\,\mathrm{d}v_x\,\mathrm{d}v_y\,\mathrm{d}v_z = 1.$$

Equation (10.29) is what is known as a *functional equation* and we need to find what forms of the functions $f(v_x)$, $f(v_y)$, $f(v_z)$ and $\phi(v)$ are consistent with (10.29). By inspection, a suitable solution is

$$f(v_x) = C\,\mathrm{e}^{Av_x^2}, \qquad f(v_y) = C\,\mathrm{e}^{Av_y^2}, \qquad f(v_z) = C\,\mathrm{e}^{Av_z^2};$$

where C and A are constants. The beauty of our exponential trial solution is that the powers are additive when the functions are multiplied together, as required by (10.29). Therefore,

$$\begin{aligned}\phi(v) &= f(v_x)f(v_y)f(v_z) = C^3\mathrm{e}^{A(v_x^2+v_y^2+v_z^2)} \\ &= C^3\mathrm{e}^{Av^2}.\end{aligned} \tag{10.30}$$

The distribution must converge as $v \to \infty$ and hence A must be negative. Maxwell wrote $A = -\alpha^{-2}$ and so

$$\phi(v) = C^3\mathrm{e}^{-v^2/\alpha^2} \tag{10.31}$$

where α is a constant to be determined. Next, we have to find the normalisation constant. For the velocity component v_x, say, the total probability that the molecule has some value of v_x must be unity, and so

$$\int_{-\infty}^{\infty} C\,\mathrm{e}^{-v_x^2/\alpha^2}\,\mathrm{d}v_x = 1.$$

This is a standard integral,

$$\int_{-\infty}^{\infty} \mathrm{e}^{-x^2}\,\mathrm{d}x = \pi^{1/2},$$

and hence

$$C = \frac{1}{\alpha\pi^{1/2}}. \tag{10.32}$$

This leads directly to four conclusions, which we quote in Maxwell's words (but using our notation).

(i) *The number of particles whose velocity, resolved in a certain direction, lies between v_x and $v_x + \mathrm{d}v_x$ is*

$$N\frac{1}{\alpha\pi^{1/2}}\mathrm{e}^{-v_x^2/\alpha^2}\,\mathrm{d}v_x. \tag{10.33}$$

We have just proved this result.

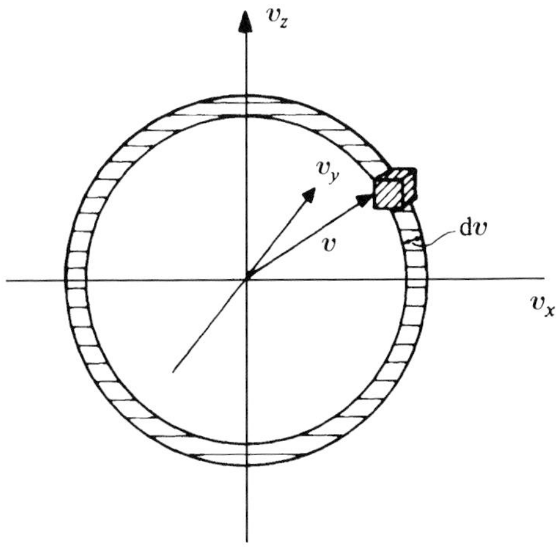

Figure 10.3: Conversion of the volume integral over $dv_x dv_y dv_z$ in velocity space to one over the speed v. The volume of the spherical shall in velocity space is $4\pi v^2\, dv$.

(ii) *The number whose actual velocity [speed] lies between v and v + dv is*

$$N\frac{4}{\alpha^3\pi^{1/2}}v^2 e^{-v^2/\alpha^2}\,dv. \tag{10.34}$$

This expression is established by transforming from Cartesian to polar coordinates in velocity space as follows. From (10.30) and (10.33), the number of particles with speed v and velocity components lying in the 'volume element' $dv_x\, dv_y\, dv_z$ at (v_x, v_y, v_z) is

$$\begin{aligned} N\phi(v)\,dv_x\,dv_y\,dv_z &= Nf(v_x)f(v_y)f(v_z)\,dv_x\,dv_y\,dv_z \\ &= N\frac{1}{\alpha^3\pi^{3/2}}e^{-v^2/\alpha^2}dv_x\,dv_y\,dv_z. \end{aligned} \tag{10.35}$$

The total number of particles with speeds in the range v to $v + dv$ is now found by summing over all 'volume elements' such as $dv_x\, dv_y\, dv_z$ between distances v and $v + dv$ from the origin in velocity space (Fig. 10.3):

$$\begin{aligned} N\phi(v)\,4\pi v^2\,dv &= N\frac{1}{\alpha^3\pi^{3/2}}\,e^{-v^2/\alpha^2}4\pi v^2\,dv \\ &= N\frac{4}{\alpha^3\pi^{1/2}}\,v^2 e^{-v^2/\alpha^2}dv, \end{aligned}$$

as stated by Maxwell.

(iii) *To find the mean value of v, add the velocities of all the particles together and divide by the number of particles; the result is*

$$\textit{mean velocity} = \frac{2\alpha}{\pi^{1/2}}. \tag{10.36}$$

This is a straightforward calculation, using the probability distribution (10.34) and writing it as $N\Phi\,\mathrm{d}v$. Then

$$\overline{v} = \frac{\int_0^\infty v\Phi(v)\,\mathrm{d}v}{\int_0^\infty \Phi(v)\,\mathrm{d}v} = \int_0^\infty \frac{4}{\alpha^3\pi^{1/2}}\,v^3\,\mathrm{e}^{-v^2/\alpha^2}\,\mathrm{d}v.$$

This involves another standard integral, $\int_0^\infty x^3\mathrm{e}^{-x^2}\,\mathrm{d}x = 1/2$. Therefore, changing variables to $x = v/\alpha$, we find

$$\overline{v} = \frac{4}{\alpha^3\pi^{1/2}}\frac{\alpha^4}{2} = \frac{2\alpha}{\pi^{1/2}}.$$

(iv) *To find the mean value of v^2, add all the values together and divide by N:*

$$\textit{mean value of } v^2 = \tfrac{3}{2}\alpha^2.$$

To find this result, take the variance of v by the usual procedure:

$$\overline{v^2} = \int_0^\infty v^2\Phi(v)\,\mathrm{d}v = \int_0^\infty \frac{4}{\alpha^3\pi^{1/2}}v^4\mathrm{e}^{-v^2/\alpha^2}\,\mathrm{d}v. \tag{10.37}$$

This can be reduced to a standard form by integration by parts. Changing variable to $x = v/\alpha$, the integral (10.37) becomes

$$\overline{v^2} = \tfrac{3}{2}\alpha^2. \tag{10.38}$$

Maxwell immediately went on to note

> that the velocities are distributed among the particles according to the same law as the errors are distributed among the observations in the theory of the *method of least squares*.

Thus, in this very first paper, the direct relation to errors and statistical procedures was established.

To put Maxwell's distribution into its standard form, we compare the result (10.38) with the value for $\overline{v^2}$ deduced from the kinetic theory of gases (10.4) and find

$$\overline{v^2} = \frac{3\alpha^2}{2} = \frac{3kT}{m}, \qquad \text{and so} \qquad \alpha = \sqrt{\frac{2kT}{m}}. \tag{10.39}$$

We can therefore write Maxwell's distribution, (10.34), in its definitive form:

$$\Phi(v)\,\mathrm{d}v = 4\pi\left(\frac{m}{2\pi kT}\right)^{3/2} v^2 \exp\left(-\frac{mv^2}{2kT}\right)\mathrm{d}v. \tag{10.40}$$

This is quite an astonishing achievement. It is an extraordinarily simple derivation, which scarcely acknowledges that we are dealing with molecules and gases at all. It is in striking contrast to the derivation which proceeds from the Boltzmann distribution. The distribution is shown in Fig. 10.4 in dimensionless form.

Maxwell then applied this law in a variety of circumstances and, in particular, he showed that if two different types of molecule are present in the same volume then the mean *vis viva*, or kinetic energy, must be the same for each of them.

In the final section of the paper, Maxwell addressed the problem of accounting for the ratio of the specific heats of gases, which for many molecular gases had been measured to be 1.4. Clausius had already noted that some additional means of storing *vis viva* was needed

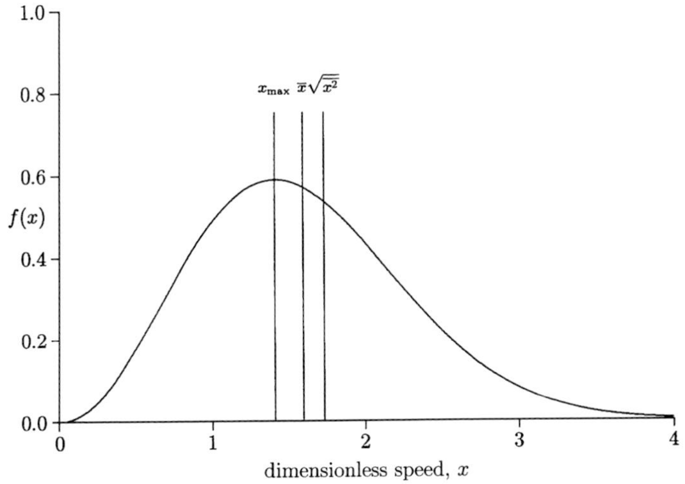

Figure 10.4: The Maxwell distribution plotted in dimensionless form, $f(x)\,dx = \sqrt{2/\pi}\,x^2\,e^{-x^2/2}\,dx$, where $x = v/\sigma$ and $\sigma = \sqrt{kT/m}$ is the standard deviation of the distribution. The mean, root mean squared and maximum values of the distribution occur at $x = \sqrt{8/\pi}$, $\sqrt{3}$ and $\sqrt{2}$ respectively.

and Maxwell proposed that it was stored in the kinetic energy of rotation of the molecules, which he modelled as rough particles. He found that, in equilibrium, as much energy could be stored in rotational as in translational motion, $\frac{1}{2}I\omega^2 = \frac{1}{2}mv^2$, and so he could derive a value for the ratio of specific heats. Instead of $U = \frac{3}{2}RT$, he found $U = 3RT$, since as much energy is stored in rotational as translational motion, and consequently, following the analysis of equations (10.24) to (10.26),

$$\gamma = \tfrac{4}{3} = 1.333.$$

This value is almost as bad as the value 1.667 which follows if only translational motion is considered. This profoundly depressed Maxwell. The last sentence of his great paper reads:

> Finally, by establishing a necessary relation between the motions of translation and rotation of all particles not spherical, we proved that a system of such particles could not possibly satisfy the known relation between the two specific heats of all gases.

His inability to explain the value $\gamma = 1.4$ was a grave disappointment to Maxwell, who, in his report to the British Association for the Advancement of Science of 1860, stated that this discrepancy 'overturned the whole hypothesis'.[9]

The second key concept which resulted directly from this paper is the principle of the equipartition of energy. This was to prove a highly contentious issue until it was finally resolved by Einstein's application of quantum concepts to the average energy of an oscillator (see Section 14.4). The principle states that in equilibrium energy is shared equally among all the independent modes in which it can be stored by molecules. The problem of accounting for values of $\gamma = 1.4$ was aggravated by the discovery of spectral lines in the spectra of gases, which were interpreted as resonances associated with the internal structure of molecules. There were so many lines that if, each of them were to acquire its share of the internal energy of the gas, the ratio of specific heats of gases would tend to unity. These two fundamental problems cast doubt upon the general validity of the principle of equipartition of energy and in consequence upon the whole idea of a kinetic theory of gases, despite its success in accounting for the perfect gas law.

These issues were at the centre of much of Maxwell's subsequent work. In 1867, he presented another derivation of the velocity distribution from the perspective of its evolution as a result of molecular collisions.[10] He considered the outcome of collisions between molecules which have velocities in the range $\boldsymbol{v}_1$ to $\boldsymbol{v}_1 + \mathrm{d}\boldsymbol{v}_1$ with those of the same mass with velocities in the range $\boldsymbol{v}_2$ to $\boldsymbol{v}_2 + \mathrm{d}\boldsymbol{v}_2$. The conservation of momentum requires

$$\boldsymbol{v}_1 + \boldsymbol{v}_2 \rightarrow \boldsymbol{v}_1' + \boldsymbol{v}_2'.$$

Because of the conservation of energy,

$$\tfrac{1}{2}mv_1^2 + \tfrac{1}{2}mv_2^2 = \tfrac{1}{2}mv_1'^2 + \tfrac{1}{2}mv_2'^2. \tag{10.41}$$

Since the assembly of molecules is assumed to have reached an equilibrium state, the joint probability distributions for the velocities before and after the collisions must satisfy the relation

$$\phi(\boldsymbol{v}_1)\,\mathrm{d}\boldsymbol{v}_1\,\phi(\boldsymbol{v}_2)\,\mathrm{d}\boldsymbol{v}_2 = \phi(\boldsymbol{v}_1')\,\mathrm{d}\boldsymbol{v}_1'\,\phi(\boldsymbol{v}_2')\,\mathrm{d}\boldsymbol{v}_2'. \tag{10.42}$$

To appreciate the force of this argument, it is simplest to inspect Maxwell's velocity distribution in the Cartesian form (10.35). Conservation of kinetic energy ensures that the exponential terms on either side of (10.42) remain the same:

$$\exp\left[-\frac{m}{2kT}\left(v_1^2 + v_2^2\right)\right] = \exp\left[-\frac{m}{2kT}\left(v_1'^2 + v_2'^2\right)\right]. \tag{10.43}$$

In addition, Maxwell showed that the elements of phase space, that is, the joint velocity space, before and after the collisions are related by

$$(\mathrm{d}v_x\,\mathrm{d}v_y\,\mathrm{d}v_z)_1 \times (\mathrm{d}v_x\,\mathrm{d}v_y\,\mathrm{d}v_z)_2 = (\mathrm{d}v_x'\,\mathrm{d}v_y'\,\mathrm{d}v_z')_1 \times (\mathrm{d}v_x'\,\mathrm{d}v_y'\,\mathrm{d}v_z')_2.$$

We can recognise that this is a special case of *Liouville's theorem*, according to which if a certain group of molecules occupies a volume of phase space $\mathrm{d}v_x\,\mathrm{d}v_y\,\mathrm{d}v_z\,\mathrm{d}x\,\mathrm{d}y\,\mathrm{d}z$ then, as they change their velocities and positions under the laws of classical mechanics, the magnitude of the volume in phase space is conserved. This result was known to Boltzmann as he struggled to establish the principles of statistical mechanics. Thus, Maxwell's velocity distribution can be retrieved from the requirement that the distribution is stationary in the above sense, following collisions between the molecules of the gas.

There is an intriguing footnote to this story. The exponential factor in Maxwell's velocity distribution is just $\exp(-E/kT)$, where $E = \frac{1}{2}mv^2$ is the kinetic energy of the molecules. Let us write down the ratio of the numbers of molecules in the speed intervals v_i to $v_i + \mathrm{d}v_i$ and v_j to $v_j + \mathrm{d}v_j$. From (10.40)

$$\frac{N(E_i)}{N(E_j)} = \frac{\exp(-E_i/kT)\,4\pi v_i^2\mathrm{d}v_i}{\exp(-E_j/kT)\,4\pi v_j^2\mathrm{d}v_j}. \tag{10.44}$$

This expression can be interpreted in terms of the probability that a molecule has energy E at temperature T and the volume of velocity space available to such a molecule, with speed

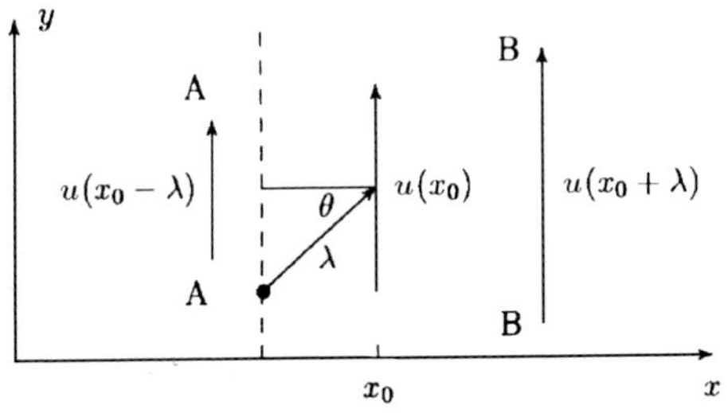

Figure 10.5: Illustrating the geometry used to evaluate the coefficient of viscosity from the kinetic theory of gases. The horizontal separation of neighbouring vertical arrows is λ.

in the range v to $v + dv$. This is an aspect of the much more general result

$$\frac{N(E_i)}{N(E_j)} = \frac{g_i}{g_j} \frac{\exp(-E_i/kT)}{\exp(-E_j/kT)}, \tag{10.45}$$

where the g_i and g_j describe the total numbers of available states in which the molecules would have energies E_i and E_j. In the language of statistical physics, g_i describes the *degeneracy* of the state i, meaning the number of different states with the same energy E_i. In thermal equilibrium, the number of molecules with energy E depends not only upon the factor $p(E) \propto \exp(-E/kT)$, the Boltzmann factor, but also upon the number of available states with energy E. We will return to this key result in Case Study V.

Despite the problems with the kinetic theory, Maxwell never seriously doubted its validity or that of the velocity distribution he had derived. Among the most compelling of his reasons were the results of his studies of the viscosity of gases.

10.5 The viscosity of gases

Maxwell immediately applied the kinetic theory of gases to their transport properties – diffusion, thermal conductivity and viscosity. His calculation of the coefficient of viscosity of gases was of special importance and so let us repeat it. The standard diagram for the transport of viscous stresses is shown in Fig. 10.5.

Gas flows in the positive y-direction at speed $u(x)$, which is assumed to be much slower than the average speed $\bar{v}$ of the molecules. There is a velocity gradient in the x-direction, as indicated by the relative lengths of the vertical arrows at AA, x_0 and BB. For illustrative purposes, the (yz)-planes through the three arrows have been separated by one mean free path λ. The object of the calculation is to work out the shear stress τ_{x_0}, that is, the net force per unit area in the y-direction acting on the gas in the yz-plane at x_0. Experimentally, this stress is found to depend upon the velocity gradient according to the expression

$$\tau_{x_0} = \eta \left(\frac{du}{dx}\right)_{x_0}, \tag{10.46}$$

where η is the viscosity of the gas. At the molecular level, the stress is the net rate at which momentum is delivered to the interface per unit area per second by the random motions of the molecules on either side of the plane at x_0.

The molecules make many collisions with each other with mean free path λ, which is related to their collision cross-section σ by $\lambda = (n\sigma)^{-1}$. In each collision it is a good approximation to assume that the angles though which the molecules are scattered are random. Therefore the momentum transferred to the yz-plane at x_0 can be considered to have originated from a distance λ at some angle θ to the normal to the plane, as indicated in Fig. 10.5. From (10.10), the flux of molecules with speed v arriving in the angular range θ to $\theta + \mathrm{d}\theta$ per second at area $\mathrm{d}A$ is

$$J(v, \theta)\,\mathrm{d}\theta = \tfrac{1}{2}n(v)\,v\,\mathrm{d}A \sin\theta\cos\theta\,\mathrm{d}\theta.$$

Each molecule brings with it the component of momentum in the y-direction which it acquired at perpendicular distance $\lambda\cos\theta$ from the plane, $mu(x_0 - \lambda\cos\theta)$, and so the rate of transfer of momentum from these molecules per unit area is

$$mu(x_0 - \lambda\cos\theta) \times \tfrac{1}{2}n(v)\,v\sin\theta\cos\theta\,\mathrm{d}\theta$$
$$= m\left[u(x_0) - \lambda\cos\theta\left(\frac{\mathrm{d}u}{\mathrm{d}x}\right)_{x_0}\right] \times \tfrac{1}{2}n(v)\,v\,\sin\theta\cos\theta\,\mathrm{d}\theta, \qquad (10.47)$$

where we have used a Taylor expansion in deriving the second equality. A pleasant aspect of this calculation is that we can now integrate over angles from $\theta = \pi$ to $\theta = 0$ and so take account of the random arrival at x_0 of molecules from both sides of the yz-plane. We then arrive at the rate of momentum transfer per unit area by molecules with speed of random motion v:

$$\tfrac{1}{2}mvn(v)\left[u(x_0)\int_\pi^0 \sin\theta\cos\theta\,\mathrm{d}\theta - \lambda\left(\frac{\mathrm{d}u}{\mathrm{d}x}\right)_{x_0}\int_\pi^0 \sin\theta\cos^2\theta\,\mathrm{d}\theta\right]. \qquad (10.48)$$

The first integral in the brackets is zero and the second equals $-2/3$. Therefore, the net rate of transport of momentum per unit area at x_0 by these molecules is

$$\tau_{x_0}(v) = \tfrac{1}{3}mvn(v)\lambda\left(\frac{\mathrm{d}u}{\mathrm{d}x}\right)_{x_0}. \qquad (10.49)$$

We now integrate over the speeds of the molecules, finally obtaining

$$\tau_{x_0} = \tfrac{1}{3}m\lambda\left(\frac{\mathrm{d}u}{\mathrm{d}x}\right)_{x_0}\int vn(v)\,\mathrm{d}v = \tfrac{1}{3}nm\overline{v}\lambda\left(\frac{\mathrm{d}u}{\mathrm{d}x}\right)_{x_0}, \qquad (10.50)$$

where $\overline{v}$ is their mean speed of random motion.

This can be compared with the expression (10.46) for the viscous stress in the presence of a velocity gradient in a fluid. Therefore, the expression for the *coefficient of viscosity* is

$$\eta = \tfrac{1}{3}\lambda\overline{v}nm = \tfrac{1}{3}\lambda\overline{v}\rho, \qquad (10.51)$$

where ρ is the density of the gas.

It is worth noting that, in the literature on fluid dynamics, two coefficients of viscosity are used. The coefficient η we have used here is known as the *viscosity* or the *dynamic* or *absolute viscosity*. Fluid dynamicists also use the quantity η/ρ, known as the *kinematic viscosity*, and the reader should be careful to use the right one when looking up tables of viscosities.

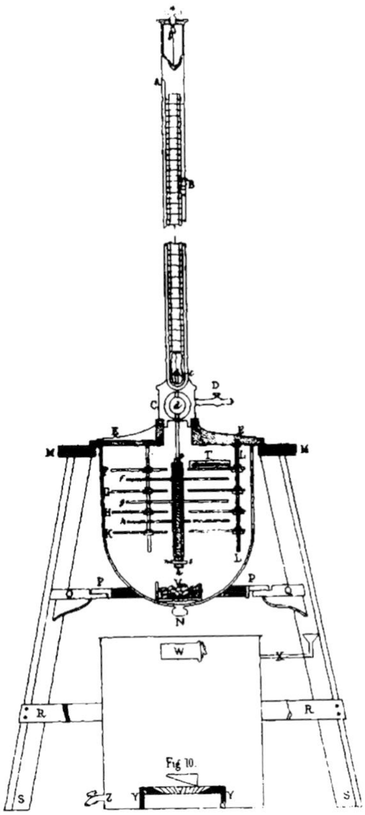

Figure 10.6: Maxwell's apparatus for measuring the viscosity of gases. The gas fills the chamber and the glass discs oscillate as a torsion pendulum. The viscosity of the gas is found by measuring the rate of decay of the oscillations of the torsion balance. The oscillations of the torsion balance were measured by reflecting light from the mirror attached to the suspension. The pressure and temperature of the gas could be varied. The oscillations were started magnetically since the volume of the chamber had to be perfectly sealed.

We can now work out how the coefficient of viscosity η is expected to change with pressure and temperature. Substituting $\lambda = (n\sigma)^{-1}$ into (10.51), we find

$$\eta = \tfrac{1}{3}\lambda\overline{v}nm = \tfrac{1}{3}m\overline{v}\sigma^{-1}. \tag{10.52}$$

Maxwell was astonished to find that the coefficient of viscosity is *independent of the pressure*, since there is no dependence upon n in the last expression in (10.52). The reason is that, although there are fewer molecules per unit volume as n decreases, the mean free path increases as n^{-1}, enabling a larger momentum increment to be transported to the plane at x_0 per molecule.

Equally remarkably, as the temperature of the gas increases, $\overline{v}$ increases as $T^{1/2}$. Therefore, the viscosity of a gas should increase with temperature, unlike the behaviour of liquids. This somewhat counter-intuitive result was the subject of a brilliant set of experiments which Maxwell carried out from 1863 to 1865 (Fig. 10.6). He fully confirmed the prediction of the kinetic theory that the viscosity of gases is independent of the pressure. He fully expected to discover a $T^{1/2}$ law, but in fact, he found a stronger dependence, $\eta \propto T$.

In his paper[10] of 1867, he interpreted this result as indicating that there must be a weak repulsive force between the molecules which varied with their spacing r as r^{-5}. This was a profound discovery since it meant that there was no longer any reason to consider the molecules to be 'elastic spheres of definite radius'. The repulsive force, proportional to r^{-5}, meant that encounters between molecules would take the form of deflections through different angles, depending upon the impact parameter (see section A4.3). Maxwell showed that it was much more appropriate to think in terms of a *relaxation time*, roughly the time it would take the molecule to be deflected through 90° as a result of random encounters with other molecules. According to Maxwell's analysis, the molecules could be replaced by the concept of centres of repulsion, or, in his words, 'mere points, or pure centres of force endowed with inertia' – it was no longer necessary to make any special assumption about molecules as hard, elastic spheres.

10.6 The statistical nature of the second law of thermodynamics

In 1867, Maxwell first presented his famous argument in which he demonstrated how, as a consequence of the kinetic theory of gases, it is possible to transfer heat from a colder to a hotter body without doing any work. He considered a vessel divided into two halves, A and B, the gas in A being hotter than that in B. Maxwell proposed that a small hole be drilled in the partition between A and B and a 'finite being' watch the molecules as they approach the hole. The finite being has a shutter which can close the hole and adopts the strategy of allowing only fast molecules to pass from B to A and slow molecules from A into B. By this means, the fast molecules in the tail of the Maxwell distribution of the cool gas in B heat the hot gas in A and the slow molecules in the low-energy side of the distribution of hot gas in A cool the cool gas in B. The finite being thus enables the system to violate the second law of thermodynamics. Thomson referred to the finite being as 'Maxwell's demon', a name to which Maxwell objected, asking Tait to

> 'call him no more a demon but a valve'.[11]

Maxwell's last remark illuminates a key point about the statistical nature of the second law of thermodynamics. Quite independently of finite beings or demons, there is a small but finite probability that, from time to time, exactly what Maxwell describes does indeed occur. Whenever one fast molecule moves from B to A, heat is transferred from the colder to the hotter body without the influence of any external agency. Now, it is overwhelmingly more likely that hot molecules move from A to B and in this process heat flows from the hotter to the colder body with the consequence that the entropy of the combined system increases. There is, however, no question but that, according to the kinetic theory of gases, there is a very small but finite probability that the reverse happens spontaneously and entropy decreases in this natural process. Maxwell was quite clear about the significance of his argument. He remarked to Tait that his argument was designed to show that the second law of thermodynamics has only a statistical certainty.[11]

This is a brilliant and compelling argument, but it does depend upon the validity of the kinetic theory of gases.

The essentially statistical nature of the second law was emphasised by another argument of Maxwell's. In the late 1860s, both Clausius and Boltzmann attempted to derive the second law of thermodynamics from mechanics, an approach known as the dynamical interpretation of the second law. In this approach, the dynamics of individual molecules were followed in the hope that this would ultimately lead to an understanding of the origin of the second law. Maxwell refuted this approach as a matter of principle because of the simple but powerful argument that Newton's laws of motion, and indeed Maxwell's equations for the electromagnetic field, are completely time reversible. Consequently the irreversibility implicit in the second law cannot be explained by a dynamical theory. The second law could only be understood as a statement based upon a statistical analysis of an immense number of molecules. The controversy about the nature of the second law will reappear in Planck's attack on the problem of understanding the spectrum of black-body radiation.

Boltzmann originally belonged to the dynamical school of thought, but was fully aware of Maxwell's work. Among the most significant of his contributions during these years was a reworking of Maxwell's analysis of the equilibrium distribution of velocities in a gas, including the presence of a potential term $\phi(r)$ which describes the potential energy of a molecule in the field. The conservation of energy requires that

$$\tfrac{1}{2}mv_1^2 + \phi(\boldsymbol{r}_1) = \tfrac{1}{2}mv_2^2 + \phi(\boldsymbol{r}_2),$$

and the corresponding probability distribution has the form

$$f(v) \propto \exp\left[-\frac{\frac{1}{2}mv^2 + \phi(\boldsymbol{r})}{kT}\right].$$

We recognise the appearance of the Boltzmann factor $\exp(-E/kT)$ in this analysis.

Eventually, Boltzmann accepted Maxwell's doctrine concerning the statistical nature of the second law and set about working out the formal relationship between entropy and probability,

$$S = C \ln W, \tag{10.53}$$

where S is the entropy of a particular equilibrium state and W is the probability that the system will be found in that state, according to the rules which will be developed below and in Chapter 13. The value of the constant of proportionality was not known. Boltzmann's analysis was of considerable mathematical complexity and this was one of the problems which prevented the scientists of his day fully appreciating the deep significance of what he had achieved.

10.7 Entropy and probability

The formal relationship between entropy and probability involves a number of tricky issues and it would take us too far afield to look into the many subtleties of the subject. John Waldram's book *The Theory of Thermodynamics*[12] can be thoroughly recommended as an exposition of the fundamentals of statistical thermodynamics with no punches pulled – it is salutory to be reminded of the care needed in setting up the underlying infrastructure of statistical thermodynamics. The objective of the remainder of this chapter is to make plausible the formal relation between entropy and probability and to illuminate its close relationship to *information theory*.

The law of the increase in entropy tells us that systems evolve in such a way as to become more uniform. To put it another way, the molecules which make up a system become more randomised – the structure is less organised and more disordered. We have already met examples in Chapter 9: when bodies at different temperatures are brought together, heat is exchanged in such a way that they come to the same temperature – temperature inequalities are evened out. In a Joule expansion, the gas expands to fill a greater volume and thus produces uniformity throughout a greater volume of space. In both cases, there is an increase in entropy.

Maxwell's arguments strongly suggest that the entropy increase is a statistical phenomenon, although he did not attempt to quantify the nature of the relation between the second law and statistics. The advance made by Boltzmann was to describe the degree of 'disorder' of any particular state of a system in terms of the probability for the state to arise by chance and to relate this probability to the entropy of the system.

Suppose we have two systems in different equilibrium states and p_1 and p_2 are the respective probabilities for these states to arise independently by chance. The probability that they both occur is the product of the two probabilities $p = p_1 p_2$. Let us associate entropies S_1 and S_2 respectively with system 1 and system 2. Entropies are additive, since entropy is an extensive quantity, and consequently the total entropy must be

$$S = S_1 + S_2. \tag{10.54}$$

Thus, if entropy and probability are related then there must be a logarithmic relation of the form $S = C \ln p$ between them, where C is some constant.

Joule expansion. Let us apply this statistical definition of entropy to the Joule expansion of a perfect gas. Consider the Joule expansion of 1 mole of gas from volume V to $2V$, noting that the system is in equilibrium states at the beginning and end of the expansion. What is the probability that, if the molecules were allowed to move freely throughout the volume $2V$, they would all occupy only a volume V? For each molecule, the probability of occupying the volume V is one-half. Therefore, if there were two molecules, the probability would be $(\frac{1}{2})^2$, if three molecules $(\frac{1}{2})^3$, if four molecules $(\frac{1}{2})^4 \ldots$ and, for one mole of gas with $N_0 \approx 6 \times 10^{23}$ molecules, Avogadro's constant, the probability is $(\frac{1}{2})^{N_0}$ – this last probability is very small indeed. Notice that this number represents the relative probability of the two equilibrium states since it is assumed that at both the start and finish the distributions of the particles are uniform.

Let us adopt the definition $S = C \ln p$ to find the entropy change associated with $p_2/p_1 = 2^{N_0}$. The entropy change is then

$$S_2 - S_1 = \Delta S = C \ln(p_2/p_1) = C \ln 2^{N_0} = C N_0 \ln 2. \tag{10.55}$$

We have, however, already worked out the entropy change for a Joule expansion according to classical thermodynamics, (9.53), so that for $V_2 = 2V_1$ we obtain

$$\Delta S = R \ln(V_2/V_1) = R \ln 2.$$

It immediately follows that

$$R = CN \qquad \text{or} \qquad C = R/N = k.$$

We have recovered Boltzmann's fundamental relation between entropy and probability, with the bonus of having determined the constant relating S and $\ln p$, Boltzmann's constant k:

$$\Delta S = S_2 - S_1 = k \ln(p_2/p_1). \tag{10.56}$$

The sign has been chosen so that the more likely event corresponds to the greater entropy.

Partitioned gas. Let us perform another calculation which illustrates how the statistical approach works. Consider two equal volumes, each $V/2$, containing gas at different pressures p_1 and p_2 separated by a partition. On one side of the partition, there are r moles of gas ($r \leq 1$) and on the other $1 - r$ moles.

Now let us find the change in entropy when the partition is removed. For the sake of argument suppose that $\frac{1}{2} > r > 0$. Then, after removal of the partition, effectively the r moles will have contracted to occupy a volume rV, and the $1 - r$ moles will have expanded to occupy a volume $(1 - r)V$. We can regard these two processes as taking place separately.

What is the entropy change? If the gas is perfect then from (9.53) the entropy change of the r moles will be

$$\Delta S_1 = rR \ln \frac{rV}{V/2} = rR \ln 2r, \tag{10.57}$$

and the entropy change of the $1 - r$ moles will be

$$\Delta S_2 = (1 - r)R \ln \frac{(1 - r)V}{V/2} = (1 - r)R \ln[2(1 - r)]. \tag{10.58}$$

Since, as mentioned above, entropies are additive, the total change in entropy on removal of the partition is

$$\begin{aligned} \Delta S = \Delta S_1 + \Delta S_2 &= R\{r \ln 2r + (1 - r) \ln[2(1 - r)]\} \\ &= R\{r \ln r + (1 - r) \ln(1 - r) + \ln 2\}. \end{aligned} \tag{10.59}$$

Note that if $r = \frac{1}{2}$ then $\Delta S = 0$ as expected: removal of the partition causes no macroscopic change in the gas. If, however, $r = 0$ then we revert to the Joule expansion discussed above, for which $\Delta S = R \ln 2$.

Statistical approach. Now let us look at the same problem from a statistical perspective. We can use simple statistical procedures to work out the number of different ways in which N objects can be distributed between two boxes. The necessary tools are developed in Section 13.2. Referring to that section, the number of ways of choosing m objects from N, that is, the number of choices with m objects in one box and $N - m$ in the other is

$$g(N, m) = \frac{N!}{(N-m)!m!}.$$

Let us write $m = \frac{1}{2}N + x$, so that for the case of equal numbers of objects in each box $x = 0$; then

$$g(N, x) = \frac{N!}{[(N/2) - x]![(N/2) + x]!}.$$

Now this is not a probability but the exact number of microscopic ways in which we can end up with $[(N/2) - x]$ objects in one box and $[(N/2) + x]$ in the other. The total number of ways of distributing the objects between the two boxes is 2^N. Therefore the probability corresponding to $g(N, x)$ is $p = g(N, x)/2^N$. Therefore, adopting Boltzmann's expression for the entropy, $S = k \ln p$, we obtain

$$\begin{aligned} S = k \ln[g(N, x)/2^N] = k\{&\ln N! - \ln[(N/2) - x]! \\ &- \ln[(N/2) + x]! - N \ln 2\}. \end{aligned} \tag{10.60}$$

Now approximate the logarithms using Stirling's approximation in the form

$$\ln M! \approx M \ln M - M. \tag{10.61}$$

After a bit of simple manipulation of (10.60), we find

$$S = -Nk\left[\left(\frac{1}{2} - \frac{x}{N}\right)\ln\left(\frac{1}{2} - \frac{x}{N}\right) + \left(\frac{1}{2} + \frac{x}{N}\right)\ln\left(\frac{1}{2} + \frac{x}{N}\right) + \ln 2\right]. \tag{10.62}$$

Now, reverting to the original problem, N is the number of molecules in 1 mole, Avogadro's constant N_0. Therefore,

$$r = \frac{1}{2} - \frac{x}{N_0}, \quad 1 - r = \frac{1}{2} + \frac{x}{N_0} \tag{10.63}$$

and from (10.63)

$$S = -N_0 k[r \ln r + (1 - r)\ln(1 - r) + \ln 2]. \tag{10.64}$$

When the system is in equilibrium $r = 1/2$ and so $S = 0$. Therefore (10.64) gives the entropy of the partitioned gas as compared with the unpartitioned gas. It corresponds exactly to the result for the same experiment obtained from classical thermodynamics, (10.59), with the correct constant in front of the logarithms, $k = R/N_0$. Thus, the definition $S = k \ln p$ can account rather beautifully for the classical result by statistical arguments.

Probability of small deviations x. Let us now work out the probability of small deviations from equal numbers of molecules in each volume. To do this, expand the function $g(N_0, x)$

for small values of x. From (10.62),

$$\frac{S}{k} = \ln p(N_0, x) = -\frac{N_0}{2}\left[\left(1 - \frac{2x}{N_0}\right)\ln\left(1 - \frac{2x}{N_0}\right) + \left(1 + \frac{2x}{N_0}\right)\ln\left(1 + \frac{2x}{N_0}\right)\right].$$

Expanding the logarithms to order x^2,

$$\ln(1 + x) = x - \tfrac{1}{2}x^2,$$

the expression for $\ln p(N_0, x)$ reduces to

$$\begin{aligned}\ln p(N_0, x) &= -\frac{N_0}{2}\left[\left(1 - \frac{2x}{N_0}\right)\left(-\frac{2x}{N_0} - \frac{1}{2}\frac{4x^2}{N_0^2}\right) + \left(1 + \frac{2x}{N_0}\right)\left(\frac{2x}{N_0} - \frac{1}{2}\frac{4x^2}{N_0^2}\right)\right] \\ &= -\frac{2x^2}{N_0} + \cdots \end{aligned} \tag{10.65}$$

and so the probability distribution is

$$p(N_0, x) = \exp\left(-\frac{2x^2}{N_0}\right), \tag{10.66}$$

that is, a Gaussian distribution having mean value 0 with probability unity! We note that the standard deviation of the distribution about the value $x = 0$ is $(N_0/2)^{1/2}$, that is, to order of magnitude, $N_0^{1/2}$. Thus, the likely deviation about the value $x = 0$ corresponds to very tiny fluctuations indeed: since $N_0 \sim 10^{24}$, $N_0^{1/2} \sim 10^{12}$ or, in terms of the fractional fluctuation, $N_0^{1/2}/N_0 \sim 10^{-12}$. Our analysis of subsection 15.2.1 shows that this is no more than the statistical fluctuation about the mean value.

This example illustrates how statistical mechanics works. We deal with huge ensembles of molecules, $N_0 \sim 10^{24}$, and hence, although it is true that there are statistical deviations from the mean behaviour, in practice, they are very, very tiny. Thus, it is possible for the statistical entropy to decrease spontaneously but the likelihood of this happening is negligibly small because N_0 is so large.

Let us show briefly how entropies depend upon the volume of the system, not only in real three-dimensional space but also in *velocity* or *phase space* as well. The entropy change of a perfect gas in passing from coordinates V_0, T_0 to V, T is

$$\Delta S(T, V) = C_V \ln(T/T_0) + R \ln(V/V_0). \tag{10.67}$$

Now express the first term on the right-hand side in terms of molecular velocities rather than temperatures. According to kinetic theory, $T \propto \overline{v^2}$ and hence, since $C_V = \frac{3}{2}R$,

$$\begin{aligned}\Delta S(T, V) &= \tfrac{3}{2}R \ln\left(\overline{v^2}/\overline{v_0^2}\right) + R \ln(V/V_0) \\ &= R\left[\ln\left(\overline{v^2}/\overline{v_0^2}\right)^{3/2} + \ln(V/V_0)\right].\end{aligned} \tag{10.68}$$

We can interpret this formula as indicating that when we change both T and V, there are two contributions to the entropy change. The volume of real space changes in the ratio V/V_0. In addition, the first term in the brackets indicates that the molecules occupy a larger volume of velocity or phase space, by the factor $(\overline{v^2}/\overline{v_0^2})^{3/2} \sim v^3/v_0^3$. Thus, we can interpret the formula for the entropy increase in terms of an increase in the accessible volumes of both real and velocity space.

10.8 Entropy and the density of states

Let us take the story one step further. You will have noticed that, in the reasoning which led to (10.60), rather than working in terms of probabilities we could have used equally well the total number of different ways in which m objects can be allocated to one box and $N - m$ to the other. This leads naturally to the theory of thermodynamics as formulated by John Waldram in his careful presentation,[12] which has strongly influenced the approach taken in this section. Let us begin with one of his examples.

Harmonic oscillators. Just as in the route pioneered by Boltzmann, John Waldram's formulation of thermodynamics begins by assuming the existence of discrete energy levels, as in any quantum system; for simplicity, it is convenient to consider the system to be a set of identical, but spatially localised and therefore *distinguishable*, harmonic oscillators, which have the helpful feature of possessing equally spaced energy levels. For present purposes, we can neglect the zero-point energy $\frac{1}{2}\hbar\omega$ of each oscillator. Suppose that we have a system consisting of 100 oscillators, each in its ground state, and that we can add energy quanta to the oscillators, the energy ϵ of each quantum being equal to the spacing of the energy levels. We add Q quanta to the system of oscillators and ask, 'In how many different ways W can we distribute the Q quanta among the $N = 100$ oscillators?' If there were no quanta, $Q = 0$, there would be only one way of distributing them and so $W = 1$. If $Q = 1$, we can place the one quantum in any of the 100 ground states and so $Q = 100$. If $Q = 2$, we can place the two quanta in different ground states in $(100 \times 99)/2 = 4950$ ways, plus the 100 ways in which the two quanta can be allocated to the same oscillator. Hence, in total, $W = 5050$. Notice that in carrying out this sum, we need to avoid duplications – it is the total number of *different* ways which is enumerated.

The extension of this procedure to distributing Q quanta among N oscillators is carried out in Section 13.3, with the result that the total number of different ways of doing so, or *configurations*, is

$$W = \frac{(Q + N - 1)!}{Q!(N - 1)!}. \qquad (10.69)$$

We can now create a table (Table 10.1) which shows how rapidly W increases with increasing Q for the case of 100 oscillators.

We notice that $W \sim 10^Q$ as Q becomes large. It is left as an exercise to the reader to show that, using Stirling's approximation, if $Q = 10^{23}$, then $W \sim 10^{10^{23}}$. Thus, the number of ways of obtaining a total energy Q increases very rapidly indeed with increasing Q. We say that the number of ways of arranging the quanta, or the number of configurations, or the *density of states*, W, increases very rapidly with energy. This is a crucial concept.

Now in reality one cannot count all the states with total energy exactly $E = Q\epsilon$. In practice we always deal with the number of states within a finite range of energies, say, from E to $E + \mathrm{d}E$, which is $g(E)\,\mathrm{d}E$, where $g(E)$, the density of states, is a continuous function with dimensions $[\text{energy}]^{-1}$.

Table 10.1. *The number of different ways W in which Q quanta can be distributed among N oscillators*

Q	W
0	1
1	100
2	5050
3	171 700
4	4421 275
5	92 846 775
10	$4.509\,3338 \times 10^{13}$
20	$2.807\,3848 \times 10^{22}$

Statistical equilibrium. To convert these ideas into a theory of statistical thermodynamics, we need to make some assumptions. Central to the subject is the *principle of equal equilibrium probabilities*. This is the statement that

> In statistical equilibrium, all quantum states accessible to the system are equally likely to be occupied.

Thus, if we share three quanta among 100 oscillators, there are 171 700 ways of sharing out this energy and, in equilibrium, the system is equally likely to be found in any one of them. There is no point in disguising the fact that there are a number of problems with this assumption – for example, does the system really explore all possible distributions in the age of the Universe? Likewise, we need to assume that there are equal probabilities for a transition between any pair of states to occur in either direction. Waldram refers to this as the *principle of jump-rate symmetry*, which fortunately follows from the equality of the transition probabilities between any two quantum states according to quantum mechanics. I simply note that the principle of equal equilibrium probabilities works incredibly well and so provides its own justification.

This is the beginning of a journey which leads to the statistical definition of the law of increase of entropy and what we mean by temperature in statistical physics. John Waldram gives a very pleasing example of how this works – again, our approach here is 'naive' compared with his careful treatment of the many subtleties involved.

Density of states. Consider two large systems A and B, containing respectively $N_A = 5000$ and $N_B = 10\,000$ oscillators. All the oscillators have the same energy levels and so, as in the above example, we can work out exactly the density of states for each of them. For A,

$$g_A = \frac{(N_A + Q_A - 1)!}{Q_A!(N_A - 1)!} \approx \frac{(N_A + Q_A)!}{Q_A!N_A!}, \tag{10.70}$$

where the last equality follows from the fact that the numbers are so large that the minus-one terms are of no importance. We can now simplify this result by using Stirling's formula in an even simpler approximation $Q! \approx Q^Q$. Therefore,

$$g_A = \frac{(N_A + Q_A)^{N_A+Q_A}}{Q_A^{Q_A} N_A^{N_A}}. \tag{10.71}$$

The densities of states for A and B are therefore

$$g_A = \frac{(5000 + Q_A)^{5000+Q_A}}{Q_A^{Q_A}\ 5000^{5000}}, \qquad g_B = \frac{(10\,000 + Q_B)^{10\,000+Q_B}}{Q_B^{Q_B}\ 10\,000^{10\,000}}. \tag{10.72}$$

If we now place the two systems in thermal contact, the quanta are free to explore all the states of the combined system A + B. This means that the density of states available to the combined system is $g_A g_B$ since we can combine every state of A with every state of B. Thus, for the combined system,

$$g = g_A g_B = \frac{(5000 + Q_A)^{5000+Q_A}}{Q_A^{Q_A}\ 5000^{5000}} \times \frac{(10\,000 + Q_B)^{10\,000+Q_B}}{Q_B^{Q_B}\ 10\,000^{10\,000}}. \tag{10.73}$$

The total number of quanta $Q = Q_A + Q_B$ is fixed. It is left as an exercise to the reader to find the maximum value of g subject to this constraint. It is simplest to take the logarithm of both sides, throw away the constants and maximise by finding $\mathrm{d}(\ln g)/\mathrm{d}Q_A$. The answer you should find is that g is maximised by

$$Q_A = \frac{Q}{3}, \qquad Q_B = \frac{2Q}{3}; \tag{10.74}$$

in other words, the energy is shared exactly in proportion to the numbers of oscillators in A and B. Thus the equilibrium distribution is found by maximising the density of states available to the system, subject to the boundary conditions of the problem.

Now suppose that the system were not in its most probable state. For example, suppose that the energies were the other way round, with

$$Q_A = \frac{2Q}{3}, \qquad Q_B = \frac{Q}{3}. \tag{10.75}$$

Inserting these values into (10.73), you will find that

$$\frac{g}{g_{max}} = 10^{-738}. \tag{10.76}$$

Statistical definition of entropy. So, how does the system evolve towards equilibrium? Because of the principle of jump-rate symmetry, when the systems are placed in thermal contact there is twice the probability of a quantum jumping from A to B as from B to A, because there are twice as many states available in B as in A. Thus, the 'natural flow of heat' is from A to B. In statistical terms, heat flows in such a way that $g_A g_B$ increases, that is,

$$\mathrm{d}(g_A g_B) \geq 0. \tag{10.77}$$

We can equally well find the maximum of the logarithm of $g_A g_B$ and so

$$\mathrm{d}\,[\ln(g_A g_B)] = \mathrm{d}(\ln g_A) + \mathrm{d}(\ln g_B) \geq 0. \tag{10.78}$$

However, we recall that g_A and g_B are functions of energy, $g_A(E_A)$ and $g_B(E_B)$. Therefore, using

$$df = \frac{\partial f}{\partial x}\,dx + \frac{\partial f}{\partial y}\,dy,$$

we obtain

$$\frac{\partial(\ln g_A)}{\partial E_A}\,dE_A + \frac{\partial(\ln g_B)}{\partial E_B}\,dE_B \geq 0. \tag{10.79}$$

When energy is exchanged between A and B, $dE_A = -dE_B$. Therefore

$$\left[\frac{\partial(\ln g_A)}{\partial E_A} - \frac{\partial(\ln g_B)}{\partial E_B}\right] dE_A \geq 0. \tag{10.80}$$

This is a fundamental equation in statistical physics. The signs are important. If dE_A is positive then A gains energy and, to satisfy (10.80), we require

$$\frac{\partial(\ln g_A)}{\partial E_A} > \frac{\partial(\ln g_B)}{\partial E_B}. \tag{10.81}$$

On the contrary, if dE_A is negative then B gains energy and

$$\frac{\partial(\ln g_B)}{\partial E_B} > \frac{\partial(\ln g_A)}{\partial E_A}. \tag{10.82}$$

Thus, we see that $\partial(\ln g)/\partial E$ is an inverse measure of temperature: the heat flow takes place, according to (10.80), from the system with the smaller value of $\partial(\ln g)/\partial E$ to the system with the larger value of $\partial(\ln g)/\partial E$. This leads to the definition of *statistical temperature* T_s, namely

$$\frac{1}{kT_s} = \frac{\partial(\ln g)}{\partial E} \tag{10.83}$$

where g is a maximum.

When A and B are in equilibrium, $d\ln(g_A g_B) = 0$ and so

$$\frac{\partial(\ln g_A)}{\partial E_A} = \frac{\partial(\ln g_B)}{\partial E_B}, \qquad \text{that is,} \qquad T_s(\text{A}) = T_s(\text{B}). \tag{10.84}$$

The definition (10.83) can be compared with the corresponding equation derived from classical thermodynamics, (9.57),

$$\left(\frac{\partial S}{\partial U}\right)_V = \frac{1}{T}. \tag{10.85}$$

Evidently, we have derived a statistical definition of entropy:

$$S = k\ln g = k\ln[g(E)\,dE], \tag{10.86}$$

where we recall that g is the maximum density of states, corresponding to the equilibrium state.

Table 10.2. *Examples of different sums over probabilities*

p_i					$-\sum p_i \ln p_i$	$-\sum \ln p_i$	$-\sum p_i^2 \ln p_i$
0.2	0.2	0.2	0.2	0.2	1.609	0.322	0.322
0.1	0.3	0.2	0.3	0.1	1.504	8.622	0.327
0.49	0.01	0.01	0.01	0.48	0.840	15.263	0.340
0.001	0.001	0.996	0.001	0.001	0.032	27.635	0.004

10.9 Gibbs entropy and information

These concepts were gradually crystallised during the last two decades of the nineteenth century, largely thanks to the heroic efforts of Boltzmann and Gibbs. Gibbs, in particular, discovered the correct procedure for defining the probabilities p_i that a system would be found in state i and how these probabilities should be combined to define the entropy of the system. What Gibbs discovered turned out to be one aspect of a very general theorem in probability theory. It was only in 1948 that Claude Shannon discovered a theorem which enabled the statistical approach to be formulated in a completely general way on the basis of information theory. *Shannon's theorem*[13] is one of the most fundamental theorems in the theory of probability and can be stated as follows.

If p_i are a set of mutually exclusive probabilities, then the function

$$f(p_1, p_2, p_3, \ldots p_n) = -C \sum_1^n p_i \ln p_i \tag{10.87}$$

is a unique function, which, when maximised, gives the most likely distribution of the p_i for a given set of constraints.

We will not prove this theorem but give a simple numerical example which illustrates how it works. The important point is that the form $-\sum p_i \ln p_i$ is not at all arbitrary. Suppose that there are five possible outcomes of an experiment and that we assign probabilities p_i to these. In Table 10.2, the results of forming various sums of these probabilities are shown.

This simple demonstration indicates that the maximum value of $-\sum p_i \ln p_i$ corresponds to the case in which all the p_i are equal. In the words of the theorem, the most likely distribution of probabilities in the absence of any constraints is that all five probabilities are the same. In the case in which the system is more or less completely ordered, as in the fourth example, $-\sum p_i \ln p_i$ tends to zero. It can be seen from the table that, as the probabilities become more and more uniform, $-\sum p_i \ln p_i$ tends towards a maximum value. The correspondence with the law of increase of entropy is striking and leads directly to the definition of the *Gibbs entropy* as

$$S = -k \sum_i p_i \ln p_i, \tag{10.88}$$

so that the maximum of this function corresponds to the most probable distribution, the equilibrium state of the system. Notice that (10.88) provides a very powerful definition of the entropy of the system, which can be used even if it has not attained an equilibrium state.

Notice how this definition quantifies the relation between entropy, disorder and information. The fourth example in the table is more or less completely ordered, has very low entropy and contains the maximum amount of information about the state of the system. The distribution in the first example is completely random, the entropy is a maximum and we obtain the minimum information about which state the system is actually in. Accordingly, in information theory, the *information* is defined as follows:

$$\text{information} = k \sum_i p_i \ln p_i ; \tag{10.89}$$

the information is sometimes called the *negentropy*.

Let us now complete the analysis by relating the Gibbs definition of entropy to the results obtained in the last section. According to the principle of equal equilibrium probabilities, in equilibrium there is an equal probability of the system being found in any one of the states accessible to it, subject to a set of given constraints. The number of states having the equilibrium energy E is g, or more exactly $g(E)\,\mathrm{d}E$. Each of these g states is equally probable and so

$$p_i = \frac{1}{g} = \frac{1}{g(E)\,\mathrm{d}E},$$

where the last equality indicates how we would carry out the same analysis if the density of states were described by a continuous function. Therefore, adopting the Gibbs definition of entropy,

$$S = -k \sum_{i=1}^{i=g} \frac{1}{g} \ln \frac{1}{g}.$$

Since the sum is over all g states,

$$S = k \ln g = k \ln(\text{density of states}), \tag{10.90}$$

exactly the definition of the equilibrium entropy derived in (10.86).

To convince ourselves of the equivalence of these procedures, we can use the example shown in Table 10.2 and place one quantum in any one of the five oscillators. We showed in Section 10.8 that the density of states then corresponds to $Q = 1$ and $N = 5$. Hence

$$g = \frac{(Q + N - 1)!}{Q!(N - 1)!} = 5,$$

as expected. Therefore, $\ln g = \ln 5 = 1.609$, which is exactly the same as the value of $-\sum_i p_i \ln p_i$ shown in the first line of Table 10.2.

Finally, let us consider the Joule expansion of Section 10.7 in the same light. At the start, the particles are partitioned into half the total volume, and the probability p_i that a particle will be in that volume is equal to unity for all N_0 particles and so $\sum_i p_i \ln p_i = 0$. When the gas has expanded into the whole volume, the probability is 0.5 that the molecule will be in one volume and 0.5 that it will be in the other. In equilibrium, these probabilities are the same for all molecules and so the Gibbs entropy for the new state is

$$S = -k \sum_{N_0} \left(\tfrac{1}{2} \ln \tfrac{1}{2} + \tfrac{1}{2} \ln \tfrac{1}{2}\right) = kN_0 \ln 2. \tag{10.91}$$

Thus, the Gibbs entropy is again seen to be identical to the classical definition of the entropy change for this process, which resulted in $\Delta S = R \ln 2 = k N_0 \ln 2$.

10.10 Concluding remarks

With this introduction, the way is now clear for the development of the full statistical mechanical interpretation of classical thermodynamics. This development is traced in such standard texts as Kittel's *Thermal Physics*,[14] Mandl's *Statistical Physics*[15] and Waldram's *The Theory of Thermodynamics*.[12]

Boltzmann's great discovery of the quantitative relations between entropy and probability was appreciated by relatively few physicists at the time. The stumbling blocks were the facts that the status of the kinetic theory of gases, and in particular the equipartition theorem, was uncertain since it failed to account for the specific heats of gases and it was not clear how to incorporate the internal vibrations of molecules, as exhibited by spectral lines. Towards the end of the nineteenth century, a reaction against atomic and molecular theories of the properties of matter gained currency in some quarters in continental Europe. According to this view, one should only deal with the bulk properties of systems, that is, with classical thermodynamics, and abolish atomic and molecular concepts as unnecessary. These were profoundly discouraging developments for Boltzmann. They may well have contributed to his suicide in 1906. It is a tragedy that he was driven to such an act at the very time when the correctness of his fundamental insights were appreciated by Einstein. The problem of the specific heats was solved in Einstein's classic paper of 1906 and in the process opened up a completely new vision of the nature of elementary processes.

We will return to Boltzmann's procedures and the concepts developed in this chapter when we survey the extraordinary achievements of Planck and Einstein which resulted in the discovery of quantisation and quanta.

10.11 References

1 Clausius, R. (1857). *Annalen der Physik*, **100**, 497 (English translation in S.G. Brush ed., 1966, *Kinetic Theory*, Vol. 1, p. 111, Oxford: Pergamon Press).
2 Clausius, R. (1858). *Annalen der Physik*, **105**, 239 (English translation in S.G. Brush ed., 1966, *Kinetic Theory*, Vol. 1, p. 135. Oxford: Pergamon Press).
3 Waterston, J.J. (1843). See article by S.G. Brush, *DSB*, Vol. 14, p. 184. See also J.G. Haldane ed. (1928), *The Collected Scientific Papers of John James Waterston*. Edinburgh: Oliver and Boyd.
4 Rayleigh, Lord (1892). *Phil. Trans. Roy. Soc.*, **183**, 1.
5 Rayleigh, Lord (1892). *Op. cit*, 2.
6 Rayleigh, Lord (1892). *Op. cit*, 3.
7 Maxwell, J.C. (1860). *Phil. Mag., Series 4*, **19**, 19 and **20**, 21. See also, W.D. Niven ed. (1890), *The Scientific Papers of James Clerk Maxwell* p. 377. Cambridge: Cambridge University Press (1890).

8 Everitt, C.W.F. (1970). *DSB*, Vol. 9, p. 218.
9 Maxwell, J.C. (1860). *Report of the British Association for the Advancement of Science*, **28**, Part 2, 16.
10 Maxwell, J.C. (1867). *Phil. Trans. Roy. Soc.*, **157**, 49.
11 Maxwell, J.C. (1867). Communication to P.G. Tait, quoted by P.M. Harman, 1982, *Energy, Force and Matter, The Conceptual Development of Nineteenth-Century Physics*, p. 140, Cambridge: Cambridge University Press.
12 Waldram, J.R. (1985). *The Theory of Thermodynamics*. Cambridge: Cambridge University Press.
13 Shannon, C.E. (1948). *The Bell System Technical Journal*, **27**, 379 and 623. (See also N.J.A. Sloane and A.D. Wyner eds., 1993, *The Collected Papers of Claude Elwood Shannon*, Piscataway, New Jersey: IEEE Press.)
14 Kittel, C. (1969). *Thermal Physics*. New York: John Wiley and Sons.
15 Mandl, F. (1971). *Statistical Physics*. Chichester: John Wiley and Sons.

Case Study V
The origins of the concept of quanta

Quanta and relativity are phenomena of physics which are quite outside our everyday experience – they are also perhaps the greatest discoveries of twentieth-century physics. In the debate between those supporting the Copernican and geocentric models of the Universe, discussed in Chapter 3, one of the issues raised by opponents of the Copernican picture concerned 'the deception of the senses': if these phenomena are of such fundamental importance, why are we not aware of them in our everyday lives? Quanta and relativity were both discovered by very careful experiment, the results of which could not be accounted for within the context of Newtonian physics. Indeed, these phenomena are arguably 'non-intuitive', and yet they lie at the foundations of the whole of modern physics.

In this case study, we will study in some detail the origins of the concept of quanta. For me, this is one of the most dramatic stories in intellectual history. It is also very exciting and catches the flavour of an epoch when, within 25 years, physicists' view of nature changed totally and completely new perspectives were opened up. The story illustrates many important points about how physics and theoretical physics work in practice. We find the greatest physicists making mistakes, individuals having to struggle against the accepted views of virtually all physicists, and, most of all, a level of inspiration and scientific creativity which I find dazzling. If only everyone, and not only those who have had a number of years training as physicists or mathematicians, could appreciate the intellectual beauty of this story.

In addition to telling a fascinating and compelling story, I want to prove everything essential to it using the physics and mathematics available at the time. This provides a splendid opportunity for reviewing a number of important areas of basic physics in their historical context. We will find a striking contrast between those phenomena which can be explained classically and those which necessarily involve quantum ideas. The story will cover the years from about 1890 to the 1920s, by which date all physicists had to come to terms with a new view of physics in which all the fundamental entities are quantised.

The story will centre upon the work of two very great physicists – Planck and Einstein. Planck is properly given credit for the discovery of quantisation and we will trace how this came about. Einstein's contribution was even greater in that, long before anyone else, he inferred that all natural phenomena are quantum in nature, and he was the first to put the subject on a firm theoretical basis.

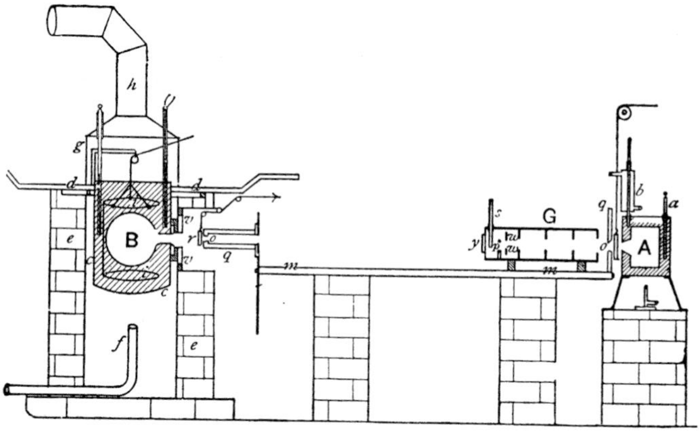

Figure V.1: Lummer and Pringsheim's apparatus of 1897 for determining experimentally the Stefan–Boltzmann law with high precision. The relative intensities of radiation from the enclosure B and the reference black body A could be found by moving the detector G along the table between them until the illumination from each source was equal. (From H.S. Allen and R.S. Maxwell, 1952, *A Text-book of Heat*, Vol. II, p. 746, London: Macmillan and Co.)

I based my original version of this story upon a set of lectures by Martin J. Klein entitled *The Beginnings of the Quantum Theory*[1] and published in the proceedings of the fifty-seventh Varenna Summer School. When I first read them, I found these lectures a revelation and felt cheated that I had not known this story before. I acknowledge fully my indebtedness to Professor Klein in inspiring what I consider in many ways to be the core of the present volume. Kindly, he sent me further materials after the publication of the first edition of this book. I have found his article 'Einstein and the wave–particle duality'[2] particularly valuable.

To my regret, having delivered lectures on these topics from 1978 to 1980, I was unaware of the important volume by Thomas S. Kuhn *Black-Body Theory and the Quantum Discontinuity 1894–1912* (1978)[3] when I sent the first edition of this book to press. Kuhn's book is in my view a masterpiece and he digs deeply into much of the material discussed in this case study. Kuhn's profound insights have changed some of my perceptions as recorded in the first edition of the present text. For those who wish to pursue these topics in more detail, his book is essential reading.

V.1 References

1 Klein, M.J. (1977). *History of twentieth century physics, Proc. International School of Physics 'Enrico Fermi'*, Course 57, p. 1. New York and London: Academic Press.
2 Klein, M.J. (1964). Einstein and the wave–particle duality, *The New Philosopher*, **3**, 1–49.
3 Kuhn, T.S. (1978). *Black-Body Theory and the Quantum Discontinuity 1894–1912*. Oxford: Clarendon Press.

11 Black-body radiation up to 1895

11.1 The state of physics in 1890

In the course of the case studies treated so far, we have been building up a picture of the state of physics and theoretical physics towards the end of the nineteenth century. The achievement had been immense. In mechanics and dynamics, the Lagrangian and Hamiltonian dynamics described briefly in Chapter 7, were well understood. In thermodynamics, the first and second laws were firmly established, largely through the efforts of Clausius and Lord Kelvin, and the full ramifications of the concept of entropy in classical thermodynamics were being elaborated. In Chapters 5 and 6, we described how Maxwell derived the equations of electromagnetism. Hertz's experiments of 1887–9 demonstrated beyond any shadow of doubt that, as predicted by Maxwell, light is a form of electromagnetic wave. This discovery provided a firm theoretical foundation for the wave theory of light, which could account for virtually all the known phenomena of optics.

The impression is sometimes given that most physicists of the 1890s believed that the combination of thermodynamics, electromagnetism and classical mechanics could account for all known physical phenomena and that all that remained to be done was to work out the consequences of these hard-won achievements. In fact, it was a period of ferment when there were still many fundamental unresolved problems which exercised the greatest minds of the period.

We have discussed the ambiguous status of the kinetic theory of gases and the equipartition theorem as expounded by Clausius, Maxwell and Boltzmann. The fact that these could not account satisfactorily for all the known properties of gases was a major barrier to their acceptance. The status of atomic and molecular theories of the structure of matter came under attack both for the technical reasons outlined above and also because of a movement away from mechanistic atomic models for physical phenomena in favour of empirical or phenomenological theories. The origin of the 'resonances' within molecules, which were presumed to be the origin of spectral lines, had no clear interpretation and was an embarrassment to supporters of the kinetic theory. Boltzmann had discovered the statistical basis of thermodynamics, but the theory had won little support, particularly in the face of a movement which denied that kinetic theories had any value, even as hypotheses. The negative result of the Michelson–Morley experiment was announced in 1887 – we will take up that story in Case Study VI on special relativity. A useful portrait of the state of physics in the last decades of the nineteenth century is provided by David Lindley in his book *Boltzmann's Atom*.[1]

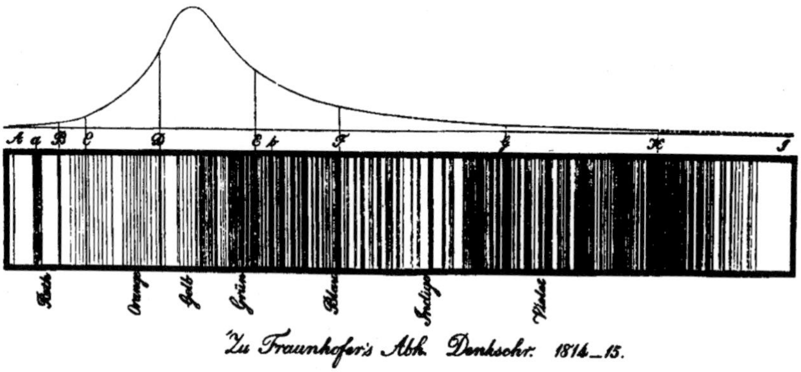

Figure 11.1: Fraunhofer's solar spectrum of 1814 showing the vast number of absorption lines. The colours of the various regions of the spectrum are shown, as well as the labels of the prominent absorption lines.

Among these unresolved problems was the origin of the spectrum of *black-body radiation*, the answer to which was to prove to be the key not only to the discovery of quantisation but also to the resolution of many other problems listed above. The discovery of quantisation and quanta was the precursor of the modern quantum theory of matter and radiation.

11.2 Kirchhoff's law of emission and absorption of radiation

The first step along the path to the understanding of the spectrum of black-body radiation was the discovery of spectral lines. In 1802, spectroscopic measurements of the solar spectrum were made by William Wollaston, who observed five strong dark lines. This observation was followed up in much greater detail by Joseph Fraunhofer, the son of a glazier and an experimenter of genius. Early in his career, Fraunhofer became one of the directors of the Benediktbreuern Glass Factory in Bavaria, one of its main functions being the production of high-quality glass components for surveying and military purposes. His objective was to improve the quality of the glasses and lenses used in these instruments and for this he needed much improved wavelength standards. The breakthrough came in 1814 with his discovery of the vast array of dark absorption features in the solar spectrum (Fig. 11.1). The ten prominent lines were labelled A, a, B, C, D, E, b, F, G, H and, in addition, there were 574 fainter lines between B and H. One of his great contributions was the invention of an instrument with which he could measure precisely the deflection of light passing through the prism. To achieve this, he placed a theodolite on its side and observed the spectrum through a telescope mounted on the theodolite's rotating ring – this device was the first high-precision *spectrograph*. Fraunhofer also measured the wavelengths of the sun's spectral lines precisely using a diffraction grating, a device invented by Thomas Young just over a decade earlier.

Fraunhofer noted that the strong absorption feature which he labelled D consisted of two lines of the same wavelengths as the strong double emission line components observed in lamp-light. Foucault was able to reproduce the D-lines in the laboratory by passing the radiation from a carbon arc through a sodium flame in front of the slit of the spectrograph. It was realised that absorption lines are the characteristic signatures of different elements; the most important experiments were carried out by Robert Bunsen and Gustav Kirchhoff. The Bunsen burner, as it came to be called, had the great advantage of enabling the spectral lines of various elements to be determined without contamination by the burning gas. In Kirchhoff's monumental series of papers of 1861–3, entitled 'Investigations of the solar spectrum and the spectra of the chemical elements',[2] the solar spectrum was compared with the spectra of 30 elements using a four-prism arrangement, designed so that the spectrum of the element and the solar spectrum could be observed simultaneously. He concluded that the cool, outer regions of the solar atmosphere contained iron, calcium, magnesium, sodium, nickel and chromium and probably cobalt, barium, copper and zinc as well.[3]

In the course of these studies, in 1859 Kirchhoff formulated his law concerning the relation between the coefficients of emission and absorption of radiation and how these can be related to the spectrum of thermal equilibrium radiation, what came to be known as *black-body radiation*. Let us begin by clarifying the definitions we will use throughout this case study.

11.2.1 Radiation intensity and energy density

Consider a region of space in which a small element of area $\mathrm{d}A$ is exposed to a radiation field. In a time $\mathrm{d}t$, the total amount of energy passing through $\mathrm{d}A$ is $\mathrm{d}E = S\,\mathrm{d}A\,\mathrm{d}t$, where S is the *total flux density* of radiation and has units W m^{-2}. We will be concerned with understanding the *spectral energy distribution* of the radiation, and so let us work in terms of the power arriving per unit area per unit frequency interval, S_ν, so that $S = \int S_\nu\,\mathrm{d}\nu$. We will refer to S_ν as the *flux density*; its units are W m^{-2} Hz^{-1}. In the case of an isotropic point source of radiation of luminosity L_ν, at distance r, placing the elementary area normal to the direction of the source, the flux density of radiation is, by the conservation of energy,

$$S_\nu = \frac{L_\nu}{4\pi r^2}. \tag{11.1}$$

The units of luminosity are W Hz^{-1}.

The flux density depends upon the orientation of the area $\mathrm{d}A$ in the radiation field. It is meaningless to think of the radiation approaching the area along a particular ray. Rather, as in Chapter 10 for a flux of molecules, we should consider the flux of radiation $\mathrm{d}S_\nu$ arriving within an element of solid angle $\mathrm{d}\Omega$ in the direction of the unit vector $\boldsymbol{i}_\theta$, which lies at some angle θ to the vector $\mathrm{d}\boldsymbol{A}$ (see Fig. 11.2(*a*)). This leads to one of the key quantities needed in our study, the intensity of radiation. If we now orient the area $\mathrm{d}\boldsymbol{A}$ normal to the direction $\boldsymbol{i}_\theta$ of the incoming rays (Fig. 11.2(*b*)), then the *intensity*, or *brightness*, of the radiation, I_ν, is defined to be the quotient of the flux density $\mathrm{d}S_\nu$ and the solid angle $\mathrm{d}\Omega$ within which this flux density originates:

$$I_\nu = \frac{\mathrm{d}S_\nu}{\mathrm{d}\Omega}. \tag{11.2}$$

Thus, the intensity is the flux density per unit solid angle – its units are W m^{-2} Hz^{-1} sr^{-1}.

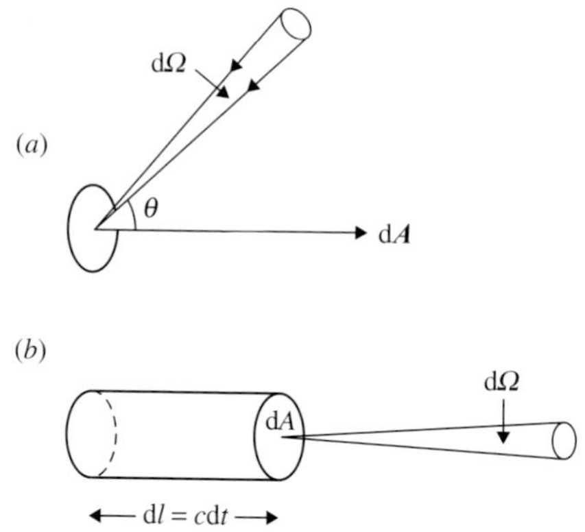

Figure 11.2: (*a*) The contribution of the flux density of radiation $\mathrm{d}S_\nu$ arriving within $\mathrm{d}\Omega$ to the total energy flux through the area $\mathrm{d}A$. Note that the flux density $\mathrm{d}S_\nu$ is defined as the power per unit frequency interval, arriving per unit area within the solid angle $\mathrm{d}\Omega$, when the area vector $\mathrm{d}A$ is parallel to the direction of propagation. Then, the intensity of radiation is defined to be $I_\nu = \mathrm{d}S_\nu/\mathrm{d}\Omega$. (*b*) Illustrating how to relate intensity I_ν to the energy density u_ν of radiation.

Just as in the case of fluid flow, we can write the energy flux through the area $\mathrm{d}A$ from the solid angle $\mathrm{d}\Omega$ at angle θ as $\boldsymbol{i}_\theta \cdot \mathrm{d}\boldsymbol{A}\, \mathrm{d}S_\nu$. Thus the energy per unit frequency interval at frequency ν flowing through $\mathrm{d}A$ in time $\mathrm{d}t$ is

$$\mathrm{d}E_\nu = \int_\Omega I_\nu \boldsymbol{i}_\theta \cdot \mathrm{d}\boldsymbol{A}\, \mathrm{d}t\, \mathrm{d}\Omega. \tag{11.3}$$

In the case of an isotropic radiation field, I_ν is independent of direction and so $I_\nu(\boldsymbol{\theta}) = I_0 = \text{constant}$. Then

$$\mathrm{d}E_\nu = I_0\, \mathrm{d}t \int_\Omega \boldsymbol{i}_\theta \cdot \mathrm{d}\boldsymbol{A}\, \mathrm{d}\Omega = I_0\, \mathrm{d}t\, |\mathrm{d}\boldsymbol{A}| \int_0^\pi \cos\theta \times 2\pi \sin\theta\, \mathrm{d}\theta = 0. \tag{11.4}$$

As expected, there is no net energy flux through any elementary area placed in an isotropic radiation field.

Let us now relate the intensity I_ν to the spectral energy density of radiation u_ν within an elementary volume $\mathrm{d}V$. Consider the power per unit area arriving normal to the elementary area $\mathrm{d}\boldsymbol{A}$ within the element of solid angle $\mathrm{d}\Omega$, which we can make as small as we like (Fig. 11.2(*b*)). Then, the radiation which passes through this area in this direction in time $\mathrm{d}t$ is contained in the volume of a cylinder of cross-sectional area $\mathrm{d}A$ and length $c\,\mathrm{d}t$ (Fig. 11.2(*b*)). By conservation of energy, as much energy flows into the near end of the cylinder at the speed of light as flows out of the other end. Therefore, the energy $\mathrm{d}U_\nu(\boldsymbol{\theta})$ contained in the volume at any time is found from

$$\mathrm{d}U_\nu(\boldsymbol{\theta})\, \mathrm{d}\Omega = I_\nu\, \mathrm{d}A\, \mathrm{d}t\, \mathrm{d}\Omega. \tag{11.5}$$

The energy density of radiation $u_\nu(\boldsymbol{\theta})\, \mathrm{d}\Omega$ is found by dividing by the volume of the cylinder $\mathrm{d}V = \mathrm{d}A\, c\, \mathrm{d}t$, that is

$$u_\nu(\boldsymbol{\theta})\, \mathrm{d}\Omega = \frac{\mathrm{d}U_\nu(\boldsymbol{\theta})}{\mathrm{d}V} \mathrm{d}\Omega = \frac{I_\nu\, \mathrm{d}\Omega}{c}. \tag{11.6}$$

To find u_ν, the energy density of radiation per unit frequency interval at frequency ν, now integrate over solid angle:

$$u_\nu = \int_\Omega u_\nu(\theta)\,\mathrm{d}\Omega = \frac{1}{c}\int I_\nu\,\mathrm{d}\Omega. \tag{11.7}$$

Thus

$$u_\nu = \frac{4\pi}{c}J_\nu, \tag{11.8}$$

where we have defined the *mean intensity* J_ν to be

$$J_\nu = \frac{1}{4\pi}\int I_\nu\,\mathrm{d}\Omega. \tag{11.9}$$

The total energy density is found by integrating over all frequencies:

$$u = \int u_\nu\,\mathrm{d}\nu = \frac{4\pi}{c}\int J_\nu\,\mathrm{d}\nu. \tag{11.10}$$

There is an important feature of the intensity of radiation from a source. Suppose the source is a sphere of luminosity L_ν and radius a at distance r. Then, the intensity of radiation measured by an observer at distance r is

$$I_\nu = \frac{S_\nu}{\Omega} = \frac{L_\nu/(4\pi r^2)}{\pi a^2/r^2} = \frac{L_\nu/(4\pi)}{\pi a^2}, \tag{11.11}$$

which is independent of the distance of the source. The intensity is just the luminosity of the source per steradian divided by its projected surface area, that is, πa^2 for a spherical source.

11.2.2 *Kirchhoff's law of emission and absorption*

Radiation in thermodynamic equilibrium with its surroundings was first described theoretically by Gustav Kirchhoff in 1859, while he was conducting the experiments described at the beginning of this section. His arguments are fundamental to the understanding of the physics of radiation processes.[4]

When radiation propagates through a medium, its intensity can decrease because of absorption by the material of the medium or increase because of its radiative properties. To simplify the argument leading to Kirchhoff's law, we assume that the emission from the material within the cylinder shown in Fig. 11.2(*b*) is *isotropic* and that we can neglect scattering of the radiation. We can then define a *monochromatic emission coefficient* j_ν such that the power radiated into the solid angle $\mathrm{d}\Omega$ from the cylindrical volume is

$$\mathrm{d}I_\nu\,\mathrm{d}A\,\mathrm{d}\Omega \equiv j_\nu\,\mathrm{d}V\,\mathrm{d}\Omega, \tag{11.12}$$

where $\mathrm{d}I_\nu$ is the intensity of the radiation from the path length $\mathrm{d}l$ of the material; j_ν has units W m^{-3} Hz^{-1} sr^{-1} and is the intensity radiated per unit solid angle per unit path length. Since the emission is assumed to be isotropic, the *volume emissivity* of the medium is $\varepsilon_\nu = 4\pi j_\nu$. We can write the volume of the cylinder $\mathrm{d}V$ as $\mathrm{d}A\,\mathrm{d}l$ and so

$$\mathrm{d}I_\nu = j_\nu\,\mathrm{d}l. \tag{11.13}$$

We now define the *monochromatic absorption coefficient* α_ν, which measures the loss of intensity from the beam in the material, by the relation

$$\mathrm{d}I_\nu\,\mathrm{d}A\,\mathrm{d}\Omega \equiv -\alpha_\nu\,I_\nu\,\mathrm{d}A\,\mathrm{d}\Omega\,\mathrm{d}l. \tag{11.14}$$

This can be regarded as a phenomenological relation based upon experiment. It can be interpreted in terms of the absorption cross-section of the atoms or molecules of the medium, but this was not necessary for Kirchhoff's argument. Equation (11.14) simplifies to

$$\mathrm{d}I_\nu = -\alpha_\nu\,I_\nu\,\mathrm{d}l. \tag{11.15}$$

Therefore, the *transfer equation* for radiation can be written, including both emission and absorption,

$$\frac{\mathrm{d}I_\nu}{\mathrm{d}l} = -\alpha_\nu\,I_\nu + j_\nu. \tag{11.16}$$

This equation enabled Kirchhoff to understand the relation between the emission and absorption properties of the lines observed in the laboratory and in solar spectroscopy. If there is no absorption then $\alpha_\nu = 0$ and the solution of the transfer equation is

$$I_\nu(l) = I_\nu(l_0) + \int_{l_0}^{l} j_\nu(l')\,\mathrm{d}l', \tag{11.17}$$

the first term representing the background emission intensity and the second emission from the medium located between l_0 and l. If there is absorption and no emission, $j_\nu = 0$, the solution of (11.16) is

$$I_\nu(l) = I_\nu(l_0)\exp\left[-\int_{l_0}^{l} \alpha_\nu(l')\,\mathrm{d}l'\right], \tag{11.18}$$

where the exponential term describes the absorption of radiation by the medium. The term in brackets is often written in terms of the *optical depth* of the medium, τ_ν, where

$$\tau_\nu = \int_{l_0}^{l} \alpha_\nu(l')\,\mathrm{d}l', \qquad \text{and so} \qquad I_\nu(l) = I_\nu(l_0)\,\mathrm{e}^{-\tau_\nu}. \tag{11.19}$$

Kirchhoff went further and determined the relation between the coefficients of emission and absorption by considering what happens when the processes of emission and absorption take place within an enclosure which has reached thermodynamic equilibrium. The first part of the argument concerns the general properties of the equilibrium spectrum. That such a unique spectrum must exist can be deduced from the second law of thermodynamics, as follows. Suppose we have two enclosures of arbitrary shape, both containing electromagnetic radiation in thermal equilibrium at temperature T, and a filter is placed between them which allows only radiation in the frequency interval ν to $\nu + \mathrm{d}\nu$ to pass between them. Then, if $I_\nu(1)\,\mathrm{d}\nu \neq I_\nu(2)\,\mathrm{d}\nu$, energy could flow spontaneously between them, violating the second law. The same type of argument can be used to show that the radiation must be isotropic. Therefore, the intensity spectrum of equilibrium radiation must be a unique function of only temperature and frequency, which can be written

$$I_\nu = B_\nu(T) \tag{11.20}$$

where $B_\nu(T)$ is a universal function of T and ν.

Now, suppose a volume of an emitting medium is maintained at temperature T. The emission and absorption coefficients will have some dependence upon temperature, which we do not know. Suppose we now place the emitting volume in an enclosure containing electromagnetic radiation in thermal equilibrium at temperature T. Then, after a very long time, the emission and absorption processes must be in balance, so that there is no change in the intensity of the radiation throughout the volume. In other words, in the transfer equation (11.16) $\mathrm{d}I_\nu/\mathrm{d}l = 0$, so that $\alpha_\nu I_\nu = j_\nu$ and the intensity spectrum is the universal equilibrium spectrum $B_\nu(T)$,

$$\alpha_\nu B_\nu(T) = j_\nu. \tag{11.21}$$

This is *Kirchhoff's law of emission and absorption*, which showed that the emission and absorption coefficients for any physical process are related by the as yet unknown spectrum of equilibrium radiation. This expression enabled Kirchhoff to understand the relation between the emission and absorption properties of flames, arcs, sparks and the solar atmosphere. In 1859, however, very little was known about the form of $B_\nu(T)$. As Kirchhoff remarked, 'It is a highly important task to find this function.'[5] This was one of the great experimental challenges for the remaining decades of the nineteenth century.

11.3 The Stefan–Boltzmann law

In 1879, Josef Stefan, Director of the Institute of Experimental Physics in Vienna, deduced the law which bears his name, primarily from experiments carried out by Tyndall at the Royal Institution of Great Britain on the radiation from platinum strip heated to different known temperatures, but also by reanalysing cooling experiments carried out by Dulong and Petit in the early years of the 19th century.[6] He found that the rate of energy emission over all wavelengths (or frequencies) is proportional to the fourth power of the absolute temperature T,

$$-\frac{\mathrm{d}E}{\mathrm{d}t} = \text{total radiant energy per second} \propto T^4. \tag{11.22}$$

In 1884, his former pupil, Ludwig Boltzmann, by then a professor at Graz, deduced this law from classical thermodynamics.[7] It is important that his analysis was entirely classical. Let us demonstrate how this can be done without taking any short cuts.

Consider a volume filled only with electromagnetic radiation in equilibrium with the walls of the container and suppose that the volume is closed by a piston so that the 'gas' of radiation can be compressed or expanded. Now add some heat $đQ$ to the 'gas' reversibly. As a result, the total internal energy increases by $\mathrm{d}U$ and work is done on the piston, so that the volume increases by $\mathrm{d}V$. Then, according to the first law of thermodynamics,

$$đQ = \mathrm{d}U + p\,\mathrm{d}V. \tag{11.23}$$

Now introduce the increase in entropy $\mathrm{d}S = đQ/T$, so that

$$T\,\mathrm{d}S = \mathrm{d}U + p\,\mathrm{d}V, \tag{11.24}$$

as was derived in Section 9.8. We convert (11.24) into a partial differential equation by dividing through by $\mathrm{d}V$ at constant T:

$$T\left(\frac{\partial S}{\partial V}\right)_T = \left(\frac{\partial U}{\partial V}\right)_T + p. \tag{11.25}$$

We now use one of Maxwell's relations (A9.12), which was derived in appendix section A9.2,

$$\left(\frac{\partial p}{\partial T}\right)_V = \left(\frac{\partial S}{\partial V}\right)_T$$

to recast relation (11.25). Therefore,

$$T\left(\frac{\partial p}{\partial T}\right)_V = \left(\frac{\partial U}{\partial V}\right)_T + p. \tag{11.26}$$

This is the relation we were seeking, because we can now find the relation between U and T if an equation of state for the gas, that is, a relation between p, V and U, can be determined.

The equation of state can be derived from Maxwell's electromagnetic theory and was proved by Maxwell in his great work *A Treatise on Electricity and Magnetism*[8] published in 1873. Let us derive the radiation pressure of a 'gas' of electromagnetic radiation by essentially the same route followed by Maxwell. If you already know the answer, $p = \frac{1}{3}u$ where u is the energy density of radiation, and you can prove it classically then you may wish to advance to subsection 11.3.3.

11.3.1 *The reflection of electromagnetic waves by a conducting plane*

The expression for radiation pressure can be derived from the fact that when electromagnetic waves are reflected by a conductor they exert a force at the interface. To understand the nature and magnitude of this force according to classical electrodynamics, we need to work out the currents which flow in a conductor under the influence of incident electromagnetic waves. Let us consider the case in which waves are normally incident on a sheet of large but finite conductivity (Fig. 11.3). This will provide excellent revision of the use of some of the tools developed in our study of electromagnetism (Chapters 5 and 6).

Figure 11.3 shows on the left a vacuum and on the right a medium with conductivity σ. The incident and reflected waves and the transmitted waves are shown by arrows. From our studies in Chapters 3 and 4, we can write down the *dispersion relations* for waves propagating in the vacuum and in the conducting medium:

in the vacuum,
$$k^2 = \omega^2/c^2; \tag{11.27}$$

in the conductor,
$$\nabla \times \boldsymbol{H} = \frac{\partial \boldsymbol{D}}{\partial t} + \boldsymbol{J}, \qquad \nabla \times \boldsymbol{E} = -\frac{\partial \boldsymbol{B}}{\partial t},$$
$$\boldsymbol{J} = \sigma \boldsymbol{E}, \qquad \boldsymbol{D} = \epsilon\epsilon_0 \boldsymbol{E}, \qquad \boldsymbol{B} = \mu\mu_0 \boldsymbol{H}. \tag{11.28}$$

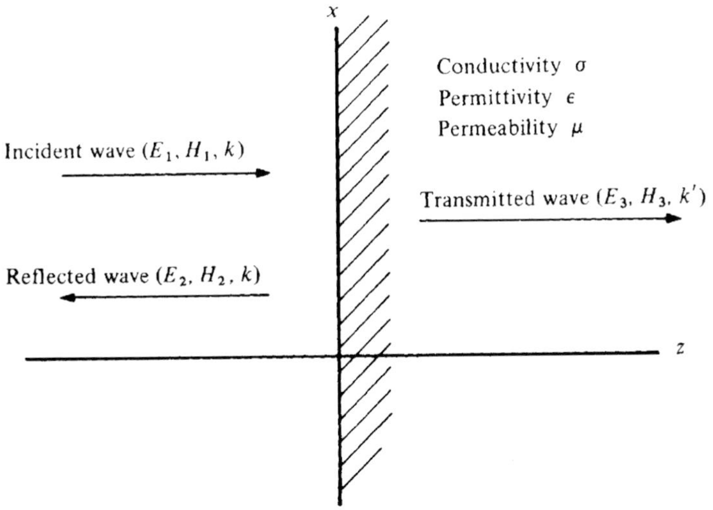

Figure 11.3: An electromagnetic wave incident at an interface between a vacuum and a highly conducting medium.

Using the relations developed in appendix section A5.6 for a travelling wave of form $\exp[\mathrm{i}(\mathbf{k}\cdot\mathbf{r}-\omega t)]$,

$$\nabla\times\to\mathrm{i}\boldsymbol{k}\,\times\qquad\text{and}\qquad\frac{\partial}{\partial t}\to-\mathrm{i}\omega,$$

we find that

$$\begin{aligned}\boldsymbol{k}\times\boldsymbol{H}&=-(\omega\epsilon\epsilon_0+\mathrm{i}\sigma)\boldsymbol{E},\\ \boldsymbol{k}\times\boldsymbol{E}&=\omega\boldsymbol{B}.\end{aligned}\tag{11.29}$$

Thus, in the conductor we obtain a dispersion relation similar to (11.27) but with $\omega\epsilon\epsilon_0$ replaced by $\omega\epsilon\epsilon_0+\mathrm{i}\sigma$, that is,

$$k^2=\epsilon\mu\frac{\omega^2}{c^2}\left(1+\frac{\mathrm{i}\sigma}{\omega\epsilon\epsilon_0}\right).\tag{11.30}$$

Let us consider the case in which the conductivity is very high, $\sigma/(\omega\epsilon\epsilon_0)\gg 1$. Then

$$k^2=\mathrm{i}\frac{\mu\omega\sigma}{\epsilon_0 c^2}.$$

Since $\mathrm{i}^{1/2}=2^{-1/2}(1+\mathrm{i})$, the solution for k is

$$k=\pm\left(\frac{\mu\omega\sigma}{2\epsilon_0 c^2}\right)^{1/2}(1+\mathrm{i}),\qquad\text{that is,}\qquad k=\pm\left(\frac{\mu\omega\sigma}{2\epsilon_0 c^2}\right)^{1/2}\mathrm{e}^{\mathrm{i}\pi/4}.\tag{11.31}$$

The phase relations between $\boldsymbol{E}$ and $\boldsymbol{B}$ in the wave can now be found from the second relation of (11.29), $\boldsymbol{k}\times\boldsymbol{E}=\omega\boldsymbol{B}$.

In *free space*, **k** is real and $\boldsymbol{E}$ and $\boldsymbol{B}$ oscillate in phase and with constant amplitude. In the conductor, however, (11.31) shows that there is a difference of $\pi/4$ between the phases of $\boldsymbol{E}$ and $\boldsymbol{B}$ and both fields decrease exponentially into the conductor, that is,

$$\begin{aligned} \boldsymbol{E} &\propto \exp\left[\mathrm{i}(kz - \omega t)\right], \\ &= \exp\left(-\frac{z}{l}\right) \exp\left[\mathrm{i}\left(\frac{z}{l} - \omega t\right)\right], \end{aligned} \tag{11.32}$$

where $l = [2\epsilon_0 c^2/(\mu\omega\sigma)]^{1/2}$. The amplitude of the wave decreases by a factor $1/\mathrm{e}$ in the length l, which is called the *skin depth* of the conductor. It is the typical depth to which electromagnetic fields can penetrate into the conductor.

There is an important general feature of the solution represented by (11.32), which is worth noting. If we trace back the steps involved in arriving at this solution, the assumption that the conductivity is high corresponds to neglecting the displacement current $\partial \boldsymbol{D}/\partial t$ in comparison with $\boldsymbol{J}$. Then, the equation we have solved has the form of a diffusion equation,

$$\nabla^2 \boldsymbol{H} = \sigma\epsilon\epsilon_0 \frac{\partial \boldsymbol{H}}{\partial t}. \tag{11.33}$$

In general, wave solutions of diffusion equations have dispersion relations of the form $k = A(1 + \mathrm{i})$; that is, the real and imaginary parts of the wave vector are equal, corresponding to waves which decay in amplitude by $\mathrm{e}^{-2\pi}$ in each complete cycle. This is true for equations such as:

(i) *the heat conduction equation,*

$$\kappa\nabla^2 T - \frac{\partial T}{\partial t} = 0, \qquad \kappa = \frac{K}{\rho C}, \tag{11.34}$$

where K is the thermal conductivity of the medium, ρ its density, C the specific heat and T the temperature;

(ii) *the diffusion equation,*

$$D\nabla^2 N - \frac{\partial N}{\partial t} = 0, \tag{11.35}$$

where D is the diffusion coefficient and N the number density of particles;

(iii) *the equation for viscous waves,*

$$\frac{\mu}{\rho}\nabla^2 \boldsymbol{u} - \frac{\partial \boldsymbol{u}}{\partial t} = 0, \tag{11.36}$$

where μ is the viscosity, ρ the density of the fluid and $\boldsymbol{u}$ the fluid velocity. This equation is derived from the Navier–Stokes equation for fluid flow in a viscous medium (see the appendix to Chapter 7).

Returning to our story (Fig. 11.3), we now match the $\boldsymbol{E}$ and $\boldsymbol{H}$ vectors of the waves at the interface between the two media. Taking z to be the direction normal to the interface, we introduce the following:

for the incident wave,

$$\begin{aligned} E_x &= E_1 \exp[\mathrm{i}(kz - \omega t)], \\ H_y &= \frac{E_1}{Z_0} \exp[\mathrm{i}(kz - \omega t)], \end{aligned} \tag{11.37}$$

where $Z_0 = (\mu_0/\epsilon_0)^{1/2}$ is the impedance of free space;

for the reflected wave,

$$E_x = E_2 \exp[-\mathrm{i}(kz + \omega t)],$$
$$H_y = -\frac{E_2}{Z_0} \exp[-\mathrm{i}(kz + \omega t)]; \tag{11.38}$$

for the transmitted wave,

$$E_x = E_3 \exp[\mathrm{i}(k'z - \omega t)],$$
$$H_y = E_3 \frac{(\mu\omega\sigma/2\epsilon_0 c^2)^{1/2}}{\omega\mu\mu_0}(1 + \mathrm{i}) \exp[\mathrm{i}(k'z - \omega t)], \tag{11.39}$$

where k' is given by the value of k in the relation (11.31). The expression for H_y has been found by substituting for k in the relation between $\boldsymbol{E}$ and $\boldsymbol{B}$, $\boldsymbol{k} \times \boldsymbol{E} = \omega \boldsymbol{B}$.

For simplicity, let us write $q = [(\mu\omega\sigma/2\epsilon_0 c^2)^{1/2}(\omega\mu\mu_0)^{-1}](1 + \mathrm{i})$. Then

$$H_y = qE_3 \exp[\mathrm{i}(k'z - \omega t)]. \tag{11.40}$$

The boundary conditions require E_x and H_y to be continuous at the interface (see Section 6.5), that is, at $z = 0$,

$$E_1 + E_2 = E_3,$$
$$\frac{E_1}{Z_0} - \frac{E_2}{Z_0} = qE_3. \tag{11.41}$$

Therefore

$$\frac{E_1}{1 + qZ_0} = \frac{E_2}{1 - qZ_0} = \frac{E_3}{2}. \tag{11.42}$$

In general, q is a complex number and hence there are phase differences between E_1, E_2 and E_3. We are, however, interested in the case in which the conductivity is very large, $|q|Z_0 \gg 1$, and hence

$$\frac{E_1}{qZ_0} = -\frac{E_2}{qZ_0} = \frac{E_3}{2}, \tag{11.43}$$

that is,

$$E_1 = -E_2 \quad \text{and} \quad E_3 = 0.$$

Therefore, at the interface, the total electric field strength $E_1 + E_2$ is zero and the magnetic field strength $H_1 + H_2 = 2H_1$.

It may seem as though we have strayed rather far from the thermodynamics of radiation, but we are now able to work out the pressure exerted by the incident wave upon the surface.

11.3.2 The formula for radiation pressure

Suppose that the radiation is confined in a box with rectangular sides and the waves bounce back and forth between the walls at $z = \pm z_1$ (Fig. 11.3). Assuming that the walls of the box are highly conducting, as in the preceding subsection, we now know the values of the electric and magnetic field strengths in the vicinity of the walls for normal incidence.

Part of the origin of the phenomenon of radiation pressure may be understood as follows. The electric field in the conductor E_x causes a current density to flow in the $+x$ direction,

$$J_x = \sigma E_x. \tag{11.44}$$

But the force per unit volume acting on this electric current in the presence of a magnetic field is

$$\boldsymbol{F} = N_q q(\boldsymbol{v} \times \boldsymbol{B}) = \boldsymbol{J} \times \boldsymbol{B}, \tag{11.45}$$

where N_q is the number of conduction electrons per unit volume and q is the electronic charge. Since $\boldsymbol{B}$ is in the $+y$ direction, this force acts in the $\boldsymbol{i}_x \times \boldsymbol{i}_y$-direction; this is the $\boldsymbol{k}$-direction, that of the incident wave. Therefore, the pressure acting on a layer of thickness $\mathrm{d}z$ in the conductor is

$$\mathrm{d}p = J_x B_y \,\mathrm{d}z. \tag{11.46}$$

However, we also know that in the conductor curl $\boldsymbol{H} = \boldsymbol{J}$, because the conductivity is very high, and hence we can relate J_x and B_y by

$$\frac{\partial H_z}{\partial y} - \frac{\partial H_y}{\partial z} = J_x.$$

Since $H_z = 0$, $-\partial H_y/\partial z = J_x$. Substituting into (11.46),

$$\mathrm{d}p = -B_y \frac{\partial H_y}{\partial z}\,\mathrm{d}z. \tag{11.47}$$

Therefore,

$$p = -\int_0^\infty B_y \frac{\partial H_y}{\partial z}\,\mathrm{d}z = \int_0^{H_0} B_y \,\mathrm{d}H_y,$$

where H_0 is the value of the magnetic field strength at the interface and, according to the analysis of subsection 11.3.1, $H \to 0$ as $z \to \infty$. For a linear medium, $B_0 = \mu\mu_0 H_0$ and hence

$$p = \tfrac{1}{2}\mu\mu_0 H_0^2. \tag{11.48}$$

Notice that this pressure is associated with induced currents flowing in the conducting medium.

Now we have to ask what other forces act at the vacuum–conductor interface. These are associated with the stresses in the electromagnetic fields themselves and are given by the appropriate components of the Maxwell stress tensor. In our simple case, we can derive the magnitude of these stresses by an argument based on Faraday's concept of lines of force. Suppose a uniform longitudinal magnetic field is confined within a long rectangular perfectly conducting tube (Fig. 11.4). If the medium is linear, so that $B = \mu\mu_0 H$, the energy per unit length of tube is

$$E = \tfrac{1}{2} BHzl,$$

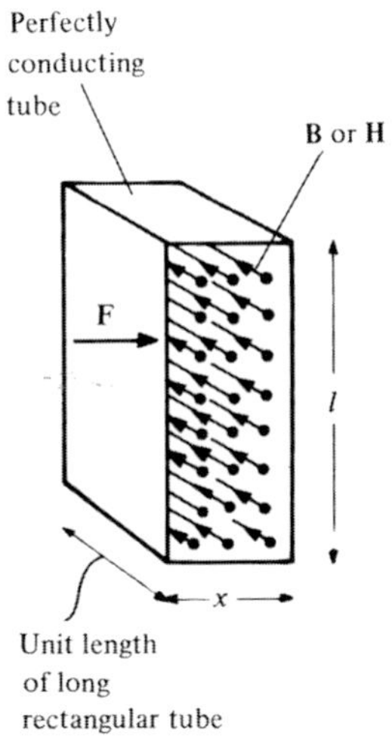

Figure 11.4: A long perfectly conducting rectangular tube enclosing a longitudinal magnetic field. When the tube is compressed by a force F in the z-direction, the magnetic flux enclosed by the tube is conserved. For more details of magnetic flux freezing, see, for example, M.S. Longair (1992), *High Energy Astrophysics*, Vol. 2, 307–12, Cambridge: Cambridge University Press.

where z is the width and l the height of the tube. Now squash the width of the tube by dz whilst maintaining the same number of lines of force through it. Then the magnetic flux density increases to $Bz/(z - \mathrm{d}z)$ because of conservation of lines of force and correspondingly, because the field is linear, H becomes $Hz/(z - \mathrm{d}z)$. Thus, the energy in the volume becomes

$$
\begin{aligned}
E + \mathrm{d}E &= \tfrac{1}{2}BHlz^2 \frac{1}{(z - \mathrm{d}z)}, \\
\mathrm{d}E &= \tfrac{1}{2}BHlz\left(1 + \frac{\mathrm{d}z}{z}\right) - \tfrac{1}{2}BHlz, \\
&= \tfrac{1}{2}BHl\,\mathrm{d}z.
\end{aligned}
$$

But this increase in energy must be the result of external work done against the magnetic field,

$$F\,\mathrm{d}z = \mathrm{d}E = \tfrac{1}{2}BHl\,\mathrm{d}z.$$

Thus, the force per unit area p_m is $F/l = \frac{1}{2}BH$; it acts *perpendicularly* to the field direction. We can write

$$p_m = \tfrac{1}{2}\mu\mu_0 H^2. \tag{11.49}$$

We can apply exactly the same argument to the electrostatic field, in which case the total force per unit area associated with the electric and magnetic fields is*

$$p = \tfrac{1}{2}\epsilon\epsilon_0 E^2 + \tfrac{1}{2}\mu\mu_0 H^2. \tag{11.50}$$

* Here E refers to the magnitude of the electric field.

This argument applies to the fields in both regions but, because the value of μ is different on either side of the interface, there is a pressure difference across it associated with the presence of the magnetic fields.

We have shown that, at the interface, $E_x = 0$ and $H_y = H_0 = 2H_1$. Hence in the vacuum adjacent to the interface $p = \frac{1}{2}\mu_0 H_0^2$. Inside the conductor the stress is $\frac{1}{2}\mu\mu_0 H_0^2$. Therefore, from (11.48), the total pressure on the conductor is

$$p = \underbrace{\tfrac{1}{2}\mu_0 H_0^2}_{\text{stress in vacuum}} - \underbrace{\tfrac{1}{2}\mu\mu_0 H_0^2}_{\text{stress in conductor}} + \underbrace{\tfrac{1}{2}\mu\mu_0 H_0^2}_{\text{force on conduction current}},$$

so that

$$p = \tfrac{1}{2}\mu_0 H_0^2. \tag{11.51}$$

The energy density in the wave propagating in the positive z-direction in the vacuum is $\frac{1}{2}(\epsilon_0 E_1^2 + \mu_0 H_1^2) = \mu_0 H_1^2$. Therefore, since $H_0 = 2H_1$,

$$p = \tfrac{1}{2}\mu_0 H_0^2 = 2\mu_0 H_1^2 = 2\varepsilon_1 = \varepsilon_0, \tag{11.52}$$

where ε_0 is the total energy density of radiation in the vacuum, being the sum of the energy densities in the incident and reflected waves.

Equation (11.52) is the relation between the pressure and energy density for a 'one-dimensional' gas of electromagnetic radiation, confined between two reflecting walls. In an isotropic-three dimensional volume, there are equal energy densities associated with radiation propagating in the three orthogonal directions, that is,

$$\varepsilon_x = \varepsilon_y = \varepsilon_z = \varepsilon_0,$$

and hence the radiation pressure p is

$$p = \tfrac{1}{3}\varepsilon, \tag{11.53}$$

where ε is the total energy density of radiation.

This somewhat lengthy demonstration shows how it is possible to derive the pressure of a gas of electromagnetic radiation entirely by classical arguments. I have purposely given this simple treatment because I prefer to use these physical arguments rather than the more mathematical treatment starting from Maxwell's equations and involving the use of the Maxwell stress tensor for the electromagnetic field. This is how Maxwell derived the expression for radiation pressure in Section 793 of the second volume of his *Treatise on Electricity and Magnetism* of 1873.[8] In his argument, the pressure of a 'gas' of electromagnetic radiation in free space was derived directly from (11.50). On averaging this expression over the period of the electromagnetic waves, the relation $p = \varepsilon_0$ for a one-dimensional gas is found directly.

11.3.3 The derivation of the Stefan–Boltzmann law

The relation (11.53) between p and ε for a gas of electromagnetic radiation leads to the Stefan–Boltzmann law. From (11.26), writing $U = \varepsilon V$

$$\frac{T}{3}\left(\frac{\partial \varepsilon}{\partial T}\right)_V = \left(\frac{\partial(\varepsilon V)}{\partial V}\right)_T + \frac{\varepsilon}{3},$$

$$\frac{T}{3}\left(\frac{\partial \varepsilon}{\partial T}\right)_V = \varepsilon + \frac{\varepsilon}{3} = \frac{4\varepsilon}{3}.$$

This relation between ε and T can be integrated immediately:

$$\frac{d\varepsilon}{\varepsilon} = 4\frac{dT}{T}, \qquad \ln \varepsilon = 4 \ln T, \qquad \varepsilon = aT^4. \tag{11.54}$$

This is the calculation which Boltzmann carried out in 1884 and his name is justly associated with the *Stefan–Boltzmann law*. The constant a is given by

$$a = \frac{8\pi^5 k^4}{15c^3h^3} = 7.566 \times 10^{-16}\ \text{J m}^{-3}\ \text{K}^{-4}. \tag{11.55}$$

We will derive this expression in Section 13.3.

The expression (11.54) for the energy density of radiation can be related to the energy emitted per unit area per second from the surface of a black body at temperature T. Suppose that the black body is placed in an enclosure maintained in thermal equilibrium at temperature T and that $N(E)$ is the number density of radiation modes of energy E in the enclosure. If the energy of a mode is E then the rate at which energy arrives at the wall per unit area per second, and consequently the rate I at which the energy must be reradiated from it, is from (10.13) $\frac{1}{4}N(E)Ec = \frac{1}{4}N\bar{E}c = \frac{1}{4}\varepsilon c$. Therefore,

$$I = \frac{\varepsilon c}{4} = \frac{ac}{4}T^4 = \sigma T^4 = 5.67 \times 10^{-8}\ T^4\ \text{W m}^{-2} \tag{11.56}$$

where σ is the Stefan-Boltzmann constant. The experimental evidence for the Stefan–Boltzmann law was not particularly convincing in 1884 and it was not until 1897 that Lummer and Pringsheim undertook very careful experiments which showed that the law was indeed correct to high precision (see Fig. V.1 at the start of this case study).

11.4 Wien's displacement law and the spectrum of black-body radiation

The spectrum of black-body radiation was not particularly well known in 1895, but there had already been important theoretical work carried out by Wilhelm Wien on the form which the radiation law should have. *Wien's displacement law*[9] was derived using a combination of electromagnetism and thermodynamics, as well as dimensional analysis. Let us show exactly what Wien did in this work, published in 1893, which was to be of central importance in the subsequent story.

First of all, Wien worked out how the properties of a 'gas' of radiation change when it undergoes a reversible adiabatic expansion. The analysis starts with the basic thermodynamic

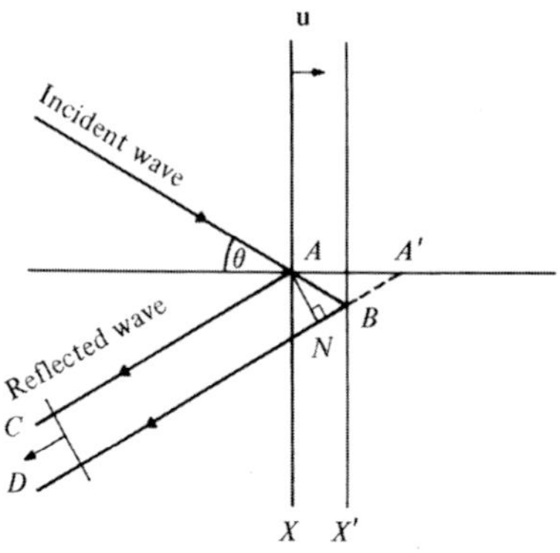

Figure 11.5: Illustrating the change in wavelength of an electromagnetic wave reflected from a perfectly conducting plane moving to the right at speed u.

relation

$$đQ = \mathrm{d}U + p\,\mathrm{d}V.$$

In an adiabatic expansion $đQ = 0$ and, for a 'gas' of radiation, $U = \varepsilon V$ and $p = \frac{1}{3}\varepsilon$. Therefore

$$\mathrm{d}(\varepsilon V) + \tfrac{1}{3}\varepsilon\,\mathrm{d}V = 0,$$
$$V\,\mathrm{d}\varepsilon + \varepsilon\,\mathrm{d}V + \tfrac{1}{3}\varepsilon\,\mathrm{d}V = 0,$$

so that

$$\frac{\mathrm{d}\varepsilon}{\varepsilon} = -\frac{4\,\mathrm{d}V}{3V}.$$

Integrating,

$$\varepsilon = \text{constant} \times V^{-4/3}. \qquad (11.57)$$

But $\varepsilon = aT^4$ and hence

$$TV^{1/3} = \text{constant}. \qquad (11.58)$$

Since V is proportional to the cube of the radius r of a spherical volume,

$$T \propto r^{-1}. \qquad (11.59)$$

The next step in deriving Wien's displacement law is to work out the relation between the wavelength of the radiation and the volume of the enclosure in a reversible adiabatic expansion. Let us carry out two simple calculations which demonstrate the answer. First of all, we determine the change in wavelength if a wave is reflected from a slowly moving mirror (Fig. 11.5). The initial position of the mirror is at X and we suppose that at that time one of the maxima of the incident waves is at A. It is then reflected along the path AC. Now suppose that by the time the next wavecrest arrives at the mirror, the latter has moved to X' and hence the maximum has to travel an extra distance ABN as compared with the first

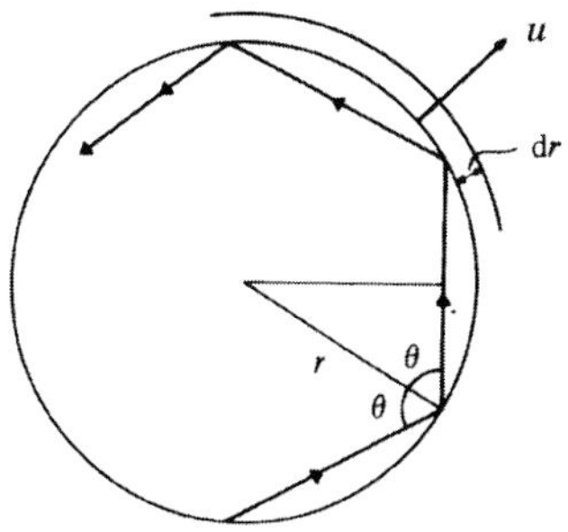

Figure 11.6: Illustrating the differential change in wavelength of an electromagnetic wave as it is reflected inside a spherical volume expanding at speed u.

maximum, that is, the distance between maxima, the wavelength, is increased by an amount $d\lambda = AB + BN$. By symmetry,

$$AB + BN = A'N = AA' \cos\theta.$$

But

$$AA' = 2 \times \text{distance moved by mirror} = 2XX' = 2uT,$$

where u is the velocity of the mirror and T is the period of the wave. Hence,

$$\begin{aligned} d\lambda &= 2uT \cos\theta, \\ &= 2\lambda \frac{u}{c} \cos\theta, \qquad \text{since} \qquad d\lambda \ll \lambda. \end{aligned} \tag{11.60}$$

Notice that this result is correct only to first order in small quantities, but that is all we need since we are interested in differential changes.

Now suppose the wave is confined within a spherical reflecting cavity, which is expanding slowly, and that its angle of incidence with respect to the normal to the sphere is θ (Fig. 11.6). We can work out the number of reflections which the wave makes with the sphere, and hence the total change in wavelength, as the cavity expands from r to $r + dr$ at speed u, assuming $u \ll c$.

If the velocity of expansion is small, θ remains the same for all reflections as the sphere expands by an infinitesimal distance dr. The time it takes for the sphere to expand a distance dr is $dt = dr/u$ and, from the geometry of Fig. 11.6, the time between reflections for a wave propagating at velocity c is $2r \cos\theta/c$. The number of reflections in time dt is therefore $c\, dt/(2r \cos\theta)$ and the change in wavelength $d\lambda$ is

$$d\lambda = \left(\frac{2u\lambda}{c} \cos\theta\right) \left(\frac{c\, dt}{2r \cos\theta}\right),$$

that is,

$$\frac{d\lambda}{\lambda} = \frac{u\, dt}{r} = \frac{dr}{r}.$$

Integrating, we find

$$\lambda \propto r. \tag{11.61}$$

Thus, the wavelength of the radiation increases linearly in proportion to the radius of the spherical volume.

We can now combine this result with (11.59), $T \propto r^{-1}$, and find that

$$T \propto \lambda^{-1}. \tag{11.62}$$

This is one aspect of *Wien's displacement law*. If radiation undergoes a reversible adiabatic expansion, the wavelength of a particular set of waves changes inversely with temperature. In other words, the wavelength of the radiation is 'displaced' as the temperature changes. In particular, if we follow the maximum of the radiation spectrum, it should follow the law $\lambda_{\text{max}} \propto T^{-1}$. This was found to be in good agreement with experiment.

Now Wien went significantly further than this and combined the two laws, the Stefan–Boltzmann law and the law $T \propto \lambda^{-1}$, to set constraints upon the form of the equilibrium radiation spectrum. His first step was to note that if any system of bodies is enclosed in a perfectly reflecting enclosure then eventually all of them come to thermal equilibrium through the emission and absorption of radiation – in this state, as much energy is radiated as is absorbed by the bodies per unit time. Thus, if we wait long enough, the radiation in the enclosure attains a black-body radiation spectrum at a certain temperature T. According to the arguments presented in subsection 11.2.2, first enunciated by Kirchhoff, the radiation must be isotropic, and the only parameter which can characterise the radiation spectrum is the temperature, T.

The next step is to note that if the black-body radiation is initially at temperature T_1 and the enclosure undergoes a reversible adiabatic expansion then by definition the expansion proceeds infinitely slowly, so that the radiation takes up an equilibrium spectrum at all stages in the expansion to a lower temperature T_2. The crucial point is that the radiation spectrum is of equilibrium black-body form, at the beginning and end of the expansion. The unknown radiation law must therefore scale appropriately with temperature.

Consider the radiation in the wavelength interval λ_1 to $\lambda_1 + \mathrm{d}\lambda_1$ and let its energy density be $\varepsilon = u(\lambda_1)\,\mathrm{d}\lambda_1$, that is, $u(\lambda)$ is the energy density per unit wavelength interval or per unit bandwidth. Then, according to Boltzmann's analysis given in subsection 11.3.3, the energy associated with any particular wavelength range $\mathrm{d}\lambda$ changes as T^4 and hence

$$\frac{u(\lambda_1)\,\mathrm{d}\lambda_1}{u(\lambda_2)\,\mathrm{d}\lambda_2} = \left(\frac{T_1}{T_2}\right)^4, \tag{11.63}$$

where λ_2 is the wavelength after the expansion. But from (11.62) $\lambda_1 T_1 = \lambda_2 T_2$, and hence $\mathrm{d}\lambda_1 = (T_2/T_1)\,\mathrm{d}\lambda_2$. Therefore,

$$\frac{u(\lambda_1)}{T_1^5} = \frac{u(\lambda_2)}{T_2^5}, \qquad \text{that is,} \qquad \frac{u(\lambda)}{T^5} = \text{constant}. \tag{11.64}$$

Since $\lambda T = \text{constant}$, (11.64) can be rewritten

$$u(\lambda)\,\lambda^5 = \text{constant}. \tag{11.65}$$

Now, the only combination of T and λ which is constant during the expansion is the product λT, and hence we can conclude that, in general, the constant in (11.65) can only be constructed out of functions involving λT. Therefore, the radiation law must have the form

$$u(\lambda)\,\lambda^5 = f(\lambda T) \tag{11.66}$$

or

$$u(\lambda)\,\mathrm{d}\lambda = \lambda^{-5} f(\lambda T)\,\mathrm{d}\lambda. \tag{11.67}$$

This is the complete form of *Wien's displacement law* and it sets important constraints upon the form of the black-body radiation spectrum.

We will find it more convenient to work in terms of frequencies rather than wavelengths, and so let us convert Wien's displacement law into frequency form:*

$$u(\lambda)\,\mathrm{d}\lambda = u(\nu)\,\mathrm{d}\nu,$$

and use

$$\lambda = c/\nu, \qquad \mathrm{d}\lambda = -\frac{c}{\nu^2}\,\mathrm{d}\nu.$$

Hence from (11.67)

$$u(\nu)\,\mathrm{d}\nu = \left(\frac{c}{\nu}\right)^{-5} f\left(\frac{\nu}{T}\right)\left(-\frac{c}{\nu^2}\,\mathrm{d}\nu\right), \tag{11.68}$$

that is,

$$u(\nu)\,\mathrm{d}\nu = \nu^3 f\left(\frac{\nu}{T}\right)\,\mathrm{d}\nu. \tag{11.69}$$

This is really rather clever. Notice how much Wien was able to deduce using only rather general thermodynamic arguments. We will see in a moment how crucial this argument proved to be in establishing the correct formula for black-body radiation.

This was all new when Planck first became interested in the problem of the spectrum of equilibrium radiation in 1895. Let us now turn to Planck's huge contributions to the understanding of the spectrum of black-body radiation.

11.5 References

1 Lindley, D. (2001). *Boltzmann's Atom*. New York: The Free Press.
2 Kirchhoff, G. (1861–3). Part 1, *Abhandl. der Berliner Akad.* (1861), p. 62, (1862), p. 227; part 2 (1863), p. 225.
3 An excellent description of these early spectroscopic studies is given by J.B. Hearnshaw in *The Analysis of Starlight* (1986), Cambridge: Cambridge University Press.
4 I particularly like the careful discussion of the fundamentals of radiative processes by G.B. Rybicki and A.P. Lightman, *Radiative Processes in Astrophysics* (1979), New York: John Wiley and Sons.
5 Kirchhoff, G. (1859). *Ber. der Berliner Akad.*, p. 662. (*Trans. Phil. Mag*, **19**, 193, 1860.)

* Here $u(\lambda)$ and $u(\nu)$ are different, though closely related, functions, the forms of which are defined by their respective arguments, likewise $f(\lambda T)$ and $f(\nu/T)$.

6 Stefan, J. (1879). *Wiener Ber. II*, **79**, 391–428.

7 Boltzmann, L. (1884). *Ann. der Physik*, **22**, 291–4.

8 Maxwell, J.C. (1873). *A Treatise on Electricity and Magnetism*. Oxford: Clarendon Press. (Reprint of the second edition of 1891 published by Oxford University Press, 1998, in two volumes.)

9 Wien, W. (1893). *Ber. der Berliner Akad.*, pp. 55–62.

12 1895–1900: Planck and the spectrum of black-body radiation

12.1 Planck's early career

Max Planck was typically straightforward and honest about his early career, as can be learned from his short scientific autobiography.[1] He studied under Helmholtz and Kirchhoff in Berlin, but in his own words

> I must confess that the lectures of these men netted me no perceptible gain. It was obvious that Helmholtz never prepared his lectures properly . . . Kirchhoff was the very opposite . . . but it would sound like a memorised text, dry and monotonous.

It is reassuring to students struggling to understand physics that even the greatest of physicists are sometimes inadequate as university teachers. In my own experience, although the best physicists are very often the most inspiring lecturers, there is considerable variation in their degree of commitment to passing on the torch to the next generation. We have all had to put up with the uneven quality of the lecturers who happen to be delivering courses.* In the end, however, we have to understand the material ourselves rather than be spoon-fed, and so perhaps it is not as disastrous as it might seem at first sight. Indeed, one might argue that a bad lecturer requires the student to think harder about the material, which is a good thing – let me hasten to add that this is no excuse for bad lecturing!

Planck's research interests were inspired by his reading of the works of Clausius and he set about investigating how the second law of thermodynamics could be applied to a wide variety of different physical problems, as well as elaborating as clearly as possible the basic tenets of the subject. He completed his dissertation in 1879. In his words,

> The effect of my dissertation on the physicists of those days was nil. None of my professors at the University had any understanding for its contents, as I learned for a fact in my conversations with them . . . Helmholtz probably did not even read my paper at all. Kirchhoff expressly disapproved of its contents, with the comment that . . . the concept of entropy . . . must not be applied to irreversible processes. I did not succeed in reaching Clausius. He did not answer my letters and I did not find him at home when I tried to see him in person in Bonn.[2]

* See the remarks about Newton's lectures in Chapter 4!

There are two interesting aspects to these statements. First of all, Kirchhoff was incorrect in stating that the concept of entropy cannot be applied to irreversible processes. Entropy is a function of state, and therefore can be determined for any given state of the system, independent of whether reversible or irreversible processes were involved in attaining that the state. Second, one should not be particularly surprised that Planck had little success in reaching the great men. We are all familiar with the fact that people are very busy and it is often difficult for even sympathetic senior scientists to have the time to devote to studies outside their immediate interests. In addition, we have to deal with real human beings and some are more approachable than others. The important lesson is not to be discouraged or give up – there is always more than one way of attaining one's goals.

Planck continued his researches on entropy and thermodynamics and, following the death of Kirchhoff in 1889, he succeeded to Kirchhoff's chair at the University of Berlin. This was a propitious advance since the University was one of the most active centres of physics in the world and German physics was at the very forefront of research.

In 1894, he turned his attention to the problem of the spectrum of black-body radiation, the topic which was to dominate his subsequent work and which led to the discovery of quantisation and quanta. It is likely that his interest in this problem was stimulated by Wien's important paper which had been published in the previous year and which we analysed in detail in Section 11.4. Wien's analysis had a strong thermodynamic flavour, which must have appealed to Planck.

In 1895, he published the first results of his work on the resonant scattering of plane electromagnetic waves by an oscillating dipole. This was the first of Planck's papers in which he diverged from his previous areas of interest, in that it appeared to be about electromagnetism rather than entropy. In the last words of the paper, however, Planck made it clear that he regarded this as a first step towards tackling the problem of the spectrum of black-body radiation. His aim was to set up a system of oscillators, in an enclosed cavity, which would radiate and interact with the radiation produced so that after a very long time the system would come into equilibrium. He could then apply the laws of thermodynamics to black-body radiation with a view to understanding the origin of its spectrum. He explained why this approach offered the prospect of providing insights into basic thermodynamics processes, as follows. When energy is lost by an oscillator by radiation, the process can be considered 'conservative' because if the radiation is enclosed in a box with perfectly reflecting walls then it will react back on the oscillator. Furthermore, the process is independent of the nature of the oscillator. In Planck's words,

> The study of conservative [rather than dissipative] damping seems to me to be of great importance, since it opens up the prospect of a possible general explanation of irreversible processes by means of conservative forces – a problem that confronts research in theoretical physics more urgently every day.[3]

It is important to remember that Planck was a great expert in thermodynamics and a follower of Clausius as an exponent of classical thermodynamics. He looked upon the second law of thermodynamics as having absolute validity – processes in which the total entropy decreased, he believed, should be strictly excluded. This was a very different point of view from that set forth by Boltzmann in his memoir of 1877, in which the second law of thermodynamics is only statistical in nature. As demonstrated in Chapter 10, according to Boltzmann there is a very high probability indeed that entropy increases in all natural processes, but there remains an extremely small but finite probability that the system will evolve to a state of lower entropy. As is clearly indicated in Kuhn's history of these years,[4] there were many technical concerns about Boltzmann's approach and Planck and his students had published papers criticising some of the steps in Boltzmann's statistical arguments.

Planck believed that, by studying classically the interaction of oscillators with electromagnetic radiation, he would be able to show that entropy increases absolutely for a system consisting of matter and radiation. This was elaborated in a series of five papers. The idea did not work, as was pointed out by Boltzmann. One cannot obtain a monotonic approach to equilibrium without some statistical assumptions about the way in which the system approaches the equilibrium state. This can be understood from Maxwell's simple but compelling argument concerning the time reversibility of the laws of mechanics, dynamics and electromagnetism (Section 10.6).

Finally, Planck conceded that statistical assumptions were necessary and introduced the concept of 'natural radiation', which corresponded to Boltzmann's assumption of 'molecular chaos'. This must have been distressing for Planck. It is a painful experience for any scientist to have the assumptions upon which their researches have been based for 20 years undermined. However, much worse was to come.

Once the assumption was made that there exists a state of 'natural radiation', Planck was able to make considerable progress with his programme. The first thing he did was to relate the energy density of radiation in an enclosure to the average energy of the oscillators within it. This is a very important result and can be derived entirely from classical arguments. Let us show how it can be done – it is a beautiful piece of theoretical physics.

12.2 Oscillators and their radiation in thermal equilibrium

Why treat oscillators rather than, say, atoms, molecules, lumps of rock, and so on? The reason is that, in thermal equilibrium, everything is in equilibrium with everything else – rocks are in equilibrium with atoms and oscillators and therefore there is no advantage in treating complicated objects. The advantage of considering simple harmonic oscillators is that the radiation and absorption laws can be calculated exactly. To express this important point in another way, Kirchhoff's law (11.21), which relates the emission and absorption coefficients, j_ν and α_ν respectively,

$$\alpha_\nu B_\nu(T) = j_\nu,$$

shows that, if we can determine these coefficients for any process, we can find the universal equilibrium spectrum $B_\nu(T)$.

12.2.1 The rate of radiation of an accelerated charged particle

Classically, charged particles emit electromagnetic radiation when they are accelerated. In the analysis which follows, we need the rather beautiful formula for the rate of emission of radiation of an accelerated charged particle. The normal derivation of this formula proceeds from Maxwell's equations through the use of retarded potentials for the field components at a large distance. We can give a much simpler derivation of the exact results of these calculations from a remarkable argument given by J.J. Thomson in his book *Electricity and Matter*,[5] the published version of his Silliman Memorial Lectures delivered in Harvard in 1903. He used this formula to derive the cross-section for the scattering of X-rays by electrons in atoms in his subsequent book *Conduction of Electricity through Gases*[6] – this was the origin of the *Thomson cross-section* for the scattering of electromagnetic radiation by free electrons. Thomson's argument indicates very clearly why an accelerated charge radiates electromagnetic radiation and also clarifies the origin of the polar diagram and polarisation properties of the radiation.

Consider a charge q stationary at the origin O of some frame of reference S at time $t = 0$. Suppose that the charge suffers a small acceleration to velocity Δv in a short interval of time Δt. Thomson visualised the resulting electric field distribution in terms of the field lines attached to the accelerated charge. After a time t, we can distinguish between the field configuration inside and outside a sphere of radius ct centred on the origin of S (Fig. 12.1(a)), recalling that Maxwell had shown that electromagnetic disturbances are propagated at the speed of light in free space. Outside the sphere, the field lines do not yet know that the charge has moved away from the origin, because information cannot travel faster than the speed of light, and therefore they are radial, centred on O. Inside this sphere, the field lines are radial about the origin of the frame of reference centred on the moving charge. Between these two regions, there is a thin shell of thickness $c\Delta t$ in which we have to join up corresponding electric field lines (see Fig. 12.1(a)). Geometrically, it is clear that there must be a component of the electric field in the circumferential direction in this shell, that is, in the $\boldsymbol{i}_\theta$-direction. This 'pulse' of electromagnetic field is propagated away from the charge at the speed of light; it represents an energy loss from the accelerated charged particle.

Let us work out the strength of the electric field in the pulse. We assume that the increment in velocity Δv is very small, that is, $\Delta v \ll c$, and therefore it is safe to assume that the field lines are radial not only at $t = 0$ but also at the time t in the frame of reference S. There will, in fact, be small aberration effects associated with the velocity Δv, but these are second order compared with the gross effects we are discussing. We may therefore consider a small cone of field lines at an angle θ with respect to the acceleration vector of the charge at $t = 0$ and a similar one at the later time t when the charge is moving at the constant velocity Δv (Fig. 12.1(b)). We now have to join up electric field lines between the two cones through the thin shell of thickness $c\Delta t$, as shown in the diagram. The strength of the E_θ-component of the field is given by the number of field lines per unit area in the $\boldsymbol{i}_\theta$-direction. From the geometry of Fig. 12.1(b), which exaggerates the discontinuities in the field lines, the E_θ-component is given by the relative sizes of the sides of the rectangle $ABCD$, that is,

$$\frac{E_\theta}{E_r} = \frac{\Delta v\, t \sin\theta}{c\Delta t}. \tag{12.1}$$

But

$$E_r = \frac{q}{4\pi\epsilon_0 r^2}, \qquad \text{where} \qquad r = ct.$$

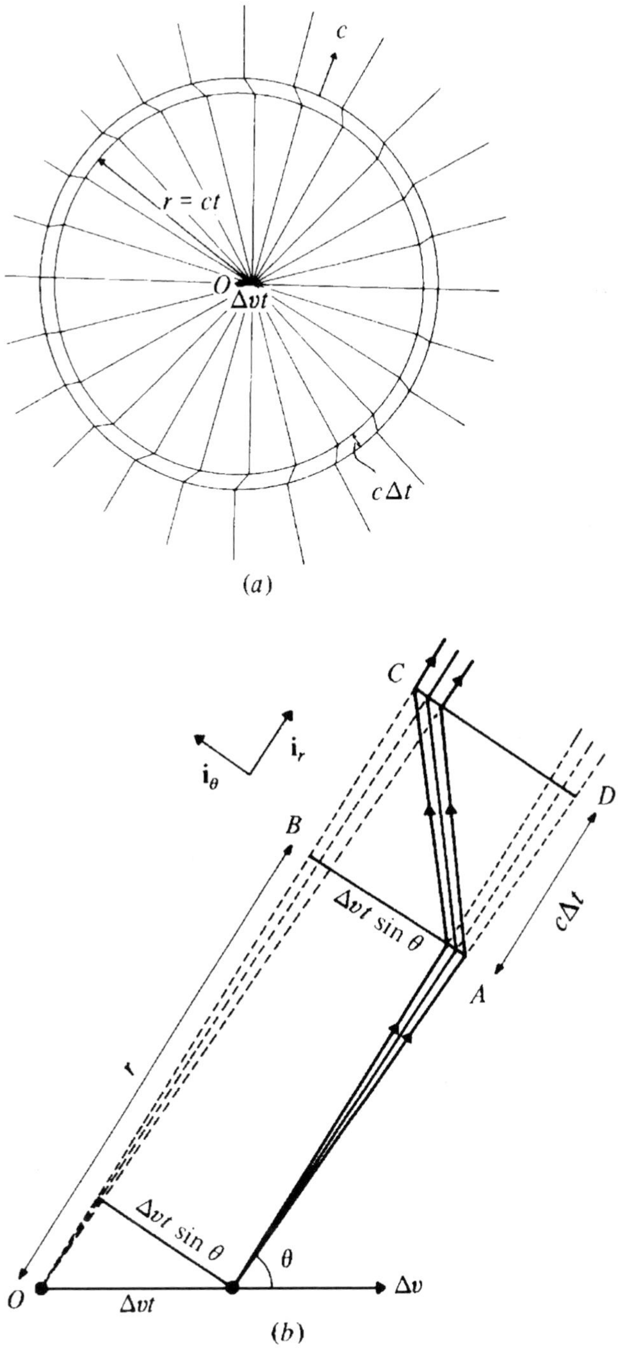

Figure 12.1: (*a*) Illustrating J.J. Thomson's method of evaluating the radiation of an accelerated charged particle. The diagram shows schematically the configuration of electric field lines at time t due to a charge accelerated to a velocity Δv in time Δt at $t = 0$. (*b*) An expanded version of part of (*a*) used to evaluate the strength of the E_θ-component of the electric field due to the acceleration of the electron. (M.S. Longair, 1997, *High Energy Astrophysics*, Vol. 1, Cambridge: Cambridge University Press: a revised and updated version of the 1992 edition.)

Therefore

$$E_\theta = \frac{q(\Delta v/\Delta t)\sin\theta}{4\pi\epsilon_0 c^2 r}.$$

$\Delta v/\Delta t$ is the acceleration $|\boldsymbol{a}|$ of the charge and hence

$$E_\theta = \frac{q\,|\boldsymbol{a}|\sin\theta}{4\pi\epsilon_0 c^2 r}. \tag{12.2}$$

Notice that the radial component of the field decreases as r^{-2}, according to Coulomb's law, but the tangential component decreases only as r^{-1}, because in the shell, as t increases, the field lines become more and more stretched in the E_θ-direction, as can be seen from (12.1). Alternatively, we can write $q\boldsymbol{a} = \ddot{\boldsymbol{p}}$, where $\boldsymbol{p}$ is the dipole moment of the charge, and hence

$$E_\theta = \frac{|\ddot{\boldsymbol{p}}|\sin\theta}{4\pi\epsilon_0 c^2 r}. \tag{12.3}$$

This electric field component represents a pulse of electromagnetic radiation, and hence the rate of energy flow per unit area per second at distance r is given by magnitude of the Poynting vector, $S = |\boldsymbol{E}\times\boldsymbol{H}| = E_0^2/Z_0$, where $Z_0 = (\mu_0/\epsilon_0)^{1/2}$ is the impedance of free space (see Section 6.12). The rate of energy flow through an area $r^2\,\mathrm{d}\Omega$ subtended by solid angle $\mathrm{d}\Omega$ at angle θ at distance r from the charge is therefore

$$Sr^2\,\mathrm{d}\Omega = \frac{|\ddot{\boldsymbol{p}}|^2\sin^2\theta}{16\pi^2 Z_0\epsilon_0^2 c^4 r^2}\,r^2\,\mathrm{d}\Omega = \frac{|\ddot{\boldsymbol{p}}|^2\sin^2\theta}{16\pi^2\epsilon_0 c^3}\,\mathrm{d}\Omega. \tag{12.4}$$

To find the total radiation rate $-\mathrm{d}E/\mathrm{d}t$, we integrate over solid angle. Because of the symmetry of the emitted intensity with respect to the acceleration vector, we can integrate over the solid angle defined by the circular strip between angles θ and $\theta + \mathrm{d}\theta$, as in Fig. 10.1; thus

$$-\frac{\mathrm{d}E}{\mathrm{d}t} = \int_0^\pi \frac{|\ddot{\boldsymbol{p}}|^2\sin^2\theta}{16\pi^2\epsilon_0 c^3}\,2\pi\sin\theta\,\mathrm{d}\theta.$$

We find the key result

$$-\frac{\mathrm{d}E}{\mathrm{d}t} = \frac{|\ddot{\boldsymbol{p}}|^2}{6\pi\epsilon_0 c^3} = \frac{q^2|\boldsymbol{a}^2|}{6\pi\epsilon_0 c^3}. \tag{12.5}$$

This result is sometimes referred to as *Larmor's formula* – precisely the same result comes out of the full theory. These formulae embody the three essential properties of the radiation of an accelerated charged particle.

(i) The total radiation rate is given by Larmor's formula (12.5). Notice that, in this formula, the acceleration is the *proper acceleration* in the relativistic sense of the charged particle and that the radiation rate is measured in the *instantaneous rest frame* of the particle.

(ii) The *polar diagram* of the radiation is of *dipolar* form, that is, the electric field strength varies as $\sin\theta$ and the power radiated per unit solid angle varies as $\sin^2\theta$, where θ is the angle with respect to the acceleration vector of the particle (Fig. 12.2). Notice that there is no radiation along the direction of the acceleration vector and the field strength is greatest at right angles to it.

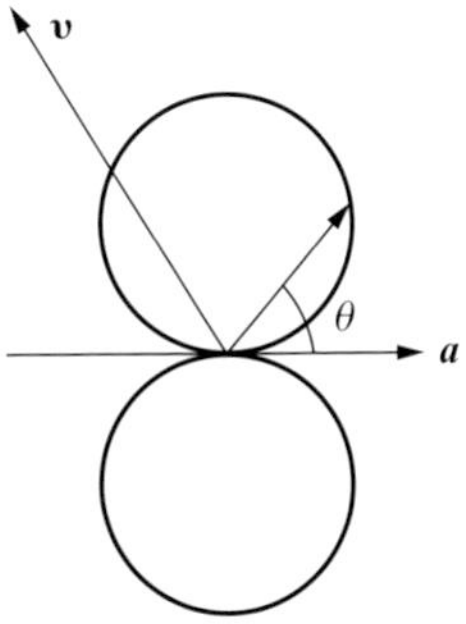

Figure 12.2: The polar diagram of the radiation field E_θ emitted by an accelerated electron, showing the magnitude of the electric field strength as a function of polar angle θ with respect to the instantaneous acceleration vector $\boldsymbol{a}$. Note that the radiation properties of the charged particle in its instantaneous rest frame are independent of the velocity vector $\boldsymbol{v}$, which in general will not be parallel to $\boldsymbol{a}$, as illustrated in the diagram. The radiation field $E_\theta \propto \sin\theta$, (12.3), corresponds to circular lobes with respect to the acceleration vector. (M.S. Longair, 1997, *High Energy Astrophysics*, Vol. 1, Cambridge: Cambridge University Press: a revised and updated version of the 1992 edition.)

(iii) The radiation is *polarised*, the electric field vector, as measured by a distant observer, always lying in the direction of the polar angle unit vector, $\boldsymbol{i}_\theta$ (see Fig. 12.2).

In fact, most radiation problems involving accelerated charged particles can be understood in terms of these simple rules. Several important examples, including the derivation of the Thomson cross-section, are given in my series of books *High Energy Astrophysics*.[7] Although Thomson's is a rather pleasing argument, you should not regard it as a substitute for the full theory, which comes out of the strict application of Maxwell's equations.

12.2.2 Radiation damping of an oscillator

We now apply the result (12.5) to the case of an oscillator performing simple harmonic oscillations with amplitude x_0 at angular frequency ω_0, $x = x_0 \exp(\mathrm{i}\omega_0 t)$. Therefore, $\ddot{x} = -\omega_0^2 x_0 \exp(\mathrm{i}\omega_0 t)$, or, taking the real part, $\ddot{x} = -\omega_0^2 x_0 \cos\omega_0 t$. The instantaneous rate of loss of energy by radiation is therefore

$$-\frac{\mathrm{d}E}{\mathrm{d}t} = \frac{\omega_0^4 e^2 x_0^2}{6\pi\epsilon_0 c^3}\cos^2\omega_0 t. \tag{12.6}$$

The average value of $\cos^2\omega_0 t$ is $\frac{1}{2}$ and hence the average rate of loss of energy in the form of electromagnetic radiation is

$$-\left(\frac{\mathrm{d}E}{\mathrm{d}t}\right)_{\text{average}} = \frac{\omega_0^4 e^2 x_0^2}{12\pi\epsilon_0 c^3}. \tag{12.7}$$

Now let us study the equation of damped simple harmonic motion, since the oscillator loses energy by virtue of emitting electromagnetic waves. It can be written in the form

$$m\ddot{x} + a\dot{x} + kx = 0$$

where m is the (reduced) mass, k the spring constant and $a\dot{x}$ the damping force. The energy associated with each of these terms is found by multiplying through by $\dot{x}$ and integrating with respect to time:

$$\int_0^t m\ddot{x}\dot{x}\,\mathrm{d}t + \int_0^t a\dot{x}^2\,\mathrm{d}t + \int_0^t kx\dot{x}\,\mathrm{d}t = 0,$$

and thus

$$\tfrac{1}{2}m\int_0^t \mathrm{d}(\dot{x}^2) + \int_0^t a\dot{x}^2\,\mathrm{d}t + \tfrac{1}{2}k\int_0^t \mathrm{d}(x^2) = 0. \tag{12.8}$$

We identify the terms in (12.8) with the kinetic energy, the damping energy loss and the potential energy of the oscillator respectively. Now evaluate each term for simple harmonic motion of the form $x = x_0 \cos\omega_0 t$. The average values of the kinetic and potential energies are as follows:

$$\begin{aligned}\text{average kinetic energy} &= \tfrac{1}{4}mx_0^2\omega_0^2,\\ \text{average potential energy} &= \tfrac{1}{4}kx_0^2.\end{aligned}$$

If the damping is very small, the natural frequency of oscillation of the oscillator is $\omega_0^2 = k/m$. Thus, as we already know, the average kinetic and potential energies of the oscillator are equal and the total energy is the sum of the two, which is

$$E = \tfrac{1}{2}mx_0^2\omega_0^2. \tag{12.9}$$

Inspection of (12.8) shows that the instantaneous rate of energy loss by radiation from the oscillator is $a\dot{x}^2$ and so, averaging over time

$$-\left(\frac{\mathrm{d}E}{\mathrm{d}t}\right)_{\text{average}} = \tfrac{1}{2}ax_0^2\omega_0^2. \tag{12.10}$$

Therefore, taking the ratio of (12.9) and (12.10), we find

$$-\left(\frac{\mathrm{d}E}{\mathrm{d}t}\right)_{\text{average}} = \frac{a}{m}E. \tag{12.11}$$

We can now compare this relation with expression (12.7). Substituting (12.9) into (12.7), we obtain

$$-\left(\frac{\mathrm{d}E}{\mathrm{d}t}\right)_{\text{average}} = \gamma E, \tag{12.12}$$

where $\gamma = \omega_0^2 e^2/(6\pi\epsilon_0 c^3 m)$. This expression becomes somewhat simpler if we introduce the *classical electron radius* $r_e = e^2/(4\pi\epsilon_0 m_e c^2)$. Then $\gamma = 2r_e\omega_0^2/(3c)$, where m_e is the mass of the electron. We can therefore obtain the correct expression for the decay constant of the oscillator by identifying γ with a/m in (12.11).

We can appreciate now why Planck believed this was a fruitful way of attacking the problem. The energy is not lost through dissipation of energy as heat, but goes into electromagnetic radiation and the constant γ depends only upon fundamental constants, if we take the oscillator to be an oscillating electron. In contrast, in frictional damping, the energy goes into heat and the loss-rate formula contains constants appropriate to the material. Furthermore, in the case of electromagnetic waves, if the oscillators and waves are confined within an enclosure with perfectly reflecting walls, energy is not lost from the system and the waves react back on the oscillators, returning energy to the latter. This is why Planck called the damping 'conservative damping'. The more common name for this phenomenon nowadays is *radiation damping*.

12.3 The equilibrium radiation spectrum of a harmonic oscillator

In the previous subsection we discussed the expression for the dynamics of an oscillator undergoing natural damping by radiation,

$$m\ddot{x} + a\dot{x} + kx = 0,$$

or

$$\ddot{x} + \gamma\dot{x} + \omega_0^2 x = 0. \tag{12.13}$$

If now an electromagnetic wave at a different angular frequency ω is incident on the oscillator, energy can be transferred to it, and so we add a forcing term to (12.13):

$$\ddot{x} + \gamma\dot{x} + \omega_0^2 x = \frac{F}{m}. \tag{12.14}$$

The oscillator will be accelerated by the E-field of the incident wave, and so we can write $F = eE_0 \exp(\mathrm{i}\omega t)$. To find the response of the oscillator, we adopt a trial solution for x of the form $x = x_0 \exp(\mathrm{i}\omega t)$. Then

$$x_0 = \frac{eE_0}{m\left(\omega_0^2 - \omega^2 + \mathrm{i}\gamma\omega\right)}. \tag{12.15}$$

Notice that there is a complex factor in the denominator, which means that the oscillator does not vibrate in phase with the incident wave. This does not matter for our calculation. We are only interested in the modulus of the amplitude, as we will show.

We are already quite far along the track to finding the amount of energy transferred to the oscillator, but let us not do that straight away. Rather, let us work out the rate of radiation of the oscillator under the influence of the incident radiation field. If we set this equal to the 'natural' radiation of the oscillator, we will have fulfilled the condition necessary for finding the equilibrium spectrum.

This procedure provides a physical picture for what we mean by the oscillator being in equilibrium with the radiation field. To put it another way, the work done by the incident radiation field is just enough to supply the energy loss per second by the oscillator. From now on the calculation is just hard work – there is only one tricky step left.

From (12.6), the rate of radiation of the oscillator is

$$-\frac{\mathrm{d}E}{\mathrm{d}t} = \frac{\omega^4 e^2 x_0^2}{6\pi\epsilon_0 c^3}\cos^2\omega_0 t,$$

where we use the value of x_0 derived in (12.15). We need the square of the modulus of x_0 and this is found by multiplying x_0 by its complex conjugate:

$$x_0^2 = \frac{e^2 E_0^2}{m^2\left[\left(\omega_0^2 - \omega^2\right)^2 + \gamma^2\omega^2\right]}.$$

Therefore, the radiation rate of the oscillator when subjected to an incident wave of angular frequency ω is

$$-\frac{\mathrm{d}E}{\mathrm{d}t}(\omega) = \frac{\omega^4 e^4 E_0^2}{6\pi\epsilon_0 c^3 m^2\left[\left(\omega_0^2 - \omega^2\right)^2 + \gamma^2\omega^2\right]}\cos^2\omega_0 t. \tag{12.16}$$

It is now convenient to average over time, and so

$$\langle\cos^2\omega t\rangle = \tfrac{1}{2}.$$

We notice that the factor that includes E_0^2 is closely related to the energy in the incident wave. The incident energy per unit area per second is given by the Poynting vector $\boldsymbol{E} \times \boldsymbol{H}$. For electromagnetic waves, $\boldsymbol{B} = \boldsymbol{k} \times \boldsymbol{E}/\omega$, $|\boldsymbol{H}| = E/(\mu_0 c)$ and $|\boldsymbol{E} \times \boldsymbol{H}| = [E_0^2/(\mu_0 c)]\cos^2\omega t$, and so the average incident energy per unit area per second is

$$\frac{1}{2\mu_0 c}E_0^2 = \frac{\epsilon_0 c E_0^2}{2}.$$

Furthermore, when we deal with a superposition of waves of random phase, as is the case for equilibrium radiation, the total incident energy is found by adding the energies of the individual waves, that is, by summing over all the terms in E^2 (see Section 15.3). Therefore, we can replace the value of E_0^2 in our formula by a sum over all the waves of the same angular frequency ω incident upon the oscillator, and we find the total *average* reradiated power at that angular frequency. From (12.16),

$$-\frac{\mathrm{d}E}{\mathrm{d}t}(\omega) = \frac{\omega^4 e^4 \times \frac{1}{2}\sum_i E_{0i}^2}{6\pi\epsilon_0 c^3 m^2\left[\left(\omega_0^2 - \omega^2\right)^2 + \gamma^2\omega^2\right]}. \tag{12.17}$$

The next step is to note that the loss rate (12.17) is in fact part of a continuum intensity distribution and also that we can write the sum in (12.17) as an incident intensity* in the frequency band ω to $\omega + \mathrm{d}\omega$,

$$I(\omega)\,\mathrm{d}\omega = \tfrac{1}{2}\epsilon_0 c \sum_i E_{0i}^2. \tag{12.18}$$

Therefore, we can write the radiation loss rate at angular frequency ω as

$$-\frac{\mathrm{d}E}{\mathrm{d}t}(\omega) = \frac{\omega^4 e^4}{6\pi\epsilon_0^2 c^4 m^2}\frac{I(\omega)\,\mathrm{d}\omega}{\left[\left(\omega_0^2 - \omega^2\right)^2 + \gamma^2\omega^2\right]}.$$

* Notice that this intensity is the *total* power per unit area, from 4π steradians. The usual definition is in terms of W m^{-2} Hz^{-1} sr^{-1}. Correspondingly, in (12.23), the relation between $I(\omega)$ and $u(\omega)$ is $I(\omega) = u(\omega)c$ rather than the usual relation $I(\omega) = u(\omega)c/(4\pi)$.

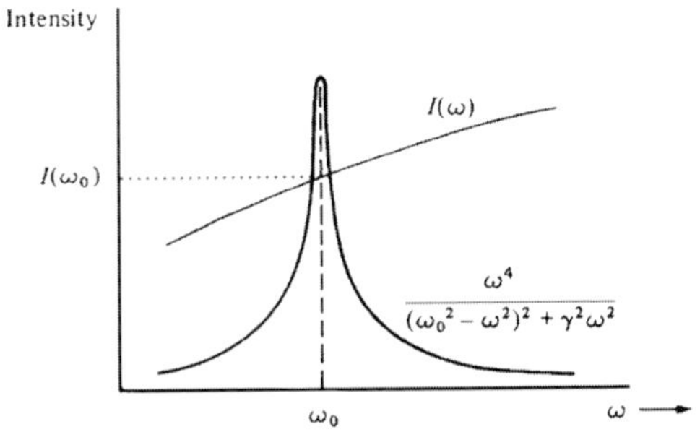

Figure 12.3: The response curve of the oscillator. The continuum spectrum $I(\omega)$ is much broader than the response function of the oscillator and can be taken as constant.

Introducing the classical electron radius, $r_e = e^2/(4\pi\epsilon_0 m_e c^2)$, we obtain

$$-\frac{dE}{dt}(\omega) = \frac{8\pi r_e^2}{3}\frac{\omega^4 I(\omega)\,d\omega}{\left[\left(\omega_0^2-\omega^2\right)^2+\gamma^2\omega^2\right]}. \qquad (12.19)$$

Now the *response curve* of the oscillator as described by the factor $\omega^4/[(\omega_0^2-\omega^2)^2+\gamma^2\omega^2]$ is very sharply peaked about the value ω_0 (see Fig. 12.3), because the radiation rate is very small in comparison with the total energy of the oscillator, that is, $\gamma \ll 1$. We can therefore make some simplifying approximations. Where ω appears in the formula on its own, we can let $\omega \to \omega_0$; also, we can write $\omega_0^2-\omega^2 = (\omega_0+\omega)(\omega_0-\omega) \approx 2\omega_0(\omega_0-\omega)$. Therefore

$$-\frac{dE}{dt}(\omega) \approx \frac{2\pi r_e^2}{3}\frac{\omega_0^2 I(\omega)\,d\omega}{\left[(\omega-\omega_0)^2+\frac{1}{4}\gamma^2\right]}. \qquad (12.20)$$

Finally, we expect $I(\omega)$ to be a slowly varying function in comparison with the sharpness of the response curve of the oscillator and so we can set $I(\omega)$ equal to a constant over the range of values of ω of interest. Then we can integrate over ω to find the total radiation loss rate:

$$-\frac{dE}{dt} = \frac{2\pi r_e^2}{3} I(\omega_0)\int_0^\infty \frac{\omega_0^2\,d\omega}{(\omega-\omega_0)^2+\frac{1}{4}\gamma^2}. \qquad (12.21)$$

The integral is easy if we set the lower limit equal to minus infinity and this is permissible since the function tends very rapidly to zero at frequencies away from that of the peak intensity. Using

$$\int_{-\infty}^{\infty}\frac{dx}{x^2+a^2} = \frac{\pi}{a},$$

the integral (12.21) becomes $2\pi/\gamma$ and so

$$-\frac{dE}{dt} = \frac{2\pi\omega_0^2 r_e^2}{3}\left(\frac{2\pi}{\gamma}\right) I(\omega_0) = \frac{4\pi^2\omega_0^2 r_e^2}{3\gamma} I(\omega_0). \qquad (12.22)$$

According to the prescription which we set out at the beginning of this section, this rate of radiation should now be set equal to the spontaneous radiation rate of the oscillator. There is only one complication. We have assumed so far that the oscillator can respond to incident radiation arriving from any direction. However, for a single oscillator with axis in, say, the x-direction, there are directions of incidence in which the oscillator does not respond to the incident wave – this occurs if the incident electric field is perpendicular to the dipole axis of the oscillator. We can get round this difficulty by a clever argument used by Richard Feynman in his analysis of this problem.[8] Suppose three oscillators are mutually at right angles. Then this system can respond like a completely free oscillator and follow any incident electric field. Therefore, we can regard (12.22) as the radiation which would be emitted by *three mutually perpendicular oscillators*, each oscillating at frequency ω_0. At last, we have arrived at the answer we have been seeking. Equating the radiation rates (12.22) and (12.12), as modified for three oscillators, we find Planck's rather spectacular result:

$$I(\omega_0)\frac{4\pi^2\omega_0^2 r_e^2}{3\gamma} = 3\gamma E, \qquad \text{where} \qquad \gamma = \frac{2r_e\omega_0^2}{3c},$$

and hence

$$I(\omega_0) = \frac{\omega_0^2}{\pi^2 c^2} E. \tag{12.23}$$

Writing (12.23) in terms of the spectral energy density $u(\omega_0)$,*

$$u(\omega_0) = \frac{I(\omega_0)}{c} = \frac{\omega_0^2}{\pi^2 c^3} E. \tag{12.24}$$

We can now drop the subscript zero on ω_0 since this result applies to all angular frequencies in equilibrium. In terms of frequency ν:

$$u(\omega)\,\mathrm{d}\omega = u(\nu)\,\mathrm{d}\nu = \frac{\omega^2}{\pi^2 c^3} E\,\mathrm{d}\omega,$$

that is,

$$u(\nu) = \frac{8\pi\nu^2}{c^3} E. \tag{12.25}$$

This is the result which Planck derived in a paper published in June 1899.[9] It is a remarkable formula. All information about the nature of the oscillator has completely disappeared from the problem. There is no mention of its charge or mass. All that remains is its average energy E. The meaning behind the relation is obviously very profound and fundamental in a thermodynamic sense. I find this an intriguing calculation: the whole analysis has proceeded through a study of the electrodynamics of oscillators, and yet the final result contains no trace of the means by which we arrived at the answer. One can imagine how excited Planck must have been when he discovered this basic result.

Even more important, if we can work out the average energy of an oscillator of frequency ν in an enclosure of temperature T, we can find immediately the spectrum of black-body radiation. The next step is therefore to find the relation between E, T and ν.

* See the footnote earlier in this section about the relation between intensity and energy density.

12.4 Towards the spectrum of black-body radiation

A surprising aspect of the subsequent development was what Planck *did not do next*. We know the answer to the problem posed in the last sentence of the previous section according to classical statistical mechanics. Classically, in thermodynamic equilibrium each degree of freedom is allotted $\frac{1}{2}kT$ of energy and hence the mean energy of a harmonic oscillator should be kT, because it has two degrees of freedom, those associated with the squared terms $\dot{x}^2$ and x^2 in the expression for its energy. Setting $E = kT$ in (12.25), we find

$$u(\nu) = \frac{8\pi\nu^2}{c^3}kT. \tag{12.26}$$

This turns out to be the correct expression for the black-body radiation law at low frequencies, the Rayleigh–Jeans law, which will be described in its proper context in a moment. Why did Planck not make this substitution for E? First of all, the equipartition theorem of Maxwell and Boltzmann is a result of statistical thermodynamics and this was a point of view which he had specifically rejected. At least, he was certainly not as familiar with statistical mechanics as he was with classical thermodynamics. In addition, as we described in Chapter 9, it was far from clear in 1899 how secure the equipartition theorem really was. Maxwell's kinetic theory could not account for the ratio of the specific heats of diatomic gases, and Planck still had reservations about Boltzmann's statistical approach.

Planck had already had his fingers burned when he failed to note the necessity of statistical assumptions in deriving the equilibrium state of black-body radiation. He therefore adopted a rather different approach. To quote his words:

> I had no alternative than to tackle the problem once again – this time from the opposite side – namely from the side of thermodynamics, my own home territory where I felt myself to be on safer ground. In fact, my previous studies of the second law of thermodynamics came to stand me in good stead now, for at the very outset I hit upon the idea of correlating not the temperature of the oscillator but its entropy with its energy ... While a host of outstanding physicists worked on the problem of the spectral energy distribution both from the experimental and theoretical aspect, every one of them directed his efforts solely towards exhibiting the dependence of the intensity of radiation on the temperature. On the other hand, I suspected that the fundamental connection lies in the dependence of entropy upon energy ... Nobody paid any attention to the method which I adopted and I could work out my calculations completely at my leisure, with absolute thoroughness, without fear of interference or competition.[10]

In March 1900, Planck[11] worked out the following relation for the change in entropy ΔS of a system which is not in equilibrium but close to it – we will not derive this result because it will be of marginal importance in the end, although it was important to Planck in obtaining the right answer at the time:

$$\Delta S = \frac{3}{5}\frac{\partial^2 S}{\partial E^2}\,\Delta E\,\mathrm{d}E. \tag{12.27}$$

This equation applies to a system whose entropy deviates from the maximum value in that an individual oscillator initially deviates by an amount ΔE from the equilibrium energy E. The entropy change ΔS occurs when the energy of the oscillator is changed by $\mathrm{d}E$. Thus, if ΔE and $\mathrm{d}E$ have opposite signs, so that the system tends to return towards equilibrium, then, because the entropy change must be positive, the function $\partial^2 S/\partial E^2$ must necessarily have a negative value. We can see by inspection that a formula of this type must be correct. A negative value of $\partial^2 S/\partial E^2$ means that there is an entropy maximum and thus if ΔE and $\mathrm{d}E$ have opposite signs, the system must approach equilibrium.

To appreciate what Planck did next, we need to review the state of experimental determination of the spectrum of black-body radiation at this time. Wien[12] had followed up his studies of the spectrum of thermal radiation by attempting to derive the radiation law from theory. We need not go into his ideas, but he derived an expression for the radiation law which was consistent with his displacement law (11.69) and which provided an excellent fit to all the data available in 1896. The displacement law is

$$u(\nu) = \nu^3 f(\nu/T),$$

and Wien's theory suggested that

$$u(\nu) = \frac{8\pi\alpha}{c^3}\nu^3 \mathrm{e}^{-\beta\nu/T}. \tag{12.28}$$

This is *Wien's law* as written in our notation, rather than Wien's. There are two unknown constants in the formula, α and β; the constant $8\pi/c^3$ has been included on the right-hand side for reasons which will become apparent in a moment. Rayleigh's comment on Wien's paper was terse:

Viewed from the theoretical side, the result appears to me to be little more than a conjecture.[13]

The importance of the formula (12.28) was that it gave an excellent account of all the experimental data available at the time and therefore could be used in theoretical studies. Typical black-body spectra are shown in Fig. 12.4. To fit them, what is needed is a function which rises steeply to a maximum and then cuts off abruptly at short wavelengths. As might be expected, the region of the maximum was well determined by the experimental data, but the uncertainties in the wings of the energy distribution were much greater.

The next step in Planck's paper was the introduction of a definition of the entropy S of an oscillator,

$$S = -\frac{E}{\beta\nu}\ln\frac{E}{\alpha\nu\mathrm{e}}. \tag{12.29}$$

E is the energy of the oscillator and α and β are constants; e is the base of natural logarithms. In fact, we know from Planck's writings that he derived this formula by working backwards from Wien's law. Let us show how this can be done.

Wien's law (12.28) can be inserted into the expression (12.25) relating $u(\nu)$ and E,

$$u(\nu) = \frac{8\pi\nu^2}{c^3}E, \qquad \text{and so} \qquad E = \alpha\nu\,\mathrm{e}^{-\beta\nu/T}. \tag{12.30}$$

Now suppose the oscillator (or a set of oscillators) is placed in a fixed volume V. If the energy of the system is E, the thermodynamic equality (9.55) can be written

$$T\,\mathrm{d}S = \mathrm{d}E + p\,\mathrm{d}V,$$

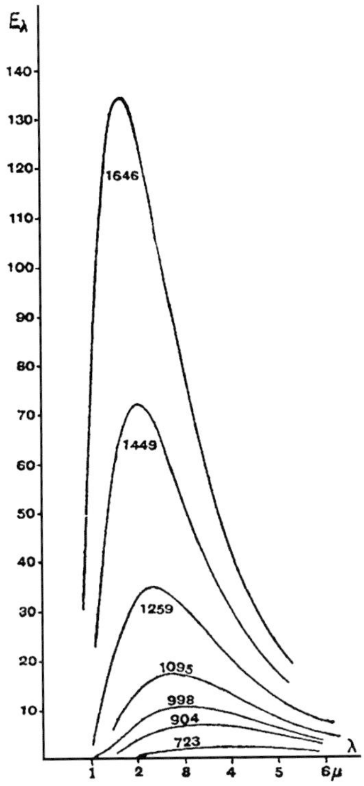

Figure 12.4: The spectrum of black-body radiation plotted on linear scales of intensity and wavelength as determined by Lummer and Pringsheim in 1899 for temperatures in degrees centigrade between 700 °C and 1600 °C. (H.S. Allen and R.S. Maxwell, 1952, *A Text-book of Heat*, Part II, p. 748, Macmillan.)

and hence

$$\left(\frac{\partial S}{\partial E}\right)_V = \frac{1}{T}. \tag{12.31}$$

E and S are additive functions of state and so (12.31) can refer to the properties of an individual oscillator, as well as to an ensemble of them. Taking the logarithm of (12.30), we obtain an expression for T and so we can relate $(\partial S/\partial E)_V$ to E, that is,

$$\frac{1}{T} = \left(\frac{\partial S}{\partial E}\right)_V = -\frac{1}{\beta\nu}\ln\left(\frac{E}{\alpha\nu}\right). \tag{12.32}$$

Now adopting Planck's definition of the entropy of the oscillator (12.29) and differentiating with respect to E, we obtain

$$\begin{aligned}\frac{\mathrm{d}S}{\mathrm{d}E} &= -\frac{1}{\beta\nu}\ln\left(\frac{E}{\alpha\nu e}\right) - \frac{E}{\beta\nu}\frac{1}{E} = -\frac{1}{\beta\nu}\left[\ln\left(\frac{E}{\alpha\nu e}\right) + 1\right] \\ &= -\frac{1}{\beta\nu}\left[\ln\left(\frac{E}{\alpha\nu}\right)\right].\end{aligned} \tag{12.33}$$

The identity of (12.32) and (12.33) demonstrates how Wien's law combined with (12.25) leads to Planck's definition of entropy (12.29).

Now we arrive at the step which was crucial for Planck. Take the next derivative of (12.33) with respect to E:

$$\frac{d^2 S}{dE^2} = -\frac{1}{\beta\nu}\frac{1}{E}. \tag{12.34}$$

Since β, ν and E are all necessarily positive quantities, d^2S/dE^2 is necessarily negative. Therefore, Wien's law (10.28) is entirely consistent with the second law of thermodynamics. Notice the remarkable simplicity of the expression for the second derivative of the entropy with respect to energy – it is proportional to the inverse of the energy. This profoundly impressed Planck, who remarked:

> I have repeatedly tried to change or generalise the equation for the electromagnetic entropy of an oscillator in such a way that it satisfies all theoretically sound electromagnetic and thermodynamic laws but I was unsuccessful in this endeavour.[14]

In a paper presented to the Prussian Academy of Sciences in May 1899 he stated:

> I believe that this must lead me to conclude that the definition of radiation entropy and therefore Wien's energy distribution law necessarily result from the application of the principle of the increase of entropy to the electromagnetic radiation theory and therefore the limits of validity of this law, in so far as they exist at all, coincide with those of the second law of thermodynamics.[15]

This is a rather dramatic conclusion and we now know that it is not correct, as will be discussed below – Planck had, however, made tremendous progress towards the correct formulation of the theory.

An interesting aspect of the above assertion is that it is often dangerous to use arguments of the form, 'I cannot think of any other function which can do the job'; in fact, any negative function of energy would have satisfied Planck's requirement so far as the second law of thermodynamics was concerned. Now, there remained only the problem of ensuring consistency with the measured spectrum of black-body radiation.

12.5 The primitive form of Planck's radiation law

These calculations were presented to the Prussian Academy of Sciences in June 1900. By October 1900, Rubens and Kurlbaum had shown beyond any doubt that Wien's law was inadequate to explain the spectrum of black-body radiation at low frequencies and high temperatures. Their experiments were carried out with the greatest of care over a wide range of temperatures. They showed that, at low frequencies and high temperatures, the intensity of radiation was proportional to temperature. This is clearly inconsistent with Wien's law because if $u(\nu) \propto \nu^3\, e^{-\beta\nu/T}$ then, for $\beta\nu/T \ll 1$, $u(\nu) \propto \nu^3$ and is independent of temperature. Wien's displacement law requires the form of the spectrum to depend only upon ν/T and so the functional dependence must be more complex than (12.28) for small values of ν/T.

Rubens and Kurlbaum showed Planck their new results before they were presented in October 1900 and he was given the opportunity to make some remarks about their implications. The result was his paper entitled 'An improvement of the Wien distribution'[16] – it was the first time that the Planck formula appeared in its primitive form. Here is what Planck did.

He now knew that he had to find a law which would result in the relation $u(\nu) \propto T$ in the limit $\nu/T \to 0$. Let us run through the relations he had already derived. Planck knew from (12.25) that

$$u(\nu) = \frac{8\pi \nu^2}{c^3} E.$$

At low frequencies, the experiments of Rubens and Kurlbaum showed that $u(\nu) \propto T$; therefore the energy E of an oscillator with frequency ν should satisfy

$$E \propto T \qquad \text{as well as} \qquad \frac{\mathrm{d}S}{\mathrm{d}E} = \frac{1}{T}.$$

Hence

$$\frac{\mathrm{d}S}{\mathrm{d}E} \propto \frac{1}{E}, \qquad \text{and} \qquad \frac{\mathrm{d}^2 S}{\mathrm{d}E^2} \propto -\frac{1}{E^2}. \tag{12.35}$$

However, Wien's law remains good for large values of ν/T and leads to

$$\frac{\mathrm{d}^2 S}{\mathrm{d}E^2} \propto -\frac{1}{E}. \tag{12.36}$$

Therefore, $\mathrm{d}^2 S/\mathrm{d}E^2$ must change its functional dependence upon E between large and small values of ν/T.

The standard technique for combining functions of the form (12.35) and (12.36) is to try an expression of the form

$$\frac{\mathrm{d}^2 S}{\mathrm{d}E^2} = -\frac{a}{E(b+E)}, \tag{12.37}$$

which has exactly the required properties for large and small values of E, that is, $E \gg b$ and $E \ll b$ respectively. The rest of the analysis is straightforward. Integrating,

$$\frac{\mathrm{d}S}{\mathrm{d}E} = -\int \frac{a}{E(b+E)}\,\mathrm{d}E = -\frac{a}{b}[\ln E - \ln(b+E)]. \tag{12.38}$$

But $\mathrm{d}S/\mathrm{d}E = 1/T$ and hence

$$\frac{1}{T} = -\frac{a}{b}\ln\left(\frac{E}{b+E}\right), \qquad \mathrm{e}^{b/(aT)} = \frac{b+E}{E},$$

$$E = \frac{b}{\mathrm{e}^{b/(aT)} - 1}. \tag{12.39}$$

Now we can find the radiation spectrum, from (12.25):

$$u(\nu) = \frac{8\pi \nu^2}{c^3} E = \frac{8\pi \nu^2}{c^3} \frac{b}{\mathrm{e}^{b/(aT)} - 1}. \tag{12.40}$$

From the high-frequency, low-temperature limit, we can compare the constants with those which appear in Wien's law (12.28):

$$u(\nu) = \frac{8\pi \nu^2 b}{c^3 \mathrm{e}^{b/aT}} = \frac{8\pi \alpha}{c^3} \frac{\nu^3}{\mathrm{e}^{\beta \nu/T}}.$$

Thus b must be proportional to the frequency ν. We can therefore write Planck's formula in its primitive form:

$$u(\nu) = \frac{A\nu^3}{e^{\beta\nu/T} - 1}. \tag{12.41}$$

Notice that this formula satisfies Wien's displacement law (11.69).

Of equal importance for the next part of the story is the fact that Planck was also able to find an expression for the entropy of an oscillator by integrating $\mathrm{d}S/\mathrm{d}E$, as we have done already. From (12.38),

$$\frac{\mathrm{d}S}{\mathrm{d}E} = -\frac{a}{b}[\ln E - \ln(b+E)] = -\frac{a}{b}\left[\ln\frac{E}{b} - \ln\left(1+\frac{E}{b}\right)\right].$$

Integrating again,

$$\begin{aligned} S &= -\frac{a}{b}\left\{\left(E\ln\frac{E}{b} - E\right) - \left[E\ln\left(1+\frac{E}{b}\right) - E + b\ln\left(1+\frac{E}{b}\right)\right]\right\} \\ &= -a\left[\frac{E}{b}\ln\frac{E}{b} - \left(1+\frac{E}{b}\right)\ln\left(1+\frac{E}{b}\right)\right], \qquad \text{with} \quad b \propto \nu. \end{aligned} \tag{12.42}$$

Planck's new formula (12.41) was remarkably elegant and could now be confronted with the experimental evidence. Before doing that, let us study the origins of another radiation formula, which had been proposed at about the same time, the *Rayleigh–Jeans law*.

12.6 Rayleigh and the spectrum of black-body radiation

Lord Rayleigh was the author of the famous text *The Theory of Sound*[17] and a leading exponent of the theory of waves in general. His interest in the theory of black-body radiation was stimulated by the inadequacies of Wien's law (12.28) in accounting for the low-frequency behaviour of black-body radiation as a function of temperature. His original paper[13] is a short and elegant exposition of the application of the theory of waves to black-body radiation and is reproduced in full in the appendix to this chapter. The reason for the inclusion of Jeans' name in what became known as the *Rayleigh–Jeans law* is that there was a numerical error in Rayleigh's analysis, which was corrected by Jeans in a paper published in *Nature* in 1906. Let us look first at the structure of Rayleigh's paper.

The first paragraph is a description of the state of knowledge of the black-body spectrum in 1900. In the second, Rayleigh acknowledges the success of Wien's formula in accounting for the maximum in the black-body spectrum, but then expresses his worries about the long-wavelength behaviour of the formula. He sets out his proposal in the third and fourth paragraphs and in the fifth describes his preferred form for the radiation spectrum, which, at the end, he hopes will be compared with experiment by 'the distinguished experimenters who have been occupied with this subject'.

Let us rework the essence of paragraphs three, four and five in present-day notation with the correct numerical factors. We begin with the problem of waves in a box. Suppose the box is a cube with sides of length L. Inside the box, all possible wave modes consistent with the boundary conditions are allowed to come into thermal equilibrium at temperature T. The general wave equation for the waves in the box is

$$\nabla^2\psi = \frac{\partial^2\psi}{\partial x^2} + \frac{\partial^2\psi}{\partial y^2} + \frac{\partial^2\psi}{\partial z^2} = \frac{1}{c_s^2}\frac{\partial^2\psi}{\partial t^2}, \tag{12.43}$$

where c_s is the speed of the waves. The walls are fixed and so the waves must have zero amplitude, $\psi = 0$, at $x, y, z = 0$ and $x, y, z = L$. The solution of this problem is well known:

$$\psi = C\mathrm{e}^{-\mathrm{i}\omega t} \sin\frac{l\pi x}{L} \sin\frac{l\pi y}{L} \sin\frac{l\pi z}{L}, \tag{12.44}$$

corresponding to standing waves which fit perfectly into the box in three dimensions, provided l, m and n are integers. Each combination of l, m and n is called a *normal mode of oscillation* of the waves in the box. What this means in physical terms is that, for any such mode, the medium within the box through which the waves propagate oscillates in phase at a particular frequency. These modes are also *orthogonal*, that is, they represent independent ways in which the medium can oscillate. In addition, the set of orthogonal modes with $0 \le l, m, n \le \infty$ is a complete set and so any pressure distribution in the medium can be described as a sum over this complete set of orthogonal functions.

We now substitute (12.44) into the wave equation (12.43) and so find the relation between the values of l, m, n and the angular frequency of the waves, ω:

$$\frac{\omega^2}{c^2} = \frac{\pi^2}{L^2}(l^2 + m^2 + n^2). \tag{12.45}$$

Writing $p^2 = l^2 + m^2 + n^2$,

$$\frac{\omega^2}{c^2} = \frac{\pi^2 p^2}{L^2}. \tag{12.46}$$

We thus find a relation between the modes as parameterised by $p^2 = l^2 + m^2 + n^2$ and their angular frequency ω. According to the Maxwell–Boltzmann equipartition theorem, energy is shared equally among each independent mode and therefore to find the total energy in the box we need to know how many modes there are in the range p to $p + \mathrm{d}p$. We find this by drawing a three-dimensional lattice in lmn-space and evaluating the number of these modes in the octant of a sphere, illustrated in two dimensions in Fig. 12.5. Notice that the number density of points in lmn-space is one per unit volume. Therefore, if p is large then the number of modes is given by the number of points (l, m, n) in an octant of the spherical shell of radius p and width $\mathrm{d}p$:

$$n(p)\,\mathrm{d}p = \tfrac{1}{8} \times 4\pi p^2\,\mathrm{d}p. \tag{12.47}$$

Expressing this result in terms of ω rather than p,

$$p = \frac{L\omega}{\pi c}, \qquad \mathrm{d}p = \frac{L}{\pi c}\,\mathrm{d}\omega, \tag{12.48}$$

and hence

$$n(p)\,\mathrm{d}p = \frac{L^3\omega^2}{2\pi^2 c^3}\,\mathrm{d}\omega. \tag{12.49}$$

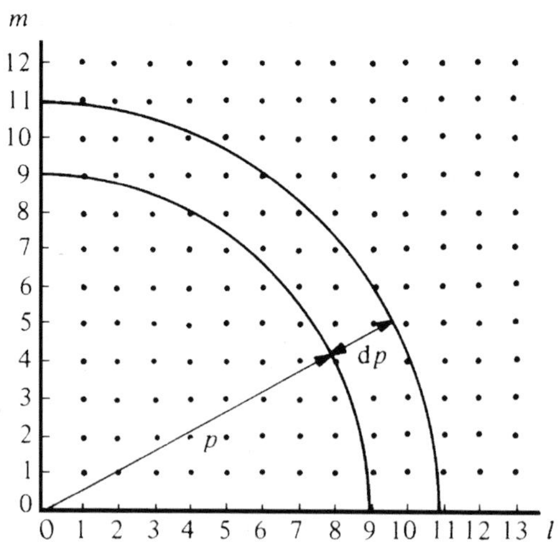

Figure 12.5: The number of modes in the interval dl dm dn can be replaced by an increment in the phase-space volume $\frac{1}{2}\pi p^2\,\mathrm{d}p$, where $p^2 = l^2 + m^2 + n^2$. The diagram shows a two-dimensional cross-section through the three-dimensional volume in l, m and n.

In the present case, the waves are electromagnetic waves and so there are two independent linear polarisations for any given mode. Therefore, there are twice as many modes as given by (12.49).

All we have to do now is to apply the Maxwell–Boltzmann doctrine of the equipartition of energy and give each mode of oscillation an energy $E = kT$ according to the classical prescription. Then, the spectral energy density of electromagnetic radiation in the box is given by:

$$u(\nu)\,\mathrm{d}\nu\, L^3 = En(p)\,\mathrm{d}p = \frac{L^3\omega^2 E}{\pi^2 c^3}\,\mathrm{d}\omega, \tag{12.50}$$

that is,

$$u(\nu) = \frac{8\pi\nu^2}{c^3}E = \frac{8\pi\nu^2}{c^3}kT. \tag{12.51}$$

The first equality is exactly the same result as that derived by Planck from electrodynamics, (12.25), but Rayleigh did not hesitate in setting $E = kT$, according to the principle of equipartition of energy. It is intriguing that two such different approaches should result in exactly the same answer.

A number of features of Rayleigh's analysis are worth noting.

(i) Rayleigh deals directly with the electromagnetic waves themselves, rather than with the oscillators which are the source of the waves and which are in equilibrium with them.

(ii) Central to the result is the doctrine of the equipartition of energy. I had great difficulty with the principle when I was first taught it as a student. It is only when one realises that one is dealing with independent, or 'orthogonal', modes of oscillation that the idea becomes clear physically. We can decompose any motion of a gas in a box into its normal modes of oscillation and these modes will be independent of one another. However, they cannot be completely independent or there would be no way of exchanging energy between them so that they can come into equipartition. There must be processes by which energy can be exchanged between the so-called independent modes of oscillation. In fact what happens is that the modes are very weakly coupled together when we study higher-order processes in the gas. Thus, if one mode gets more energy than another, there are physical processes by which the excess energy can be redistributed among the modes. What the Maxwell–Boltzmann doctrine states is that, if you leave the system long enough, irregularities in the energy distribution among the oscillations are smoothed out by these energy-interchange mechanisms. In many natural phenomena, the equilibrium distributions are set up very quickly and so there is no need to worry, but it is important to remember that there is a definite assumption about the ability of the interaction processes to bring about the equipartition distribution of energy.

(iii) Rayleigh was aware of 'the difficulties which attend the Boltzmann–Maxwell doctrine of the partition of energy'. Among these is the fact that (12.51) must break down at high frequencies because the experimental spectrum of black-body radiation does not increase as ν^2 to infinite frequency (Fig. 12.4). (This divergence of the spectrum of black-body radiation from classical statistical physics became known as the 'ultraviolet catastrophe' – we will return to this key result in Section 14.2.) Rayleigh proposed, however, that the analysis 'may apply to the graver modes', that is, to long wavelengths or low frequencies.

(iv) Then, right out of the blue, we read in the fifth paragraph 'If we introduce the exponential factor, the complete expression is [in our notation]

$$u(\nu) = \frac{8\pi\nu^2}{c^3} kT\,\mathrm{e}^{-\beta\nu/T}.\text{'} \tag{12.52}$$

Rayleigh included this factor empirically so that the radiation spectrum would converge at high frequencies and because the exponential in Wien's law provided a good fit to the data. Notice that (12.52) is consistent with Wien's displacement law (11.69).

Rayleigh's analysis was brilliant but, as we will see, it did not get its due credit in 1900.

12.7 Comparison of the laws for black-body radiation with experiment

On 25 October 1900, Rubens and Kurlbaum compared their new precise measurements of the black-body spectrum with five different predictions. These were (i) Planck's formula (12.41), (ii) Wien's relation (12.28), (iii) Rayleigh's results (12.52) and, (iv) and (v), two empirical relations which had been proposed by Thiesen and by Lummer and Jahnke. Rubens and Kurlbaum concluded that Planck's formula was superior to all the others and that it gave precise agreement with experiment. Rayleigh's proposal was found to be a poor representation of the experimental data. The functional forms of the three relations discussed above are shown in Fig. 12.6. Rayleigh was justifiably upset by the tone in which they discussed his result. When his scientific papers were republished two years later he

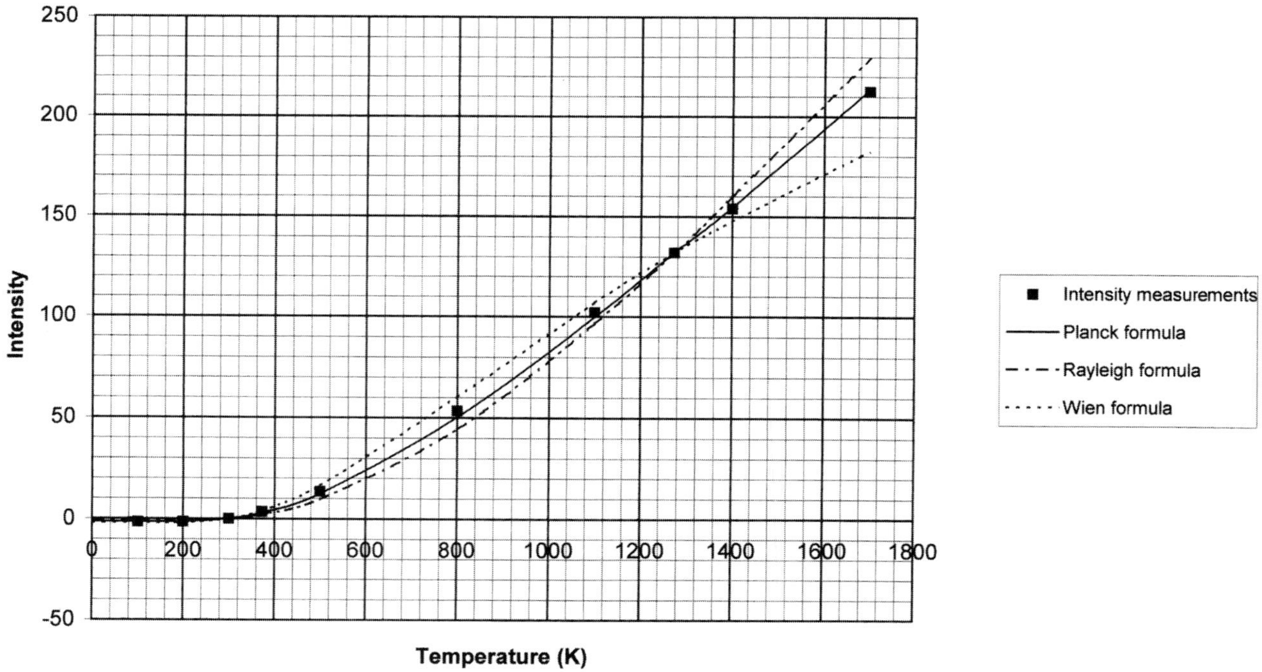

Figure 12.6: Comparison of the variation of the intensity of black-body radiation with temperature at 8.85 microns with different formulae (quartz filter). (After H. Rubens and F. Kurlbaum, 1901, *Annalen der Physik*, **4**, 649.)

remarked on his important conclusion that the intensity of radiation should be proportional to temperature at low frequencies (see the footnote in the appendix to this chapter). He pointed out:

> This is what I intended to emphasise. Very shortly afterwards the anticipation above expressed was confirmed by the important researches of Rubens and Kurlbaum who operated with exceptionally long waves.[18]

The essential theoretical point of Rayleigh's paper, that the radiation spectrum can be described accurately by the Rayleigh–Jeans formula in the wavelength region in which it is applicable, had been missed by the experimenters, who had simply compared the formulae with their experimental measurements. There is a warning to all theorists and experimenters in this tale. Experimenters sometimes use the results of theory without a full appreciation of their range of applicability. In addition, experimenters like to have lots of theories between which to discriminate experimentally, no matter how outlandish they may be. Thus, in the present example, the theories of Planck and Rayleigh are much more profound than the other three. The experimenter is, however, right to try to maintain an unbiased stance between theories.

It seems inconceivable that Planck was not aware of Rayleigh's work. Rubens had told Planck about his experiments and he must have seen the comparison of the various curves with the experiments. We can only assume that it was the statistical basis of the equipartition theorem which put Planck off, together with the fact that Rayleigh's law did not give a good account of the experimental data. Even so, he ought to have been impressed by the ease with which Rayleigh found the correct low-frequency, high-temperature relation and also by the method used to obtain a result identical to Planck's relation between the energy density of radiation and the mean energy of each mode (12.25). It has to be said that Rayleigh's paper does not exactly make this point clear.

Planck had not, however, explained anything. All he had was a formula and it had no sound theoretical basis. He immediately embarked upon this problem. The formula (12.34) was presented to the German Physical Society on 19 October 1900 and, on 14 December 1900, he presented another paper entitled 'On the theory of the energy distribution law in the normal spectrum'.[19] In his memoirs, he writes

> After a few weeks of the most strenuous work of my life, the darkness lifted and an unexpected vista began to appear.[20]

This new vista, which was to revolutionise the fundamental structure of physics, is the subject of the next chapter.

12.8 References

1 Planck, M. (1950). *Scientific Autobiography and Other Papers*, p. 15. London: Williams and Norgate.
2 Planck, M. (1950). *Op. cit.*, p. 18–19.
3 Planck M. (1896). Quoted by M.J. Klein (1977) in *History of Twentieth Century Physics, Proc. International School of Physics 'Enrico Fermi'*, Course 57, p. 3. New York and London: Academic Press.

4 Kuhn, T.S. (1978). *Black-Body Theory and the Quantum Discontinuity 1894–1912*. Oxford: Clarendon Press.
5 Thomson, J.J. (1906). *Electricity and Matter*. London: Archibald Constable and Co.
6 Thomson, J.J. (1907). *Conduction of Electricity through Gases*. Cambridge: Cambridge University Press.
7 Longair, M.S. (1997). *High Energy Astrophysics*, Vols. 1 and 2. Cambridge: Cambridge University Press (a revised and updated version of the 1992 and 1994 editions).
8 Feynman, R.P. (1963). *Feynman Lectures on Physics*, Vol. 1, eds. R.P. Feynman, R.B. Leighton and M. Sands, pp. 41–5. Redwood City, California: Addison-Wesley.
9 Planck, M. (1899). *Berl. Ber.*, pp. 440–86.
10 Planck, M. (1950). *Op. cit.*, pp. 37–8.
11 Planck, M. (1900). *Ann. der Phys.*, **1**, 719–37.
12 Wien, W. (1896). *Ann. der Phys.*, **581**, 662–9.
13 Rayleigh, Lord (1900). *Phil. Mag.*, **49**, 539. See also *Scientific Papers by John William Strutt, Baron Rayleigh*, Vol. 4, *1892–1901*, p. 483, Cambridge: Cambridge University Press.
14 Planck, M. (1958). *Physikalische Abhandlungen und Vorträge (Collected Scientific Papers)*, Vol. 1, p. 596. Braunschweig: Friedr. Vieweg und Sohn. (English trans.: see A. Hermann, 1971, *The Genesis of Quantum Theory (1899–1913)*, p. 10, Cambridge, Massachusetts: MIT Press.)
15 Planck, M. (1958). *Op. cit*, Vol. 1, p. 597. (English trans.: A. Hermann, 1971, *op. cit*, p. 10.)
16 Planck, M. (1900). *Verhandl. der Deutschen Physikal. Gesellsch.*, **2**, 202. See also *Collected Scientific Works* (1958), *op. cit.*, Vol. 1, p. 687–9. (English trans.: *Planck's Original Papers in Quantum Physics*, 1972, annotated by H. Kangro, pp. 35–7, London: Taylor and Francis.)
17 Rayleigh, Lord (1894). *The Theory of Sound*, two volumes. London: Macmillan.
18 Rayleigh, Lord (1902). *Scientific Papers by John William Strutt, Baron Rayleigh*, Vol. 4, *1892–1901*, p. 483. Cambridge: Cambridge University Press.
19 Planck, M. (1900). *Verhandl. der Deutschen Physikal. Gesellsch.*, **2**, 237–45. See also *Collected Scientific Works* (1958), *op. cit.*, Vol. 1, pp. 698–706. (English trans.: *Planck's Original Papers in Quantum Physics*, 1972, *op. cit.*, pp. 38–45.)
20 Planck, M. (1925). *A Survey of Physics*, p. 166. London: Methuen and Co.

Appendix to Chapter 12: Rayleigh's paper of 1900, with the original footnotes

REMARKS UPON THE LAW OF COMPLETE RADIATION.[18]

(From *Philosophical Magazine*, 1900, XLIX, pp. 539, 540.)

By complete radiation I mean the radiation from an ideally black-body, which according to Stewart* and Kirchhoff is a definite function of the absolute temperature θ and the

* Stewart's work appears to be insufficiently recognized upon the Continent. [See *Phil. Mag.* I. p. 98, 1901; p. 494 below.]

wavelength λ. Arguments of (in my opinion†) considerable weight have been brought forward by Boltzmann and W. Wien leading to the conclusion that the function is of the form

$$\theta^5\,\phi(\theta\lambda)\,\mathrm{d}\lambda \tag{1}$$

expressive of the energy in that part of the spectrum which lies between λ and $\lambda + \mathrm{d}\lambda$. A further specialization by determining the form of the function ϕ was attempted later.‡ Wien concludes that the actual law is

$$c_1\lambda^{-5}\,\mathrm{e}^{-c_2/\lambda\theta}\,\mathrm{d}\lambda \tag{2}$$

in which c_1 and c_2 are constants, but viewed from the theoretical side the result appears to me to be little more than a conjecture. It is, however, supported upon general thermodynamic grounds by Planck.§

Upon the experimental side, Wien's law (2) has met with important confirmation. Paschen finds that his observations are well represented, if he takes

$$c_2 = 14\,455,$$

θ being measured in centigrade degrees and λ in thousandths of a millimetre (μ). Nevertheless, the law seems rather difficult of acceptance, especially the implication that as the temperature is raised, the radiation of given wavelength approaches a limit. It is true that for visible rays the limit is out of range. But if we take $\lambda = 60\mu$, as (according to the remarkable researches of Rubens) for the rays selected by reflexion at surfaces of Sylvin, we see that for temperatures over 1000° (absolute) there would be but little further increase of radiation.

The question is one to be settled by experiment; but in the meantime I venture to suggest a modification of (2), which appears to me more probable *a priori*. Speculation upon this subject is hampered by the difficulties which attend the Boltzmann–Maxwell doctrine of the partition of energy. According to this doctrine every mode of vibration should be alike favoured; and although for some reason not yet explained the doctrine fails in general, it seems possible that it may apply to the graver modes. Let us consider in illustration the case of a stretched string vibrating transversely. According to the Boltzmann–Maxwell law, the energy should be equally divided among all the modes, whose frequencies are as 1, 2, 3, ... Hence if k be the reciprocal of λ, representing the frequency, the energy between the limits k and $k + \mathrm{d}k$ is (when k is large enough) represented by $\mathrm{d}k$ simply.

When we pass from one dimension to three dimensions, and consider for example the vibrations of a cubical mass of air, we have (*Theory of Sound*, §267) as the equation for k^2,

$$k^2 = p^2 + q^2 + r^2$$

where p, q, r are integers representing the number of subdivisions in the three directions. If we regard p, q, r as the coordinates of points forming a cubic array, k is the distance of any point from the origin. Accordingly the number of points for which k lies between k and

† *Phil. Mag.* Vol. XLV. p. 522 (1898).
‡ *Wied. Ann.* Vol. LVIII. p. 662 (1896).
§ *Wied. Ann* Vol. I. p. 74 (1900).

$k + dk$, proportional to the volume of the corresponding spherical shell, may be represented by $k^2\, dk$, and this expresses the distribution of energy according to the Boltzmann–Maxwell law, so far as regards the wave-length or frequency. If we apply this result to radiation, we shall have, since the energy in each mode is proportional to θ,

$$\theta k^2\, dk \tag{3}$$

or, if we prefer it,

$$\theta\lambda^{-4}\, d\lambda. \tag{4}$$

It may be regarded as some confirmation of the suitability of (4) that it is of the prescribed form (1).

The suggestion is that (4) rather than, as according to (2),

$$\lambda^{-5}\, d\lambda \tag{5}$$

may be the proper form when $\lambda\theta$ is great.¶ If we introduce the exponential factor, the complete expression will be

$$c_1\theta\lambda^{-4}\, e^{-c_2/\lambda\theta}\, d\lambda \tag{6}$$

If, as is probably to be preferred, we make k the independent variable, (6) becomes

$$c_1\theta k^2\, e^{-c_2 k/\theta}\, dk. \tag{7}$$

Whether (6) represents the facts of observation as well as (2) I am not in a position to say. It is to be hoped that the question may soon receive an answer at the hands of the distinguished experimenters who have been occupied with this subject.

¶ [1902. This is what I intended to emphasize. Very shortly afterwards the anticipation above expressed was confirmed by the important researches of Rubens and Kurlbaum (*Drude Ann.* IV. p. 649, 1901), who operated with exceptionally long waves. The formula of Planck given about the same time, seems best to fit the observations. According to this modification of Wien's formula, $e^{-c_2/\lambda\theta}$ is replaced by $1/(e^{c_2/\lambda\theta} - 1)$. When $\lambda\theta$ is great, this becomes $\lambda\theta/c_2$, and the complete expression reduces to (4).]

13 Planck's theory of black-body radiation

13.1 Introduction

On the very day when I formulated this law, I began to devote myself to the task of investing it with a true physical meaning. This quest automatically led me to study the interrelation of entropy and probability – in other words, to pursue the line of thought inaugurated by Boltzmann.[1]

Planck recognised that, in order to give his expression (12.41) for the spectrum of black-body radiation physical meaning, the way forward involved adopting a point of view which he had rejected in essentially all his previous work. As is apparent from his words quoted at the end of Chapter 12, Planck was working at white-hot intensity, because he was not a specialist in statistical physics. We will find that his analysis did not, in fact, follow the precepts of classical statistical mechanics. Despite the basic flaws in his argument, he discovered the essential role which *quantisation* plays in accounting for the expression (12.41) – recall that Planck's derivation of October 1900 was essentially a thermodynamic argument.

We have already discussed Boltzmann's expression for the relation between entropy and probability, $S \propto \ln W$, in Sections 10.7 and 10.8. The constant of proportionality was not known at the time and so we write $S = C \ln W$, where C is some unknown universal constant. First of all, let us describe how Planck should have proceeded, according to classical statistical mechanics.

13.2 Boltzmann's procedure in statistical mechanics

Planck knew that he had to work out the average energy $\overline{E}$ of an oscillator in thermal equilibrium in an enclosure at temperature T. Suppose there are N oscillators in such an enclosure, so that their total energy is $E = N\overline{E}$. The entropy is also an additive function, and so, if $\overline{S}$ is the average entropy of an oscillator, the entropy of the whole system is $S = N\overline{S}$.

One of the tricks of classical statistical mechanics is to begin by considering the molecules to have discrete energies, $0, \epsilon, 2\epsilon, 3\epsilon, \ldots$. This procedure is adopted so that exact probabilities can be determined statistically. At the appropriate step in the argument, the value of ϵ is allowed become infinitesimally small, whilst the total energy of the system remains finite. Then, the probability distribution changes from a discrete to a continuous distribution.

Table 13.1. *Planck's example of one possible distribution of 100 energy elements among 10 oscillators. The total energy* $E = \Sigma_k E_k = 100\epsilon$

Oscillator label, k	1	2	3	4	5	6	7	8	9	10
Energy E_k in units of ϵ	7	38	11	0	9	2	20	4	4	5

This is exactly how Planck's famous paper[2] of December 1900 begins. Suppose there exists an energy unit ϵ and a fixed amount of energy E is to be distributed among N oscillators. There are many different ways in which the $r = E/\epsilon$ energy elements can be distributed over the N oscillators. The example presented by Planck in his paper is shown in Table 13.1 – he supposed that there were $N = 10$ oscillators and $r = 100$ energy elements to be distributed among them.

There are obviously many different ways of distributing the energy among the oscillators besides this one. If Planck had followed Boltzmann's prescription, here is how he would have proceeded.

Boltzmann noted that each such distribution of energies can be represented by a set of numbers $w_0, w_1, w_2, w_3, \ldots, w_r$, which describe the number of molecules or oscillators with energies $0, 1, 2, 3, \ldots, r$ in units of ϵ. In Planck's example, the set $\{w_j\}$ would begin $\{1, 0, 1, 0, 2, 1, 0, 1, 0, 1, 0, 1\}$; this includes the w_j from $j = 0$ to $j = 11$ but would continue, with mostly zero entries, up to w_{100}. Therefore, we should work out the number of ways in which the energy elements can be distributed over the oscillators so as to result in the same distribution of energies, $w_0, w_1, w_2, w_3, \ldots, w_r$.

Let us revise some basic elements of permutation theory. The number of different ways in which n different objects can be ordered is $n!$. For example, for three objects a, b, c, these can be ordered in the following ways: $abc, acb, bac, bca, cab, cba$, that is, in $3! = 6$ ways. If m of the objects are identical then the number of different orderings is reduced, because we can interchange them and it makes no difference to the distribution. Since m objects can be ordered, when different, in $m!$ ways, the numbers of different arrangements is reduced to $n!/m!$. Thus, if $a \equiv b$ in the above example, the possible arrangements are aac, aca, caa, that is, $3!/2! = 3$ different arrangements. If a further l objects are identical, the number of different arrangements is further reduced to $n!/(m!l!)$, and so on.

Now we ask, in how many different ways can we select x from n objects? We divide the set of n objects into two groups of objects, x objects which make up the set selected and $n - x$ objects which are not selected. Using the same reasoning as in the last paragraph, the number of different ways of making this selection is $n!/[(n - x)!x!]$. This is often written $\binom{n}{x}$ or ${}^nC_x$ and will be recognised as the set of coefficients in the binomial expansion of $(1 + t)^n$, that is,

$$
\begin{aligned}
(1+t)^n &= 1 + nt + \frac{n(n-1)}{2!}t^2 + \cdots + \frac{n!}{(n-x)!x!}t^x + \cdots + t^n \\
&= \sum_{x=0}^{n} \binom{n}{x} t^x. \qquad (13.1)
\end{aligned}
$$

Let us now return to our aim of working out the number of different ways in which we can obtain the particular distribution of oscillator energies $w_0, w_1, w_2, \ldots, w_r$ over the N oscillators, recalling that $N = \sum_{j=0}^{r} w_j$. First of all, we select w_0 oscillators from the N, which can be done in $\binom{N}{w_0}$ ways. This leaves $N - w_0$, from which we can select w_1 in $\binom{N-w_0}{w_1}$ different ways. Next, we select w_2 from the remaining $N - w_0 - w_1$ in $\binom{N-w_0-w_1}{w_2}$ ways, and so on until we have accounted for all N oscillators. Therefore, the total number of different ways of arriving at the particular distribution $w_0, w_1, w_2, \ldots, w_r$ is the product of all these numbers, which is

$$
\begin{aligned}
W_i(w_0, w_1, w_2, \ldots, w_r) &= \binom{N}{w_0}\binom{N-w_0}{w_1}\binom{N-w_0-w_1}{w_2}\cdots \\
&\quad \times \binom{N-w_0-w_1-\cdots-w_{r-1}}{w_r}, \\
&= \frac{N!}{w_0!w_1!w_2!\cdots w_r!}. \qquad (13.2)
\end{aligned}
$$

We have used the facts that $N = \sum_{j=0}^{r} w_j$ and $0! = 1$. It is clear from the discussion of the last two paragraphs that $W_i(w_0, w_1, w_2, \ldots, w_r)$ is just the number of different ways of making the selection $\{w_0, w_1, w_2, \ldots, w_r\}$ of oscillator energies. According to the *principle of equal equilibrium probabilities*, each such distribution $\{w_0, w_1, w_2, \ldots, w_r\}$ is equally likely, so that the probability of finding a particular distribution is, from (13.2),

$$p_i = \frac{W_i(w_0, w_1, w_2, \ldots, w_r)}{\sum_i W_i(w_0, w_1, w_2, \ldots, w_r)}. \qquad (13.3)$$

According to the Boltzmann procedure, the equilibrium state is that which has the greatest value of p_i. This is equivalent to maximising the entropy, if it is defined as

$$S = C \ln p, \qquad (13.4)$$

that is, the state of maximum W_i also corresponds to the state of maximum entropy (see Section 10.8).

We have the choice of continuing to work with the probabilities p_i or the density of states W_i, which was introduced in Section 10.8. These only differ by an unimportant (very large) factor. For consistency with modern usage, let us work in terms of the density of states. Therefore, using $S = C \ln W_i$ and taking the logarithm of (13.2),

$$\ln W_i = \ln N! - \sum_{j=0}^{j=r} \ln w_j!. \qquad (13.5)$$

Again, we use Stirling's formula,

$$n! \approx (2\pi n)^{1/2}\left(\frac{n}{\mathrm{e}}\right)^n, \qquad \ln n! \approx n \ln n - n, \qquad (13.6)$$

if n is very large indeed. Substituting into (13.5),

$$\begin{aligned}\ln W_i &\approx N \ln N - N - \sum_j w_j \ln w_j + \sum_j w_j \\ &= N \ln N - \sum_j w_j \ln w_j,\end{aligned} \tag{13.7}$$

where we have used the fact that $N = \sum_j w_j$.

To find the state of maximum entropy, now maximise $\ln W_i$ subject to the following constraints:

$$\begin{aligned}&\text{number of oscillators,} \quad && N = \textstyle\sum_j w_j = \text{constant},\\ &\text{total energy of oscillators,} \quad && E = \textstyle\sum_j \epsilon_j w_j = \text{constant},\end{aligned} \tag{13.8}$$

where ϵ_j is the energy of the jth state, $\epsilon_j = j\epsilon$. This is a classic example of a problem to be solved by the technique of *undetermined multipliers*. In (13.7) we need preserve only terms in the variable w_j, since the number of oscillators N is fixed. Therefore, we need to find the turning values of the function

$$S(w_j) = -\sum_j w_j \ln w_j - A \sum_j w_j - B \sum_j \epsilon_j w_j, \tag{13.9}$$

where A and B are constants to be found from the boundary conditions. Maximising $S(w_j)$, we require

$$\begin{aligned}\delta[S(w_j)] &= -\delta \sum_j w_j \ln w_j - A\,\delta \sum_j w_j - B\,\delta \sum_j \epsilon_j w_j \\ &= -\sum_j [(\ln w_j\, \delta w_j + \delta w_j) + A\, \delta w_j + B\, \epsilon_j\, \delta w_j] \\ &= -\sum_j \delta w_j\, (\ln w_j + \alpha + \beta\epsilon_j) = 0.\end{aligned} \tag{13.10}$$

Since the δw_j are independent, the sum in the parenthesis must be zero for all j and therefore

$$w_j = \mathrm{e}^{-\alpha - \beta\epsilon_j}. \tag{13.11}$$

This is the primitive form of Boltzmann's distribution. The term $\mathrm{e}^{-\alpha}$ is a constant factor in front of the exponential, and so that we can write $w_j \propto \mathrm{e}^{-\beta\epsilon_j}$. There is nothing in this analysis so far to tell us what the constant β should be. Indeed, so far, we have simply carried out a statistical analysis. We have to appeal to analyses such as that which leads to Maxwell's velocity distribution to find the value of β (see Section 10.4). From (10.45) it can be seen that $\beta \propto (kT)^{-1}$ and so

$$w_j \propto \mathrm{e}^{-\epsilon_j/kT}, \tag{13.12}$$

where k is Boltzmann's constant. Finally, Boltzmann allows the energy elements ϵ to tend to zero in such a way that $E = \epsilon_j = j\epsilon$ remains finite, and so he ends up with a continuous energy distribution

$$w(E) \propto \mathrm{e}^{-E/kT}. \tag{13.13}$$

(*a*) × | × × × || × × × | × × | × × × × × | × | × × × × || ×

(*b*) × × × × | × × || × | × × × | × × | × | × × × | × × × | ×

Figure 13.1: Two ways of distributing 20 energy elements among 10 boxes.

13.3 Planck's analysis

Planck's analysis began by following Boltzmann's procedure. There is a fixed total energy E to be divided among the N oscillators and energy elements ϵ are introduced. Therefore, as above, there are $r = E/\epsilon$ energy elements to be shared among the oscillators. Rather than following the procedure outlined in the last section, however, Planck simply worked out the *total number of ways* in which r energy elements can be distributed over the N oscillators. Let us work this number out using the permutation theory outlined in the last section.

The problem can be represented by the diagrams shown in Fig. 13.1. Planck had two fixed quantities, the total number of energy elements r and the number of boxes N into which he wished to place them. In Fig. 13.1(*a*), one way of placing the 20 energy elements into 10 boxes is illustrated; note that there are two empty boxes. It can be seen that the whole problem reduces to no more than determining the total number of ways in which the elements and the walls of the boxes can be arranged between the end stops. A second example is shown in Fig. 13.1(*b*). Thus, the problem amounts to working out the number of different ways in which r elements and $N - 1$ walls can be arranged, remembering that the r elements are identical and the $N - 1$ walls are identical. We deduced the answer in the last section:

$$\frac{(N + r - 1)!}{r!(N - 1)!}. \tag{13.14}$$

This elegant argument was first presented by Ehrenfest[3] in 1914. Equation (13.14) represents the *total number of ways* of distributing an energy $E = r\epsilon$ over N oscillators. Now Planck made the crucial step in his argument. He *defined* (13.14) to be the probability which should be used in the relation

$$S = C \ln W.$$

Let us see where this leads. N and r are very large indeed and so we can use Stirling's approximation

$$n! = (2\pi n)^{1/2}\left(\frac{n}{e}\right)^n\left(1 + \frac{1}{12n} + \cdots\right). \tag{13.15}$$

We need to take the logarithm of (13.14) and so we can use an even simpler approximation than that used in (13.6): since n is very large, we write to a good approximation $n! \approx n^n$ and so

$$W = \frac{(N + r - 1)!}{r!(N - 1)!} \approx \frac{(N + r)!}{r!N!} \approx \frac{(N + r)^{N+r}}{r^r N^N}. \tag{13.16}$$

Therefore

$$S = C[(N + r)\ln(N + r) - r \ln r - N \ln N],$$

$$r = \frac{E}{\epsilon} = \frac{N\overline{E}}{\epsilon}, \tag{13.17}$$

where $\overline{E}$ is the average energy of the oscillators. Thus

$$S = C\left\{N\left(1 + \frac{\overline{E}}{\epsilon}\right)\ln\left[N\left(1 + \frac{\overline{E}}{\epsilon}\right)\right] - \frac{N\overline{E}}{\epsilon}\ln\frac{N\overline{E}}{\epsilon} - N\ln N\right\}. \tag{13.18}$$

The average entropy per oscillator $\overline{S}$ is therefore

$$\overline{S} = \frac{S_N}{N} = C\left[\left(1 + \frac{\overline{E}}{\epsilon}\right)\ln\left(1 + \frac{\overline{E}}{\epsilon}\right) - \frac{\overline{E}}{\epsilon}\ln\frac{\overline{E}}{\epsilon}\right]. \tag{13.19}$$

But this looks rather familiar. Equation (13.19) is exactly the expression for the entropy of an oscillator which Planck had derived to account for the spectrum of black-body radiation. From (12.42), we see that

$$S = a\left[\left(1 + \frac{\overline{E}}{b}\right)\ln\left(1 + \frac{\overline{E}}{b}\right) - \frac{\overline{E}}{b}\ln\frac{\overline{E}}{b}\right],$$

with the requirement that $b \propto \nu$. Thus, *the energy elements* ϵ *must be proportional to frequency*, and Planck wrote this result in the form which has persisted to this day:

$$\epsilon = h\nu, \tag{13.20}$$

where h is rightly known as Planck's constant. This is the origin of the concept of *quantisation*. According to classical statistical mechanics, we ought now to allow $\epsilon \to 0$, but evidently we cannot obtain the expression for the entropy of an oscillator unless the energy elements do *not* disappear but have finite magnitude $\epsilon = h\nu$. Furthermore, we have now determined the value of a in terms of the universal constant C. Therefore, we can write the complete expression for the energy density of black-body radiation:

$$u(\nu) = \frac{8\pi h\nu^3}{c^3}\frac{1}{e^{h\nu/CT} - 1}. \tag{13.21}$$

Finally, what about C? Planck pointed out that C must be a universal constant relating the entropy to the probability, and Boltzmann had implicitly worked out its value for a perfect gas. If C is a universal constant then a single application of any suitable law, such as the perfect gas law, that determines its value defines C for all processes. For example, we could use the results of the Joule expansion of a perfect gas, treated classically and statistically, as was demonstrated in Section 10.7. As shown in that section, the ratio $C = k = R/N_A$, where R is the gas constant and N_A is Avogadro's constant, the number of molecules per mole. We can at last write down the Planck distribution once and for all in its final form:

$$u(\nu) = \frac{8\pi h\nu^3}{c^3}\frac{1}{e^{h\nu/kT} - 1}. \tag{13.22}$$

It is straightforward to integrate this expression to find the total energy density of radiation u in the black-body spectrum,

$$u = \int_0^\infty u(\nu)\,d\nu = \frac{8\pi h}{c^3}\int_0^\infty \frac{\nu^3\,d\nu}{e^{h\nu/kT}-1}. \tag{13.23}$$

This is a standard integral, the value of which is found from

$$\int_0^\infty \frac{x^3\,dx}{e^x-1} = \frac{\pi^4}{15}.$$

Therefore

$$u = \left(\frac{8\pi^5 k^4}{15c^3h^3}\right)T^4 = aT^4. \tag{13.24}$$

We have recovered the Stefan–Boltzmann law for the total energy density of radiation u. Substituting the values of the constants,

$$a = 7.566 \times 10^{-16}\ \mathrm{J\,m^{-3}\,K^{-4}}.$$

We can relate this energy density to the energy I emitted per second from the surface of a black body maintained at temperature T, using exactly the same procedure as that described in subsection 11.3.3. As shown in that subsection, u, I and T are related by

$$I = \tfrac{1}{4}uc = \tfrac{1}{4}acT^4 = \sigma T^4 = 5.67 \times 10^{-8}T^4\ \mathrm{W\,m^{-2}}.$$

This calculation determines the Stefan–Boltzmann constant in terms of fundamental constants:

$$\sigma = \frac{ac}{4} = \frac{2\pi^5k^4}{15c^2h^3} = 5.67 \times 10^{-8}\ \mathrm{W\,m^{-2}\,K^{-4}}. \tag{13.25}$$

What is one to make of Planck's argument? There are two fundamental criticisms.

(i) Planck certainly does not follow Boltzmann's procedure for finding the equilibrium distribution. What he defines as a probability is simply the total number of ways of allocating r energy elements to N different boxes. His analysis involves no maximisation of probabilities to find the most probable state. Planck had no illusions about this. In his own words:

> In my opinion, this stipulation basically amounts to a definition of the probability W; for we have absolutely no point of departure, in the assumptions which underlie the electromagnetic theory of radiation, for talking about such a probability with a definite meaning.[4]

Einstein repeatedly pointed out this weak point in Planck's argument; for example, in 1912 he wrote

> The manner in which Mr Planck uses Boltzmann's equation is rather strange to me in that a probability of a state W is introduced without a physical definition of this quantity. If one accepts this, then Boltzmann's equation simply has no physical meaning.[5]

(ii) A second problem concerns a logical inconsistency in Planck's analysis. On the one hand the oscillators can only take energies E which are multiples of ϵ, and yet, on the other hand, a classical result has been used to work out the rate of radiation of the oscillator (subsection 9.2.1). Implicit in that analysis was the assumption that the energies of the oscillators can vary continuously rather than take only discrete values.

These were major stumbling blocks and it is fair to say that nobody quite understood the significance of what Planck had done. The theory did not in any sense gain immediate acceptance. Nonetheless, whether one likes it or not, the concept of quantisation has been introduced by the energy elements without which it is not possible to reproduce the Planck function. It was some time before Planck himself fully appreciated the profound significance of his analysis. In 1906, Einstein showed that if Planck had followed strictly Boltzmann's procedure then he would have obtained the same answer whilst maintaining the essential concept of energy quantisation. We will repeat Einstein's analysis in the next chapter.

A second point concerns the contrast between the quantum and classical parts of the derivation of Planck's formula for the black-body spectrum. Einstein was not nearly as worried as other physicists about mixing up macroscopic and microscopic physical concepts. He regarded equations such as Maxwell's equations for the electromagnetic field as being statements only about the *average values* of the quantities measured. Furthermore, the expression relating $u(\nu)$ and E might be a relation which has significance independent of electromagnetic theory, although it can be derived by that means. More important, while the relation might not be precisely true on the microscopic scale, it could still be a good representation of the average behaviour of the system as measured in laboratory experiments. This was a very advanced point of view, and was characteristic of Einstein's thinking in the crucial years of the first decade of the twentieth century. We will see in a moment how this reasoning led to Einstein's spectacular discovery of the light quantum.

13.4 Planck and 'natural units'

Why did Planck take his derivation of the black-body radiation formula so seriously? As noted by Martin Klein, Planck had a deeply held conviction that

> the search for the absolute (is) the loftiest of all scientific activity.[6]

Planck realised that there were two fundamental constants in his theory of black-body radiation, k and h. The fundamental nature of k was apparent from its appearance in the kinetic theory as the gas constant per molecule, Boltzmann's constant. Furthermore, Planck had shown that the constant C in Boltzmann's relation $S = C \ln W$ was in fact the same Boltzmann constant k. Both constants could be determined with some precision from the experimentally measured form of the spectrum of black-body radiation (13.22) and from the value of the constant in the Stefan–Boltzmann law (13.24) or (13.25). Combining the known value of the gas constant R with his new determination of k, Planck found a value for Avogadro's constant N_A of 6.175×10^{23} molecules per mole, by far the best estimate known at that time. The present adopted value is $N_A = 6.022 \times 10^{23}$ molecules per mole.

The electric charge carried by a mole of monovalent ions was obtained from electrolytic theory and is known as Faraday's constant. Knowing N_A precisely, Planck was able to derive the elementary unit of charge and found it to be $e = 4.69 \times 10^{-10}$ esu, corresponding to 1.56×10^{-19} C. Again, Planck's value was by far the best available at that time, contemporary experimental values ranging between 1.3 and 6.5×10^{-10} esu. The present standard value is 1.602×10^{-19} C.

Equally compelling for Planck was the fact that h, in conjunction with the constant of gravitation G and the velocity of light c, enabled a set of 'natural' units to be defined in terms of fundamental constants. It is worthwhile quoting Klein's eloquent words on this topic:

> 'All systems of units previously employed, owed their origins to accidents of human life on Earth', wrote Planck. The usual units of length and time derived from the size of the Earth and the period of its orbit, those of mass and temperature from the special properties of water, the Earth's most characteristic feature. Even the standardisation of length using some spectral line would be quite as arbitrary, as anthropomorphic, since the particular line, say the sodium D-line, would be chosen to suit the convenience of the physicist. The new units he was proposing would be truly 'independent of particular bodies or substances, would necessarily retain their significance for all times and for all cultures including extra-terrestrial and non-human ones', and therefore deserve the term *natural units*.[7]

From the dimensions of h, G and c, we can derive 'natural units' of time, length and mass as follows:

Unit	Defining equation	S.I. value
Time	$t_{Pl} = (Gh/c^5)^{1/2}$	10^{-43} s
Length	$l_{Pl} = (Gh/c^3)^{1/2}$	4×10^{-35} m
Mass–energy	$m_{Pl} = (hc/G)^{1/2}$	5.4×10^{-8} kg $\equiv 3 \times 10^{19}$ GeV

Often these *Planck units* are written in terms of $\hbar = h/(2\pi)$, in which case their values are 2.5 times smaller than those quoted in the table. It is evident that the natural units of time and length are very small indeed, while the mass unit is much greater than the mass of any known elementary particle. Nonetheless, Planck was absolutely correct to note that these are the fundamental dimensions involving combinations of the 'classical' constants c and G with the fundamental *quantum of action* h, as Planck referred to it.

It is striking that, a century later, these quantities play a central role in the physics of the very early Universe. They define the times, scales and masses at which we need to take seriously the quantisation of gravity.[8] There is no theory of quantum gravity, but current thinking is that many of the large-scale features of our Universe were laid down by physical processes occurring at the Planck era t_{Pl} at the enormous temperatures found by setting the natural unit of mass–energy, 3×10^{19} GeV, equal to kT. We will return to some aspects of these ideas in Chapter 19.

13.5 Planck and the physical significance of h

It was a number of years before the truly revolutionary nature of what Planck had achieved in these crucial last months of 1900 was appreciated. Little interest was shown in the physical content of his paper, largely, one suspects, because no one quite understood what he had done. Perhaps surprisingly, Planck wrote no papers on the subject of quantisation over the next five years. The next publication which casts some light on his understanding was his text *Lectures on the Theory of Thermal Radiation*[9] of 1906. Thomas Kuhn gives a detailed analysis of Planck's thoughts on quantisation through the period 1900 to 1906 on the basis of these lectures in Chapter 5 of his monograph.[10]

Kuhn's analysis is brilliant, but it is a tortured, unresolved story. What is clear from Kuhn's analysis is that Planck undoubtedly believed that the classical laws of electromagnetism were applicable to the processes of the emission and absorption of radiation, despite the introduction of the finite energy elements in his theory. In the *Lectures*, Planck describes two versions of Boltzmann's procedure in statistical physics. One is the version described in Section 13.2, in which it is supposed that the energies of the oscillators take values 0, ϵ, 2ϵ, 3ϵ, and so on. But there is also a second version in which the energies were considered to lie within the *ranges* 0 to ϵ, ϵ to 2ϵ, 2ϵ to 3ϵ, and so on. This procedure leads to exactly the same statistical probabilities as the first version. In a subsequent passage, in which the motions of the oscillators are traced in phase space, he again refers to the energies of the trajectories corresponding to certain energy ranges U to $U + \Delta U$, ΔU eventually being identified with $h\nu$. Thus, in some sense, Planck regarded quantisation as referring to the average properties of the oscillators.

In the *Lectures*, Planck had little to say about the nature of the quantum of action h, but he was well aware of its fundamental importance. In his words,

> The thermodynamics of radiation will therefore not be brought to an entirely satisfactory state until the full and universal significance of the constant h is understood.[10]

He suggested that the solution might lie in a more detailed understanding of the microphysics of the process of radiation, and this is supported by a letter to Ehrenfest of July 1905 in which he remarks that:

> It seems to me not impossible that this assumption [the existence of an elementary quantum of electricity e] offers a bridge to the existence of an elementary energetic quantum h, particularly since h has the same dimensions as e^2/c.[11]

Planck spent many years trying to reconcile his theory with classical physics, but he failed to find any physical significance for h beyond its appearance in the radiation formula. In his words:

> My futile attempts to fit the elementary quantum of action somehow into the classical theory continued for a number of years and they cost me a great deal of effort. Many of my colleagues saw in this something bordering on tragedy. But I feel differently about it. For the thorough enlightenment I thus received was all the more valuable. I now knew for a fact that the elementary quantum of action played a far more significant part in physics than I had originally been inclined to suspect and this recognition made me see clearly the need for the introduction of totally new methods of analysis and reasoning in the treatment of atomic problems.[12]

Indeed it was not until after about 1908 that Planck fully appreciated the quite fundamental nature of quantisation, which has no counterpart in classical physics. His original view was that the introduction of energy elements was

> . . . a purely formal assumption and I really did not give it much thought except that no matter what the cost, I must bring about a positive result.[13]

This quotation is from a letter by Planck to R.W. Wood written in 1931, 30 years after the events described in this chapter. I find it a rather moving letter and it is worthwhile reproducing it in full.

> 7 October 1931
> My dear colleague,
> You recently expressed the wish, after our fine dinner in Trinity Hall, that I should describe from a psychological viewpoint the considerations which had led me to propose the hypothesis of energy quanta. I shall attempt herewith to respond to your wish.
>
> Briefly summarised, what I did can be described as simply an act of desperation. By nature I am peacefully inclined and reject all doubtful adventures. But by then I had been wrestling unsuccessfully for six years (since 1894) with the problem of equilibrium between radiation and matter and I knew that this problem was of fundamental importance to physics; I also knew the formula that expresses the energy distribution in the normal spectrum. A theoretical interpretation therefore had to be found at any cost, no matter how high. It was clear to me that classical physics could offer no solution to this problem and would have meant that all energy would eventually transfer from matter into radiation. In order to prevent this, a new constant is required to assure that energy does not disintegrate. But the only way to recognise how this can be done is to start from a definite point of view. This approach was opened to me by maintaining the two laws of thermodynamics. The two laws, it seems to me, must be upheld under all circumstances. For the rest, I was ready to sacrifice every one of my previous convictions about physical laws. Boltzmann had explained how thermodynamic equilibrium is established by means of a statistical equilibrium, and if such an approach is applied to the equilibrium between matter and radiation, one finds that the continuous loss of energy into radiation can be prevented by assuming that energy is forced at the outset to remain together in certain quanta. This was purely a formal assumption and I really did not give it much thought except that no matter what the cost, I must bring about a positive result.
>
> I hope that this discussion is a satisfactory response to your inquiry. In addition I am sending you as printed matter the English version of my Nobel lecture on the same topic. I cherish the memory of my pleasant days in Cambridge and the fellowship with our colleagues.
> With kind regards,
> Very truly yours
> M. Planck.[13]

It is intriguing that Planck was prepared to give up the whole of physics, *except the 'two laws of thermodynamics'*, in order to understand the black-body radiation spectrum.

There is undoubtedly a sense of incompleteness about Planck's epoch-making achievements. They are very great indeed and the sense of a titanic intellectual struggle is vividly conveyed. This tortuous struggle is in sharp contrast to the virtuosity of the next giant steps taken by Albert Einstein. Before moving on to this new phase of development, let us conclude the story of Planck's statistical mechanics.

13.6 Why Planck found the right answer

Why did Planck find the correct expression for the radiation spectrum, despite the fact that the statistical procedures he used were more than a little suspect? There are two answers, one methodological, the other physical. The first is that it seems quite likely that Planck worked backwards. It was suggested by Rosenfeld, and endorsed by Klein on the basis of an article by Planck of 1943, that he started with the expression for the entropy of an oscillator (12.42) and worked backwards to find W from $\exp(S/k)$. This results in the permutation formula on the right in (13.16), which is more or less exactly the same as (13.14) for large values of N and r. Equation (13.14) was a well-known formula in permutation theory, which appeared in the works of Boltzmann in his patient exposition of the fundamentals of statistical physics. Planck then regarded (13.14) as the definition of entropy according to statistical physics. If this is indeed what happened, it in no sense diminishes Planck's achievement, in my view. Physicists need to be pragmatic in the face of such crises.

The second answer is that Planck had stumbled by accident upon one of the correct methods of evaluating the statistics of indistinguishable particles according to quantum mechanics, a concept which was unheard of at the time. These procedures were first demonstrated by the Indian mathematical physicist Satyendra Nath Bose in a manuscript entitled 'Planck's law and the hypothesis of light quanta',[14] which he sent to Einstein in 1924. Einstein immediately appreciated its deep significance, translated it into German himself and arranged for it to be published in the journal *Zeitschrift für Physik*. Bose's paper and his collaboration with Einstein led to the establishment of the method of counting indistinguishable particles in quantum mechanics known as *Bose–Einstein statistics*, which differ radically from classical Boltzmann statistics. Equally remarkable is the fact that the new statistical procedures were formulated *before* the discovery of quantum mechanics. Einstein went on to apply these new procedures to the statistical mechanics of an ideal gas.[15]

Bose was not really aware of the profound implications of his derivation of the Planck spectrum. To paraphrase Pais's account of the originality of his paper, Bose introduced three new features into statistical physics:

(i) Photon number is not conserved.
(ii) Bose divides phase space into coarse-grained cells and works in terms of the numbers of particles per cell. The counting explicitly requires that, because the photons are taken to be identical, each possible distribution of states should be counted only once. Thus, Boltzmann's axiom of the distinguishability of particles is gone.
(iii) Because of this way of counting, the statistical independence of particles has gone.

These very profound differences from the classical Boltzmann approach were to find an explanation in quantum mechanics and are associated with the symmetries of the wavefunctions for particles of different spins. It was only with Dirac's discovery of relativistic quantum mechanics that the significance of these deep symmetries was to be fully understood. As Pais remarks,

> The astonishing fact is that Bose was correct on all three counts. (In his paper, he commented on none of them.) I believe there had been no such successful shot in the dark since Planck introduced the quantum in 1900.[16]

The full exposition of these topics would take us far from the main development of this chapter, but let us show how (13.22) can derived according to Bose–Einstein statistics and why Plancks's expression for the probability and hence the entropy turned out to be correct in this case. An excellent concise treatment of these topics is given by Huang in his book *Introduction to Statistical Physics*.[17] The approach is different from the procedure discussed above in that the argument starts by dividing the volume of phase space into elementary cells. Consider one of these cells, which we label k and which has energy ϵ_k and degeneracy g_k, the latter meaning the number of available states with the same energy ϵ_k within that cell. Now suppose there are n_k particles to be distributed over these g_k states and that the particles are identical. Then, the number of different ways in which the n_k particles can be distributed over these states is

$$\frac{(n_k + g_k - 1)!}{n_k!(g_k - 1)!}. \tag{13.26}$$

using the exactly the same logic which led to (13.14). This is the key step in the argument and differs markedly from the corresponding Boltzmann result (13.2). The key point is that in (13.2) all possible ways of distributing the particles over the energy states are included in the statistics, whereas in (13.26), duplications of the same distribution are eliminated. At the quantum level, the explanation for this distinction is neatly summarised by Huang who remarks: 'The classical way of counting in effect accepts all wave functions regardless of their symmetry properties under the interchange of coordinates. The set of acceptable wave functions is far greater than the union of the two quantum cases [the Fermi–Dirac and Bose–Einstein cases].'[17]

The result (13.26) refers only to a single cell in phase space and we need to extend it to all the cells which make up the phase space. The total number of ways of distributing the particles over the cells is the product of all numbers such as (13.26), that is,

$$W = \prod_k \frac{(n_k + g_k - 1)!}{n_k!(g_k - 1)!}. \tag{13.27}$$

To find the equilibrium distribution of the $N = \sum_k n_k$ particles among the cells we ask as before, 'What is the distribution which results in the maximum value of W?' At this point, we return to the Boltzmann procedure. First, Stirling's theorem is used to simplify $\ln W$:

$$\ln W = \ln \prod_k \frac{(n_k + g_k - 1)!}{n_k!(g_k - 1)!} \approx \sum_k \ln \frac{(n_k + g_k)^{n_k+g_k}}{(n_k)^{n_k}(g_k)^{g_k}}. \tag{13.28}$$

Now we maximise W subject to the constraints $\sum_k n_k = N$ and $\sum n_k\epsilon_k = E$. Using the method of undetermined multipliers as before,

$$\delta(\ln W) = 0 = \sum_k \delta n_k\{[\ln(g_k + n_k) - \ln n_k] - \alpha - \beta\epsilon_k\}$$

so that

$$[\ln(g_k + n_k) - \ln n_k] - \alpha - \beta\epsilon_\kappa = 0,$$

and finally

$$n_k = \frac{g_k}{\mathrm{e}^{\alpha+\beta\epsilon_k} - 1}. \tag{13.29}$$

This is known as the *Bose–Einstein distribution* and is the correct statistics for counting the indistinguishable particles known as *bosons*. According to quantum mechanics, bosons are particles with even spin. For example, photons are spin-1 particles, gravitons are spin-2 particles, and so on.

In the case of black-body radiation, we do not need to specify the number of photons present. We can see this from the fact that for black-body radiation the distribution is determined solely by one parameter – the total energy, or the temperature, of the system. Therefore, in the method of undetermined multipliers, we can drop the restriction on the total number of particles. The distribution will automatically readjust to the total amount of energy present, and so $\alpha = 0$. Therefore,

$$n_k = \frac{g_k}{\mathrm{e}^{\beta\epsilon_k} - 1}. \tag{13.30}$$

By inspecting the low-frequency, high-temperature behaviour of the Planck spectrum, we can show that $\beta = 1/kT$, as in the classical case.

Finally, the degeneracy g_k of the cells in phase space for radiation in the frequency interval ν to $\nu + \mathrm{d}\nu$ has already been worked out in our discussion of Rayleigh's approach to the origin of the black-body spectrum (Section 12.6). One of the reasons for Einstein's enthusiasm for Bose's paper was that Bose had derived this factor entirely by considering the phase space available to the photons rather than by appealing to Planck's or Rayleigh's approaches, which relied upon results from classical electromagnetism. In Planck's analysis, the derivation of his expression (12.25) was entirely electromagnetic, and Rayleigh's argument proceeded by fitting electromagnetic waves into a box with perfectly conducting walls. Bose considered the photons to have momenta $p = h\nu/c$ and so the volume of momentum, or phase, space for photons in the energy range $h\nu$ to $h(\nu + d\nu)$ is, using the standard procedure,

$$\mathrm{d}V_p = V\mathrm{d}p_x\,\mathrm{d}p_y\,\mathrm{d}p_z \to V \times 4\pi p^2\,\mathrm{d}p = \frac{4\pi h^3\nu^2\,\mathrm{d}\nu}{c^3}V \tag{13.31}$$

where V is the volume of real space. Now Bose considered this volume of phase space to be divided into elementary cells of volume h^3, following ideas first stated by Planck in his 1906 lectures, and so the number of cells in phase space in the energy range $h\nu$ to $h(\nu + \mathrm{d}\nu)$ is

$$\mathrm{d}N_\nu = \frac{4\pi\nu^2\,\mathrm{d}\nu}{c^3}V. \tag{13.32}$$

He needed to take account of the two polarisation states of the photon and so Rayleigh's result was recovered,

$$dN_\nu = \frac{8\pi\nu^2}{c^3}\,d\nu \qquad \text{with} \qquad \epsilon_k = h\nu. \tag{13.33}$$

We immediately find that the spectral energy density $u(\nu)$ of the radiation is given by

$$u(\nu)\,d\nu = \frac{8\pi h\nu^3}{c^3}\frac{1}{e^{h\nu/kT} - 1}\,d\nu. \tag{13.34}$$

This is Planck's expression for black-body radiation and it has been derived using Bose–Einstein statistics for indistinguishable particles. The statistics are applicable not only to photons but also to integral-spin particles of all types.

We have now run far ahead of our story. None of this was known in 1900 and it was in the following years that Einstein made his revolutionary contributions to modern physics.

13.7 References

1 Planck, M. (1950). *Scientific Autobiography and Other Papers*, p. 41. London: Williams and Norgate.

2 Planck, M. (1900). *Verhandl. der Deutschen Physikal. Gesellsch.*, **2**, 237–45. See also *Collected Scientific Works* (1958), *Physikalische Abhandlungen und Vorträge (Collected Scientific Papers)*, Vol. 1, p. 698, Braunschweig: Friedr. Vieweg und Sohn. (English trans.: *Planck's Original Papers in Quantum Physics*, 1972, annotated by H. Kangro, pp. 38–45, London: Taylor and Francis.)

3 Ehrenfest, P. and Kammerlingh Onnes, H. (1914). *Proc. Acad. Amsterdam*, **17**, 870.

4 Planck, M. (1900). Quoted by M.J. Klein (1977), *History of Twentieth Century Physics, Proc. International School of Physics 'Enrico Fermi'*, Course 57, p. 17. New York and London: Academic Press.

5 Einstein, A. (1912). In *The Theory of Radiation and Quanta, Trans.* First Solvay Conference, eds. P. Langevin and M. de Broglie, p. 115, Gautier-Villars, Paris. (Translation of quotation: see A. Hermann, 1971, *The Genesis of Quantum Theory (1899–1913)*, p. 20, Cambridge, Massachusetts: MIT Press.)

6 Planck, M. (1950). *Scientific Autobiography and Other Papers*, p. 35. London: Williams and Norgate.

7 Klein, M.J. (1977). *History of Twentieth Century Physics, Proc. International School of Physics 'Enrico Fermi'*, Course 57, pp. 13–14. New York and London: Academic Press.

8 Frolov, V.I. and Novikov, I.D. (1998). *Black Hole Physics*. Dordrecht: Kluwer Academic Publishers.

9 Planck, M. (1906). *Vorlesungen über die Theorie der Wärmstrahlung*, first edition. Leipzig: Barth.

10 Kuhn, T.S. (1978). *Black-Body Theory and the Quantum Discontinuity 1894–1912*. Oxford: Clarendon Press.

11 Planck, M. (1905). Quoted by Kuhn (1978), *op. cit.*, 132.

12 Planck, M. (1950). *Op. cit.*, pp. 44–5.
13 Planck, M. (1931). Letter from M. Planck to R.W. Wood. See Hermann, A. (1971), *op. cit.*, pp. 23–4, Cambridge, Massachusetts: MIT Press.
14 Bose, S.N. (1924). *Zeitschrift für Physik*, **26**, 178.
15 Einstein, A. (1924). *Sitz. Preuss. Akad. Wissenschaften*, p. 261; (1925), *ibid.*, p. 3.
16 Pais, A. (1982). *Subtle is the Lord . . . The Science and Life of Albert Einstein*. Oxford: Clarendon Press.
17 Huang, K. (2001). *Introduction to Statistical Physics*. London: Taylor and Francis.

14 Einstein and the quantisation of light

14.1 1905 – Einstein's *annus mirabilis*

Up to 1905, Planck's work had made little impression, and he was no further forward in understanding the profound implications of what he had done. As discussed in Chapter 13, he expended a great deal of unsuccessful effort in trying to find a classical interpretation for the 'quantum of action' h, which he correctly recognised had fundamental significance for understanding the spectrum of black-body radiation. The next great steps were taken by Albert Einstein, and it is no exaggeration to state that he was the first person to appreciate the full significance of quantisation and the reality of quanta. He showed that this is a fundamental aspect of all physical phenomena, rather than just a 'formal assumption' for accounting for the Planck distribution. From 1905 onwards, he never deviated from his belief in the reality of quanta – it was some considerable time before the great figures of the day conceded that Einstein was indeed correct. He came to this conclusion in a series of brilliant papers of dazzling scientific virtuosity.

Einstein completed what we would now call his undergraduate studies in August 1900. Between 1902 and 1904, he wrote three papers on the foundations of Boltzmann's statistical mechanics. Once again, notice how a deep understanding of thermodynamics and statistical physics provided the starting point for the investigation of basic problems in theoretical physics. As was explained in Case Study IV, thermodynamics and statistical physics do not deal with specific physical processes, which might not be particularly well understood; rather, they deal with the overall properties of physical systems and provide general rules about the expected behaviour.

14.1.1 Einstein's three great papers of 1905

In 1905, Einstein was 26 and was employed as 'technical expert, third class' at the Swiss patent office in Bern. In that year, he completed his doctoral dissertation 'A new determination of molecular dimensions', which he presented to the University of Zurich on 20 July 1905. In the same year, he published three papers which are among the greatest classics in the literature of physics. Any one of them would have ensured that his name would remain a permanent fixture in the scientific literature. These papers are:

(i) 'On the motion of small particles suspended in stationary liquids required by the molecular-kinetic theory of heat'.[1]

(ii) 'On the electrodynamics of moving bodies'.[2]

(iii) 'On an heuristic point of view concerning the production and transformation of light'.[3]

These papers are conveniently available in English translations in *Einstein's Miraculous Year*,[4] edited and introduced by John Stachel. The above translations are taken from *The Collected Papers of Albert Einstein*: Vol. 2, *The Swiss Years: Writings, 1900–1909*,[5] which can also be strongly recommended.

The first paper is more familiarly known by the title of a subsequent paper published in 1906 entitled 'On the theory of Brownian motion'[6] and is a reworking of some of the results of his doctoral dissertation. Brownian motion is the irregular motion of microscopic particles in fluids and had been studied in detail in 1828 by the botanist Robert Brown, who had noted the ubiquity of the phenomenon. The motion results from the statistical effect of very large numbers of collisions between molecules of the fluid and microscopic particles. Although each impact is very small, the net result of a very large number of them randomly colliding with a particle is a 'drunken man's walk'. Einstein quantified this problem by relating the diffusion of the particles to the properties of the molecules responsible for the collisions. In a beautiful analysis, he derived the formula for the mean squared distance travelled by the particle in time t,

$$\langle r^2 \rangle = \frac{kTt}{3\pi\eta a}, \tag{14.1}$$

where a is the radius of the particle, T is the temperature, η is the viscosity of the fluid and k is Boltzmann's constant. Crucially, Einstein had discovered the relation between the molecular properties of fluids and the observed diffusion of macroscopic particles. In his estimates of the magnitude of the effect for particles 1 μm in diameter, he needed a value for Avogadro's constant N_A and he used the values which Planck and he had found from their studies of the spectrum of black-body radiation (see Section 14.2 below). He predicted that such particles would diffuse about 6 μm in one minute. In the last paragraph of the paper, Einstein states:

> Let us hope that a researcher will soon succeed in solving the problem presented here, which is so important for the theory of heat!

Remarkably, Einstein wrote this paper 'without knowing that observations concerning Brownian motion were already long familiar', as he remarked in his *Autobiographical Notes*[7] – this accounts for the absence of the term 'Brownian motion' in the title of his first paper on the subject.

Einstein's concern about the importance of this calculation for the theory of heat was well founded. The battle to establish the kinetic theory and Boltzmann's statistical procedures against the energeticists, led by Wilhelm Ostwald and Georg Helm, who denied the existence of atoms and molecules, was not yet won. In defence of their position, it has to be recognised that, at that time, there was no microscopic evidence that heat is indeed associated with the random motions of atoms and molecules. Ernst Mach was hostile to the concept of atoms and molecules since they are not accessible directly to our senses, although he admitted that atomism was a useful conceptual tool. Many scientists regarded the concepts of atoms and molecules simply as useful working hypotheses. Einstein realised that the agitational motion of the particles observed in Brownian motion is heat – the macroscopic particles reflect the motion of the molecules on the microscopic scale.

Precise observations of Brownian motion were difficult at that time, but in 1908 Jean Perrin[8] carried out a meticulous series of brilliant experiments which confirmed in detail all Einstein's predictions. This work convinced everyone, even the sceptics, of the reality of molecules. In Perrin's words,

> I think it is impossible that a mind free from all preconception can reflect upon the extreme diversity of the phenomena which thus converge to the same result without experiencing a strong impression, and I think it will henceforth be difficult to defend by rational arguments a hostile attitude to molecular hypotheses.[9]

The second of Einstein's great papers of 1905 is his famous paper on special relativity, and its contents will be the subject of Chapter 16. While listing Einstein's papers of 1905, it is worth remarking that there is another, much shorter, paper on special relativity, entitled 'Does the inertia of a body depend on its energy content?'[10] This paper is an explicit statement that inertial mass and energy are the same thing, a forceful reminder that the expression $E = mc^2$ can be read forwards or backwards.

The third great paper of 1905 is often referred to as Einstein's paper on the photoelectric effect. This is a gross misrepresentation of the profundity of this paper. As Einstein wrote to his friend Conrad Habicht in May 1905:

> I promise you four papers . . . the first of which I could send you soon, since I will soon receive the free reprints. The paper deals with radiation and the energetic properties of light and is very revolutionary . . .[4]

What Einstein was referring to is the theoretical content of the paper. It really is revolutionary and we study it in some detail in the next section.

14.1.2 Einstein in 1905

The year 1905 is rightly referred to as Einstein's *annus mirabilis* – the three papers listed in subsection 14.1.1 changed the face of physics. The term *annus mirabilis* was first used in physics in connection with Newton's extraordinary achievements of 1665–7, which were discussed in Section 4.4. Einstein's achievement certainly ranks with Newton's, although they were very different characters with quite different outlooks on physics and mathematics. Stachel[4] gives a revealing comparison of the characters of these physicists of genius.

Newton's genius was not only in physics, but also in mathematics and, whilst the seeds of his great achievements were sown when he was only 22, it was many years before his ideas were crystallised into the forms we celebrate today. Einstein confessed that he was no mathematician, and the mathematics needed to understand his papers of 1905 is no more than is taught in the first couple of years of an undergraduate physics course. His genius lay in his extraordinary physical intuition, which enabled him to see deeper into physical problems than his contemporaries. The three great papers were the result of a sudden burst of creativity following almost a decade of deep pondering about three fundamental areas of physics, which suddenly came to fruition almost simultaneously in 1905. Despite their obvious differences, the three papers have a striking commonality of approach. In each case, Einstein stands back from the specific problem at hand and studies carefully the underlying physical principles. Then, with dazzling virtuosity, he reveals the underlying physics in a new light. Nowhere is this genius more apparent than in his stunning papers on quanta and quantisation of 1905 and 1906.

14.2 'On an heuristic viewpoint concerning the production and transformation of light'

Let us reproduce the opening paragraphs of Einstein's third great paper of 1905 – they are revolutionary and startling. They demand attention, like the opening of a great symphony.

> There is a profound formal difference between the theoretical ideas which physicists have formed concerning gases and other ponderable bodies and Maxwell's theory of electromagnetic processes in so-called empty space. Thus, while we consider the state of a body to be completely defined by the positions and velocities of a very large but finite number of atoms and electrons, we use continuous three-dimensional functions to determine the electromagnetic state existing within some region, so that a finite number of dimensions is not sufficient to determine the electromagnetic state of the region completely . . .
>
> The undulatory theory of light, which operates with continuous three-dimensional functions, applies extremely well to the explanation of purely optical phenomena and will probably never be replaced by any other theory. However, it should be kept in mind that optical observations refer to values averaged over time and not to instantaneous values. Despite the complete experimental verification of the theory of diffraction, reflection, refraction, dispersion and so on, it is conceivable that a theory of light operating with continuous three-dimensional functions will lead to conflicts with experience if it is applied to the phenomena of light generation and conversion.[3]

In other words, there may well be circumstances under which Maxwell's theory of the electromagnetic field cannot explain all electromagnetic phenomena and Einstein specifically gives as examples the spectrum of black-body radiation, photoluminescence and the photoelectric effect. His proposal is that, for some purposes, it may be more appropriate to consider light to be

> discontinuously distributed in space. According to the assumption considered here, in the propagation of a light ray emitted from a point source, the energy is not distributed continuously over ever-increasing volumes of space, but consists of a finite number of energy quanta localised at points in space that move without dividing, and can be absorbed and generated only as complete units.

He ends by hoping that

> the approach to be presented will prove of use to some researchers in their investigations.

Einstein was really asking for trouble. The full implications of Maxwell's theory were still being worked out and here he was proposing to replace these hard-won achievements by light corpuscles. To the physicists of the time, this must have looked like a re-run of the controversy between the wave picture of Huyghens and the particle, or corpuscular, picture of Newton, and everyone knew which theory had won the day. The proposal must have seemed particularly untimely when Maxwell's discovery of the electromagnetic nature of light had been so completely validated only 15 years previously by Heinrich Hertz (see Section 5.4).

Notice carefully how Einstein's proposal differs from Planck's approach. Planck found that the 'energy element' $\epsilon = h\nu$, from which the energy of his oscillators are composed, must not vanish. These oscillators are the source of the electromagnetic radiation in the black-body spectrum, but Planck had absolutely nothing to say about the radiation emitted by them. He firmly believed that the waves emitted by the oscillators were simply the classical electromagnetic waves of Maxwell. In contrast, Einstein proposes that the radiation field itself should be quantised.

Like the other Einstein papers, the article is beautifully written and very clear. It all looks so very simple and obvious that it can be forgotten just how revolutionary its contents are. For clarity of exposition, we will continue to use the same modern notation which we have used up till now, rather than Einstein's notation.

After the introduction, Einstein states Planck's formula (12.25), which we have now seen many times, relating the average energy of an oscillator to the energy density of black-body radiation in thermodynamic equilibrium. Einstein does not hesitate, however, to set the average energy of the oscillator equal to kT, according to the kinetic theory, so that

$$u(\nu) = \frac{8\pi\nu^2}{c^3}kT, \tag{14.2}$$

and then writes the total energy in the black-body spectrum in the following provocative form:

$$\text{total energy} = \int_0^\infty u(\nu)\,\mathrm{d}\nu = \frac{8\pi kT}{c^3}\int_0^\infty \nu^2\,\mathrm{d}\nu = \infty. \tag{14.3}$$

This is exactly the problem which had been pointed out by Rayleigh in 1900 and which led to his arbitrary introduction of the exponential factor to prevent the spectrum diverging at high frequencies (see Section 12.6). This phenomenon was later called the *ultraviolet catastrophe* by Paul Ehrenfest.

Einstein next agrees that, despite the high-frequency divergence of the expression (14.2), it is a very good description of the black-body radiation spectrum at low frequencies and high temperatures, and hence the value of Boltzmann's constant k can be derived from that part of the spectrum. Einstein's value of k agreed precisely with Planck's estimate, which Einstein interpreted as meaning that Planck's estimate was, in fact, independent of the details of the theory he had developed to account for the black-body spectrum.

Now we come to the heart of the paper. We have already emphasised the central role which entropy plays in the thermodynamics of radiation. Einstein now rederives a suitable form for the entropy of black-body radiation, using only thermodynamics and the observed form of the radiation spectrum. Entropies are additive, and since, in thermal equilibrium, we may consider the radiation of different wavelengths to be independent, we can write the entropy of the radiation enclosed in volume V as

$$S = V\int_0^\infty \phi[u(\nu), \nu]\,\mathrm{d}\nu. \tag{14.4}$$

The function ϕ is the entropy of the radiation per unit frequency interval per unit volume. The aim of the calculation is to find an expression for ϕ in terms of the spectral energy density $u(\nu)$ and the frequency ν. No other quantities besides the temperature T can be involved in the expression for the equilibrium spectrum, as was shown by Kirchhoff according to the argument reproduced in subsection 11.2.2. The problem had already been solved by Wien but, for completeness, Einstein gives an elegant proof of the result as follows.

The function ϕ must be such that, in thermal equilibrium, the entropy is a maximum for a fixed value of the total energy, and so the problem can be written in terms of the calculus of variations:

$$\delta S = \delta \int_0^\infty \phi[u(\nu), \nu]\, \mathrm{d}\nu = 0, \tag{14.5}$$

subject to the constraint

$$\delta E = \delta \int_0^\infty u(\nu)\, \mathrm{d}\nu = 0, \tag{14.6}$$

where E is the total energy. Using the method of undetermined multipliers, the equation for the function $\phi(u)$ is given by

$$\int_0^\infty \left(\frac{\partial \phi}{\partial u} \delta u\, \mathrm{d}\nu - \lambda\, \delta u\, \mathrm{d}\nu \right) = 0,$$

where the undetermined multiplier λ is independent of frequency. The integrand must be zero to ensure that the integral is zero and so

$$\frac{\partial \phi}{\partial u} = \lambda.$$

Now suppose that the temperature of unit volume of black-body radiation is increased by an infinitesimal amount $\mathrm{d}T$, keeping the volume of the enclosure constant; we use differentials here to emphasis that we are now considering an infinitesimal change between *equilibrium* states. Then, the increase in entropy is

$$\mathrm{d}S = \int_{\nu=0}^{\nu=\infty} \frac{\partial \phi}{\partial u}\, \mathrm{d}u\, \mathrm{d}\nu.$$

But, we have just shown that $\partial\phi/\partial u = \lambda$ is independent of frequency and hence

$$\mathrm{d}S = \frac{\partial \phi}{\partial u}\, \mathrm{d}E,$$

where

$$\mathrm{d}E = \int_{\nu=0}^{\nu=\infty} \mathrm{d}u\, \mathrm{d}\nu. \tag{14.7}$$

But $\mathrm{d}E$ is just the energy added to cause the infinitesimal increase in temperature, and so we can use the thermodynamic relation (9.57) to write

$$\left(\frac{\partial S}{\partial E} \right)_V = \frac{1}{T}. \tag{14.8}$$

Therefore,

$$\frac{\partial\phi}{\partial u} = \frac{1}{T}. \tag{14.9}$$

This is the equation we have been seeking. Notice the pleasing symmetry between the following relations:

$$\begin{aligned} S &= \int_0^\infty \phi\,\mathrm{d}\nu & E &= \textstyle\int_0^\infty u(\nu)\,\mathrm{d}\nu, \\ \left(\frac{\partial S}{\partial E}\right)_V &= \frac{1}{T} & \frac{\partial\phi}{\partial u} &= \frac{1}{T}. \end{aligned} \tag{14.10}$$

Einstein now uses (14.9) to work out the entropy of black-body radiation. Rather than use Planck's formula, he uses Wien's formula since, although it is not correct for low frequencies and high temperatures, it is the correct law in the region in which the classical theory breaks down. Therefore, analysis of the entropy associated with the high-frequency part of the spectrum is likely to give insight into where the classical calculation has gone wrong.

First, Einstein writes down the form of Wien's law as derived from experiment. In the notation of (12.28),

$$u(\nu) = \frac{8\pi\alpha}{c^3}\frac{\nu^3}{\mathrm{e}^{\beta\nu/T}}. \tag{14.11}$$

From (14.11), we immediately find an expression for $1/T$, which, using (14.9), gives us

$$\frac{1}{T} = \frac{1}{\beta\nu}\ln\frac{8\pi\alpha\nu^3}{c^3 u(\nu)} = \frac{\partial\phi}{\partial u}. \tag{14.12}$$

An expression for ϕ is found by integration:

$$\frac{\partial\phi}{\partial u} = -\frac{1}{\beta\nu}\left(\ln u + \ln\frac{c^3}{8\pi\alpha\nu^3}\right),$$

and so

$$\begin{aligned} \phi &= -\frac{u}{\beta\nu}\left(\ln u - 1 + \ln\frac{c^3}{8\pi\alpha\nu^3}\right), \\ &= -\frac{u}{\beta\nu}\left(\ln\frac{uc^3}{8\pi\alpha\nu^3} - 1\right). \end{aligned} \tag{14.13}$$

Now Einstein makes clever use of this formula for ϕ. Consider the energy density of radiation in the spectral range ν to $\nu + \mathrm{d}\nu$ and suppose it has energy $E = Vu\,\mathrm{d}\nu$, where V is the volume. Then the entropy associated with this radiation is

$$S = V\phi\,\Delta\nu = -\frac{E}{\beta\nu}\left(\ln\frac{Ec^3}{8\pi\alpha\nu^3 V\,\mathrm{d}\nu} - 1\right). \tag{14.14}$$

Suppose the volume changes from V_0 to V, while the total energy remains constant. Then, the entropy change is

$$S - S_0 = \frac{E}{\beta\nu}\ln\frac{V}{V_0}. \tag{14.15}$$

But this formula looks familiar. Einstein recalls that this entropy change is exactly the same as that found in the Joule expansion of a perfect gas according to elementary statistical mechanics. Let us repeat the analysis of Section 10.7, which led to (10.55). Boltzmann's relation can be used to work out the entropy difference $S - S_0$ between the initial and final states, $S - S_0 = k \ln(W/W_0)$, where W_0 and W are the probabilities of these states. In the initial state, the system has volume V_0 and the particles move randomly throughout this volume. At any time, the probability that a single particle occupies a given smaller volume V is V/V_0 and hence the probability that all N end up together in the volume V is $(V/V_0)^N$. Therefore, the entropy difference for a gas of N particles is

$$S - S_0 = kN \ln(V/V_0). \tag{14.16}$$

Einstein notes that (14.15) and (14.16) are formally identical. He immediately concludes that the radiation behaves thermodynamically as if it consisted of discrete particles, their number N being equal to $E/(k\beta\nu)$. In Einstein's own words,

> Monochromatic radiation of low density (within the limits of validity of Wien's radiation formula) behaves thermodynamically as though it consisted of a number of independent energy quanta of magnitude $k\beta\nu$.

Rewriting this result in Planck's notation, since $\beta = h/k$ the energy of each quantum is $h\nu$.

Einstein then works out the average energy of the quanta according to Wien's formula for the black-body spectrum. The energy in the frequency interval ν to $\nu + \mathrm{d}\nu$ is E and the number of quanta is $E/(k\beta\nu)$. Therefore, the average energy is

$$\overline{E} = \frac{\int_0^\infty (8\pi\alpha/c^3)\nu^3\, \mathrm{e}^{-\beta\nu/T}\, \mathrm{d}\nu}{\int_0^\infty (8\pi\alpha/c^3)[\nu^3/(k\beta\nu)]\, \mathrm{e}^{-\beta\nu/T}\, \mathrm{d}\nu} = k\beta \frac{\int_0^\infty \nu^3\, \mathrm{e}^{-\beta\nu/T}\, \mathrm{d}\nu}{\int_0^\infty \nu^2\, \mathrm{e}^{-\beta\nu/T}\, \mathrm{d}\nu}.$$

Integrating the denominator by parts,

$$\int_0^\infty \nu^2\, \mathrm{e}^{-\beta\nu/T}\, \mathrm{d}\nu = \left[\frac{\nu^3}{3}\, \mathrm{e}^{-\beta\nu/T}\right]_0^\infty + \frac{\beta}{3T}\int_0^\infty \nu^3\, \mathrm{e}^{-\beta\nu/T}\, \mathrm{d}\nu, \tag{14.17}$$

and hence

$$\overline{E} = k\beta \times \frac{3T}{\beta} = 3kT. \tag{14.18}$$

The average energy of the quanta is closely related to the mean kinetic energy per particle in the black-body enclosure, $\frac{3}{2}kT$. This also is a highly suggestive result.

So far, Einstein has stated that the radiation 'behaved as though' it consisted of a number of independent particles. Is this meant to be taken seriously, or is it just another 'formal device'? The last sentence of Section 6 of his paper leaves the reader in no doubt:

> The next obvious step is to investigate whether the laws of emission and transformation of light are also of such a nature that they can be interpreted or explained by considering light to consist of such energy quanta.

In other words, 'Yes, let us assume that they are real particles and see whether we can understand other phenomena.'

Einstein considers three phenomena which cannot be explained by classical electromagnetic theory: Stokes' rule of photoluminescence, the photoelectric effect and the ionisation of gases by ultraviolet light.

Stokes' rule is the experimental observation that the frequency of photoluminescent emission is less than the frequency of the incident light. This is now explained as a consequence of the conservation of energy. If the incoming quanta each have energy $h\nu_1$ then a re-emitted quantum can at most have this energy. If some of the energy of a quantum is absorbed by the material before re-emission then the re-emitted quantum has energy $h\nu_2 \leq h\nu_1$.

The photoelectric effect. This is probably the most famous result of the paper because Einstein makes a definite quantitative prediction on the basis of the theory expounded above. Ironically, the photoelectric effect had been discovered by Heinrich Hertz in 1887 in the same experiments which fully validated Maxwell's equations. Perhaps the most remarkable feature of the effect was Lénard's discovery that the energies of the electrons emitted from the metal surface are independent of the intensity of the incident radiation.

Einstein's proposal provided an immediate solution to this problem. Radiation of a given frequency consists of quanta of the same energy $h\nu$. If one of these is absorbed by the material, the electron may receive sufficient energy to remove it from the surface against the forces which bind it to the material. If the intensity of the light is increased then more electrons are ejected, but their energies remain unchanged. Einstein wrote this result in the following form. The maximum kinetic energy which the ejected electron can have, E_k, is

$$E_\mathrm{k} = h\nu - W, \tag{14.19}$$

where W is the amount of work necessary to remove the electron from the surface of the material, what is called the *work function* of the material. Experiments to estimate the magnitude of the work function involve placing the photocathode in an opposing potential so that, when the potential reaches some value V, the ejected electrons can no longer reach the anode and the photoelectric current falls to zero. This occurs at the potential at which $E_\mathrm{k} = eV$. Therefore,

$$V = \frac{h}{e}\nu - \frac{W}{e}. \tag{14.20}$$

In Einstein's words,

> If the formula derived is correct, then V must be a straight line function of the frequency of the incident light, when plotted in Cartesian coordinates, whose slope is independent of the nature of the substance investigated.

Thus, the quantity h/e, the ratio of Planck's constant to the electronic charge, can be found directly from the slope of this relation. These were remarkable predictions because nothing was known at that time about the dependence of the photoelectric effect upon the frequency of the incident radiation. After ten years of difficult experimentation, all aspects of Einstein's equation were fully confirmed experimentally. In 1916, Millikan was able to summarise the results of his extensive experiments:

> Einstein's photoelectric equation has been subjected to very searching tests and it appears in every case to predict exactly the observed results.[12]

Photoionisation of gases. The third piece of experimental evidence discussed in Einstein's paper was the fact that the energy of each photon has to be greater than the ionisation potential of the gas if photoionisation is to take place. He showed that the smallest energy quanta for the ionisation of air were approximately equal to the ionisation potential determined independently by Stark. Once again, the quantum hypothesis is in agreement with experiment.

At this point, the paper ends. It is one of the great papers in physics and is the work described in Einstein's Nobel Prize citation.

14.3 The quantum theory of solids

In 1905, Planck and Einstein had somewhat different views about the role of quantisation and quanta. Planck had quantised the oscillators and there is no mention of this approach in Einstein's paper. Indeed, it seems that Einstein was not at all clear that they were actually describing the same phenomenon. In 1906, however, he showed that the two approaches were, in fact, the same.[13] Then, in a paper submitted to *Annalen der Physik* in November 1906 and published in 1907,[14] he came to the same conclusion by a different argument, and went on to extend the idea of quantisation to solids.

In the first of these papers, Einstein asserts that he and Planck are actually describing the same phenomena of quantisation.

> At that time [1905] it seemed to me as though Planck's theory of radiation formed a contrast to my work in a certain respect. New considerations which are given in the first section of this paper demonstrate to me, however, that the theoretical foundation on which Planck's radiation theory rests differs from the foundation that would result from Maxwell's theory and the electron theory and indeed differs exactly in that Planck's theory implicitly makes use of the hypothesis of light quanta just mentioned.[13]

These arguments are developed further in the paper of 1907.[14] Einstein demonstrated that if Planck had followed the proper Boltzmann procedure then he would have still obtained the correct formula for black-body radiation, whilst maintaining the assumption that the oscillators can only take definite energies, $0, \epsilon, 2\epsilon, 3\epsilon, \ldots$. Let us repeat Einstein's argument, which now appears in all the standard textbooks.

If Planck had followed Boltzmann's procedures, he would have ended up with the Boltzmann expression for the probability that a state of energy $E = r\epsilon$ is occupied, even though the limit $\epsilon \to 0$ is not taken, as we demonstrated in Section 13.2:

$$p(E) \propto \mathrm{e}^{-E/kT}.$$

Einstein assumed that the energy of an oscillator is quantised in units of ϵ. Thus, if there are N_0 oscillators in the ground state, the number in the $r = 1$ state is $N_0\,\mathrm{e}^{-\epsilon/kT}$, the number in the $r = 2$ state is $N_0\mathrm{e}^{-2\epsilon/kT}$ and so on. Therefore, the average energy of the oscillator is

$$\overline{E} = \frac{N_0 \times 0 + \epsilon N_0 \,\mathrm{e}^{-\epsilon/kT} + 2\epsilon N_0 \,\mathrm{e}^{-2\epsilon/kT} + \cdots}{N_0 + N_0 \,\mathrm{e}^{-\epsilon/kT} + N_0 \,\mathrm{e}^{-2\epsilon/kT} + \cdots}$$

$$= \frac{N_0 \epsilon \,\mathrm{e}^{-\epsilon/kT}\left[1 + 2\mathrm{e}^{-\epsilon/kT} + 3(\mathrm{e}^{-\epsilon/kT})^2 + \cdots\right]}{N_0\left[1 + \mathrm{e}^{-\epsilon/kT} + (\mathrm{e}^{-\epsilon/kT})^2 + \cdots\right]}. \tag{14.21}$$

We recall the following series:

$$\frac{1}{1-x} = 1 + x + x^2 + x^3 + \cdots,$$
$$\frac{1}{(1-x)^2} = 1 + 2x + 3x^2 + \cdots, \tag{14.22}$$

and hence we can see that the mean energy of the oscillator is

$$\overline{E} = \frac{\epsilon \,\mathrm{e}^{-\epsilon/kT}}{1 - \mathrm{e}^{-\epsilon/kT}} = \frac{\epsilon}{\mathrm{e}^{\epsilon/kT} - 1}. \tag{14.23}$$

Thus, using the proper Boltzmann procedure, Planck's relation for the mean energy of an oscillator is recovered, provided that the energy element ϵ does not vanish. Einstein's approach indicates clearly the origin of the departure from the classical result. The mean energy $\overline{E} = kT$ is recovered from (14.23) in the classical limit $\epsilon \to 0$. Notice that, by allowing $\epsilon \to 0$, the averaging takes place over a continuum of energies which the oscillator might take. To put it another way, it is assumed that equal volumes of phase space are given equal weights in the averaging process, and this is the origin of the classical equipartition theorem. Einstein shows that Planck's formula requires that this assumption is wrong. Rather, only those volumes of phase space with energies $0, \epsilon, 2\epsilon, 3\epsilon \ldots$ * should have non-zero weights and these should all be equal.

Einstein then relates this result directly to his previous paper on light quanta:

> We must assume that for ions which can vibrate at a definite frequency and which make possible the exchange of energy between radiation and matter, the manifold of possible states must be narrower than it is for the bodies in our direct experience. We must in fact assume that the mechanism of energy transfer is such that the energy can assume only the values $0, \epsilon, 2\epsilon, 3\epsilon \ldots$.[15]

But this is only the beginning of the paper. Much more is to follow – Einstein puts it beautifully.

> I now believe that we should not be satisfied with the result. For the following question forces itself upon us. If the elementary oscillators that are used in the theory of the energy exchange between radiation and matter cannot be interpreted in the sense of the present kinetic molecular theory, must we not also modify the theory for the other oscillators that are used in the molecular theory of heat? There is no doubt about the answer, in my opinion. If Planck's theory of radiation strikes to the heart of the matter, then we must also expect to find contradictions between the present kinetic molecular theory and experiment in other areas of the theory of heat, contradictions that can be resolved by the route just traced. In my opinion, this is actually the case, as I try to show in what follows.[16]

* According to quantum mechanics, the oscillator energies are each increased by an amount $\frac{1}{2}\epsilon = \frac{1}{2}h\nu$, the zero-point energy; this does not affect the argument.

This is the paper often referred to as Einstein's application of the quantum theory to solids, but, just as in the case of his paper on the photoelectric effect, this paper is very much deeper than this name suggests and strikes right at the heart of the quantum nature of matter and radiation.

The problem discussed by Einstein concerns the heat capacities of solids. According to the *Dulong and Petit law*, the heat capacity per mole of a solid is about $3R$. This result can be derived simply from the equipartition theorem. The model of the solid consists of N_A atoms per mole and it is supposed that they can all vibrate in the three independent directions, x, y, z. According to the equipartition theorem, the internal energy per mole of the solid should therefore be $3N_A kT$, since each independent mode of vibration is awarded a total energy kT. The heat capacity per mole follows directly by differentiation: $C = \partial U/\partial T = 3N_A kT = 3R$.

It was known that some materials do not obey the Dulong and Petit law, in that they have significantly smaller heat capacities than $3R$ – this was particularly true for light elements such as carbon, boron and silicon. In addition, by 1900, it was known that the heat capacities of some elements change rapidly with temperature and only attain the value $3R$ at high temperatures.

The problem is readily solved if Einstein's quantum hypothesis is adopted. For oscillators, the classical formula for the average energy of the oscillator, kT, should be replaced by the quantum formula

$$\overline{E} = \frac{h\nu}{e^{h\nu/kT} - 1}.$$

Now atoms are complicated systems, but let us suppose for simplicity that, for a particular material, they all vibrate at the same frequency, the *Einstein frequency* ν_E, and that these vibrations are independent. Since each atom has three independent modes of vibration, the internal energy is then

$$U = 3N_A \frac{h\nu_E}{e^{h\nu_E/kT} - 1} \tag{14.24}$$

and the heat capacity is

$$\begin{aligned} \frac{dU}{dT} &= 3N_A h\nu_E \frac{1}{(e^{h\nu_E/kT} - 1)^2} e^{h\nu_E/kT} \frac{h\nu_E}{kT^2} \\ &= 3R \left(\frac{h\nu_E}{kT}\right)^2 \frac{e^{h\nu_E/kT}}{(e^{h\nu_E/kT} - 1)^2}. \end{aligned} \tag{14.25}$$

This expression turns out to provide a remarkably good fit to the measured variation of the heat capacity with temperature. In his paper, Einstein compared the experimentally determined variation of the heat capacity of diamond with (14.25), with the results shown in Fig. 14.1. The decrease in the heat capacity at low temperatures is apparent, although the experimental points lie slightly above the predicted relation at low temperatures.

We can now understand why light elements have smaller heat capacities than the heavier elements. Presumably, the lighter atoms have higher vibrational frequencies than heavier atoms and hence, at a given temperature, ν_E/T is larger and the heat capacity is smaller. To

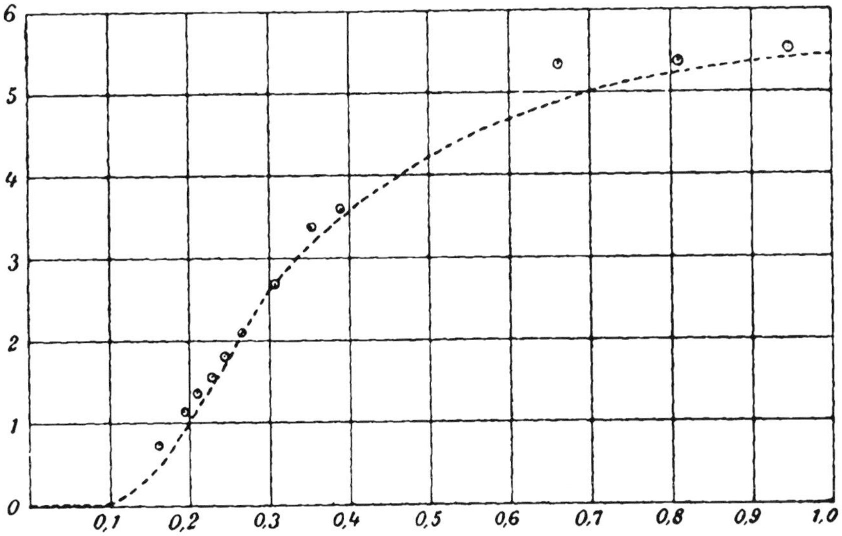

Figure 14.1: The variation in the heat capacity of diamond with temperature, compared with the prediction of Einstein's quantum theory. The abscissa is T/θ_{E}, where $k\theta_{\mathrm{E}} = h\nu_{\mathrm{E}}$, and the ordinate is the molar heat capacity in calories mole^{-1}. This diagram appears in Einstein's paper of 1907[14] and uses the results of H.F. Weber, which were listed in the tables of Landolt and Börnstein.

account for the experimental data for diamond, shown in Fig. 14.1, the frequency ν_{E} must lie in the infrared waveband. As a result, all vibrations at higher frequencies make only a vanishingly small contribution to the heat capacity. As expected, there is strong absorption in diamond at infrared wavelengths, corresponding to frequencies $\nu \approx \nu_{\mathrm{E}}$. Einstein compares his estimates of ν_{E} with the strong absorption features observed in a number of materials and finds remarkable agreement, granted the simplicity of the model.

Perhaps the most important prediction of the theory was that the heat capacities of all solids should decrease to zero at low temperatures, as indicated in Fig. 14.1. This was of key importance from the point of view of furthering the acceptance of Einstein's ideas, because, at about this time, Walther Nernst began a series of experiments to measure the heat capacities of solids at low temperatures. Nernst's motivation for undertaking these experiments was to test his *heat theorem*, or the *third law of thermodynamics*, which he had developed theoretically in order to understand the nature of chemical equilibria. The heat theorem enabled the calculation of chemical equilibria to be carried out precisely and also led to the prediction that the heat capacities of all materials should tend to zero at low temperatures. As recounted by Frank Blatt,

> ...shortly after Einstein had assumed a junior faculty position at the University of Zurich [in 1909], Nernst paid the young theorist a visit so that they could discuss problems of common interest. The chemist George Hevsey ... recalls that among his colleagues it was this visit by Nernst that raised Einstein's reputation. He had come as an unknown man to Zurich. Then Nernst came, and the people at Zurich said, 'This Einstein must be a clever fellow, if the great Nernst comes so far from Berlin to Zurich to talk to him'.[17]

14.4 Debye's theory of specific heats

Although Einstein took little interest in the heat capacities of solids after 1907, his ideas of quantisation were taken significantly further by Peter Debye in an important paper of 1912.[18] Einstein was well aware that the assumption that the atoms of a solid vibrate independently was a crude approximation. Debye took the opposite approach, of returning to a continuum picture almost identical to that developed by Rayleigh in his treatment of the spectrum of black-body radiation (Section 12.6). Debye realised that the collective modes of vibration of a solid could be represented by the complete set of normal modes which follows from fitting waves into a box, as described by Rayleigh. Each independent mode of oscillation of the solid as a whole should be awarded an average energy

$$\overline{E} = \frac{\hbar\omega}{\exp(\hbar\omega/kT) - 1}, \tag{14.26}$$

according to Einstein's prescription; ω is the angular frequency of vibration of the mode and $\hbar = h/(2\pi)$. The number of modes $\mathrm{d}\mathcal{N}$ in the angular frequency range $\mathrm{d}\omega$ had been evaluated by Rayleigh according to the procedure described by Fig. 12.5 and was shown to be

$$\mathrm{d}\mathcal{N} = \frac{L^3\omega^2}{2\pi^2 c_s^3}\,\mathrm{d}\omega, \tag{14.27}$$

where c_s is the speed of propagation of the waves in the material. Just as in the case of electromagnetic radiation, we need to determine how many independent polarisation states there are for the waves. In this case, there are two transverse modes and one longitudinal mode, corresponding to the independent directions in which the material can be stressed by the wave, and so in total there are $3\,\mathrm{d}\mathcal{N}$ modes in the internal $\mathrm{d}\omega$, each of which is awarded the energy (14.26). Debye makes the assumption that these modes have the same speed of propagation, which is independent of the frequency of the modes. Therefore, the total internal energy of the material is

$$\begin{aligned} U &= \int_0^{\omega_{\max}} \frac{\hbar\omega}{\exp(\hbar\omega/kT) - 1}\, 3\,\mathrm{d}\mathcal{N} \\ &= \frac{3}{2\pi^2}\left(\frac{kTL}{\hbar c_s}\right)^3 \int_0^{x_{\max}} \frac{x^3}{e^x - 1}\,\mathrm{d}x, \end{aligned} \tag{14.28}$$

where $x = \hbar\omega/kT$.

The problem is now to determine the value of $x_{\max}$. Debye introduced the idea that there must be a limit to the total number of modes in which energy could be stored. In the high-temperature limit, he argued that the total energy should not exceed that given by the classical equipartition theorem, namely $3N_A kT$ for one mole of the material, where N_A is Avogadro's constant. Since each mode of oscillation has energy kT in this limit, there should be a maximum of $3N_A$ modes in which energy is stored. Therefore, recalling that there are $3\mathrm{d}\mathcal{N}$ modes in the angular frequency range $\mathrm{d}\omega$, Debye's condition can be found by integrating (14.27):

$$3N_A = 3\int_0^{\omega_{\max}} \mathrm{d}\mathcal{N} = 3\int_0^{\omega_{\max}} \frac{L^3\omega^3}{2\pi^2 c_s^3}\,\mathrm{d}\omega,$$

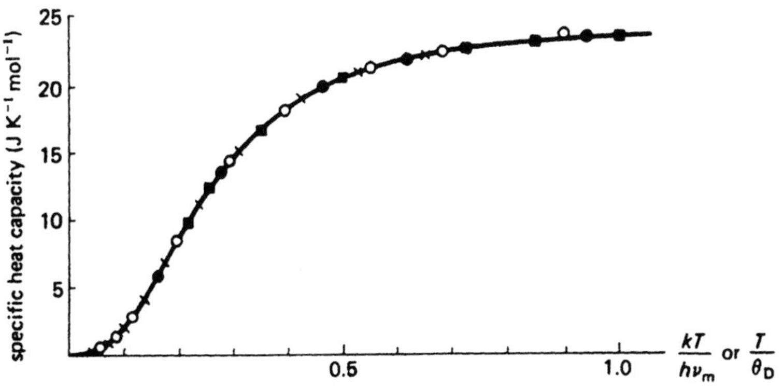

Figure 14.2: The molar heat capacity for various solids as a function of $kT/\hbar\omega_D$, where ω_D is the Debye angular frequency, compared with the predictions of the Debye theory. θ_D is the Debye temperature. The substances plotted are copper (open circles, $\theta_D = 315$ K), silver (solid circles, $\theta_D = 215$ K), lead (solid squares, $\theta_D = 88$ K) and carbon (crosses, $\theta_D = 1860$ K). (From D. Tabor, 1991, *Gases, Liquids and Solids and Other States of Matter*, p. 236, Cambridge: Cambridge University Press.)

yielding

$$\omega_{\max}^3 = \frac{6\pi^2 N_A}{L^3} c_s^3. \tag{14.29}$$

It is conventional to write $x_{\max} = \hbar\omega_{\max}/kT = \theta_D/T$, where θ_D is known as the *Debye temperature*. Therefore, the expression for the total internal energy of the material (14.28) can be rewritten

$$U = 9RT\left(\frac{T}{\theta_D}\right)^3 \int_0^{\theta_D/T} \frac{x^3}{e^x - 1}\,dx \tag{14.30}$$

for one mole of the material. This is the famous expression derived by Debye for the internal energy per mole of the solid.

To find the heat capacity, it is simplest to consider an infinitesimal increment dU associated with a single frequency ω and then integrate over x as before:

$$C = \frac{dU}{dT} = 9R\left(\frac{T}{\theta_D}\right)^3 \int_0^{\theta_D/T} \frac{x^4}{(e^x - 1)^2}\,dx. \tag{14.31}$$

This integral cannot be written in closed form, but it provides an even better fit to the data on the heat capacity of solids than Einstein's expression (14.25), as can be seen from Fig. 14.2.

As pointed out by Tabor,[19] there are two big advantages of the Debye theory. First, the value of the Debye temperature can be worked out from the bulk properties of the material using (14.29), whereas the Einstein temperature is an arbitrary parameter to be found by fitting to the data. Second, the expression (14.31) for the heat capacity gives a much improved fit to the data at low temperatures. If $T \ll \theta_D$, the upper limit to the integral in (14.31) can be set to infinity, and then the integral has the value $4\pi^4/15$. Therefore, at low temperatures $T \ll \theta_D$, the heat capacity is

$$C = \frac{dU}{dT} = \frac{12\pi^4}{5} R \left(\frac{T}{\theta_D}\right)^3 : \tag{14.32}$$

rather than decreasing exponentially at low temperatures, the heat capacity varies as T^3.

Two other points are of interest. Firstly, for many solids it turns out that $\theta_E \approx 0.75\theta_D$, $\nu_E \approx 0.75\nu_D$. Secondly, there is a simple interpretation of ω_{max} in terms of wave propagation in the solid. From (14.29), the maximum frequency is

$$\nu_{max} = \left(\frac{3}{4\pi}\right)^{1/3} \left(\frac{N_A}{L^3}\right)^{1/3} c_s, \tag{14.33}$$

for one mole of the solid, where N_A is Avogadro's constant. But $L/N_A^{1/3}$ is just the typical interatomic spacing, a. Therefore, (14.33) states that $\nu_{max} \approx c_s/a$, that is, the minimum wavelength of the waves $\lambda_{min} = c_s/\nu_{max} \approx a$. This makes a great deal of sense physically. On scales less than the interatomic spacing a, the concept of collective vibration of the atoms of the material ceases to have any meaning.

14.5 The specific heats of gases revisited

Einstein's quantum theory solves the problems of accounting for the specific heats of diatomic gases, which had caused Maxwell such distress (see Section 10.4). According to Einstein, all energies associated with atoms and molecules should be quantised, and then the distribution of particles among the quantised energy states depends upon the temperature of the gas. Einstein's analysis, given in Sections 14.2 and 14.3, showed that the average energy of an oscillator of frequency ν at temperature T is

$$\overline{E} = \frac{h\nu}{e^{h\nu/kT} - 1}. \tag{14.34}$$

If $kT \gg h\nu$, the average kinetic energy is kT, as expected from kinetic theory. If however $kT \ll h\nu$, the average energy tends to $h\nu\, e^{-h\nu/kT}$, which is negligibly small.

Let us therefore look again at the various modes by which energy can be stored by the molecules of a gas.

14.5.1 *Linear motion*

For atoms or molecules confined to a volume of, say, one cubic metre, the energy levels are quantised but their spacing is absolutely tiny. To anticipate de Broglie's brilliant insight of 1923, a wavelength λ can be associated with the momentum p of a particle, such that $p = h/\lambda$, where λ is known as the *de Broglie wavelength* and h is Planck's constant. Therefore, quantum effects only become important on the metre scale at energies $\epsilon = p^2/(2m) \sim h^2/(2m\lambda^2) = 3 \times 10^{-41}$ J. Setting ϵ equal to $\frac{3}{2}kT$, we find $T = 10^{-18}$ K. Thus, for all accessible temperatures $kT \gg \epsilon$ and we need not worry about the effects of quantisation on the linear motion of molecules. Therefore, for the atoms or molecules of a gas, there are always the classical three degrees of freedom associated with their linear motion.

14.5.2 Molecular rotation

Energy can also be stored in the rotational motion of a molecule. The amount of energy which can be stored can be found from a basic result of quantum mechanics, that orbital angular momentum J is quantised such that only discrete values are allowed according to the relation

$$J = [j(j+1)]^{1/2}\hbar,$$

where $j = 0, 1, 2, \ldots$ and $\hbar = h/2\pi$. We recall that, if the particles have *intrinsic angular momentum*, or *spin*, then j can take half-integral values as well, but we will restrict our arguments to bulk rotational motion. The rotational energy of the molecule is given by the analogue of the classical relation between energy and angular momentum

$$E = \frac{J^2}{2I},$$

where I is the moment of inertia of the molecule about the rotation axis. Therefore

$$E = \frac{j(j+1)}{2I}\hbar^2. \tag{14.35}$$

If we write the rotational energy of the molecule as $E = \frac{1}{2}I\omega^2$ then the angular frequency of rotation ω is

$$\omega = \frac{[j(j+1)]^{1/2}\hbar}{I}.$$

The first rotational state of the molecule corresponds to $j = 1$ and so the first excited state has energy $E = \hbar^2/I$. Therefore, the condition that rotational motion contributes to the storage of energy at temperature T is

$$kT \geq E = \frac{\hbar^2}{I},$$

which is equivalent to $kT \geq \hbar\omega$.

Let us consider the cases of hydrogen and chlorine molecules and work out the temperatures at which the energy E_1 of the first excited state is equal to kT. Table 14.1 shows the results of these calculations.

Thus, for hydrogen, at room temperature energy can be stored in its rotational degrees of freedom, but at low temperatures, $T \ll 173$ K, the rotational degrees of freedom are 'frozen out', or 'quenched', and do not contribute to the heat capacity of the gas. For the case of chlorine, the temperature at which $E_1 = kT$ is so low that the rotational degrees of freedom contribute to the heat capacity at all temperatures at which it is gaseous.

There is one further key issue. For a rotating *diatomic molecule*, which of its three rotational degrees of freedom are actually excited? Figure 14.3 shows rotational motion

Table 14.1. *Rotational properties of the diatomic molecules hydrogen and chlorine, including the first excitation energy* E_1

Property	Hydrogen	Chlorine
Mass of atom m in amu	1	35
Bond length d in nm	0.0746	0.1988
Moment of inertia		
$I = \frac{1}{2}md^2$ in kg m^2	4.65×10^{-48}	1.156×10^{-45}
Energy E_1 in J	2.30×10^{-21}	9.62×10^{-24}
$T = E_1/k$	173 K	0.697 K

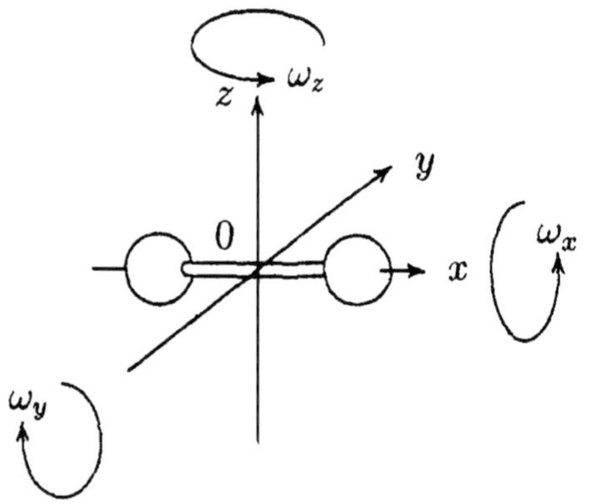

Figure 14.3: Illustrating the modes by which angular momentum can be stored in a linear diatomic molecule.

about the three perpendicular axes, with angular frequencies ω_x, ω_y and ω_z. The calculation performed above refers to the moments of inertia about the y- and z- axes. It is certainly *not* the correct calculation, however, for the moment of inertia about the x-axis. All the mass of the atoms is concentrated in their nuclei and so the moment of inertia about the x-axis is absolutely tiny compared with those about the y- and z- axes. In fact, it is about 10^{10} times smaller than that about the other two axes. Therefore, the temperature needed to excite rotation about the x-axis is about 10^{10} times greater than the others and so this mode is not excited at the temperatures at which the molecules can exist.

Thus, for diatomic molecules such as hydrogen and oxygen, there are two rotational degrees of freedom which can be excited at room temperatures. Therefore, five degrees of freedom f in total are excited, three translational and two rotational, and the ratio of heat capacities is expected to be $\gamma = (f + 2)/f = 1.4$ and $C_V = 2.5R$. This solves Maxwell's problem.

This argument can be extended to molecules which have finite moments of inertia about all three axes and then there would be three rotational degrees of freedom and $\gamma = 1.33$, as found by Maxwell.

14.5.3 *Molecular vibration*

To complete the analysis, consider the quantisation of the vibrational modes of molecules. A diatomic molecule behaves like a linear oscillator and so has two degrees of freedom, one associated with its kinetic energy and the other with its potential energy. As in the case of rotational motion, however, energy can only be stored in these degrees of freedom at a given temperature if the vibrational states of the molecule can be excited. Thus, we need to know the vibrational frequencies of the molecules in their ground states. Let us consider again the cases of hydrogen and chlorine molecules.

In the case of hydrogen, its vibrational frequency is about 2.6×10^{14} Hz and so $\epsilon = h\nu = 1.72 \times 10^{-19}$ J. Setting $h\nu = kT$, we see that $T = 12\,500$ K and so, for hydrogen, the vibrational modes cannot contribute to the heat capacity at room temperature. In fact, the vibrational modes scarcely come into play at all, because hydrogen molecules are dissociated at about 2000 K.

In the case of chlorine, the atoms are much heavier and the ground-state vibrational frequency is about an order of magnitude smaller. Thus, the vibrational modes come into play at temperatures above about 600 K. At high temperatures, the degrees of freedom of the chlorine molecule which come into play are as follows:

$$\begin{aligned} &\text{three translational} + \text{two rotational} + \text{two vibrational} \\ &\qquad = \text{seven degrees of freedom.} \end{aligned}$$

Thus, the heat capacity of chlorine at very high temperatures tends to $\frac{7}{2}R$.

14.5.4 *The cases of hydrogen and chlorine*

We are now in a position to understand the details of Fig. 14.4, which shows the variation in the heat capacities of hydrogen and chlorine with temperature. For hydrogen at low temperatures, only the translational degrees of freedom can store energy and the heat capacity is $\frac{3}{2}R_0$. Above about 170 K, the rotational modes can store energy and so there are five degrees of freedom in play. Above 2000 K the vibrational modes begin to have a role, but the molecules are dissociated at about 2000 K before these modes can come completely into play.

For chlorine, even at low temperatures, both the translational and rotational modes are excited and so the heat capacity is $\frac{5}{2}R_0$, although, as noted in the diagram, chlorine begins to liquefy between 200 and 300 K. Above 600 K, the vibrational modes begin to have a role and at the highest temperatures, the heat capacity becomes equal to $\frac{7}{2}R_0$. The corresponding values for the ratios of heat capacities are found from the rule $\gamma = (C_V + R)/C_V$.

Thus, Einstein's quantum theory can account in detail for the properties of ideal gases and solve completely the problems which had frustrated Maxwell and Boltzmann.

14.6 Conclusion

Einstein's revolutionary ideas on quanta and quantisation predate the discovery of wave and quantum mechanics by Schrödinger and Heisenberg in the 1920s by about 20 years.

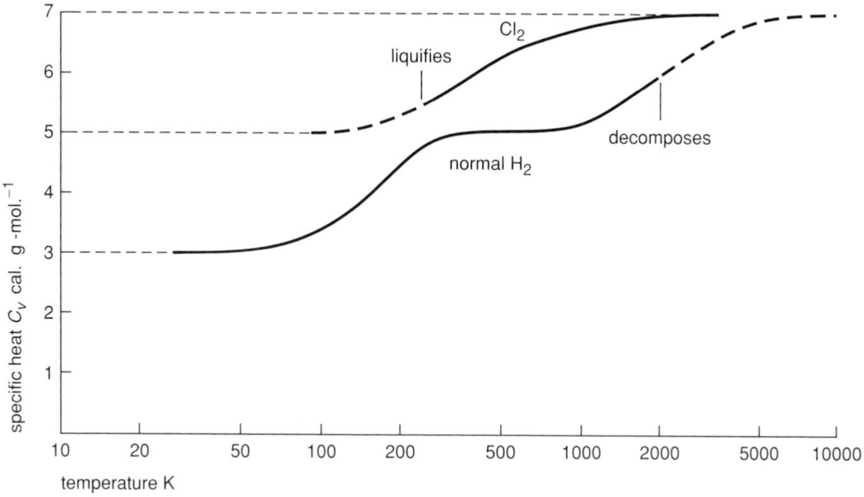

Figure 14.4: The variation of the molar heat capacity with temperature for molecular hydrogen and chlorine. The units on the ordinate are calories per mole. One calorie = 4.184 J. Therefore, $\frac{3}{2} R_0 = \frac{3}{2} \times 8.31$ J mole^{-1} = 12.465 J mole^{-1} = 3.0 calories mole^{-1}. (See D. Tabor, 1991, *Gases, Liquids and Solids and Other States of Matter*, p. 110. Cambridge: Cambridge University Press.)

Consequently, to physicists today, Einstein's arguments appear schematic, rather than formally precise in the sense of being derived from a fully developed theory of quantum processes. New concepts and approaches were needed to provide a complete description of quantum mechanical systems. Despite these technical problems, Einstein's fundamental insights were indeed revolutionary in opening up completely new realms for physics – all physical processes are basically quantum in nature, despite the fact that classical physics is successful in explaining so much.

14.7 References

1 Einstein, A. (1905a). *Ann. Phys.*, **17**, 547.
2 Einstein, A. (1905b). *Ann. Phys.*, **17**, 891.
3 Einstein, A. (1905c). *Ann. Phys.*, **17**, 132.
4 Stachel, J. ed. (1998). *Einstein's Miraculous Year: Five Papers That Changed the Face of Physics*. Princeton: Princeton University Press.
5 Einstein, A. (1900–1909). *The Collected Papers of Albert Einstein*: Vol. 2, *The Swiss Years: Writings, 1900–1909*, eds. J. Stachel and D.C. Cassidy (English transl. supplement). Princeton: Princeton University Press.
6 Einstein, A. (1906). *Ann. Phys.*, **19**, 371.
7 Einstein, A. (1979). *Autobiographical Notes*, translated and edited by P.A. Phillips, pp. 44–5. La Salle, Illinois: Open Court.
8 Perrin, J. (1909). *Ann. Chem. Phys.*, **18**, 1.

9 Perrin, J. (1910). *Brownian Movement and Molecular Reality* (translated by F. Soddy). London: Taylor and Francis.
10 Einstein, A. (1905d). *Annalen der Physik*, **17**, 639.
11 Einstein, A. (1995). *The Collected Papers of Albert Einstein:* Vol. 5., p. 20 (English trans. supplement). Princeton: Princeton University Press.
12 Millikan, R.A. (1916). *Phys. Rev.*, **7**, 18.
13 Einstein, A. (1906). *Ann. Phys.*, **20**, 199.
14 Einstein, A. (1907). *Ann. Phys.*, **22**, 180.
15 Einstein, A. (1907). *op. cit.*, pp. 183–4.
16 Einstein, A. (1907). *op. cit.*, p. 184.
17 Blatt, F.J. (1992). *Modern Physics*. New York: McGraw-Hill.
18 Debye, P. (1912). *Annalen der Physik*, **39**, 789.
19 Tabor, D. (1991). *Gases, Liquids and Solids and Other States of Matter*. Cambridge: Cambridge University Press.

15 The triumph of the quantum hypothesis

15.1 The situation in 1909

In no sense were Einstein's startling new ideas on light quanta immediately accepted by the scientific community at large. Most of the major figures in physics rejected the idea that light could be considered to be made up of discrete quanta. In a letter to Einstein in 1907, Planck wrote:

> I look for the significance of the elementary quantum of action (light quantum) not in vacuo but rather at points of absorption and emission and assume that processes in vacuo are accurately described by Maxwell's equations. At least, I do not yet find a compelling reason for giving up this assumption which for the time being seems to be the simplest.[1]

Planck continued to reject the light-quantum hypothesis as late as 1913. In 1909, Lorentz, who was generally regarded as the leading theoretical physicist in Europe and whom Einstein held in the highest esteem, wrote:

> While I no longer doubt that the correct radiation formula can only be reached by way of Planck's hypothesis of energy elements, I consider it highly unlikely that these energy elements should be considered as light quanta which maintain their identity during propagation.[2]

Einstein never deviated from his conviction concerning the reality of quanta and continued to find other ways in which the experimental features of black-body radiation would lead inevitably to the conclusion that light consists of quanta. In one of his most beautiful papers, written in 1909,[3] he showed how fluctuations in the intensity of the black-body radiation spectrum provide further evidence on the quantum nature of light. I consider this to be a paper of the very highest inspiration. The subject of fluctuations often causes problems for students and so let us first revise some elementary ideas on the statistical fluctuations of particles and waves before studying Einstein's paper.

15.2 Fluctuations of particles in a box

Let us deal first of all with the problem of particles in a box. The box is divided into N cells and a large number of particles n is distributed randomly among them. The numbers of particles in each cell are then counted. If n is very large, the number of particles in each cell is roughly the same, but there is a real scatter about the mean value because of statistical fluctuations.

Let us recall how the exact expression for these variations is derived. In the simplest case of tossing coins, we ask, 'What is the probability of obtaining x heads (or successes) in n throws?' In each throw, the probability of success $p = \frac{1}{2}$ and that of failure $q = \frac{1}{2}$, so that $p + q = 1$. If we throw two coins, the possible outcomes of the experiment are

HH, HT, TH, TT

Since the order is not important, the frequencies of occurrence of the different combinations are:

HH	H T	TT
1	2	1

giving probabilities $\frac{1}{4}, \frac{1}{2}, \frac{1}{4}$ for two coins. Similarly, for three coins, the possible combinations are

HHH, HHT, HTH, THH, HTT, THT, TTH, TTT

giving

HHH	HH T	H TT	TTT
1	3	3	1

that is, probabilities $\frac{1}{8}, \frac{3}{8}, \frac{3}{8}, \frac{1}{8}$. It is well known that these probabilities correspond to the coefficients of t in the binomial expansions of $(p + qt)^2$ and $(p + qt)^3$ with $p = q = \frac{1}{2}$.

There is an alternative way of considering these probabilities. Suppose we ask, 'What is the probability of tossing one head a 'success' and then two tails ('failures')?' The answer is plainly $pq^2 = \frac{1}{8}$ for this particular ordering. However, if the order does not matter, then we need to know the number of ways in which we could have obtained one success and two failures. This is exactly the same problem that we discussed in connection with the derivation of the Boltzmann distribution (Section 13.2). The number of different ways of arranging n objects, x of which are identical to each other and another y of which are also identical to each other, is $n!/(x!y!)$. In the present case, $y = n - x$ and hence the answer is $n!/x!(n - x)!$. Thus, in the second example above, the number of ways is $3!/(1!2!) = 3$, that is, the total probability is $\frac{3}{8}$, as can readily be seen by writing out the possibilities.

It immediately follows that, in general, the probability of x successes out of n events is $n!/[x!(n - x)!]\, p^x q^{n-x}$, which is just the coefficient of the term in t^x in the expansion of $(q + pt)^n$. Thus, if we write this probability as $P_n(x)$,

$$P_n(x) = \frac{n!}{(n - x)!x!} p^x q^{n-x}, \tag{15.1}$$

then we can write

$$(q + pt)^n = P_n(0) + P_n(1)t + P_n(2)t^2 + \cdots + P_n(x)t^x + \cdots + P_n(n)t^n. \tag{15.2}$$

Setting $t = 1$, we obtain

$$1 = P_n(0) + P_n(1) + P_n(2) + \cdots + P_n(x) + \cdots + P_n(n), \tag{15.3}$$

showing that the total probability is 1, as it must be.

Now let us take the derivative with respect to t of (15.2):

$$\begin{aligned} np(q + pt)^{n-1} &= P_n(1) + 2P_n(2)t + \cdots + xP_n(x)t^{x-1} \\ &\quad + \cdots + nP_n(n)t^{n-1}. \end{aligned} \tag{15.4}$$

Setting $t = 1$,

$$pn = \sum_{x=0}^{n} xP_n(x). \tag{15.5}$$

But the quantity on the right-hand side is just the average value of x, that is, $\bar{x} = pn$, which agrees with our intuition.

Let us repeat the procedure to find the variance of the distribution. Taking the derivative of (15.4) with respect to t,

$$\begin{aligned} p^2 n(n-1)(q + pt)^{n-2} &= 2P_n(2) + \cdots + x(x-1)P_n(x)t^{x-2} + \cdots \\ &\quad + n(n-1)P_n(n)t^{n-2}. \end{aligned} \tag{15.6}$$

Again setting $t = 1$, we find

$$\begin{aligned} p^2 n(n-1) &= \sum_{x=0}^{n} x(x-1)P_n(x) = \sum_{x=0}^{n} x^2 P_n(x) - \sum_{x=0}^{n} xP_n(x) \\ &= \sum_{x=0}^{n} x^2 P_n(x) - np, \end{aligned} \tag{15.7}$$

that is,

$$\sum_{x=0}^{n} x^2 P_n(x) = np + p^2 n(n-1). \tag{15.8}$$

Now $\sum_{x=0}^{n} x^2 P_n(x)$ is a measure of the variance of the distribution of x, but it is measured with respect to the origin rather than the mean. Fortunately, there is a rule which tells us how to measure the variance with respect to the mean:

$$\sigma^2 = \sum_{x=0}^{n} x^2 P_n(x) - \bar{x}^2. \tag{15.9}$$

Therefore

$$\begin{aligned} \sigma^2 &= \sum_{x=0}^{n} x^2 P_n(x) - (pn)^2 = np + p^2 n(n-1) - (pn)^2 \\ &= np(1-p) = npq. \end{aligned} \tag{15.10}$$

Finally, we convert (15.1) from a discrete to a continuous distribution by a procedure which is carried in all the standard textbooks on statistics. The result is the normal, or

Gaussian, distribution $p(x)\,\mathrm{d}x$, which can be written

$$p(x)\,\mathrm{d}x = \frac{1}{(2\pi\sigma^2)^{1/2}} \exp\left(-\frac{x^2}{2\sigma^2}\right)\mathrm{d}x. \tag{15.11}$$

For this distribution, σ^2 is the variance and has the value npq; x is measured with respect to the mean value np. We have already met this distribution in the context of the Maxwell–Boltzmann velocity distribution (10.40).

Equation (15.10) now gives us the answer we have been seeking. If the box is divided into distinct N cells, the probability of a particle being in a particular cell in one experiment is $p = 1/N$, giving $q = 1 - 1/N$. The total number of particles is n. Therefore, the average number of particles per sub-box is n/N and the variance about this mean value, that is, the mean squared statistical fluctuation about the mean, is

$$\sigma^2 = \frac{n}{N}\left(1 - \frac{1}{N}\right). \tag{15.12}$$

If N is large, $\sigma^2 = n/N$ and is just the average number of particles in each cell, that is, $\sigma = (n/N)^{1/2}$. Notice the well-known result that, for large values of N, the mean is equal to the variance.

This is the origin of the useful rule that the fractional fluctuation about the average value is $1/M^{1/2}$, where M is the number of discrete objects counted. This is the statistical behaviour we would expect of particles in a box.

15.3 Fluctuations of randomly superposed waves

The random superposition of waves is different in important ways. Suppose the electric field $\boldsymbol{E}$ at some point in space is the random superposition of the electric fields from N sources, where N is very large. For simplicity, we consider only propagation in the z-direction and electric fields polarised in the x-direction. We also assume that the frequencies ν and the amplitudes ξ of all the waves are the same, the only difference being their random phases. Then, the quantity $E_x^* E_x = |\boldsymbol{E}|^2$ is proportional to the Poynting vector flux density in the z-direction and so is proportional to the energy density u of the radiation; E_x^* is the complex conjugate of E_x. Summing all the waves and assuming they have random phases, $E_x^* E_x$ can be written

$$E_x^* E_x = \left(\xi \sum_k \mathrm{e}^{\mathrm{i}\varphi_k}\right)^* \left(\xi \sum_j \mathrm{e}^{\mathrm{i}\varphi_j}\right) = \xi^2 \left(\sum_k \mathrm{e}^{-\mathrm{i}\varphi_k}\right)\left(\sum_j \mathrm{e}^{\mathrm{i}\varphi_j}\right) \tag{15.13}$$

$$= \xi^2 \left(N + \sum_{j\neq k} \mathrm{e}^{\mathrm{i}(\varphi_j - \varphi_k)}\right)$$

$$= \xi^2 \left[N + 2\sum_{j>k} \cos(\varphi_j - \varphi_k)\right]. \tag{15.14}$$

The average of the term in $\cos(\varphi_j - \varphi_k)$ is zero, since the phases of the waves are random, and therefore

$$\langle E_x^* E_x \rangle = N\xi^2 \propto u. \tag{15.15}$$

This is a familiar result: for incoherent radiation, that is, for waves with random phases, the total energy density is given by the sum of the energies in all the waves.

Let us now work out the fluctuations in the average energy density of the waves. We need to work out the difference between the quantity $\langle (E_x^* E_x)^2 \rangle$ and the square of the mean value (15.15). We recall that $\langle \Delta n^2 \rangle = \langle n^2 \rangle - \langle \overline{n} \rangle^2$ and hence

$$\Delta u^2 \propto \langle (E_x^* E_x)^2 \rangle - \langle E_x^* E_x \rangle^2. \tag{15.16}$$

Now

$$\begin{aligned}\langle (E_x^* E_x)^2 \rangle &= \xi^4 \left(N + \sum_{j \neq k} \mathrm{e}^{\mathrm{i}(\varphi_j - \varphi_k)} \right)^2 \\ &= \xi^4 \left(N^2 + 2N \sum_{j \neq k} \mathrm{e}^{\mathrm{i}(\varphi_j - \varphi_k)} + \sum_{l \neq m} \mathrm{e}^{\mathrm{i}(\varphi_l - \varphi_m)} \sum_{j \neq k} \mathrm{e}^{\mathrm{i}(\varphi_j - \varphi_k)} \right).\end{aligned} \tag{15.17}$$

The second term on the right-hand side again averages to zero because the phases are random. In the double sum, most of the constituent terms average to zero, because the phases are random, but not all of them: those terms for which $l = k$ and $m = j$ do not vanish. Recalling that $l = m$ is excluded from the summation, the matrix of l, m values that give non-vanishing contributions $(E_x^* E_x)$ is

	$l \rightarrow$					
m	–	2, 1	3, 1	4, 1	...	N, 1
$\downarrow$	1, 2	–	3, 2	4, 2	...	N, 2
	1, 3	2, 3	–	4, 3	...	N, 3
	1, 4	2, 4	3, 4	–	...	N, 4
	⋮	⋮	⋮	⋮	⋮	⋮
	1, N	2, N	3, N	4, N	...	–

There are clearly $N^2 - N$ terms and therefore we find that

$$\langle (E_x^* E_x)^2 \rangle = \xi^4 [N^2 + N(N-1)] \approx 2N^2 \xi^4, \tag{15.18}$$

since N is very large. Therefore, from (15.15) and (15.16), we find

$$\Delta u^2 \propto \langle (E_x^* E_x)^2 \rangle = 2N^2 \xi^4 - N^2 \xi^4 = N^2 \xi^4.$$

We have derived the important result,

$$\Delta u^2 = u^2 \tag{15.19}$$

that is, *the fluctuations in the energy density are of the same magnitude as the energy density of the radiation field itself.* This is a remarkable property of electromagnetic radiation and is the reason why phenomena such as interference and diffraction occur even for incoherent radiation. Despite the fact that the radiation measured by a detector is a superposition of a large number of waves with random phases, the fluctuations in the fields are as large as the magnitude of the total intensity.

The physical meaning of this calculation is clear. The matrix of non-vanishing contributions to $(E_x^* E_x)^2$ is no more than the sum of all pairs of waves added separately. Every pair of waves of frequency ν interferes to produce fluctuations in intensity of the radiation $\Delta u \approx u$, that is, multiplying out an arbitrary pair of terms of (15.13) with $j \neq k$

$$\xi^2 \sin(kz - \omega t)\sin(kx - \omega t + \varphi) = \frac{\xi^2}{2}\{\cos\varphi - \cos[2(kz - \omega t) + \varphi]\}.$$

Notice that this analysis refers to waves of random phase φ and of a particular angular frequency ω, that is, what we would refer to as waves all corresponding to a single mode.

Let us now study Einstein's analysis of fluctuations in black-body radiation.

15.4 Fluctuations in black-body radiation

Einstein begins his paper of 1909 by inverting Boltzmann's relation between entropy and probability:

$$W = \exp\left(\frac{S}{k}\right). \tag{15.20}$$

Consider only the radiation energy E in the frequency interval ν to $\nu + d\nu$. As before, we write $E = Vu(\nu)\,d\nu$. Now divide the volume V into a large number of cells and suppose that ΔE_i is the fluctuation within the ith cell. Then the entropy of this cell is

$$S_i = S_i(0) + \left(\frac{\partial S}{\partial E}\right)\Delta E_i + \frac{1}{2}\left(\frac{\partial^2 S}{\partial E^2}\right)(\Delta E_i)^2 + \ldots. \tag{15.21}$$

But, averaging over all cells, we know that there is no net fluctuation, $\sum_i \Delta E_i = 0$, and therefore, to second order,

$$S = \sum_i S_i = S(0) + \frac{1}{2}\left(\frac{\partial^2 S}{\partial E^2}\right)\sum_i (\Delta E_i)^2. \tag{15.22}$$

Therefore, using (15.20), the probability distribution for the fluctuations is

$$W \propto \exp\left[\frac{1}{2}\left(\frac{\partial^2 S}{\partial E^2}\right)\frac{\sum(\Delta E_i)^2}{k}\right]. \tag{15.23}$$

This is just the sum of a set of normal distributions which, for any individual cell, can be written

$$W_i \propto \exp\left[-\frac{(\Delta E_i)^2}{2\sigma^2}\right], \tag{15.24}$$

where the variance of the distribution is

$$\sigma^2 = -\frac{k}{\partial^2 S/\partial E^2}. \tag{15.25}$$

Notice that we have obtained a physical interpretation for the second derivative of the entropy with respect to energy, which had played a prominent part in Planck's original analysis.

Let us now derive σ^2 for the black-body spectrum

$$u(\nu) = \frac{8\pi h\nu^3}{c^3} \frac{1}{e^{h\nu/kT} - 1}. \tag{15.26}$$

Inverting (15.26),

$$\frac{1}{T} = \frac{k}{h\nu} \ln\left(\frac{8\pi h\nu^3}{c^3 u} + 1\right). \tag{15.27}$$

We now express this result in terms of the total energy in the cavity in the frequency interval ν to $\nu + d\nu$, $E = Vu\, d\nu$. As before, $\partial S/\partial E = 1/T$. Therefore

$$\frac{\partial S}{\partial E} = \frac{k}{h\nu} \ln\left(\frac{8\pi h\nu^3}{c^3 u} + 1\right) = \frac{k}{h\nu} \ln\left(\frac{8\pi h\nu^3 V\, d\nu}{c^3 E} + 1\right),$$

$$\frac{\partial^2 S}{\partial E^2} = -\frac{k}{h\nu} \frac{1}{\left(\dfrac{8\pi h\nu^3 V\, d\nu}{c^3 E} + 1\right)} \times \frac{8\pi h\nu^3 V\, d\nu}{c^3 E^2},$$

$$\frac{k}{\partial^2 S/\partial E^2} = -\left(h\nu E + \frac{c^3}{8\pi\nu^2 V\, d\nu} E^2\right) = -\sigma^2. \tag{15.28}$$

In terms of the square of the fractional fluctuation,

$$\frac{\sigma^2}{E^2} = \frac{h\nu}{E} + \frac{c^3}{8\pi\nu^2 V\, d\nu}. \tag{15.29}$$

Einstein noted that the two terms on the right-hand side have quite specific meanings. The first originates from the high-frequency, Wien part of the spectrum and, if we suppose that the radiation consists of photons each of energy $h\nu$, we see that it corresponds to the statement that the fractional fluctuation in the intensity is just $1/N^{1/2}$, where N is the number of photons, that is,

$$\Delta N/N = 1/N^{1/2}. \tag{15.30}$$

As we have shown in subsection 15.2.8, this is exactly the result expected if light consists of discrete particles.

Let us now look more closely at the second term in (15.29). It originates from the Rayleigh–Jeans part of the spectrum. We ask, 'How many independent modes are there in the box in the frequency range ν to $\nu + d\nu$?' We have already shown in Section 12.6 that there are $N_{\text{modes}} = 8\pi\nu^2 V\, d\nu/c^3$ modes (see (12.50)). We have also shown in Section 15.3 that the energy density fluctuations associated with each wave mode have magnitude $\Delta u^2 = u^2$. When we add together randomly the N_{modes} independent modes in the frequency interval ν to $\nu + d\nu$, the energy density totals $E = N_{\text{modes}}\, u$, and the fluctuation in the randomly superimposed modes is $\Delta E = \sqrt{N_{\text{modes}}}u$. Thus

$$\frac{\Delta E^2}{E^2} = \frac{1}{N_{\text{modes}}} = \frac{c^3}{8\pi \nu^2 V \,d\nu},$$

which is exactly the same as the second term on the right-hand side of (15.29).

Thus, the two parts of the fluctuation spectrum correspond to *particle* and *wave* statistics, the former corresponding to the Wien part of the spectrum and the latter to the Rayleigh–Jeans part. The amazing aspect of this formula for the fluctuations is that when we add together the variances due to independent causes, the equation

$$\frac{\sigma^2}{E^2} = \frac{h\nu}{E} + \frac{c^3}{8\pi \nu^2 V \,d\nu} \tag{15.31}$$

states that *we should add independently the 'wave' and 'particle' aspects of the radiation field to find the total magnitude of the fluctuations*. I regard this as a quite miraculous piece of theoretical physics.

15.5 The first Solvay conference

Einstein published these remarkable results in 1909, but still obtained little support for the concept of light quanta. Among those who were persuaded of the importance of quanta was Walther Nernst, who at that time was measuring the low-temperature heat capacities of various materials. As recounted in Section 14.3, Nernst visited Einstein in Zurich in March 1910 and they compared Einstein's theory with Nernst's recent experiments. These experiments showed that Einstein's predictions of the low-temperature variation of the specific heat with temperature (14.25) gave a good description of the experimental results. As Einstein wrote to his friend Jakob Laub after the visit,

> I consider the quantum theory certain. My predictions with respect to specific heats seem to be strikingly confirmed. Nernst, who has just been here, and Rubens are eagerly occupied with experimental tests, so that people will soon be informed about this matter.[4]

By 1911, Nernst was convinced not only of the importance of Einstein's results but also of the theory underlying them. The outcome of the meeting with Einstein was dramatic. The number of papers on quanta began to increase rapidly, as Nernst popularised the results of the quantum theory of solids.

Nernst was a friend of the wealthy Belgian industrialist Ernest Solvay and he persuaded him to sponsor a meeting of a select group of physicists to discuss the issues of quanta and radiation. The idea was first mooted in 1910, but Planck urged that the meeting should be postponed for a year. As he wrote,

> My experience leads me to the opinion that scarcely half of those you envisage as participants have a sufficiently lively conviction of the pressing need for reform to be motivated to attend the conference... Of the entire list you name, I believe that besides ourselves [only] Einstein, Lorentz, W. Wien and Larmor are deeply interested in the topic.[5]

Figure 15.1: The participants in the first Solvay conference on physics, Brussels 1911. (From *La théorie du rayonnement et les quanta*, 1912, eds. P. Langevin and M. De Broglie, Gautier-Villars, Paris.)

By the following year, matters were very different. This was the first, and perhaps the most significant, of the famous series of Solvay conferences. The eighteen official participants met on 29 October 1911 at the Hotel Metropole in Brussels and the meeting took place between the 30th of that month and 3rd November (Fig. 15.1). By this time, the majority of the participants were supporters of the quantum hypothesis. Here is how they lined up. Two were definitely against quanta – Jeans and Poincaré. Rayleigh was invited but did not attend – his views were essentially the same as those of Jeans. Five were initially neutral – Rutherford, Brillouin, Marie Curie, Perrin, Knudsen. The eleven others were basically pro-quanta – Lorentz (chairman), Nernst, Planck, Rubens, Sommerfeld, Wien, Warburg, Langevin, Einstein, Hasenohrl and Onnes. The secretaries were Goldschmidt, De Broglie and Lindemann; Solvay, who hosted the conference, was there, as well as his collaborators Herzen and Hostelet. Rayleigh and van der Waals were unable to attend.

As a result of the meeting, Poincaré was converted to the notion of quanta; the physicists who had taken a neutral position had done so because they were unfamiliar with the arguments. The conference had a profound effect in that it provided a forum at which all the arguments could be presented. In addition, all the participants wrote their lectures for publication beforehand and these were then discussed in detail. These discussions were recorded and the full proceedings published within a year of the event in the important volume *La théorie du rayonnement et les quanta: rapports et discussions*

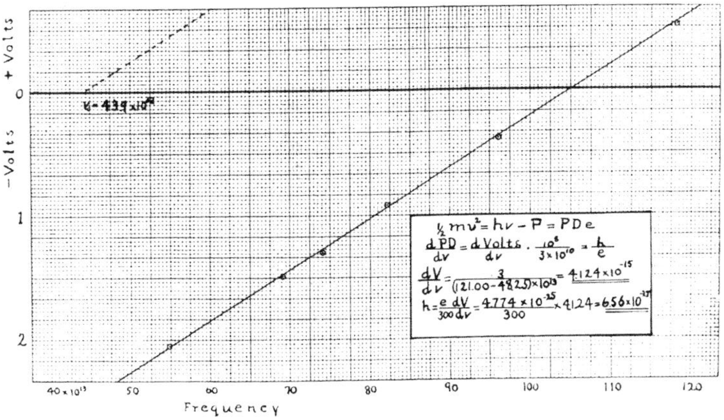

Figure 15.2: Millikan's results on the photoelectric effect compared with the predictions of Einstein's quantum theory. (R.A. Millikan, 1916, *Phys. Rev.*, **7**, 355.)

de la réunion tenue à Bruxelles, du 30 octobre au 3 novembre 1911.[6] Thus, all the important issues were made available to the scientific community in one volume. As a result, the next generation of students became fully familiar with the arguments and many of them set to work immediately to tackle the problems of quanta. Furthermore, these problems began to be appreciated beyond the central-European German-speaking scientific community.

It would be wrong, however, to believe that everyone was suddenly convinced of the existence of quanta. In his paper of 1916 on his great series of experiments in which he verified the dependence of the photoelectric effect upon frequency (Fig. 15.2), Millikan stated:

> We are confronted however by the astonishing situation that these facts were correctly and exactly predicted nine years ago by a form of quantum theory which has now been generally abandoned.[7]

Millikan refers to Einstein's 'bold, not to say reckless, hypothesis of an electromagnetic light corpuscle of energy $h\nu$ which flies in the face of the thoroughly established facts of interference'.[7]

15.6 Bohr's theory of the hydrogen atom

In parallel with these major developments in the theory of quanta, great advances had been made in understanding the nature of atoms, largely as a result of the brilliant experiments associated with the names of J.J. Thomson and Ernest Rutherford.

15.6.1 *Thomson and Rutherford*

The discovery of the electron in 1897 is traditionally attributed to J.J. Thomson on the basis of his famous series of experiments, in which he established that the charge-to-mass ratio,

e/m_e, of cathode rays is about two-thousand times that of hydrogen ions. At the same time, several continental physicists had been hot on the trail.

(i) In 1896, Pieter Zeeman discovered the broadening of spectral lines when a sodium flame is placed between the poles of a strong electromagnet. Lorentz interpreted this result in terms of the splitting of the spectral lines due to the motion of the 'ions' in the atoms about the magnetic field direction – he found a lower limit of 1000 for the value of e/m_e.

(ii) In January 1897, Emil Wiechert used the magnetic deflection technique to obtain a measurement of e/m_e for cathode rays and concluded that these particles had mass between 2000 and 4000 times smaller than that of hydrogen, assuming their electric charge to be the same in magnitude as that of hydrogen ions. He obtained only an upper limit to the speed of the particles, since it was assumed that the kinetic energy of the cathode rays was $E_{\rm kin} = eV$, where V is the accelerating voltage of the discharge tube.

(iii) Walter Kaufmann's experiment was similar to Thomson's. He found the same value of e/m_e no matter which gas filled the discharge tube, a result which puzzled him; the value was 1000 times greater than that for hydrogen ions. He concluded that 'the hypothesis of cathode rays as emitted particles is by itself inadequate for a satisfactory explanation of the regularities I have observed'.

J.J. Thomson was the first of these pioneers to interpret the experiments in terms of a sub-atomic particle. In his words, written in 1897, cathode rays constituted

> . . . a new state, in which the subdivision of matter is carried very much further than in the ordinary gaseous state.[8]

In 1899, Thomson used one of C.T.R. Wilson's early cloud chambers to measure the charge of the electron. He counted the total number of droplets formed and their total charge. From these, he estimated $e = 2.2 \times 10^{-19}$ C, compared with the present standard value of 1.602×10^{-19} C. This experiment was the precursor of the famous Millikan oil drop experiment, in which the water-vapour droplets were replaced by fine drops of a heavy oil, which did not evaporate during the course of the experiment. Thomson also carried out an important series of experiments in which he demonstrated that the β-particles emitted in radioactive decays and those ejected in the photoelectric effect had the same charge-to-mass ratio as the cathode rays.

Thomson pursued a much more sustained and detailed campaign than the other physicists in establishing the universality of what became known as *electrons*, the name coined for cathode rays by Johnstone Stoney in 1891. It seems fair to regard Thomson as the discoverer of the first sub-atomic particle.

It was inferred that there must be electrons within the atom, but it was not clear how many there must be. There might have been as many as 2000 electrons, if they were to make up the mass of the hydrogen atom. The answer was provided by a brilliant series of experiments carried out by Thomson and his colleagues, who studied the scattering of X-rays by thin films. X-rays are scattered from the electrons in atoms by *Thomson scattering*, the theory of which was worked out by Thomson[9] using the classical expression for the radiation of an accelerated electron (see subsection 12.2.1). It is straightforward to show[10] that the cross-section for the scattering of a beam of incident radiation by an electron is

$$\sigma_T = \frac{e^4}{6\pi\epsilon_0^2 m_e^2 c^4} = \frac{8\pi r_e^2}{3} = 6.653 \times 10^{-29}\ \mathrm{m}^2, \tag{15.32}$$

the *Thomson cross-section*, where $r_e = e^2/(4\pi\epsilon_0 m_e c^2)$ is the classical electron radius.

Thomson found that the intensity of the X-rays scattered from thin films indicated that there were not thousands of electrons in each atom, but a much smaller number. In collaboration with Charles Barkla, he showed that, except for hydrogen, the number of electrons was roughly half the atomic weight. The picture which emerged was one in which the number of electrons and, consequently, the amount of positive charge in the atom increased by units of the electronic charge. Furthermore, most of the mass of the atom had to be associated with the positive charge. A key question was, 'How are the electrons and the positive charge distributed inside atoms?' In the picture favoured by Thomson, the positive charge was distributed throughout the atom and, within this sphere, the negatively charged electrons were placed on carefully chosen orbits – this rather clever model became known by a somewhat pejorative name, the 'plum-pudding' model of the atom. We will return to the nasty problems Thomson was trying to solve in the next subsection.

The discovery of the nuclear structure of atoms resulted from a brilliant series of experiments carried out by Ernest Rutherford and his colleagues Hans Geiger and Ernest Marsden in the period 1909–12. Rutherford had taken up the chair of physics at Manchester University in 1907 and, in the following year, demonstrated convincingly that α-particles are helium nuclei.[11] He had also been impressed by the fact that α-particles could pass through thin films rather easily, suggesting that much of the volume of atoms is empty space, although there was clear evidence of small-angle scattering. Rutherford persuaded Marsden, who was still an undergraduate, to investigate whether α-particles were deflected through large angles on being fired at a thin gold foil. To Rutherford's astonishment, a few particles were deflected by more than 90°, and a very small number almost returned along the direction of incidence. In Rutherford's words:

> It was quite the most incredible event that has ever happened to me in my life. It was almost as incredible as if you fired a 15-inch shell at a piece of tissue paper and it came back and hit you.[12]

Rutherford realised that it required a very considerable force to send the α-particle back along its track. It was only in 1911 that he hit upon the idea that if all the positive charge were concentrated in a compact nucleus, the scattering could be attributed to the repulsive electrostatic force between the incoming α-particle and the positive nucleus. Rutherford was no theorist, but he used his knowledge of central orbits in inverse-square-law fields of force to work out the properties of what became known as *Rutherford scattering*.[13] As was shown in the appendix to Chapter 4, the orbit of the α-particle is a *hyperbola*, the angle of deflection φ being given by

$$\cot\frac{\varphi}{2} = \left(\frac{4\pi\epsilon_0 m_\alpha}{2Ze^2}\right) p_0 v_0^2, \tag{15.33}$$

where p_0 is the collision parameter, v_0 is the initial velocity of the α-particle and Z the nuclear charge. It is straightforward to work out the probability that the α-particle is scattered through an angle φ. The result is

$$p(\varphi) \propto \frac{1}{v_0^4} \operatorname{cosec}^4 \frac{\varphi}{2}, \tag{15.34}$$

the famous $\operatorname{cosec}^4(\varphi/2)$ law derived by Rutherford, which was found to explain precisely the observed distribution of scattering angles of the α-particles.[14]

Rutherford had, however, achieved much more. The fact that the scattering law was obeyed so precisely, even for large angles of scattering, meant that the inverse-square law of electrostatic repulsion held good to very small distances indeed. Rutherford and his colleagues found that the nucleus had to have a size less than about 10^{-14} m, very much less than the sizes of atoms, which are typically about 10^{-10} m.

This is one of the great experiments of physics and the papers by Rutherford, Geiger and Marsden from 1909 to 1913 are classics of twentieth-century physics. Rutherford attended the first Solvay Conference in 1911, but made no mention of his remarkable experiments, which led directly to his nuclear model of the atom. Remarkably, this key result for understanding the nature of atoms made little impact upon the physics community at the time and it was not until 1914 that Rutherford was thoroughly convinced of the necessity of adopting the nuclear model of the atom. Before that time, however, someone else did adopt it – Niels Bohr, the first theorist to apply quantum concepts to the structure of atoms successfully.

15.6.2 *The structure of atoms and the Bohr model of the hydrogen atom*

The construction of atomic models was a major industry in the early years of the twentieth century, particularly in England – the splendid survey by Heilbron[15] can be thoroughly recommended. The model builders, of whom Thomson was a leading practitioner, faced two major problems. The first was the dynamical stability of the distribution of electrons within atoms and the second, and potentially even greater, problem was the fact that an electron accelerated in the field of the positive charge radiates electromagnetic radiation, and so would spiral very rapidly to the centre of the atom.

However the electrons are distributed in the atom, they cannot be stationary because of *Earnshaw's theorem*, which states that any static distribution of electric charges is mechanically unstable, in that they either collapse or disperse to infinity under the action of the electrostatic forces. The alternative is to place the electrons in orbits, what is often called the planetary, or Saturnian, model of the atom. The most famous of these early models was that due to the Japanese physicist Nagaoka, who attempted to associate the spectral lines of atoms with small vibrational perturbations of the electrons about their equilibrium orbits. The problem with this model was that perturbations in the plane of the electron's orbit are unstable, leading to instability of the atom as a whole.

The radiative instability of the electron can be understood from the following calculation. Suppose the electron is in a circular orbit of radius a. Then, equating the centripetal force to the electrostatic force of attraction between the electron and the nucleus of charge Ze,

$$\frac{Ze^2}{4\pi\epsilon_0 a^2} = \frac{m_e v^2}{a} = m_e \ddot{r} \qquad (15.35)$$

where $\ddot{r}$ is the centripetal acceleration. The rate at which the electron loses energy by radiation is given by (12.5). The kinetic energy of the electron is $E = \frac{1}{2}m_e v^2 = \frac{1}{2}m_e a\ddot{r}$. Therefore, the time it takes the electron to lose all its kinetic energy by radiation is

$$T = \frac{E}{|\mathrm{d}E/\mathrm{d}t|} = \frac{2\pi a^3}{\sigma_T c}, \qquad (15.36)$$

where σ_T is the Thomson cross-section. Taking the radius of the atom to be $a = 10^{-10}$ m, the time it takes the electron to lose all its energy is about 3×10^{-10} s. Something is profoundly wrong. Moreover, as the electron loses energy, it moves into an orbit of smaller radius, loses energy more rapidly and spirals into the centre.

The pioneer atom-model builders were well aware of this problem. Their solution was to place the electrons in orbits such that there is no net acceleration when the acceleration vectors of all the electrons in the atom are added together. This requires, however, the electrons to be very well ordered in their orbits about the nucleus. For example, in the case of hydrogen, the analysis which leads to (15.36) applies. If there are two electrons in the atom, they can be placed in the same circular orbit on opposite sides of the nucleus and so, to first order, there is no net dipole moment as observed at infinity, and hence no dipole radiation. There is, however, a finite electric quadrupole moment and hence radiation at the level $(\lambda/a)^2$ relative to the intensity of dipole radiation is emitted. Since $\lambda/a \sim 10^{-3}$ the radiation problem is significantly relieved. By adding more electrons to the orbit, the quadrupole moment can be cancelled out as well and so, by adding sufficient electrons to each orbit, the radiation problem can be reduced to manageable proportions. It is, however, achieved at the expense of requiring each orbit to be densely populated with a very well-ordered system of electrons. This was the basis of Thomson's 'plum-pudding' model. The 'plums', or electrons, were most certainly not randomly located within the sphere of positive charge, but were precisely located to ensure that the radiative instability did not occur. The radiative problem remained, however, particularly for hydrogen, which possesses only one electron.

Niels Bohr completed his doctorate on the electron theory of metals in 1911. Even at that stage, he had convinced himself that this theory was seriously incomplete and required further mechanical constraints on the motion of electrons at the microscopic level. He spent the following year in England, working for seven months with J.J. Thomson at the Cavendish Laboratory in Cambridge and four months with Ernest Rutherford in Manchester. Bohr was immediately struck by the significance of Rutherford's model of the nuclear structure of the atom and began to devote all his energies to understanding atomic structure on that basis. He quickly appreciated the distinction between the chemical properties of atoms, which are associated with the orbiting electrons, and radioactive processes, which are associated with activity in the nucleus. On this basis, he could understand the nature of the isotopes of a particular chemical species. Bohr also realised from the outset that the structure of atoms could not be understood on the basis of classical physics. The obvious way forward was to incorporate the quantum concepts of Planck and Einstein into the models of atoms. Einstein's statement, quoted in Section 14.3,

> ... for ions which can vibrate with a definite frequency, ... the manifold of possible states must be narrower than it is for bodies in our direct experience.[16]

was precisely the type of constraint which Bohr was seeking. Such a mechanical constraint was essential to understand how atoms could survive the inevitable instabilities according to classical physics. How could these ideas be incorporated into the models for atoms?

Bohr was not the first physicist to introduce quantum concepts into the construction of atomic models. In 1910, a Viennese doctoral student, A.E. Haas, realised that if Thomson's sphere of positive charge were uniform an electron would perform simple harmonic motion through the centre of the sphere, since the restoring force at radius r from the centre would be, according to Gauss's theorem in electrostatics,

$$f = m_e \ddot{r} = -\frac{eQ(\leq r)}{4\pi\epsilon_0 r^2} = -\left(\frac{eQ}{4\pi\epsilon_0 a^3}\right) r, \tag{15.37}$$

where a is the radius of the atom and Q the total positive charge. For a hydrogen atom, for which $Q = e$, the frequency of oscillation of the electron would be

$$\nu = \frac{1}{2\pi}\left(\frac{e^2}{4\pi\epsilon_0 m_e a^3}\right)^{1/2}. \tag{15.38}$$

Haas argued that the energy of oscillation of the electron, $E = e^2/(4\pi\epsilon_0 a)$, should be quantised and set equal to $h\nu$. It followed that,

$$h^2 = \frac{\pi m_e e^2 a}{\epsilon_0}. \tag{15.39}$$

Haas used (15.39) to show how Planck's constant could be related to the properties of atoms, taking for ν the short-wavelength limit of the Balmer series, that is, setting $n \to \infty$ in the Balmer formula (15.44). Haas's efforts[17] were discussed by Lorentz at the 1911 Solvay conference, but they did not attract much attention. According to Haas's approach, Planck's constant was simply a property of atoms defined by (15.39), whereas those already converted to quanta preferred to believe that h had much deeper significance.

In the summer of 1912, Bohr wrote an unpublished memorandum[18] for Rutherford, in which he made his first attempt at quantising the energy levels of the electrons in atoms. He proposed relating the kinetic energy T of the electron to the frequency $\nu' = v/(2\pi a)$ of its orbit about the nucleus through the relation

$$T = \tfrac{1}{2} m_e v^2 = K\nu', \tag{15.40}$$

where K is a constant which he expected would be of the same order of magnitude as Planck's constant h. Bohr believed there must be some such non-classical constraint in order to guarantee the stability of atoms. Indeed, his criterion (15.40) absolutely fixed the kinetic energy of the electron about the nucleus. For a bound circular orbit,

$$\frac{mv^2}{a} = \frac{Ze^2}{4\pi\epsilon_0 a^2} \tag{15.41}$$

where Z is the positive charge of the nucleus in units of the charge of the electron e. As is well known, the binding energy of the electron is

$$E = T + U = \frac{m_e v^2}{2} - \frac{Ze^2}{4\pi \epsilon_0 a} = -T = \frac{U}{2}, \tag{15.42}$$

where U is the electrostatic potential energy. The quantisation condition (15.40) enables both v and a to be eliminated from the expression for the kinetic energy of the electron. A straightforward calculation shows that

$$T = \frac{m Z^2 e^2}{32 \epsilon_0^2 K^2}, \tag{15.43}$$

which was to prove to be of great significance for Bohr. His memorandum containing these ideas was principally about issues such as the number of electrons in atoms, atomic volumes, radioactivity, the structure and binding of diatomic molecules and so on. There is no mention of spectroscopy, which he and Thomson considered too complex to provide useful information.

The next clue was provided by the work of the Cambridge physicist John William Nicholson,[19] who arrived at the concept of the quantisation of angular momentum by a quite different route. Nicholson had shown that, although the Saturnian model of the atom is unstable for perturbations in the plane of the orbit, perturbations perpendicular to the plane are stable for orbits containing up to five electrons – he assumed that the unstable modes in the plane of the orbit were suppressed by some unspecified mechanism. The frequencies of the stable oscillations were multiples of the orbital frequency and he compared these with the frequencies of the lines observed in the spectra of bright nebulae, particularly with the 'nebulium' and 'coronium' lines. Performing the same exercise for ionised atoms with one fewer orbiting electron, further matches to the astronomical spectra were obtained. The frequency of the orbiting electrons remained a free parameter, but when he worked out the angular momentum associated with an electron Nicolson found that it turned out to be a multiple of $h/2\pi$. When Bohr returned to Copenhagen later in 1912, he was perplexed by the success of Nicholson's model, which seemed to provide a successful, quantitative, model for the structure of atoms and which could account for the spectral lines observed in astronomical spectra.

The breakthrough came in early 1913, when H.M. Hansen told Bohr about the Balmer formula for the wavelengths, or frequencies, of the spectral lines in the spectrum of hydrogen,

$$\frac{1}{\lambda} = \frac{\nu}{c} = R\left(\frac{1}{2^2} - \frac{1}{n^2}\right), \tag{15.44}$$

where $R = 1.097 \times 10^7\ \mathrm{m}^{-1}$ is known as the *Rydberg constant* and $n = 3, 4, 5, \ldots$. As Bohr recalled much later,

> As soon as I saw Balmer's formula, the whole thing was clear to me.[20]

He realised immediately that this formula contained within it the crucial clue for the construction of a model of the hydrogen atom, which he took to consist of a single negatively charged electron orbiting a positively charged nucleus. He went back to his memorandum on the quantum theory of the atom, in particular, to his expressions (15.42) and (15. 43) for the total energy and the kinetic energy of the electron. He realised that he could determine the value of his constant K from the expression for the Balmer series. The running term in $1/n^2$ can be associated with (15.43) if we write for hydrogen with $Z = 1$

$$E_n = -T_n = -\frac{m_e e^2}{32\epsilon_0 n^2 K^2}. \quad (15.45)$$

Then, when the electron changes from an orbit with quantum number n to that with $n = 2$, the energy of the emitted radiation would be the difference in the total energies $E_n - E_2$ of the two states. Applying Einstein's quantum hypothesis, this energy should be equal to $h\nu$. Inserting the numerical values of the constants into (15.45), Bohr found that the constant K was exactly $h/2$. Therefore, the energy of the state with quantum number n is

$$E_n = -\frac{m_e e^2}{8\epsilon_0 n^2 h^2}. \quad (15.46)$$

The angular momentum J of the state could be found immediately by writing $T = \frac{1}{2} I\omega'^2 = nh\nu'/2$ from (15.40). It immediately follows that

$$J = I\omega' = \frac{nh}{2\pi}. \quad (15.47)$$

This is how Bohr arrived at the quantisation of angular momentum according to the 'old' quantum theory. In the first paper of his famous trilogy of 1913,[21] Bohr acknowledged that Nicholson had discovered the quantisation of angular momentum in his papers of 1912. These results were the inspiration for what became known as the *Bohr model of the atom.*

In the first paper of the trilogy of 1913, Bohr noted that a formula similar to (15.44) could account for the *Pickering series*, which had been discovered in 1896 by Edward Pickering in the spectra of stars. In 1912, Alfred Fowler discovered the same series in laboratory experiments. Bohr argued that singly ionised helium atoms would have exactly the same spectrum as hydrogen but the wavelengths of the corresponding lines would be four times shorter, as observed in the Pickering series. Fowler objected, however, that the ratio of the Rydberg constants for singly ionised helium and hydrogen was not 4, but 4.00163. Bohr realised that the problem arose from neglecting the contribution of the mass of the nucleus to the computation of the moments of inertia of the hydrogen atom and the helium ion. If the angular velocity of the electron and the nucleus about their centre of mass is ω, the condition for the quantisation of angular momentum is

$$\frac{nh}{2\pi} = \mu\omega R^2, \quad (15.48)$$

where $\mu = m_e m_N/(m_e + m_N)$ is the *reduced mass* of the atom or ion. This takes into account the contributions of both the electron and the nucleus to the angular momentum; R is their separation. Therefore, the ratio of the Rydberg constants for ionised helium and hydrogen should be

$$\frac{R_{He^+}}{R_H} = 4\left(\frac{1 + m_e/M}{1 + m_e/4M}\right) = 4.00160, \quad (15.49)$$

where M is the mass of the hydrogen atom. Thus, precise agreement was found between the theoretical and laboratory estimates of the ratio of Rydberg constants for hydrogen and ionised helium.

The Bohr theory of the hydrogen atom was a quite remarkable achievement and the first convincing application of the quantum theory to atoms. Bohr's dramatic results were persuasive evidence for many scientists that Einstein's quantum theory really had to be taken seriously for processes occurring on the atomic scale. The 'old' quantum theory was, however, fundamentally incomplete and constituted an uneasy mixture of classical and quantum ideas without any self-consistent theoretical underpinning.

From Einstein's point of view, the results provided further strong support for his quantum picture of elementary processes. In his biography of Bohr, Pais[22] tells the story of Hevesy's encounter with Einstein in September 1913. When Einstein heard of Bohr's analysis of the Balmer series of hydrogen, Einstein remarked cautiously that Bohr's work was very interesting, and important if right. When Hevesy told him about the helium results, Einstein responded,

> This is an enormous achievement. The theory of Bohr must then be right.

15.7 Einstein (1916) 'On the quantum theory of radiation'

During the years 1911 to 1916, Einstein was preoccupied with the formulation of general relativity, which will be discussed in Chapter 17. By 1916, the pendulum of scientific opinion was beginning to swing in favour of the quantum theory, particularly following the success of Bohr's theory of the hydrogen atom. These ideas fed back into Einstein's thinking about the problems of the emission and absorption of radiation and resulted in his famous derivation of the Planck spectrum through the introduction of what are now called *Einstein's A and B coefficients*.

Einstein's great paper[23] of 1916 begins by noting the formal similarity between the Maxwell–Boltzmann distribution for the velocity distribution of the molecules in a gas and Planck's formula for the black-body spectrum. Einstein shows how these distributions can be reconciled through a derivation of the Planck spectrum, which gives insight into what he refers to as the 'still unclear processes of emission and absorption of radiation by matter'. The paper begins with a description of a quantum system consisting of a large number of molecules which can occupy a discrete set of molecular states $Z_1, Z_2, Z_3, \ldots$ with corresponding energies $\epsilon_1, \epsilon_2, \epsilon_3, \ldots$. According to classical statistical mechanics, the relative probabilities W_n that these states are occupied in thermodynamic equilibrium at temperature T are given by Boltzmann's relation

$$W_n = g_n \exp\left(-\frac{\epsilon_n}{kT}\right), \tag{15.50}$$

where the g_n are the *statistical weights*, or *degeneracies*, of the states Z_n. As Einstein remarks forcefully in his paper

> [Equation (15.50)] expresses the farthest-reaching generalisation of Maxwell's velocity distribution law.

Consider two quantum states of the gas molecules, Z_m and Z_n, with energies ϵ_m and ϵ_n respectively such that $\epsilon_m > \epsilon_n$. Following the precepts of the Bohr model, it is assumed that a quantum of radiation is emitted if the molecule changes from the state Z_m to Z_n, the energy of the quantum being $h\nu = \epsilon_m - \epsilon_n$. Similarly, when a photon of energy $h\nu$ is absorbed the molecule changes from the state Z_n to Z_m.

The quantum description of these processes follows by analogy with the classical processes of the emission and absorption of radiation.

Spontaneous emission. Einstein notes that a classical oscillator emits radiation in the absence of excitation by an external field. The corresponding process at the quantum level is called *spontaneous emission*, and the probability of its taking place in the time interval $\mathrm{d}t$ is

$$\mathrm{d}W = A_m^n \,\mathrm{d}t, \tag{15.51}$$

similar to the law of radioactive decay.

Induced emission and absorption. By analogy with the classical case, if the oscillator is excited by waves of the same frequency as the oscillator then it either gains or loses energy, depending upon the phase of the wave relative to that of the oscillator, that is, the work done on the oscillator can be either positive or negative. The magnitude of the positive or negative work done is proportional to the energy density of the incident waves. The quantum mechanical equivalents of these processes are those of *induced absorption*, in which the molecule is excited from the state Z_n to Z_m, and *induced emission*, in which the molecule emits a photon under the influence of the incident radiation field. The probabilities of these processes are:

$$\begin{aligned} \text{for induced absorption,}\quad \mathrm{d}W &= B_n^m \rho \,\mathrm{d}t; \\ \text{for induced emission,}\quad \mathrm{d}W &= B_m^n \rho \,\mathrm{d}t. \end{aligned}$$

The lower indices refer to the initial state and the upper indices to the final state; ρ is the energy density of radiation with frequency ν. B_n^m and B_m^n are constants for a particular physical process, and are referred to as 'coefficients for changes of state by induced absorption or emission'.

We now seek the spectrum of the energy density of radiation $\rho(\nu)$ in thermal equilibrium. The relative numbers of molecules with energies ϵ_m and ϵ_n in thermal equilibrium are given by the Boltzmann relation (15.50) and so, in order to leave the equilibrium distribution unchanged under the processes of spontaneous and induced emission and induced absorption of radiation, the probabilities must balance, that is,

$$\underbrace{g_n \mathrm{e}^{-\epsilon_n/kT} B_n^m \rho}_{\text{absorption}} = \underbrace{g_m \mathrm{e}^{-\epsilon_m/kT} \left(B_m^n \rho + A_m^n\right)}_{\text{emission}}. \tag{15.52}$$

Now, in the limit $T \to \infty$ the radiation energy density $\rho \to \infty$, and the induced processes dominate the equilibrium. Therefore, allowing $T \to \infty$ and $A_m^n = 0$, (15.52) becomes

$$g_n B_n^m = g_m B_m^n. \tag{15.53}$$

Reorganising (15.52), the equilibrium radiation spectrum ρ can be written

$$\rho = \frac{A_m^n}{B_m^n} \frac{1}{\exp\left(\dfrac{\epsilon_m - \epsilon_n}{kT}\right) - 1}, \tag{15.54}$$

which is Planck's radiation law. Suppose we consider only Wien's law, which is known to be the correct expression in the frequency range in which light should be considered as consisting of photons. Then, in the limit $(\epsilon_m - \epsilon_n)/kT \gg 1$,

$$\rho = \frac{A_m^n}{B_m^n} \exp\left(-\frac{\epsilon_m - \epsilon_n}{kT}\right) \propto \nu^3 \exp\left(-\frac{h\nu}{kT}\right). \tag{15.55}$$

Comparing the factors in (15.55) we find the key relations:

$$\frac{A_m^n}{B_m^n} \propto \nu^3, \tag{15.56}$$

$$\epsilon_m - \epsilon_n = h\nu. \tag{15.57}$$

The constant of proportionality in (15.56) can be found from the Rayleigh–Jeans limit of the black-body spectrum, $(\epsilon_m - \epsilon_n)/kT \ll 1$. From (12.51),

$$\rho(\nu) = \frac{8\pi\nu^2}{c^3} kT = \frac{A_m^n}{B_m^n} \frac{kT}{h\nu},$$

thus

$$\frac{A_m^n}{B_m^n} = \frac{8\pi h\nu^3}{c^3}. \tag{15.58}$$

The importance of these relations between the A and B coefficients is that they are associated with atomic processes at the microscopic level. Once A_m^n or B_m^n or B_n^m is known, the other coefficients can be found from (15.53) and (15.58) immediately.

It is intriguing that this important analysis occupies only the first three sections of Einstein's paper. The main thrust of the paper for Einstein was that he could now use these results to determine how the motions of molecules would be affected by the emission and absorption of quanta. The analysis was similar to his earlier studies of Brownian motion but was now applied to the case of quanta interacting with molecules. The quantum nature of the processes of emission and absorption were essential features of his argument. We will not go into the details of this remarkable calculation, except to quote the key result that, when a molecule emits or absorbs a quantum $h\nu$, there must be a positive or negative change in the momentum of the molecule of magnitude $|h\nu/c|$, even in the case of spontaneous emission. In Einstein's words,

> There is no radiation of spherical waves. In the spontaneous emission process, the molecule suffers a recoil of magnitude $h\nu/c$ in a direction that, in the present state of the theory, is determined only by 'chance'.[23]

Direct experimental evidence for the correctness of Einstein's conclusion was provided by Arthur Holly Compton's beautiful X-ray scattering experiments of 1923.[24] These showed that photons could undergo collisions with nearly free electrons in which they behaved like particles, the *Compton effect* or *Compton scattering*. It is a standard result of elementary special relativity that the increase in wavelength of the photon in collision with a stationary electron is

$$\lambda' - \lambda = \frac{hc}{m_e c^2}(1 - \cos\theta), \tag{15.59}$$

where θ is the angle through which the photon is scattered.[25] Implicit in this calculation is the conservation of relativistic three-momentum in which the momentum of the photon is $h\nu/c$. In June 1929, Werner Heisenberg wrote in a review article, entitled 'The development of the quantum theory 1918–1928',

> At this time [1923] experiment came to the aid of theory with a discovery which would later become of great significance for the development of quantum theory. Compton found that with the scattering of X-rays from free electrons, the wavelength of the scattered rays was measurably longer than that of the incident light. This effect, according to Compton and Debye, could easily be explained by Einstein's light-quantum hypothesis; the wave theory of light, on the contrary, failed to explain this experiment. With that result, the problems of radiation theory which had hardly advanced since Einstein's works of 1906, 1909 and 1917 were opened up.[26]

Compton's remark in his reminisciences in the year before he died is revealing:

> These experiments were the first to give, at least to physicists in the United States, a conviction of the fundamental validity of the quantum theory.[27]

Returning to the Einstein A and B coefficients, it is interesting to relate them to the emission and absorption coefficients appearing in Kirchhoff's law, which was introduced in subsection 11.2.2. Again, Einstein's profound insight opened up entirely new ways of understanding the interaction between matter and radiation. The transfer equation for radiation (11.16) is

$$\frac{\mathrm{d}I_\nu}{\mathrm{d}l} = -\alpha_\nu I_\nu + j_\nu,$$

where l is the path length through the medium. Suppose the upper and lower states are labelled 2 and 1 respectively. It is natural to associate the emission coefficient j_ν with the spontaneous rate of emission of quanta per unit solid angle, that is,

$$j_\nu = \frac{h\nu}{4\pi} n_2 A_{21}, \tag{15.60}$$

where n_2 is the number density of molecules in the upper state. In the same way, the energy absorbed by the molecules by induced absorption is

$$\frac{h\nu}{4\pi} n_1 B_{12} I_\nu, \tag{15.61}$$

so that the absorption coefficient is

$$\alpha_\nu = \frac{h\nu}{4\pi} n_1 B_{12}. \tag{15.62}$$

We still have to take account of induced or stimulated emission. Although it might seem that this term should be included in the emission coefficient, it is preferable to consider the term to represent *negative absorption* since it depends upon the intensity of radiation along the direction of the ray, as does the induced absorption term. Therefore, we can write the absorption coefficient *corrected for stimulated emission* as

$$\alpha_\nu = \frac{h\nu}{4\pi}(n_1 B_{12} - n_2 B_{21}) = \frac{h\nu}{4\pi} n_1 B_{12}\left(1 - \frac{g_1 n_2}{g_2 n_1}\right). \tag{15.63}$$

The transfer equation for radiation then becomes

$$\frac{\mathrm{d}I_\nu}{\mathrm{d}l} = -\frac{h\nu}{4\pi} n_1 B_{12}\left(1 - \frac{g_1 n_2}{g_2 n_1}\right) I_\nu + \frac{h\nu}{4\pi} n_2 A_{21}. \tag{15.64}$$

In the case in which the populations of the states are given by the Boltzmann distribution and the system has reached thermodynamic equilibrium, $\mathrm{d}I_\nu/\mathrm{d}l = 0$, the Planck distribution is recovered. Equation (15.64) is a remarkably powerful expression. In many physical situations, the distribution of energies among the molecules is maintained in thermal equilibrium by particle collisions, but the radiation is not in thermodynamic equilibrium with the matter – the matter is said to be in a state of *local thermodynamic equilibrium*. In such a circumstance,

$$\frac{n_2}{n_1} = \frac{g_2}{g_1}\exp\left(-\frac{h\nu}{kT}\right), \qquad \frac{n_2 g_1}{n_1 g_2} = \exp\left(-\frac{h\nu}{kT}\right) < 1. \tag{15.65}$$

When the inequality in (15.65) is satisfied, even if the distribution of particle energies is not Maxwellian the populations of the states are said to be *normal*. It is, however, possible to violate this condition by overpopulating the upper state, in which case the absorption coefficient (15.63) can become *negative* and the intensity of radiation is amplified along the path of the rays. In this case, the populations of the states of the molecules are said to be *inverted*. This is the principle behind the operation of *masers* and *lasers*.

One aspect of the introduction of Einstein's A and B coefficients was crucial in understanding the equilibrium between matter and radiation. In his analyses of 1909 Einstein had considered the thermodynamic equilibrium between oscillators and the radiation field, but Lorentz[28] argued that the same equilibrium spectrum should be obtained if, rather than oscillators, free electrons and radiation were maintained in thermal contact within an enclosure at temperature T. The process by which energy is exchanged between photons and electrons is the *Compton scattering* mechanism discussed above. Photons with energy $h\nu$ collide with electrons with momentum $\boldsymbol{p}$, resulting in an increase (or decrease) in the momentum of the electron to $\boldsymbol{p}'$, while the photon leaves with decreased (or increased) energy $h\nu'$. In 1923, Wolfgang Pauli[29] showed that the mean energy of the electrons is $\frac{1}{2}m\overline{v^2} = \frac{3}{2}kT$ and that the Planck spectrum is obtained, *provided* induced processes are taken into account in determining the probability of scattering into the direction $h\nu'$. The full analysis of the problem is non-trivial, but the essence of it is that if the energy density of radiation at frequency ν is ρ and the energy density at frequency ν' is ρ' then the probability of scattering from ν to ν' is given, in the notation used by Pauli, by an expression of the form $(A\rho + B\rho\rho')\,\mathrm{d}t$, where the term in A is the spontaneous scattering and B the induced scattering. The remarkable part of this expression is the term in B, which includes the energy density of the radiation *after* scattering and so corresponds to *induced Compton scattering*. In fact, analysing the origin of these two terms, the A-term corresponds to the normal Compton scattering process, in which the radiation can be considered as consisting of photons, while the B-term corresponds to the wave properties of the radiation. If both terms are included then the mean energy of the electrons is $\frac{3}{2}kT$.

The relation to Einstein's paper of 1916 was elucidated a few months after Pauli's paper by Einstein and Paul Ehrenfest,[30] who argued that the process of Compton scattering could be considered to consist of the absorption of a photon of frequency ν and the emission of a photon of frequency ν'. Then, the concepts of induced absorption and spontaneous and induced emission can be applied to the process. The probability of scattering into a particular state with frequency ν' is proportional to the product of the probabilities of absorption and emission and an expression of the form $B\rho(A' + B'\rho')$ is found, where A' and B' represent the spontaneous and induced emission terms respectively.

15.8 The story concluded

The big problem for all physicists was to discover a theory which could reconcile the apparently contradictory wave and particle properties of light. During the year 1905 to 1925 the fundamentals of physics were progressively undermined. The necessity of adopting quantum concepts became increasingly compelling with Planck's, Einstein's and Bohr's deep insights into the deficiencies of classical physics at the microscopic level. The 'old' quantum theory, which has been the subject of the last five chapters, was a patchwork of ideas and, as the theory developed in the hands of theorists such as Sommerfeld, became more and more *ad hoc* and unsatisfactory. The breakthrough came with the discoveries of wave and matrix mechanics by Erwin Schrödinger and Werner Heisenberg in 1925–6, and the relativistic wave equation for the electron by Paul Dirac in 1928. These established a new set of physical principles and a new mathematical framework for understanding the wave–particle duality which had caused such distress to physicists of the first two decades of the twentieth century. The resolution of the problem is nicely summarised by Stephen Hawking:

> The theory of quantum mechanics is based on an entirely new type of mathematics that no longer describes the real world in terms of particles and waves; it is only observations of the world that may be described in these terms.[31]

The story which has unfolded over the last five chapters is one of the most exciting and dramatic intellectual developments in theoretical physics, and indeed in the whole of science. The contributions of many very great physicists were crucial in reaching this new level of understanding. In the course of revising this text, I have read again many of Einstein's papers and have been absolutely astounded by the imagination and fertility of his thinking. Not only the quality but also the sheer quantity of papers spanning the whole of physics is quite staggering. What comes through is a sustained clarity of thinking and a sureness of intuition in tackling head-on the most challenging problems facing physicists in

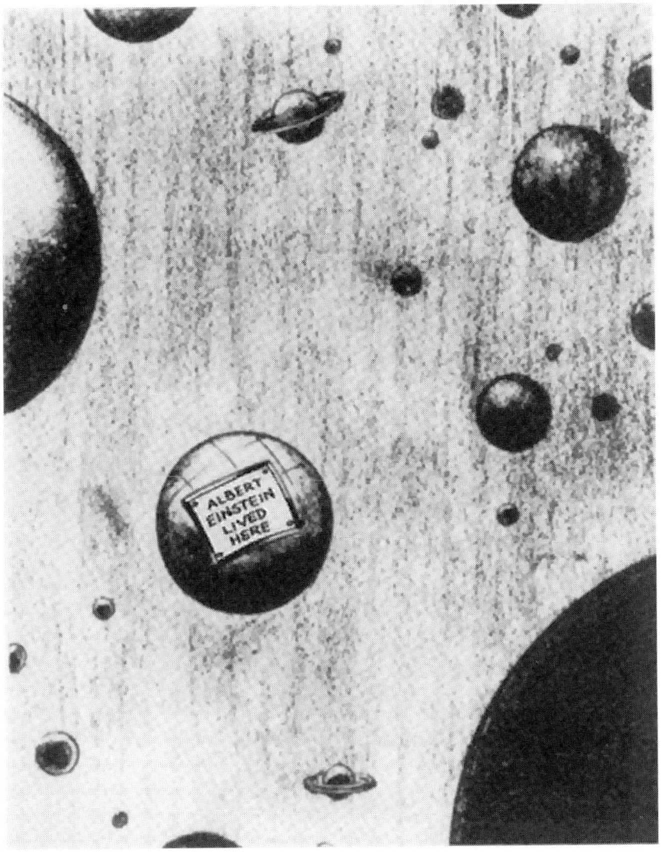

Figure 15.3: Cartoon by Herblock in memory of Albert Einstein, who died on 15 April 1955. (From Herblock (Herbert L. Block), 1995, *Here and Now*, Simon and Schuster.)

the period 1905 to 1925. If relativity and quantum mechanics were the greatest discoveries of twentieth-century physics, the person at the heart of these developments was undoubtedly Albert Einstein.

I have been asked by non-scientists whether Einstein deserves his reputation as the outstanding figure of modern science as he is portrayed in the popular press (Fig. 15.3). The case study of the last five chapters and those to come on special and general relativity should leave the reader in no doubt that Einstein's contributions transcend those of virtually all other physicists, the only physicists who can seriously be mentioned in the same breath being Newton and Maxwell. I am reminded of an occasion in the 1970s when I had dinner in Moscow with academician V.L. Ginzburg. He recounted how Landau, the greatest Russian theoretical physicist of the twentieth century, put physicists into leagues, some belonging to the first division, others to the second division and so on. The first division contained names such as Newton and Maxwell; he rated his own contribution rather modestly. But there was a separate division, the zero division, in which there was only one physicist – and that physicist was Einstein.

15.9 References

1 Planck, M. (1907). Letter to Einstein of 6 July 1907, Einstein Archives, Princeton, New Jersey, quoted by A. Hermann (1971) in *The Genesis of Quantum Theory (1899–1913)*, p. 56, Cambridge, Massachusetts: MIT Press.

2 Lorentz, H.A. (1909). Letter to W. Wein of 12 April 1909, quoted by A. Hermann (1971) in *The Genesis of Quantum Theory (1899–1913)*, p. 56, Cambridge, Massachusetts: MIT Press.

3 Einstein, A. (1909). *Phys. Zeitschrift*, **10**, 185.

4 Eintein, A. (1910). Letter to Laub, 16 March 1910. See Einstein Archives, *op. cit.* and Kuhn, T.S. (1978), *Black-body Theory and the Quantum Discontinuity 1894–1912*, p. 214. Oxford: Clarendon Press.

5 Planck, M. (1910). Letter to W. Nernst of 11 June 1910. See Kuhn (1978), *Black-body Theory and the Quantum Discontinuity 1894–1912*, p. 230.

6 Langevin, P. and De Broglie, M. eds. (1912). *La Théorie du rayonnement et les quanta: rapports et discussions de la réunion tenue à Bruxelles, du 30 Octobre au 3 Novembre 1911.* Paris: Gautier-Villars.

7 Millikan, R.A. (1916). *Phys. Rev.*, **7**, 355.

8 Thomson, J.J. (1897). *Phil. Mag.*, **44**, 311.

9 Thomson, J.J. (1907). *Conduction of Electricity through Gases*. Cambridge: Cambridge University Press.

10 Longair, M.S. (1997). *High Energy Astrophysics*, Vol. 1. Cambridge: Cambridge University Press (a revised and updated version of the 1992 edition).

11 Rutherford, E. and Royds, T. (1909). *Phil. Mag.*, **17**, 281.

12 See Andrade, E.N. da C. (1964). *Rutherford and the Nature of the Atom*, p. 111. New York: Doubleday.

13 Rutherford, E. (1911). *Phil. Mag.*, **21**, 669.

14 Geiger, H. and Marsden, E. (1913). *Phil. Mag.*, **25**, 604.

15 Heilbron, J. (1978). *History of Twentieth Century Physics*, *Proc. International School of Physics 'Enrico Fermi'*, Course 57, p. 40. New York and London: Academic Press.

16 Einstein, A. (1907). *Ann. Phys.*, **22**, p. 184.

17 Haas, A.E. (1910). *Wiener Berichte IIa*, **119**, 119; *Jahrb. der Radioakt. und Elektr.*, **7**, 261; *Phys. Zeitschr.*, **11**, 537.

18 Bohr, N. (1912). Bohr's memorandum is reproduced in *On the Constitution of Atoms and Molecules*, ed. L. Rosenfeld, Copenhagen (1963).

19 Nicholson, J.W. (1911), *Phil. Mag.*, **22**, 864; (1912) *Mon. Not. Roy. Astron. Soc.*, **72**, 677. See also R. McCormach (1966), The atomic theory of John William Nicholson, *Arch. Hist. Exact Sci.*, **2**, 160.

20 See Heilbron, J. (1978). *Op. cit.*, 70.

21 Bohr, N. (1913). *Phil. Mag.*, **26**, 1, 476, 857.

22 Pais, A. (1991). *Niels Bohr's times, in Physics, Philosophy and Polity*, p. 154. Oxford: Clarendon Press.

23 Einstein, A. (1916). *Phys. Gesell. Zürich. Mitteil.*, **16**, 47. See also *Deutsche Phys. Gesell. Verhandl.*, **18**, 318 and *Zeit. Phys.*, **18**, 121 (1917).

24 Compton, A.H. (1923). *Phys. Rev.*, **21**, 483.
25 See, for example, M.S. Longair (1997), *op. cit.*, p. 96 *et seq.*
26 Heisenberg, W. (1929). The development of the quantum theory 1918–1928, *Naturwiss.*, **17**, 491. (See translation of the quotation in R.H. Steuwer, 1975, *The Compton Effect*, p. 287, New York: Science History Publications.)
27 Compton, A.H. (1961). *Am. J. Phys.*, **29**, 817.
28 Lorentz, H.A. (1912). See P. Langevin and M. De Broglie eds. (1912), *op. cit.*, pp. 35–9.
29 Pauli, W. (1923). *Zeit. Phys.*, **18**, 272.
30 Einstein, A. and Ehrenfest, P. (1923). *Zeit. Phys.*, **19**, 301.
31 Hawking, S.W. (1988). *A Brief History of Time: From the Big Bang to Black Holes*, p. 56. London: Bantam Books.

Appendix to Chapter 15: The detection of signals in the presence of noise

We can use the tools developed in the last two chapters to study the problem of the detection of signals in the presence of noise. In many cases, very faint signals need to be detected and the noise may arise from a number of different sources. For example, the signal itself is of finite magnitude and therefore only determined to a certain statistical precision, and it may have to be detected in the presence of thermal noise in the detector; there may also be unwanted background radiation incident on the detector. One very useful result, before we consider the general problem, is an expression for the electrical noise in a resistor due to thermal fluctuations. This general result is known as *Nyquist's theorem*.

A15.1 *Nyquist's theorem and Johnson noise*

Nyquist's theorem is a beautiful application of Einstein's expression for the mean energy per mode (14.23) to the case of the spontaneous fluctuations in the energy present in a resistor at temperature T. This is a very important result for the design of electronic amplifiers and electric circuits used in association with low-noise receivers.

Let us derive the expression for the noise power generated in a transmission line which is terminated at either end by matched resistors R, that is, the impedance of the transmission line Z_0 for the propagation of electromagnetic waves is equal to R (Fig. A15.1). Suppose that the circuit is placed in an enclosure in thermal equilibrium at temperature T. Energy is shared equally among all the modes present in the system according to Einstein's prescription (14.23) that, per mode, the average energy is

$$\overline{E} = \frac{h\nu}{e^{h\nu/kT} - 1}, \tag{A15.1}$$

where ν is the frequency of the mode. Therefore, we need to know how many modes there are associated with the transmission line and award each of them energy $\overline{E}$ according to Einstein's version of the Maxwell–Boltzmann doctrine of equipartition.

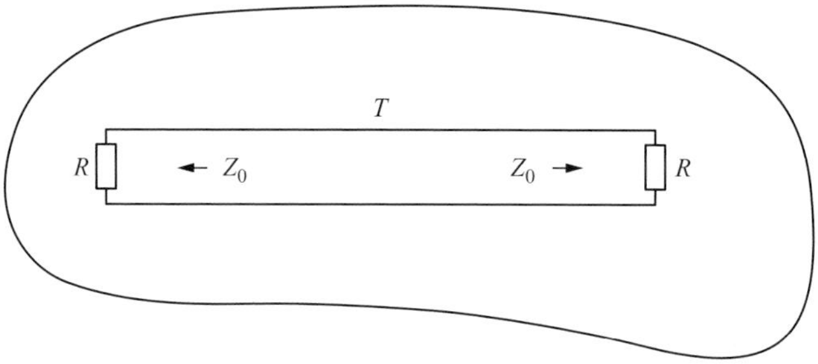

Figure A15.1: A transmission line of wave impedance Z_0, terminated at each end with matched resistors R in thermodynamic equilibrium within an enclosure at temperature T.

In effect, we perform the same calculation as Rayleigh carried out for the normal modes of oscillation of waves in a box, but now we are dealing with a one-dimensional, rather than a three-dimensional, problem. The one-dimensional case of a line of length L is easily derived from the relation (12.45),

$$\frac{\omega}{c} = \frac{n\pi}{L}, \tag{A15.2}$$

where n takes only integral values, $n = 1, 2, 3, \ldots$ and c is the speed of light. Standard analysis of the nature of these modes starting from Maxwell's equations shows that there is only one polarisation associated with each value of n. In thermodynamic equilibrium, each mode is awarded energy $\overline{E}$; hence, to work out the energy per unit frequency range we need to know the number of modes in the frequency range $d\nu$. The number is found by differentiating (A15.2):

$$dn = \frac{L}{\pi c}\, d\omega = \frac{2L}{c}\, d\nu, \tag{A15.3}$$

and hence the energy per unit frequency range is $2L\overline{E}/c$.

One of the basic properties of standing waves such as those represented by (A15.2) is that they correspond precisely to the superposition of waves of equal amplitude propagating at velocity c in opposite directions along the line. For each value of n, these travelling waves correspond to the only two permissible solutions of Maxwell's equations for propagation along the transmission line. A certain amount of power is therefore emitted and absorbed by the matched resistors R at either end of the line and, since they are matched, all the energy of the incident waves is absorbed and then re-emitted by them. Therefore, from (A15.3), equal amounts of energy $L\overline{E}/c$ per unit frequency interval propagate in either direction and travel along the wire into the resistors in a travel time $t = L/c$. The power P delivered to each resistor per unit frequency interval is

$$P = \frac{L\overline{E}/c}{L/c} = \overline{E}$$

in units of W Hz^{-1}. In thermal equilibrium, however, the same amount of power must be returned to the transmission line, or else the resistors will heat up above the equilibrium temperature T. This proves the fundamental result that the noise power delivered by a resistor at temperature T is

$$P = \frac{h\nu}{e^{h\nu/kT} - 1}. \tag{A15.4}$$

At low frequencies, $h\nu \ll kT$, which is normally the case for radio and microwave receivers, (A15.4) reduces to

$$P = kT. \tag{A15.5}$$

This is *Nyquist's theorem* applied to the thermal noise of a resistor at low frequencies. The power available in the frequency range ν to $\nu + d\nu$ is $P\,d\nu = kT\,d\nu$. In the opposite limit, $h\nu \gg kT$, the noise power decreases exponentially, as $P = h\nu \exp(-h\nu/kT)$.

The result (A15.5) was confirmed experimentally by J.B. Johnson in 1928 and the low-frequency noise that it describes is known as *Johnson noise*. For a variety of resistors, he found exactly the relation predicted by expression (A15.5) and derived an estimate of k within 8% of the present-day standard value.

Equation (A15.5) provides a convenient way of describing the performance of a radio receiver which delivers a certain noise power P_n. We define an *equivalent noise temperature* T_n for the performance of the receiver at frequency ν by the relation

$$T_n = P_n/k. \tag{A15.6}$$

For low frequencies, we note another key feature of the result (A15.5). Per unit frequency range per second, the electrical noise power kT corresponds exactly to the energy of a single mode in thermal equilibrium. Since this energy is in the form of electrical signals, or waves, the fluctuations in this mode correspond to

$$\Delta E/E = 1, \tag{A15.7}$$

as was shown in Section 15.3. Therefore, we expect the amplitude of the noise fluctuations per unit frequency interval to be kT. We will use this result in Section A15.3.

A15.2 The detection of photons in the presence of background noise

Let us consider first the case in which $h\nu \gg kT$. The results of Sections 15.4 and A15.1 show that in this case we can consider light to consist of independent particles, that is, photons. The statistical properties are described by the relations developed in Section 15.2. If there is no background radiation, the precision with which the signal is measured is

$$\frac{\Delta I}{I} = \frac{1}{N^{1/2}},$$

where N is the number of photons detected from the source.

Often faint sources of radiation are observed in the presence of a much greater background signal. If the number of background photons is $N_b \gg N$, the uncertainty in the measurements is largely determined by the uncertainty with which the background can be determined:

$$\frac{\Delta I}{I} = \frac{1}{(N + N_b)^{1/2}} \approx \frac{1}{N_b^{1/2}}.$$

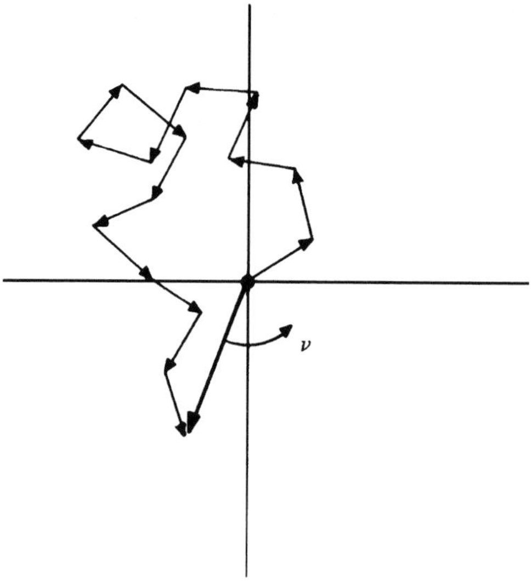

Figure A15.2: Illustrating the random superposition of electromagnetic waves in the frequency range ν to $\nu + \mathrm{d}\nu$ on an Argand diagram. The large vector is the sum of all the electric fields oscillating in this frequency range.

In this case, to measure the intensity of the source, we first measure the source plus background:

$$(N + N_\mathrm{b}) \pm (N + N_\mathrm{b})^{1/2} \approx (N + N_\mathrm{b}) \pm N_\mathrm{b}^{1/2}.$$

We next measure the background on its own, obtaining

$$N_\mathrm{b} \pm N_\mathrm{b}^{1/2}.$$

We then subtract these measures to produce an estimate of N, but in the subtraction we have to add the variances of the error estimates, so that the best estimate is

$$N \pm (N + 2N_\mathrm{b})^{1/2} \approx N \pm (2N_\mathrm{b})^{1/2}.$$

A15.3 The detection of electromagnetic waves in the presence of noise

In the case $h\nu \ll kT$, black-body radiation behaves like classical electromagnetic radiation. The signal induces a current or voltage in the detector, which can be modelled by the resistor considered in Section A15.1. Suppose the signal is received within a waveband ν to $\nu + \mathrm{d}\nu$. The detected signal is the vector sum of all the electric fields of the waves as displayed on an Argand diagram (Fig. A15.2), which shows the amplitude and phase of the contributions of each field to the total signal. This diagram rotates at a rate ν, the frequency of the waves. If all the waves were of exactly the same frequency, the pattern would continue to rotate in this configuration forever. However, because there is a spread in frequencies the vectors change their phase relations so that after a time $T \sim 1/\Delta\nu$ the vector sum of the waves is different. The time T is called the *coherence time* of the waves; it is roughly the time during which the vector sum of the waves provides a particular estimate of the resultant amplitude.

Whereas with photons we obtain an independent piece of information every time a photon arrives, in the case of waves we obtain a new estimate once per coherence time $T = 1/\Delta\nu$. Thus, if a source is observed for a time t we obtain $t/T = t\Delta\nu$ independent estimates of the intensity of the source. These results tell us how often we need to sample the signal.

Often we are interested in measuring very weak signals in the presence of noise in the receiver. The noise power itself fluctuates with amplitude $\Delta E/E = 1$ per mode per second, according to our analysis of Section A15.1. We can therefore reduce the amplitude of the fluctuations by integrating over long periods or by increasing the bandwidth of the receiver. In both cases, the number of independent estimates of the strength of the signal is $\Delta\nu\, t$ and hence the amplitude of the power fluctuations after time t is reduced to

$$\Delta P = \frac{kT}{(\Delta\nu t)^{1/2}}.$$

Thus, if we are prepared to integrate over a long enough time interval and/or use large enough bandwidths, very weak signals can be detected.

Case Study VI
Special relativity

Special relativity is a subject in which intuition is a dangerous tool. Statements such as 'moving clocks go slow' or 'moving rulers are length contracted' are common currency but, in my view, they can hinder rather than help an understanding of special relativity. In such non-intuitive areas, it is important to have at hand a robust formalism which can be relied upon to give correct answers. After a number of years of teaching the subject, I have concluded that the best approach is to introduce the Lorentz transformations and four-vectors as quickly as possible. These provide a simple set of rules which enable most reasonably straightforward problems in special relativity to be tackled with confidence. This approach involves constructing a four-vector algebra and calculus with rules essentially identical to those of three-vectors.

On researching the history of special relativity, it became apparent that many of my favourite tricks were well known to the pioneers, and so this case study begins with a modest historical introduction. I would emphasise that this is not intended as a course in special relativity, in which every 'i' is dotted and every 't' crossed. Rather the emphasis is upon the creation of a mathematical structure which enables relativistic calculations to be carried out as simply as possible.

Throughout this book, I have suppressed any detailed discussion of what I call the *genius of experiment*, the exception being in Chapter 18, in which I give full rein to my enthusiasm for the experimental and observation foundations of cosmology. Nonetheless, I cannot resist including a mention of the famous Michelson–Morley experiment in these introductory paragraphs. Albert Abraham Michelson was an experimenter of genius and his little book *Studies in Optics*[1] deserves to be on the shelf of every physicist. He was the first American citizen to win the Nobel Prize in science. His genius was to recognise the power of optical interferometry in measuring extremely small path differences between light rays. The object of the experiment was to use this extremely sensitive technique to measure the motion of the Earth through the hypothetical stationary aether, the medium through which electromagnetic phenomena were supposed to propagate, even in a vacuum. Figure VI.1 shows the famous Michelson–Morley experiment of 1887. In Michelson's own words,

> . . . the interferometer was mounted on a block of stone 1.5 m square and 0.25 m thick resting on an annular wooden ring which floated the whole apparatus on mercury.

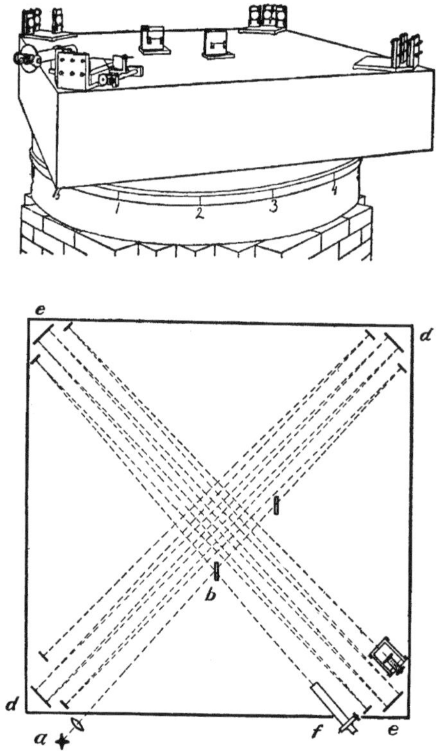

Figure VI.1: The Michelson–Morley experiment of 1887. (From A.A. Michelson, 1927, *Studies in Optics*, Chicago: University of Chicago Press.)

The lower diagram shows the multiple reflections used to increase the optical path lengths of the arms of the interferometer.

The experiment was performed by observing the movement of the central fringe of the interferometer as the apparatus was rotated 'fairly uniformly and continuously' through 360° and measurements made every one-sixteenth of a revolution. In Fig. VI.2, the mean displacements of the central fringe during rotation through 360° (solid lines) are compared with one eighth of the sinusoidal variation expected if the Earth moved through a stationary aether at 30 km s^{-1} (broken lines). Again, in Michelson's words,

It must be concluded that the experiment shows no evidence of a displacement greater than 0.01 fringe . . . With $V/c = 1/10\,000$, this gives an expected displacement of 0.4 fringes. The actual value is certainly less than one-twentieth of this actual amount and probably less than one-fortieth.

The experiment was repeated at different times of the year by Morley and Miller, in case the motion of the Earth about the Sun was cancelled out by the drift of the aether, but the same null result was found.

In the 1962 reprint of Michelson's book, Harvey B. Lemon writes in his introduction:

To the complete astonishment and mystification of the scientific world this refined experiment also yielded absolutely negative results. Again must we note at this point the universal confidence with

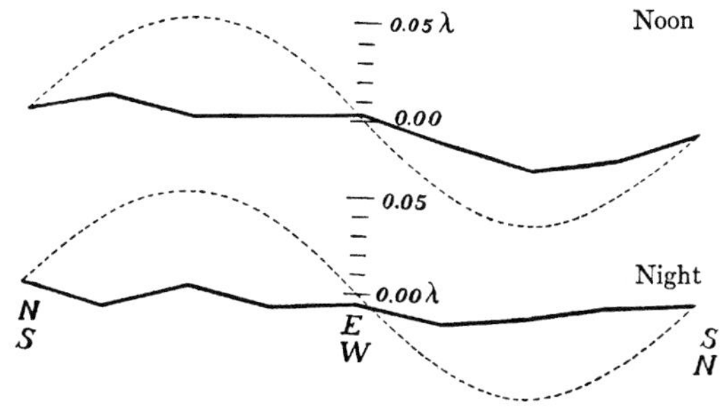

Figure VI.2: The null result of the Michelson–Morley experiment. The solid lines show the average movement of the central fringe as the apparatus was rotated through 360°. The broken lines show one eighth of the sinusoidal variation expected if the Earth moved through a stationary aether at 30 km s^{-1}. (From A.A. Michelson, 1927, *Studies in Optics*, Chicago: University of Chicago Press.)

which any experimental fact announced by Michelson was instantly accepted. Not for twenty years did anyone have the temerity to challenge his conclusion.

This may well be part of the reason that there is no explicit reference to the Michelson–Morley experiment in Einstein's great paper of 1905 – Michelson's null result instantly became one of the established facts in the thorny problem of understanding the nature of the aether.

VI.1 Reference

1 Michelson, A.A. (1927). *Studies in Optics*. Chicago: University of Chicago Press (first Phoenix edition 1962).

16 Special relativity – a study in invariance

16.1 Introduction

There is a compelling logic about the standard route to the Lorentz transformations and relativistic dynamics, which can be summarised as follows.

- Bradley's observations of stellar aberration of 1727–8 imply that the Earth moves through a stationary aether.
- The null result of the Michelson–Morley experiment of 1887 indicates that there is no detectable motion of the Earth through the aether.
- Einstein's second postulate of special relativity states that the speed of light is the same for an observer in any inertial frame of reference.
- Derivation of the Lorentz transformations and relativistic kinematics follows.
- The invariance of all the laws of physics under Lorentz transformation is invoked, according to Einstein's first postulate of relativity.
- Relativistic dynamics can now be derived.
- A consequence of the latter is $E = mc^2$.
- The invariance of Maxwell's equations follows automatically from the properties of the Lorentz transformations.

It is a splendid story, but is not at all how the theory came about historically. As with many of the most dramatic turning points in the history of physics, the route to the final theory was tortuous and involved many blind alleys. The origins of special relativity have been the subject of a great deal of study by historians of science and, fortunately, there are two excellent and accessible surveys of the key events by Stachel.[1,2] More details of the physical arguments which led to the special theory of relativity are contained in Chapters 6–8 of the important study by Pais, *Subtle is the Lord . . . the Science and Life of Albert Einstein*.[3] I will use these sources extensively in this section.

The emphasis of this case study is more on the concept of invariance, rather than on the history of relativity, but some of the events which led up to Einstein's great papers of 1905 illuminate the terminology and set them in context.

16.1.1 Maxwell's equations and Voigt's paper (1887)

As discussed in Section 5.5, it was some time after 1865 before Maxwell's formulation of the laws of electromagnetism, as encapsulated in the compact set of four equations (5.45),was fully accepted. Maxwell's version of the theory was much more cumbersome than this familiar set of equations, which were put into their definitive form by the efforts of Heaviside, Helmholtz and Hertz. Hertz's demonstration that electromagnetic phenomena are propagated at the speed of light and that electromagnetic waves have all the properties of light waves were persuasive evidence that Maxwell's equations indeed encapsulated in compact form the known phenomena of electromagnetism and unified the physics of electromagnetism and light.

This triumph came, however, at a cost. At the time, all known wave phenomena resulted from the perturbation of some material medium and so the nature of the medium through which electromagnetic waves were thought to be propagated, the aether, became a central concern of late nineteenth-century physics.[4] Bradley's observations of stellar aberration had seemed to show that the Earth must be moving through a stationary aether. To account for these observations, the Galilean expression for the addition of velocities was used; this follows from the Galilean transformations under which Newton's laws of motion are form-invariant. A problem for nineteenth-century physicists was that, unlike Newton's laws of motion, Maxwell's equations are not form-invariant under Galilean transformation.

To demonstrate this, the Galilean transformations can be used to derive expressions which describe how partial derivatives transform between inertial frames of reference S and S'. Thus, if we write

$$\begin{aligned} t' &= t, \\ x' &= x - Vt, \\ y' &= y, \\ z' &= z, \end{aligned} \tag{16.1}$$

then using the chain rule we find, for example,

$$\frac{\partial}{\partial t} = \frac{\partial t'}{\partial t}\frac{\partial}{\partial t'} + \frac{\partial x'}{\partial t}\frac{\partial}{\partial x'} + \frac{\partial y'}{\partial t}\frac{\partial}{\partial y'} + \frac{\partial z'}{\partial t}\frac{\partial}{\partial z'} = \frac{\partial}{\partial t'} - V\frac{\partial}{\partial x'} \tag{16.2}$$

from (16.1). The transforms of the partial derivatives are thus

$$\begin{aligned} \frac{\partial}{\partial t} &\to \frac{\partial}{\partial t'} - V\frac{\partial}{\partial x'}, \\ \frac{\partial}{\partial x} &\to \frac{\partial}{\partial x'}, \\ \frac{\partial}{\partial y} &\to \frac{\partial}{\partial y'}, \\ \frac{\partial}{\partial z} &\to \frac{\partial}{\partial z'}. \end{aligned} \tag{16.3}$$

The problem with Maxwell's equations can be appreciated by transforming them from the laboratory frame to a frame moving at velocity V along the positive x-axis. Let us write

Maxwell's equations *in free space* in Cartesian coordinates:

$$\begin{aligned}\frac{\partial E_z}{\partial y} - \frac{\partial E_y}{\partial z} &= -\frac{\partial B_x}{\partial t},\\ \frac{\partial E_x}{\partial z} - \frac{\partial E_z}{\partial x} &= -\frac{\partial B_y}{\partial t},\\ \frac{\partial E_y}{\partial x} - \frac{\partial E_x}{\partial y} &= -\frac{\partial B_z}{\partial t};\end{aligned} \tag{16.4}$$

$$\begin{aligned}\frac{\partial B_z}{\partial y} - \frac{\partial B_y}{\partial z} &= \frac{1}{c^2}\frac{\partial E_x}{\partial t},\\ \frac{\partial B_x}{\partial z} - \frac{\partial B_z}{\partial x} &= \frac{1}{c^2}\frac{\partial E_y}{\partial t},\\ \frac{\partial B_y}{\partial x} - \frac{\partial B_x}{\partial y} &= \frac{1}{c^2}\frac{\partial E_z}{\partial t};\end{aligned} \tag{16.5}$$

$$\frac{\partial E_x}{\partial x} + \frac{\partial E_y}{\partial y} + \frac{\partial E_z}{\partial z} = 0; \tag{16.6}$$

$$\frac{\partial B_x}{\partial x} + \frac{\partial B_y}{\partial y} + \frac{\partial B_z}{\partial z} = 0. \tag{16.7}$$

Substituting for the partial derivatives in (16.4) and (16.5), and using (16.6) and (16.7) as transformed into the S' frame of reference, (16.4) and (16.5) become

$$\frac{\partial(E_z + VB_y)}{\partial y'} - \frac{\partial(E_y - VB_z)}{\partial z'} = -\frac{\partial B_x}{\partial t'}, \tag{16.8}$$

$$\frac{\partial E_x}{\partial z'} - \frac{\partial(E_z + VB_y)}{\partial x'} = -\frac{\partial B_y}{\partial t'}, \tag{16.9}$$

$$\frac{\partial(E_y - VB_z)}{\partial x'} - \frac{\partial E_x}{\partial y'} = -\frac{\partial B_z}{\partial t'}, \tag{16.10}$$

and

$$\frac{\partial(B_z - VE_y)}{\partial y'} - \frac{\partial(B_y + VE_z)}{\partial z'} = \frac{1}{c^2}\frac{\partial E_x}{\partial t'}, \tag{16.11}$$

$$\frac{\partial B_x}{\partial z'} - \frac{\partial(B_z - VE_y)}{\partial x'} = \frac{1}{c^2}\frac{\partial E_y}{\partial t'}, \tag{16.12}$$

$$\frac{\partial(B_y + VE_z)}{\partial x'} - \frac{\partial B_x}{\partial y'} = \frac{1}{c^2}\frac{\partial E_z}{\partial t'}. \tag{16.13}$$

We can now seek definitions for the components of $\boldsymbol{E}$ and $\boldsymbol{B}$ in the moving frame S'. For example, in the first term of (16.8) and the second term of (16.9) we might write $E'_z = E_z + VB_y$, but then we would run into trouble with the term in (16.13) involving E_z. Similarly, although we might be tempted to write $B'_y = B_y + VE_z$ in the second term of (16.11) and the first term of (16.13), this runs into serious trouble with the last term of (16.9).

It can be seen that the only frame of reference in which the Galilean transformations result in a self-consistent set of equations is that for which $V = 0$. For these reasons, it was believed that Maxwell's equations would hold true only in the frame of reference of a stationary aether – they were considered to be *non-relativistic* under Galilean transformations.

Remarkably, in 1887 Woldmar Voigt[5] noticed that Maxwell's wave equation for electromagnetic waves

$$\nabla^2 \boldsymbol{H} - \frac{1}{c^2}\frac{\partial^2 \boldsymbol{H}}{\partial^2 t} = 0 \tag{16.14}$$

is form-invariant under the transformation

$$\begin{aligned} t' &= t - \frac{Vx}{c^2}, \\ x' &= x - Vt, \\ y' &= y/\gamma, \\ z' &= z/\gamma \end{aligned} \tag{16.15}$$

where $\gamma = (1 - V^2/c^2)^{-1/2}$. Except for the fact that the transformations on the right-hand side have been divided by the Lorentz factor γ, the set of equations (16.15) is the Lorentz transformation. Voigt derived this expression using the invariance of the phase of a propagating electromagnetic wave, exactly the approach we will take in Section 16.2. This work was unknown to Lorentz when he derived what we now know as the Lorentz transformations.

16.1.2 Fitzgerald's paper (1889)

The Irish physicist George Francis Fitzgerald is rightly credited with the remarkable insight that the null result of the Michelson–Morley experiment could be explained by supposing that moving objects are length contracted in their direction of motion. His brief paper entitled 'The aether and the Earth's atmosphere', published in *Science* in 1889, reads as follows.

> I have read with much interest Messrs Michelson and Morley's wonderfully delicate experiment attempting to decide the important question as to how far the aether is carried along by the Earth. Their result seems opposed to other experiments showing that the aether in the air can . . . be carried along only to an inappreciable extent. I would suggest that almost the only hypothesis that can reconcile this opposition is that the length of material bodies changes, according as they are moving through the aether or across it, by an amount depending on the square of the ratio of their velocity to that of light. We know that electric forces are affected by the motion of electrified bodies relative to the aether, and it seems a not improbable supposition that the molecular forces are affected by the motion, and that the size of the body alters consequently . . .[6]

The above quotation is more than 60% of his brief note. Remarkably, this paper, by which Fitzgerald is best remembered, was not included in his complete works edited by Larmor in 1902. Lorentz knew of the paper in 1894, but Fitzgerald was uncertain as to whether it had been published, when Lorentz wrote to him. The reason was that *Science* went bankrupt in 1889 and was only refounded in 1895. Notice that Fitzgerald's proposal was only qualitative and that he was proposing a real physical contraction of the body in its direction of motion because of interaction with the aether.

16.1.3 The Lorentz transformation

Hendrik Lorentz had agonised over the null result of the Michelson–Morley experiment and in 1892 came up with same suggestion as Fitzgerald, but with a quantitative expression for the length contraction. In his words,

> This experiment has been puzzling me for a long time, and in the end I have been able to think of only one means of reconciling it with Fresnel's theory. It consists in the supposition that the line joining two points of a solid body, if at first parallel to the direction of the Earth's motion, does not keep the same length when subsequently turned through 90°.[7]

Lorentz worked out that the length contraction had to amount to

$$l = l_0\left(1 - \frac{V^2}{2c^2}\right), \tag{16.16}$$

which is just the low-velocity limit of the expression

$$l = \frac{l_0}{\gamma}, \qquad \text{where} \qquad \gamma = \left(1 - \frac{V^2}{c^2}\right)^{-1/2}.$$

Subsequently, this phenomenon has been referred to as the *Fitzgerald–Lorentz contraction*.

In 1895, Lorentz[8] tackled the problem of finding transformations which would result in the form invariance of Maxwell's equations and derived the following relations (which we express in SI notation):

$$\begin{aligned} &x' = x - Vt, \qquad y' = y, \qquad z' = z,\\ &t' = t - \frac{Vx}{c^2},\\ &\boldsymbol{E}' = \boldsymbol{E} + \boldsymbol{V} \times \boldsymbol{B},\\ &\boldsymbol{B}' = \boldsymbol{B} - \frac{\boldsymbol{V} \times \boldsymbol{E}}{c^2},\\ &\boldsymbol{P}' = \boldsymbol{P} \end{aligned} \tag{16.17}$$

where $\boldsymbol{P}$ is the polarisation. Under this set of transformations, Maxwell's equations are form-invariant to first order in V/c. Notice that time is no longer absolute. Lorentz apparently considered this simply to be a convenient mathematical tool in order to ensure form invariance to first order in V/c. He called t the *general time* and t' the *local time*. In order to account for the null result of the Michelson–Morley experiment, he had to include an additional second-order compensation factor, the Fitzgerald–Lorentz contraction $(1 - V^2/c^2)^{-1/2}$, into the theory. One important innovation of this paper was the *assumption* that the force on an electron should be given by the first-order expression

$$\boldsymbol{f} = e(\boldsymbol{E} + \boldsymbol{V} \times \boldsymbol{B}). \tag{16.18}$$

This is the origin of the expression for the *Lorentz force* for the joint action of electric and magnetic fields on a charged particle.

Einstein knew of Lorentz's paper of 1895 but was unaware of his subsequent work. In 1899, Lorentz[9] established the invariance of the equations of electromagnetism to all orders in V/c, through a new set of transformations:

$$x' = \epsilon\gamma(x - Vt), \qquad y' = \epsilon y, \qquad z' = \epsilon z,$$
$$t' = \epsilon\gamma\left(t - \frac{Vx}{c^2}\right). \tag{16.19}$$

These are the *Lorentz transformations* in a form that includes the scale factor ϵ. By this means, he was able to incorporate length contraction into the transformations. Almost coincidentally, in 1898 Joseph Larmor wrote his prize-winning essay *Aether and Matter*,[10] in which he derived the standard form of the Lorentz transformations and showed that they included the Fitzgerald–Lorentz contraction.

In his major paper of 1904, entitled 'Electromagnetic phenomena in a system moving with any velocity smaller than light',[11] Lorentz presented the transformations (16.19) with $\epsilon = 1$. However, as expressed by Gerald Holten, the apparatus used to arrive at this result contained a number of 'fundamental assumptions'. In Holten's words,

> [The theory] in fact contained eleven *ad hoc* hypotheses: restriction to small ratios of velocities V to light velocity c, postulation *a priori* of the transformation equations . . . , assumption of a stationary aether, assumption that the stationary electron is round, that its charge is uniformly distributed, that all mass is electromagnetic, that the moving electron changes one of its dimensions precisely in the ratio $(1 - V^2/c^2)^{1/2}$ to one, ...[12]

Thus, in Lorentz's approach, the transformations were part of the theory of the electron and were postulated *a priori* as part of the aether theory of electromagnetism. Einstein was not aware of this paper in 1905.

16.1.4 Poincaré's contribution

Henri Poincaré made a number of crucial contributions to the development of special relativity. In a remarkable paper of 1898, entitled 'La mesure du temps',[13] he identified the problem of what is meant by simultaneity and the measurement of time intervals. In the conclusion of his paper, he writes

> The simultaneity of two events or the order of their succession, as well as the equality of two time intervals, must be defined in such a way that the statement of the natural laws be as simple as possible. In other words, all rules and definitions are but the result of unconscious opportunism.

In 1904, Poincaré surveyed the current problems in physics and included the statement:

> . . . the principle of relativity, according to which the laws of physical phenomena should be the same whether for an observer fixed or for an observer carried along by uniform movement or translation.[14]

Notice that this is no more than a restatement of Galilean relativity. However, he concluded by remarking

> Perhaps likewise, we should construct a whole new mechanics, . . . where, inertia increasing with the velocity, the velocity of light would become an impassible limit.

16.1.5 *Einstein before 1905*

Albert Einstein had been wrestling with exactly these problems since 1898. He was certainly aware of Poincaré's writings and these were the subject of intense discussion with his scientific colleagues in Bern. According to Einstein, in a letter of 25 April 1912 to Paul Ehrenfest,

> I knew that the principle of the constancy of the velocity of light was something quite independent of the relativity postulate and I weighted which was the more probable, the principle of the constancy of c, as required by Maxwell's equations, or the constancy of c exclusively for an observer located at the light source. I decided in favour of the former.[15]

In 1924, he stated:

> After seven years of reflection in vain (1898–1905), the solution came to me suddenly with the thought that our concepts of space and time can only claim validity insofar as they stand in a clear relation to our experiences; and that experience could very well lead to the alteration of these concepts and laws. By a revision of the concept of simultaneity into a more malleable form, I thus arrived at the special theory of relativity.[15]

Once he had discovered the concept of the relativity of simultaneity, it took him only five weeks to complete his great paper, 'On the electrodynamics of moving bodies'.[16] As we will discuss below, this paper of Einstein contains the two postulates of special relativity, in addition to which he made only four assumptions, one concerning the isotropy and homogeneity of space, the others concerning three logical properties involved in the definition of the synchronisation of clocks. Thus, Einstein's approach not only simplified greatly many of the ideas current at the time, but completely revolutionised our understanding of the nature of space and time.

Einstein did not regard what he had done as particularly revolutionary. In his own words,

> With respect to the theory of relativity, it is not at all a question of a revolutionary act, but a natural development of a line which can be pursued through the centuries.[17]

It is noteworthy that, of the three great papers published in 1905, Einstein regarded that on quanta as the really revolutionary paper of the three.

16.1.6 *Reflections*

Relativity is a subject in which the results are not at all intuitively obvious on the basis of our everyday experience. The number of books on the subject is immense, but, above all others, pride of place must go to Einstein's original paper of 1905.[16] It is as clear as any of the subsequent papers in expounding the basis of the theory. It is one of the miracles of theoretical physics that Einstein produced such a complete and elegant exposition with such profound implications for our understanding of the nature of space and time in this single paper.

As I have already noted, in this case study I want to look at special relativity from the point of view of *invariance*. This is very much in the spirit of Einstein. I do not want to go into the formalities of the subject but rather to concentrate on deriving the formulae of

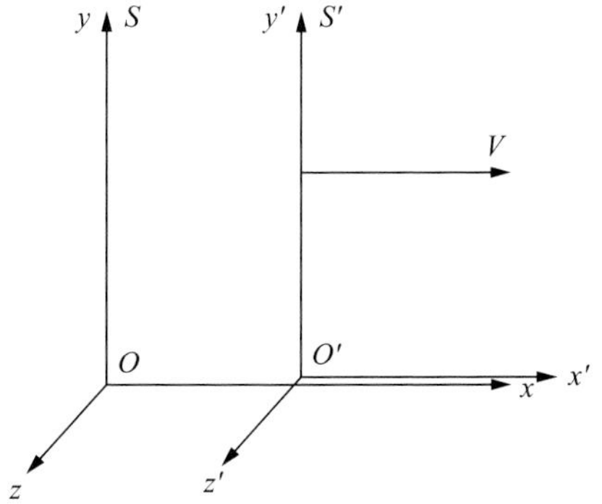

Figure 16.1: Inertial frames of reference S and S' in standard configuration.

special relativity with the minimum number of assumptions and exposing clearly where we have to be guided by experiment. The approach will be similar to that of Rindler in his excellent textbook *Relativity: Special, General and Cosmological.*[18] I find Rindler's exposition one of the most satisfying of all books on relativity.

16.2 Geometry and the Lorentz transformation

Let us begin by deriving the Lorentz transformations as directly as possible from Einstein's postulates of relativity. As in the case of Galilean relativity, we begin with two inertial frames of reference moving with relative velocity V. We call S the laboratory frame of reference, or the local rest frame, and S' the moving frame, which we take to have a velocity V in the positive x-direction with respect to S. We adopt rectangular Cartesian coordinate systems in S and S' with their axes parallel – frames of reference S and S' in this orientation are referred to as being in 'standard configuration' (Fig. 16.1). We will not go into the formalities of showing that it is indeed possible to set up such a standard configuration – this is done by Rindler – but we will assume that it can be done.

The special theory of relativity as formulated by Einstein in his great paper of 1905 contains two basic postulates. Let us quote the words of Einstein himself.

> The same laws of electrodynamics and optics are valid for all frames of reference for which the equations of mechanics hold good. We will raise this conjecture (the purport of which will hereafter be called the 'Principle of Relativity') to the status of a postulate and also introduce another postulate which is only apparently irreconcilable with the former, namely that light is always propagated in empty space with a definite velocity c which is independent of the state of motion of the emitting body.

Rephrasing Einstein's postulates, the first is that the laws of physics are the same in all inertial frames of reference and, second, that the speed of light should have the same value c in all inertial frames. It is the second postulate which is crucial in leading to the special theory of relativity.

Suppose the origins of the inertial frames S and S' are coincident at $t = 0$ in S and $t' = 0$ in S'. It is always possible to arrange this to be the case at any point in space by resetting the clocks in S and S'. Now let us send out an electromagnetic wave from the origin at $t = 0$, $t' = 0$. Because of Einstein's second postulate, we can immediately write down equations for the motion of a point on the wavefront in the two frames of reference:

$$\begin{aligned} &\text{in } S, && c^2t^2 - x^2 - y^2 - z^2 = 0, \\ &\text{in } S', && c^2t'^2 - x'^2 - y'^2 - z'^2 = 0, \end{aligned} \tag{16.20}$$

guaranteeing that in both frames the speed of light is c. We will use essentially the same argument as that used by Voigt to derive his version of the transformations (16.15) in 1887. The *four-vectors* $[ct, x, y, z]$ and $[ct', x', y', z']$ specify an *event* in S and S' respectively. We seek a set of transformations between the coordinate sets x, y, z, ct and x', y', z', t' that satisfies (16.20). An important way of rephrasing this statement is that we seek transformations which will leave $(c^2t^2 - x^2 - y^2 - z^2)$ *form-invariant* between inertial frames of reference.

We see immediately that there is a formal analogy between (16.20) and the properties of the components of a three-vector under rotation: if two Cartesian frames of reference have the same origin and one is rotated through an angle θ with respect to the other, the norm (or the square of the magnitude) of a three-vector is the same in both frames, that is,

$$R^2 = x^2 + y^2 + z^2 = x'^2 + y'^2 + z'^2. \tag{16.21}$$

Let us consider the problem of transforming (16.20) in the same way. For simplicity, suppose the wave is propagated along the positive x-direction, so that the transform we are seeking must result in

$$c^2t^2 - x^2 = c^2t'^2 - x'^2 = 0. \tag{16.22}$$

We first convert (16.22) into a form similar to (16.21) by writing $c\tau = \mathrm{i}ct$, $c\tau' = \mathrm{i}ct'$. Thus, we have already introduced *imaginary time* τ, a concept which has caused so much grief to lay readers of Stephen Hawking's *A Brief History of Time*[19] – it is no more than a change of variable. Then

$$c^2\tau^2 + x^2 = c^2\tau'^2 + x'^2 = 0. \tag{16.23}$$

By comparison with (16.21), we can see that the necessary transformation formulae are just the rotation formulae for vectors in two dimensions:

$$\begin{aligned} c\tau' &= -x \sin\theta + c\tau' \cos\theta, \\ x' &= x\cos\theta + c\tau \sin\theta. \end{aligned} \tag{16.24}$$

Notice that it is assumed here that the 'time coordinate' corresponds to the 'y-direction'. Now τ is imaginary and so the angle θ must be imaginary as well. To convert into real quantities, write $\theta = \mathrm{i}\varphi$, where φ is real. Because

$$\begin{aligned} \cos \mathrm{i}\varphi &= \cosh\varphi, \\ \sin \mathrm{i}\varphi &= \mathrm{i}\sinh\varphi, \end{aligned}$$

we find

$$\begin{aligned} ct' &= -x \sinh\varphi + ct \cosh\varphi, \\ x' &= x \cosh\varphi - ct \sinh\varphi. \end{aligned} \tag{16.25}$$

If we had known that this pair of equations leaves $x^2 - c^2t^2$ form-invariant, we could have avoided introducing imaginary time.

We now need to determine the constant φ. We note that at time t in S the origin of S', $x' = 0$, has moved to $x = Vt$ in S. Substituting into the second relation of (16.25), we find

$$0 = x \cosh\varphi - ct \sinh\varphi, \qquad \tanh\varphi = x/(ct) = V/c.$$

Then, because

$$\cosh\varphi = (1 - V^2/c^2)^{-1/2} \qquad \text{and} \qquad \sinh\varphi = (V/c)(1 - V^2/c^2)^{-1/2},$$

we find from (16.25)

$$\begin{aligned} ct' &= \gamma\,(ct - Vx/c)\,, \\ x' &= \gamma(x - ct), \end{aligned} \tag{16.26}$$

where $\gamma = (1 - V^2/c^2)^{-1/2}$ is often referred to as the *Lorentz factor* – it appears in virtually all calculations in special relativity. We also find that $y' = y$, $z' = z$ from the symmetry of the transformations between S and S'.

We have therefore derived the complete set of Lorentz transformations using the idea of form-invariance and Einstein's second postulate, which is a statement about the invariance of the speed of light between inertial frames of reference. Let us write the Lorentz transformations in the standard form that we will employ throughout this exposition:

$$\begin{aligned} ct' &= \gamma\,(ct - Vx/c)\,, \\ x' &= \gamma(x - ct), \\ y' &= y, \\ z' &= z, \\ \gamma &= \left(1 - V^2/c^2\right)^{-1/2}. \end{aligned} \tag{16.27}$$

It should be stressed that x, y, z, t and x', y', z', t' are the coordinates of the same *event* as measured in the two frames S and S'. In the case of (16.20) the event is the arrival at (x, y, z) of the wavefront of the light emitted at $t = 0$.

In the standard exposition of special relativity, one now goes on to explore the remarkable consequences of the set of transformations (16.27) and the so-called paradoxes which are supposed to arise. Let me state immediately that there are no paradoxes – only some rather non-intuitive features of space–time which result from the Lorentz transformations. The origin of these phenomena is what Einstein recognised as the key feature of special relativity, the *relativity of simultaneity*. This concept elucidates the origin of some of the difficulties which students have in understanding relativity. As we have already stated, at any point in space we can reset the clocks in S and S' so that they read the same time at that instant.

In the above example, we have arranged that $x' = 0$, $x = 0$, $t' = 0$, $t = 0$. This event, the coincidence of the origins of the two sets of axes in Fig. 16.1, is simultaneous in the two frames of reference. For an event at any other value of x, the observers in S and S' disagree about the time at which the event occurs, since

$$ct' = \gamma\,(ct - Vx/c)\,, \tag{16.28}$$

and hence if $x \neq 0$, $t' \neq t$. In other words, although observers in S and S' can agree on simultaneity at one point in space–time, they will disagree *at all other points*. This is the origin of the phenomena of time dilation, length contraction, the twin paradox etc. It is the fundamental difference between Galilean and special relativity. In Galilean relativity, observers in S and S' always agree everywhere about simultaneity. This is clear from the Newtonian limit of (16.28). If $V/c \to 0$, $\gamma \to 1$ and $t' = t$ everywhere.

16.3 Three-vectors and four-vectors

Three-vectors provide a compact notation for writing the laws of physics in a form which is independent of the choice of reference frame. No matter how we rotate or translate the frame of reference, or change the system of coordinates used, the vector relations remain form-invariant. For example:

(i) Vector addition is preserved:

$$\text{if}\quad \boldsymbol{a} + \boldsymbol{b} = \boldsymbol{c} \qquad \text{then} \qquad \boldsymbol{a}' + \boldsymbol{b}' = \boldsymbol{c}'.$$

(ii) Equally,

$$\text{if}\quad \boldsymbol{a}\cdot(\boldsymbol{b} + \boldsymbol{c}) \qquad \text{then} \qquad \boldsymbol{a}'\cdot(\boldsymbol{b}' + \boldsymbol{c}').$$

(iii) The *norm* or *magnitude* of the three-vector is invariant with respect to rotations and displacements:

$$\boldsymbol{a}^2 = |\boldsymbol{a}|^2 = a_1^2 + a_2^2 + a_3^2 = a_1'^2 + a_2'^2 + a_3'^3.$$

(iv) Scalar products are invariant:

$$\boldsymbol{a}\cdot\boldsymbol{b} = a_1 b_1 + a_2 b_2 + a_3 b_3 = a_1' b_1' + a_2' b_2' + a_3' b_3'.$$

In other words, these vector relations express truths which are independent of the frame of reference in which we perform our calculation.

Our objective is to find quantities similar to three-vectors which remain form-invariant under Lorentz transformation. These *four-vectors* are objects which enable us to write down the law of physics in relativistic form. The components of the four-vectors should transform like the components of the primitive four-vector $[ct, x, y, z]$, which we met in the previous section; we need to find expressions for these components which relate them to physical quantities measurable in the laboratory.

There remains the thorny question of notation. I will use the Lorentz transformations in the form (16.27) and define the components of four-vectors as quantities which transform like ct and x, y, z. The components of the four-vectors are written in brackets and I use bold italic capital letters to represent a four-vector. Thus

$$\boldsymbol{R} \equiv [ct, x, y, z]. \tag{16.29}$$

To work out the *norm* of $\boldsymbol{R}$ we write

$$\boldsymbol{R}^2 = |\boldsymbol{R}|^2 = c^2t^2 - x^2 - y^2 - z^2, \tag{16.30}$$

that is, there is a plus sign in front of c^2t^2 and minus signs in front of x^2, y^2 and z^2; in the language of relativity, the *signature* of the metric is $[1, -1, -1, -1]$. The time component has a different status from the spatial components, and so I will use the rule that the time component has subscript 0.

Thus, if a four-vector is written as $\boldsymbol{A} = [A_0, A_1, A_2, A_3]$ then its norm is

$$|\boldsymbol{A}|^2 = A_0^2 - A_1^2 - A_2^2 - A_3^2, \tag{16.31}$$

exactly equivalent to $R^2 = x^2 + y^2 + z^2$ for three-vectors. The transformation rules for the components of $\boldsymbol{A}$ can be found from (16.27), according to the identifications

$$\begin{array}{ll} ct' \to A_0', & ct \to A_0; \\ x' \to A_1', & x \to A_1; \\ y' \to A_2', & y \to A_2; \\ z' \to A_3', & z \to A_3. \end{array}$$

In the rest of this section we will be concerned with finding four-vectors for *kinematic* quantities, that is, quantities which *describe* motion. We have already introduced our primitive four-vector $\boldsymbol{R} \equiv [ct, x, y, z]$. Let us proceed now to the *displacement four-vector*.

16.3.1 Displacement four-vector

The four-vector $[ct, x, y, z]$ transforms between inertial frames of reference according to the Lorentz transformation and therefore so does the four-vector $[ct + c\Delta t, x + \Delta x, y + \Delta y, z + \Delta z]$. From the linearity of the Lorentz transformations, it follows that $[c\Delta t, \Delta x, \Delta y, \Delta z]$ also transforms like $[ct, x, y, z]$. We therefore define the quantity

$$\Delta \boldsymbol{R} = [c\Delta t, \Delta x, \Delta y, \Delta z] \tag{16.32}$$

to be the *displacement four-vector*. To express it another way, it must be a four-vector by virtue of being the difference of two four-vectors.

Evidently, the time interval between two events depends upon the inertial frame of reference in which the observer is located. Of special importance is the case in which two events take place at the same spatial location in some frame S', that is, for which $\Delta x' = \Delta y' = \Delta z' = 0$. The *proper time* Δt_0 is the time interval between events in this special frame of reference. It is straightforward to show that it is the *shortest time interval* measured in any frame of reference, as follows. Taking the norms of the displacement four-vector in the frames of reference S and S',

$$c^2\Delta t^2 - \Delta x^2 - \Delta y^2 - \Delta z^2 = c^2\Delta t'^2 - \Delta x'^2 - \Delta y'^2 - \Delta z'^2.$$

In the frame S', $\Delta x' = \Delta y' = \Delta z' = 0$ and $\Delta t' = \Delta t_0$, and so

$$c^2 \Delta t_0^2 = c^2 \Delta t^2 - \Delta x^2 - \Delta y^2 - \Delta z^2; \tag{16.33}$$

since Δx^2, Δy^2, Δz^2 are necessarily positive, Δt_0 to must be the minimum time measured in any inertial frame of reference.

An important aspect of the proper time interval Δt_0 is that it is the only invariant time interval upon which all observers can agree for a pair of events separated by $c\Delta t$, Δx, Δy, Δz. Observers in different inertial frames make different measurements of $c\Delta t$, Δx, Δy, Δz and, although they all measure different values of Δt for the time interval between the same two events, they all agree about the value of Δt_0 when they measure Δx, Δy and Δz as well and insert their values into (16.33). As indicated above, the proper time is the only invariant time interval on which they can all agree.

Let us work out the relation between the proper time interval Δt_0 measured in S' and the time interval between the same two events measured in the laboratory frame of reference S. The four-vectors associated with the proper time interval Δt_0 and the laboratory frame are

$$[c\Delta t_0, 0, 0, 0], \qquad [c\Delta t, \Delta x, \Delta y, \Delta z]. \tag{16.34}$$

Equating the norms of these four-vectors,

$$c^2 \Delta t_0^2 = c^2 \Delta t^2 - \Delta x^2 - \Delta y^2 - \Delta z^2 = c^2 \Delta t^2 - \Delta \boldsymbol{r}^2. \tag{16.35}$$

But $\Delta \boldsymbol{r}$ is just the displacement of the origin of S' in the time Δt. In other words, the velocity V of the origin of S' is $\Delta \boldsymbol{r}/\Delta t$ and hence from (16.35)

$$\Delta t_0^2 = \Delta t^2 \left(1 - \frac{V^2}{c^2}\right), \tag{16.36}$$

that is,

$$\Delta t_0 = \frac{\Delta t}{\gamma}. \tag{16.37}$$

Since γ is always greater than unity, this calculation demonstrates that the proper time interval Δt_0 is the shortest time between two events. We will use (16.37) repeatedly in what follows.

The relation (16.37) explains the observation of muons at sea-level. Cosmic rays are very high energy protons and nuclei which originate in violent explosions of stars and make their way to the Solar System through the interstellar medium. Muons are created in the upper layers of the atmosphere by collisions between cosmic rays, mostly protons, and the nuclei of atoms of the atmospheric gases. A typical interaction is

$$p + p \to n\pi^+ + n\pi^- + n\pi^0 + p + p + \cdots,$$

that is, a shower of positive, negative and neutral pions is created in the collision. In these collisions, roughly equal numbers of the three types of pion are produced. The positive and negative pions have very short lifetimes, $\tau = 2.551 \times 10^{-8}$ s, and decay into muons:

$$\pi^+ \to \mu^+ + \nu_\mu, \qquad \pi^- \to \mu^- + \bar{\nu}_\mu;$$

the ν_μ and $\bar{\nu}_\mu$ are muon neutrinos and antineutrinos respectively. The muons decay after a short lifetime $\Delta t_0 = 2.2 \times 10^{-6}$ s into electrons, positrons, neutrinos, muon neutrinos and so on. Since they are created at the top of the atmosphere, at a height of about 10 km, according to Galilean relativity they should travel a typical distance of only $c\Delta t_0$ before decaying, that is, a distance of $3 \times 10^5 \times 2.2 \times 10^{-6}$ km $= 660$ m. Therefore, we would expect very few of them to reach the surface of the Earth. However, intense fluxes of relativistic muons are in fact observed at the surface of the Earth. These high-energy muons have Lorentz factors $\gamma > 20$. Because of the effects of *time dilation*, (16.37), the observer on the Earth measures a decay half-life $\Delta t = \gamma \Delta t_0$, an effect which is often confusingly referred to as 'moving clocks run slow', meaning that Δt_0 is the shortest possible time interval between the muon's creation and decay. The muon itself acts as a moving clock. According to the observer on the surface of the Earth, the muon has a half life of $\gamma \Delta t_0$, in which time it can travel a distance of 13 km and so easily reach the surface of the Earth.

16.3.2 *The velocity four-vector*

To find the velocity four-vector, we need a quantity which transforms like $\Delta \boldsymbol{R}$, (16.32), and has the form of a velocity. The only time which is Lorentz invariant is the proper time Δt_0. As we discussed above, an observer in any inertial frame of reference can measure $c\Delta t, \Delta x, \Delta y, \Delta z$ for two events and compute from these $c^2 \Delta t_0^2 = c^2 \Delta t^2 - \Delta x^2 - \Delta y^2 - \Delta z^2$. An observer in any other inertial frame of reference would find exactly the same value, $c^2 \Delta t_0^2$. Therefore, let us define the *velocity four-vector* using the Lorentz-invariant quantity Δt_0:

$$\boldsymbol{U} = \frac{\Delta \boldsymbol{R}}{\Delta t_0}. \tag{16.38}$$

But, from the analysis which led to (16.37), the proper time interval is related to the time interval Δt measured in the laboratory frame S by $\Delta t_0 = \Delta t / \gamma$. We can therefore write $\boldsymbol{U}$ as

$$\begin{aligned}\boldsymbol{U} = \frac{\Delta \boldsymbol{R}}{\Delta t_0} &= \gamma \left[c\frac{\Delta t}{\Delta t}, \frac{\Delta x}{\Delta t}, \frac{\Delta y}{\Delta t}, \frac{\Delta z}{\Delta t} \right], \\ &= \left[\gamma c, \gamma u_x, \gamma u_y, \gamma u_z \right], \\ &= \left[\gamma c, \gamma \boldsymbol{u} \right]. \end{aligned} \tag{16.39}$$

In this relation $\boldsymbol{u}$ is the three-velocity of the particle in the frame S and γ is the corresponding Lorentz factor; u_x, u_y, u_z, are the components of $\boldsymbol{u}$. Notice the procedure we use to work out the components of the four-velocity. In the frame S we measure the three-velocity $\boldsymbol{u}$ and hence γ. We then form the quantities γc and $\gamma u_x, \gamma u_y, \gamma u_z$ and know that they will transform exactly as ct and x, y, z.

Let us use this procedure to add two velocities relativistically. If the relative velocity of the two frames of reference is V in standard configuration and $\boldsymbol{u}$ is the velocity of the particle in the frame S then what is its velocity in the frame S'? Call the Lorentz factor associated with the relative motion of S and S' γ_V. First, we write down the velocity four-vectors for the particle in S and S':

$$\text{in } S, \quad [\gamma c, \gamma \boldsymbol{u}] \equiv [\gamma c, \gamma u_x, \gamma u_y, \gamma u_z], \qquad \gamma = (1 - u^2/c^2)^{-1/2};$$
$$\text{in } S', \quad [\gamma' c, \gamma' \boldsymbol{u}'] \equiv [\gamma' c, \gamma' u'_x, \gamma' u'_y, \gamma' u'_z], \qquad \gamma' = (1 - u'^2/c^2)^{-1/2}.$$

We can relate the components of the four-vectors through the following identities:

$$\begin{aligned} ct' &\to \gamma' c, & ct &\to \gamma c, \\ x' &\to \gamma' u'_x, & x &\to \gamma u_x, \\ y' &\to \gamma' u'_y, & y &\to \gamma u_y, \\ z' &\to \gamma' u'_z, & z &\to \gamma u_z. \end{aligned}$$

Therefore, applying the Lorentz transformation (16.27), we find

$$\begin{aligned} \gamma' c &= \gamma_V \left(\gamma c - V \gamma u_x / c\right), \\ \gamma' u'_x &= \gamma_V (\gamma u_x - V \gamma), \\ \gamma' u'_y &= \gamma u_y, \\ \gamma' u'_z &= \gamma u_z. \end{aligned} \tag{16.40}$$

The first relation gives

$$\frac{\gamma \gamma_V}{\gamma'} = \frac{1}{1 - V u_x / c^2},$$

and therefore, from the spatial terms of (16.40), we find

$$\begin{aligned} u'_x &= \frac{u_x - V}{(1 - V u_x / c^2)}, \\ u'_y &= \frac{u_y}{\gamma_V (1 - V u_x / c^2)}, \\ u'_z &= \frac{u_z}{\gamma_V (1 - V u_x / c^2)}. \end{aligned} \tag{16.41}$$

These are the standard expressions for the addition of velocities in special relativity. They have many pleasing features. For example, if any of u_x, u_y or u_z is separately equal to c and, necessarily, the others are zero then the magnitude of the total velocity u' in S' also equals c, as required by Einstein's second postulate.

Note that the norm of the four-velocity (16.39) is

$$|\boldsymbol{U}^2| = c^2\gamma^2 - \gamma^2 u_x^2 - \gamma^2 u_y^2 - \gamma^2 u_z^2 = \gamma^2 c^2 (1 - u^2/c^2) = c^2, \tag{16.42}$$

which is an invariant as expected.

16.3.3 The acceleration four-vector

We can now repeat the procedure of subsection 16.3.2 to find the acceleration four-vector. First, we form the increment of the four-velocity $\Delta \boldsymbol{U} \equiv [c\Delta\gamma, \Delta(\gamma \boldsymbol{u})]$, which is necessarily a four-vector. Then we define the only invariant acceleration-like quantity we can form, by dividing $\Delta \boldsymbol{U}$ by the proper time interval Δt_0:

$$\boldsymbol{A} = \frac{\Delta \boldsymbol{U}}{\Delta t_0} = \left[\gamma c \frac{\Delta \gamma}{\Delta t}, \gamma \frac{\Delta}{\Delta t}(\gamma \boldsymbol{u})\right] = \left[\gamma c \frac{\mathrm{d}\gamma}{\mathrm{d}t}, \gamma \frac{\mathrm{d}}{\mathrm{d}t}(\gamma \boldsymbol{u})\right], \tag{16.43}$$

in the limit $\Delta t \to 0$. This is the *acceleration four-vector*. Let us convert it into a more useful form for practical applications. First, note how to differentiate $\gamma = (1 - u^2/c^2)^{-1/2}$:

$$\begin{aligned}\frac{\mathrm{d}\gamma}{\mathrm{d}t} &= \frac{\mathrm{d}}{\mathrm{d}t}\left(1 - \frac{u^2}{c^2}\right)^{-1/2} = \frac{\mathrm{d}}{\mathrm{d}t}\left(1 - \frac{\boldsymbol{u}\cdot\boldsymbol{u}}{c^2}\right)^{-1/2} \\ &= \frac{\boldsymbol{u}\cdot\boldsymbol{a}}{c^2}\left(1 - \frac{u^2}{c^2}\right)^{-3/2} = \gamma^3\left(\frac{\boldsymbol{u}\cdot\boldsymbol{a}}{c^2}\right).\end{aligned} \tag{16.44}$$

Then

$$\frac{\mathrm{d}}{\mathrm{d}t}(\gamma \boldsymbol{u}) = \gamma^3\left(\frac{\boldsymbol{u}\cdot\boldsymbol{a}}{c^2}\right)\boldsymbol{u} + \gamma \boldsymbol{a}$$

and so

$$\boldsymbol{A} = \frac{\mathrm{d}\boldsymbol{U}}{\mathrm{d}t} = \left[\gamma^4\left(\frac{\boldsymbol{u}\cdot\boldsymbol{a}}{c}\right), \gamma^4\left(\frac{\boldsymbol{u}\cdot\boldsymbol{a}}{c^2}\right)\boldsymbol{u} + \gamma^2\boldsymbol{a}\right]. \tag{16.45}$$

Notice what this means. In some frame of reference, say S, we measure at a particular instant the three-velocity $\boldsymbol{u}$ and the three-acceleration $\boldsymbol{a}$ of a particle. From these quantities we form the scalar quantity $\gamma^4(\boldsymbol{u}\cdot\boldsymbol{a})/c$ and the three-vector $\gamma^4(\boldsymbol{u}\cdot\boldsymbol{a}/c^2)\boldsymbol{u} + \gamma^2\boldsymbol{a}$. Because $\boldsymbol{A}$ is a four-vector, we then know from (16.29) that these quantities transform exactly as ct and $\boldsymbol{r}$ between inertial frames of reference.

It is often convenient in relativistic calculations to transform into the frame of reference in which the particle is instantaneously at rest. The particle may be accelerated, but this does not matter since there is no dependence upon acceleration in the Lorentz transformations. In the *instantaneous rest frame*, $\boldsymbol{u} = 0$ and $\gamma = 1$ and hence the four-acceleration in this frame is

$$\boldsymbol{A} = [0, \boldsymbol{a}] \equiv [0, \boldsymbol{a}_0], \tag{16.46}$$

where $\boldsymbol{a}_0$ is the *proper acceleration* of the particle. In the same frame, the four-velocity of the particle is

$$\boldsymbol{U} = [\gamma c, \gamma \boldsymbol{u}] = [c, 0].$$

Therefore, using the definition of the scalar product as for three-vectors, $\boldsymbol{A}\cdot\boldsymbol{U} = 0$. Since four-vector relations are true in any frame of reference, this means that, no matter which frame we care to work in, the scalar product of the velocity and acceleration four-vectors is always zero, that is, the velocity and acceleration four-vectors are orthogonal. If you are suspicious, it is a useful exercise to take the scalar product of the expressions (16.39) and (16.45) for the four-vectors $\boldsymbol{A}$ and $\boldsymbol{U}$ and show that it is indeed zero.

16.4 Relativistic dynamics – the momentum and force four-vectors

So far, we have been dealing with *kinematics*, that is, the description of motion, but now we have to tackle the problems of *dynamics*. This means introducing the concepts of momentum and force and how they are related.

16.4.1 The momentum four-vector

First of all, let us go through the purely formal exercise of defining suitable momentum and force four-vectors and then finding out if they make physical sense. Let us introduce a four-vector quantity $\boldsymbol{P}$, which has the dimensions of momentum:

$$\boldsymbol{P} = m_0\boldsymbol{U} = [\gamma m_0 c, \gamma m_0 \boldsymbol{u}], \tag{16.47}$$

where $\boldsymbol{U}$ is the velocity four-vector and m_0 is the mass of the particle, which is taken to be a scalar invariant – it will be identified with the particle's rest mass. We note the following immediate consequences:

(i) $m_0\boldsymbol{U}$ is certainly a four-vector since m_0 is an invariant;
(ii) the space components of $\boldsymbol{P}$ reduce to the Newtonian formula for momentum if $u \ll c$, that is, $m_0\gamma\boldsymbol{u} \to m_0\boldsymbol{u}$ as $u \to 0$.

Therefore, *if this agrees with experiment*, we have found a suitable form for the momentum four-vector.

Notice that we can also define a *relativistic three-momentum* $\boldsymbol{p} = \gamma m_0\boldsymbol{u}$ from the spatial components of the four-vector. The quantity $m = \gamma m_0$ is defined to be the *relativistic inertial mass*. As will be discussed below, I do not like this term, but it is commonly found in the literature. We need to show that these definitions result in a self-consistent set of dynamics.

Let us find a relation between the relativistic momentum and the relativistic mass of the particle. Equating the norms of the momentum four-vector in the laboratory frame, $\boldsymbol{P} \equiv [\gamma m_0 c, \gamma m_0\boldsymbol{u}]$, and that in the rest frame of the particle, $\boldsymbol{P} \equiv [m_0 c, 0]$, we obtain

$$\begin{aligned} m_0^2c^2 &= \gamma^2 m_0^2 c^2 - \gamma^2 m_0^2 u^2, \\ &= m^2c^2 - p^2, \end{aligned}$$

or

$$p^2 = m^2c^2 - m_0^2c^2. \tag{16.48}$$

We will rewrite this expression in terms of energies in Section 16.5.

16.4.2 The force four-vector

Following the logic of the previous sections, the natural four-vector generalisation of Newton's second law of motion is

$$\boldsymbol{F} = \frac{\mathrm{d}\boldsymbol{P}}{\mathrm{d}t_0}, \tag{16.49}$$

where $\boldsymbol{F}$ is the *four-force* and $\mathrm{d}t_0$ is the differential of proper time. We now relate the force which we measure in the laboratory to the four-force. The best approach is through the quantity we have called the relativistic three-momentum. Why did we adopt the above definition? Consider a collision between two particles which initially have momentum four-vectors $\boldsymbol{P}_1$ and $\boldsymbol{P}_2$. After the collision, these become $\boldsymbol{P}'_1$ and $\boldsymbol{P}'_2$. The conservation of four-momentum can then be written

$$\boldsymbol{P}_1 + \boldsymbol{P}_2 = \boldsymbol{P}'_1 + \boldsymbol{P}'_2. \tag{16.50}$$

In terms of the components of the four-vectors, this equation implies

$$\begin{aligned} m_1 + m_2 &= m'_1 + m'_2, \\ \boldsymbol{p}_1 + \boldsymbol{p}_2 &= \boldsymbol{p}'_1 + \boldsymbol{p}'_2, \end{aligned} \tag{16.51}$$

where the m's are the relativistic masses. Thus, implicit in this formalism is the requirement that the relativistic three-momentum is conserved and so, for relativistic particles, $\gamma m_0 \boldsymbol{u}$ plays the role of momentum. The corresponding force equation is suggested by Newton's second law:

$$\boldsymbol{f} = \frac{\mathrm{d}\boldsymbol{p}}{\mathrm{d}t} = \frac{\mathrm{d}}{\mathrm{d}t}(\gamma m_0 \boldsymbol{u}), \tag{16.52}$$

where $\boldsymbol{f}$ is the normal three-force of Newtonian dynamics.

Are these definitions self-consistent? We have to be a bit careful. On the one hand, we could just argue, 'Let's look at experiment and see if it works', and in many ways this is more or less all that the argument amounts to. We can do a little better. In point collisions of particles, the relativistic generalisation of Newton's third law should apply, that is, $\boldsymbol{f} = -\boldsymbol{f}$. We can only consider point collisions or else we would get into trouble with action at a distance in relativity – recall the relativity of simultaneity in different frames of reference. For a point collision, $\boldsymbol{f} = -\boldsymbol{f}$ is true if we adopt the definition of $\boldsymbol{f}$ given above, because we have already argued that relativistic three-momentum is conserved. That is,

$$\Delta \boldsymbol{p}_1 = -\Delta \boldsymbol{p}_2, \qquad \frac{\Delta \boldsymbol{p}_1}{\Delta t} = -\frac{\Delta \boldsymbol{p}_2}{\Delta t}, \qquad \boldsymbol{f}_1 = -\boldsymbol{f}_2.$$

However, we cannot be absolutely sure that we have made the correct choice without appealing to experiment. We faced the same sort of logical problem when we tried to understand the meaning of Newton's laws of motion (Section 7.1). They ended up being a set of definitions which give results consistent with experiment. Similarly, relativistic dynamics cannot come out of pure thought but can be put into a logically self-consistent mathematical structure which is consistent with experiment.

We therefore adopt the definition (16.52) of $\boldsymbol{f}$ as the three-force, in the same sense as in Newtonian dynamics, but now the particle may be moving relativistically and the relativistic three-momentum should be used for $\boldsymbol{p}$. Within this framework, we can derive a number of pleasing results.

16.4.3 $F = m_0 A$

This follows directly from the definition of F,

$$F = \frac{dP}{dt_0} = m_0 \frac{dU}{dt_0} = m_0 A. \tag{16.53}$$

In addition, because $A \cdot U = 0$, it follows that

$$F \cdot U = 0;$$

that is, the force and velocity four-vectors are orthogonal.

16.4.4 The relativistic generalisation of $f = dp/dt$

Let us write out the four-vector form of Newton's second law in terms of its components:

$$F = [f_0, f_1, f_2, f_3] = \frac{dP}{dt_0} = \left[\gamma \frac{d(\gamma m_0 c)}{dt}, \gamma \frac{dp}{dt}\right], \tag{16.54}$$

where $p = \gamma m_0 u$. Since we have argued that the relativistic form of Newton's second law in three-vector form is $f = dp/dt$, it follows that we must write the four-force in the form

$$F = [f_0, \gamma f] = \left[\gamma \frac{d(\gamma m_0 c)}{dt}, \gamma \frac{dp}{dt}\right] = m_0 A. \tag{16.55}$$

Equating the spatial components of $[f_0, \gamma f] = m_0 A$ and using (16.45), we find

$$f = m_0 \gamma^3 \left(\frac{u \cdot a}{c^2}\right) u + m_0 \gamma a. \tag{16.56}$$

This is the relativistic generalisation of the Newtonian expression $f = m_0 a$.

Now let us analyse the 'time' component of the force four-vector, that is, of

$$F = \frac{dP}{dt_0} = \left[\gamma \frac{d(\gamma m_0 c)}{dt}, \gamma \frac{d}{dt}(\gamma m_0 u)\right].$$

From (16.44), we can write

$$\gamma \frac{d(\gamma m_0 c)}{dt} = m_0 \gamma^4 \left(\frac{u \cdot a}{c}\right),$$

or, in terms of the relativistic mass m,

$$\frac{dm}{dt} = \gamma^3 m_0 \left(\frac{u \cdot a}{c^2}\right). \tag{16.57}$$

Now inspect the quantity $(f \cdot u)/c^2$. Substituting for f from (16.56),

$$\begin{aligned} \frac{f \cdot u}{c^2} &= m_0 \gamma^3 \left(\frac{u \cdot a}{c^2}\right) \frac{u^2}{c^2} + m_0 \gamma \left(\frac{u \cdot a}{c^2}\right), \\ &= m_0 \gamma^3 \left(\frac{u \cdot a}{c^2}\right) \left(\frac{u^2}{c^2} + \frac{1}{\gamma^2}\right), \\ &= m_0 \gamma^3 \left(\frac{u \cdot a}{c^2}\right). \end{aligned} \tag{16.58}$$

Therefore, (16.57) becomes

$$\frac{\mathrm{d}m}{\mathrm{d}t} = \frac{\boldsymbol{f} \cdot \boldsymbol{u}}{c^2}, \tag{16.59}$$

or

$$\frac{\mathrm{d}(mc^2)}{\mathrm{d}t} = \frac{\mathrm{d}(\gamma m_0 c^2)}{\mathrm{d}t} = \boldsymbol{f} \cdot \boldsymbol{u}. \tag{16.60}$$

This is one of the amazing results of Einstein's paper of 1905. The quantity $\boldsymbol{f} \cdot \boldsymbol{u}$ is just the rate at which work is done on the particle, that is, its rate of increase of energy. Thus mc^2 is identified with the total energy of the particle. This is the formal proof of perhaps the most famous equation in physics,

$$E = mc^2. \tag{16.61}$$

Notice the profound implication of equation (16.61): there is a certain amount of inertial mass associated with the energy produced when work is done. It does not matter what the form of the energy is – electrostatic, magnetic, kinetic, elastic, etc. All energies are the same thing as inertial mass. Likewise, reading the equation backwards, inertial mass is energy. Nuclear power stations and nuclear explosions are vivid demonstrations of the identity of inertial mass and energy.

16.4.5 *A mild polemic*

Sometimes references are made to *longitudinal* and *transverse* masses in textbooks. These arise from an unhappy use of the relativistic form of Newton's second law of motion. If $\boldsymbol{u} \parallel \boldsymbol{a}$ then the force law (16.56) reduces to

$$\begin{aligned} \boldsymbol{f} &= m_0\gamma^3 \left(\frac{u^2}{c^2} + \frac{1}{\gamma^2}\right)\boldsymbol{a} \\ &= \gamma^3 m_0 \boldsymbol{a}. \end{aligned}$$

The quantity $\gamma^3 m_0$ is called the *longitudinal* mass. However, if $\boldsymbol{u} \perp \boldsymbol{a}$ then we find that

$$\boldsymbol{f} = \gamma m_0 \boldsymbol{a},$$

and γm_0 is called the *transverse* mass. I dislike this introduction of different sorts of mass along with 'relativistic' forms of Newton's second law of motion, in the form $\boldsymbol{f} =$ 'm'$\boldsymbol{a}$. It is preferable to stick with the correct generalisation of Newton's second law in the form $\boldsymbol{f} = \mathrm{d}\boldsymbol{p}/\mathrm{d}t$, where $\boldsymbol{p} = \gamma m_0 \boldsymbol{u}$, and abolish all masses except the rest mass m_0, for the reasons discussed in the next section.

16.5 The relativistic equations describing motion

Consider a particle of rest mass m_0 travelling at velocity $\boldsymbol{u}$ in some inertial frame of reference S. The particle has Lorentz factor γ given by the standard expression $\gamma = \left(1 - u^2/c^2\right)^{-1/2}$. Then:

- The *relativistic three-momentum* of the particle is $\gamma m_0 \boldsymbol{u}$.
- Its *total energy* is $E = mc^2 = \gamma m_0 c^2$.

- Its *rest mass energy* is $E_0 = m_0c^2$.
- Its *kinetic energy* is the difference between its total and rest mass energies,

$$E_{\rm kin} = E - E_0 = \gamma mc^2 - m_0c^2 = (\gamma - 1)m_0c^2. \tag{16.62}$$

The last definition follows from the calculation which led to (16.60), in which we showed that, when we do work on a particle, we increase the quantity γm_0c^2. For $u \ll c$, (16.62) reduces to the Newtonian non-relativistic expression for the kinetic energy,

$$E_{\rm kin} = (\gamma - 1)m_0c^2 = \left[\left(1 - \frac{u^2}{c^2}\right)^{-1/2} - 1\right] m_0c^2 \approx \tfrac{1}{2}m_0u^2. \tag{16.63}$$

- The relativistic form of Newton's second law of motion is

$$\boldsymbol{f} = \frac{\mathrm{d}\boldsymbol{p}}{\mathrm{d}t}. \tag{16.64}$$

- The conservation laws of energy and momentum are subsumed into the single law of conservation of four-momentum,

$$\boldsymbol{P}_1 + \boldsymbol{P}_2 = \boldsymbol{P}_1' + \boldsymbol{P}_2', \tag{16.65}$$

where the components of the four vectors are given by

$$\boldsymbol{P} \equiv [p_0,\ p_x,\ p_y,\ p_z] = [\gamma m_0c,\ \gamma m_0u_x,\ \gamma m_0u_y,\ \gamma m_0u_z]. \tag{16.66}$$

- Equating the spatial components in the equation of conservation of four-momentum, we find the law of *conservation of relativistic three-momentum*,

$$\boldsymbol{p}_1 + \boldsymbol{p}_2 = \boldsymbol{p}_1' + \boldsymbol{p}_2'. \tag{16.67}$$

- From the 'time' component of the equation of conservation of four-momentum, we find the *relativistic law of conservation of energy*; we can express this in three ways:

$$\begin{aligned} p_{01} + p_{02} &= p_{01}' + p_{02}', \\ E_1 + E_2 &= E_1' + E_2', \\ \gamma_1 m_{01} + \gamma_2 m_{02} &= \gamma_1' m_{01} + \gamma_2' m_{02}, \end{aligned} \tag{16.68}$$

where the E's are the total energies of the particles.

16.5.1 *Massless particles*

There are real advantages in rewriting the relations we have just derived in terms of the various energies defined above. Let us rewrite the list entirely in terms of energies:

- total energy $E = \gamma m_0c^2$;
- rest mass energy $E_0 = m_0c^2$;
- relativistic momentum $\boldsymbol{p} = \gamma m_0\boldsymbol{u} = (E/c^2)\boldsymbol{u}$;
- kinetic energy $E_{\rm kin} = E - E_0$.

Thus, we can write the four-momentum of a particle as follows:

$$\boldsymbol{P} = m_0 \boldsymbol{U} = m_0[\gamma c,\ \gamma u_x,\ \gamma u_y,\ \gamma u_z] = \left[\frac{E}{c},\ \frac{E}{c^2}\boldsymbol{u}\right]. \tag{16.69}$$

In Case Study V, we demonstrated in considerable detail that light consists of massless particles, *photons*, which travel at the speed of light. Their energies are related to their frequencies ν by $E = h\nu$, where h is Planck's constant. For these particles, the rest mass energy $E_0 = 0$ and $\boldsymbol{u} = \boldsymbol{c}$. Thus, for massless particles, such as the photon, we find directly

$$\boldsymbol{p} = \frac{E}{c^2}\boldsymbol{c} \qquad \text{and} \qquad E_{\text{kin}} = E, \tag{16.70}$$

and the momentum four-vector for such particles is

$$\boldsymbol{P} = \left[\frac{E}{c},\ \frac{E}{c^2}\boldsymbol{c}\right]. \tag{16.71}$$

Notice that we have obtained directly another expression for the momentum of a photon:

$$\boldsymbol{p} = \frac{E}{c^2}\boldsymbol{c} = \frac{h\nu}{c^2}\boldsymbol{c}. \tag{16.72}$$

It can be seen that there is a real advantage in writing the four-momenta in terms of energies since then we can treat particles and photons on exactly the same basis. Notice also that all masses have been eliminated from the expressions.

To complete this rewrite of the expressions, let us perform the same change of variable for the expression relating the masses and momenta of the particles in their rest and moving frames (16.48). Equating the norms of the momentum four-vector (16.69) in the laboratory frame and the rest frame of the particle,

$$\left(\frac{E}{c}\right)^2 - \left(\frac{E}{c^2}\boldsymbol{u}\right)^2 = \left(\frac{E_0}{c}\right)^2,$$

or

$$E^2 - \boldsymbol{p}^2 c^2 = E_0^2 = \text{constant}. \tag{16.73}$$

This equation is a great help in solving problems in relativity. We can work out the total energy $E = \sum_i E_i$ and total momentum $\boldsymbol{p} = \sum_i \boldsymbol{p}_i$ in some frame of reference and then we know that the quantity $E^2 - \boldsymbol{p}^2 c^2 = E_0^2$ is an invariant in all inertial frames of reference. This is particularly useful in working out threshold energies for the creation of particles in high-energy collisions. One final note is that, in the extreme high-energy limit $E \gg E_0$, (16.73) becomes

$$E = |\boldsymbol{p}|c. \tag{16.74}$$

Thus, in the ultrarelativistic limit, particles behave like photons.

16.5.2 *Relativistic particle dynamics in a magnetic field*

Consider next the dynamics of a relativistic particle in a magnetic field. The expression for the *Lorentz force* is

$$\boldsymbol{f} = e(\boldsymbol{E} + \boldsymbol{u} \times \boldsymbol{B})$$

and hence, if $\boldsymbol{E} = 0$,

$$\boldsymbol{f} = e(\boldsymbol{u} \times \boldsymbol{B}).$$

Therefore, from (16.56),

$$e(\boldsymbol{u} \times \boldsymbol{B}) = m_0\gamma^3 \left(\frac{\boldsymbol{u} \cdot \boldsymbol{a}}{c^2}\right) \boldsymbol{u} + \gamma m_0 \boldsymbol{a}. \tag{16.75}$$

The left-hand side of this expression and the first term on the right are perpendicular vectors, because $\boldsymbol{u} \perp (\boldsymbol{u} \times \boldsymbol{B})$, and hence we require simultaneously that

$$e(\boldsymbol{u} \times \boldsymbol{B}) = \gamma m_0 \boldsymbol{a} \qquad \text{and} \qquad \boldsymbol{u} \cdot \boldsymbol{a} = 0, \tag{16.76}$$

that is, the acceleration impressed by the magnetic field is perpendicular to both $\boldsymbol{B}$ and the velocity $\boldsymbol{u}$. This is the origin of the circular or spiral motion of charged particles in magnetic fields.

16.6 The frequency four-vector

Finally, let us derive the *frequency four-vector* from the *scalar product rule*: namely, if $\boldsymbol{A}$ is a four-vector and $\boldsymbol{A} \cdot \boldsymbol{B}$ is an invariant then $\boldsymbol{B}$ must also be a four-vector. The simplest approach is to consider the phase of a wave. If we write the expression for the wave in the form $\exp[\mathrm{i}(\boldsymbol{k} \cdot \boldsymbol{r} - \omega t)]$ then the quantity $\boldsymbol{k} \cdot \boldsymbol{r} - \omega t$ is the phase of the wave, in other words, it defines how far through the cycle of the wave from 0° to 360° one happens to be at coordinates $[ct, \boldsymbol{r}]$. This is an invariant scalar quantity, whatever frame of reference one happens to be in. Thus, in any inertial frame of reference,

$$\boldsymbol{k} \cdot \boldsymbol{r} - \omega t = \text{invariant}.$$

But $[ct, \boldsymbol{r}]$ is a four-vector and therefore the quantity

$$\boldsymbol{K} = [\omega/c, \boldsymbol{k}] \tag{16.77}$$

must also be a four-vector; it is called the *frequency four-vector*.

We can also derive this four-vector for photons by deriving the appropriate momentum four-vector. Thus, from (16.71),

$$\boldsymbol{P} \equiv \left[\frac{E}{c}, \frac{E}{c^2}\boldsymbol{c}\right] \equiv \left[\frac{h\nu}{c}, \frac{h\nu}{c}\boldsymbol{i}_k\right]$$

where $h\nu/c$ is the momentum of the photon and $h\nu$ its energy; $\boldsymbol{i}_k$ is the unit vector in the direction of propagation of the photon. Writing

$$\frac{h\nu}{c^2} = \frac{\hbar\omega}{c^2}, \qquad \frac{h\nu}{c} = \frac{\hbar\omega}{c} = \hbar|\boldsymbol{k}|,$$

the momentum four-vector of the photon becomes

$$\boldsymbol{P} \equiv \hbar\left[\frac{\omega}{c}, \boldsymbol{k}\right], \tag{16.78}$$

that is,

$$\boldsymbol{P} = \hbar\boldsymbol{K}.$$

16.7 Lorentz contraction and the origin of magnetic fields

We have stated that special relativistic phenomena are outside our normal experience, but there is one remarkable example in which these second-order effects in V/c are omnipresent in everyday life – they are the origin of magnetic fields, which are associated with electric current. We noted in subsection 16.1.1 the interchangeability of electric and magnetic fields depending upon the inertial frame of reference in which the measurements are made. We need to repeat the analyses of subsection 16.1.1 to work out the transformation laws for $\boldsymbol{E}$ and $\boldsymbol{B}$ according to special relativity. This involves replacing (16.3) by the Lorentz transformations for partial derivatives; these are

$$\begin{aligned}
\frac{\partial}{\partial t} &\rightarrow \gamma\left(\frac{\partial}{\partial t'} - V\frac{\partial}{\partial x'}\right), \\
\frac{\partial}{\partial x} &\rightarrow \gamma\left(\frac{\partial}{\partial x'} - \frac{V}{c^2}\frac{\partial}{\partial t'}\right), \\
\frac{\partial}{\partial y} &\rightarrow \frac{\partial}{\partial y'}, \\
\frac{\partial}{\partial z} &\rightarrow \frac{\partial}{\partial z'}.
\end{aligned} \tag{16.79}$$

It is left to the reader to show that Maxwell's equations are form-invariant under these transformations and that the transformed components of $\boldsymbol{E}$ and $\boldsymbol{B}$ are

$$\begin{aligned}
E'_x &= E_x, \\
E'_y &= \gamma(E_y - VB_z), \\
E'_z &= \gamma(E_z - VB_y),
\end{aligned} \tag{16.80}$$

$$\begin{aligned}
B'_x &= B_x, \\
B'_y &= \gamma\left(B_y - \frac{V}{c^2}E_z\right), \\
B'_z &= \gamma\left(B_z - \frac{V}{c^2}E_y\right).
\end{aligned} \tag{16.81}$$

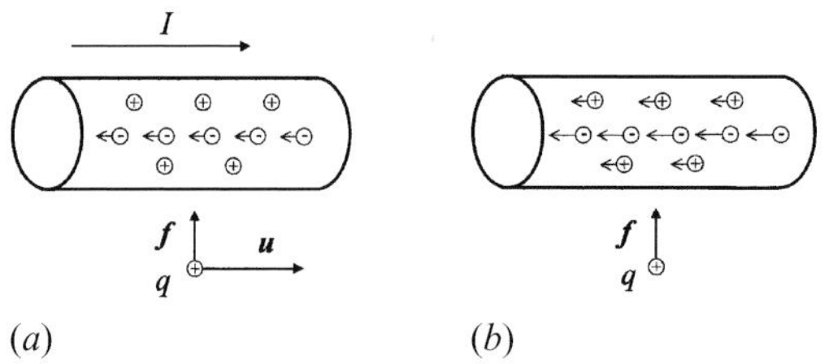

Figure 16.2: Illustrating the origin of the force on a charge moving with velocity $\boldsymbol{u}$ in the vicinity of a current-carrying wire. (*a*) There is no net charge on the wire in the lab frame S. (*b*) In a frame S' moving with the charge q, there is a net negative charge on the wire.

If, for example, there is a uniform magnetic field in some frame of reference then it can be transformed to zero in some other frame, in which its effects would be attributed to the action of an electric field $\boldsymbol{E}'$. Let us consider the case of the magnetic field of a current-carrying wire.

Suppose a current I flows in the wire as observed in the laboratory frame of reference S, the electrons drifting at speed v while the ions remain stationary. The current is $I = en_e v$, where n_e is the number of electrons per unit length and is equal to n_i, the number of ions per unit length, so that charge neutrality is preserved in S; e is the charge of the electron. Applying Ampère's law (5.10), the magnetic flux density at radial distance r from the wire is

$$\oint \boldsymbol{B} \cdot \mathrm{d}\boldsymbol{s} = \mu_0 I, \qquad B = \frac{\mu_0 e n_e v}{2\pi r}. \tag{16.82}$$

If a charge q moves at velocity $\boldsymbol{u}$ parallel to the wire in the positive x-direction, as shown in Fig. 16.2, the Lorentz force is

$$\boldsymbol{f} = q(\boldsymbol{u} \times \boldsymbol{B}) = q\frac{\mu_0 e n_e u v}{2\pi r} \tag{16.83}$$

towards the wire, if $q > 0$. Notice that the force direction is also given by the rule that like currents attract each other.

We now repeat the calculation in the frame of reference S' of the moving charge q. The ions now move in the negative x'-direction at speed u and the electrons have speed v', which is the relativistic sum of v and u and is also in the negative x'-direction. The corresponding Lorentz factors are

$$\gamma_i' = \frac{1}{(1 + u^2/c^2)^{1/2}}, \qquad \gamma_e' = \frac{1 + uv/c^2}{(1 - u^2/c^2)^{1/2}(1 - v^2/c^2)^{1/2}},$$

where the first relation of (16.40) has been used to find the Lorentz factor for the electrons – notice that we add together the speeds relativistically although the speeds of the ions and electrons are very small, $u \ll c$, $v \ll c$. The charge densities per unit length in S' are given by the length-contracted values for the ions and electrons. For the ions, which are stationary in S, the number per unit length increases to

$$n_i' = n_i \gamma_i'.$$

For the electrons, the number per unit length in S is already increased relative to the number density in their own rest frame by the factor γ_e. Therefore, the number per unit length when moving at speed v', relative to its value in S, is

$$n'_e = n_e \frac{\gamma'_e}{\gamma_e}.$$

Therefore, the net charge on the wire per unit length in S' is

$$en' = e(n'_i - n'_e). \tag{16.84}$$

Preserving quantities to second order in v/c, we find

$$en' = -\frac{n_e euv}{c^2}. \tag{16.85}$$

In S' there is, therefore, a net negative charge on the wire, which results in an *electrostatic attractive force* on the stationary charge q in S'. The field due to a line charge ρ_e per unit length is found from Gauss's law for electrostatics to be $E_r = \rho_e/(2\pi\epsilon_0 r)$, and so the force acting on the charge q in S' is

$$f_E = q\frac{en_e uv}{2\pi\epsilon_0 c^2 r}\gamma = q\frac{\mu_0 en_e uv}{2\pi r}, \tag{16.86}$$

which is exactly the same as (16.83). Thus, we can understand the origin of the magnetic field as being due to the second-order effect of length contraction upon the charges flowing in the wire.

It is remarkable just how small the effect is. Using the figures adopted by French,[20] suppose a current of 10 A flows in a copper wire of cross-section $\sigma = 1$ mm^2. In solid copper, there are $n_e \sim 10^{20}$ conduction electrons mm^{-3}. Therefore, since the current is $I = n_e ev\sigma$, we find $v \sim 0.6$ mm s^{-1}. This corresponds to $v/c \sim 2 \times 10^{-12}$ – the electrons are scarcely moving at all!

The reason the effect is so tiny is that without the almost perfect cancellation of the charges of the ions the electrostatic forces would be enormous. To see this, we can compare the electrostatic force due to the ions alone, which would be $f_i = qn_i e/(2\pi\epsilon_0 r)$, with the expression (16.85):

$$\frac{f_E}{f_i} = \frac{q\mu_0 en_e uv/(2\pi r)}{qn_i e/(2\pi\epsilon_0 r)} = \mu_0\epsilon_0 uv = \frac{uv}{c^2}. \tag{16.87}$$

Notice that $\mu_0\epsilon_0$ is just the ratio of the strengths of the magnetostatic and electrostatic forces for unit poles and charges (see Section 5.1).

16.8 Reflections

I will end this case study here. It is a beautiful example of how an appropriate mathematical structure can clarify the underlying symmetries of the theory. I could have gone through the whole exercise as a piece of mathematics and the points at which I had to look at the real world would hardly have been apparent. In fact, there are assumptions present, as we have shown, just as there are in Newtonian theory. It so happens that in this case the mathematical structure is very elegant indeed – it might not have turned out that way.

From the practical point of view, the use of four-vectors greatly simplifies all calculations in relativity. It is only necessary to remember a single set of transformations and then understand how to derive the relevant four-vectors. There are great advantages in having a prescription such as this which can be trusted for carrying out relativistic calculations. I am very suspicious of arguments which try to provide intuitive ways of looking at relativistic problems. Phenomena such as length contraction and time dilation should be treated cautiously and the simplest way of ensuring that confusion does not reign is to write down the four-vectors associated with events and then use the Lorentz transforms to relate coordinates.

16.9 References

1 Stachel, J. (1995). The history of relativity, in *Twentieth Century Physics*, Vol. 1, eds. L.M. Brown, A. Pais and A.B. Pippard, 249–356. Bristol: Institute of Physics Publishing and New York: American Institute of Physics Press.
2 Stachel, J. (1998). *Einstein's Miraculous Year: Five Papers that Changed the Face of Physics*. Princeton: Princeton University Press.
3 Pais, A. (1982). *Subtle is the Lord ... the Science and Life of Albert Einstein*. Oxford: Clarendon Press.
4 Whittaker, E.T. *History of the Theories of the Aether and Electricity*: Vol. 1 (1910), London: Longmans, Green and Co. (revised and enlarged, New York: Nelson and Sons, 1951); Vol. 2 (1953), New York: Nelson and Sons. This history is regarded as the standard work on the subject, with the one caveat that for some reason Whittaker gives the credit for special relativity to Lorentz and Poincaré rather than Einstein. As will be apparent from this case study, this scarcely matches the facts as must have been known at the time.
5 Voigt, W. (1887). *Goett. Nachr.*, 41–51.
6 Fitzgerald, G.F. (1889). *Science*, **13**, 390.
7 Lorentz, H.A. (1892). *Versl. K. Ak. Amsterdam*, **1**, 74. (See A. Pais, 1982, *op. cit.*, pp. 123–4.)
8 Lorentz, H.A. (1895). *Versuch einer Theorie der Elektrischen und Optischen Erschweinungen in bewegten Körpern*. Leiden: Brill. (See A. Pais, 1982, *op. cit.*, pp. 124–5.)
9 Lorentz, H.A. (1899). *Versl. K. Ak. Amsterdam*, **10**, 793. (See A. Pais, 1982, *op. cit.*, pp. 125–6.)
10 Larmor, J. (1900). *Aether and Matter*. Cambridge: Cambridge University Press.
11 Lorentz, H.A. (1904). *Proc. K. Ak. Amsterdam*, **6**, 809. (See A. Pais, 1982, *op. cit.*, p. 126.)
12 Holten, G. (1988). *The Thematic Origins of Scientific Thought: Kepler to Einstein*, pp. 169–70. Cambridge, Massachusetts: Harvard University Press.
13 Poincaré, H. (1898). *Rev. Métaphys. Morale*, **6**, 1.
14 Poincaré, H. (1904). *L'Etat actuel et l'avenir de la physique mathématique*, Bulletin des Sciences Mathématiques.
15 Stachel, J. (1998). *Op. cit.*, p. 112.

16 Einstein, A. (1905). *Annalen der Physik*, **17**, 891. (See transl. in J. Stachel, 1998, *op. cit.*)
17 Holten, G. (1988). *Op. cit.*, p. 176.
18 Rindler, W. (2001). *Relativity: Special, General and Cosmological.* Oxford: Oxford University Press.
19 Hawking, S.W. (1988). *A Brief History of Time.* London: Bantam Press.
20 French, A.P. (1968). *Special Relativity.* London: Chapman and Hall.

Case Study VII

General relativity and cosmology

It is sometimes a matter of dispute whether general relativity and cosmology should appear in undergraduate syllabuses at all. The criticism is often made that they are included simply to add glamour to physics courses, to act as a lure to attract students into 'real hard-core' physics. I take a much more positive view of their inclusion as an integral part of physics.

General relativity follows on rather naturally from special relativity and results in even more profound changes to our concepts of space and time than those which follow from special relativity. The idea that the geometry of space–time is influenced by the distribution of matter, which then moves along paths in curved space–time is one of the fundamental concepts of modern physics. Indeed, the bending of space–time is now regularly observed as the gravitational lens effect in deep astronomical images (Fig. VII.1). Unfortunately, general relativity is technically complex, in the sense that the necessary mathematics goes beyond what can normally be introduced at the undergraduate level, and it would be wrong to underestimate these technical difficulties. Nonetheless, a great deal can be achieved without the use of advanced techniques, provided the reader is prepared to accept a few results which will simply be made plausible, the exact results then being quoted. This seems a small price to pay for some insight into the intimate connection between space–time geometry and gravitation and for understanding some of the more remarkable phenomena which are expected to be observed in strong gravitational fields, such as those found in the vicinity of black holes.

It might be thought even more perverse to introduce cosmology into this case study. My own view is that the physics of the large-scale structure and geometry of the Universe are just as much integral parts of physics as the structure of the Universe on the microscopic scale, the scale of atoms, nuclei and elementary particles. Therefore, I consider the astrophysical and cosmological sciences to be just as much a part of physics as condensed matter physics, statistical mechanics, and so on.

Another reason for including these topics is that observational and astrophysical cosmology has been revolutionised over the last forty years, and we now have a clear physical picture of the processes involved in the evolution of the Universe from the time it was less than a second old to the present time. This story is full of splendid physics and illustrates how laws established in the laboratory can be applied successfully on the scale

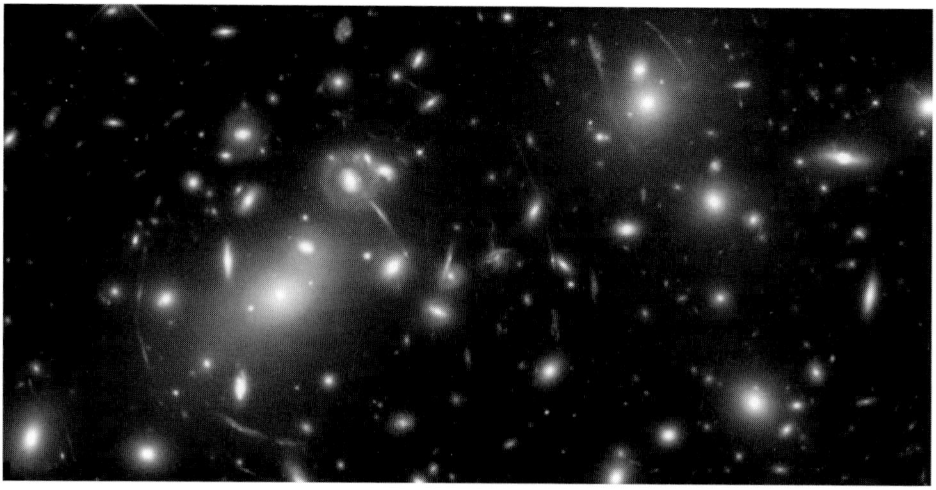

Figure VII.1: The rich cluster of galaxies Abell 2218 showing the remarkable arcs, coincident with the centre of the cluster, due to the gravitational lensing of very distant background galaxies. (Courtesy of NASA and the STScI.)

of the very large, as well as the very small. This new understanding has also posed a number of fundamental problems for physicists and cosmologists and these will undoubtedly become part of the new physics of the twenty-first century. Fortunately, the essential physics necessary to understand these issues can be developed rather precisely using the tools of an undergraduate physics course combined with ideas from elementary general relativity, which will be discussed in Chapter 17.

By ending with cosmology in Chapter 19, we close the circle begun in Case Study I, where the route to the discovery of Newton's laws of motion and gravity was described. I have allowed myself one indulgence, which is to counterbalance the highly theoretical contents of Chapters 17 and 19 with an interlude on the *technology of cosmology* in Chapter 18. Hard-line theorists may wish to proceed directly from Chapter 17 to Chapter 19, but that would be a pity; Chapter 18 describes the struggles of physicists, technologists and observers to provide the observational evidence which is the foundation for our present understanding of the origin and evolution of our Universe. As in all the physical sciences, the experimental and observational validation of theory remains the great technical challenge for the future and the only way in which theory can be constrained.

17 An introduction to general relativity

17.1 Introduction

Einstein's great paper of 1905 on the special theory of relativity appears effortless, and as discussed in Chapter 16, he did not regard the formulation of the theory as a particularly 'revolutionary act'. The route to the discovery of the general theory of relativity was very different. Whereas others had come close to elucidating the essential features of special relativity, Einstein was on his own and went far beyond all his contemporaries in his discovery of the general theory. How he arrived at the theory is one of the great stories of theoretical physics and involved the very deepest physical insight, imagination, intuition and sheer doggedness. It would lead to concepts barely conceivable even by a genius like Einstein – the phenomenon of black holes and the possibility of testing theories of gravity in the strong-field limit through the observation of relativistic stars. General relativity also provided for the first time a relativistic theory of gravity which could be used to construct fully self-consistent models of the Universe as a whole.

The history of the discovery of general relativity is admirably told by Abraham Pais in his scientific biography of Einstein, *Subtle is the Lord... the Science and Life of Albert Einstein,*[1] which discusses many of the technical details of the papers published in the period 1907 to 1915. Equally to be recommended is the survey by John Stachel of the history of the discovery of both theories of relativity.[2] The thrust of our case study is much more towards the content of the theory than its history, but a brief chronology of the discovery in this intellectual *terra incognita* is worth recording.

17.1.1 1907

It is simplest to quote Einstein's own words from his Kyoto address[3] of December 1922.

> In 1907, while I was writing a review of the consequences of special relativity... I realised that all the natural phenomena could be discussed in terms of special relativity except for the law of gravitation. I felt a deep desire to understand the reason behind this... It was most unsatisfactory to me that, although the relation between inertia and energy is so beautifully derived [in special relativity], there is no relation between inertia and weight. I suspected that this relationship was inexplicable by means of special relativity.

In the same lecture, he remarks

I was sitting in a chair in the patent office in Bern when all of a sudden a thought occurred to me: 'If a person falls freely he will not feel his own weight.' I was startled. This simple thought made a deep impression upon me. It impelled me towards a theory of gravitation.

In his comprehensive review of relativity[4] published in 1907, Einstein devoted the whole of the last section, Section V, to 'The principle of relativity and gravitation'. In the very first paragraph, he raised the question,

Is it conceivable that the principle of relativity also applies to systems that are accelerated relative to one another?

He had no doubt about the answer and stated the *principle of equivalence* explicitly for the first time:

... in the discussion that follows, we shall therefore assume the complete physical equivalence of a gravitational field and a corresponding acceleration of the reference system.

From this postulate, he derived the time-dilation formula in a gravitational field

$$\mathrm{d}t = \mathrm{d}\tau \left(1 + \frac{\Phi}{c^2}\right), \tag{17.1}$$

where Φ is the gravitational potential, τ is the proper time and t is the time measured at zero potential. Then, applying Maxwell's equations to the propagation of light in a gravitational potential, he found that the equations are form-invariant provided the speed of light varies in the radial direction as

$$c(r) = c\left[1 + \frac{\Phi(r)}{c^2}\right], \tag{17.2}$$

recalling that Φ is always negative. Einstein realised that, as a result of Huyghens' principle or, equivalently, Fermat's principle of least time, light rays are bent in a non-uniform gravitational field. He was disappointed to find that the effect was too small to be detected in any terrestrial experiment.

17.1.2 1911

Einstein published nothing on gravity and relativity until 1911, although he was undoubtedly wrestling with these problems through the intervening period. In his paper[5] of that year, he reviewed his earlier ideas but noted that a gravitational dependence of the speed of light would result in the deflection of the light of background stars by the Sun. Applying Huyghens' principle to the propagation of light rays with a variable speed of light, he found the standard 'Newtonian' result that the angular deflection of light by a mass M would amount to

$$\Delta\theta = \frac{2GM}{pc^2}, \tag{17.3}$$

where p is the collision, or impact, parameter (see Fig. A4.5). For the Sun, this deflection amounts to 0.87 arcsec, although Einstein estimated 0.83 arcsec. Einstein urged astronomers to attempt to measure the deflection. Intriguingly, (17.3) had been published by Johann Soldner[6] in 1804 on the basis of the Newtonian corpuscular theory of light.

17.1.3 *1912–15*

Following the Solvay conference of 1911, Einstein returned to the problem of incorporating gravity into the theory of relativity and, from 1912 to 1915, his efforts were principally devoted to formulating the relativistic theory of gravity. It was to prove to be a titanic struggle.

During 1912, he realised that he needed more general space–time transformations than those of special relatively. Two quotations will illustrate the evolution of his throught.

> The simple physical interpretation of the space–time coordinates will have to be forfeited, and it cannot yet be grasped what form the general space–time transformations could have.[7]

> If all accelerated systems are equivalent, then Euclidean geometry cannot hold in all of them.[8]

Towards the end of 1912, he realised that what was needed was non-Euclidean geometry. From his student days, he vaguely remembered Gauss's theory of surfaces, which had been taught to him by Carl Friedrich Geiser. Einstein consulted his old school friend, the mathematician Marcel Grossmann, about the most general forms of transformation between frames of reference for metrics of the form

$$ds^2 = g_{\mu\nu}\,dx^\mu\,dx^\nu.^* \qquad (17.4)$$

Although this was outside Grossmann's field of expertise, he soon came back with the answer that the most general transformation formulae were the Riemannian geometries, but that they had the 'bad feature' that they are non-linear. Einstein instantly recognised that, on the contrary, this was a great advantage since any satisfactory theory of relativistic gravity must be non-linear, as will be discussed in subsection 17.2.4.

The collaboration between Einstein and Grossmann was crucial in elucidating the features of Riemannian geometry essential for the development of the theory, Einstein fully acknowledging the central role which Grossmann had played. At the end of the introduction to his first monograph on general relativity, Einstein wrote

> Finally, grateful thoughts go at this place to my friend the mathematician Grossmann, who by his help not only saved me the study of the relevant mathematical literature but also supported me in the search for the field equations of gravitation.[9]

The Einstein–Grossmann paper of 1913 was the first exposition of the role of Riemannian geometry in the search for a relativistic theory of gravity.[10] The details of Einstein's struggles over the next three years are fully recounted by Pais. It was a huge and exhausting intellectual endeavour, which culminated in the presentation of the theory in its full glory in November 1915. In that month, Einstein discovered that he could account precisely for

* Occasionally in this chapter we shall need to use this superfix notation to label the components of vectors or tensors.

the perihelion shift of Mercury, discovered by Le Verrier in 1859, as a natural consequence of his general theory of relativity. He now knew that the theory must be right.

In the next section, we will put flesh on the various arguments which led Einstein to the conclusion that the relativistic theory of gravity had to involve the complexities of Riemannian geometry and on the tools needed to relate that structure to the properties of matter and radiation.

17.1.4 The plan of attack

The aim of this chapter is to make plausible the basic concepts of space, time and gravity as embodied in the general theory of relativity. Then we will study one specific solution of the theory, the Schwarzschild solution, to illustrate how Einstein's relativistic gravity differs from Newtonian gravity.

My recommended approach would be to begin with Berry's *Principles of Cosmology and Gravitation*[11] and Rindler's *Relativity: Special, General and Cosmological*,[12] both of which are excellent introductory texts. To deepen appreciation of the theory, Weinberg's *Gravitation and Cosmology*[13] shows clearly why general relativity has to be as complex as it is. What I particularly like about his approach is that the physical content of the theory is clearly described at each stage of its development. Another strong recommendation is d'Inverno's *Introducing Einstein's Relativity*,[14] which is particularly clear on the geometric aspects of the theory and goes much further than Weinberg's text in studying black hole solutions of Einstein's field equations. Finally, the real enthusiast should not be without the classic, mammoth volume *Gravitation*[15] by Misner, Thorne and Wheeler.

17.2 Essential features of the relativistic theory of gravity

17.2.1 The principle of equivalence

Einstein's insight that 'a person in free fall will not feel his own weight' is embodied in the *principle of equivalence*. The starting point is the comparison of Newton's law of gravity,

$$\boldsymbol{f} = -\frac{Gm_1m_2}{r^2}\,\boldsymbol{i}_r, \tag{17.5}$$

with Newton's second law of motion,

$$\boldsymbol{f} = \frac{\mathrm{d}}{\mathrm{d}t}(m\boldsymbol{v}) = m\ddot{\boldsymbol{r}}, \tag{17.6}$$

for the case of point masses. Three different types of mass appear in these formulae. The mass which appears in (17.6) is the *inertial mass* of the body m_i, meaning the constant which appears in the relation between force and acceleration. It describes the resistance, or inertia, of the body to changes in its state of motion, quite independently of the nature of the impressed force $\boldsymbol{f}$.

The masses appearing in (17.5) are *gravitational masses*, but again care is needed. Suppose we ask, 'What is the gravitational force acting on the mass m_2 due to the presence of the mass m_1?' The answer is in two parts. First, the mass m_1 is the source of a gravitational field $\boldsymbol{g}_2$ at the location of the mass m_2,

$$\boldsymbol{g}_2 = -\frac{Gm_1}{r^2}\boldsymbol{i}_r.$$

The mass m_1, the source of the field, is called the *active gravitational mass*. When the mass m_2 is placed in this field, it experiences a force $m_2\boldsymbol{g}_2$, describing its response to the pre-existing field, and so m_2 is called the *passive gravitational mass*. The proportionality of the active and passive gravitational masses follows directly from Newton's third law of motion $\boldsymbol{f}_1 = -\boldsymbol{f}_2$: that is,

$$Gm_{1a}m_{2p} = Gm_{2a}m_{1p} \qquad \text{and so} \qquad \frac{m_{1a}}{m_{2a}} = \frac{m_{1p}}{m_{2p}},$$

where the subscripts a and p refer to the active and passive gravitational masses respectively. Since the ratios of the masses of the two bodies are the same, the active and passive gravitational masses can be taken to be identical by a suitable choice of units.

The relation between gravitational and inertial mass has to be found from experiment. The gravitational mass which appears in Newton's law of gravity is akin to the electric charge in electrostatics – they both describe the force associated with a particular physical property of the bodies involved. In the case of electrostatics, there is a clear distinction between the electric charge of a body and its inertial mass. In fact, it would be more helpful to refer to the gravitational mass as a 'gravitational charge', to emphasise the physical distinction between it and inertial mass.

In deriving his law of gravity, Newton was well aware of the fact that he had assumed that gravitational and inertial mass are the same thing – this was implicit in his analysis demonstrating that the planets move in elliptical orbits about the Sun. Newton tested this assumption by comparing the periods of pendulums having identical dimensions but made of different materials. Preserving the distinction between gravitational and inertial mass, the equation of simple harmonic motion for a simple pendulum of length l is

$$\frac{\mathrm{d}^2\theta}{\mathrm{d}t^2} = -\left(\frac{m_\mathrm{g}}{m_\mathrm{i}}\right)\left(\frac{g}{l}\right)\theta. \tag{17.7}$$

Newton showed that the inertial and gravitational mass are proportional to each other to better than one part in 1000.

The expression for centripetal force also involves the inertial mass and so the same type of test can be carried out for bodies in which the centripetal force is provided by the gravitational force. This approach was subject to increasingly accurate measurements by Eötvös in 1880s, by Dicke in 1964 and by Braginski in 1971. In the version of Dicke's experiment carried out by Braginski and Panov, the linearity of the relation between gravitational and inertial mass was established to about one part in 10^{12}. These experiments and many others in gravitational physics are carefully reviewed by Clifford Will in his excellent book *Theory and Experiment in Gravitational Physics*.[16] The precise linearity of the relation between the two masses enables rather subtle tests to be carried out. For example, the inertial mass $m = E/c^2$ associated with the total energy $E = \gamma m_0 c^2 = mc^2$ of electrons moving at relativistic speeds in atoms behaves exactly as a gravitational mass of the same magnitude. There is a need to continue to improve the precision with which the linearity of this relation is established. This is the objective of the NASA/ESA space mission known as STEP, the Satellite Test of the Equivalence Principle[17] – the project would be able to detect any non-linearity at a level of one part in 10^{18}. Any deviation from exact proportionality would have a profound impact upon fundamental physics.

The null result of these experiments is the experimental basis for Einstein's *principle of equivalence*, although he was notoriously vague about the precise experimental evidence for many of his greatest insights. Rindler quotes Einstein's statement of the principle as follows:

> All local, freely falling, non-rotating laboratories are fully equivalent for the performance of all physical experiments.[18]

Einstein expressed this somewhat more *in extenso* in an unpublished review of the general theory written in January 1920, now in the Pierpoint Morgan Library in New York City:

> ...*for an observer falling freely from the roof of a house there exists* – at least in his immediate surroundings – *no gravitational field* [Einstein's italics]. Indeed, if the observer drops some bodies, then these remain relative to him in the same state of rest or of uniform motion, independent of their particular chemical or physical nature.... The observer therefore has the right to interpret his state as 'at rest'.[19]

Another statement of the principle is as follows:

> At any point in a gravitational field, in a frame of reference moving with the free-fall acceleration at that point, all the laws of physics have their usual special-relativistic form except for the force of gravity, which disappears locally.[20]

A freely falling frame of reference is one that is accelerated at the gravitational acceleration at that point in space: $\boldsymbol{a} = \boldsymbol{g}$.

These statements formally identify inertial and gravitational mass, since the force acting on a particle in a gravitational field depends upon the particle's *gravitational mass*, whereas its acceleration depends upon its *inertial mass*.

Einstein's first statement of the principle, quoted by Rindler, is a natural extension of the postulates which form the basis of special relativity. In particular, any two inertial frames in relative motion are 'freely falling' frames in zero gravity. Hence, the interval between events must be given by the metric of special relativity, the *Minkowski metric*

$$\mathrm{d}s^2 = \mathrm{d}t^2 - \frac{1}{c^2}\,\mathrm{d}l^2. \tag{17.8}$$

But the principle of equivalence goes far beyond this. Inside a 'freely falling laboratory', there is no gravitational field. By transforming into such a frame of reference, gravity is replaced at that point in space by an accelerated frame of reference. The principle of equivalence asserts that the laws of physics should be identical in all such accelerated frames of reference. Thus, Einstein's great insight was to 'abolish' gravity and replace it by appropriate transformations between accelerated frames of reference.

To reinforce the idea of the equivalence of gravity and accelerations locally, it is useful to consider pictures of laboratories in free fall, in gravitational fields and also accelerated

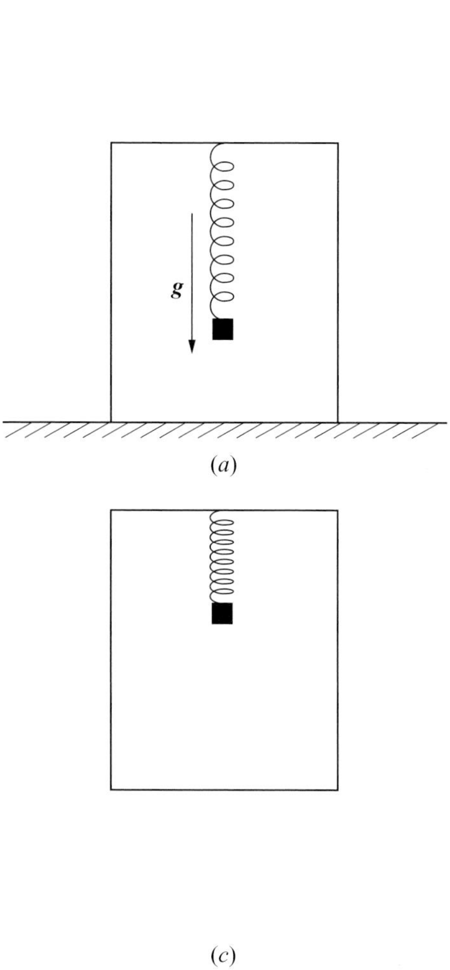

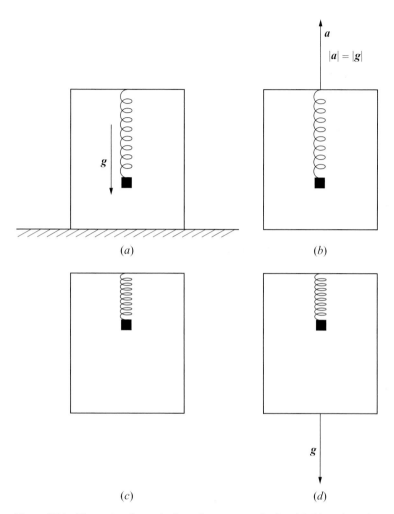

Figure 17.1: Illustrating the equivalence between gravitational fields and accelerated reference frames. In each diagram the laboratory has a spring suspended from the ceiling with a weight attached. (*a*) A static gravitational field $\boldsymbol{g}$, and (*b*) the equivalent accelerated frame of reference. In (*a*) and (*b*), the springs are stretched by the same amount. (*c*) The same laboratory in the absence of gravitational fields and accelerations; (*d*) a laboratory in free fall in a gravitational field $\boldsymbol{g}$. In (*c*) and (*d*), there is no extension of the spring.

so as to mimic the effects of gravity. In Figs. 17.1(*a*) and (*b*), the equivalence of a static gravitational field and an accelerated laboratory is demonstrated by a spring balance. For comparison, a laboratory in a 'region of zero gravitational field' and one in free fall in a gravitational field of field strength $\boldsymbol{g}$ are illustrated in Figs. 17.1(*c*) and (*d*) respectively.

Although the *acceleration* due to gravity can be eliminated at a particular point in space, the same transformation cannot completely eliminate the effects of gravity in the vicinity of that point. This is most easily seen by considering the gravitational field at distance r in the vicinity of a mass M (Fig. 17.2). It is apparent that different freely falling laboratories are needed at different points in space in order to eliminate gravity everywhere. Thus, if

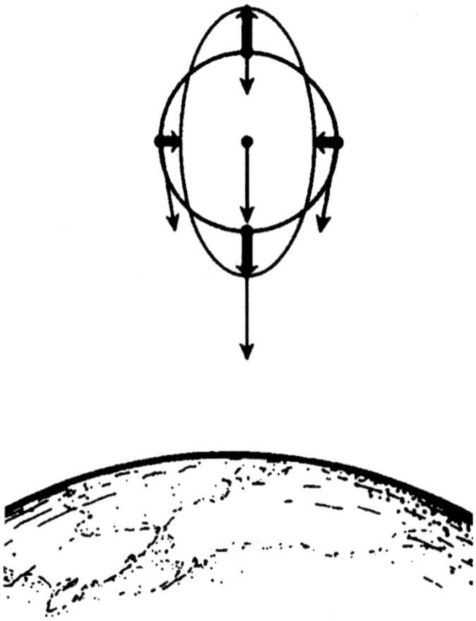

Figure 17.2: Illustrating the 'tidal forces' which cannot be eliminated when the acceleration due to gravity $\boldsymbol{g}$ is replaced by an accelerated reference frame at a particular point in space. In this example, the observer, at the centre of the sphere, is in centrifugal equilibrium in the field of a mass M. Initially, test masses are located on a sphere about the observer. At a later time, the sphere is distorted into an ellipsoid because of the tidal forces which cannot be eliminated by transforming away the gravitational field at the observer. (From R. Penrose, 1997, *The Large, the Small and the Human Mind*, Cambridge: Cambridge University Press.)

very precise measurements are made in the vicinity of the mass M, neighbouring particles are observed to move under the influence of the components of gravity which were not precisely eliminated by transforming to a single accelerated reference frame. For example, consider a standard Euclidean x-, y-, z-coordinate frame inside an orbiting space station, the z-coordinate being taken to lie in the radial direction. It is a useful exercise to show that, if two test particles are released from rest, with an initial separation vector ξ, the separation vector varies with time as

$$\frac{\mathrm{d}^2}{\mathrm{d}t^2}\begin{bmatrix}\xi^x \\ \xi^y \\ \xi^z\end{bmatrix} = \begin{bmatrix} -\dfrac{GM}{r^3} & 0 & 0 \\ 0 & -\dfrac{GM}{r^3} & 0 \\ 0 & 0 & +\dfrac{2GM}{r^3}\end{bmatrix}\begin{bmatrix}\xi^x \\ \xi^y \\ \xi^z\end{bmatrix}. \tag{17.9}$$

A pleasing result of this analysis is that the uncompensated forces vary as r^{-3}. This is exactly the same dependence as the 'tidal forces' which cause Earth–Moon and Earth–Sun tides.

We therefore need a theory which reduces locally to Einstein's special relativity in a freely falling reference frame and which transforms correctly into another freely falling frame when we move to a slightly different point in space. There is no such thing as a global Lorentz frame in the presence of a non-uniform gravitational field.

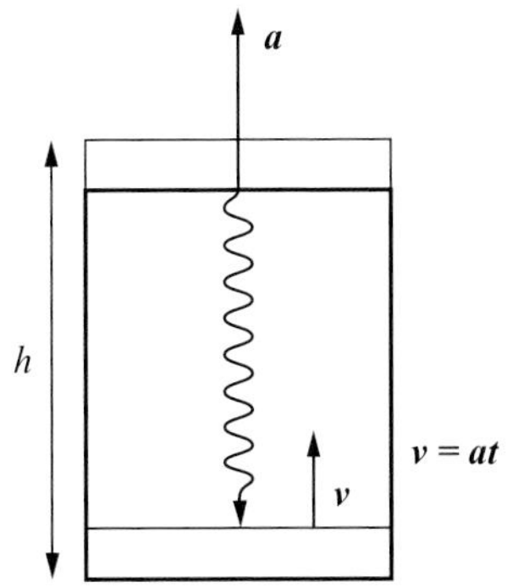

Figure 17.3: Illustrating the gravitational redshift. The laboratory is accelerated to velocity $\boldsymbol{at}$ after time t.

17.2.2 The gravitational redshift and time dilation in a gravitational field

Let us repeat Einstein's argument of 1907 concerning the redshift of electromagnetic waves in a gravitational field. Consider a light wave of frequency ν propagating from the ceiling to the floor of a lift in a gravitational field $\boldsymbol{g}$. The gravitational force can be mimicked by accelerating the lift upwards so that $\boldsymbol{g} = -\boldsymbol{a}$ (Fig. 17.3). If the height of the lift is h, the light signal travels from the roof to the floor in a time $t = h/c$. In this time, the floor is accelerated relative to an external observer to a speed $v = at = |\boldsymbol{g}|t$ and hence

$$v = \frac{|\boldsymbol{g}|h}{c}. \tag{17.10}$$

Therefore, the light wave will be observed with a higher frequency when it arrives at the floor of the lift, because of the Doppler effect. If v is small compared with c then, to first order in v/c, the observed frequency ν' is

$$\nu' = \nu\left(1 + \frac{v}{c}\right) = \nu\left(1 + \frac{|\boldsymbol{g}|h}{c^2}\right). \tag{17.11}$$

Let us now express this result in terms of the change in gravitational potential ϕ between the ceiling and floor of the lift. Since $\boldsymbol{g} = -\mathrm{grad}\,\phi$, we can write

$$|\boldsymbol{g}| = -\frac{\Delta\phi}{h}. \tag{17.12}$$

Notice that, because of the attractive nature of the gravitational force, ϕ is more negative at $h = 0$ than at the ceiling, and so $\Delta\phi$ is negative. Equation (17.11) then becomes

$$\nu' = \nu\left(1 - \frac{\Delta\phi}{c^2}\right). \tag{17.13}$$

This leads to expression for the *gravitational redshift* z_{g} in the 'Newtonian' limit. The redshift is defined to be

$$z = \frac{\lambda_{\mathrm{obs}} - \lambda_{\mathrm{em}}}{\lambda_{\mathrm{em}}} = \frac{\nu - \nu'}{\nu}, \tag{17.14}$$

and so

$$z_g = \frac{\Delta\phi}{c^2} \tag{17.15}$$

for small changes in ϕ such that $|\Delta\phi|/c^2 \ll 1$. Thus, the frequency of the waves depends upon the *gravitational potential* in which the light waves are propagated.

The first observation which demonstrated the reality of the gravitational redshift was proposed by Arthur Eddington in 1924 and involved measuring the shift in the lines in the spectrum of the white dwarf star Sirius B. The success of this observation by Walter Adams is described in Section. 18.5. Laboratory experiments to measure gravitational redshift were carried out by Pound and Rebka in 1960 and by Pound and Snider in 1965. They measured the difference in redshift of γ-ray photons moving up and then down a tower 22.5 m high at Harvard University using the Mössbauer effect. In this effect, the recoil effects of the emission and absorption of the γ-ray photons are effectively zero, since the momentum is absorbed by the atomic lattice as a whole. The γ-ray resonance is very sharp indeed and only tiny Doppler shifts are needed to move away from resonance absorption. In the Harvard experiment, the theoretical difference in redshifts for γ-ray photons moving up and down the tower was

$$z_{\rm up} - z_{\rm down} = \frac{2gh}{c^2} = 4.905 \times 10^{-15}.$$

The measured value was $(4.900 \pm 0.037) \times 10^{-15}$, providing excellent agreement within about 1%.

Notice the important point that the gravitational redshift is incompatible with special relativity. According to that theory, observers at the top and bottom of the tower are considered to be at rest in the same inertial frame of reference.

Suppose we now write (17.13) in terms of the period T of the waves. Then

$$T' = T\left(1 + \frac{\Delta\phi}{c^2}\right). \tag{17.16}$$

This expression is exactly the same as the time dilation formula relating time intervals in two inertial frames in special relativity, only now the expression refers to different locations in the gravitational field. This expression for the time dilation is exactly what would be evaluated for any time interval, and so we can write in general

$$\mathrm{d}t' = \mathrm{d}t\left(1 + \frac{\Delta\phi}{c^2}\right). \tag{17.17}$$

Let us take the gravitational potential to be zero at infinity and then measure its value at any point in the field relative to that value. Still assuming the weak-field limit, in which changes in the gravitational potential are small, at any point in the gravitational field we can write

$$\mathrm{d}t'^2 = \mathrm{d}t^2\left[1 + \frac{\phi(r)}{c^2}\right]^2, \tag{17.18}$$

where $\mathrm{d}t$ is the time interval measured at $\phi = 0$, that is, at $r = \infty$. Since $\phi(r)/c^2$ is small, (17.18) can be written

$$\mathrm{d}t'^2 = \mathrm{d}t^2\left[1 + \frac{2\phi(r)}{c^2}\right]. \tag{17.19}$$

Adopting the Newtonian expression for the gravitational potential of a point mass M, $\phi(r) = -GM/r$, we find

$$\mathrm{d}t'^2 = \mathrm{d}t^2\left(1 - \frac{2GM}{rc^2}\right). \tag{17.20}$$

Let us now introduce (17.20) for the time interval into the Minkowski metric of special relativity, (17.8), where the quantities on the right-hand side refer to the position r in the field:

$$\mathrm{d}s^2 = \mathrm{d}t'^2 - \frac{1}{c^2}\,\mathrm{d}l^2, \tag{17.21}$$

where $\mathrm{d}l$ is the differential element of proper distance. The metric of space–time in the vicinity of the point mass can therefore be written

$$\mathrm{d}s^2 = \mathrm{d}t^2\left(1 - \frac{2GM}{rc^2}\right) - \frac{1}{c^2}\,\mathrm{d}l^2. \tag{17.22}$$

This calculation shows how the metric coefficients become more complicated than those of Minkowski space when we attempt to construct a relativistic theory of gravity. It is important to appreciate that, in the above analysis, the time intervals $\mathrm{d}t$ and $\mathrm{d}t'$ are *proper times*. Thus, if we consider the observer to be located at zero potential at $r = \infty$, $\mathrm{d}t$ is the proper time measured on that clock at a fixed point in space, that is, the interval $\mathrm{d}s$ is a pure time interval and hence is the proper time. In exactly the same way, the observer at rest at gravitational potential $\Delta\phi$ measures a proper time interval $\mathrm{d}t'$. Since $\mathrm{d}t \neq \mathrm{d}t'$, there is a synchronisation problem – the rate at which proper time runs depends upon the gravitational potential.

We have stumbled across a 'general relativistic' version of the twin paradox. If one twin goes on a cosmic journey through a deeper gravitational potential than the twin who stays at home, less proper time passes on the clock of the travelling twin than on the clock of the less adventurous twin. This is no more than time dilation in a gravitational field.

The clock synchronisation problem is resolved if we decide to refer all time intervals to *coordinate time*, which takes the same value at all points in space. The natural coordinate time interval to use is $\mathrm{d}t$, the proper time measured at infinity, which can then be related to that measured in a gravitational potential $\phi(r)$ by (17.19).

The difference between $\mathrm{d}t$ and $\mathrm{d}t'$ as expressed by (17.17) enables us to understand Einstein's result of 1907 that the speed of light can be considered to depend upon the gravitational potential ϕ. At any point in the field, light propagates at the speed of light, that is, $\mathrm{d}l/\mathrm{d}t' = c$. When we measure the speed of light as observed at infinity, however, the observer measures $\mathrm{d}t$ and not $\mathrm{d}t'$. Hence, the observer at infinity measures the speed of light in the radial direction $c(r)$ to be less than c by the factor appearing in (17.17), that is

$$c(r) = c\left(1 + \frac{\Delta\phi}{c^2}\right)$$

remembering that $\Delta\phi$ in negative. Note that this is not the complete story, because we have not yet dealt with the space component of the metric (17.22). As Einstein realised, this expression for $c(r)$ is convenient for estimating time-lag effects when an electromagnetic wave passes through a varying gravitational potential and also for working out the deflection of the waves.

One final comment about the gravitational redshift (17.13) is worth making. Notice that it is no more than an expression of the conservation of energy in a gravitational field. In the final form (17.58) of the metric, this property will be built into the metric coefficients. Thus, we begin to see how bits of real physics can be built into the space–time metric.

17.2.3 *Space curvature*

Let us illustrate how the expression for the spatial component of the metric (17.22), dl, has to change as well. Consider again the propagation of light rays in our lift, but now travelling perpendicular to the gravitational acceleration. Again, we use the principle of equivalence to replace the gravitational field in a stationary laboratory by an accelerated lift in free space (Fig. 17.4).

In the time a light ray propagates across the lift, a distance d, the lift moves upwards a distance $\frac{1}{2}|\boldsymbol{g}|t^2$. Therefore, in a frame accelerated by $\boldsymbol{a} = -\boldsymbol{g}$ in free space, and also in the stationary frame in the gravitational field, the light ray follows a parabolic path. Let us approximate this light path by a circular arc of radius R. The length of the chord l across the circle is then found from

$$l^2 = \tfrac{1}{4}|\boldsymbol{g}|^2t^4 + d^2. \tag{17.23}$$

From the geometry of the diagram, it can be seen that $\varphi \approx |\boldsymbol{g}|t^2/d$. Hence, since $R\varphi \approx l$,

$$R^2 = \frac{l^2}{\varphi^2} = \frac{d^2}{4} + \frac{d^4}{|\boldsymbol{g}|^2t^4}. \tag{17.24}$$

Now, $\frac{1}{2}|\boldsymbol{g}|t^2 \ll d$, $d = ct$ and so

$$R = \frac{d^2}{|\boldsymbol{g}|t^2} = \frac{c^2}{|\boldsymbol{g}|}. \tag{17.25}$$

Thus, the radius of curvature R of the path of the light rays depends only upon the local gravitational acceleration $|\boldsymbol{g}|$. Since $\boldsymbol{g}$ is determined by the gradient of the gravitational potential, it follows that the curvature of the paths of light rays depends upon the mass distribution.

This simple argument indicates that, according to the principle of equivalence, the geometry along which light rays propagate in an accelerated laboratory, which we recall is the frame in which the local gravity has been abolished, is not flat but curved space,

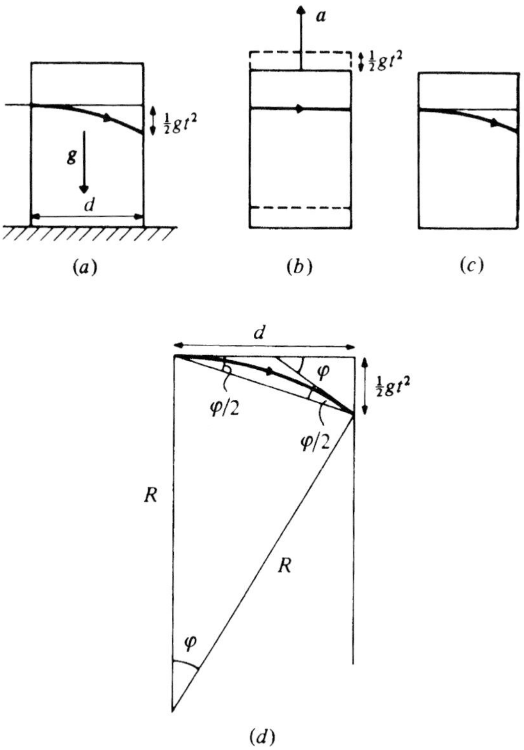

Figure 17.4: Illustrating the gravitational deflection of light according to the principle of equivalence. (*a*) The path of a light ray in a gravitational field $\mathbf{g}$. (*b*) The path of the light ray in free space, as viewed by a stationary external observer. The broken lines show the position of the accelerated laboratory after time t. (*c*) The path of the light ray as observed in the accelerated frame of reference. (*d*) The geometry of the light deflection in the accelerated frame of reference.

the spatial curvature depending upon the local value of the gravitational acceleration $\mathbf{g}$. It was arguments of this nature which convinced Einstein that he had to take non-Euclidean geometries seriously in order to derive a relativistic theory of gravity. Thus, the appropriate geometry for 'dl^2' in the metric (17.22) must, in general, refer to curved space. But, we have just shown, in subsection 17.2.2, that the time component of the metric also depends upon the gravitational potential and so, in the full theory, the metric must refer to four-dimensional curved space–time.

17.2.4 *The non-linearity of relativistic gravity*

To complicate matters further, any relativistic theory of gravity must be non-linear. This follows from Einstein's mass–energy relation $E = mc^2$ as applied to the gravitational field. The gravitational field of some mass distribution has a certain local energy density at each point in the field. Since $E = mc^2$, it follows that there is a certain inertial mass density in the gravitational field that is itself a source of gravitational field. This property contrasts with, say, the case of an electric field, which possesses a certain amount of electromagnetic field energy and which consequently has a certain inertial mass, but the latter does not generate additional electrostatic charge. Thus, relativistic gravity is intrinsically a non-linear theory and this accounts for a great deal of its complexity.

This was the aspect of the Riemannian geometries which appealed to Einstein in 1912, when he was informed by Grossmann of this 'bad feature' of these geometries.

17.3 Isotropic curved spaces

Before outlining how the general theory of relativity is formulated, it is helpful to consider the properties of isotropic curved spaces, because they illuminate some aspects of the interpretation of the Schwarzschild metric, which will be studied in detail in Section 17.5, and because these spaces are central to the development of the standard world models in Chapter 19.

In *flat* space, the distance between two points separated by $\mathrm{d}x$, $\mathrm{d}y$, $\mathrm{d}z$ is

$$\mathrm{d}l^2 = \mathrm{d}x^2 + \mathrm{d}y^2 + \mathrm{d}z^2.$$

Let us write this as in (17.4), in the form of a *Riemannian metric*:

$$\mathrm{d}s^2 = g_{\mu\nu}\,\mathrm{d}x^\mu\,\mathrm{d}x^\nu.$$

Then

$$g_{\mu\nu} = \begin{bmatrix} 1 & 0 & 0 \\ 0 & 1 & 0 \\ 0 & 0 & 1 \end{bmatrix}.$$

Now consider the simplest two-dimensional isotropic *curved* space, the surface of a sphere. This 'two-space' is isotropic because the radius R of the sphere is the same at all points on the surface. An orthogonal frame of reference can be set up at each point locally on the surface of the sphere. Let us choose spherical polar coordinates to describe the positions of points on the surface of the two-sphere, as indicated in Fig. 17.5. In this case, the natural orthogonal coordinates x^1 and x^2 (where the superfixes 1 and 2 number the coordinates, as in the above metric) are the angular coordinates θ and φ, and hence

$$\mathrm{d}x^1 = \mathrm{d}\theta, \qquad \mathrm{d}x^2 = \mathrm{d}\varphi.$$

From the geometry of the sphere, the length $\mathrm{d}l$ of the line element defined by $\mathrm{d}\theta$ and $\mathrm{d}\varphi$ is given by

$$\mathrm{d}l^2 = R^2\,\mathrm{d}\theta^2 + R^2\sin^2\theta\,\mathrm{d}\varphi^2. \tag{17.26}$$

Therefore, the metric tensor in this case is

$$g_{\mu\nu} = \begin{bmatrix} R^2 & 0 \\ 0 & R^2\sin^2\theta \end{bmatrix}. \tag{17.27}$$

The key point is that $g_{\mu\nu}$ contains information about the intrinsic geometry of the two-space. We now need a prescription which enables us to find the intrinsic geometry of the surface from the components of the metric tensor. In the case of a sphere, this may appear hardly

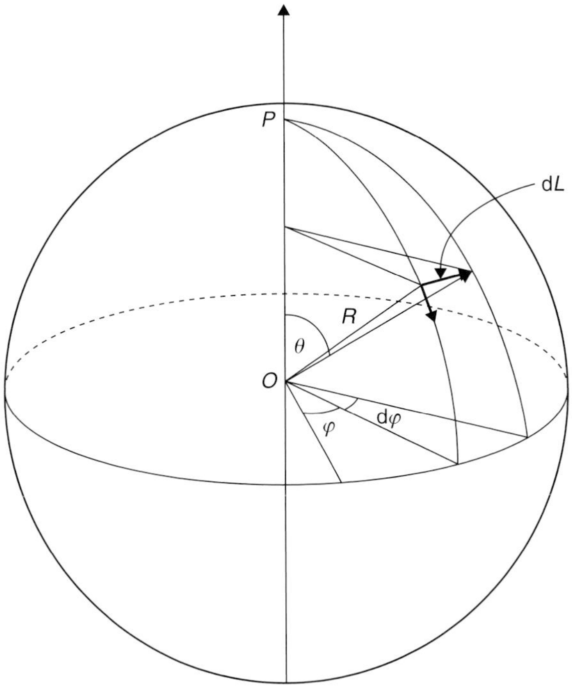

Figure 17.5: The surface of a sphere as an example of a two-dimensional isotropic curved space. The lengths of the small displacements in the $\boldsymbol{i}_\theta$ and $\boldsymbol{i}_\varphi$ directions are $R\,\mathrm{d}\theta$ and $\mathrm{d}L = R\sin\theta\,\mathrm{d}\varphi$ respectively.

necessary, but in the discussion above we could easily have chosen some strange set of coordinates which would have obscured the intrinsic geometry of the space.

This is where the memory of Gauss's theory of surfaces played a key role in Einstein's thinking in 1912. For the case of two-dimensional metric tensors which can be reduced to diagonal form, that is, those with $g_{12} = g_{21} = 0$, Gauss showed that the curvature of the surface is given by the formula

$$\kappa = \frac{1}{2g_{11}g_{22}}\left\{-\frac{\partial^2 g_{11}}{\partial(x^2)^2} - \frac{\partial^2 g_{22}}{\partial(x^1)^2} + \frac{1}{2g_{11}}\left[\frac{\partial g_{11}}{\partial x^1}\frac{\partial g_{22}}{\partial x^1} + \left(\frac{\partial g_{11}}{\partial x^2}\right)^2\right]\right.$$
$$\left. + \frac{1}{2g_{22}}\left[\frac{\partial g_{11}}{\partial x^2}\frac{\partial g_{22}}{\partial x^2} + \left(\frac{\partial g_{22}}{\partial x^1}\right)^2\right]\right\}. \tag{17.28}$$

A proof of this theorem is outlined in Berry's book[11] and the general case for two-spaces is quoted by Weinberg.[13] With $g_{11} = R^2$, $g_{22} = R^2\sin^2\theta$ and $x^1 = \theta$, $x^2 = \varphi$, it is straightforward to show that $\kappa = 1/R^2$, that is, the space is one of constant curvature, meaning that it is independent of θ and φ. In the appendix to this chapter, we prove that the only isotropic two-spaces are those for which the curvature κ is a constant, which can be positive, negative or zero. The value $\kappa = 0$ corresponds to flat, Euclidean space. If κ is negative, the two-space is hyperbolic, the radii of curvature being in opposite senses at all points of the two-space, as illustrated schematically in Fig. 17.6. For these spaces, trigonometric

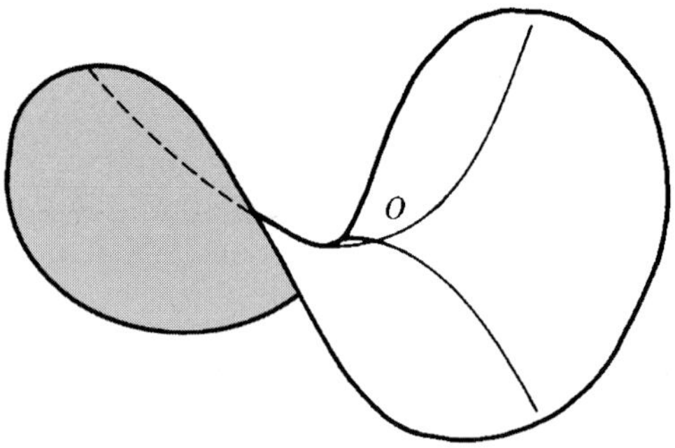

Figure 17.6: Illustrating the geometry of an isotropic hyperbolic two-space. The principal radii of curvature of the surface are equal in magnitude, but have opposite signs in orthogonal directions.

functions such as $\sin\theta$ and $\cos\theta$ are replaced by their hyperbolic counterparts, $\sinh\theta$ and $\cosh\theta$.

In isotropic curved spaces κ is a constant everywhere, but in the case of a *general two-space* the curvature is a function of the spatial coordinates.

The extension to isotropic three-spaces is straightforward, but it is no longer possible to envisage the three-space geometrically since it would have to be embedded in an isotropic four-space. We can proceed in a straightforward manner however, when it is realised that a two-dimensional section through an isotropic three-space must itself be an isotropic two-space, for which we have already worked out the metric tensor.

Suppose we want to work out the angle $\mathrm{d}\varphi$ subtended at P by great circles from the ends of the length $\mathrm{d}L$ at polar angle θ, as illustrated in Fig. 17.5. In the notation of that figure, a simple application of spherical geometry shows that

$$\mathrm{d}L = R\,\sin\theta\,\mathrm{d}\varphi. \tag{17.29}$$

If x is the distance round the great circle from P to $\mathrm{d}L$, that is, a 'geodesic distance' on the surface of the sphere, $\theta = x/R$ and hence

$$\mathrm{d}L = R\,\sin(x/R)\,\mathrm{d}\varphi. \tag{17.30}$$

By the *geodesic distance* we mean the shortest distance between two points in a two-space and, in the isotropic case, this is along the great circle joining P and $\mathrm{d}L$.

To extend the argument to isotropic three-spaces, suppose we want to work out the solid angle subtended at P by a circular area perpendicular to the line of sight from P in the isotropic three-space. This elementary area is defined by its diameter $\mathrm{d}L$ and another diameter perpendicular to $\mathrm{d}L$. To find this second diameter, we take a two-sphere section through the curved space perpendicular to the previous direction, as illustrated schematically in Fig. 17.7. Because of the isotropy of the space, the relation between $\mathrm{d}L$ and $\mathrm{d}\varphi$ is the same in the perpendicular direction and hence the area of the circle of diameter $\mathrm{d}L$ is

$$\mathrm{d}A = \frac{\pi\,\mathrm{d}L^2}{4} = \frac{\pi R^2}{4}\sin^2\frac{x}{R}\,\mathrm{d}\varphi^2 = R^2\sin^2\frac{x}{R}\,\mathrm{d}\Omega, \tag{17.31}$$

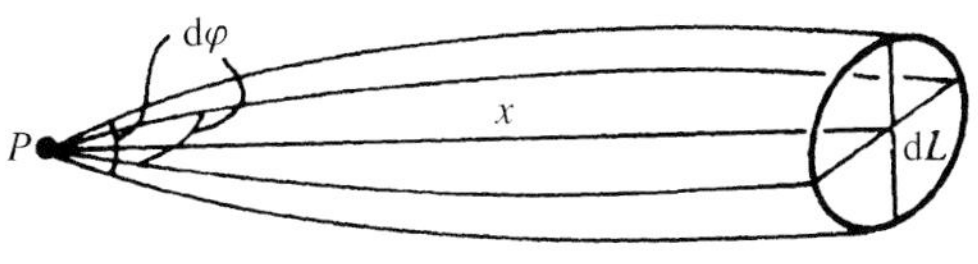

Figure 17.7: Illustrating how a surface area $\mathrm{d}A = \pi\, \mathrm{d}l^2/4$ can be measured in an isotropic three-space by taking orthogonal sections through the space, each of which is an isotropic two-space.

where $\mathrm{d}\Omega = (\pi/4)\mathrm{d}\varphi^2$ is the solid angle subtended at P by the area $\mathrm{d}A$ in the isotropic three-space. Notice that, if $R \gg x$, (17.31) reduces to $A = x^2\, \mathrm{d}\Omega$, the Euclidean result.

This example illustrates an important feature of the idea of 'distance' in non-Euclidean geometries. If x is taken to be the geodesic distance then the formula for surface area involves something more complex than the Euclidean formula. We can, however, make the formulae for areas look like the Euclidean relation by writing $\mathrm{d}A = r^2\, \mathrm{d}\Omega$, defining a new distance coordinate r. In this example, we would write

$$r = R\sin(x/R). \tag{17.32}$$

Now the formula for a spatial interval lying in a spherical two-space is, generalising (17.26),

$$\mathrm{d}l^2 = \mathrm{d}x^2 + R^2\sin^2(x/R)\,\mathrm{d}\varphi^2. \tag{17.33}$$

Let us rewrite (17.33) in terms of r rather than x. Differentiating (17.32), we find

$$\mathrm{d}r = \cos(x/R)\,\mathrm{d}x, \tag{17.34}$$

and so

$$\begin{aligned}\mathrm{d}r^2 &= \cos^2(x/R)\,\mathrm{d}x^2\\ &= [1 - \sin^2(x/R)]\,\mathrm{d}x^2 = \left(1 - r^2/R^2\right)\,\mathrm{d}x^2,\end{aligned}$$

that is,

$$\mathrm{d}x^2 = \frac{\mathrm{d}r^2}{1 - \kappa r^2} \tag{17.35}$$

where $\kappa = 1/R^2$ is the curvature of the two-space. The metric can therefore be written

$$\mathrm{d}l^2 = \frac{\mathrm{d}r^2}{1 - \kappa r^2} + r^2\,\mathrm{d}\varphi^2. \tag{17.36}$$

The metrics (17.33) and (17.36) are exactly equivalent but notice the different meanings of x and r: x is a *geodesic distance* in the three-space, whereas r is a distance coordinate, the *angular diameter distance*, which gives the correct answer for distances normal to the geodesic according to the relation $\mathrm{d}L = r\,\mathrm{d}\varphi$. Notice how conveniently the metric (17.36) takes care of spherical, flat and hyperbolic spaces, depending on the sign of κ.

17.4 The route to general relativity

The considerations of Sections 17.2 and 17.3 make it clear why general relativity is a theory of considerable technical complexity. Space–time has to be defined by a general curved metric tensor $g_{\mu\nu}$, which is a function of space–time coordinates. It is apparent from the calculations carried out in the preceding sections that the $g_{\mu\nu}$ are analogous to gravitational potentials, their variation from point to point in space–time defining the local curvature of space. From the purely geometrical point of view, we have to give mathematical substance to the principle of equivalence, that is, to devise transformations in Riemannian space which can relate the values of $g_{\mu\nu}$ at different points in space–time. From the dynamical point of view, we have to be able to relate the $g_{\mu\nu}$ to the distribution of matter in the Universe.

We are led to four-dimensional curved spaces which locally reduce to a metric of the form (17.8). As we have already seen, the appropriate choice of the metric, in general, is a *Riemannian metric*

$$ds^2 = g_{\mu\nu}\, dx^\mu\, dx^\nu, \tag{17.37}$$

where x^μ and x^ν are coordinates defining points in the four-dimensional space. Thus, according to (17.37), ds^2 is given by a homogeneous quadratic differential form in these coordinates. The indices μ and ν on the dx's are written as superscripts for the technical reason that in the full theory of four-dimensional spaces we need to distinguish covariant quantities (carrying subscripts) and contravariant quantities (carrying superscripts). We will not need to worry about this distinction in our approach.

As mentioned above, locally these spaces reduce to the Minkowski metric (17.8), for which

$$g_{\mu\nu} = \begin{bmatrix} 1 & 0 & 0 & 0 \\ 0 & -\dfrac{1}{c^2} & 0 & 0 \\ 0 & 0 & -\dfrac{1}{c^2} & 0 \\ 0 & 0 & 0 & -\dfrac{1}{c^2} \end{bmatrix}, \tag{17.38}$$

where $[x^0, x^1, x^2, x^3] = [t, x, y, z]$. The *signature* of the metric is $[+, -, -, -]$ and this property is preserved under general Riemannian transformations.

An important conceptual point is that the coordinates should be considered as 'labels' for the points in the space rather than 'distances', as is conventional in Euclidean space. It so happens that x, y and z are real distances in Euclidean space. In general, however, the coordinates may not have this intuitively obvious meaning.

17.4.1 *Four-tensors in special relativity*

All the laws of physics, with the exception of the law of gravitation, can be written in Lorentz-invariant form, that is, they are *form-invariant* under Lorentz transformations. *Four-vectors*, introduced in Chapter 16, are designed to be objects which are form-invariant under Lorentz transformations. Just for this section, we will adopt the notation, used by professional relativists, according to which the velocity v is measured in units of the speed of light, so that we set $c = 1$. Then, the time coordinate $x^0 = t$ and the spatial components are $x = x^1$, $y = x^2$ and $z = x^3$. The transformation of a four-vector V^α between two inertial frames of reference in standard configuration is then written

$$V^\alpha \rightarrow V'^\alpha = \Lambda^\alpha_\beta V^\beta, \tag{17.39}$$

where the matrix Λ^α_β is the Lorentz transformation

$$\Lambda^\alpha_\beta = \begin{bmatrix} \gamma & -\gamma v & 0 & 0 \\ -\gamma v & \gamma & 0 & 0 \\ 0 & 0 & 1 & 0 \\ 0 & 0 & 0 & 1 \end{bmatrix}, \tag{17.40}$$

with $\gamma = (1 - v^2/c^2)^{-1/2}$. The convention of summing over identical indices is adopted in (17.39).

Many physical quantities are naturally described in terms of *tensors* rather than vectors. Therefore, the natural extension of the idea of four-vectors in relativity is to *four-tensors*, which are objects that transform according to the rule

$$T^{\alpha\gamma} \rightarrow T'^{\alpha\gamma} = \Lambda^\alpha_\beta \Lambda^\gamma_\delta T^{\beta\delta}. \tag{17.41}$$

As an example of how this works, consider what relativists call 'dust', by which they mean matter without any internal pressure. The *energy–momentum tensor* for dust is $T^{\alpha\beta} = \rho_0 u^\alpha u^\beta$, where ρ_0 is the proper mass density of the dust, meaning the density measured by an observer moving with the flow, what is called a *comoving observer*; u^α is the velocity four-vector. $T^{\alpha\beta}$ can therefore be written

$$T^{\alpha\beta} = \rho_0 \begin{bmatrix} 1 & u^{x_1} & u^{x_2} & u^{x_3} \\ u^{x_1} & (u^{x_1})^2 & u^{x_1}u^{x_2} & u^{x_1}u^{x_3} \\ u^{x_2} & u^{x_1}u^{x_2} & (u^{x_2})^2 & u^{x_2}u^{x_3} \\ u^{x_3} & u^{x_1}u^{x_3} & u^{x_2}u^{x_3} & (u^{x_3})^2 \end{bmatrix}. \tag{17.42}$$

The simplest case is the transformation of the T^{00} component of this four-tensor, $T'^{00} = \Lambda^0_\beta \Lambda^0_\delta T^{\beta\delta}$. Performing this double sum, the result is $\rho' = \gamma^2 \rho_0$, which has a natural interpretation in special relativity: the observed density of the dust ρ' is increased by two powers of the Lorentz factor γ over the proper value ρ_0. One of these is associated with the formula for the relativistic momentum of the dust $p = \gamma m v$ and the other with length contraction in the direction of motion of the flow, $l = l_0/\gamma$.

For a perfect fluid, as opposed to dust, the pressure p cannot be neglected and then the energy–momentum tensor becomes $T^{\alpha\beta} = (\rho_0 + p)u^\alpha u^\beta - p g^{\alpha\beta}$, where $g^{\alpha\beta}$ is the metric tensor.* It is a pleasant exercise to show that the equation

$$\frac{\partial T^{\alpha\beta}}{\partial x^\beta} \equiv \partial_\beta T^{\alpha\beta} = 0 \tag{17.43}$$

* Here the metric tensor is written in terms of its contravariant components.

expresses both the laws of conservation of momentum and energy in relativity. Thus the operator ∂_β takes the form

$$[\partial/\partial x^0, \partial/\partial x^1, \partial/\partial x^2, \partial/\partial x^3]. \qquad (17.44)$$

These aspects of the energy–momentum tensor $T^{\mu\nu}$ are elegantly described by Rindler.[12]

Maxwell's equations in a vacuum can be written in compact form in terms of the antisymmetric electromagnetic field tensor $F^{\alpha\beta}$:

$$F^{\alpha\beta} \equiv \begin{bmatrix} 0 & E_x & E_y & E_z \\ -E_x & 0 & B_z & -B_y \\ -E_y & -B_z & 0 & B_x \\ -E_z & B_y & -B_x & 0 \end{bmatrix}, \qquad (17.45)$$

and the current density four-vector $j^\alpha = [\rho_e, \boldsymbol{j}]$. I apologise for deviating from my normal practice of strictly using SI units; the above form of Maxwell's equations is written in Heaviside–Lorentz units, for which $c = 1$. The equation of continuity becomes

$$\partial_\alpha j^\alpha = 0. \qquad (17.46)$$

Maxwell's equations for the relations between electric and magnetic fields and their sources become

$$\partial_\beta F^{\alpha\beta} = j^\alpha. \qquad (17.47)$$

Thus, four-tensors provide the natural language for expressing the laws of physics in a form which guarantees that they transform correctly according to the Lorentz transformations.

17.4.2 What Einstein did

Einstein's objective was to find the relation between the $g_{\mu\nu}$ and the mass–energy distribution, that is, the analogue of Poisson's equation in Newtonian gravity, which involves second-order partial differential equations. For example, for the metric (17.22), we rationalised from (17.19) that the g_{00} component should have the form

$$g_{00} = 1 + \frac{2\phi}{c^2}. \qquad (17.48)$$

Poisson's equation for gravity is

$$\nabla^2\phi = 4\pi G\rho, \qquad (17.49)$$

and hence, using (17.48) and the rest-frame value for $T_{00} = \rho$, we find

$$\nabla^2 g_{00} = \frac{8\pi G}{c^2} T_{00}.^* \qquad (17.50)$$

* For this simple case $T_{00} = T^{00}$.

This simple, and somewhat crude, calculation shows why it is reasonable to expect a close relation between the derivatives of $g_{\mu\nu}$ and the energy–momentum tensor $T_{\mu\nu}$. In fact, it turns out that the constant in front of T_{00} is the same as in the full theory.

The tensor equivalent of Poisson's equation involves the differentiation of tensors and this is where the complications begin; the partial differentiation of tensors does not generally yield other tensors. As a result, the operations corresponding to grad, div and curl are more complicated for tensors than for vectors.

What is needed is a tensor which involves the metric tensor $g_{\mu\nu}$ and its first and second derivatives and which is linear in its second derivatives. It turns out that there is a unique answer to this problem, the fourth-rank tensor known as the *Riemann–Christoffel tensor* $R^{\lambda}_{\mu\nu\kappa}$. Other tensors can be formed from this tensor by contraction, the most important of these being the *Ricci tensor*

$$R_{\mu\kappa} = R^{\lambda}_{\mu\lambda\kappa}, \tag{17.51}$$

and the *curvature scalar*

$$R = g^{\mu\kappa} R_{\mu\kappa}. \tag{17.52}$$

Einstein's stroke of genius was to relate these tensors to the energy–momentum tensor in the following way:

$$R_{\mu\nu} - \frac{g_{\mu\nu} R}{2} = -\frac{8\pi G}{c^2} T_{\mu\nu}. \tag{17.53}$$

This is the key relation showing how the components of the metric tensor $g_{\mu\nu}$ are related to the mass–energy distribution in the Universe.

We will go no further along this route, except to note that Einstein realised that he could add an additional term to the left-hand side of (17.53). This is the origin of the famous *cosmological constant* Λ, which was introduced in order to construct a static closed model for the Universe (Section 19.6). Equation (17.53) then becomes

$$R_{\mu\nu} - \frac{g_{\mu\nu} R}{2} + \Lambda g_{\mu\nu} = -\frac{8\pi G}{c^2} T_{\mu\nu}. \tag{17.54}$$

The cosmological constant will reappear in Chapter 19.

We still need a rule which tells us how to find the path of a particle or light ray in space–time from the metric (17.37). In Euclidean three-space, the answer is that we seek that path which minimises the distance ds between the points A and B. We need to find the corresponding result for space–time. If we join the points A and B by a large number of possible routes through space–time, the path corresponding to free fall between A and B must be the shortest path. From consideration of the twin paradox in general relativity, the free-fall path between A and B must correspond to the maximum proper time between A and B, that is, we require $\int_A^B \mathrm{d}s$ to be a *maximum* for the shortest path. In terms of the calculus of variations, this can be written

$$\delta \int_A^B \mathrm{d}s = 0. \tag{17.55}$$

It is interesting that, for our 'Newtonian' metric (17.22), this condition is exactly equivalent to Hamilton's principle in mechanics and dynamics, as we now show. The 'action integral' derived from (17.22) is

$$\begin{aligned}\int_A^B \mathrm{d}s &= \int_{t_1}^{t_2} \frac{\mathrm{d}s}{\mathrm{d}t}\,\mathrm{d}t, \\ &= \int_{t_1}^{t_2} \left[\left(1+\frac{2\phi}{c^2}\right) - \frac{v^2}{c^2}\right]^{1/2} \mathrm{d}t, \end{aligned} \tag{17.56}$$

since $\mathrm{d}l/\mathrm{d}t = v$. For weak fields, $2\phi/c^2 - v^2/c^2 \ll 1$, ignoring the constant term (17.56) reduces to

$$\int_A^B \mathrm{d}s = \int_{t_1}^{t_2} (U - T)\,\mathrm{d}t, \tag{17.57}$$

where $U = m\phi$ is the potential energy of the particle and $T = \frac{1}{2}mv^2$ is its kinetic energy. Thus, the constraint that $\mathrm{d}s$ must be a maximum results in Hamilton's principle in dynamics (see Section 7.3).

17.5 The Schwarzschild metric

The solution of Einstein's field equations for the gravitational field of a point mass was discovered by Karl Schwarzschild[22] in 1916, only two months after the publication of Einstein's theory in its definitive form. Schwarzschild was a brilliant astronomer, and director of the Potsdam Astrophysical Observatory, who had volunteered for military service in 1914 at the beginning of the First World War. In 1916, while serving on the Russian front, he wrote two papers on his exact solution of Einstein's field equations for a stationary point mass. Tragically, he contracted a rare skin disease at the front and died in May 1916 after a short illness. The *Schwarzschild metric* for a point mass M has the form

$$\mathrm{d}s^2 = \left(1 - \frac{2GM}{rc^2}\right)\mathrm{d}t^2 - \frac{1}{c^2}\left[\frac{\mathrm{d}r^2}{1 - \dfrac{2GM}{rc^2}} + r^2\left(\mathrm{d}\theta^2 + \sin^2\theta\,\mathrm{d}\varphi^2\right)\right]. \tag{17.58}$$

From the analyses of Sections 17.2 and 17.3, we can understand the meanings of the various coordinates and terms in the metric. We recall that at any point in space (17.58) reduces to the local Minkowski metric.

(i) The *time coordinate* in (17.58) is written in a form such that clocks can be synchronised everywhere, that is, the interval of proper time for a *stationary observer* at radial distance r is*

$$\mathrm{d}\tau = \left(1 - \frac{2GM}{rc^2}\right)^{1/2} \mathrm{d}t, \tag{17.59}$$

* Notice that here we use τ rather t' as in Section 17.2.

whereas the corresponding interval of coordinate time measured at infinity is $\mathrm{d}t$. Thus, $\mathrm{d}t$ keeps track of time at infinity and the proper time at any point in the field is found from (17.59). Note that, whereas the factor $[1 - 2GM/(rc^2)]^{1/2}$ in the metric (17.22) was only approximate, (17.59) is *exact* in general relativity for all values of the parameter $2GM/(rc^2)$.

(ii) The *angular coordinates* in (17.58) have been written in terms of spherical polar coordinates centred on the point mass. By inspection, it can be seen that the radial coordinate r is such that proper distances perpendicular to the radial coordinate are given by

$$\mathrm{d}l_\perp^2 = r^2(\mathrm{d}\theta^2 + \sin^2\theta\,\mathrm{d}\varphi^2). \tag{17.60}$$

Hence, the distance measure r can be thought of as an *angular-diameter distance* and is distinct from the *geodesic distance* x from the mass to the point labelled r. The square of the increment of proper, or geodesic, distance $\mathrm{d}x$ in the radial direction is the first component of the spatial part of the metric (17.58) and so

$$x = \int_0^r \mathrm{d}x = \int_0^r \frac{\mathrm{d}r}{\left(1 - \dfrac{2GM}{rc^2}\right)^{1/2}}. \tag{17.61}$$

The combination of (17.59) and (17.61) enables us to derive the correct version of Einstein's argument concerning the apparent variability of the speed of light in the radial direction according to an observer at infinity, for the case of the gravitational potential of a point mass. The local speed of light is $c = \mathrm{d}x/\mathrm{d}\tau$ and hence, for a distant observer, the speed of light in terms of coordinate time t and the distance measure r is

$$c(r) = \frac{\mathrm{d}r}{\mathrm{d}t} = c\left(1 - \frac{2GM}{rc^2}\right). \tag{17.62}$$

This is a convenient expression for working out time delays due to time dilation in the gravitational field of a point mass.

(iii) The *metric* (17.58) describes how the gravity of the point mass bends space–time. The effects of time dilation were described in (i) but, in addition, the space sections are curved. For an isotropic curved two-space, we showed that the spatial component of the metric could be written, (17.36), as

$$\frac{\mathrm{d}r^2}{1 - \kappa r^2} + r^2\mathrm{d}\varphi^2.$$

In an isotropic curved three-space the distance element in the plane perpendicular to the line of sight is now

$$\mathrm{d}l_\perp^2 = r^2(\mathrm{d}\theta^2 + \sin^2\theta\,\mathrm{d}\varphi^2),$$

so that the spatial terms of the metric are

$$\frac{\mathrm{d}r^2}{1 - \kappa r^2} + r^2\left(\mathrm{d}\theta^2 + \sin^2\theta\,\mathrm{d}\varphi^2\right). \tag{17.63}$$

Comparing (17.58) with (17.63), it can be seen that the local space curvature is found by equating κr^2 and $2GM/(rc^2)$:

$$\kappa = \frac{2GM}{r^3 c^2}. \tag{17.64}$$

This provides a measure of the curvature of space produced by a point mass M at distance coordinate r. Notice that the space is isotropic at all points, but the curvature changes as r^{-3}. Equation (17.64) has the attractive features that κ tends to zero as r tends to infinity and that κ is proportional to M: writing the curvature κ as $1/R^2$, we find

$$\frac{r^2}{R^2} = \frac{2GM}{rc^2}. \tag{17.65}$$

This is a measure of the curvature of space relative to the distance from the point mass. Notice that $2GM/(rc^2)$ is the 'relativistic factor' appearing in the time and radial components of the metric. Some typical values of this parameter are as follows.

At the surface of the Earth,	$2GME_{\mathrm{E}}/(rc^2) = 1.4 \times 10^{-8}$
At the surface of the Sun,	$2GM_{\odot}/(rc^2) = 4 \times 10^{-6}$
At the surface of a neutron star,	$2GM_{\mathrm{ns}}/(rc^2) \approx 0.3$

Thus, within the solar system, the effects of space curvature are very small and require measurements of extreme precision in order to be detectable at all. The first example indicates the magnitude of the factor by which the sum of the angles of a triangle described by light beams differ from 180° in measurements made at the surface of the Earth. Gauss is reputed to have carried out this experiment in the Harz mountains in the 1820s. Using a theodolite, he is said to have found that the angles of the triangle formed by three tall mountains added up to 180°, but we now understand how small the effects of space curvature are expected to be. What is remarkable is that he had the insight to realise that Euclid's fifth postulate, which asserts that parallel lines only meet at infinity or that the sum of the angles of a triangle is 180°, need not necessarily be correct over large enough distances. At the surface of neutron stars, for example, where the gravitational fields are strong, the effects of general relativity become significant and play an important role in their stability.

(iv) Finally, note that something 'funny' must happen at the radial coordinate $2GM/(rc^2) = 1$. The radius defined by this relation, $r_{\mathrm{g}} = 2GM/c^2$, is known as the *Schwarzschild radius* and plays a particularly important role in the study of black holes, which will be described later in this chapter.

17.6 Particle orbits about a point mass

Let us now analyse the motion of a test particle in the vicinity of a point mass according to the Schwarzschild metric. This is the simplest solution of Einstein's field equations – the point mass has no angular momentum and is uncharged. To simplify the analysis, we consider orbits in the equatorial plane $\theta = \pi/2$. There is no loss of generality in making this simplifiation, since we can always rotate the coordinate system in such a way that $d\theta = 0$. The Schwarzschild metric (17.58) can therefore be written

$$ds^2 = \alpha\, dt^2 - \frac{1}{c^2}(\alpha^{-1}\, dr^2 + r^2\, d\varphi^2). \tag{17.66}$$

In this notation, φ is the angle in the equatorial plane and $\alpha = 1 - 2GM/(rc^2)$. We now form the 'action integral'

$$s_{AB} = \int_A^B ds = \int_A^B G(x^\mu, \dot{x}^\mu)\, ds, \tag{17.67}$$

where the function $G(x^\mu, \dot{x}^\mu)$ is found by dividing (17.66) by ds^2:

$$G(x^\mu, \dot{x}^\mu) = \left(\alpha \dot{t}^2 - \frac{1}{c^2}\alpha^{-1}\dot{r}^2 - \frac{r^2}{c^2}\dot{\varphi}^2\right)^{1/2}. \tag{17.68}$$

This is the function to be inserted into the Euler–Lagrange equation (7.26) for the independent coordinates $x^\mu = t, r, \varphi$.

Notice what we are doing in following this procedure. The dots refer to differentiation with respect to ds, that is, with respect to proper time in the case of the motion of particles. Note also that $G(x^\mu, \dot{x}^\mu) = 1$. The important point is that, when we insert $G(x^\mu, \dot{x}^\mu)$ into the Euler–Lagrange equation, it is only the functional dependence of G upon x^μ and $\dot{x}^\mu$ which is important. Exactly the same procedure occurs in Hamiltonian mechanics, in which the Hamiltonian is a constant, but it is the dependence upon the coordinates which determines the dynamics. To reinforce this point, consider the case of flat space, in which $G(x^\mu, \dot{x}^\mu)$ has the form

$$G(x^\mu, \dot{x}^\mu) = \left[\dot{t}^2 - \frac{1}{c^2}\left(\dot{x}^2 + \dot{y}^2 + \dot{z}^2\right)\right]^{1/2}, \tag{17.69}$$

where again the dots mean differentiation with respect to proper time. It is apparent that (17.69) is the magnitude of the velocity four-vector

$$\boldsymbol{U} = \left[\dot{t}, \frac{\dot{x}}{c}, \frac{\dot{y}}{c}, \frac{\dot{z}}{c}\right] = \left[\gamma, \gamma\frac{\boldsymbol{u}}{c}\right], \tag{17.70}$$

where $\boldsymbol{u}$ is the three-velocity, and so $|G| = 1$.

We notice that the expression (17.68) for G does not depend upon either t or φ and so the Euler–Lagrange equation,

$$\frac{d}{d\tau}\left(\frac{\partial G}{\partial \dot{x}^\mu}\right) - \frac{\partial G}{\partial x^\mu} = 0, \tag{17.71}$$

is greatly simplified for these coordinates. Considering first the t-coordinate,

$$\frac{d}{d\tau}\left(\frac{\partial G}{\partial \dot{t}}\right) = 0, \qquad \frac{\partial G}{\partial \dot{t}} = \text{constant}. \tag{17.72}$$

Taking the partial derivative of (17.68) with respect to $\dot{t}$ and recalling that $G = 1$,

$$\alpha \dot{t} = k, \tag{17.73}$$

where k is a constant. This is a very important result: we have derived an expression for a constant of the motion and we guess that it is equivalent to the *conservation of energy* in the gravitational field. What our procedure has done is to find a property of the motion which is stationary with respect to proper time – in classical mechanics, this is the total energy for a conservative field of force. We will return to the interpretation of (17.73) in a moment.

We now carry out the same analysis for the φ-coordinate. Then

$$\frac{\mathrm{d}}{\mathrm{d}\tau}\left(\frac{\partial G}{\partial \dot{\varphi}}\right) = 0, \qquad \frac{\partial G}{\partial \dot{\varphi}} = \text{constant}. \tag{17.74}$$

Taking the partial derivative of (17.68) with respect to $\dot{\varphi}$ and recalling that $G = 1$,

$$r^2 \dot{\varphi} = h, \tag{17.75}$$

where h is a constant. This is another important result. This time, our procedure has found a constant of the motion for variations in the φ-direction. The Newtonian analogue of this result is the *conservation of angular momentum*, where h is the *specific angular momentum*, that is, the angular momentum per unit mass.

The equation for the r-coordinate is somewhat more complicated. It is simplest to substitute for $\dot{t}$ and $\dot{\varphi}$ into (17.68) in two stages. First, substituting for $\dot{t}$ from (17.73) into expression (17.68) gives

$$\begin{aligned} G^2 &= \alpha \dot{t}^2 - \frac{1}{c^2}\alpha^{-1}\dot{r}^2 - \frac{r^2}{c^2}\dot{\varphi}^2, \\ &= \frac{k^2}{\alpha} - \frac{1}{\alpha c^2}\dot{r}^2 - \frac{r^2}{c^2}\dot{\varphi}^2 = 1. \end{aligned} \tag{17.76}$$

Equation (17.76) can now be reorganised as follows:

$$\dot{r}^2 + \alpha (r\dot{\varphi})^2 - \frac{2GM}{r} = c^2(k^2 - 1). \tag{17.77}$$

Finally, multiplying through by half the mass of a test particle m, we obtain

$$\tfrac{1}{2} m\dot{r}^2 + \tfrac{1}{2}\alpha m (r\dot{\varphi})^2 - \frac{GMm}{r} = \frac{mc^2}{2}(k^2 - 1). \tag{17.78}$$

This calculation shows that the constant k is indeed related to the conservation of energy. All the terms on the left-hand side look superficially like those appearing in the Newtonian expression for the conservation of energy in bound orbits, but there are *important differences*.

- The three terms on the left-hand side look formally the same as the Newtonian expression for conservation of energy, except for the inclusion of the factor α in front of the term $\frac{1}{2}m(r\dot{\varphi})^2$ corresponding to motion in the circumferential direction.
- The dots refer to differentiation with respect to *proper time* $\mathrm{d}\tau$, not to coordinate time.
- The r-coordinate is an angular-diameter distance rather than a geodesic distance, but we are now familiar with the care needed in the use of distance measures in curved space–time.

17.6.1 Energy in general relativity

The definition of energy in general relativity is discussed by Shapiro and Teukolsky.[23] The simplest approach is to adopt the Lagrangian formulation of mechanics, in which the total energy is defined as the momentum conjugate to the time coordinate through the relation

$$E \equiv p_t \equiv \frac{\partial \mathcal{L}}{\partial \dot{t}} \tag{17.79}$$

(see subsection 7.5.1). Shapiro and Teukolsky show that $\mathcal{L} \propto G^2$ and so the definition of E can be found from (17.68),

$$E \propto \alpha \dot{t} = k. \tag{17.80}$$

In fact, the reasoning which led to (17.73) performed precisely the same operation. To find the constant relating E to k, it is simplest to consider the case of a stationary particle at $r = \infty$, in which case (17.78) shows that $k = 1$. In this case the total energy is the rest mass energy mc^2 and hence, in general, the total energy must take the form

$$E = kmc^2 \qquad \text{or} \qquad k = \frac{E}{mc^2}. \tag{17.81}$$

Notice that, by taking the partial derivative with respect to $\dot{t}$ in (17.79), we are finding the energy according to coordinate time t, that is, the energy as measured at infinity rather than at a general point in the field.

Let us illustrate this in another way. The four-vector velocity of the point mass m can be written down from (17.68), by analogy with the flat-space case (17.70). The four-momentum is

$$\boldsymbol{P} = \left[mc\alpha^{1/2}\dot{t},\ m\alpha^{-1/2}\dot{r},\ 0,\ mr\dot{\varphi}\right] \tag{17.82}$$

for the case of motion in the $r\varphi$-plane. The second and fourth terms correspond to the radial and tangential components of the particle's momentum, and the first to its total energy at that point in the field. Therefore, we can write the *local* energy of the particle in the field as

$$E_{\text{local}} = mc^2\alpha^{1/2}\dot{t} = k\alpha^{-1/2}mc^2. \tag{17.83}$$

The energy as measured at infinity is, however, subject to the same time-dilation factor as the time component of the metric, since they are both the 'time' components of four-vectors. If we jump forward and use (17.126), the full general relativistic formula for time dilation, the energy measured by the observer at infinity is

$$E = \alpha^{1/2}E_{\text{local}} = \alpha^{1/2}k\alpha^{-1/2}mc^2 = kmc^2 \tag{17.84}$$

(see also Shapiro and Teukolsky's book[23]). This result will prove to be important in working out the maximum amount of energy which can be extracted from matter forming an accretion disc about a black hole.

17.6.2 Particle orbits in Newtonian gravity and in general relativity

Let us now contrast the orbits of particles about a point mass M according to Newtonian gravity and general relativity. Substituting for $\dot{\varphi}$ in (17.78) using (17.75),

$$m\dot{r}^2 + \frac{\alpha m h^2}{r^2} - \frac{2GMm}{r} = mc^2(k^2 - 1). \tag{17.85}$$

Using the definition of α, as in (17.66), this becomes

$$m\dot{r}^2 + \frac{mh^2}{r^2} - \frac{2GMm}{r} - \frac{2GMmh^2}{r^3c^2} = mc^2(k^2 - 1). \tag{17.86}$$

The substitution resulting in (17.85) incorporates the law of conservation of angular momentum into the law of conservation of energy according to general relativity. For Newtonian gravity, the equivalent expression is

$$m\dot{r}^2 + \frac{mh^2}{r^2} - \frac{2GMm}{r} = m\dot{r}_\infty^2 \tag{17.87}$$

where $\dot{r}_\infty$ is the radial velocity of the test particle at infinity. If the particle does not reach infinity, $\dot{r}_\infty$ is imaginary.

Although (17.86) and (17.87) look similar, the meanings of the coordinates are quite different, as noted in the three bullet points just before the previous subsection. The most important difference is the presence of the term $-2GMmh^2/(r^3c^2)$ on the left-hand side of (17.86). The origin of this term can be traced to the α-factor in front of the term representing rotational motion about the point mass; this term has profound implications for the dynamics of particles about the central point mass.

The radial component of the velocity of a test particle in Newtonian theory can be represented in terms of the variation of the different terms in (17.87) with radius. Rewriting (17.87) for unit mass in terms of the Schwarzschild radius $r_g = 2GM/c^2$ and a dimensionless specific angular momentum $\eta = h^2/(r_g^2c^2)$, we obtain

$$\text{for the Newtonian case} \qquad \dot{r}^2 = -c^2\left[\frac{\eta}{(r/r_g)^2} - \frac{1}{r/r_g}\right] + \dot{r}_\infty^2. \tag{17.88}$$

The terms in the brackets act as a potential Φ such that

$$\Phi = \frac{\eta}{(r/r_g)^2} - \frac{1}{r/r_g}, \tag{17.89}$$

where the first term is a 'centrifugal' potential and the second the Newtonian gravitational potential. The variation of $-\Phi$ with r/r_g is shown in Fig. 17.8, in which the contributions of the two terms are displayed separately. We now discuss various cases.

- If the particle has zero radial velocity at infinity, that is, it starts from A', the radial velocity at different radii can be found from the bold curve. When the particle reaches the radial coordinate at A, the radial component of its velocity is zero and this is the closest the particle can approach to $r = 0$, since $\dot{r}$ becomes imaginary for smaller values of r. At A, all the kinetic energy of the particle is in its tangential motion.

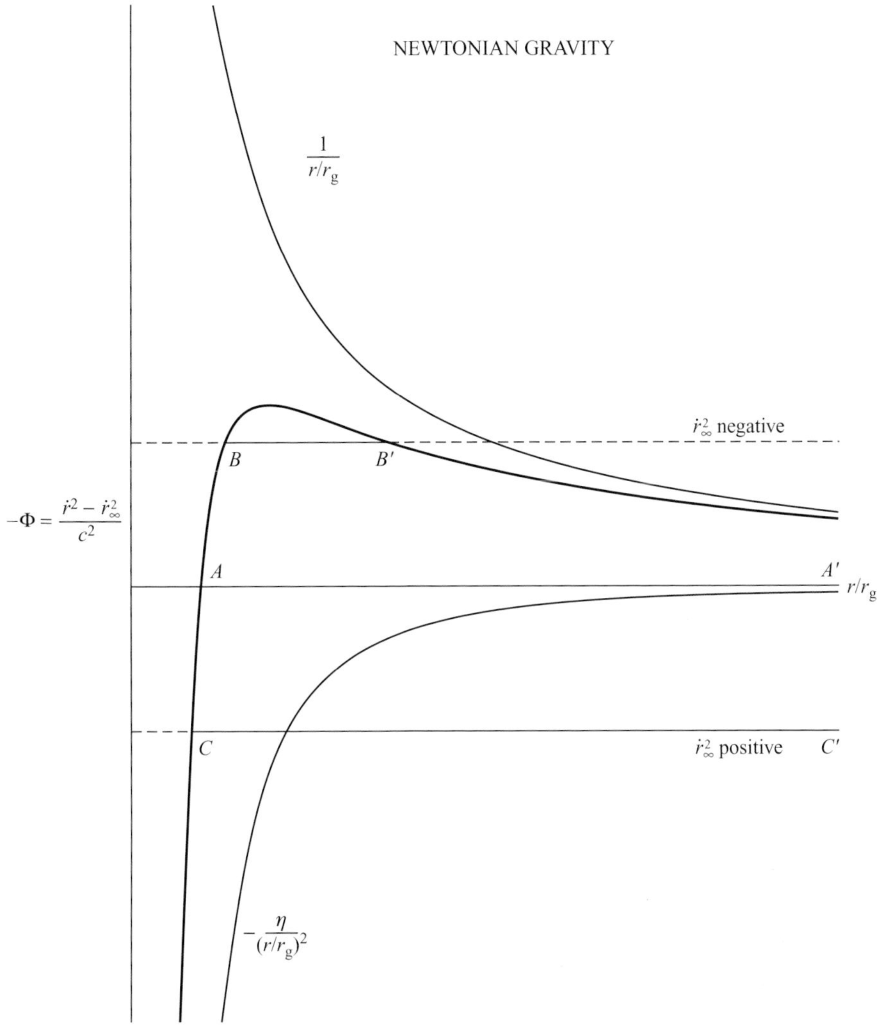

Figure 17.8: The bold curve shows the potential $-\Phi$ in Newtonian gravity, in which there are two contributions to the potential, plotted against r/r_g.

- Bound orbits are found if $\dot{r}^2_\infty$ is negative. Then, the particle moves in an elliptical orbit between B and B'.
- Hyperbolic orbits of unbound particles correspond to positive values of $\dot{r}^2_\infty$ and are represented by loci such as CC'. C is the distance of closest approach and is found from the solution of (17.86) at which $\dot{r} = 0$. Notice that the particle can only reach $r = 0$ if $\eta = 0$. This is an important result in that even if η is tiny it is always sufficient to prevent the test mass reaching the origin, according to classical dynamics.

In the corresponding analysis for general relativity, the expression for $\dot{r}^2$ becomes

$$\text{for the GR case} \qquad \dot{r}^2 - c^2(k^2 - 1) = -c^2\left[\frac{\eta}{(r/r_g)^2} - \frac{1}{r/r_g} - \frac{\eta}{(r/r_g)^3}\right] \qquad (17.90)$$

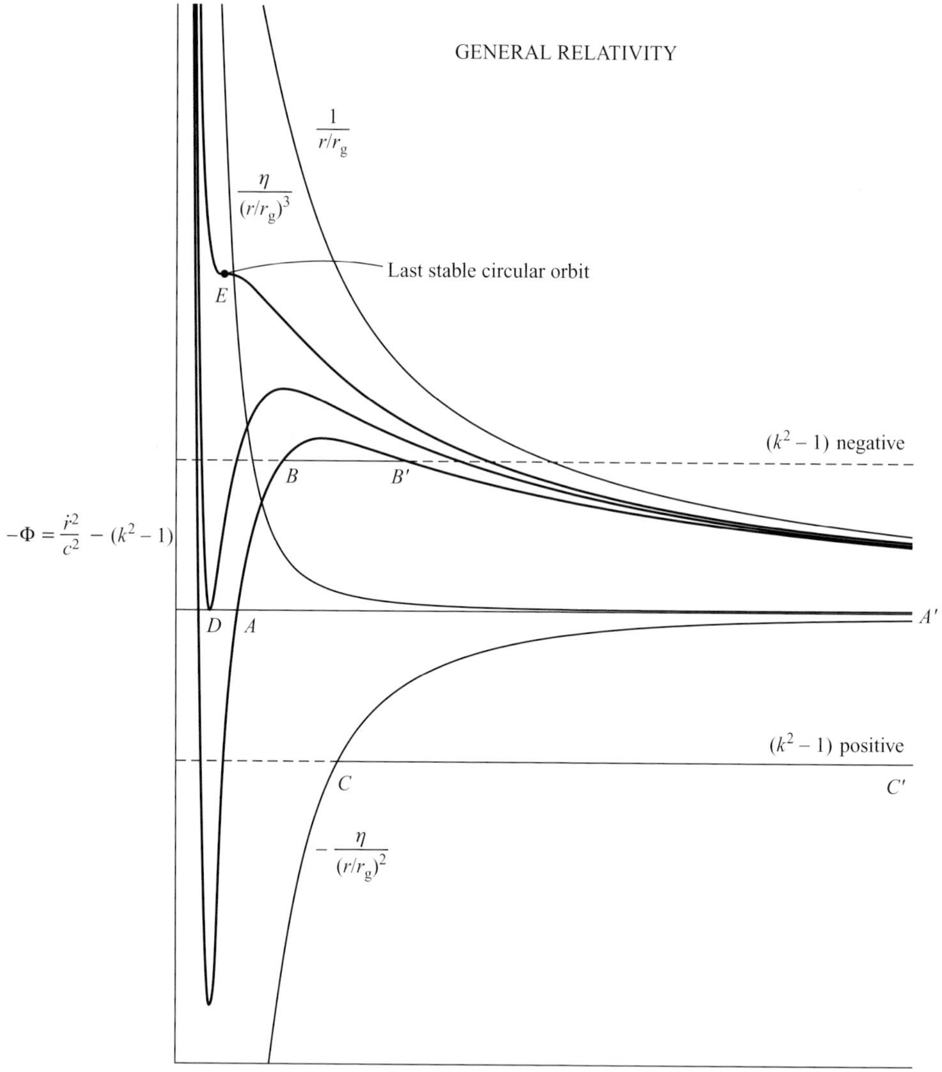

Figure 17.9: The variation of the potential $-\Phi$ acccording to general relativity. There is an attractive potential associated with the term $\eta(r/r_g)^{-3}$, in addition to the two terms found in Newtonian gravity. This term dominates the dynamics close to the origin.

so that now

$$-\Phi = \frac{\eta}{(r/r_g)^2} - \frac{1}{r/r_g} - \frac{\eta}{(r/r_g)^3}. \tag{17.91}$$

Much of the analysis is similar to the Newtonian case, but now we have to include the specifically general relativistic term $\eta(r/r_g)^{-3}$, which acts as a negative potential. The key point is that its magnitude increases even more rapidly with decreasing r than does the centrifugal term. The corresponding energy diagram for the general relativistic case is shown in Fig. 17.9.

The presence of the term in $(r/r_g)^{-3}$ introduces a strong attractive potential, which becomes dominant if the particle can penetrate to small enough values of r/r_g. The behaviour of the particle depends upon the value of η. If η is large, the particle behaves more or less as in the Newtonian case and attains zero velocity at the point A. For small enough η, however, the behaviour illustrated by DA' is found. At the point D, the negative general relativistic potential term becomes sufficient to balance the centrifugal potential and the particle can fall into $r = 0$. If $k^2 - 1$ is negative, bound orbits such as BB' are possible but eventually even these become impossible. So long as the chord BB' is of finite length, a range of elliptical and circular orbits is possible, but eventually η decreases to such a value that B and B' coincide at the point E, which corresponds to the *last stable circular orbit* about the point mass M.

It is straightforward to work out from (17.91) the ranges of η in which these different types of behaviour are found. The limiting case D corresponds to the orbit of a particle which falls from infinity with finite angular momentum but nonetheless can reach $r = 0$. This point occurs when the roots of $\Phi = 0$, that is, the roots of

$$\frac{1}{(r/r_g)^3}\left[\eta\left(\frac{r}{r_g}\right) - \left(\frac{r}{r_g}\right)^2 - \eta\right] = 0$$

are equal. Solving this equation, the limiting value occurs at $\eta = 4$, $r/r_g = 2$.

The last stable circular orbit E occurs when the turning points of the potential function coincide and $\mathrm{d}^2\Phi/\mathrm{d}r^2 = 0$. The turning points coincide when $\eta = 3$, the corresponding value of r/r_g being 3. This is a particularly important result for the dynamics of particles in the vicinity of black holes, in particular, for understanding the structure of their accretion discs. Circular orbits at $r < 3r_g$ are not stable and the particle spirals rapidly in towards $r = 0$.

These results illustrate the physical origin of *black holes* in general relativity. The potential barrier created by the centrifugal potential in Newtonian theory is modified in general relativity so that particles can fall into $r = 0$ despite the fact that they possess finite angular momentum.

17.7 Advance of perihelia of planetary orbits

The detailed analysis of the orbit of a test mass m proceeds from the Schwarzchild metric (17.58) via the energy equation (17.86),

$$m\dot{r}^2 + \frac{mh^2}{r^2} - \frac{2GMm}{r} - \frac{2GMmh^2}{r^3c^2} = mc^2(k^2 - 1). \tag{17.92}$$

Here $\dot{r}$ means $\mathrm{d}r/\mathrm{d}\tau$, and the dependence upon τ can be eliminated by replacing it with the dependence upon φ, in order to obtain an expression for the orbit in the polar coordinates r and φ:

$$\dot{r} = \frac{\mathrm{d}r}{\mathrm{d}\tau} = \frac{\mathrm{d}r}{\mathrm{d}\varphi}\frac{\mathrm{d}\varphi}{\mathrm{d}\tau} = \frac{\mathrm{d}r}{\mathrm{d}\varphi}\dot{\varphi} = \frac{h}{r^2}\frac{\mathrm{d}r}{\mathrm{d}\varphi}, \tag{17.93}$$

using (17.75). Therefore (17.92) becomes

$$\left(\frac{h}{r^2}\frac{\mathrm{d}r}{\mathrm{d}\varphi}\right)^2 + \frac{h^2}{r^2} - \frac{2GM}{r} - \frac{2GMh^2}{r^3c^2} = c^2(k^2 - 1). \tag{17.94}$$

There is a standard substitution, $u = 1/r$, which simplifies the analysis. Then $\mathrm{d}r = -(1/u^2)\,\mathrm{d}u$ and, dividing by h^2,

$$\left(\frac{\mathrm{d}u}{\mathrm{d}\varphi}\right)^2 + u^2 - \frac{2GM}{h^2}u - \frac{2GM}{c^2}u^3 = \frac{c^2}{h^2}(k^2 - 1). \tag{17.95}$$

Now differentiating with respect to φ,

$$\frac{\mathrm{d}^2u}{\mathrm{d}\varphi^2} + u = \frac{GM}{h^2} + \frac{3GM}{c^2}u^2. \tag{17.96}$$

This is the equation we have been seeking. It is illuminating to compare it with the corresponding Newtonian result. The Newtonian law of conservation of energy is

$$\tfrac{1}{2}mv^2 - \frac{GMm}{r} = \tfrac{1}{2}m(\dot{r}^2 + r^2\dot{\varphi}^2) - \frac{GMm}{r} = \text{constant}. \tag{17.97}$$

Because of the conservation of angular momentum, $v_\varphi r = \dot{\varphi}r^2 = h = \text{constant}$, where h is the specific angular momentum, and hence

$$m\dot{r}^2 + \frac{mh^2}{r^2} - \frac{2GMm}{r} = \text{constant}. \tag{17.98}$$

Performing the same types of substitution involved in deriving (17.96), namely, dividing through by m, setting $u = 1/r$ and then converting the time derivative into a derivative with respect to φ by writing

$$\frac{\mathrm{d}r}{\mathrm{d}t} = \frac{\mathrm{d}r}{\mathrm{d}\varphi}\frac{\mathrm{d}\varphi}{\mathrm{d}t} = \frac{h}{r^2}\frac{\mathrm{d}r}{\mathrm{d}\varphi}, \tag{17.99}$$

we find the result

$$\left(\frac{\mathrm{d}u}{\mathrm{d}\varphi}\right)^2 + u^2 - \frac{2GM}{h^2}u = \text{constant}. \tag{17.100}$$

Differentiating with respect to φ, we find in the Newtonian case

$$\frac{\mathrm{d}^2u}{\mathrm{d}\varphi^2} + u = \frac{GM}{h^2}. \tag{17.101}$$

The expressions (17.96) and (17.101) are similar, the important difference being the presence of the term $(3GM/c^2)u^2$ in (17.96). We already know the solutions of (17.101); for bound orbits, these are *ellipses*. Notice that, in the limit $c \to \infty$, the term $(3GM/c^2)u^2$ vanishes, enabling us to recover the Newtonian expression (17.101). Within the Solar System this term is a very small correction to the Newtonian solution. To see this, we can compare the magnitudes of the two 'gravitational' terms on the right-hand side of (17.96) for a circular orbit, $u = \text{constant}$:

$$\frac{(3GM/c^2)u^2}{GM/h^2} = \frac{3u^2h^2}{c^2} = \frac{3v^2}{c^2}. \tag{17.102}$$

Thus, the correction to the orbit is second order in v/c. Since the speed of the Earth about the Sun is 30 km s^{-1}, the correction amounts to only about three parts in 10^8.

Although the effect is very small, it results in a measurable change in the orbits of the planets. The largest effect within the solar system is found for Mercury's orbit about the Sun, which is somewhat elliptical, having ellipticity $e = 0.2$. The position of closest approach of a planet to the Sun is known as its *perihelion* and the effect of the small general relativistic correction is to cause the perihelion of Mercury's orbit to advance by a small amount each orbit. Let us determine the magnitude of the advance for a *circular orbit*.

For a circular orbit, $u =$ constant, $\mathrm{d}^2u/\mathrm{d}\varphi^2 = 0$ and so, in the Newtonian approximation,

$$u = \frac{GM}{h^2}. \tag{17.103}$$

We can insert this solution into equation (17.96) and solve for a small perturbation $g(\varphi)$ to the orbital phase:

$$u = \frac{GM}{h^2} + g(\varphi). \tag{17.104}$$

Preserving terms to first order in $g(\varphi)$ gives

$$\frac{\mathrm{d}^2 g}{\mathrm{d}\varphi^2} + g\left[1 - \left(\frac{3GM}{c^2}\right)\left(\frac{2GM}{h^2}\right)\right] = \frac{3GM}{c^2}\left(\frac{GM}{h^2}\right)^2. \tag{17.105}$$

This is a harmonic equation for g and, in the non-relativistic limit in which $3GM/c^2 \to 0$, it becomes

$$\frac{\mathrm{d}^2 g}{\mathrm{d}\varphi^2} + g = 0. \tag{17.106}$$

The harmonic solutions have period 2π, in other words, correspond to a perfect circular orbit.

Because of the relativistic perturbation term, the phase of the planet in its orbit is slightly different from 2π per revolution. From the expression (17.105), we find

$$\omega^2 = \left[1 - \left(\frac{3GM}{c^2}\right)\left(\frac{2GM}{h^2}\right)\right], \tag{17.107}$$

and the period T of the orbit is

$$T = \frac{2\pi}{\left[1 - \left(\frac{3GM}{c^2}\right)\left(\frac{2GM}{h^2}\right)\right]^{1/2}} \approx 2\pi\left(1 + \frac{3G^2M^2}{c^2h^2}\right). \tag{17.108}$$

Thus, the orbit closes up slightly later each orbit. The fractional change of phase per orbit is $\mathrm{d}T/T = \mathrm{d}\varphi/(2\pi)$, where

$$\frac{\mathrm{d}\varphi}{2\pi} \approx \frac{3G^2M^2}{c^2h^2} = \frac{3}{4}\left(\frac{2GM}{hc}\right)^2 = \frac{3}{4}\left(\frac{c}{v}\right)^2\left(\frac{r_{\mathrm{g}}}{r}\right)^2, \tag{17.109}$$

since $h = rv$. For elliptical orbits, the exact answer is

$$\frac{\mathrm{d}\varphi}{2\pi} = \frac{3}{4}\left(\frac{c}{v}\right)^2\left(\frac{r_g}{r}\right)^2\frac{1}{1-e^2}, \tag{17.110}$$

where e is the ellipticity. For Mercury, $r = 5.8 \times 10^{10}$ m, $T = 88$ days and $\epsilon = 0.2$, and for the Sun $r_g = 3$ km. The advance of the perihelion of Mercury is therefore predicted to amount to 43 arcsec per century. This was the extraordinary resolution of an unsolved problem in nineteenth-century celestial mechanics. Once the effects of the perturbations of the other planets upon Mercury's orbit about the Sun had been taken into account, Le Verrier found in 1859 that there remained an unaccounted advance of its perihelion of 43 arcsec per century. In his great paper on general relativity of November 1915, Einstein triumphantly showed that this was precisely accounted for by the general theory of relativity.

The effect is very small within the Solar System, but it is very much larger in the case of compact close binary star systems. The most important example is the binary pulsar PSR 1913+16, which consists of two neutron stars orbiting their common centre of mass. The orbital period is only 7.75 hours, and the ellipticity of the orbits of the neutron stars is $e = 0.617$. In this case, the rate of advance of the perihelia of their elliptical orbits provides important information about the masses of the neutron stars as well as providing very sensitive tests of general relativity itself.[24]

17.8 Light rays in Schwarzschild space–time

Einstein realised as early as 1907 that light rays must be deflected by the presence of matter and in 1911 urged astronomers to attempt to measure the deflection of light rays by the Sun, four years before the general theory was presented in its definitive form. The predicted deflection can be determined from the Schwarzschild metric for a point mass. Light rays propagate along paths in space–time for which $\mathrm{d}s = \mathrm{d}\tau = 0$; these are known as *null geodesics*. This has consequences for the energy and angular momentum conservation relations, (17.73) and (17.75) respectively, which can be written

$$\alpha\dot{t} = \left(1 - \frac{2GM}{rc^2}\right)\frac{\mathrm{d}t}{\mathrm{d}s} = k, \qquad r^2\dot{\varphi} = r^2\frac{\mathrm{d}\varphi}{\mathrm{d}s} = h. \tag{17.111}$$

Since $\mathrm{d}s = \mathrm{d}\tau = 0$, both k and h are infinite although their ratio remains finite. The propagation equation for light rays can be found by inserting the value $h = \infty$ into (17.96), so that

$$\frac{\mathrm{d}^2u}{\mathrm{d}\varphi^2} + u = \frac{3GM}{c^2}u^2. \tag{17.112}$$

The term on the right-hand side represents the influence of curved space–time upon the propagation of the light ray. Notice that there is an interesting limiting case in which photons propagate in a circular orbit about a black hole. In this case, the first term on the left-hand side of (17.112) is zero, and since $u = 1/r$ the radius of the photon orbit is

$$r = \frac{3GM}{c^2} = \frac{3r_g}{2}. \tag{17.113}$$

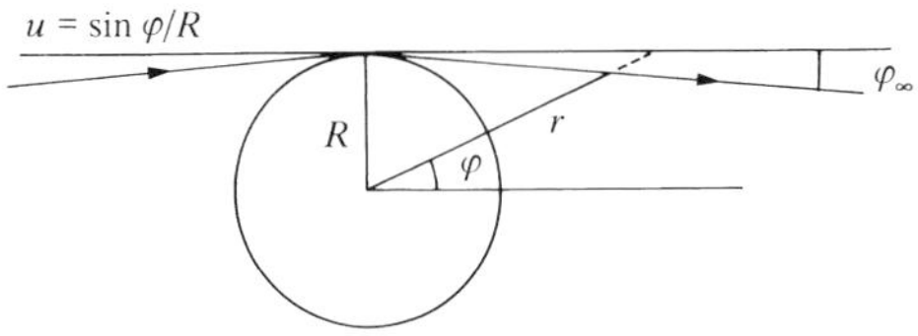

Figure 17.10: The coordinate system used to work out the deflection of light rays by the Sun; $r = 1/u$.

Returning to the much less extreme case of the deflection of light rays by the Sun, let us first work out the path of a light ray in the absence of the term $(3GM/c^2)u^2$. The propagation equation becomes

$$\frac{d^2u}{d\varphi^2} + u = 0. \tag{17.114}$$

A suitable solution of this equation is $u_0 = \sin\varphi/R$, where R is the distance of closest approach of the light ray, as illustrated in Fig. 17.10. This solution corresponds to a straight line, tangent to the (empty) sphere at radius R.

To find the path of the light ray in the next order of approximation, we seek solutions of (17.112) with $u = u_0 + u_1$. Then, we find

$$\frac{d^2u_1}{d\varphi^2} + u_1 = \frac{3GM}{c^2R^2}\sin^2\varphi. \tag{17.115}$$

By inspection, a suitable trial solution is $u_1 = A + B\cos 2\varphi$. Differentiating and equating coefficients results in the following solution for u:

$$u = u_0 + u_1 = \frac{\sin\varphi}{R} + \frac{3GM}{2c^2R^2}\left(1 + \frac{\cos 2\varphi}{3}\right). \tag{17.116}$$

The angle φ is expected to be very small and so we can use the approximations $\sin\varphi \approx \varphi$ and $\cos 2\varphi \approx 1$. In the limit $r \to \infty$, $u \to 0$, we find

$$u = \frac{\varphi_\infty}{R} + \frac{3GM}{2c^2R^2}\left(1 + \frac{1}{3}\right) = 0,$$

so that

$$\varphi_\infty = -\frac{2GM}{Rc^2}. \tag{17.117}$$

From the geometry of Fig. 17.10, the total deflection is twice φ_∞, and so

$$\Delta\varphi = \frac{4GM}{Rc^2}. \tag{17.118}$$

For light rays just grazing the limb of the Sun, the deflection amounts to 1.75 arcsec.

Historically, this was a very important result. The photon has a certain momentum $p = \hbar k$ and so the Rutherford scattering formula can be used to work out the angular deviation of the light path. The answer can be derived from the analysis of appendix section A4.3 and is $\Delta\varphi_{\text{Newton}} = 2GM/(Rc^2) = 0.87$ arcsec, the result found by Soldner in 1801. These predictions led to the famous eclipse expeditions of 1919 led by Eddington and Crommelin to measure precisely the angular deflections of the positions of stars observed close to the limb of the Sun during a solar eclipse. One expedition went to Sobral in Northern Brazil and the other to the island of Principe, off the coast of West Africa. The Sobral result was 1.98 ± 0.12 arcsec and the Principe result 1.61 ± 0.3 arcsec. These were technically demanding observations and there was some controversy about the accuracy of the results. The latter were, however, clearly consistent with Einstein's prediction and from that time on Einstein's name became synonymous with scientific genius in the public imagination. The matter was laid to rest in the 1970s when radio interferometry found exactly the expected result.

The deflection of the images of background stars and galaxies by intervening masses has developed dramatically over the last 15 years, and the subject of *gravitational lensing* has become a major and very exciting astronomical industry. Figure VII.1 in the introduction to this case study shows the remarkable arcs observed about the centre of a rich cluster of galaxies, the mass of the cluster acting as a gravitational lens which magnifies and distorts the images of very distant background galaxies. Images such as these enable the mass distribution in the cluster to be determined in considerable detail and has produced convincing evidence for dark matter in the haloes of the galaxies and within the cluster as a whole.

17.9 Particles and light rays near black holes

Let us now consider a particle falling radially from rest at infinity to $r = 0$. We begin with the energy equation (17.86) with $h = 0$,

$$m\dot{r}^2 - \frac{2GMm}{r} = mc^2(k^2 - 1). \tag{17.119}$$

If the particle is at rest at infinity, the total energy of the particle is its rest mass energy $E = mc^2$; hence, according to (17.81), $k = 1$ and so

$$\dot{r}^2 = \left(\frac{dr}{d\tau}\right)^2 = \frac{2GM}{r}. \tag{17.120}$$

This equation describes the dynamics of the particle in terms of the *proper time* measured at its point in space. Performing the integral

$$\int_{\tau_1}^{\tau_2} d\tau = -\int_{r_1}^{0} \frac{r^{1/2}}{(2GM)^{1/2}}\, dr,$$

we obtain

$$\tau_2 - \tau_1 = \left(\frac{2}{9GM}\right)^{1/2} r_1^{3/2}. \tag{17.121}$$

Thus, the particle falls from r_1 to $r = 0$ in a *finite proper time*, and nothing strange takes place at the Schwarzschild radius r_g. The observer at infinity has, however, a different view

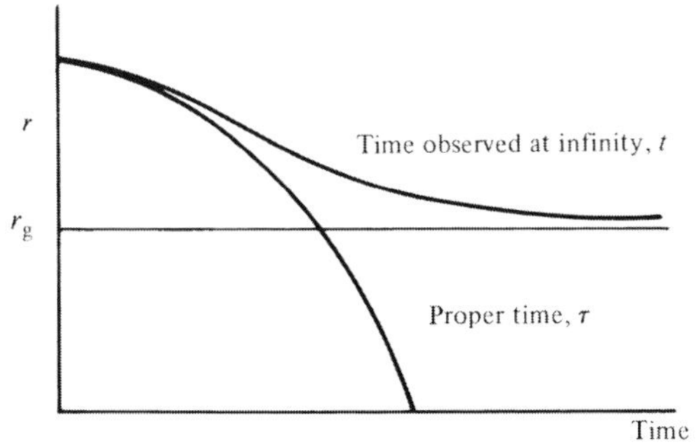

Figure 17.11: Comparison of the proper time τ and the time observed at infinity t for a test particle falling radially from rest at infinity to $r = 0$.

of the world. That observer measures $\mathrm{d}t$ and not $\mathrm{d}\tau$. The metric (17.66) with $\mathrm{d}\varphi = 0$,

$$\mathrm{d}s^2 = \mathrm{d}\tau^2 = \alpha\, \mathrm{d}t^2 - \frac{1}{c^2}\alpha^{-1}\, \mathrm{d}r^2, \tag{17.122}$$

can be written as

$$\mathrm{d}t^2 = \alpha^{-1}\mathrm{d}\tau^2 + \frac{1}{(\alpha c)^2}\, \mathrm{d}r^2. \tag{17.123}$$

We can now substitute for $\mathrm{d}\tau^2$ using (17.120) to find the relation between $\mathrm{d}t$ and $\mathrm{d}r$. After some reorganisation, we find:

$$\mathrm{d}t = -\frac{\mathrm{d}r}{c\left(\dfrac{2GM}{rc^2}\right)^{1/2}\left(1 - \dfrac{2GM}{rc^2}\right)} = -\frac{\mathrm{d}r}{c\left(\dfrac{r_\mathrm{g}}{r}\right)^{1/2}\left(1 - \dfrac{r_\mathrm{g}}{r}\right)}. \tag{17.124}$$

We can now integrate to find the coordinate time t for collapse from r_1 to $r = 0$,

$$t = \int_{t_1}^{t_2} \mathrm{d}t = -\frac{1}{c r_\mathrm{g}^{1/2}} \int_{r_1}^{0} \frac{r^{3/2}}{r - r_\mathrm{g}}\, \mathrm{d}r. \tag{17.125}$$

Thus the coordinate time t diverges, as measured by the distant observer, as r tends to r_g: although the particle takes a finite proper time to collapse to $r = 0$, it takes an infinite time to reach the Schwarzschild radius r_g according to the external observer. These different types of behaviour are illustrated in Fig. 17.11.

Correspondingly, light signals detected by a distant observer are progressively redshifted as $r \to r_\mathrm{g}$. The proper time interval for signals emitted from some point in the field is, from the metric (17.58),

$$\mathrm{d}s = \mathrm{d}\tau = \left(1 - \frac{2GM}{rc^2}\right)^{1/2} \mathrm{d}t. \tag{17.126}$$

Hence, in terms of frequencies,

$$\begin{aligned}\nu_{obs} &= \left(1 - \frac{2GM}{rc^2}\right)^{1/2} \nu_{em} \\ &= \left(1 - \frac{r_g}{r}\right)^{1/2} \nu_{em},\end{aligned} \tag{17.127}$$

where ν_{em} and ν_{obs} are the emitted and observed frequencies respectively. This leads to the correct expression for the *gravitational redshift* z_g according to general relativity, which can be written

$$z_g = \frac{\Delta\lambda}{\lambda} = \left(1 - \frac{r_g}{r}\right)^{-1/2} - 1. \tag{17.128}$$

Thus, the gravitational redshift tends to infinity as $r \to r_g$; in other words, as photons are emitted closer and closer to r_g the observed frequency tends to zero. As a result, we cannot observe light signals originating from radii $r < r_g$. Light signals can certainly travel inwards to $r = 0$, but those within r_g cannot propagate outwards from within r_g.

We have uncovered a number of the key properties of *black holes*. They are *black* because no radiation can travel outwards through r_g; they are *holes* because matter inevitably falls into a singularity at $r = 0$ if it approaches too close to the Schwarzschild radius r_g. Similar properties are found in the more general case of rotating, or Kerr, black holes.

Various singularities appear in the standard form of the Schwarzschild metric. The singularity at $r = r_g$ is now understood not to be a real physical singularity, but to be associated with the choice of coordinate system. In other coordinate systems, for example, in Kruskal or Finkelstein coordinates, this singularity disappears. More details of these features of coordinate systems in general relativity are discussed by Rindler in his book *Relativity: Special General and Cosmological.*[12] In contrast, the singularity at $r = 0$ appears to be a real physical singularity and so there is something missing from the theory. The obvious guess is that we need a quantum theory of gravity to eliminate the singularity, but no such theory yet exists.

17.10 Circular orbits about Schwarzschild black holes

To complete this brief introduction to general relativity, let us use the Schwarzschild metric to work out some of the properties of particles in circular orbits about black holes. These studies are of central importance for high-energy astrophysics, in particular, in the study of accretion discs about black holes. Matter falling towards a black hole is most unlikely to possess zero angular momentum and, consequently, collapse occurs along the rotation axis of the infalling material, resulting in the formation of an *accretion disc* normal to this axis. I have discussed some elementary features of thin accretion discs in *High Energy Astrophysics,* Vol. 2.[24] Matter in the accretion disc moves in essentially Keplerian orbits about the black hole. In order that matter can arrive at the last stable orbit, at $r = 3r_g$, it has to lose angular momentum and this takes place through the action of viscous forces in the differentially rotating disc. Viscosity results in the transport of angular momentum outwards, enabling the material to spiral slowly in towards the last stable orbit at $r = 3r_g$ and at the same time heating up the disc to a high temperature. Let us use the tools we have developed to understand how much energy can be released as the matter drifts in towards the last stable orbit.

We showed in subsection 17.6.1 that the total energy of a test mass orbiting a black hole, as observed at infinity, is $E = kmc^2$. From (17.86),

$$m\dot{r}^2 + \alpha m(r\dot{\varphi})^2 - \frac{2GMm}{r} = mc^2(k^2 - 1). \tag{17.129}$$

For circular orbits, $\dot{r} = 0$ and $r^2\dot{\varphi} = h$ is an invariant of the motion. Therefore

$$\frac{\alpha m h^2}{r^2} - \frac{2GMm}{r} = mc^2(k^2 - 1). \tag{17.130}$$

The value of h can be found from (17.96). For a circular orbit, $\mathrm{d}^2u/\mathrm{d}\varphi^2 = 0$, and so

$$u = \frac{GM}{h^2} + \frac{3GM}{c^2}u^2.$$

Therefore,

$$h^2 = \frac{GMr}{1 - \dfrac{3GM}{rc^2}}. \tag{17.131}$$

In the non-relativistic case, (17.131) reduces to the Newtonian result for the specific angular momentum, $h^2 = GMr$. In the case of the last stable orbit about the black hole at $r = 3r_g$, $h = \sqrt{12}GM/c$. As can be seen from (17.82), the component of the velocity four-vector in the circumferential direction is

$$\gamma v = r\frac{\mathrm{d}\varphi}{\mathrm{d}\tau} = r\dot{\varphi} = \frac{h}{r} \tag{17.132}$$

and hence for the last stable orbit $\gamma v = \sqrt{3}$ and $v = 0.5c$: particles on the last stable orbit move at half the speed of light.

Substituting (17.131) into (17.130), we obtain the important result

$$k = \frac{1 - \dfrac{2GM}{rc^2}}{\left(1 - \dfrac{3GM}{rc^2}\right)^{1/2}}. \tag{17.133}$$

In the Newtonian limit, this result reduces to

$$E = kmc^2 = mc^2 - \frac{GM}{2r}. \tag{17.134}$$

This represents the sum of the particle's rest mass energy mc^2 and the *binding energy* of its circular orbit about the point mass. We recall that, for a bound circular orbit, the kinetic energy T is one half the magnitude of the gravitational potential energy $|U|$, what is known as the *virial theorem* in stellar dynamics,

$$T = \tfrac{1}{2}|U|. \tag{17.135}$$

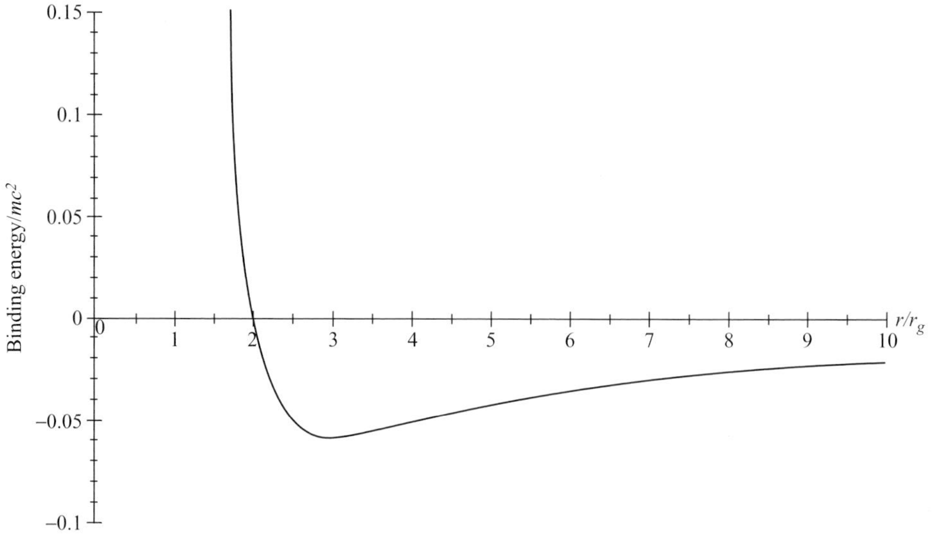

Figure 17.12: The variation of the binding energy of a circular orbit in units of mc^2 with distance coordinate r in units of r_g for the Schwarzschild metric. The last stable orbit occurs at $r/r_g = 3$ and the Schwarzschild radius at $r/r_g = 1$.

Recall, however, that the gravitational potential energy is negative and so the total energy of the particle is

$$E_{\text{orbit}} = T + U = \frac{GM}{2r} - \frac{GM}{r} = -\frac{GM}{2r}. \tag{17.136}$$

According to Newtonian gravity, if a particle is released from rest at infinity, as it falls towards the point mass. $T = |U|$ and so, if it does not lose kinetic energy, it will return back to infinity. In order to attain a bound orbit, however, half the particle's kinetic energy has to be lost, either by dissipation as heat or by some other physical process. Thus, the expression for the *total energy* tells us how much energy has to be released in order that the particle can attain a bound orbit at that radius. This is why it is often referred to as the *binding energy* of the particle's orbit – this energy would have to be supplied to return the particle to $r = \infty$.

The corresponding result in general relativity follows directly from (17.133). The expression for the binding energy becomes

$$E_{\text{orbit}} = E - mc^2 = (k-1)mc^2 = \left\{ \frac{1 - \dfrac{2GM}{rc^2}}{\left(1 - \dfrac{3GM}{rc^2}\right)^{1/2}} - 1 \right\} mc^2, \tag{17.137}$$

which describes how much energy is released by infall of matter from infinity to a circular orbit at radius r about the black hole. According to the Newtonian version of this argument, (17.136), the amount of energy available is unbounded; as $r \to 0$ the binding energy tends to infinity. This does not occur in general relativity, as shown by (17.137). The plot of the binding energy as a function of radius in Fig. 17.12 shows that only a finite amount of binding energy can be made available for powering binary X-ray sources, active galactic nuclei and quasars.

The maximum binding energy occurs at $r = 3r_g$, the last stable orbit, and is

$$E_{\text{orbit}} = -\left[1 - \left(\frac{8}{9}\right)^{1/2}\right] mc^2 = -0.0572mc^2. \quad (17.138)$$

This is a very important result for the following reasons:

- To reach the last stable orbit, 5.72% of the rest mass energy of the particle has to be released. This is almost an order of magnitude greater than the energy which can be liberated by nuclear reactions – for example, if four hydrogen nuclei are combined to form a helium nucleus, only 0.7% of the rest mass energy of the hydrogen atoms is released.
- The radius of the last stable orbit represents the most compact scale from which an object of mass M can liberate this energy. If time variations on a time scale τ are observed from any source, causality requires the size of the source to be less than $r = c\tau$. The radius of the last stable orbit is the smallest dimension from which energy can be emitted from a object of mass M.
- Even greater efficiences of energy release can be obtained by accretion onto rotating black holes. Solutions of Einstein's field equations for a rotating black hole were discovered by Roy Kerr in 1962. For a maximally rotating black hole, up to 42% of the rest mass energy of the infalling matter can be released, providing an enormous source of energy for powering extreme astrophysical objects such as quasars and radio galaxies.

Notice an interesting feature of (17.137). The binding energy is zero for an orbit with $r = 2r_g$, although such an orbit is unstable. This corresponds precisely to the locus passing though the point D in Fig. 17.9, which shows the binding energy as a function of radius r. Recall that a particle which reaches the point D along the locus from A can fall in from infinity to $r = 0$ despite the fact that it has a finite specific angular momentum. That orbit has zero binding energy.

17.11 References

1 Pais, A. (1982). *Subtle is the Lord . . . the Science and Life of Albert Einstein*. Oxford: Clarendon Press.

2 Stachel, J. (1995). The history of relativity, in *Twentieth Century Physics,* Vol. 1, eds. L.M. Brown, A. Pais and A.B. Pippard, pp. 249–356. Bristol: Institute of Physics Publishing and New York: American Institute of Physics Press.

3 Einstein, A. (1922), his Kyoto address of December 1922. J. Ishiwara (1977), *Einstein Köen-Roku*, Tokyo: Tokyo-Tosho.

4 Einstein, A. (1907). *Jahrbuch Radioaktiv. Elektronik*, **4**, 411–62.

5 Einstein, A. (1911). *Ann. Phys.*, **35**, 898–908.

6 Soldner, J.G. von (1804). *Berliner Astr. Jahrbuch*, p. 161.
7 Einstein, A. (1912). *Ann. Phys.*, **38**, 1059.
8 Einstein, A. (1922). In Ishiwara (1977), *op. cit.*
9 Einstein, A. (1916). *Die Grundlage der Allgemeinen Relativitätstheorie*, p. 6. Leipzig: J.A. Barth.
10 Einstein, A. and Grossmann, M. (1913). *Z. Math. Physik*, **62**, 225.
11 Berry, M. (1995). *Principles of Cosmology and Gravitation*. Cambridge: Cambridge University Press.
12 Rindler, W. (2001). *Relativity: Special, General and Cosmological*. Oxford: Oxford University Press.
13 Weinberg, S. (1972). *Gravitation and Cosmology*. New York: John Wiley and Sons.
14 d'Inverno, R. (1992). *Introducing Einstein's Relativity*. Oxford: Clarendon Press.
15 Misner, C.W., Thorne, K.S. and Wheeler, J.A. (1973). *Gravitation*. San Francisco: Freeman and Co.
16 Will, C.M. (1993). *Theory and Experiment in Gravitational Physics*. Cambridge: Cambridge University Press.
17 The Satellite Test of the Equivalence Principle (STEP) is described at http://www.sstd.rl.ac.uk/fundphys/step/.
18 Rindler, W. (2001). *op. cit.*, p. 19.
19 Einstein, A. (1920). Unpublished manuscript, now in the Pierpoint Morgan Library, New York. See Pais (1982), *op. cit.*, p. 178.
20 This statement is due to Prof. Anthony Lasenby in his lectures on gravitation and cosmology.
21 Eddington, A.S. (1927). *Stars and Atoms*, p. 52. Oxford: Clarendon Press.
22 Schwarzschild, K. (1916). *Sitz. Preuss. Akad. Wissenschaften* 189.
23 Shapiro, S.L. and Teukolsky, S.A. (1983). *Black Holes, White Dwarfs and Neutron Stars: The Physics of Compact Objects*. New York: John Wiley and Sons.
24 Longair, M.S. (1997). *High Energy Astrophysics*, Vol. 2. Cambridge: Cambridge University Press (corrected reprint of 1994 edition).

Appendix to Chapter 17: Isotropic curved spaces

A17.1 *A brief history of non-Euclidean geometries*

During the late eighteenth century, non-Euclidean spaces began to be taken seriously by mathematicians who realised that Euclid's fifth postulate, that parallel lines meet only at infinity, might not be essential for the construction of self-consistent geometries. The first suggestions that the global geometry of space might not be Euclidean were discussed by Lambert and Saccheri. In 1786, Lambert noted that, if space were hyperbolic rather than flat, the radius of curvature of space could be used as an absolute measure of distance. In 1816, Gauss repeated this proposal in a letter to Gerling and is reputed to have tested whether space is locally Euclidean by measuring the sum of the angles of a triangle between three peaks in the Harz mountains.

The fathers of non-Euclidean geometry were Nicolai Ivanovich Lobachevski in Russia and János Bolyai in Hungary. In his book *On the Principles of Geometry*, published in

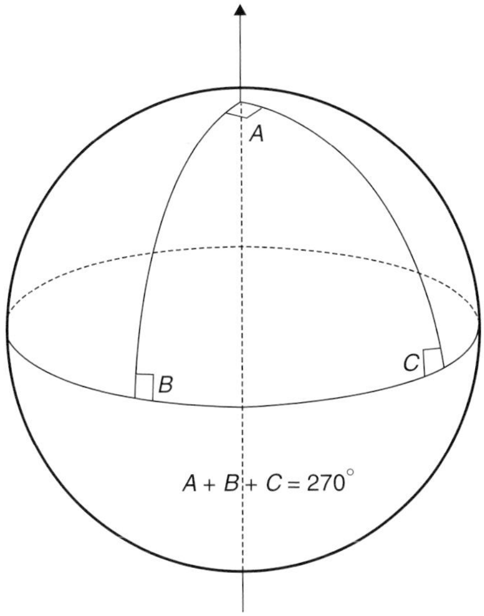

Figure A17.1: Illustrating the sum of the angles of a triangle on the surface of a sphere.

1829, Lobachevski at last solved the problem of the existence of non-Euclidean geometries and showed that Euclid's fifth postulate could not be deduced from the other postulates. Non-Euclidean geometry was placed on a firm theoretical basis by the studies of Bernhard Riemann, and the English-speaking world was introduced to these ideas through the works of Clifford and Cayley.

Einstein's monumental achievement was to combine special relativity and the theory of gravity through the use of Riemannian geometry and tensor calculus to create the general theory of relativity. Within a couple of years of formulating the theory, Einstein realised that he now had the tools with which to construct fully self-consistent models of the Universe as a whole. In Einstein's model, to be discussed in Chapter 19, the Universe is static and closed and has isotropic, spherical geometry. The Friedman solutions, published in 1922 and 1924, were also isotropic models, but they were expanding solutions and included geometries which were spherical, flat and hyperbolic.

It turns out that it is not necessary to become embroiled in the details of Riemannian geometry to appreciate the geometrical properties of isotropic curved spaces. We can demonstrate simply why the only isotropic curved spaces are those in which the two-dimensional curvature of any space section, κ, is constant throughout the space and can only take positive, zero or negative values. The essence of the following argument was first shown to me by Dr Peter Scheuer.

A17.2 Parallel transport and isotropic curved spaces

Consider first of all the simplest two-dimensional curved geometry, the surface of a sphere (Fig. A17.1). In the diagram, a triangle is shown consisting of two lines drawn from the north pole down to the equator, the triangle being completed by the line drawn along the equator. For simplicity, we consider first the case in which the angle between the lines AB and AC at the 'north pole' is 90°. The three sides of the triangle are all segments of great circles on the sphere and so are the shortest distances between the three corners of the triangle. The three lines are *geodesics* in the curved geometry.

We need a procedure for working out how non-Euclidean the curved geometry is. The way this is done in general is by the procedure known as the *parallel displacement* or *parallel transport* of a vector which makes a complete circuit around a closed figure such as the triangle in Fig. A17.1. Suppose we start with a little vector perpendicular to AC at the pole and lying in the surface of the sphere as shown. We then transport that vector from A to C, keeping it perpendicular to AC. At C, we rotate the vector through 90° so that it is now perpendicular to CB. We then transport the vector, keeping it perpendicular to BC, to the corner B. We make a further rotation through 90° to rotate the vector perpendicular to BA and then transport it back to A. At that point, we make a final rotation through 90° to bring the vector back to its original direction. Thus, the total rotation of the vector is 270°. Clearly, the surface of the sphere is a non-Euclidean space. This procedure illustrates how we can work out the geometrical properties of any two-space entirely by making measurements within the two-space.

Another simple calculation illustrates an important feature of parallel transport on the surface of a sphere. Suppose the angle at A is not 90°, but some arbitrary angle θ. Then, if the radius of the sphere is R_c, the surface area of the curved triangle ABC is θR_c^2. Thus, if $\theta = 90^\circ$, the area is $\pi R_c^2/2$ and the sum of the angles of the triangle is 270°; if $\theta = 0^\circ$, the area is zero and the sum of the angles of the triangle is 180°. This illustrates that the difference between the sum of the angles of the triangle and 180° is proportional to the area of the triangle, that is

$$\text{sum of angles of triangle} - 180^\circ \quad \propto \quad \text{area of triangle}. \tag{A17.1}$$

This result is a general property of isotropic curved spaces.

Let us again take a vector around a loop but now in some general two-space. Define a small area $ABCD$ by light rays, which travel along null-geodesics in the two-space. Consider two light rays, OAD and OBC, originating from O, crossed by the light rays AB and DC as shown in Fig. A17.2(a). The light rays AB and DC are chosen so that they each cross the light ray OAD at right angles. We start at A with the vector to be transported round the loop parallel to AD as shown. We then move the vector by parallel transport along AB until it reaches B. At this point, it has to be rotated clockwise through an angle $90^\circ - \beta$ in order to lie perpendicular to BC at B. It is then transported along BC to C, where it is rotated clockwise through an angle $90^\circ + (\beta + \mathrm{d}\beta)$, as shown in the diagram, in order to lie perpendicular to CD at C. It can be seen that the extra rotations β at B and C have opposite signs. Since CD and AB are perpendicular to the light ray AD, after C the subsequent rotations of the vector on parallel transport amount to 180° and so the total rotation round the loop is $360^\circ + \mathrm{d}\beta$. Thus, $\mathrm{d}\beta$ is a measure of the departure of the two-space from Euclidean geometry, for which $\mathrm{d}\beta = 0$.

Now, the rotation of the vector $\mathrm{d}\beta$ must depend upon the area of such a loop. In the case of an isotropic space, we would obtain the same rotation wherever we placed the loop in

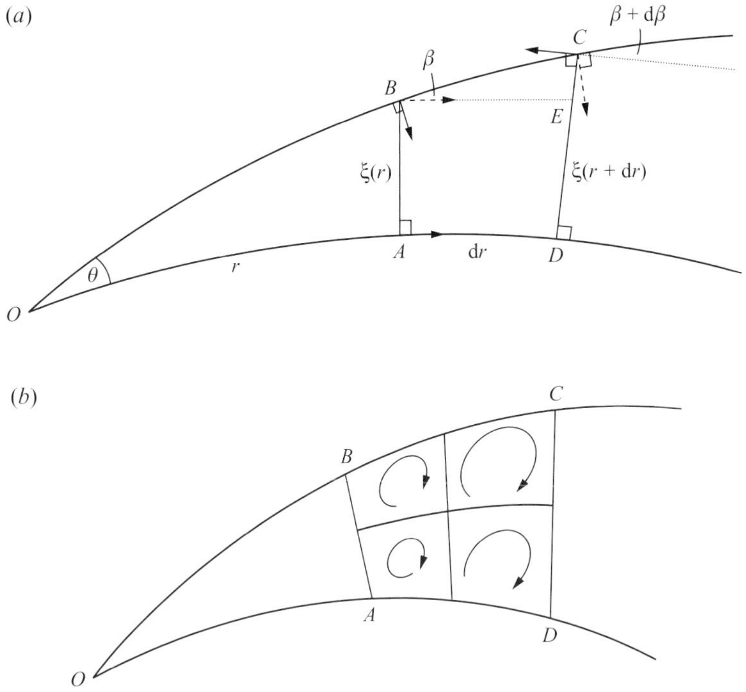

Figure A17.2: (*a*) Illustrating the parallel transport of a vector around the small loop $ABCD$ formed by the light rays OBC, OAD, AB and DC. (*b*) Illustrating how the sum of the rotations around the sub-loops must add up linearly to a total rotation $d\beta$.

the two-space. Furthermore, if we were to split the loop into a number of sub-loops then the rotations around the separate sub-loops would add up linearly to the total rotation $d\beta$ (Fig. A17.2(*b*)). Thus, in an isotropic two-space, the rotation $d\beta$ would be proportional to the area of the loop $ABCD$. We conclude that the constant relating the rotation $d\beta$ to the area of the loop must be a constant everywhere in the curved two-space, just as was found in the particular case of a spherical surface in Fig. A17.1.

The complication is that, since the space is non-Euclidean, we do not know how to relate the length AB to the geodesic distance $OA = r$ along the light path and the angle θ subtended by AB at the origin. Therefore, we have to write the distance AB as an unknown function of r, $\xi(r)$ (Fig. A17.2(*a*)), such that

$$\theta = \frac{\xi(r)}{r}. \tag{A17.2}$$

The angle β between the geodesic OBC and a line perpendicular to AB can now be found by determining how ξ changes when we move a distance dr along the geodesic OBC (Fig. A17.2(*a*)):

$$\xi(r + dr) = \xi(r) + \frac{d\xi}{dr}\, dr. \tag{A17.3}$$

From the geometry of the small triangle BCE we can write

$$\beta = \frac{\mathrm{d}\xi}{\mathrm{d}r}. \tag{A17.4}$$

At a distance Δr further along the geodesic OBC, the angle becomes

$$\beta + \mathrm{d}\beta = \frac{\mathrm{d}\xi}{\mathrm{d}r} + \left(\frac{\mathrm{d}^2\xi}{\mathrm{d}r^2}\right)\Delta r, \tag{A17.5}$$

and so the net rotation $\mathrm{d}\beta$ is

$$\mathrm{d}\beta = \left(\frac{\mathrm{d}^2\xi}{\mathrm{d}r^2}\right)\Delta r. \tag{A17.6}$$

But, we have argued that the net rotation around the loop must be proportional to the area of the loop, $\mathrm{d}A = \xi\,\Delta r$; thus

$$\left(\frac{\mathrm{d}^2\xi}{\mathrm{d}r^2}\right)\Delta r = -\kappa\,\xi\,\Delta r \qquad \text{and so} \qquad \left(\frac{\mathrm{d}^2\xi}{\mathrm{d}r^2}\right) = -\kappa\xi \tag{A17.7}$$

where κ is a constant, the minus sign being chosen for convenience. This is the equation of simple harmonic motion, which has solution

$$\xi = \xi_0 \sin \kappa^{1/2} r. \tag{A17.8}$$

The expression for ξ_0 can be found from the value of ξ for very small values of r, which must reduce to the Euclidean expression $r\theta$. Therefore, $\xi_0 = \theta/\kappa^{1/2}$ and so

$$\xi = \frac{\theta}{\kappa^{1/2}} \sin \kappa^{1/2} r. \tag{A17.9}$$

The constant κ is defined to be the *curvature* of the two-space and can be either positive, negative or zero. If it is negative, we can write $\kappa = -\kappa'$, where κ' is positive, and then the circular functions become hyperbolic functions:

$$\xi = \frac{\theta}{\kappa'^{1/2}} \sinh \kappa'^{1/2} r. \tag{A17.10}$$

In the case $\kappa = 0$, we find the Euclidean result

$$\xi = \theta r. \tag{A17.11}$$

The results include all possible isotropic curved two-spaces. The constant κ can be positive, negative or zero corresponding to spherical, hyperbolic and flat spaces respectively. In geometric terms, $R_c = \kappa^{-1/2}$ is the radius of curvature of a two-dimensional section through the isotropic curved space (Fig. 17.5) and has the same value at all points and in all orientations within the plane. It is often convenient to write the expression for ξ in the form

$$\xi = \theta R_c \sin \frac{r}{R_c}, \tag{A17.12}$$

where R_c is real for closed spherical geometries, imaginary for open hyperbolic geometries and infinite for the case of Euclidean geometry.

The simplest examples of such spaces are the spherical geometries, in which R_c is just the radius of the sphere illustrated in Fig. A17.1. The hyperbolic spaces are more difficult to envisage. The fact that R_c is imaginary can be interpreted in terms of the principal radii of curvature of the surface having opposite sign. The geometry of a hyperbolic two-space can be represented by a saddle-shaped figure (Fig. 17.6), just as a two-sphere provides an visualisation of the properties of a spherical two-space.

18 The technology of cosmology

18.1 Introduction

This chapter is very different from all the others in this book. I have a firmly held conviction that astrophysical cosmology is an observational, if not experimental, science and that the quality of the astrophysics is only as good as the data available to validate cosmological and astrophysical theories. The objective of this chapter is to survey some aspects of the technologies which have enabled astrophysical cosmology to be placed on a firm observational basis. In the telling of this story, we will encounter a number of heroic figures who deserve as much honour, in my view, as the rather better-known theorists and physicists who have played such a startling role in the development of cosmological understanding.

My reason for including this chapter is to bring home to even the most hard-line of theorists the essential role which experimental genius plays in the development of theory. In many ways, this chapter is an attempt to do for cosmology what Peter Galison[1] achieved in his splendid book *Image and Logic* for particle physics, but at a very much more modest level. Without the imaginative development of novel technology, with which to address the challenges presented by theory, theoretical physics lacks experimental validation.

To oversimplify greatly, the revolution in twentieth-century astrophysics and cosmology can be traced to three great technical developments, which were the heritage of the nineteeth century: (i) the invention of astronomical spectroscopy, (ii) the measurement of the parallaxes, and hence the distances, of stars and (iii) the invention of the photographic process. For the purposes of this review, I will regard an understanding of the stars as an integral part of astrophysical cosmology – without an understanding of stellar astrophysics, astrophysical cosmology would not be in good shape. Many more details of this remarkable story can be found in references 2 to 10 in Section 18.10.

18.2 Joseph Fraunhofer

We have already described in Section 11.2 some of the major technical contributions made by Joseph Fraunhofer in manufacturing optical components of very high quality. These led to his discovery of the multitude of absorption lines in the solar spectrum. In turn, this discovery led directly to the remarkable work of Kirchhoff, Bunsen and others in identifying these absorption lines as the signatures of different elements in the solar atmosphere. From the point of view of optical technology, Fraunhofer was able to use these spectral lines as wavelength standards with which to characterise the chromatic properties of glasses and lenses quantitatively and precisely. This resulted in much improved glasses, as well as much improved polishing and testing methods. These technical improvements gave rise to the best astronomical telescopes then available. Fraunhofer's masterpiece was the 24-cm Dorpat telescope built for Wilhelm Struve at the Dorpat, now Tartu, Observatory in Estonia.

Equally important was the 16-cm heliometer which Fraunhofer built for Friedrich Bessel at Koenigsberg. The heliometer consisted of a lens cut in half to form two D-shapes, after a design by John Dolland. The images of separated stars could be brought together and their separation measured precisely from the reading on a micrometer screw. This telescope was used by Bessel to measure the motion of the nearby star 61 Cygni. In 1838, he published its parallax, meaning its change in position on the sky due to the motion of the Earth about the Sun, and so its distance could be found. The measured parallax amounted to about one third of an arcsecond and was the first direct measurement of the distance of any star other than the Sun. This was a key observation and direct confirmation of the generally held view at the time that the stars are objects similar to the Sun. Parallax measurement remained a demanding business, however, and by 1900 only about 100 parallaxes had been measured.

18.3 The invention of photography

The third great contribution to the development of astronomy and astrophysics was the invention of the photographic process by Daguerre and Fox Talbot. Louis-Jacques-Mandé Daguerre was a remarkable character who began life as an inland revenue official and then became a scene painter at the opera. The search for methods of recording images by what was to become the photographic process began with the discovery that some natural compounds are rendered insoluble when they are exposed to light. In the course of his experiments, Daguerre discovered that iodine-treated silver paper was also sensitive to light. By 1835, he had made the important discovery of the *latent image* which is recorded on sensitised paper even if the light is not bright enough to darken the paper. This latent image could be developed by exposure to mercury vapour and fixed by a strong salt solution. The use of the latent image meant that the exposure could be reduced to 20 to 30 minutes. Remarkably, the announcement of the discovery of what was called the *Daguerreotype process* was made by an astronomer, Arago, the director of the Paris Observatory, on 9 January 1839.

A similar announcement was made almost simultaneously by William Henry Fox Talbot in England. The first astronomical images were taken in the succeeding years, but the photographic process was slow. One of the earliest and, for me, most moving images is the photograph taken by John Herschel of his father's 40-foot telescope in February 1839 through the window of their house at Slough. In a two-hour exposure, the structure supporting the large tube of the telescope can be clearly seen – the telescope was dismantled in the following year (Fig. 18.1). John Herschel had a passionate interest in photography and invented much of its terminology, including the terms 'photography', 'positive', 'negative', and so on.

The next important advance was the invention of the *wet collodion process* by Frederick Scott Archer in 1851. This process resulted in finely detailed negatives, and exposure times

Figure 18.1: John Herschel's photograph of 1839 of part of the supporting structure of his father's 40-foot telescope seen looking out through the window of their house at Slough. (From R. Learner, 1981, *Astronomy through the Telescope*, London: Evans Brothers. Courtesy of the Science Museum / Science and Society Picture Library.)

for astronomical images were reduced to about 10 to 15 minutes. Here is the procedure involved in taking wet collodion photographs:

- cover a clean glass plate with a mixture of collodion (gun-cotton or cellulose nitrate) and potassium iodide dissolved in ether;
- allow the ether to evaporate and, while still tacky, immerse in a solution of silver nitrate, later to be improved by mixing with silver bromide;
- allow the silver nitrate to react with the potassium iodide to precipitate insoluble silver iodide;
- expose the plate, but do not let it dry out. Once exposed, developed and dried, a permanent negative image is created;
- finally, at leisure, print the positive on albumen-coated paper.

The net result was faster, fine-grained plates, which quickly superseded the daguerreotype process. These inventions sparked an enormous popular interest in photography in the 1850s and many commercial photographic studios were set up. The wet-collodion process was used by Julia Margaret Cameron in her spectacular portraits of great nineteenth-century figures, including the famous images of the aged John Herschel.

The search for improved photographic materials continued throughout the remaining years of the century. The boom in photography meant that there was no lack of plates for astronomical use. At the same time, telescope design needed to be considerably improved, since a telescope designed to record photographic images had to be able to track and guide with much improved precision as compared with one used only visually – for visual observations, the length of the 'exposure' is determined by the response time of the eye, which is only about a tenth of a second. Let us first complete the story of the development of photographic techniques.

The first photographic spectrum taken using the wet-collodion process was that obtained for the bright star Vega by Henry Draper in 1872. The spectrum showed the Hα and Hβ lines of hydrogen, as well as the first detections of the next seven lines in this hydrogen series. These were to be used by Balmer in his remarkable little papers of 1886, in which he proposed the Balmer formula for the spectrum of hydrogen, the first quantum formula to be discovered.

The dry-collodion process was invented in the mid-1870s and made the photographic process much more straightforward. The search for improved materials continued and culminated in the discovery of emulsions consisting of silver salts suspended in gelatin by R.L. Maddox and Charles Bennett in 1879. Over the next few years, superb astronomical images were obtained of star clusters and nebulae, revealing unambiguously the remarkable power of photography for astronomy. Further improvements were made by exposing the plates to heat or by the addition of ammonia – this was the beginning of the dark art of hypersensitising photographic plates to increase their quantum efficiencies, which eventually reached about 1%–2%.

It is striking that the photographic pioneers developed their techniques on small telescopes – the larger telescopes were still used for the traditional pursuits of astronomers, the accurate measurement of time and stellar positions. As expressed by Richard Learner,

> The lessons of photography and spectroscopy, where astronomy of the highest class had been carried out by observers with very modest telescopes . . . were not learned by the astronomical establishment. To them, Urania, the muse of astronomy, was cold and distant, concerned with the smooth and silent motions of the stars, not a grubby figure in an apron, standing at the laboratory sink and doing the washing up.[7]

18.4 The new generation of telescopes

The requirements of accurate tracking and guiding for long astronomical exposures required significant improvements in telescope design. The heroes of this story are Lewis Morris Rutherfurd, also famous for his pioneering spectroscopic observations of bright stars, and John Draper, the father of Henry Draper. Rutherfurd invented a clockwork drive for his photographic telescope and, during the 1850s and 1860s, took some excellent astronomical photographs. John Draper devoted huge efforts to optimising telescopes for photographic purposes, which included continuing the improvement of the quality of the lenses and mirrors. Over a three-year period he devised a series of seven grinding and polishing machines. These endeavours, all carried out on small telescopes, were to pave the way for the spectacular burst of telescope construction in the late nineteenth-century and the beginning of the twentieth-century.

Refracting telescopes, similar in principle to Galileo's telescopes, were the preferred choice for astrometric studies, but this development reached the end of the line with the construction of the 1-metre (40-inch) refractor at the Yerkes observatory. Refractors had

Figure 18.2: Lord Rosse's 72-inch telescope at Birr Castle following refurbishment of the instrument in the 1990s. (Birr Scientific and Heritage Foundation, courtesy of the Earl of Rosse.)

outstanding capabilities for the visual determination of parallaxes and for the detection of double stars. In the latter case, the observer simply waited until there was a period of good astronomical 'seeing' and then by direct observation noted whether the stellar image was single or double. The problem was that these telescopes were designed with large f-numbers to obtain good image quality and so were very long, with the consequence that they tended to bend under their own weight and so required large telescope domes and buildings. It was also difficult to obtain high optical quality for such large lenses.

Several large reflecting telescopes, based on Newton's design for an all-reflecting telescope, had been constructed earlier in the century. The largest of these was the great 'Leviathan', Lord Rosse's 1.8-metre (72-inch) reflector, which he had built at his home at Birr Castle in Ireland (Fig. 18.2). The biggest problems lay with the large primary reflecting mirror, which was made of speculum metal. This material is an alloy of tin and copper with a pinch of arsenic, resulting in a polished surface which is 50% reflective. The problem was that speculum metal is a very brittle material and difficult to handle. When the mirror tarnished it had to be repolished, which was a hazardous procedure with the potential for destroying the mirror. Rosse acquired two speculum mirrors so that one could be installed on the telescope while the other was being repolished.

Despite almost insuperable problems, including the inclement Irish weather, Lord Rosse was able to make good visual observations of diffuse nebulae, perhaps his greatest achievement being the discovery of spiral arms in nebulae such as M51 (Fig. 18.3(a)).

(*a*)
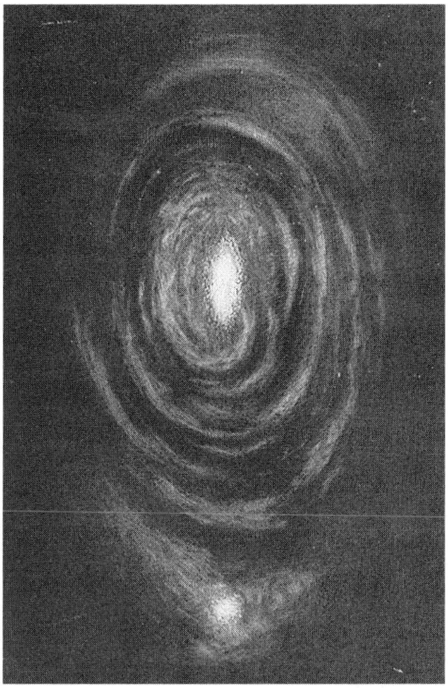

(*b*)

Figure 18.3: (*a*) A drawing of M51 made by Lord Rosse from visual observations with his 72-inch reflecting telescope at Birr Castle. (Birr Scientific and Heritage Foundation, courtesy of the Lord of Rosse). (*b*) A photograph of the galaxy M51, taken by Keeler and his colleagues during the commissioning of the Crossley 91-cm reflector. (Courtesy of the Lick Observatory, University of California at Santa Cruz.)

The solution to the problem of constructing primary mirrors for large reflectors was to be found in an exhibit at the Great Exhibition of 1851, held at the Crystal Palace in London. Glass-makers had on show decorative exhibits in which silver was chemically deposited on glass, resulting in the production of excellent mirrors. The telescope builders quickly realised that this was the solution to the problem. The film of silver could be made thin and uniform, with the result that, when the silver tarnished, rather than repolish the mirror the layer of silver could be removed chemically and a new surface laid down. The chemical process by which the silver layer was laid down was reinvented by Justus von Leibig a few years later, and the first reflectors using silvered mirrors were built by Steinheil in 1856, a 10-cm reflector, and by Foucault in 1857, a 33-cm reflector.

The problems of constructing large reflectors were still considerable, not least because the reflector design is much more susceptible to flexure and to vibrational and temperature effects. The challenge was taken up by Andrew A. Common, the telescope designer and astronomer, and George Calver, the mirror-maker. During the 1870s they made a major effort to overcome the problems inherent in the design of the reflecting telescope and introduced a number of innovations, which were to be incorporated into the next generation of instruments. The principal innovations employed in the construction of their 91-cm reflector were the relief of the weight on the telescope bearings, achieved by submerging a hollow steel float in mercury, and the introduction of an adjustable plate-holder. The result was that the tracking and guiding of the telescope were very smooth. The adjustable plate holder had the great advantage that a guide-star could be selected outside the field of view of the photographic plate and continuously monitored to ensure that precisely the same field was exposed within the limits of the seeing disc. The 90-minute exposure of the Orion Nebula by Common and Calver won the Gold Medal of the Royal Astronomical Society.

The next key step came through the generosity of the English amateur astronomer Edward Crossley. In 1895, he presented his 91-cm reflector, built to the Calver–Common design, to the Lick Observatory (Fig. 18.4). An important innovation was that the observatory was to be located on an excellent high site, on Mount Hamilton in California, where the transparency and stability of the atmosphere were very good and there was a large percentage of clear nights. The mirror was repolished by Howard Grubb and the mounting of the telescope stiffened by J.E. Keeler. The results were spectacular. Among the first plates taken by Keeler in 1900 were magnificant images of spiral nebulae (Fig. 18.3(*b*)), including the detection of large numbers of faint spiral nebulae. It was not difficult to infer that, if these objects were similar to the Andromeda Nebula, they must lie at very great distances.

The next step in increased telescope aperture followed George Ellery Hale's appointment as the Director of the Mount Wilson Observatory in California. He persuaded his father to buy the 1.5-metre blank for a 60-inch reflecting telescope. The design was an enlarged version of the Calver–Common design for the 91-cm reflector at the Lick Observatory. Before the 60-inch telescope was completed, however, Hale had persuaded J.D. Hooker to fund an even bigger telescope, the 100-inch telescope to be built on Mount Wilson. The technological challenges were proportionally greater, the mass of the telescope being 100 tonnes, but the basic Calver–Common design was retained. The tracking was provided by a mechanism very much like that of a grandfather clock, but with a 2-ton driving weight,

Figure 18.4: The Crossley 91-cm reflector at the Lick Observatory. (Courtesy of the Mary Lea Shane Archives of the Lick Observatory, University of California at Santa Cruz.)

which had to be wound up at the beginning of each night's observing. The optics were the responsibility of George Ritchey, an optical designer of genius, who was to invent the optical configuration known as the Ritchey–Chrétien design, which enabled excellent imaging to be achieved over a wide field of view. The 100-inch telescope was to be at the heart of observational cosmology through the key years from 1918 until 1950, when the 200-inch telescope was commissioned. In addition to his skill as an entrepreneur, Hale had the foresight to appoint Harlow Shapley and Edwin Hubble as staff astronomers at the Mount Wilson Observatory.

Hale was well aware of the importance of advanced instrumentation, and one of the instruments which he commissioned was an interferometer from Albert A. Michelson to be located on the top ring of the 100-inch telescope (Fig. 18.5). The principal objective was to detect double stars below the seeing limit of the 100-inch telescope. The story goes that Michelson was uncertain about the choice of length of the interferometer arm. Arthur Eddington was aware that the instrument was under construction and used his recently completed theory of the structure of the envelopes of red giant stars to predict that Betelgeuse should have an angular diameter of about 0.05 arcsec. With a 6-metre baseline, Michelson measured an angular diameter of 0.047 arcsec, the first triumph for Eddington's theory of stellar structure. This was a notable achievement, since Eddington had assumed that the core of the star was a fluid sphere and that only the extended red giant envelope was gaseous. This success led to his application of the same astrophysics to the internal structures of main sequence stars, including their cores, and to his remarkable explanation of the

Figure 18.5: Michelson's stellar interferometer mounted on the top ring of the Hooker 100-inch telescope at Mount Wilson. This instrument was used to measure the angular diameter of Betelgeuse in 1919. (A.A. Michelson and F.G. Pease, 1921, *Astrophys. J.*, **53**, 249.)

mass–luminosity relation for main-sequence stars. These were key steps in the growing understanding of the astrophysics of stars.

The 100-inch telescope was the premier instrument for studies of diffuse nebulae and the story of Hubble's and Shapley's pioneering studies of the nature of the spiral nebulae and the structure of our Galaxy is well known. The major contributions of V.M. Slipher to the unravelling of that story should be emphasised. He realised that, for the spectroscopy of low-surface-brightness objects such as the spiral nebulae, the crucial factor was not the size of the telescope but the speed, or small f-ratio, of the spectrograph camera. His heroic observations of the spectra of spiral nebulae were taken with a long-focus 24-inch refractor, often involving exposures as long as 20, 40 or even 80 hours. By 1925, 44 radial velocities of spiral nebulae had been measured and 39 of them were obtained by Slipher. As Hubble acknowledged in his book *The Realm of the Nebulae*, most of the radial velocities used in his famous paper of 1929, in which the velocity–distance relation for galaxies was first presented, were due to Slipher. The sparsity of the data on which Hubble's discovery of what is often termed the 'expansion of the Universe' was based is remarkable. He used the radial velocities of only 24 nearby galaxies and employed three different methods of distance estimation. In fact, even then, he had in his possession data which extended the relation more securely to larger distances and greater recession velocities.

The work of Hubble's which I find particularly impressive is his paper of 1926, immediately following the announcement of his secure determination of the distance of the Andromeda Nebula, which showed that the spiral nebulae are distant extragalactic systems.

In this paper, the properties of the extragalactic nebulae are described in detail, the statistics of different types of galaxies are given and the mean mass density in galaxies in the Universe as a whole derived for the first time. Hubble then compared this value with the mean density of the Einstein universe and concluded that the observations already extended to about $1/600$ of the radius of this closed-model universe. In the last paragraph of his paper, he concluded by remarking that

> . . . with reasonable increases in the speed of plates and sizes of telescopes, it may become possible to observe an appreciable fraction of the Einstein Universe.[11]

It is no surprise that Hale began his campaign to raise funds for a 200-inch telescope in 1928. By the end of the year, he had received the promise of a grant of \$6000 000 from the Rockefeller Foundation. The design of the 200-inch telescope represented the culmination of the classical tradition of large telescope design and included a number of major advances, which were built into succeeding generations of large telescopes.

- The Corning Company cast the primary 200-inch mirror from Pyrex, because of its low coefficient of expansion.
- The primary mirror was mass-reduced by creating a hexagonal cellular structure within the Pyrex.
- The f-ratio of the telescope was reduced to $f/3.3$ to reduce the length of the telescope tube and so reduce the size and cost of the huge building and dome needed to protect the telescope from the elements.
- The weight of the telescope was supported on oil-pads rather than floated in mercury.
- Serrurier trusses were used to maintain the separation of the primary and secondary mirrors. In this design, the primary and secondary mirrors remain parallel and accurately aligned even when the telescope tube bends under gravity.
- The mirrors were coated with aluminium rather than silver, with the result that the mirror needed to be recoated less frequently.

Technologically, the 200-inch telescope is a masterpiece of engineering, and it stretched mirror and telescope technology to the limit. Anyone who has used it will have noticed the extraordinary attention to detail which went into the design – it was the first telescope to have a large enough prime-focus cage to house the observer. The 200-inch telescope dominated classical observational cosmology from the time it was commissioned in the late 1940s until the 1980s, when 4-metre-class telescopes on much better sites provided astronomers with superior observing facilities.

18.5 The funding of astronomy

The huge investment made by the Rockefeller Foundation in the 200-inch telescope was only the latest in a remarkable history of private funding for astronomy, which placed the United States of America in an undoubted position of leadership in observational cosmology. Before the 1880s, astronomy was supported by National Observatories, the principal function of which were the accurate measurement of time and latitude. Otherwise, astronomy was carried out by dedicated enthusiasts or as a rich man's hobby.

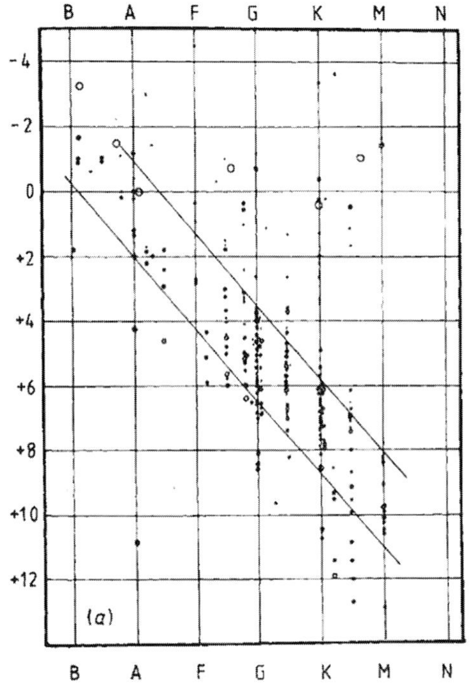

Figure 18.6: Henry Norris Russell's first luminosity–spectral-type diagram. The luminosities on the ordinate are astronomical absolute magnitudes and are related to luminosities L by $M = -2.5 \log L +$ constant. The spectral classes OBAFGKMN are measures of the surface temperature of the stars, O (not shown) being the hottest and N the coolest. (H.N. Russell, 1914, *Nature*, **93**, 227, 252, 281.)

After the American Civil War, a number of individuals became fabulously wealthy and proved to be remarkably generous patrons of astronomy. The organisation and management of the vast Harvard Observatory surveys of the apparent magnitudes (or flux densities) and spectra of very large numbers of stars by Edward Pickering provide good examples of the generosity of private patrons in supporting astronomy. Over the years when the great surveys were undertaken, Pickering obtained several \$100 000 from the Henry Draper Fund, \$400 000 from the Paine Fund in 1886, \$230 000 from the Boyden Fund in 1887 and \$50 000 from the Bruce Fund. It is estimated that Pickering's personal contribution to the project amounted to about \$100 000. At today's prices, the cost of the Harvard surveys would amount to many tens of millions of dollars.

One way of looking at these figures is that they represent the cost of providing the data needed to place the astrophysics of the structure and evolution of stars on a firm observational foundation. Let us look briefly at the fruits of this investment in observational astronomy from the perspective of the technology involved and the organisation of such projects. One of the great achievements, directly attributable to these enormous efforts, was the discovery of the Hertzsprung–Russell diagram, which displays the relation between the luminosities of stars and their spectral types; we now know that the latter are measures of surface temperature (Fig. 18.6).

Pickering and his remarkable team of lady 'computers' devoted an enormous effort to the classification of stellar spectra on the basis of the presence or absence of different lines in a spectrum. The data were acquired using telescopes optimised for wide-field photographic astronomy, the spectral types being inferred from the features appearing on dispersion-prism images, which enabled large numbers of stellar spectra to be acquired in a single observation. By 1912, almost 5000 stars had been classified spectrally over the whole sky according to the famous Harvard spectral sequence OBAFGKM..., which was formulated by Annie Cannon and which could encompass the vast majority of stars. In 1911, she began the spectral classification of 225 300 stars and, in an extraordinary feat of concentrated effort, completed the task four years later. The results of the survey were published as the Henry Draper (HD) Catalogue in nine volumes of the *Annals of the Astronomical Observatory of Harvard College*.

Henry Norris Russell published the first Hertzsprung–Russell diagram in 1914 based on his studies of about 300 stars. He had selected these stars for study whilst at the Cambridge observatories in England from 1902 to 1905. Crucial to the success of this programme was a mastery of the techniques of measuring photographic parallaxes. In collaboration with Arthur Hinks, astrometric photographs were taken of the star fields with the 12-inch Sheepshanks refractor. In their papers of 1905 and 1906, they described in some detail the necessary procedures and precautions to be taken in order to obtain precision photographic parallaxes.

On Russell's return to Princeton, Pickering offered to provide him with apparent magnitudes (or flux densities) and stellar classifications for the 300 stars in the parallax programme. This was the origin of the famous diagram published by Russell in *Nature* in 1914 (Fig. 18.6). The key discovery was the realisation that not all possible combinations of luminosity and spectral type are found among the stars. Rather, the stars fell along 'sequences' or 'branches' in the luminosity–spectral-type diagram, most of them lying along the 'main sequence', which runs from bottom right to top left in Fig. 18.6. These discoveries were to lead to the spectacular theoretical studies of Eddington in understanding the astrophysics of the stars.

Another crucial discovery to result from the Harvard surveys of stellar spectra was the existence of white dwarf stars. Their discovery is charmingly told in a reminiscence of Henry Norris Russell delivered at a colloquium in Princeton in 1954. In 1910, Russell had suggested to Pickering that it would be useful to obtain the spectra of stars for which parallaxes had been measured. Russell's reminiscence continues:

> Pickering said 'Well, name one of these stars'. Well, said I, for example, the faint component of Omicron Eridani. So Pickering said, 'Well, we make rather a speciality of being able to answer questions like that'. And so we telephoned down to the office of Mrs Fleming and Mrs Fleming said, yes, she'd look it up. In half an hour she came up and said, I've got it here, unquestionably spectral type A. I knew enough, even then, to know what that meant. I was flabbergasted. I was really baffled trying to make out what it meant... Well, at that moment, Pickering, Mrs Fleming and I were the only people in the world who knew of the existence of white dwarfs.[12]

The remarkable feature of the faint companion of o-Eridani was that it was a very-low-luminosity star and yet it had the type of spectrum associated with hot stars on the upper part of the main sequence. Russell included it without comment in his first 'Russell' diagram (Fig. 18.6), as a single A star lying roughly 10 astronomical magnitudes below typical main sequence A stars. Walter Adams drew attention to its remarkable properties in 1914, and discovered another example in the following year, the faint companion of Sirius A known as Sirius B.

Eddington realised that these observations implied that white dwarf stars had to be very dense indeed. Their masses could be determined since both examples are members of binary star systems and their radii could be estimated from Planck's radiation formula and the luminosities of the stars. Their mean densities were found to be about 10^8 kg m^{-3}. Eddington argued that there was nothing inherently implausible about such large densities. Matter at high temperatures inside stars would be completely ionised and so there was no reason at that time why the matter could not be compressed to very much higher densities than typical terrestrial densities. In his paper of 1924, he estimated the gravitational redshift which would be expected from such a compact star, according to general relativity, and found that it corresponded to a Doppler shift of the spectral lines to longer wavelengths of about 20 km s^{-1}. Adams made very careful spectroscopic observations of Sirius B with the 100-inch telescope in 1925 and, once account was taken of the orbital motions of the binary stars, a shift of 19 km s^{-1} was measured. Eddington was jubilant:

> Prof. Adams has thus killed two birds with one stone. He has carried out a new test of Einstein's theory of general relativity, and he has shown that matter at least 2000 times denser than platinum is not only possible, but actually exists in the stellar Universe.[13]

The theory of white dwarfs was one of the first triumphs of the new quantum theory of statistical mechanics as applied to astrophysics.

One final example is the work of Henrietta Leavitt, another of Pickering's remarkable team of assistants, who is best remembered for her discovery of 1777 variable stars in the Magellanic Clouds, the closest dwarf galaxies to our own Galaxy. The catalogue of variables included 25 Cepheid variables, from which she derived the famous luminosity–period relation for Cepheids, which was to be crucial in establishing the extragalactic nature of the spiral nebulae. To find the variable stars, she had positive plates made which could be compared with the negative from the first set of observations. By superimposing the positive and negative plates, she was able to distinguish rather easily those stars which were variable and to estimate their light curves, meaning the variation of luminosity with phase for periodic variables.

Her main work for many years was, however, the establishment of the North Polar Sequence, the accurate determination of the apparent-magnitude scale for stars in a region of sky which would always be accessible to observers in the Northern Hemisphere. By the time of her death in 1921, she had extended the North Polar Sequence from magnitude 2.7 to magnitude 21, with an error less than 0.1 magnitude. To achieve this, she used observations from 13 telescopes ranging from 0.5 to 60 inches in diameter and compared her scale using five different photographic photometric techniques. This work was continued by F.H. Seares at Mount Wilson through the 1920s and 1930s. Although these studies may lack the glamour of the discovery of the Hertzsprung–Russell diagram, or of white dwarfs or of the period–luminosity relation for Cepheid variables, this painstaking and time-consuming work lies at the very foundation of quantitative observational astrophysics and is of singular importance for the advance of the subject.

These achievements were hard won and are just a few examples of the remarkable discoveries which resulted from the enlightened sponsorship of American benefactors of astronomy. The lesson from these examples is that, even in the late nineteenth and early twentieth centuries, astronomy was already 'big science' and relied upon the infusion of resources on a considerable scale to maintain progress. Other examples include the generosity of James Lick, a successful manufacturer and retailer of pianos and enthusiast for astronomy. On his death in 1876, he left a bequest of $700 000 to build 'a powerful telescope, superior to and more powerful than any telescope ever yet made . . . and also a suitable observatory connected therewith'. The observatory was constructed on Mount Hamilton and officially opened in 1888 with the completion of the 36-inch telescope, under which James
L [illegible] the terms of the bequest. Another remarkable benefactor was
th [illegible] ne in the oil and steel industries and
w [illegible] 1907, following a visit to Mount
W [illegible] l $10 million to the endowment of
th [illegible] benefaction be used to enable the
w [illegible] pidly as possible.

18.6 T [illegible]

A [illegible] n 1885–7 by Heinrich Hertz, it was
n [illegible] to make an impact upon astronomy
v [illegible] hotomultiplier tubes which made a
r [illegible] dimir Zworykin at the RCA (Radio
C [illegible] ary use of photomultiplier tubes was
i [illegible] icture industry. The device was first
u [illegible] 60-inch telescope at Mount Wilson.
T [illegible] photon causes a secondary electron
c [illegible] that each detected photon results in
a [illegible] iciency of detection of photons was
limited by the quantum [illegible] photon detection. These devices had the advantage of having a linear response over a wide dynamic range and so enabled the calibration of the magnitudes of stars and galaxies to be carried out much more effectively. These devices became the preferred means of calibrating apparent magnitude scales after the Second World War.

The next step in the application of advanced electronics to optical astronomy came with the development of image intensifiers. These were off-shoots of the television industry and, in particular, found application as low-level light detectors for military purposes during the 1960s and 1970s. The principle of these devices was that each photon detected by the photocathode resulted in an electron cascade, as in the photomultiplier, but now the electron beam was focussed onto a photo-emitting screen which was scanned by a television camera. The arrival of each detected photon was registered and the image reconstructed by photon

Figure 18.7: Willard S. Boyle (left) and George E. Smith, the inventors of the charge-coupled device (CCD) in 1974. (From I.S. McLean, 1997, *Electronic Imaging in Astronomy – Detectors and Instrumentation*, Chichester: John Wiley and Sons. Courtesy of Praxis Publishing.)

counting. These systems, which included Vidicons and image photon counting systems, completely transformed the spectroscopy of faint objects during the 1970s. They were ideal for faint objects since the counting rate is limited to about one photon per pixel during the time it takes the television system to register the arrival of the photon. The Faint Object Camera of the Hubble Space Telescope used this technology for deep imaging in the ultraviolet region of the spectrum.

In 1969, the charge-coupled device or CCD was invented by Willard Boyle and George Smith at the Bell Telephone Laboratories at Murray Hill, New Jersey (Fig. 18.7). Their objective was to develop the technology for a 'picturephone', which would enable telephone callers to see each other. The semiconducting materials, which detect the photons, can have very high quantum efficiencies, the quantum effeciency being the fraction of the incident radiation which is registered by the detector. The ejected electrons are stored in potential wells within the semiconductor material. The problem was how to extract the signals without undue losses. This is where the concept of charge-coupling plays a key role. Once the signals are accumulated on the chip, the electrons are shuffled along the rows of potential wells of the detector array and read out through a single amplifier at the end of the row. The patent for the device was obtained in 1974. The development of these devices for astronomy received an enormous boost as a result of their selection as the preferred detectors for the Wide-Field Camera of the Hubble Space Telescope in 1977. Since then, CCDs have dominated optical astronomy in providing direct digital images with very high quantum efficiency. Rather than the 1%–2% achievable by photographic plates, the CCDs can have quantum efficiencies up to about 80%.

A similar story could be told about the development of infrared detectors for astronomy. Again, much of the stimulus for the development of these technologies came from military sources, in particular, from the development of cameras which could 'see in the dark'. The problem for the astronomers was that the photon fluxes in astronomy were many orders of magnitude less than those emitted by the relatively bright objects detected by night-vision cameras. The breakthrough for imaging and spectroscopy in the 1 to 5 μm waveband came in the mid to late 1980s with the release to the astronomical community of indium antimonide arrays with direct electron readouts. The original motivation for the construction of these arrays was as guidance devices for cruise missiles, but they were ideal for astronomy, once devices with low dark current became available. These arrays and their close relatives for longer infrared wavelengths are now the standard detectors for infrared astronomy.

18.7 The impact of the Second World War

The importance of commercial and military technology for the advance of astronomy has been alluded to in previous sections. By far the most important influence for many different aspects of astronomy was the Second World War, in particular, for the growth of the new astronomies which flourished after the war.

Radio signals of astronomical origin were discovered by accident by Carl Jansky in 1933. He was working for the Bell Telephone Company at Holmdel, New Jersey, and was assigned the task of detecting sources of radio noise which would interfere with telecommunication signals. With his array of aerials, he first detected radio emission from the plane of our Galaxy. In the late 1930s, Grote Reber, a radio engineer and keen amateur astronomer, constructed the first fully steerable radio telescope in his back garden and made the first radio maps of the Milky Way, which were published in the *Astrophysical Journal* in 1944. These discoveries did not raise much interest in the astronomical community.

Radio technology made enormous strides during the Second World War, stimulated by the urgent need to develop powerful radar systems for the detection of enemy aircraft and to construct low-noise radio receivers. After the war, those deeply involved in these activities turned their attention to understanding more about the astronomical signals which had been detected in the course of the development of radar. Martin Ryle at Cambridge, Bernard Lovell at Manchester and J.L. Pawsey at Sydney were the pioneers of low-frequency radio astronomy, all of them coming from a background of radar and electronics from the war effort. Radio interferometry was very much in their blood and they understood fully how relatively simple antenna systems could provide accurate positions at low radio frequencies. The ability to preserve phase in low-frequency observations was crucial to the development of the technique of aperture synthesis, which enabled high-angular-resolution radio maps to be created by combining in phase the signals from separated telescopes. Crucial to the success of this approach to imaging in radio astronomy was the availability of high-speed computers. The very first generation of computers available in Cambridge, EDSAC, was

Table 18.1. *Contributions to the total measured radio signal in the experiments of Penzias and Wilson at 4.08 GHz.* (Penzias, A. and Wilson, R. (1965). *Astrophys. J.*, **142**, 419.)

Signal	Noise signal T/K
Total zenith noise temperature	6.7 ± 0.3
Atmospheric emission	2.3 ± 0.3
Ohmic losses	0.8 ± 0.4
Backlobe response	≤ 0.1
Cosmic background radiation	3.5 ± 1.0

essential to the reconstruction of the aperture synthesis radio maps made by Ryle and his colleagues in the late 1950s and early 1960s.

The experience in the USA was different, in that the US astronomers began to explore the radio sky at somewhat higher frequencies. One of the great fruits of this endeavour was the discovery of the cosmic microwave background radiation by Arno Penzias and Robert Wilson in 1964. This story has been splendidly told by Wilson in his article in the volume *Modern Cosmology in Retrospect.*[2] At the Bell Laboratories at Holmdel, New Jersey, a 20-foot horn antenna had been constructed for satellite communications at centimetre wavelengths. Penzias and Wilson had the responsibility of calibrating the antenna for use at these frequencies, for which they had access to a 7.3-cm cooled maser receiver. The understanding was that the telescope could be used for astronomical observations for some fraction of the observing time. Wherever they pointed the telescope on the sky, they found an excess of about 3 K in antenna temperature, which could not be accounted for by noise sources in the telescope or receiver system. A list of contributions to the total detected signal is given in Table 18.1. The rest of the story is history. The cosmic microwave background radiation is the cooled remnant of the hot early phases of expansion of our Universe and undoubtedly the most important discovery in astrophysical cosmology since Hubble's discovery of the velocity–distance relation for galaxies. We will discuss the consequences of this discovery in more detail in Chapter 19. Notice that the primary motivation for the experiments was clearly commercial, rather than astronomical, interest.

There was, in addition, a major change in attitude among the scientists who had worked on physics-related problems through the Second World War. To quote Sir Bernard Lovell, they adopted an approach to research which was

> . . . utterly different from that deriving from the pre-war environment. The involvement with massive operations had conditioned them to think and behave in ways which would have shocked the pre-war university administrators. All these facts were critical in the large-scale development of astronomy.[14]

It is often said that the First World War was dominated by chemistry, whereas the Second World War was most concerned with physics. The significance of advances in physics was not lost upon the authorities, particularly in the USA. There were massive increases in funding for all sciences, with the US military in the vanguard of those supporting basic science. In the UK, this led the Wilson government to promote the 'white heat of technological revolution'. The astronomers rode this wave of investment in the fundamental sciences, but generally these initiatives were in a national or international context, rather than a result of sponsorship by private institutions, as had occurred earlier in the USA.

18.8 Ultraviolet, X-ray and γ-ray astronomy

Immediately after the Second World War, those physicists who aspired to undertake ultraviolet, X-ray and γ-ray astronomy took the first tentative steps into space astronomy. The atmosphere is opaque to all radiation with wavelengths shorter than about 330 nm and so ultraviolet, X-ray and γ-ray astronomy had to be conducted from above the Earth's atmosphere. The German V-2 rocket programme had made enormous strides in rocket technology during the war and many of the German scientists who had built these rockets, led by Werner von Braun, went to the USA; they also took with them 300 box cars full of V-2 parts. Thus was formed the core of the US army's rocket programme. The rockets were also made available for scientific research. The technological challenge was to make the scientific instruments as compact and lightweight as possible. This was a particular challenge for the higher-energy wavebands, in which the detectors were small-scale versions of the devices used with ground-based particle accelerators.

Among the earliest beneficiaries of the opening up of space for astronomy were the ultraviolet astronomers. The central figure in this story is Lyman Spitzer, who in 1946 wrote a report on the utilisation of space for astronomical purposes for the RAND project of the US air force. One of the prime targets of the early rocket experiments was the ultraviolet and X-ray emission of the Sun, which, it was surmised, might be responsible for the ionisation of the Earth's ionosphere. The first successful rocket ultraviolet observations of the Sun were made in October 1946 by Herbert Friedman's group at the Naval Research Laboratory. In the following year, they made the first successful X-ray observations of the Sun, confirming the expectation that the Sun's corona is very hot.

The flights of Sputniks 1 and 2 in late 1957 and Yuri Gagarin's orbital flight in 1961 came as a profound shock to the US administration, which realised that the USA had fallen behind the USSR in space technology. The US response was to set up the National Aeronautics and Space Administration (NASA) in July 1958 as a civilian organisation to begin the process of catching up with the USSR. As part of that endeavour, the American Science and Engineering group (AS & E) was set up in association with the Massachusetts Institute of Technology to carry out military and civilian contracts. The AS & E group was responsible for the first successful flight dedicated to searches for extraterrestrial X-rays, which took place in June 1962 (Fig. 18.8). In the five minutes of observing time, during which the rocket payload was above the Earth's atmosphere, Giacconi and his colleagues discovered an intense discrete source of emission in the constellation of Scorpius, which became known as Sco X-1. In addition, an intense background of X-rays was observed, which was remarkably uniformly distributed over the sky.

Spitzer and his colleagues planned a series of three space observatories, to be known as the Orbiting Astrophysical Observatories (OAO), which were to be dedicated to spectroscopy in the ultraviolet waveband between 90 and 330 nm. Unlike the other new astronomical

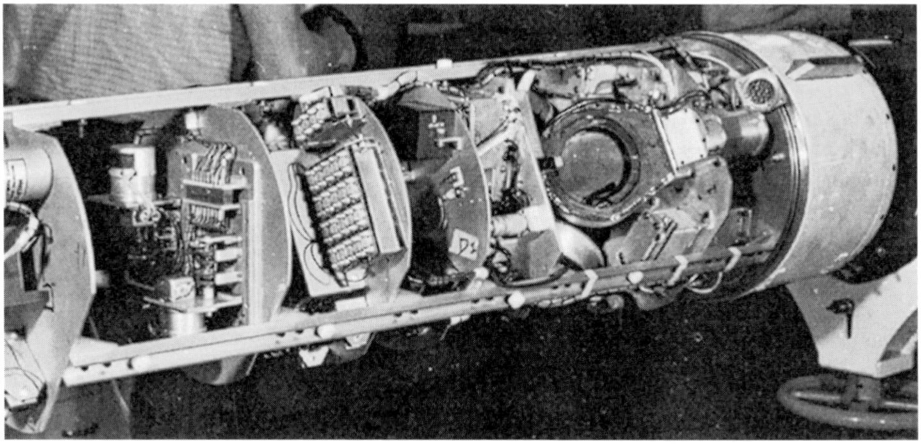

Figure 18.8: The payload of the rocket containing the X-ray detectors which made the first observation of the discrete X-ray source Sco X-1 and the X-ray background radiation in June 1962. The payload was constructed by the AS & E group (American Science and Engineering). (From W. Tucker and R. Giacconi, 1985, *The X-ray Universe*, Cambridge, Massachusetts: Harvard University Press.)

wavebands, the astrophysical objectives of ultraviolet astronomy were very well defined. The resonance transitions of essentially all the common elements were known to lie in the ultraviolet, rather than the optical, waveband. Access to wavelengths shorter than Lyman-α was of special significance because among the many resonance lines are those of deuterium, which is of great cosmological importance. The Copernicus satellite (OAO-3), launched in 1972, was the great success of the series, measuring the abundances of the common elements, as well as deuterium, in the interstellar medium for the first time. The OAO observatories led in turn to the launch of the International Ultraviolet Explorer (IUE) in 1978, a joint UK–European Space Agency–NASA project. The IUE was in turn the precursor of the Hubble Space Telescope.

Until 1970, all X-ray astronomy was carried out from rockets, which provided tantalising glimpses of what was there, but the picture was confused. These problems were resolved with the launch in December 1970 of the UHURU X-ray observatory, the first satellite dedicated to X-ray astronomy. This observatory conducted the first survey of the X-ray sky and revealed the true nature of the X-ray population – X-ray binaries, supernova remnants, young radio pulsars, active galactic nuclei and the hot intergalactic gas in clusters of galaxies were all detected.

In the early 1960s, there were already γ-ray detectors in space, monitoring the atmospheric nuclear test-ban treaties concluded between the USA and the USSR. The Vela series of satellites was launched for this purpose in the 1960s, but there was no intention that they should have any astronomical role. Cosmic γ-rays were first detected in observations made by the Explorer II satellite in 1965, but this experiment did little more than show that there exist γ-rays which originate from beyond the Earth's atmosphere. The first important astronomical observations were made by the third Orbiting Solar Observatory (OSO-III), launched in March 1967. The prime discovery of this mission was the detection of γ-rays with energies $E_\gamma > 100$ MeV from the general direction of the Galactic centre. This γ-ray flux was convincingly interpreted as the γ-ray emission associated with the decay of neutral pions created in collisions between relativistic protons and the cold interstellar gas.

The Small Astronomical Satellite, SAS-2, was launched in November 1972 and included an array of spark chambers to detect the electron–positron pairs created when an incoming γ-ray is converted into a pair within the instrument. Although it operated for only eight months and detected only about 8000 γ-rays of cosmic origin, it confirmed the existence of a general concentration of γ-rays towards the plane of the galaxy as well as two sources associated with the pulsars in the Crab and Vela supernova remnants, and it provided evidence for diffuse extragalactic γ-ray background radiation. The SAS-2 mission was followed in 1975 by the equally successful COS-B satellite, lauched by a European consortium. It also consisted of an array of spark chambers sensitive to γ-rays with energies greater than about 70 MeV, resulting in a detailed map of the Galactic plane as well as evidence for 24 discrete γ-ray sources.

To everyone's surprise, astronomical γ-ray bursts were discovered by the Vela satellites, each burst lasting typically less than a minute. During that time, each burst is the brightest source in the sky. The first of these was detected by the Vela satellite in 1967 but they were not reported in the scientific literature until 1973. Observations with the Compton Gamma-ray Observatory, launched in April 1991, have shown that these bursts occur about once per day, apparently randomly over the sky. Only in 1999 was it established beyond any doubt that these extremely high energy events take place in galaxies at cosmological distances.

18.9 Reflections

I have intentionally not brought this exposition of the technology of cosmology right up to date, since recent developments are well known but the underlying theme would remain the same. The development of new technologies, very often quite independently of astrophysical goals, has played an essential role in enabling astrophysical and cosmological theories to be confronted with observational data. The enabling technologies have often come from commercial or military initiatives, where much larger research and development resources have been available than would be realisable for any pure science project.

Another lesson is the way in which new technologies can change how astrophysical and cosmological problems are tackled. Often, when one technology reaches the end of the road, a completely new way of addressing the same problems is discovered through technological innovation. A splendid example is the present generation of 8–10 metre optical–infrared telescopes. The key technological factor was the realisation that it is more cost effective to spend money on computers rather than on steel, and to build control systems which can correct in real time for the thermal and mechanical distortions of a large telescope structure and mirror.

It is important to encourage the next generation of astrophysicists and cosmologists to take a real interest in these issues, so that they can provide leadership not only in theory and interpretation but also in the technology and instrumentation which will ultimately lead to the new insights of the astrophysics and cosmology of the twenty-first century.

I emphasised at the outset that this chapter is very different from all the others in that it has concentrated upon the technological issues which influence the development of theory. The same type of story could be told about any other area of physics. There is no better introduction to these extraordinarily compelling stories than various chapters of the three-volume work *20th Century Physics*.[15]

18.10 References

The information on which this chapter is based is derived from a very large number of sources. The principal references, which provide good introductions to many of these topics and to more detailed references, are references 2–10. This chapter began as a contribution to the conference proceedings *Historical Development of Modern Cosmology*,[16] which contains many excellent papers.

1 Galison, P. (1997). *Image and Logic: A Material Culture of Microphysics*. Chicago: Chicago University Press.

2 Bertotti, B., Balbinot, R., Bergia, S.A. and Messina, A. eds. (1990). *Modern Cosmology in Retrospect*. Cambridge: Cambridge University Press.

3 Gingerich, O.J. ed. (1984). *The General History of Astronomy* (general editor M. Hoskin), Vol. 4, *Astrophysics and Twentieth-Century Astronomy to 1950*. Cambridge: Cambridge University Press.

4 Hearnshaw, J.B. (1986). *The Analysis of Starlight – One Hundred and Fifty Years of Astronomical Spectroscopy*. Cambridge: Cambridge University Press.

5 Hearnshaw, J.B. (1996). *The Measurement of Starlight – Two Centuries of Astronomical Photometry*. Cambridge: Cambridge University Press.

6 Lang, K.R. and Gingerich, O. eds. (1979). *A Source Book in Astronomy and Astrophysics, 1900–1975*. Cambridge, Massachusetts: Harvard University Press.

7 Learner, R. (1981). *Astronomy through the Telescope*. London: Evans Brothers.

8 Leverington, D. (1996). *A History of Astronomy: From 1890 to the Present*. Berlin: Springer-Verlag.

9 Longair, M.S. (1995). Astrophysics and cosmology, in *20th Century Physics*, eds. L.M. Brown, A. Pais and A.B. Pippard, p. 1691. Bristol: Institute of Physics Publications; New York: American Institute of Physics.

10 Longair, M.S. *The Cosmic Century*. Cambridge: Cambridge University Press (in preparation).

11 Hubble, E.P. (1926). *Astrophys. J.*, **64**, 321.

12 See Davis Philip, A.G. and DeVorkin, D.H. eds. (1977). In memory of Henry Norris Russell. *Dudley Observatory Report* No. 13, pp. 90, 107.

13 Eddington, A.S. (1927). *Stars and Atoms*. Oxford: Clarendon Press. See also A.V. Douglas (1956), *The Life of Arthur Stanley Eddington*, pp. 75–8, London: Nelson.

14 Lovell, A.C.B. (1987). *Q. J. Roy. Astr. Soc.*, **28**, 8.

15 Brown, L.M., Pais, A. and Pippard, A.B. eds. (1995). *20th Century Physics*. Bristol: Institute of Physics Publications; New York: American Institute of Physics.

16 Martinez, V.J., Trimble, V. and Pons-Bordieria, M.-J. eds. (2001). *Historical Development of Modern Cosmology*, Vol. CS-252. San Francisco: ASP Conference Proceedings.

19 Cosmology

19.1 Cosmology and physics

By cosmology, I mean the application of the laws of physics to the Universe as a whole. As a result, the validation of theory depends upon observation rather than experiment, placing us at one stage further removed from our 'apparatus' than is the case in laboratory physics. Yet, throughout history, astronomical observations have played their role in establishing new physics, which has been rapidly assimilated into the mainstream of established science. In Chapter 2, Tycho Brahe's observations of the motions of the planets, which led to Newton's law of gravitation, were discussed. From observation of the eclipses of the satellites of Jupiter, Ole Rømer showed conclusively that the speed of light is finite and in 1676 estimated it from the time it takes light to travel across the Earth's orbit about the Sun.

To construct self-consistent cosmological models a relativistic theory of gravity is needed, and most of the tests of general relativity involve the use of astronomical objects. The discovery of the binary pulsar PSR 1913+16 has proved to be of particular importance for physics. The pulsar is a magnetised, rotating, neutron star, which has mass about 1.4 times the mass of the Sun, radius $r \approx 10$ km and general relativistic parameter $2GM/(rc^2) \approx 0.3$. Its companion star is another neutron star of similar mass and they orbit their common centre of mass every 7.75 hours. The pulsar emits a sharp pulse of radio emission once per rotation period of the neutron star, in the case of PSR 1913+16 once every 0.059 seconds. This system is a ideal gift to general relativists, since the pulsar is essentially a perfect 'clock' in a rotating frame of reference, enabling extremely subtle tests of general relativistic effects to be carried out by very precise timing of the arrival of the radio pulses. General relativity has passed all these tests with flying colours. Among the most important of these observations has been the speeding up of the binary orbit as a result of the emission of gravitational waves by the binary star system. This gives us confidence that general relativity provides an excellent description of relativistic gravity.

Thus, astronomy has played, and continues to play, a major role in fundamental physics. It is striking how the very best of laboratory physics enables us to understand the properties of celestial objects in a wholly convincing way. Conversely, astronomical observations enable us to test and extend the laws of physics in physical circumstances which are unattainable in the laboratory.

There is, however, a serious problem, which besets the application of the laws of physics to the Universe as a whole. We have only one Universe to observe, and physicists are rightly cautious about accepting the results of any one-off experiment. In cosmology, however, there is no prospect of doing any better. Despite this limitation, the application of the laws of physics to cosmological problems turns out to be a remarkably successful programme and leads to the standard *Big Bang* model. This picture also leads to a number of fundamental problems, which will be developed in what follows. The general consensus is that these problems can only be addressed by new types of physics applicable at the ultra-high temperatures and densities encountered in the very early Universe. These conditions far exceed those at which the laws have been tested in terrestrial laboratories. The numerous successes of the Big Bang picture have, however, led theorists to use the very earliest phases of the Big Bang as a laboratory for physics at these extreme energies.

19.2 Basic cosmological data

This is not the place to go into the many details of the observations which provide the basis for the standard Big Bang picture. I have summarised the relevant material in my book *Galaxy Formation*,[1] to which the reader is referred for details. The observational foundations of cosmology will therefore be treated quite briefly. The models which describe the large-scale structure of the Universe are based upon two key observations, (i) the isotropy and homogeneity of the Universe on the large scale and (ii) Hubble's law. These observations on their own enable us to set up the theoretical infrastructure needed to undertake cosmological studies. Then we can incorporate the physical processes needed to carry out more detailed investigations.

19.2.1 The isotropy and homogeneity of the Universe

The Universe is highly anisotropic and inhomogeneous on a small scale but, as we observe larger and larger scales in the Universe, it becomes more and more isotropic and homogeneous. Galaxies are the building blocks which define the large-scale distribution of visible matter in the Universe. On the small scale, they are clustered into groups and clusters of galaxies, but, when viewed on a large enough scale, these irregularities are smoothed out. Figure. 19.1 shows the distribution of over two million galaxies in a region of sky covering about one tenth of the celestial sphere, the image being derived from the Cambridge APM survey.

It is apparent from Fig. 19.1 that, looked at on a large enough scale, one region of the Universe looks very much like another, which is *prime facie* evidence that, in the simplest approximation, the Universe is isotropic on a large enough scale. Despite the overall evidence for isotropy, however, it is apparent from Fig. 19.1 that there are non-random features in the galaxy distribution.

There appear to be 'walls' of galaxies, as well as regions of significant under-density. These are now known to be real features of the large-scale distribution of galaxies. Major galaxy surveys are under way to measure the redshifts, or distances, of large samples of the galaxies seen in Fig. 19.1, and these show that the distribution of galaxies exhibits a 'cellular' structure. The simplest analogy for the distribution of galaxies is to think of it as resembling a sponge. Where the galaxies are located corresponds to the material of the

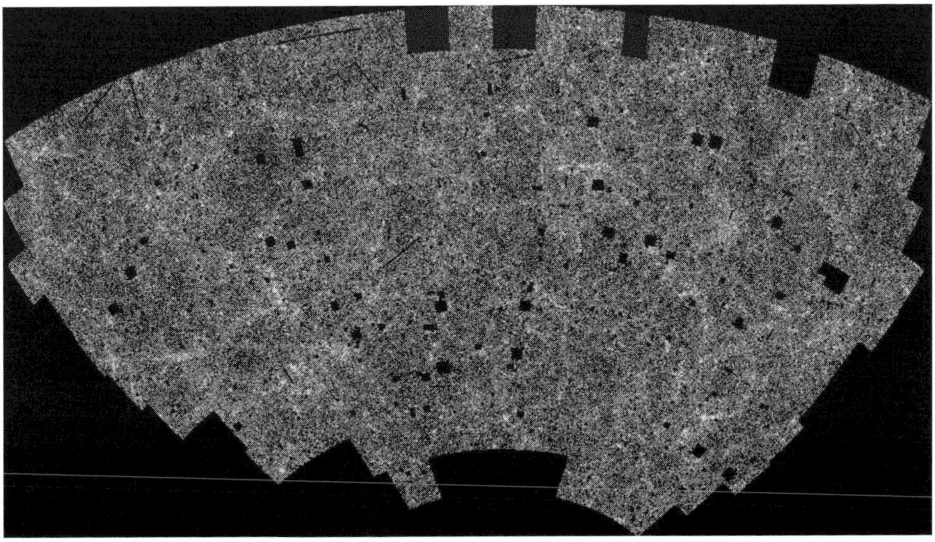

Figure 19.1: The distribution of galaxies with apparent astronomical magnitudes $17 \leq b_j \leq 20.5$ shown in an equal-area projection centred on the Southern Galactic Pole. This image was reconstructed from machine scans of 185 UK Schmidt plates by the Cambridge APM measuring machine. There are over two million galaxies in this image. The small empty patches are regions which have been excluded around bright stars, nearby dwarf galaxies, globular clusters and step wedges. (From S.J. Maddox, G. Efstathiou, W.G. Sutherland and J. Loveday 1990, *MNRAS*, **242**, p. 43.)

sponge, and the holes correspond to the 'holes' or 'voids' in the distribution of galaxies. The sizes of the holes can be very large indeed, some of them being as large as about 50 times the typical size of a cluster of galaxies. Once the distance information is included in the analysis of Fig. 19.1, it can be shown that the same 'spongy' distribution persists from the local Universe out to the limit of the APM survey. This is direct evidence not only for the isotropy of the Universe but also for its homogeneity – although the distribution of galaxies is irregular, it displays the same degree of irregularity as observations are extended to larger and larger distances.

The evidence of the distribution of galaxies is compelling enough, but even more spectacular have been observations of the cosmic microwave background radiation, the discovery of which was described in Section 18.7. Soon after its discovery in 1965, it was established that the spectrum of the radiation is of black-body form and is remarkably isotropic over the sky. These features of the cosmic microwave background radiation were studied in detail by the Cosmic Background Explorer, COBE, which was launched by NASA in September 1989. The spectrum of the radiation is that of a perfect black body at a radiation temperature $T = 2.728$ K, deviations from this spectrum amounting to less than 0.03% of its maximum intensity in the wavelength interval $2.5 > \lambda > 0.5$ mm. The distribution of the radiation away from the plane of our Galaxy is quite extraordinarily isotropic, deviations from isotropy being less than about one part in 10^5 on all angular scales greater than 10° (Fig. 19.2). As we will show, this radiation is the cooled remnant of the very hot early phases of the Big Bang; it was last scattered when the Universe was only about 300 000 years old.

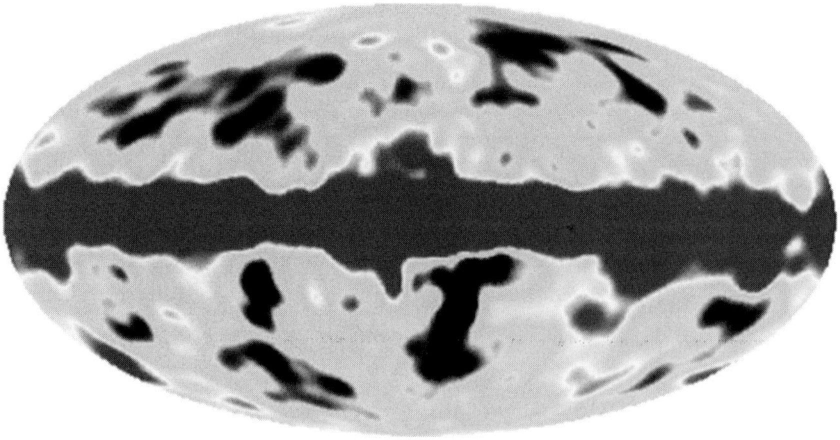

Figure 19.2: A map of the whole sky in galactic coordinates as observed by the COBE satellite at a wavelength of 5.7 mm (53 GHz). The centre of our Galaxy is in the centre of the image and the plane of our Galaxy is the broad band across the centre of the image. The intensity fluctuations away from the Galactic plane are real fluctuations in the intensity of the background radiation and correspond to an rms temperature fluctuation of only 35 ± 2 μK, compared with a total brightness temperature of 2.728 K (From C.L. Bennett *et. al.*, 1996, *Astrophys. J.*, **464**, p. L1.)

This statement begs a number of important questions, which will be discussed in detail in what follows. At this point in our story, the relevant aspect of the cosmic background radiation is that it demonstrates that *something* is extraordinarily isotropic in our Universe. We can be certain that the isotropy refers to the Universe as a whole because 'shadows' of very distant clusters of galaxies, containing hot gas, are observed in the background radiation as a result of the Sunyaev–Zeldovich effect.[2] Thus, we can be certain that, on large scales, the Universe is isotropic to better than one part in 100 000 – this is a quite astounding cosmological result and will simplify a great deal of the theoretical analysis.

19.2.2 Hubble's law

A brief history of Edwin Hubble's discovery of the velocity–distance relation for galaxies in 1929 was given in Section 18.4. Hubble used the radial velocities of 24 galaxies, most of which had been measured painstakingly by V.M. Slipher. All the galaxies had distances within 2 Mpc of our own Galaxy and it is remarkable that Hubble found this relation on the basis of the fragmentary evidence he had concerning their distances and in spite of what we now know, that the distribution of galaxies is highly inhomogeneous on these scales. Hubble's discovery was that the recession velocity v of a galaxy from our Galaxy is proportional to its distance r, that is, $v = H_0 r$ where H_0 is known as Hubble's constant and the relation as *Hubble's law*.

In modern studies of the velocity–distance relation, accurate velocities of galaxies can be readily derived from their *redshifts*. The *redshift* is defined to be a shift of spectral lines to longer wavelengths, in the cosmological case because of the recessional velocity of the galaxy. If λ_e is the wavelength of the line as emitted and λ_o the observed wavelength, the redshift z is

$$z = \frac{\lambda_o - \lambda_e}{\lambda_e}. \tag{19.1}$$

According to special relativity, the radial velocity inferred from the redshift is given by the relation

$$1 + z = \left(\frac{1 + v/c}{1 - v/c}\right)^{1/2} \tag{19.2}$$

if the galaxy is moving radially away from the observer at velocity v. In the limit of small redshifts, $v/c \ll 1$,

$$v = cz. \tag{19.3}$$

This is the type of velocity used in Hubble's law. It is, in fact, a misfortune that recessional velocities are used at all. The redshift is a much better dimensionless parameter and appears naturally in the development of the cosmological relations, as will be described below.

It is much more difficult to estimate the distances of galaxies and other extragalactic objects. The standard procedure is to find some class of astronomical objects, the members of which all have the same intrinsic luminosity L. Then the observed flux density S of the object is given by the inverse square law, $S = L/(4\pi r^2)$. Astronomers use a logarithmic scale of flux densities known as *apparent magnitudes* m such that $m = -2.5 \log S +$ constant. Therefore, if the intrinsic luminosity of the distance indicator is independent of distance, we expect that

$$m = -2.5 \log S + \text{constant} = 5 \log r + \text{constant}. \tag{19.4}$$

Figure. 19.3 shows the velocity–distance relation for the brightest galaxies in rich clusters of galaxies. It can be seen that there is a near-perfect correlation between apparent magnitude and the logarithm of the redshift, with exactly the slope expected if recessional velocity were precisely proportional to distance. The same relation is found for all classes of extragalactic objects.

A major programme for observational cosmologists has been the precise determination of the value of Hubble's constant H_0. Thanks to the efforts of many astronomers and a large programme of observations using the Hubble Space Telescope, there is now general agreement that the value of H_0 is

$$H_0 = (72 \pm 8) \text{ km s}^{-1} \text{ Mpc}^{-1} = (2.3 \pm 0.3) \times 10^{-18} \text{ s}, \tag{19.5}$$

where the error is at the $\pm 1\sigma$ level.

19.2.3 The local expansion of the distribution of galaxies

The combination of the isotropy of the Universe and Hubble's law shows that the Universe as a whole is expanding uniformly at the present time. In a uniform expansion, the ratio of

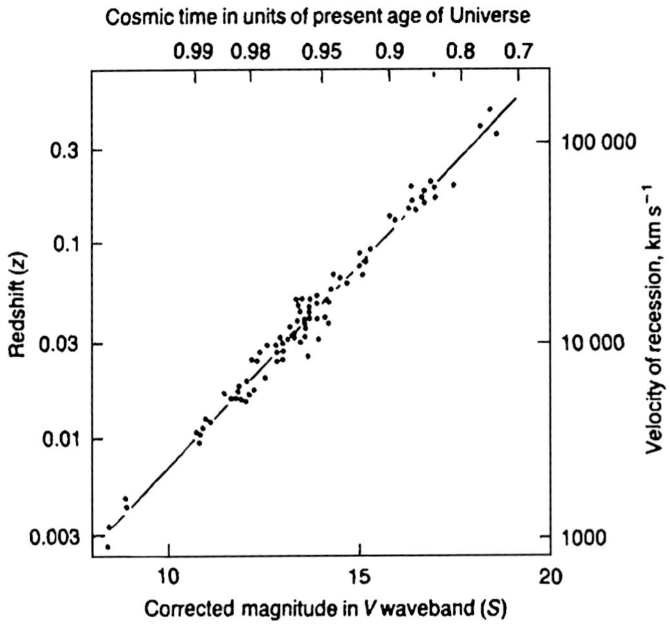

Figure 19.3: A modern version of the velocity–distance relation, for the brightest galaxies in rich clusters of galaxies. This correlation indicates that the brightest galaxies in clusters have remarkably standard properties and that their velocities of recession from our Galaxy are proportional to their distances. (From A.R. Sandage, 1968, *Observatory*, **88**, 91.)

the distances between any two points increases by the same factor in a given time interval, that is,

$$\frac{r_1(t_2)}{r_1(t_1)} = \frac{r_2(t_2)}{r_2(t_1)} = \cdots = \frac{r_n(t_2)}{r_n(t_1)} = \cdots = \alpha = \text{constant}, \tag{19.6}$$

in a time interval $t_2 - t_1$. The recession velocity of galaxy 2 from galaxy 1, which we can choose as the origin, is therefore

$$\begin{aligned} v_2 &= \frac{r_2(t_2) - r_2(t_1)}{t_2 - t_1} = \frac{r_2(t_1)}{t_2 - t_1}\left[\frac{r_2(t_2)}{r_2(t_1)} - 1\right], \\ &= \frac{\alpha - 1}{t_2 - t_1} r_2(t_1) = H_0 r_2(t_1). \end{aligned} \tag{19.7}$$

Similarly, for galaxy n,

$$v_n = \frac{\alpha - 1}{t_2 - t_1} r_n(t_1) = H_0 r_n(t_1). \tag{19.8}$$

Thus, a uniformly expanding distribution of galaxies automatically results in a velocity–distance relation of the form $v = H_0 r$. Notice that it does not matter which galaxy we choose as the origin. All observers rightly believe that they are at the centre of an expanding Universe with the same Hubble relation, if the observations are made at the same cosmic time.

19.3 The Robertson–Walker metric

The observational evidence discussed in Section 19.2 indicates that the natural starting point for the construction of cosmological models is the assumption that the Universe is isotropic, homogeneous and expanding uniformly at the present epoch. One of the problems facing the pioneers of relativistic cosmology was the interpretation of the space and time coordinates to be used in these calculations. The problem was solved independently by H.P. Robertson and A.G. Walker in 1935, who derived the form of the metric of space–time for *all* isotropic, homogeneous, uniformly expanding models of the Universe. The form of the metric is independent of the assumption that the large-scale dynamics of the Universe is described by general relativity; whatever the physics of the expansion, the space–time metric must be of Robertson–Walker form, because of the assumptions of isotropy and homogeneity.

A key step in the development of these models was the introduction by Hermann Weyl in 1923 of what is known as *Weyl's postulate*. To eliminate the arbitrariness in the choice of coordinate frames, Weyl introduced the idea that, in the words of Hermann Bondi,

> The particles of the substratum (representing the nebulae) lie in space–time on a bundle of geodesics diverging from a point in the (finite or infinite) past.[3]

The most important aspect of this statement is the postulate that the geodesics, which represent the world-lines of galaxies, do not intersect except at a singular point in the finite, or infinite, past. Again, it is extraordinary that Weyl introduced this postulate *before* Hubble's discovery of the recession of the nebulae. By the term 'substratum', Bondi meant an imaginary medium which can be thought of as a fluid that defines the overall kinematics of the system of galaxies. A consequence of Weyl's postulate is that there is only one geodesic passing through each point in space-time, except at the origin. Once this postulate is adopted, it becomes possible to assign a notional observer to each world line and these are known as *fundamental observers*. Each fundamental observer carries a standard clock and time measured on that clock, which was set to zero at the origin, is called *cosmic time*.

One further assumption is needed before we can derive the framework for the standard models. This is the assumption known as the *cosmological principle*:

> We are not located at any special location in the Universe.

A corollary of this statement is that we are located at a *typical* position in the Universe and that any other fundamental observer located anywhere in the Universe at the same cosmic epoch would observe the same large-scale features which we observe. Thus, we assert that every fundamental observer at the same cosmic epoch observes the same Hubble expansion of the distribution of galaxies, the same isotropic cosmic microwave background radiation, the same large-scale spongy structure in the distribution of galaxies and voids and so on.

The *Robertson–Walker metric* is the metric, in the sense of special relativity, of all isotropically expanding universes. Because of the observed large-scale isotropy of the Universe, the obvious starting point is to consider models in which we smooth out all structure, that is, a uniformly expanding isotropic, homogeneous Universe. Since the curvature of space is an intrinsic property of space–time, it follows that the curvature of space must also be isotropic and so the geometry must be one of the isotropic curved spaces which we considered in some detail in Section 17.3.

19.3.1 *Isotropic curved spaces again*

A key result derived in Section 17.3 was that all possible isotropic curved three-spaces are described by a spatial increment of the form

$$dl^2 = dx^2 + R_c^2 \sin^2\left(\frac{x}{R_c}\right)[d\theta^2 + \sin^2\theta \, d\varphi^2], \tag{19.9}$$

in terms of a fundamental observer's spherical polar coordinates x, θ, φ: where R_c is the radius of curvature of the geometry of the isotropic three-space and is imaginary in open, hyperbolic, geometries. If instead we write the spatial increment in terms of ρ, θ, φ, we find the result (17.63),

$$dl^2 = \frac{d\rho^2}{1 - \kappa\rho^2} + \rho^2[d\theta^2 + \sin^2\theta \, d\varphi^2], \tag{19.10}$$

where $\kappa = R_c^2$ is the curvature of the three-space. I have changed the notation for the 'angular-diameter distance' from r to ρ for reasons which will become apparent in a moment – ρ will have this meaning only in the present section. Recall that the distance coordinate x describes the *metric distance* in the radial direction, while the ρ-coordinate ensures that distances perpendicular to the radial direction are given by $dl = \rho \, d\theta$. The *Minkowski metric* for any isotropic three-space is given by

$$ds^2 = dt^2 - \frac{1}{c^2} dl^2, \tag{19.11}$$

where dl is given by either of the above forms of the spatial increment.

19.3.2 *The Robertson–Walker metric*

To develop the Robertson–Walker metric, we need to incorporate the *cosmological principle* into the construction of isotropic cosmological models. As discussed above, we introduce the concept of *fundamental observers*, who move in such a way that the Universe always appears to be isotropic to them – they all partake in the uniform Hubble flow. The next important concept is *cosmic time*, which is defined to be *proper time* as measured by a fundamental observer with a standard clock. According to the Weyl hypothesis, all the clocks can be synchronised by assuming that the world lines of all the fundamental observers diverged from some point in the distant past and we set $t = 0$ at that origin.

There is a problem in using the metric (19.9), or (19.10), which can be illustrated by a simple space–time diagram (Fig. 19.4). Since light travels at a finite speed, all astronomical objects are observed along the *past light cone* of the fundamental observer, which in our case is centred on the Earth at the present epoch t_0. Therefore, distant objects are observed at earlier epochs than the present, when the Universe was homogeneous and isotropic but the distances between fundamental observers were smaller and the spatial curvature different.

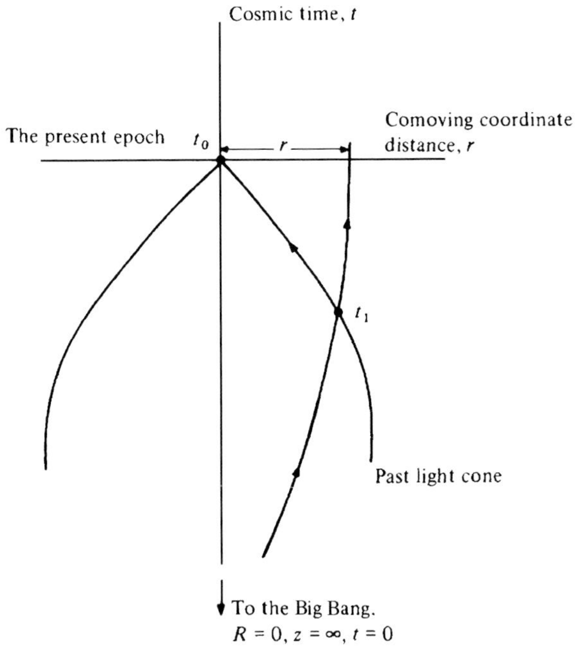

Figure 19.4: A space–time diagram showing the past light cone along which a fundamental observer observes all objects in the Universe. The curve intersecting the light cone at t_1 is the world line of a galaxy observed at epoch t_1 and redshift z.

The problem is that we can only apply the metric (19.9) to an isotropic curved space *at a single epoch*.

To derive a distance measure which can be included in the metric, we carry out a thought experiment in which we line up a set of fundamental observers between the Earth and a distant galaxy. The observers are all instructed to measure the geodesic distance dx to the next observer *at a fixed cosmic time t*, which they read on their own clocks. By adding together all the dx's, they measure a geodesic distance x at a single epoch, which can be used in the metric (19.9). Notice that x is a *fictitious distance*: distant galaxies can only be observed as they were at some epoch earlier than the present and we do *not* know how to project their positions relative to us forward to the present epoch without a knowledge of the kinematics of the expanding Universe. In other words, *the distance measure x depends upon the choice of cosmological model*, which we do not know.

The definition of a uniform expansion is that between two cosmic epochs, t_1 and t_2, the distances of any two fundamental observers, i and j, from some arbitrary fundamental observer change in such a way that

$$\frac{x_i(t_1)}{x_j(t_1)} = \frac{x_i(t_2)}{x_j(t_2)} = \text{constant}, \tag{19.12}$$

that is,

$$\frac{x_i(t_1)}{x_i(t_2)} = \frac{x_j(t_1)}{x_j(t_2)} = \cdots = \text{constant} = \frac{R(t_1)}{R(t_2)}. \tag{19.13}$$

$R(t)$ is defined to be the *scale factor* and describes how the relative distance between any two observers changes with cosmic time t. We will set $R(t) = 1$ at the present epoch t_0 and let the value of x at t_0 be r. Therefore

$$x(t) = R(t)r. \tag{19.14}$$

Thus r becomes a *distance label* attached to a galaxy for all time and the variation of proper distance in the expanding Universe is taken care of by the scale factor $R(t)$; r is called the *comoving radial distance coordinate* – it is very important.

Proper distances perpendicular to the line of sight must also change by a factor R between the epochs t and t_0, because of the isotropy of the world model:

$$\frac{\Delta l(t)}{\Delta l(t_0)} = R(t) \tag{19.15}$$

and hence, from the components of the metric (19.9) in the θ-direction at epochs t and t_0,

$$R(t) = \frac{R_c(t)\sin\dfrac{x}{R_c(t)}\,\mathrm{d}\theta}{R_c(t_0)\sin\dfrac{r}{R_c(t_0)}\,\mathrm{d}\theta}.$$

Reorganising this expression, we see that

$$\frac{R_c(t)}{R(t)}\sin\frac{R(t)r}{R_c(t)} = R_c(t_0)\sin\frac{r}{R_c(t_0)}.$$

This is only true if

$$R_c(t) = R_c(t_0)R(t), \tag{19.16}$$

that is, the radius of curvature of the spatial sections is proportional to $R(t)$. In order to preserve isotropy and homogeneity, *the curvature of space must change as $\kappa = R_c^{-2} \propto R^{-2}$ as the Universe expands.*

Let us call the radius of curvature at the present epoch $\Re$. Then

$$R_c(t) = \Re R(t). \tag{19.17}$$

Substituting (19.14) and (19.17) into the metric (19.11) and using (19.9) we obtain the result we have been seeking:

$$\mathrm{d}s^2 = \mathrm{d}t^2 - \frac{R^2(t)}{c^2}\left[\mathrm{d}r^2 + \Re^2\sin^2\left(\frac{r}{\Re}\right)(\mathrm{d}\theta^2 + \sin^2\theta\,\mathrm{d}\varphi^2)\right]. \tag{19.18}$$

This is the *Robertson–Walker metric* in the form we will use in analysing observations in cosmology. It contains one unknown function $R(t)$, the scale factor, and an unknown constant, the curvature of space at the present epoch $\kappa = \Re^{-2}$.

The metric can be written in different ways. If we use a *comoving angular diameter distance* $r_1 = \Re\sin(r/\Re)$ then the metric becomes

$$\mathrm{d}s^2 = \mathrm{d}t^2 - \frac{R^2(t)}{c^2}\left[\frac{\mathrm{d}r_1^2}{1-\kappa r_1^2} + r_1^2(\mathrm{d}\theta^2 + \sin^2\theta\,\mathrm{d}\varphi^2)\right], \tag{19.19}$$

where $\kappa = 1/\Re^2$. By a suitable rescaling of the r_1-coordinate so that $\kappa r_1^2 = r_2^2$, the metric can equally well be written

$$ds^2 = dt^2 - \frac{R_1^2(t)}{c^2}\left[\frac{dr_2^2}{1 - kr_2^2} + r_2^2(d\theta^2 + \sin^2\theta\, d\varphi^2)\right], \tag{19.20}$$

with $k = +1, 0$ and -1 for universes with spherical, flat and hyperbolic geometries respectively. In this rescaling, the value of $R_1(t_0)$ at the present epoch is $\Re$ and not unity.

The Robertson–Walker metric enables us to define the invariant interval ds^2 between events at any epoch or location in the expanding Universe. Note that:

- t is cosmic time;
- $R(t)\,dr = dx$ is the element of proper (or geodesic) distance in the radial direction;
- $R(t)[\Re \sin(r/\Re)]\,d\theta = R(t)r_1\,d\theta$ is the element of proper distance perpendicular to the radial direction subtended by the angle $d\theta$ at the origin;
- Similarly, $R(t)[\Re \sin(r/\Re)] \sin\theta\, d\varphi = R(t)r_1 \sin\theta\, d\varphi$ is the element of proper distance in the φ-direction.

Note the following features of the above arguments.

- The Robertson–Walker metric (19.18) has been derived using only special relativity and the postulates of isotropy and homogeneity – there is nothing explicitly general relativistic about the argument.
- We have specified nothing about the physics which determines the dynamics of the expanding Universe – this has all been absorbed into the function $R(t)$.

19.4 Observations in cosmology

Many results useful for relating the intrinsic properties of distant objects to their observed properties are independent of the specific cosmological model. In this section we provide a list of important results, which can be applied to the cosmology of your choice.

19.4.1 Redshift

Consider a wave packet of frequency ν_1 emitted between cosmic times t_1 and $t_1 + \Delta t_1$ from a distant galaxy. This wave packet is received by an observer at the present epoch in the cosmic time interval t_0 to $t_0 + \Delta t_0$. The signal propagates along null cones, or null geodesics, for which $ds^2 = 0$, and so, considering radial propagation from source to observer, $d\theta = 0$, $d\varphi = 0$, the metric (19.18) reduces to

$$dt = -\frac{R(t)}{c}dr, \qquad \frac{c\,dt}{R(t)} = -dr. \tag{19.21}$$

The minus sign appears because the origin of the r-coordinate is at the observer. Considering first the leading edge of the wave packet, we integrate (19.21) from the source to observer,

$$\int_{t_1}^{t_0} \frac{c\,\mathrm{d}t}{R(t)} = -\int_r^0 \mathrm{d}r. \tag{19.22}$$

The end of the wave packet must travel the same comoving coordinate distance, since the r-coordinate is fixed to the galaxy for all time. Therefore

$$\int_{t_1+\Delta t_1}^{t_0+\Delta t_0} \frac{c\,\mathrm{d}t}{R(t)} = -\int_r^0 \mathrm{d}r.$$

Hence

$$\int_{t_1}^{t_0} \frac{c\,\mathrm{d}t}{R(t)} + \frac{c\,\Delta t_0}{R(t_0)} - \frac{c\,\Delta t_1}{R(t_1)} = \int_{t_1}^{t_0} \frac{c\,\mathrm{d}t}{R(t)}. \tag{19.23}$$

Since $R(t_0) = 1$,

$$\Delta t_0 = \frac{\Delta t_1}{R(t_1)}. \tag{19.24}$$

This is the cosmological expression for the phenomenon of *time dilation*. When distant galaxies are observed, $R(t_1) < 1$ and so phenomena are observed to take longer in the observer's frame of reference than they do in that of the source.

The result also provides an expression for the *redshift*. If $\Delta t_1 = \nu_1^{-1}$ is the period of the emitted waves and $\Delta t_0 = \nu_0^{-1}$ the observed period then

$$\nu_0 = \nu_1 R(t_1). \tag{19.25}$$

Rewriting (19.25) in terms of the redshift z,

$$z = \frac{\lambda_\mathrm{o} - \lambda_\mathrm{e}}{\lambda_\mathrm{e}} = \frac{\lambda_\mathrm{o}}{\lambda_\mathrm{e}} - 1 = \frac{\nu_1}{\nu_0} - 1,$$

that is,

$$1 + z = \frac{1}{R(t_1)}. \tag{19.26}$$

Thus, *redshift is a measure of the scale factor of the Universe at the epoch when the source emitted the radiation*. For example, when a galaxy is observed with redshift $z = 1$, the scale factor of the Universe when the light was emitted was $R(t) = 0.5$, that is, the separation between fundamental observers, or galaxies, was half its present value. We obtain no information, however, about *when* the light was emitted. Notice that, in cosmology, cosmological redshift does not really have anything to do with recessional velocities.

We can now find an expression for the comoving radial distance coordinate r,

$$r = \int_{t_1}^{t_0} \frac{c\,\mathrm{d}t}{R(t)} = \int_{t_1}^{t_0} (1+z)c\,\mathrm{d}t. \tag{19.27}$$

Thus, once we know how $R(t)$ varies with cosmic time t, we can find r by integration. Note that this emphasises the somewhat artificial nature of r – we need to know the kinematics of the Universe as described by $R(t)$ between the times when the light was emitted and received before we can work out r.

19.4.2 Hubble's law

In terms of proper distances, Hubble's law can be written $v = H_0 x$ and so

$$\frac{\mathrm{d}x}{\mathrm{d}t} = H_0 x. \tag{19.28}$$

Substituting from (19.14) $x = R(t)r$,

$$r\frac{\mathrm{d}R(t)}{\mathrm{d}t} = H_0 R(t) r, \tag{19.29}$$

that is,

$$H_0 = \dot{R}/R. \tag{19.30}$$

Since we measure Hubble's constant at the present epoch, $t = t_0$, $R = 1$,

$$H_0 = (\dot{R})_{t_0}. \tag{19.31}$$

Hubble's constant H_0 describes the present expansion rate of the Universe. Its value changes with cosmic epoch and can be defined at any cosmic time through the more general relation

$$H(t) = \dot{R}/R. \tag{19.32}$$

19.4.3 Angular diameters

The spatial element of the metric in the $\mathrm{d}\theta$-direction is the *proper length* d. Using (19.18), the proper length is related to the angular diameter $\Delta\theta$ by

$$d = R(t) D\,\Delta\theta = \frac{D\Delta\theta}{1+z},$$

so that

$$\Delta\theta = \frac{d(1+z)}{D}, \tag{19.33}$$

where we have introduced a *distance measure* $D = \Re \sin(r/\Re)$. For small redshifts, $z \ll 1, r \ll \Re$, (19.33) reduces to the Euclidean relation $d = r\Delta\theta$. Equation (19.33) can also be written

$$\Delta\theta = \frac{d}{D_{\mathrm{A}}}, \tag{19.34}$$

so that the relation between d and $\Delta\theta$ looks like the standard Euclidean relation; to do this, we have introduced another distance measure, $D_{\mathrm{A}} = D/(1+z)$, which is known as the *angular-diameter distance* and is often used in the literature.

19.4.4 Apparent intensities

Suppose a source at redshift z has luminosity $L(\nu_1)$, measured in W Hz^{-1}; $L(\nu_1)$ is the total energy emitted over 4π steradians per unit time per unit frequency interval. What is the flux density $S(\nu_0)$ of the source at the observing frequency ν_0, by which we mean the energy received per unit time, per unit area and per unit bandwidth (W m^{-2} Hz^{-1}), where $\nu_0 = R(t_1)\nu_1 = \nu_1/(1+z)$?

Suppose that in the proper-time interval Δt_1 the source emits $N(\nu_1)$ photons of energy $h\nu_1$ in the bandwidth $\Delta\nu_1$. Then, the luminosity of the source is

$$L(\nu_1) = \frac{N(\nu_1)h\nu_1}{\Delta\nu_1\,\Delta t_1}. \tag{19.35}$$

These photons are distributed over a 'sphere' centred on the source at epoch t_1, and when this 'shell' of photons propagates to the observer at the present epoch t_0, a certain fraction of them is intercepted by the telescope. The photons are observed at the present epoch t_0 with frequency $\nu_0 = R(t_1)\nu_1$ in a proper time interval $\Delta t_0 = \Delta t_1/R(t_1)$ and in the waveband $\Delta\nu_0 = R(t_1)\,\Delta\nu_1$.

Finally, we need to relate the diameter of our telescope Δl to the angular diameter which it subtends at the source at epoch t_1. The metric (19.18) provides the answer. The proper distance Δl refers to the present epoch at which $R(t_0) = 1$ and hence

$$\Delta l = D\,\Delta\theta \tag{19.36}$$

where $\Delta\theta$ is the angle measured by a fundamental observer located at the source. Notice the difference between the results (19.33) and (19.36). They are angular diameters measured in opposite directions along the light cone. The factor $1+z$ difference between them is part of a more general relation concerning angular-diameter measures along light cones, which is known as the *reciprocity theorem*.

The surface area of the telescope is $\pi\,\Delta l^2/4$ and the solid angle subtended by this area at the source is $\Delta\Omega = \pi\,\Delta\theta^2/4$. The number of photons incident upon the telescope in time Δt_0 is therefore $N(\nu_1)\,\Delta\Omega/4\pi$, but they are observed with frequency ν_0. Therefore, the flux density of the source, that is, the energy received per unit time, per unit area and per unit bandwidth, is

$$S(\nu_0) = \frac{N(\nu_1)h\nu_0\,\Delta\Omega}{4\pi\,\Delta t_0\,\Delta\nu_0(\pi/4)\Delta l^2}. \tag{19.37}$$

We can now relate the quantities in (19.37) to the properties of the source using the above relations and (19.24), (19.25) and (19.35):

$$S(\nu_0) = \frac{L(\nu_1)R(t_1)}{4\pi D^2} = \frac{L(\nu_1)}{4\pi D^2(1+z)}, \tag{19.38}$$

where $D = \Re \sin(r/\Re)$. If the spectra of sources are of power-law form, $L(\nu) \propto \nu^{-\alpha}$, this relation becomes

$$S(\nu_0) = \frac{L(\nu_0)}{4\pi D^2 (1+z)^{1+\alpha}}. \tag{19.39}$$

We can repeat the analysis for *bolometric* luminosities and flux densities. In this case, we consider the total energy emitted in a *finite* bandwidth $\Delta\nu_1$ and received in the bandwidth $\Delta\nu_0$, that is

$$L_{\rm bol} = L(\nu_1)\,\Delta\nu_1 = 4\pi D^2 (1+z)^2 S_{\rm bol}, \tag{19.40}$$

where the bolometric flux density is $S_{\rm bol} = S(\nu_0)\Delta\nu_0$. Therefore,

$$S_{\rm bol} = \frac{L_{\rm bol}}{4\pi D^2 (1+z)^2} = \frac{L_{\rm bol}}{4\pi D_{\rm L}^2}, \tag{19.41}$$

where $D_{\rm L}$ is defined below. The bolometric luminosity $L_{\rm bol}$ can be summed over any suitable bandwidth, so long as the corresponding redshifted bandwidth is used to measure the bolometric flux density at the present epoch, that is,

$$\sum_{\nu_0} S(\nu_0)\,\Delta\nu_0 = \frac{\sum_{\nu_1} L(\nu_1)\,\Delta\nu_1}{4\pi D^2 (1+z)^2} = \frac{\sum_{\nu_1} L(\nu_1)\,\Delta\nu_1}{4\pi D_{\rm L}^2}. \tag{19.42}$$

The quantity $D_{\rm L} = D(1+z)$ is called the *luminosity distance*, since this definition makes the relation between $S_{\rm bol}$ and $L_{\rm bol}$ look like an inverse square law.

We can rewrite the expression for the flux density of the source $S(\nu_0)$ in terms of the luminosity of the source at the observing frequency ν_0 as follows:

$$S(\nu_0) = \frac{L(\nu_0)}{4\pi D_{\rm L}^2} \left[\frac{L(\nu_1)}{L(\nu_0)} (1+z) \right]. \tag{19.43}$$

The last term in square brackets is known as the *K-correction*. It was introduced by the pioneer optical cosmologists in the 1930s in order to 'correct' the apparent magnitude of distant galaxies for the effects of redshift on the spectrum, when observations are made at a fixed observing frequency ν_0 and in a fixed bandwidth $\Delta\nu_0$. Taking logarithms and multiplying by -2.5, the term in brackets can be converted into a correction to the apparent magnitude of the galaxy,

$$K(z) = -2.5 \log_{10} \left[\frac{L(\nu_1)}{L(\nu_0)} (1+z) \right]. \tag{19.44}$$

Notice that this form of the K-correction is correct for *monochromatic* flux densities and luminosities. In the case of observations in the optical waveband, in which magnitudes are measured through broad standard filters, averages have to be taken over the spectral energy distributions of the objects within the appropriate spectral windows in the emitted and observed wavebands.

19.4.5 Number densities

The numbers of sources in the redshift range z to $z + \mathrm{d}z$ can be readily found using r because r is the radial proper distance defined at the present epoch, and hence the number we need is just the number of objects in the interval of radial comoving coordinate distance r to $r + \mathrm{d}r$ at the present day. The volume of a spherical shell of thickness $\mathrm{d}r$ at comoving radial distance coordinate r is the product of the surface area of a sphere in curved three-space $4\pi\Re^2 \sin^2(r/\Re)$, as can be seen from the θ-component of the metric (19.18), times the element $\mathrm{d}r$ of comoving coordinate distance, both measured at the present epoch, $R(t) = 1$.

$$\mathrm{d}V = 4\pi\Re^2 \sin^2(r/\Re)\,\mathrm{d}r \equiv 4\pi D^2\,\mathrm{d}r. \tag{19.45}$$

Therefore, if N_0 is the present space density of objects and their number is conserved as the Universe expands,

$$\mathrm{d}N = 4\pi N_0 D^2\,\mathrm{d}r. \tag{19.46}$$

Notice that, because of the way in which the comoving radial distance coordinate is defined, the expansion of the system of galaxies is automatically taken into account.

19.4.6 The age of the Universe

The differential relation needed to work out the age of the Universe is (19.21),

$$-\frac{c\,\mathrm{d}t}{R(t)} = \mathrm{d}r. \tag{19.47}$$

Therefore,

$$T_0 = \int_0^{t_0} \mathrm{d}t = \int_0^{r_{\max}} \frac{R(t)\mathrm{d}r}{c}, \tag{19.48}$$

where $r_{\max}$ is the comoving distance coordinate corresponding to $R = 0, z = \infty$.

19.4.7 Summary

The above results can be used to work out the relations between the intrinsic properties of distant objects and observables for any isotropic, homogeneous world model. Let us summarise the procedure as follows:

(i) First work out from theory the function $R(t)$ and the curvature of the space at the present epoch $\kappa = \Re^{-2}$.

(ii) Next, work out the *comoving radial distance coordinate* r as a function of redshift z using the integral

$$r = \int_{t_1}^{t_0} \frac{c\,\mathrm{d}t}{R(t)}. \tag{19.49}$$

Notice what this expression means – the proper distance interval $c\,\mathrm{d}t$ at epoch t is projected forward to the present epoch t_0 by dividing by the scale factor $R(t)$.

(iii) Now, find the *distance measure* D as a function of redshift z:

$$D = \Re \sin \frac{r}{\Re}. \tag{19.50}$$

(iv) If so desired, the *luminosity distance* $D_{\mathrm{L}} = D(1+z)$ and *angular diameter distance* $D_{\mathrm{A}} = D/(1+z)$ can be introduced.

(v) The number of objects $\mathrm{d}N$ in the redshift interval $\mathrm{d}z$ and solid angle Ω can be found from the expression

$$\mathrm{d}N = \Omega N_0 D^2\, \mathrm{d}r, \tag{19.51}$$

where N_0 is the number density *at the present epoch* of objects which partake in the uniform expansion of the Universe.

19.5 Historical interlude – steady state theory

To construct cosmological models we need to determine the function $R(t)$, and the standard procedure is to use general relativity as the starting point. Alternatively, one might attempt to use the properties of the Universe to determine the function $R(t)$ directly from observation. In the 1940s and 1950s, there was an attempt to derive the function $R(t)$ from very general considerations of what was known about the Universe at that time, and this resulted in the *steady state theory* of Hermann Bondi, Thomas Gold and Fred Hoyle. Although the theory now has few advocates, it is intriguing historically and illustrates the power of the tools which were developed in the last section.

The original motivation for the theory was the fact that, in the 1940s and early 1950s, the value of Hubble's constant H_0 was estimated to be very much larger than the values favoured today. Hubble had found values of $H_0 \approx 500$ km s^{-1} Mpc^{-1} and this resulted in a significant time-scale problem. We will show below that all the standard world models with zero cosmological constant have ages less than $T = H_0^{-1}$. For $H_0 \approx 500$ km s^{-1} Mpc^{-1}, the age T turned out to be 2×10^9 years. This was less than the age of the Earth, which is known to be about 4.6×10^9 years from radioactive dating.

Bondi, Hoyle and Gold came up with an ingenious solution to the time-scale problem by replacing the cosmological principle (Section 19.3) with what they called the *perfect cosmological principle*. This is the statement that the large-scale properties of the Universe as observed today should be *the same for all fundamental observers at all epochs*. In other words, the Universe should present an unchanging appearance for all fundamental observers for all time. This postulate immediately determines the kinematics of the Universe, since Hubble's constant becomes a fundamental constant of nature. Thus, from (19.32),

$$\dot{R}/R = H_0, \qquad \text{and hence} \qquad R \propto \exp H_0 t. \tag{19.52}$$

Normalising to the value $R(t_0) = 1$ at the present epoch t_0,

$$R(t) = \exp[H_0(t - t_0)], \tag{19.53}$$

where t is cosmic time.

Since according to this model matter is continually dispersing, matter must continously appear out of nothing in order that the Universe preserve the same appearance at all epochs. As a consequence, the theory is sometimes referred to as the theory of *continuous creation*. The proponents of the theory showed that the rate at which matter would be created in order to replace the dispersing matter was unobservably small, only one particle per m^3 every 300 000 years, in any terrestrial experiment.

Other features of the model follow immediately from the formalism we have already derived. Equation (19.17) states that $R_c(t) = \Re R(t)$ and so the curvature $\kappa = \Re^{-2}$ must be zero, the only value which does not change with time. Equally straightforward is the relation between the comoving radial distance coordinate r and redshift z. Recalling that $R(t) = (1+z)^{-1}$, the expression for r becomes

$$\begin{aligned} r &= \int_t^{t_0} \frac{c\,\mathrm{d}t}{R(t)} = \int_t^{t_0} \frac{c\,\mathrm{d}t}{\exp[H_0(t - t_0)]} \\ &= \frac{c}{H_0}\left[\frac{1}{R(t)} - 1\right] = \frac{cz}{H_0}. \end{aligned} \tag{19.54}$$

Remarkably, this relation is precisely Hubble's law, but now it holds good for all redshifts. Therefore, the metric for steady state cosmology is

$$\mathrm{d}s^2 = \mathrm{d}t^2 - \frac{R^2(t)}{c^2}[\mathrm{d}r^2 + r^2(\mathrm{d}\theta^2 + \sin^2\theta\,\mathrm{d}\varphi^2)], \tag{19.55}$$

with $R(t) = \exp[H_0(t - t_0)]$ and $r = cz/H_0$. Finally, the Universe is of infinite age, resolving the time-scale problem, and there is no initial singularity, unlike in the standard Big Bang scenario.

In the mid-1950s, Walter Baade showed that Hubble's constant had been significantly overestimated, and Humason, Mayall and Sandage derived a revised value of H_0, 180 km s^{-1} Mpc^{-1}, giving $T = H_0^{-1} = 5.6 \times 10^9$ years, which is comfortably greater than the age of the Earth. Subsequent studies have reduced the value of H_0 considerably further, as discussed in subsection 19.2.2.

Despite the fact that one of the strongest motivations for the steady state theory had been undermined, its simplicity and elegance was a profound attraction for a number of theorists, who saw in it a unique cosmological model, as opposed to the standard models which, as we will show, depend upon the mass density of the Universe and the cosmological constant and possess a singularity at $t = 0$.

The uniqueness of the theory made it highly testable. One of the first pieces of evidence against it was derived from the number counts of faint radio sources, which suggested that there were many more radio sources per unit comoving volume in the past than there are at the present day – such a result violates the basic premise of the steady state theory that the overall properties of the Universe should be unchanging with time. The really fatal blow to the theory was, however, the cosmic microwave background radiation. There is no natural origin for this radiation in the steady state picture – there do not exist sources which could mimic its black-body spectrum and account for its enormous energy density. However, these properties find a natural explanation in the Big Bang picture.

The story of the steady state theory is intriguing from the point of view of methodology in cosmology. The elegance of the model was achieved at a heavy price – the introduction of totally new physics involving the continuous creation of matter. For some scientists, this step was so objectionable that they ruled the theory out immediately. Even after the time-scale problem was resolved, others were prepared to investigate whether the theory could survive observational validation despite the introduction of new physics.

When the isotropic, homogeneous steady state model ran into conflict with the observations, alternatives were proposed according to which a Universe with continuous creation of matter could mimic an evolutionary Universe. In other words, the perfect cosmological principle was dropped but the concept of the creation of matter was maintained, with the possibility that the rate of creation of mass varies with time or location in the Universe. In my view, this takes the theory into very much more speculative territory. Logically, such a modified picture is quite feasible, but it has lost much of the attraction of the original picture, which was its simplicity and uniqueness. There comes a point when even an open-minded physicist has to make the value judgement as to whether the theory has strayed too far from conventional physics for the ideas to maintain credibility. The problem is particularly acute in cosmology, because we only have one Universe to observe.

Hoyle's approach to the steady state theory was somewhat different from that of Bondi and Gold, in that he believed that the concept of the continuous creation of matter was the fundamental aspect of the theory and he sought to give it a proper theoretical foundation. There is a delightful sequel to this story. On the occasion of his 80th birthday in 1995, I invited Hoyle to lecture to the Cavendish Physical Society. Although there had been a somewhat acrimonious dispute between Hoyle and Martin Ryle, the head of the Radio Astronomy Group in the Cavendish Laboratory, over the evidence of the radio source counts, Hoyle was delighted to accept the invitation, since he had given his first lecture on the theory at the Cavendish Physical Society in 1948. In order to describe the creation of matter, Hoyle introduced what he called the C-field, which had the property of corresponding to a negative-energy equation of state. As we will show in Section 19.7, this results in the exponential expansion of the Universe. Hoyle remarked wryly that his only mistake had been to call the field C rather than ψ. In the popular inflationary model of the early Universe, there is a phase of exponential increase of the scale factor $R(t)$ under the influence of a vacuum scalar field ψ, which performs exactly the same function dynamically as Hoyle's C-field.

19.6 The standard world models

In 1917, Einstein realised that in general relativity he had discovered a theory which enabled fully self-consistent models for the Universe as a whole to be constructed for the first time. The simplest models are those in which it is assumed that the Universe is isotropic, homogeneous and uniformly expanding. These assumptions result in an enormous simplification of Einstein's field equations. All the standard text-books show that they reduce to

$$\ddot{R} = -\frac{4\pi G R}{3}\left(\rho + \frac{3p}{c^2}\right) + \frac{1}{3}\Lambda R, \tag{19.56}$$

$$\dot{R}^2 = \frac{8\pi G\rho}{3}R^2 - \frac{c^2}{\Re^2} + \frac{1}{3}\Lambda R^2, \tag{19.57}$$

where $R \equiv R(t)$ is the scale factor as before, ρ is the inertial mass density of the matter and radiation content of the Universe and p is its pressure. Notice that the pressure term in (19.56) is a relativistic correction to the inertial mass density. Unlike normal pressure forces, which depend upon the gradient of the pressure, this term depends linearly on the pressure. $\Re$ is the radius of curvature of the geometry of the world model at the present epoch, and so the term $-c^2/\Re^2$ in (19.57) is simply a constant of integration.

The *cosmological constant* Λ has been included in (19.56) and (19.57). This term has had a chequered history. It was introduced by Einstein in 1917 in order to produce static solutions of the field equations, more than 10 years before Hubble showed that, in fact, the Universe is expanding uniformly. We will discuss its role in modern cosmology in Section 19.7. Let us first discuss the models with $\Lambda = 0$.

19.6.1 The standard dust models – the Friedman world models with $\Lambda = 0$

By *dust*, cosmologists mean a pressureless fluid, $p = 0$. It is convenient to refer the density of matter to its value at the present epoch ρ_0. Because of the conservation of mass, $\rho = \rho_0 R^{-3}$ and so (19.56) and (19.57) become, for $\Lambda = 0$,

$$\ddot{R} = -\frac{4\pi G\rho_0}{3R^2}, \tag{19.58}$$

$$\dot{R}^2 = \frac{8\pi G\rho_0}{3R} - \frac{c^2}{\Re^2}. \tag{19.59}$$

In 1934, Milne and McCrea first showed how this relation can be derived using Newtonian dynamics. Consider a galaxy at distance x from the Earth (Fig. 19.5). Its deceleration is due to the gravitational attraction of the matter inside the uniform sphere of radius x and density ρ centred on the Earth. According to Gauss's theorem for gravity, we can replace the mass $M = (4\pi/3)\rho x^3$ by a point mass at the centre of the sphere. Therefore, the deceleration of the galaxy is

$$m\ddot{x} = -\frac{GMm}{x^2} = -\frac{4\pi x\rho m}{3}. \tag{19.60}$$

The mass of the galaxy m cancels out on either side of the equation, showing that the deceleration refers to the dynamics of the Universe as a whole, rather than to any particular galaxy. Now, replace x by the comoving radial distance coordinate r, $x = Rr$ (equation (19.14)),

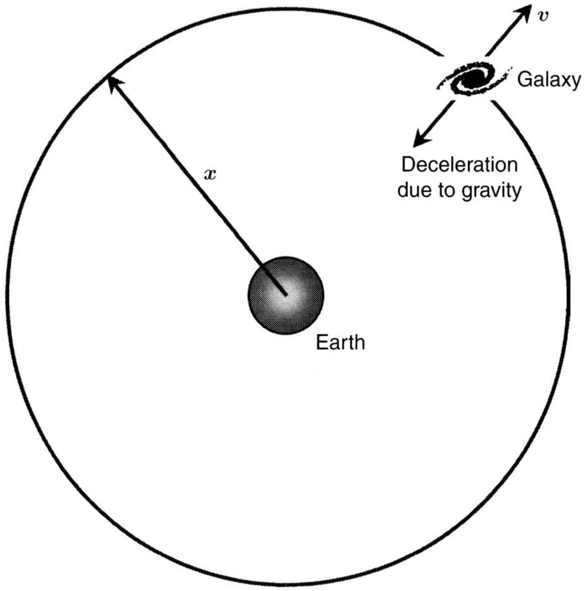

Figure 19.5: Illustrating the dynamics of isotropic world models. A galaxy at distance x is decelerated by the gravitational attraction of the matter within distance x of our Galaxy. Because of the assumption of isotropy, a fundamental observer on any galaxy participating in the uniform expansion would carry out the same calculation.

and express the density in terms of its value at the present epoch, $\rho = \rho_0 R^{-3}$. Therefore

$$\ddot{R} = -\frac{4\pi G\rho_0}{3R^2}, \tag{19.61}$$

which is identical to equation (19.58). Multiplying equation (19.61) by $\dot{R}$ and integrating,

$$\dot{R}^2 = \frac{8\pi G\rho_0}{3R} + \text{constant}. \tag{19.62}$$

This result is identical to (19.59) if we identify the constant with $-c^2/\Re^2$.

This analysis illustrates a number of important features of the world models of general relativity. First of all, note that because of the assumption of isotropy local physics is also global physics, which is why the simple Newtonian argument of Milne and McCrea works. The physics which defines the local behaviour of matter also defines its behaviour on the largest scales. For example, the curvature of space within one cubic metre is exactly the same as that on the scale of the Universe itself. A second point is that, although we might appear to have placed the Earth in a rather special position, a fundamental observer located on any galaxy would perform exactly the same calculation.

Third, notice that at no point in the argument did we ask over what physical scale the calculation was to be valid. In fact, this calculation describes correctly the dynamics of the Universe on scales that are greater than the *horizon scale*, which we take to be $r = ct$, the maximum distance between points that can be causally connected at the epoch t; the reason is the same as above – local physics is also global physics and so, if the Universe is

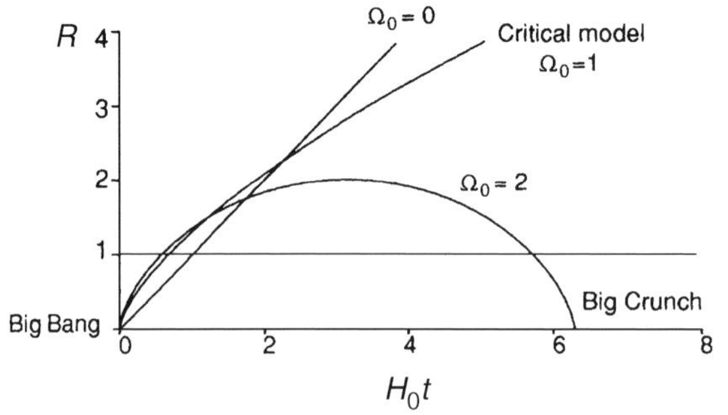

Figure 19.6: Comparison of the dynamics of different Friedman models with $\Lambda = 0$ characterised by the density parameter $\Omega_0 = \rho_0/\rho_c$; t is cosmic time and H_0 is Hubble's constant. The scale factor $R = 1$ at the present epoch. In this format, all the models have the same value of Hubble's constant at the present day.

set up in such a way that it has uniform density on scales far exceeding the horizon scale, the dynamics on these very large scales will be exactly the same as the local dynamics.

The solutions of Einstein's field equations were presented by A.A. Friedman in two papers published in 1922 and 1924. He died of typhoid during the civil war in Leningrad in 1925 and did not live to see the standard models of the Universe bear his name. Georges Lemaître rediscovered Friedman's solutions in 1927, and brought them to the notice of astronomers and cosmologists during the 1930s. The solutions of the field equations are appropriately referred to as the *Friedman models* of the Universe.

It is convenient to express the densities of the world models in terms of a *critical density* ρ_c, which is defined as $3H_0^2/(8\pi G)$, and then to refer the density of any specific model at the present epoch ρ_0 to this value through a *density parameter* $\Omega_0 = \rho_0/\rho_c$. Thus,

$$\Omega_0 = \frac{8\pi G\rho_0}{3H_0^2}. \tag{19.63}$$

The subscript 0 has been attached to Ω because the critical density ρ_c changes with cosmic epoch, as does ρ. Equation (19.59) therefore becomes

$$\dot{R}^2 = \frac{\Omega_0 H_0^2}{R} - \frac{c^2}{\Re^2}. \tag{19.64}$$

Several important results can be deduced from (19.64). At the present epoch, $t = t_0$, $R = 1$,

$$\Re = \frac{c/H_0}{(\Omega_0 - 1)^{1/2}} \quad \text{and so} \quad \kappa = \frac{\Omega_0 - 1}{(c/H_0)^2}. \tag{19.65}$$

Equation (19.65) shows that there is a one-to-one relation between the density of the Universe and its spatial curvature, one of the most beautiful results of the standard world models with $\Lambda = 0$. The solutions of (19.64) shown in Fig. 19.6 are the well-known relations between the dynamics and geometry of the Friedman world models, namely:

(i) The models with $\Omega_0 > 1$ have closed, spherical geometry and they collapse to an infinite density in a finite time, an event sometimes referred to as the 'big crunch'.

(ii) The models having $\Omega_0 < 1$ have open, hyperbolic geometries and expand forever. The velocity of expansion remains non-zero as the separation of galaxies tends to infinity.

(iii) The model with $\Omega_0 = 1$ separates the open from the closed models, that is, the collapsing models from those which expand forever. This model is often referred to as the *Einstein–de Sitter model* or the *critical model*. The velocity of expansion tends to zero as R tends to infinity. The model has a particularly simple variation of $R(t)$ with cosmic epoch,

$$R(t) = \left(\frac{3H_0 t}{2}\right)^{2/3}, \qquad \kappa = 0. \tag{19.66}$$

Another useful result is the function $R(t)$ for the empty world model $\Omega_0 = 0$: $R(t) = H_0 t$, $\kappa = -(H_0/c)^2$. This model is sometimes referred to as the *Milne model*. It is an interesting exercise to understand why it is that, in an empty world model, the global geometry of the Universe is hyperbolic. The reason is that, in this model, the test particles partaking in the universal expansion are undecelerated and any particle always has the same velocity relative to a given fundamental observer. Therefore, the cosmic times measured in different frames of reference are related by the Lorentz transform $t' = \gamma(t - vr/c^2)$, where $\gamma = (1 - v^2/c^2)^{-1/2}$. The key point is that the conditions of isotropy and homogeneity have to be applied at constant cosmic time t' in the frames of all fundamental observers. The Lorentz transform shows that this cannot be achieved in flat space, but it is uniquely satisfied in hyperbolic space with $\kappa = -(H_0/c)^2$ (see the appendix to this chapter).

The general solutions of (19.64) can be conveniently written in parametric form, as follows:

$$\text{for } \Omega_0 > 1, \qquad R = a(1 - \cos\theta), \qquad t = b(\theta - \sin\theta); \tag{19.67}$$

$$\text{for } \Omega_0 < 1, \qquad R = a(\cosh\varphi - 1), \qquad t = b(\sinh\varphi - \varphi). \tag{19.68}$$

Here

$$\text{for } \Omega_0 > 1, \qquad a = \frac{\Omega_0}{2(\Omega_0 - 1)}, \qquad b = \frac{\Omega_0}{2H_0(\Omega_0 - 1)^{3/2}}; \tag{19.69}$$

$$\text{for } \Omega_0 < 1, \qquad a = \frac{\Omega_0}{2(1 - \Omega_0)}, \qquad b = \frac{\Omega_0}{2H_0(1 - \Omega_0)^{3/2}}. \tag{19.70}$$

All the models tend towards the dynamics of the critical model at early times, but with a different constant; thus for $\theta \ll 1$, $\varphi \ll 1$,

$$R = \Omega_0^{1/3}\left(\frac{3H_0 t}{2}\right)^{2/3}. \tag{19.71}$$

Just as Hubble's constant H_0 measures the local expansion rate of the distribution of galaxies, so we can define the local deceleration of the Universe at the present epoch, $\ddot{R}(t_0)$. It is conventional to define the *deceleration parameter* q_0 to be the dimensionless deceleration at the present epoch through the expression

$$q_0 = -\left(\frac{\ddot{R}}{\dot{R}^2}\right)_{t_0}. \tag{19.72}$$

Substituting into (19.58) and including the definition of Ω_0, we find that the deceleration parameter q_0 is directly proportional to the density parameter Ω_0,

$$q_0 = \Omega_0/2, \tag{19.73}$$

provided that the cosmological constant Λ is zero.

An important result for many aspects of cosmology is the relation between redshift z and cosmic time t. From (19.64) and (19.65), and recalling that $R = (1+z)^{-1}$, we find

$$\frac{\mathrm{d}z}{\mathrm{d}t} = -H_0(1+z)^2(\Omega_0 z + 1)^{1/2}. \tag{19.74}$$

Cosmic time t measured from the Big Bang follows immediately by integration,

$$t = \int_0^t \mathrm{d}t = -\frac{1}{H_0}\int_\infty^z \frac{\mathrm{d}z}{(1+z)^2(\Omega_0 z + 1)^{1/2}}. \tag{19.75}$$

It is a straightforward calculation to evaluate this integral to find the present age of the universe from the Big Bang:

for $\Omega_0 > 1$,

$$t_0 = \frac{\Omega_0}{H_0(\Omega - 1)^{3/2}}\left[\sin^{-1}\left(\frac{\Omega_0 - 1}{\Omega_0}\right)^{1/2} - \frac{(\Omega_0 - 1)^{1/2}}{\Omega_0}\right];$$

for $\Omega_0 = 1$,

$$t_0 = \frac{2}{3H_0};$$

for $\Omega_0 < 1$,

$$t_0 = \frac{\Omega_0}{H_0(1 - \Omega_0)^{3/2}}\left[\frac{(1 - \Omega_0)^{1/2}}{\Omega_0} - \sinh^{-1}\left(\frac{1 - \Omega_0}{\Omega_0}\right)^{1/2}\right].$$

Thus, for the critical model $\Omega_0 = 1$ the present age of the Universe is $(2/3)H_0^{-1}$ and for the empty model, $\Omega_0 = 0$, it is H_0^{-1}. The above results justify the statement made in Section 19.5 that the age of the Universe for all world models with $\Lambda = 0$ is $\leq H_0^{-1}$.

Just as it is possible to define Hubble's constant at any epoch by $H = \dot{R}/R$, so we can define a density parameter Ω at any epoch through the definition $\Omega = 8\pi G\rho/(3H^2)$. Since $\rho = \rho_0 R^{-3} = \rho_0(1+z)^3$, it follows that

$$\Omega H^2 = \frac{8\pi G}{3}\rho_0(1+z)^3. \tag{19.76}$$

It is a straightforward exercise to show that this relation can be rewritten

$$1 - \frac{1}{\Omega} = (1+z)^{-1}\left(1 - \frac{1}{\Omega_0}\right). \tag{19.77}$$

This is an important result because it shows that, whatever the value of Ω_0 now, Ω tends to the value 1 in the distant past, because $(1+z)^{-1}$ becomes very small at large redshifts. It is remarkable that at the present day the Universe is close to the value $\Omega_0 = 1$ (see Section 19.9). If the value of Ω had been significantly different from 1 in the very distant past, then Ω_0 would be very widely different indeed from 1 now, as can be seen from equation (19.77). The fact that the curvature of space κ is close to zero now results in what is referred to as the *flatness problem*. Our Universe must have been *very finely tuned indeed* to the value $\Omega = 1$ in the distant past, for it to have ended up close to $\Omega_0 = 1$ now. Some cosmologists argue that because it is so remarkable that our Universe is close to $\Omega_0 = 1$ now the only reasonable value for Ω_0 is precisely 1. Proponents of the inflationary picture of the early Universe have good reasons why this should be the case. We will revisit this problem in subsection 9.7.6 once we have completed the discussion of models with $\Lambda \neq 0$ and radiation-dominated models.

We can now find expressions for the comoving radial coordinate distance r and the 'distance measure' D. We recall that

$$\mathrm{d}r = \frac{c\,\mathrm{d}t}{R(t)} = -c\,\mathrm{d}t(1+z) = \frac{c\,\mathrm{d}z}{H_0(1+z)(\Omega_0 z+1)^{1/2}}. \tag{19.78}$$

Integrating from redshifts 0 to z, the expression for r is

$$r = \frac{2c}{H_0(\Omega_0 - 1)^{1/2}}\left[\tan^{-1}\left(\frac{\Omega_0 z + 1}{\Omega_0 - 1}\right)^{1/2} - \tan^{-1}(\Omega_0 - 1)^{-1/2}\right]. \tag{19.79}$$

Finally, D is found by evaluating $D = \Re \sin(r/\Re)$, where $\Re$ is given by the expression (19.65). It is a straightforward calculation to show that

$$D = \frac{2c}{H_0\Omega_0^2(1+z)}\left\{\Omega_0 z + (\Omega_0 - 2)[(\Omega_0 z + 1)^{1/2} - 1]\right\}. \tag{19.80}$$

This is the famous formula first derived by Mattig in 1959. Although it has been derived for spherical geometry, it has the great advantage of being correct for all values of Ω_0. It can be used in all the expressions listed in Section 19.4 for relating intrinsic properties to observables.

19.6.2 Models with non-zero cosmological constant

These models have sprung into prominence over recent years since there is now convincing evidence, from studies of the power spectrum of fluctuations in the cosmic microwave background radiation and from the magnitude–redshift relation for Type Ia supernovae, that the cosmological constant Λ is non-zero.

Let us consider Einstein's field equation (19.56), including the cosmological constant, for the case of dust-filled universes:

$$\ddot{R} = -\frac{4\pi G R\rho}{3} + \frac{\Lambda R}{3} = -\frac{4\pi G\rho_0}{3R^2} + \frac{\Lambda R}{3}. \tag{19.81}$$

Inspection of (19.81) gives insight into the meaning of the cosmological constant: even in an empty universe, $\rho = p = 0$, there is still a net force acting on a test particle. In the words of Ya.B. Zeldovich, if Λ is positive, the term may be thought of as the 'repulsive effect of the vacuum', the repulsion being relative to an absolute geometrical frame of reference. There is no obvious interpretation of this term in terms of classical physics.

There is, however, a natural interpretation in the context of quantum field theory. In the modern picture of a vacuum, there are zero-point fluctuations associated with the zero-point energies of all quantum fields. The stress–energy tensor of a vacuum has a negative-energy equation of state, $p = -\rho c^2$. This pressure may be thought of as a *tension*, rather than a pressure. When such a vacuum expands, the work done $p\,\mathrm{d}V$ in expanding from V to $V + \mathrm{d}V$ is $-\rho c^2\,\mathrm{d}V$ so that, during the expansion, the mass density of the negative energy field remains constant. Carroll, Press and Turner[4] show how the theoretical value of Λ can be evaluated using simple concepts from quantum field theory, and they find the mass density of the repulsive field to be $\rho_\mathrm{v} = 10^{95}$ kg m^{-3}. This is a problem. This mean mass density associated with the cosmological constant is about 10^{120} times greater than the best values derived from observation, which correspond to $\rho_\mathrm{v} \approx 6 \times 10^{-27}$ kg m^{-3}.

This is an enormous discrepancy and constitutes one of the major unsolved problems of theoretical cosmology. Its importance for the modern picture of the early universe can be understood as follows. In the inflationary model of the very early Universe, forces of exactly this type are believed to cause the exponential expansion. If the inflationary picture is adopted, we have to explain why ρ_v decreased by a factor of at least 10^{120} at the end of the inflationary era, since it turns out that the value of ρ_v observed today is of the same order of magnitude as ρ_0; there is, however, no convincing physical explanation either for the smallness of ρ_v or for the near equality of the two density parameters.

Nevertheless, it is now quite natural to believe that there are forces in nature which provide Zeldovich's 'repulsion of the vacuum', and to associate a certain mass density ρ_v with the energy density of the vacuum at the present epoch. The term *dark energy* is often used to describe the vacuum energy density associated with ρ_v. It is convenient to rewrite the dynamical equations in terms of a density parameter Ω_v associated with ρ_v as follows. We begin with (19.56) but include the density ρ_v and pressure p_v of the vacuum fields in place of the Λ term:

$$\ddot{R} = -\frac{4\pi GR}{3}\left(\rho_\mathrm{m} + \rho_\mathrm{v} + \frac{3p_\mathrm{v}}{c^2}\right), \tag{19.82}$$

where the density of ordinary mass is ρ_m. Since $p_\mathrm{v} = -\rho_\mathrm{v} c^2$,

$$\ddot{R} = -\frac{4\pi GR}{3}(\rho_\mathrm{m} - 2\rho_\mathrm{v}). \tag{19.83}$$

But, as was shown above, as the Universe expands $\rho_\mathrm{m} = \rho_0/R^3$ while ρ_v remains constant, and so

$$\ddot{R} = -\frac{4\pi G\rho_0}{3R^2} + \frac{8\pi G\rho_\mathrm{v} R}{3}. \tag{19.84}$$

Equations (19.81) and (19.84) have exactly the same dependence of the 'cosmological term' upon the scale factor R, and so we can formally identify the cosmological constant with the vacuum mass density by

$$\Lambda = 8\pi G\rho_\mathrm{v}. \tag{19.85}$$

At the present epoch, $R = 1$ and so

$$\ddot{R}(t_0) = -\frac{4\pi G\rho_0}{3} + \frac{8\pi G\rho_v}{3}. \tag{19.86}$$

Introducing a density parameter Ω_v associated with ρ_v, we obtain

$$\Omega_v = \frac{8\pi G\rho_v}{3H_0^2}, \qquad \text{and so} \qquad \Lambda = 3H_0^2\Omega_v. \tag{19.87}$$

We can therefore find a new relation between the deceleration parameter q_0, Ω_0 and Ω_v, from (19.86) and (19.87):

$$q_0 = \frac{\Omega_0}{2} - \Omega_v. \tag{19.88}$$

The dynamical equations (19.56) and (19.57) can be rewritten

$$\ddot{R} = -\frac{\Omega_0 H_0^2}{2}\frac{1}{R^2} + \Omega_v H_0^2 R, \tag{19.89}$$

$$\dot{R}^2 = \frac{\Omega_0 H_0^2}{R} - \frac{c^2}{\Re^2} + \Omega_v H_0^2 R^2. \tag{19.90}$$

Substituting the values $R = 1$ and $\dot{R} = H_0$ at the present epoch into (19.90), the curvature of space is related to Ω_0 and Ω_v by

$$\frac{c^2}{\Re^2} = H_0^2[(\Omega_0 + \Omega_v) - 1], \tag{19.91}$$

$$\kappa = \frac{1}{\Re^2} = \frac{(\Omega_0 + \Omega_v) - 1}{c^2/H_0^2}. \tag{19.92}$$

Thus, the condition that the spatial sections are flat Euclidean space is

$$\Omega_0 + \Omega_v = 1. \tag{19.93}$$

We recall that the radius of curvature R_c of the model changes with scale factor as $R_c = R\Re$, and so if the space curvature κ is zero now, it must have been the same at all times in the past. The same flatness problem as that discussed in subsection 19.6.1 applies in these models as well since, at large redshifts, $R \ll 1$, the dynamics are dominated by the term in Ω_0 in (19.89). A complete discussion is given in subsection 19.7.6.

Let us now investigate the dynamics of the models with $\Lambda \neq 0$.

Models with $\Lambda < 0$; $\Omega_v < 0$. Models with negative cosmological constant are not of great interest, because the net effect is to incorporate an attractive force, in addition to gravity, which slows down the expansion of the Universe. The one difference from the models with $\Lambda = 0$ is that, no matter how small the values of Λ and Ω_0, the universal expansion is eventually reversed, as may be seen from inspection of (19.89).

Models with $\Lambda > 0$; $\Omega_v > 0$. These models are much more interesting, because the positive cosmological constant leads to a repulsion which opposes the attractive force of gravity. In each of these models, there is a minimum rate of expansion, which occurs at scale factor

$$R_{\min} = \left(\frac{4\pi G\rho_0}{\Lambda}\right)^{1/3} = \left(\frac{\Omega_0}{2\Omega_v}\right)^{1/3}. \tag{19.94}$$

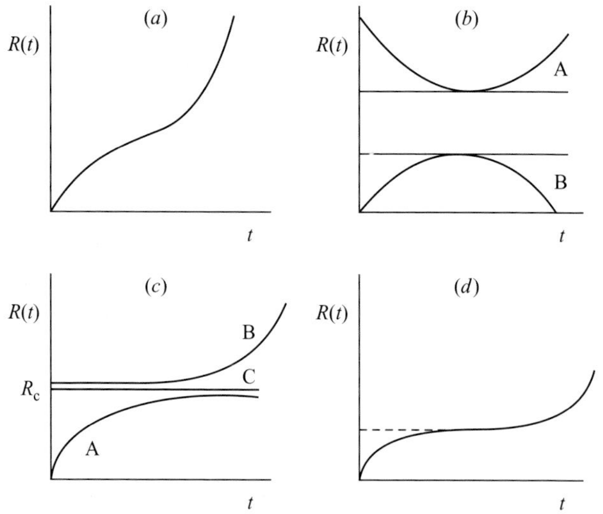

Figure 19.7: Examples of the dynamics of world models with $\Lambda \neq 1$ (Bondi 1960[3]). Models (*a*) and (*d*) are referred to as Lemaître models. Model (*c*) is the Eddington–Lemaître model. In the Einstein static model $R(t)$ is constant for all time. The other models are described in the text.

The minimum rate of expansion is, from (19.90),

$$\dot{R}^2_{\rm min} = \Lambda^{1/3}\left(\frac{3\Omega_0 H_0^2}{2}\right)^{2/3} - \frac{c^2}{\Re^2} = 3H_0^2\left(\frac{\Omega_{\rm v}\Omega_0^2}{4}\right)^{1/3} - \frac{c^2}{\Re^2}. \tag{19.95}$$

If the right-hand side of (19.95) is greater than zero then the dynamical behaviour shown in Fig. 19.7(*a*) is found. For large values of R, the dynamics become those of the de Sitter universe,

$$R(t) \propto \exp\left[(\Lambda/3)^{1/2}\, t\right] = \exp\left(\Omega_{\rm v}^{1/2} H_0 t\right). \tag{19.96}$$

If the right-hand side of (19.95) is less than zero, there exists a range of scale factors for which no solution exists, and it can be shown readily that the function $R(t)$ then has two branches, as illustrated in Fig. 19.7(*b*). For the branch labelled A, the dynamics are dominated by the Λ term and the asymptotic solutions are described by

$$R(t) \propto \exp\left[\pm(\Lambda/3)^{1/2}\, t\right]. \tag{19.97}$$

For branch B, the Universe never expands to sufficiently large values of R for the repulsive effect of the Λ-term to prevent the Universe collapsing.

The most interesting cases are those for which $\dot{R}_{\rm min} \approx 0$. The case $\dot{R}_{\rm min} = 0$ is known as the *Eddington–Lemaître model* and is illustrated in Fig. 19.7(*c*). The interpretation of these models is either A, the Universe expanded from an origin at some finite time in the past, and will eventually attain a stationary state in the infinite future, or B, the Universe is expanding away from a stationary solution in the infinite past. The stationary state C is unstable because, if it is perturbed, the Universe moves either onto branch B, or onto the collapsing variant of branch A. In Einstein's static Universe of 1917, the stationary phase occurs at the present day.

In general, the value of Λ corresponding to $\dot{R}_{\min} = 0$ is

$$\Lambda = \frac{3\Omega_0 H_0^2}{2}(1+z_c)^3, \qquad \text{so that} \qquad \Omega_v = \frac{\Omega_0}{2}(1+z)^3, \tag{19.98}$$

where z_c is the redshift at the stationary state.

The Eddington–Lemaître models can have ages much greater than H_0^{-1}; in fact, in the extreme Eddington–Lemaître model with $\dot{R}_{\min} = 0$ in the infinite past, the Universe is infinitely old. A closely related set of models, also with ages greater than H_0^{-1}, are known as the *Lemaître models*, and have Λ-terms such that the value of $\dot{R}_{\min}$ is just greater than zero. An example of this type of model is shown in Fig. 19.7(*d*). It has a long 'coasting phase' when the velocity of expansion of the Universe was very small.

19.6.3 Λ-models with zero curvature

Recent observations of the power spectrum of fluctuations in the cosmic microwave background radiation have strongly suggested that the spatial curvature of the Universe is very close to zero. In general, the expressions for the dynamics, the comoving radial distance coordinate and the age of models with non-zero cosmological constant have to be determined by numerical integration, but there are convenient solutions in the zero-curvature case. Just as in subsection 19.6.1, we need the expression for $\mathrm{d}z/\mathrm{d}t$. From (19.90), the expression which replaces (19.74) is

$$\frac{\mathrm{d}z}{\mathrm{d}t} = -H_0(1+z)\left[(1+z)^2(\Omega_0 z + 1) - \Omega_v z(z+2)\right]^{1/2}. \tag{19.99}$$

Cosmic time t measured from the Big Bang follows immediately by integration,

$$t = \int_0^t \mathrm{d}t = -\frac{1}{H_0}\int_\infty^z \frac{\mathrm{d}z}{(1+z)\left[(1+z)^2(\Omega_0 z+1) - \Omega_v z(z+2)\right]^{1/2}}. \tag{19.100}$$

For models with zero curvature $\Omega_0 + \Omega_v = 1$, and hence (19.99) becomes

$$\frac{\mathrm{d}z}{\mathrm{d}t} = -H_0(1+z)[\Omega_0(1+z)^3 + (1-\Omega_0)]^{1/2}; \tag{19.101}$$

from this an expression for $\mathrm{d}t$ can be substituted into the integral (19.27) to find the radial comoving distance coordinate, since $\kappa = 0$. The cosmic-time–redshift relation is found by integration of the zero-curvature form of (19.100):

$$t = \int_0^t \mathrm{d}t = -\frac{1}{H_0}\int_\infty^z \frac{\mathrm{d}z}{(1+z)[\Omega_0(1+z)^3 + (1-\Omega_0)]^{1/2}}. \tag{19.102}$$

The parametric solution of (19.102) is

$$t = \frac{2}{3H_0\Omega_v^{1/2}}\ln\left(\frac{1+\cos\theta}{\sin\theta}\right), \qquad \tan\theta = \left(\frac{\Omega_0}{\Omega_v}\right)^{1/2}(1+z)^{3/2}. \tag{19.103}$$

The present age of the Universe for $\kappa = 0$ is then found by setting $z = 0$:

$$t_0 = \frac{2}{3H_0\Omega_v^{1/2}}\ln\left[\frac{1+\Omega_v^{1/2}}{(1-\Omega_v)^{1/2}}\right]. \tag{19.104}$$

This relation illustrates how it is possible to find Friedman models with ages greater than H_0^{-1} with flat spatial sections. For example, if $\Omega_v = 0.9$ and $\Omega_0 = 0.1$, the age of the world model is $1.28 H_0^{-1}$.

19.7 The thermal history of the Universe

Once the dynamical framework of the world models has been set up, the thermal history of the Universe can be found by application of the laws of thermodynamics. This analysis leads to important tests of the standard picture. Let us first consider the dynamics of radiation-dominated universes.

19.7.1 *Radiation-dominated universes*

For a gas of photons, massless particles or a relativistic gas in the ultrarelativistic limit $E \gg mc^2$, the pressure p is related to the energy density u by $p = \frac{1}{3}u$. For the case of radiation, the inertial mass density ρ_{rad} is related to u by $u = \rho_{\text{rad}} c^2$.

If $N(\nu)$ is the number density of photons of energy $h\nu$ then the energy density of radiation is found by summing over all frequencies:

$$u = \sum_{\nu} h\nu N(\nu). \tag{19.105}$$

The number density of photons varies as $N = N_0 R^{-3} = N_0(1+z)^3$, and the energy of each photon changes by the redshift factor $h\nu = h\nu_0(1+z)$. Therefore, the variation in the energy density of radiation with cosmic epoch is

$$u = \sum_{\nu_0} h\nu_0 N_0(\nu_0)(1+z)^4 = u_0(1+z)^4 = u_0 R^{-4}. \tag{19.106}$$

For black-body radiation, the energy density is given by the Stefan–Boltzmann law, $u = aT^4$, and its spectral energy density by the Planck distribution

$$u(\nu)\,d\nu = \frac{8\pi h\nu^3}{c^3} \frac{1}{e^{h\nu/kT} - 1}\,d\nu. \tag{19.107}$$

It immediately follows that, for black-body radiation, the temperature T varies with redshift as $T = T_0(1+z)$ and its spectrum as

$$\begin{aligned} u(\nu_1)\,d\nu_1 &= \frac{8\pi h\nu_1^3}{c^3} \frac{1}{e^{h\nu_1/kT_1} - 1}\,d\nu_1, \\ &= \frac{8\pi h\nu_0^3}{c^3} \frac{1}{e^{h\nu_0/kT_0} - 1}(1+z)^4\,d\nu_0, \\ &= (1+z)^4 u(\nu_0)\,d\nu_0. \end{aligned} \tag{19.108}$$

Thus, a black-body spectrum preserves its form as the Universe expands, but the radiation temperature changes as $T = T_0(1+z)$ and the frequency of each photon as $\nu = \nu_0(1+z)$.

Another way of thinking about these results is in terms of the adiabatic expansion of a gas of photons. The ratio of specific heats for radiation, and for a relativistic gas in the ultrarelativistic limit, is $\gamma = 4/3$. As shown in subsection 9.3.3, in an adiabatic expansion,

$$T \propto V^{-(\gamma-1)} = V^{-1/3} \propto R^{-1} = (1+z),$$

which is exactly the same as the above result.

The variations of p and ρ with R can now be substituted into Einstein's field equations (19.56) and (19.57). Setting the cosmological constant $\Lambda = 0$,

$$\ddot{R} = -\frac{8\pi G u_0}{3c^2}\frac{1}{R^3}, \qquad \dot{R}^2 = \frac{8\pi G u_0}{3c^2}\frac{1}{R^2} - \frac{c^2}{\Re^2}. \tag{19.109}$$

At early epochs, the constant term $c^2/\Re^2$ can be neglected and so, integrating,

$$R = \left(\frac{32\pi G u_0}{3c^2}\right)^{1/4} t^{1/2} \quad \text{or} \quad u = u_0 R^{-4} = \left(\frac{3c^2}{32\pi G}\right) t^{-2}. \tag{19.110}$$

The dynamics of the radiation-dominated models, $R \propto t^{1/2}$, depend only upon the total inertial mass density present in relativistic or massless forms. Notice that we need to add together the contributions to u of all forms of relativistic matter and radiation at the appropriate epochs.

19.7.2 The matter and radiation content of the Universe

The cosmic microwave background radiation provides by far the largest contribution to the energy density of radiation in intergalactic space. Comparing the inertial mass density in this radiation and in the matter in the Universe, we find

$$\frac{\rho_{\text{rad}}}{\rho_{\text{m}}} = \frac{aT^4(z)}{\Omega_0 \rho_c (1+z)^3 c^2} = \frac{2.6 \times 10^{-5}(1+z)}{\Omega_0 h^2}, \tag{19.111}$$

where we have written $h = H_0/(100 \text{ km s}^{-1} \text{ Mpc}^{-1})$ to take account of uncertainties in the precise value of Hubble's constant H_0. The Universe has therefore been *matter dominated* since redshift $z = 4 \times 10^4\, \Omega_0 h^2$ with dynamics described by $R \propto t^{2/3}$ provided $\Omega_0 z \gg 1$. At redshifts $z \geq 4 \times 10^4\, \Omega_0 h^2$, the Universe was *radiation dominated*, with dynamics described by (19.110), $R \propto t^{1/2}$.

The present *photon-to-baryon ratio* is another key cosmological parameter. Since the microwave background radiation contains the bulk of the photon energy, adopting $T = 2.728$ K, we find

$$\frac{N_\gamma}{N_{\text{B}}} = \frac{3.6 \times 10^7}{\Omega_{\text{B}} h^2}, \tag{19.112}$$

where N_{B} and Ω_{B} refer to the number density and density parameter of baryons respectively. If photons are neither created nor destroyed during the expansion of the Universe, this ratio is an invariant and is a measure of the factor by which photons outnumber the baryons in the Universe at the present epoch. It is also proportional to the specific entropy per baryon during the radiation-dominated phases of the expansion.

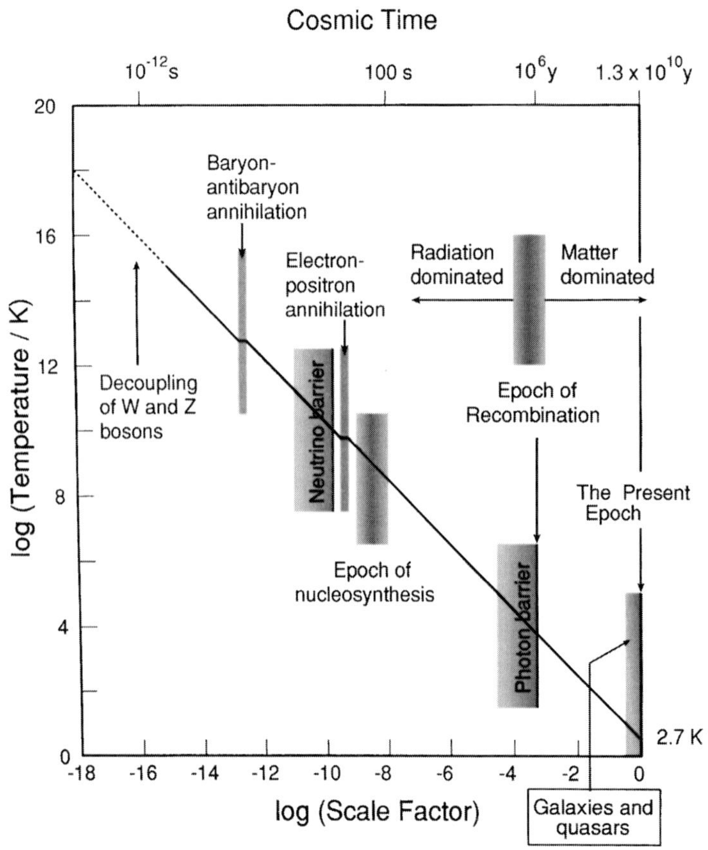

Figure 19.8: The thermal history of the Universe. The radiation temperature decreases as $T \propto R^{-1}$, except for small jumps as different particle–antiparticle pairs annihilate. Various important epochs in the standard Big Bang model are indicated. An approximate time scale is shown along the top of the diagram. The photon and neutrino barriers are shown. In the standard picture, the Universe is opaque for electromagnetic radiation and neutrinos respectively, prior to these epochs.

The resulting thermal history of the Universe is shown in Fig. 19.8. Certain epochs are of special significance and we now deal with these in a little more detail.

19.7.3 The epoch of recombination

At a redshift $z \approx 1500$, the temperature of the cosmic microwave background radiation was $T \approx 4000$ K and there were then sufficient photons with energies $h\nu \geq 13.6$ eV in the Wien region of the Planck distribution to ionise all the neutral hydrogen in the intergalactic medium. Let us demonstrate why this took place at such a relatively low temperature.

I leave it as an exercise to show that the number density of photons in the Wien region of the Planck distribution, $h\nu \gg kT$, with energies $h\nu \geq E$ is

$$n(\geq E) = \int_{E/h}^{\infty} \frac{8\pi\nu^2}{c^3} \frac{\mathrm{d}\nu}{\mathrm{e}^{h\nu/kT}} = \frac{1}{\pi^2}\left(\frac{2\pi kT}{hc}\right)^3 \mathrm{e}^{-x}(x^2 + 2x + 2), \tag{19.113}$$

where $x = E/kT$. The total number density of photons in a black-body spectrum at temperature T is

$$N = 0.244 \left(\frac{2\pi kT}{hc} \right)^3 \ \mathrm{m}^{-3}, \tag{19.114}$$

and so the fraction of the photons with energies greater than E is

$$\frac{n(\geq E)}{N} = \frac{\mathrm{e}^{-x}(x^2 + 2x + 2)}{0.244\pi^2}. \tag{19.115}$$

In the simplest approximation the intergalactic gas is ionised if there are as many ionising photons as hydrogen atoms; that is, from (19.112), for every $3.6 \times 10^7/(\Omega_\mathrm{B} h^2)$ photons in the cosmic microwave background radiation there need only be one with energy greater than 13.6 eV to ionise the gas.

For illustrative purposes, let us take the ratio (19.112) to be one part in 10^9. Then, the solution of

$$\frac{1}{10^9} = \frac{\mathrm{e}^{-x}(x^2 + 2x + 2)}{0.244\pi^2} \tag{19.116}$$

is $x = E/kT \approx 26.5$. This is an important result. There are so many photons relative to hydrogen atoms that the temperature of the radiation can be 26.5 times less than that found by setting $E = kT$ and yet there will still be sufficient photons with energy $E \geq 13.6$ eV to ionise the gas. Since $kT = 13.6$ eV at 150 000 K, the intergalactic gas is expected to be fully ionised at temperature $T \approx 150\,000/26.5 \approx 5600$ K. Since the present temperature of the cosmic microwave background radiation is 2.728 K, the hydrogen in the pregalactic gas was fully ionised at a scale factor $R \approx 2.728/5600 = 5 \times 10^{-4}$, that is, a redshift $z \approx 2000$.

It is intriguing that the same type of calculation appears in a number of different guises in astrophysics – the nuclear reactions which power the Sun take place at a much lower temperature than expected, the temperature at which regions of ionised hydrogen become fully ionised is only about 10 000 K, light nuclei are destroyed in the early Universe at much lower temperatures than would be expected. In all these cases, the high-energy tails of the Planck and Maxwell distributions contain large numbers of particles or photons with energies very much greater than the mean.

Detailed calculations show that the pregalactic gas was 50% ionised at a redshift $z_\mathrm{r} \approx 1500$, and this epoch is referred to as the *epoch of recombination*. The reason for this name is that the pregalactic gas was fully ionised prior to this epoch and so, when the clocks are run forward, the universal plasma recombined at this epoch.

The most important consequence of this calculation is that the Universe was opaque to Thomson scattering by free electrons at early epochs. Detailed calculations show that the optical depth of the intergalactic gas for Thomson scattering was unity at a redshift close to 1000 and then became very large as soon as the intergalactic hydrogen was fully ionised at redshift z_r. As a result, the Universe beyond $z \approx 1000$ is unobservable, because any photons originating from larger redshifts were scattered many times before they propagated to the Earth. Consequently, there is a *photon barrier* at a redshift of 1000 beyond which we cannot obtain information directly, using photons. If there was no further scattering of the photons between the epoch of recombination and the present epoch, the redshift of about 1000 was the *last scattering surface* for the cosmic microwave background radiation. The very small spatial fluctuations in the image of the microwave sky seen in Fig. 19.2 were imprinted on this last scattering surface at $z \approx 1000$.

19.7.4 The horizon problem

The *particle horizon* is defined to be the maximum distance over which causal communication could have taken place by a particular epoch t. In other words, it is the distance a light signal could have travelled from the origin of the Big Bang at $t = 0$ by the epoch t. The radial comoving distance coordinate r corresponding to the distance travelled by a light signal from the origin of the Big Bang to the epoch t is, (19.21),

$$r = \int_0^t \frac{c\,\mathrm{d}t}{R(t)} = \int_\infty^z (1+z)c\,\mathrm{d}t. \qquad (19.117)$$

To find the horizon scale at the epoch corresponding to redshift z, we simply scale the comoving radial distance coordinate r by the scale factor $R(t) = (1+z)^{-1}$. Thus, the particle horizon $r_\mathrm{H}(t)$ at epoch t, corresponding to redshift z, is

$$r_\mathrm{H}(t) = R(t)\int_0^t \frac{c\,\mathrm{d}t}{R(t)} = \frac{1}{1+z}\int_\infty^z (1+z)c\,\mathrm{d}t. \qquad (19.118)$$

At the epoch of recombination, it is safe to use the matter-dominated solutions with $\Lambda = 0$ and neglect the curvature term. Then, all the Friedman models tend towards the dynamics of the critical model:

$$R(t) = \frac{1}{1+z} = \Omega_0^{1/3}\left(\frac{3H_0 t}{2}\right)^{2/3}. \qquad (19.119)$$

Therefore, the particle horizon is

$$r_\mathrm{H}(t) = \frac{1}{1+z}\int_\infty^z (1+z)c\,\mathrm{d}t = \frac{2c}{H_0\Omega_0^{1/2}}(1+z)^{-3/2}. \qquad (19.120)$$

Using (19.119), this result can be written in terms of cosmic time t, $r_\mathrm{H}(t) = 3ct$. This result makes physical sense since it is expected that the typical distance which light could travel by epoch t would be of order ct. The factor 3 takes account of the fact that the fundamental observers were closer together at earlier epochs and so greater distances than ct could be causally connected then.

We can use these results to illustrate the origin of the *horizon problem* for the standard Friedman models with $\Omega_\mathrm{v} = 0$. Let us work out the angle θ_H which the particle horizon subtends at the last scattering surface according to an observer at the present epoch. At redshift $z = 1000$ we can safely use the standard matter-dominated solutions of Friedman's equation in the limit $\Omega_0 z \gg 1$. From (19.80), it can be seen that, in the limit of very large redshifts, the distance measure D tends to the constant value $2c/(H_0\Omega_0)$. Therefore, from (19.33) the angular scale of the horizon length at $z = 1000$ is

$$\theta_\mathrm{H} = \frac{r_\mathrm{H}(t)(1+z)}{D} = \frac{\Omega_0^{1/2}}{(1+z)^{1/2}} = 1.8\Omega_0^{1/2} \quad \text{degrees}. \qquad (19.121)$$

This result means that, according to the standard Friedman picture, regions of the Universe separated by an angle of more than $1.8\Omega_0^{1/2}$ degrees on the sky could not have been in causal contact on the last scattering surface. Why then is the cosmic microwave background radiation so uniform over the whole sky to a precision of about one part in 10^5? This is the *horizon problem* and, in the standard picture, it has to be assumed that the remarkable isotropy of the Universe was an aspect of the initial conditions from which the Universe expanded.

19.7.5 The epoch of equality of matter and radiation inertial mass densities

It follows from (19.111) that matter and radiation made equal contributions to the inertial mass density at redshift $z = 4 \times 10^4\ \Omega_0 h^2$ and that at larger redshifts the Universe was radiation dominated. After the intergalactic gas recombined at redshifts $z \leq 1000$, there was negligible coupling between the neutral matter and the photons of the microwave background radiation. Since the matter and radiation were not thermally coupled, they cooled independently, the ratio of specific heats γ being 5/3 for the gas and 4/3 for the radiation. The result was adiabatic cooling, which depends upon the scale factor R as $T_{\rm m} \propto R^{-2}$ and as $T_{\rm rad} \propto R^{-1}$ for the matter and radiation respectively. We therefore expect that the matter cooled much more rapidly than the radiation during the post-recombination epochs. This was not the case prior to recombination, however, because then the matter and radiation were strongly coupled by *Compton scattering*: the optical depth of the intergalactic plasma for Thomson scattering was so large that the small energy transfers which took place between the photons and the electrons by Compton scattering were sufficient to maintain the matter at the same temperature as the radiation.

Once the thermal history of the intergalactic gas has been determined, the variation of the speed of sound with cosmic epoch can be found. The adiabatic speed of sound $c_{\rm s}$ is given by

$$c_{\rm s}^2 = \left(\frac{\partial p}{\partial \rho}\right)_S,$$

where the subscript S means 'at constant entropy'. The complication is that, from the epoch when the energy densities of matter and radiation were equal to beyond the epoch of recombination and the subsequent neutral phase, the dominant contributions to p and ρ changed dramatically as the Universe changed from being radiation dominated to matter dominated; the coupling between the matter and the radiation became weaker and finally the plasma recombined at redshifts $z \approx 1000$. For the epochs when the matter and radiation were closely coupled by Compton scattering, the square of the sound speed can be written

$$c_{\rm s}^2 = \frac{(\partial p/\partial T)_{\rm rad}}{(\partial \rho/\partial T)_{\rm rad} + (\partial \rho/\partial T)_{\rm m}}, \tag{19.122}$$

where the partial derivatives are taken at constant entropy. This expression reduces to the following result:

$$c_{\rm s}^2 = \frac{c^2}{3}\left(\frac{4\rho_{\rm rad}}{4\rho_{\rm rad} + 3\rho_{\rm m}}\right). \tag{19.123}$$

Thus, in the radiation-dominated phases, $z \geq 4 \times 10^4\ \Omega_0 h^2$, the speed of sound tended to the relativistic sound speed, $c_s = c/\sqrt{3}$. At smaller redshifts, the sound speed decreased as the contribution of the inertial mass density of the thermal matter became more important.

19.7.6 The flatness problem revisited

In subsection 19.6.1, we gave a simple treatment of the *flatness problem*, but we now give a more complete treatment, including both the vacuum energy and the radiation in the Friedman equations. Let us insert these additional energy densities into (19.57) and (19.90), obtaining

$$\dot{R}^2 = \frac{8\pi G R^2}{3}(\rho_{\rm m} + \rho_{\rm rad} + \rho_{\rm v}) - \frac{c^2}{\Re^2}. \tag{19.124}$$

Each of these mass densities can be written in terms of the density parameters at any epoch as follows:

$$\Omega_{\rm m} = \frac{8\pi G\rho_{\rm m}}{3H^2}, \qquad \Omega_{\rm rad} = \frac{8\pi G\rho_{\rm rad}}{3H^2}, \qquad \Omega_{\rm v} = \frac{8\pi G\rho_{\rm v}}{3H^2} \tag{19.125}$$

where $H = \dot{R}/R$. Each of these terms has a different dependence upon scale factor as the Universe evolves:

$$\rho_{\rm m} = \rho_{\rm m}(0)\, R^{-3}, \qquad \rho_{\rm rad} = \rho_{\rm rad}(0)\, R^{-4}, \qquad \rho_{\rm v} = \rho_{\rm v}(0), \tag{19.126}$$

where the zero in brackets means at the present epoch. Thus, the complete dynamical equation can be written

$$\dot{R}^2 = H_0^2 R^2 \left[\Omega_{\rm m}(0)\, R^{-3} + \Omega_{\rm rad}(0)\, R^{-4} + \Omega_{\rm v}(0)\right] - \frac{c^2}{\Re^2}. \tag{19.127}$$

Let us now repeat the analysis which led to (19.77), but including the $\Omega_{\rm rad}$ and $\Omega_{\rm v}$ terms. Dividing (19.124) through by $\dot{R}^2$, we find

$$1 = \Omega_{\rm m} + \Omega_{\rm rad} + \Omega_{\rm v} - \frac{c^2}{\Re^2 \dot{R}^2}, \tag{19.128}$$

$$1 = \Omega - \frac{c^2}{\Re^2 H^2 R^2}, \tag{19.129}$$

$$H^2 R^2(\Omega - 1) = \frac{c^2}{\Re^2}, \tag{19.130}$$

where $\Omega = \Omega_{\rm m} + \Omega_{\rm rad} + \Omega_{\rm v}$. At the present epoch, this expression reduces to

$$H_0^2[\Omega(0) - 1] = \frac{c^2}{\Re^2}. \tag{19.131}$$

To determine how Ω changes with cosmic epoch, we first need to know how $H(t)$ changes. For the matter-dominated era, (19.76) leads to (19.77) and the same result applies in the present case as well. In the radiation-dominated era, the inertial mass density is dominated by relativistic matter and fields, for which the energy and mass densities vary as R^{-4},

and so

$$\Omega = \frac{8\pi G\rho}{H^2} = \frac{8\pi G\rho(0)}{H^2}R^{-4} = \Omega(0)\frac{H_0^2}{H^2}R^{-4}. \tag{19.132}$$

Equating (19.130) and (19.131), and substituting for H^2/H_0^2 from (19.132), we find

$$1 - \frac{1}{\Omega} = \frac{1}{(1+z)^2}\left[1 - \frac{1}{\Omega(0)}\right]. \tag{19.133}$$

The same *flatness problem* has recurred. If the density parameter Ω had differed by even a tiny amount from unity at very large redshifts, its value today would certainly not be close to $\Omega(0) = 1$.

19.7.7 Early epochs

To complete this brief thermal history of the Universe, let us summarise the physical processes which were important at earlier epochs.

- Extrapolating back to redshifts $z \approx 10^8$, the radiation temperature was $T \approx 3 \times 10^8$ K and the background photons attained γ-ray energies, $\epsilon = kT = 25$ keV. At this high temperature, the photons in the Wien region of the Planck spectrum were energetic enough to dissociate light nuclei such as helium and deuterium. At earlier epochs, all nuclei were dissociated.
- At redshifts $z > 10^9$, as a result of electron–positron pair production from the thermal background radiation the Universe was flooded with electron–positron pairs, one pair for every pair of photons present in the Universe today. When the clocks are run forward from an earlier epoch, the electrons and positrons annihilate at $z \approx 10^9$ and their energy is transferred to the photon field – this accounts for a small discontinuity at $R = 10^{-9}$ in the temperature history, as shown in Fig. 19.8.
- At a slightly earlier epoch, $z \approx 10^{9.5}$, the opacity of the Universe for weak interactions was unity. This results in a *neutrino barrier*, similar to the photon barrier at $z \sim 1000$.
- We can extrapolate even further back in time to $z \approx 10^{12}$, when the temperature of the background radiation was sufficiently high for baryon–antibaryon pair production to take place from the thermal background. Just as in the case of the epoch of electron–positron pair production, the Universe was flooded with baryons and antibaryons, one pair for every pair of photons present in the Universe now. Again, there is a small discontinuity in the temperature history at this epoch.

The process of extrapolation further and further back into the mists of the early Universe continues until we run out of physics which has been established in the laboratory. Most particle physicists would probably agree that the standard model of elementary particles has been tried and tested to energies of at least 100 GeV, and so we can probably trust laboratory physics back to epochs as early as 10^{-6} s, although the more conservative among us might be happier to accept 10^{-3} s. How far back one is prepared to extrapolate beyond these times is largely a matter of taste. The most ambitious theorists have no hesitation in extrapolating back to the very earliest Planck eras, $t_P \sim (Gh/c^5)^{1/2} = 10^{-43}$ s, when the relevant physics was certainly very different from the physics of the Universe from redshifts of about 10^{12} to the present day. These ideas are motivated by the need to understand some of the fundamental problems associated with the standard Big Bang.

19.8 Nucleosynthesis in the early Universe

One of the reasons why the standard Big Bang model is taken so seriously is its remarkable success in accounting for the observed abundances of the light elements by *primordial nucleosynthesis*. The details of the arguments given below are described in more detail in my book *Galaxy Formation*.[1]

Consider a particle of mass m at very high temperatures, $kT \gg mc^2$. If the time scales of the interactions which maintain this species in thermal equilibrium with all the other species present at temperature T are shorter than the age of the Universe at that epoch, the equilibrium number densities of the particle and its antiparticle are given by statistical mechanics:

$$N = \overline{N} = \frac{4\pi g}{h^3} \int_0^\infty \frac{p^2 \, \mathrm{d}p}{\mathrm{e}^{E/kT} \pm 1}, \tag{19.134}$$

where g is the statistical weight of the particle, p its momentum and the $\pm$ sign depends upon whether the particles are fermions (+) or bosons (−). For (i) photons, (ii) nucleons and electrons and (iii) neutrinos respectively, these are:

$$\begin{aligned}
&\text{(i)} \quad g = 2, \qquad N = 0.244 \left(\frac{2\pi kT}{hc}\right)^3 \mathrm{m}^{-3}, \qquad u = aT^4;\\
&\text{(ii)} \quad g = 2, \qquad N = 0.183 \left(\frac{2\pi kT}{hc}\right)^3 \mathrm{m}^{-3}, \qquad u = \tfrac{7}{8} aT^4;\\
&\text{(iii)} \quad g = 1, \qquad N = 0.091 \left(\frac{2\pi kT}{hc}\right)^3 \mathrm{m}^{-3}, \qquad u = \tfrac{7}{16} aT^4.
\end{aligned}$$

To find the total energy density ε, we add all the equilibrium energy densities:

$$\varepsilon = \chi(T) a T^4. \tag{19.135}$$

This is the expression for the energy density which should be included in (19.110) to determine the dynamics of the early Universe.

When the particles become non-relativistic, $kT \ll mc^2$, and the abundances of the different species are still maintained by interactions between the particles, the non-relativistic limit of (19.134) gives an equilibrium number density

$$N = g \left(\frac{mkT}{h^2}\right)^{3/2} \exp\left(-\frac{mc^2}{kT}\right). \tag{19.136}$$

Thus, once the particles become non-relativistic they no longer contribute to the inertial mass density which determines the rate of expansion of the Universe.

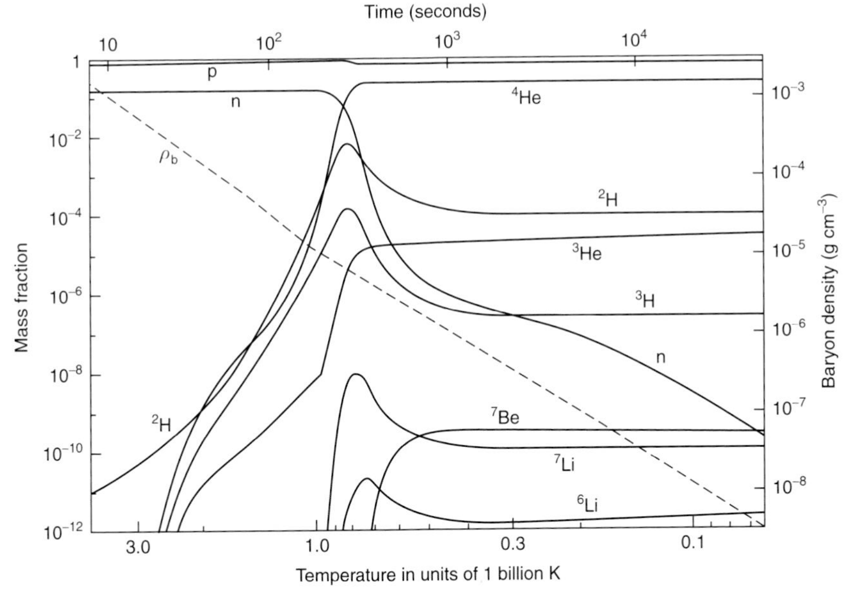

Figure 19.9: An example of the time and temperature evolution of the abundances of the light elements in the early evolution of the Big Bang model of the Universe.[5] Before 10 seconds, no significant synthesis of light elements took place because deuterium ^{2}H is destroyed by hard γ-rays in the Wien region of the black-body radiation spectrum. As the temperature decreases, more and more deuterium survives and synthesis of the light elements becomes possible through the reactions (19.140). The synthesis of elements such as D (^{2}H), ^{3}He, ^{4}He, ^{7}Li and ^{7}Be was completed after about 15 minutes.

At redshifts $z < 10^{12}$, the neutrons and protons are non-relativistic, $kT \ll mc^2$, and their equilibrium abundances are maintained by the electron–neutrino weak interactions

$$e^+ + n \to p + \bar{\nu}_e, \qquad \nu_e + n \to p + e^- . \tag{19.137}$$

For the neutrons and protons, the values of g are the same and so the relative abundance of neutrons and protons is

$$\left[\frac{n}{p}\right] = \exp\left(-\frac{\Delta mc^2}{kT}\right), \tag{19.138}$$

where Δmc^2 is the mass–energy difference between the neutron and the proton.

This abundance ratio 'freezes out' when the neutrino interactions can no longer maintain the equilibrium abundances of neutrons and protons. The condition for freezing out is that the time scale of the weak interactions becomes greater than the age of the Universe. Detailed calculations show that the time scales for the expansion of the Universe and the decoupling of the neutrinos were the same when the Universe was almost precisely one second old, at a temperature of 10^{10} K. At that time, the neutron fraction, as determined by (19.138), was

$$\left[\frac{n}{n+p}\right] = 0.21. \tag{19.139}$$

The neutron fraction decreased very slowly after this time. Detailed calculations show that after 300 s the neutron fraction had fallen to 0.123.[6] At this epoch, the bulk of the formation of the light elements took place. Almost all the neutrons combined with protons to form ^{4}He nuclei, so that for every pair of neutrons a helium nucleus was formed. The reactions involved are

$$
\begin{array}{ll}
\mathrm{p}+\mathrm{n} \rightarrow \mathrm{D}+\gamma, & \mathrm{p}+\mathrm{D} \rightarrow {}^3\mathrm{He}+\gamma, \\
\mathrm{n}+\mathrm{D} \rightarrow {}^3\mathrm{H}+\gamma, & \mathrm{p}+{}^3\mathrm{H} \rightarrow {}^4\mathrm{He}+\gamma, \\
\mathrm{n}+{}^3\mathrm{He} \rightarrow {}^4\mathrm{He}+\gamma, & \mathrm{d}+\mathrm{d} \rightarrow {}^4\mathrm{He}+\gamma, \\
{}^3\mathrm{He}+{}^3\mathrm{He} \rightarrow {}^4\mathrm{He}+2\mathrm{p}. &
\end{array}
\qquad (19.140)
$$

The results of detailed computations of the evolution of the light element abundances during the first few hours of the Universe are shown in Fig. 19.9. Most of the nucleosynthesis took place at temperatures less than about 1.2×10^9 K since, at higher temperatures, the deuterons would have been destroyed by the γ-rays of the background radiation. The binding energy of deuterium is $E_\mathrm{B} = 2.23$ MeV and this energy is equal to kT at $T = 2.6 \times 10^{10}$ K. However, just as in the case of the recombination of the intergalactic gas (see subsection 19.7.3), the photons far outnumbered the nucleons and it is only when the temperature of the expanding gas had decreased to about 26 times less than this temperature that the number of dissociating photons became fewer than the number of nucleons. Although some of the neutrons had decayed spontaneously by this time, the bulk of them survived and so, according to the above calculation, the predicted helium-to-hydrogen mass ratio is just twice the neutron fraction: $[^4\mathrm{He}/\mathrm{H}] \approx 0.25$. Detailed studies show that, in addition to ^{4}He, which is always produced with an abundance of about 23% to 25%, there are traces of the light elements deuterium (D), helium-3 (^{3}He) and lithium-7 (^{7}Li) (see Fig. 19.9).

These are quite remarkable results. It had been a great mystery why the abundance of helium is so high wherever it can be observed in the Universe. Its chemical abundance always appears to be greater than about 23%. In addition, there had always been a problem in understanding how deuterium could have been synthesised. It is a very fragile nucleus and is destroyed rather than created in stellar interiors. The same argument applies to the isotope ^{3}He of helium and to ^{7}Li. Precisely these elements, however, were synthesised in the early stages of the Big Bang. In stellar interiors, nucleosynthesis takes place in roughly thermodynamic equilibrium conditions over very long time scales, whereas in the early stages of the Big Bang the 'explosive' nucleosynthesis is all over in a few minutes.

The deuterium and ^{3}He abundances provide strong constraints upon the present baryon density of the Universe. The observed deuterium abundance relative to hydrogen is always found to be $[\mathrm{D}/\mathrm{H}] \approx 1.5 \times 10^{-5}$. Therefore, since after the epochs corresponding to $z = 10^8$ we only know of ways of destroying deuterium rather than creating it, this figure provides a firm lower limit to the amount of deuterium which could have been produced by primordial nucleosynthesis. Wagoner's simulations show that the predicted abundance of deuterium is a strong function of the present baryon density of the Universe, whereas the abundance of helium is remarkably constant (Fig. 19.10).

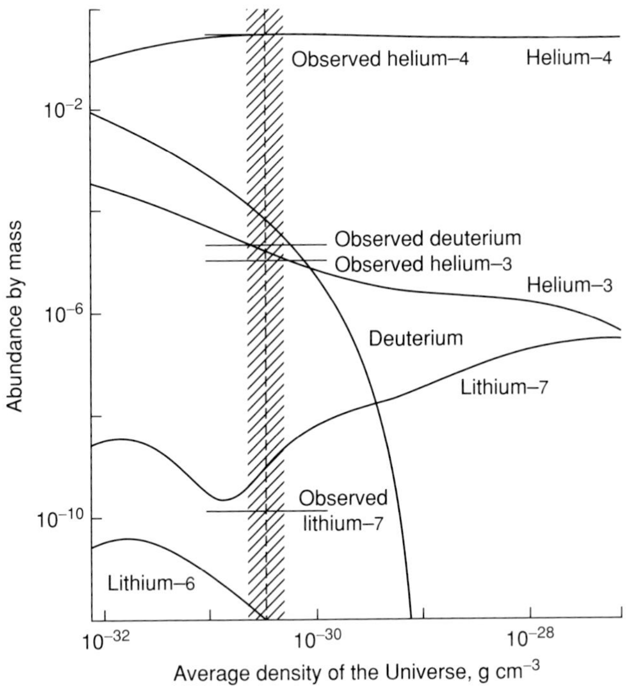

Figure 19.10: The predicted primordial abundances of the light elements compared with their observed values. The horizontal marker for 'Observed helium-4' largely coincides with the line showing the variation in helium-4 abundance. The present density of baryonic matter in the world model is plotted along the abscissa. The shaded band shows the range for Ω_B, or ρ_B, that could explain the abundances of the light elements. The observed abundances of the elements are in good agreement with a model having $\Omega_B \approx 0.015h^{-2}$.

The reasons for this can be understood as follows. The amount of helium synthesised is determined by the equilibrium abundances of protons and neutrons as the Universe cools down, which is primarily determined by the thermodynamics of the expanding radiation-dominated Universe. However, the abundance of deuterium depends upon the number density of nucleons: if the Universe has a high baryon number density then essentially all the deuterons were converted into ^{4}He, whereas, if the Universe has a low density then not all the deuterions were converted. The same argument applies to ^{3}He. Thus, the deuterium and ^{3}He abundances set an upper limit to the present baryon density of the Universe. The figure which results from recent analyses is

$$\Omega_B h^2 \leq 0.015 \tag{19.141}$$

where h is the value of Hubble's constant in units of 100 km s^{-1} Mpc^{-1}. Thus, even adopting a small value of $h = 0.5$, $\Omega_B < 0.06$ and so *baryonic matter cannot close the Universe.* Indeed, as will be discussed in the next section, it cannot even account for the amount of dark matter present in the Universe.

One of the more remarkable results of these studies is that they have enabled limits to be set to the number of neutrino species which could have been present during the epochs when the light elements were synthesised. If there had been more than three species of neutrino present, they would have contributed to the inertial mass density of massless particles and so would have speeded up the early expansion of the Universe (see (19.110)). The decoupling of the neutrinos would have taken place at a higher temperature, resulting in an overproduction of helium. From this type of cosmological argument, it was shown that there cannot be more than three families of neutrinos, a result subsequently confirmed from the width of the decay spectrum of $W^{\pm}$ and Z^0 bosons measured by LEP at CERN.

19.9 The best-buy cosmological model

The analyses of the preceding sections lead to four independent pieces of evidence, all of which are consistent with the standard Big Bang scenario:

- Hubble's law describing the linearity of the relation between the recession velocities of galaxies and their distances;
- the remarkable isotropy and black-body spectrum of the cosmic microwave background radiation;
- the formation of the light elements deuterium, ^{3}He, ^{4}He and ^{7}Li in the early phases of the Big Bang;
- the similarity of the expansion time scale of the Universe and the ages of the oldest objects we can measure in the Universe. These studies involve the radioactive dating of long-lived isotopes, according to the science of nucleocosmochronology, and age estimates for the oldest stellar systems known to us, the globular clusters.

These arguments have persuaded cosmologists that the standard Big Bang model provides the most convincing framework for studies of the origin and evolution of the Universe and its contents. The elaboration of that story requires a whole book in itself. To the above list of successes of the model, I would now add the reconciliation of the large-scale structure of the Universe as observed in the distribution of galaxies with the observed fluctuation power spectrum of the cosmic microwave background radiation, which originated at the last scattering surface at a redshift $z \approx 1000$. This argument shows convincingly that large-scale structures in the Universe must have been formed by the gravitational collapse of small density perturbations which originated in the very early Universe.

The upshot of a vast amount of observational and theoretical effort is that there is now a remarkable consensus among cosmologists about the preferred values of the cosmological parameters introduced in the course of this study. In summary, a set of cosmological parameters which would be widely accepted is as follows.

- The value of *Hubble's constant* from the Hubble Space Telescope key project, which used Cepheid variables to measure the cosmic distance scale, is $H_0 = 70 \pm 7$ km s^{-1} Mpc^{-1}, where the error is at the 1σ level. This result is consistent with independent estimates, many of which find values in the range 60–75 km s^{-1} Mpc^{-1}.
- The *curvature of space* can be measured from the first maximum in the power spectrum of the fluctuations in the cosmic microwave background radiation, which is expected on an angular scale of about 1°. This maximum is associated with the horizon scale on

the last scattering surface at $z = 1000$, which acts as an excellent 'rigid rod' at a known redshift (see subsection 19.7.4). It has been detected convincingly in the Boomerang, Maxima and DASI experiments. The most recent result from the DASI experiment is $\Omega_0 + \Omega_v = 1.04 \pm 0.06$; we recall that $\Omega_0 + \Omega_v$ is unity for flat space.

- The average *mass density in the Universe* at the present epoch is dominated by the dark matter, the nature of which is still unknown but which is certainly present in galaxies and clusters of galaxies. Recent studies of the average amount of dark matter estimated from the infall of galaxies into large-scale structures have found values of $\Omega_m \approx 0.3$, consistent with studies of individual clusters of galaxies. The result from the DASI experiment is $\Omega_m h^2 = 0.14 \pm 0.04$.
- Studies of distant supernovae of Type 1A have shown that they are the best 'standard candles' yet discovered for extending the apparent-magnitude–redshift relation from small to large redshifts. These have provided convincing evidence that the deceleration parameter of the Universe is negative. Values of $\Omega_v \approx 0.7$ and $\Omega_m \approx 0.3$ provide a good fit to the observations.
- The value of the density parameter in baryonic matter found from studies of primordial nucleosynthesis, $\Omega_B h^2 \approx 0.010$–$0.015$, is in encouraging agreement with the values found from studies of the fluctuation spectrum of the cosmic microwave background radiation; the DASI experiment found a value $\Omega_B h^2 = 0.022 \pm 0.004$.

The upshot of these studies is that we seem to be living in an accelerating Universe, in which the expansion is driven by the energy of the vacuum fields Ω_v. The mass density of the dark matter is about one third of the critical cosmological density and that of the baryonic matter about an order of magnitude smaller than that of the dark matter. This new orthodoxy brings with it a new range of problems which are stretching the imaginations of cosmologists and astrophysicists. These include the following.

- Why do we live in a Universe with zero curvature, when the Universe could have been set up with any value? This is referred to as the *flatness problem*.
- Why is the Universe so uniform on the very largest scales? As shown in subsection 19.7.4, this uniformity leads to the *horizon problem* – according to the standard picture, there was not time in the Universe for one region of the sky to communicate with another on scales greater than 1° on the last scattering surface.
- Why is the Universe made out of matter rather than an equal mixture of matter and antimatter? This is the *baryon asymmetry problem*.
- What is the nature of the dark matter?
- Why does the vacuum density parameter take the value $\Omega_v \approx 0.7$? The natural value which comes from quantum field theory is about 10^{120} greater than this value.
- Why is the density parameter Ω_0 of the same magnitude as the vacuum density parameter Ω_v at the present epoch, when they have very different dependences upon the scale factor R?
- What is the origin of the fluctuations from which large-scale structures in the Universe formed?

The most popular solution to a number of these problems is to assume that the Universe went through a phase of very rapid expansion in the very early Universe, the *inflationary model of the early Universe*. Let us assume that the scale factor, R, increased exponentially

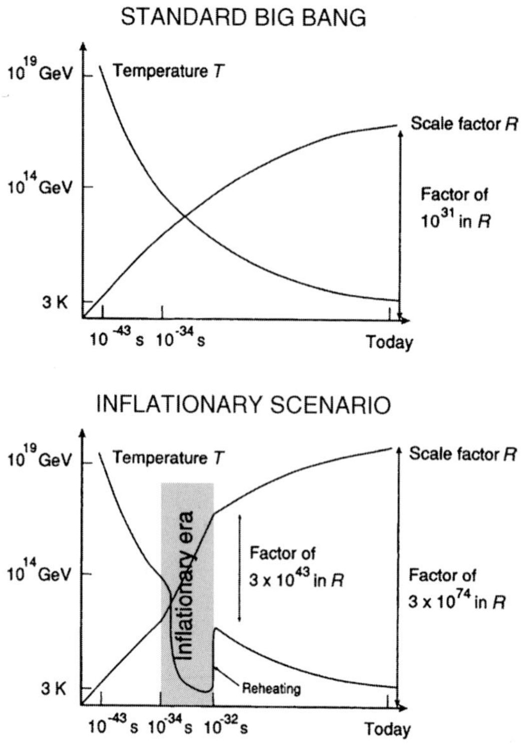

Figure 19.11: Comparison of the evolution of the scale factor and temperature in the standard Big Bang and inflationary cosmologies.

with time, as $R \propto e^{t/T}$; the exponential expansion continued for a certain time and then the Universe switched over to the standard radiation-dominated phase. Consider a tiny region of the early Universe expanding under the influence of this exponential expansion. Particles within the region are initially very close together and so are in causal communication with each other. Suppose that before the inflationary expansion begins the region has a physical scale less than the particle horizon, and so there has been time for the small region to attain a uniform, homogeneous state. The region then expands exponentially so that neighbouring parts of the region are driven to such distances that they can no longer communicate with each other by light signals – causally connected regions are swept beyond their local horizons by the inflationary expansion.

In the most popular version of the inflationary scenario for the very early Universe, the exponential expansion is associated with the breaking of symmetries of grand unified theories of elementary particles at very high energies. According to these theories, at high enough energies the strong and electroweak forces are unified and it is only at lower energies that they appear as distinct forces. The characteristic energy at which the grand unification phase transition is expected to take place is $E \sim 10^{14}$ GeV, only about 10^{-34} seconds after the Big Bang, which is also the characteristic e-folding time for the exponential expansion. This energy scale is commonly referred to as the GUT scale. In a typical realisation of the inflation picture, the exponential inflationary expansion took place from this time until the Universe was about 100 times older. At the end of this period, there was an enormous release of energy associated with the phase transition and this heated up the Universe to a very high temperature; the dynamics then became those of the standard radiation-dominated Big Bang.

Let us put some figures into this calculation. Over the interval from 10^{-34} seconds to 10^{-32} seconds, the scale factor of the Universe increased exponentially by a factor of about $e^{100} \approx 10^{43}$. The horizon scale at the beginning of this period was only $r \approx ct \approx 3 \times 10^{-26}$ m and this was inflated to a dimension 3×10^{17} m by the end of the period of inflation. This dimension then scaled as $t^{1/2}$, as in the standard radiation-dominated Universe, so that the region would have expanded to a size of 3×10^{42} m by the present day – this dimension far exceeds the present scale of the Universe, which is about 10^{26} m. Thus, our present Universe would have arisen from a tiny region in the very early Universe which was much smaller than the horizon scale at that time. This guarantees that our present Universe would be isotropic on the large scale, resolving the horizon problem. This history of the inflationary Universe is compared with the standard Friedman picture in Fig. 19.11.

A further consequence of the exponential expansion is that it straightens out the geometry of the early Universe, however complicated it may have been to begin with. The radius of curvature of the geometry scales as $R_c(t) = \Re R(t)$, and so the radius of curvature of the geometry within the tiny region is inflated to dimensions vastly greater than the present size of the Universe – the geometry of the inflated region is driven towards flat Euclidean geometry. This process ends at the end of the inflationary era. When the Universe transforms from the exponentially expanding inflationary state into the radiation-dominated phase, the geometry is Euclidean, $\kappa = 0$ and consequently the Universe must have $\Omega = 1$.

These are just some of the issues facing the present generation of cosmologists, both observers and theorists. They complete the circle in a very pleasing way. Our story began with the contribution that astronomy played in understanding our place in the Universe, which led to the understanding of the laws of gravity, and we end with the cosmological challenges facing physicists at the dawn of the twenty-first century. These are problems worthy of successors to Tycho Brahe, Kepler, Galileo and Newton.

19.10 References

1 Longair, M.S. (1997). *Galaxy Formation*. Berlin: Springer-Verlag.
2 The physics of the Sunyaev–Zeldovich effect and its applications in cosmology are described in R.A. Sunyaev and Ya.B. Zeldovich (1980), *Ann. Rev. Astron. Astrophys.*, **18**, 537–60.
3 Bondi, H. (1960). *Cosmology*. Cambridge: Cambridge University Press.
4 Carroll, S.M., Press, W.H. and Turner, E.L. (1992). *Ann. Rev. Astr. Astrophys.*, **30**, 499.
5 Wagoner, R.V. (1973). *Astrophys. J.*, **148**, 3.
6 Weinberg, S. (1972). *Gravitation and Cosmology*. New York: John Wiley and Co.

Appendix to Chapter 19
The Robertson–Walker metric for an empty universe

The empty world model, $\Omega_0 = 0$, $\Omega_v = 0$, is often referred to as the *Milne model*, because it can be developed purely from kinematics: Milne's major contribution to cosmology was in elucidating the meaning of time and kinematics in cosmology, and he developed a particular approach to the construction of cosmological models which are known as *kinematic*

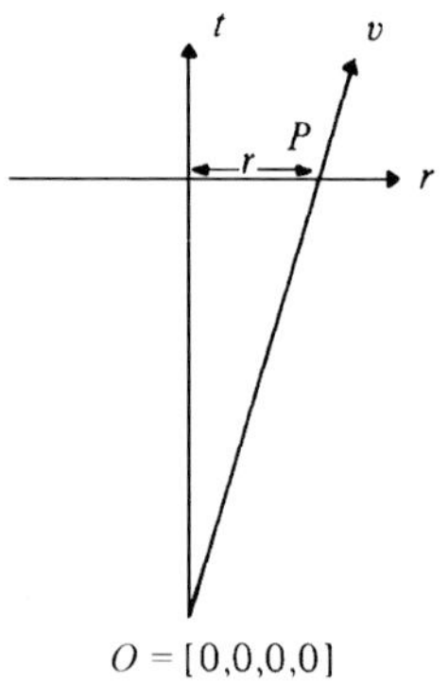

Figure A19.1: A space–time diagram for an empty universe

cosmologies. The most famous of these is the empty model in which there is no gravity and so particles move apart at constant velocity from $t = 0$ to $t = \infty$. The Robertson–Walker metric, Section 19.3, can be derived for this special case using only special relativity. This is an instructive exercise since it reveals clearly some of the problems which arise in the more general treatment using general relativity.

The origin O of the uniform expansion is taken to be $[0, 0, 0, 0]$ and the world lines of particles diverge from this point, each point maintaining constant velocity with respect to the others. The space–time diagram for this Universe is shown in Fig. A19.1. Our own world line is the t-axis, and that of particle P, which has constant velocity v with respect to us, is shown. The problem becomes apparent as soon as we attempt to define a suitable *cosmic time* for ourselves and for the fundamental observer moving with the particle P. At time t, the distance of P from us is r and, since P's velocity is constant, $r = vt$ in our reference frame. Because of the relativity of simultaneity, the observer at P measures a different time τ. From the Lorentz transformation,

$$\tau = \gamma\left(t - \frac{vr}{c^2}\right), \qquad \gamma = \left(1 - \frac{v^2}{c^2}\right)^{-1/2}.$$

Since $r = vt$,

$$\tau = t\left(1 - \frac{r^2}{c^2 t^2}\right)^{1/2}. \tag{A19.1}$$

The problem is now apparent: t is the proper time for the observer at O and for nobody else. We need to be able to define surfaces of constant cosmic time τ, because it is only on these surfaces that we can impose the conditions of isotropy and homogeneity, in accordance with the cosmological principle. The appropriate surface for which $\tau =$ constant is given by those points which satisfy

$$\tau = t\left(1 - \frac{r^2}{c^2 t^2}\right)^{1/2} = \text{constant}. \tag{A19.2}$$

Locally, at each point in the space–time, this surface must be normal to the world line of a fundamental observer.

The next requirement is to define the local element of radial distance $\mathrm{d}l$ at the point P on the surface $\tau =$ constant.* The interval $\mathrm{d}s^2 = \mathrm{d}t^2 - (1/c^2)\,\mathrm{d}r^2$ is an invariant. Since $\tau =$ constant over the surface, $\mathrm{d}s^2 = -(1/c^2)\,\mathrm{d}l^2$ and hence

$$\mathrm{d}l^2 = \mathrm{d}r^2 - c^2\,\mathrm{d}t^2. \tag{A19.3}$$

Thus τ and $\mathrm{d}l$ define locally the proper time and proper distance of events at P and are identical to the cosmic time t and an element of the radial distance coordinate x introduced in Section 19.3. It is only in τ- and l-coordinates that the cosmological principle can be applied.

Now let us transform from the frame S of the observer at O to the frame S' of the world line at P, by moving at radial velocity v. Distances perpendicular to the radial coordinate remain unaltered under Lorentz transformation and, therefore, if in S

$$\mathrm{d}s^2 = \mathrm{d}t^2 - \frac{1}{c^2}\left(\mathrm{d}r^2 + r^2\,\mathrm{d}\theta^2\right) \tag{A19.4}$$

then in S',

$$\mathrm{d}s^2 = \mathrm{d}\tau^2 - \frac{1}{c^2}\left(\mathrm{d}l^2 + r^2\,\mathrm{d}\theta^2\right). \tag{A19.5}$$

Now all we need do is to express r in terms of l and τ to complete the transformation to (τ, l) coordinates.

We have already shown that, along the surface of constant τ, $\mathrm{d}l^2 = \mathrm{d}r^2 - c^2\,\mathrm{d}t^2$. In addition, since $\mathrm{d}l$ is measured at fixed time τ in S', the relation between $\mathrm{d}r$ and $\mathrm{d}t$ can be found from the Lorentz transform of $\mathrm{d}\tau$,

$$\mathrm{d}\tau = \gamma\left(\mathrm{d}t - \frac{v}{c^2}\,\mathrm{d}r\right) = 0.$$

Therefore

$$\mathrm{d}t^2 = \frac{v^2}{c^4}\,\mathrm{d}r^2, \tag{A19.6}$$

and, from (A19.3),

$$\mathrm{d}l^2 = \mathrm{d}r^2\left(1 - \frac{v^2}{c^2}\right) = \mathrm{d}r^2\left(1 - \frac{r^2}{c^2t^2}\right) = \mathrm{d}r^2 - \frac{r^2\,\mathrm{d}r^2}{c^2t^2}. \tag{A19.7}$$

Notice that this makes sense since, in the rest frame at the point P, the recession speed is the same as observed in S', $v = \mathrm{d}l/\mathrm{d}\tau = \mathrm{d}r/\mathrm{d}t$.

Finally, we can substitute for $\mathrm{d}r/t$ in the last term of (A19.7) using $\mathrm{d}l/\tau = \mathrm{d}r/t$, so that

$$\mathrm{d}l^2 = \frac{\mathrm{d}r^2}{\left(1 + \dfrac{r^2}{c^2\tau^2}\right)}. \tag{A19.8}$$

Integrating, using the substitution $r = c\tau\,\sinh x$, the solution is

$$r = c\tau\,\sinh(l/c\tau). \tag{A19.9}$$

* The meanings of $\mathrm{d}l$ and $\mathrm{d}r$ are somewhat different from our earlier usage in this chapter.

The metric (A19.5) can therefore be written

$$ds^2 = d\tau^2 - \frac{1}{c^2}\left[dl^2 + c^2\tau^2 \sinh^2 \frac{l}{c\tau}\, d\theta^2\right]. \tag{A19.10}$$

This corresponds precisely to the expression (19.9) for an isotropic curved space with hyperbolic geometry, the radius of curvature of the geometry $\Re$ being $c\tau$. This explains why an empty universe has hyperbolic spatial sections. The conditions (A19.1) and (A19.8) are the key relations which indicate why we can only define a consistent cosmic time and radial distance coordinate in hyperbolic rather than flat space, for an empty Universe.

20 Epilogue

We have come to the end of our story. To leave it on the brink of quantum mechanics and its subsequent development is tantalising, but to take it further would result in a book aimed at a different audience and would need more advanced mathematical tools. How far I have succeeded in the many aims I set myself at the outset is not for me to judge. I can only state that, in preparing the original lectures and revising and amplifying them for publication, I have learned so much that I wish I had known long ago. My overwhelming impression, after 40 years of studying physics, astrophysics and cosmology, is an enhanced appreciation and admiration for the physicists and mathematicians who worked out the basic laws in the first place. These are outstanding intellectual achievements with flashes of genius and insight that raised the whole discipline to a new level of understanding.

On reading many of the original papers, I was impressed over and over again by the clarity of many of the great papers in the development of classical and modern physics. To be honest, I have found reading the original papers by physicists such as Maxwell, Rayleigh and Einstein easier than many modern textbooks. In these great works, I find a clarity of thought and exposition, which results from a clear understanding of the relation between our physical world and the mathematics we need to describe it. It is this deep appreciation of the basic laws of physics and theoretical physics which is the springboard for new insights and discoveries. This understanding cannot be gained by taking short cuts – if it were possible, you can be quite sure that they would have been discovered by now. It requires a great deal of hard work and experience of using the laws before the richness of their content can be fully appreciated.

Although the working out of any particular problem may be complicated, the basic physical ideas and mathematical structures are simple. Once these are understood, it is basically a matter of technique to apply them to a specific problem. I remember vividly a story told to me by former supervisor and colleague, the late Peter Scheuer, about an incident which involved his colleague, John Hunter Thomson, during his period of army service. Being a physicist John was put into a section working on radio receivers, and one day he asked a sergeant a question about some aspect of one of the circuits. The sergeant's reply is engraved on my memory: 'All you need to know is Ohm's law, but you need to know Ohm's law bloody well!' It is a statement which could be made about any of the laws of physics.

Index

The titles of original papers and books quoted in the main text are shown in italic type.

Those topics which I consider to be particularly valuable for revision purposes are shown in bold type. The page numbers indicated in bold type refer to the principal theoretical discussion of the topic.